Handbook of
SUSTAINABLE APPAREL PRODUCTION

Handbook of
SUSTAINABLE APPAREL PRODUCTION

Edited by
Subramanian Senthilkannan Muthu

CRC Press
Taylor & Francis Group
Boca Raton London New York

CRC Press is an imprint of the
Taylor & Francis Group, an **informa** business

First published in paperback 2024

First published 2015
by CRC Press
2385 NW Executive Center Drive, Suite 320, Boca Raton FL 33431

and by CRC Press
4 Park Square, Milton Park, Abingdon, Oxon, OX14 4RN

CRC Press is an imprint of Taylor & Francis Group, LLC

© 2015, 2024 Taylor & Francis Group, LLC

Reasonable efforts have been made to publish reliable data and information, but the author and publisher cannot assume responsibility for the validity of all materials or the consequences of their use. The authors and publishers have attempted to trace the copyright holders of all material reproduced in this publication and apologize to copyright holders if permission to publish in this form has not been obtained. If any copyright material has not been acknowledged please write and let us know so we may rectify in any future reprint.

Except as permitted under U.S. Copyright Law, no part of this book may be reprinted, reproduced, transmitted, or utilized in any form by any electronic, mechanical, or other means, now known or hereafter invented, including photocopying, microfilming, and recording, or in any information storage or retrieval system, without written permission from the publishers.

For permission to photocopy or use material electronically from this work, access www.copyright.com or contact the Copyright Clearance Center, Inc. (CCC), 222 Rosewood Drive, Danvers, MA 01923, 978-750-8400. For works that are not available on CCC please contact mpkbookspermissions@tandf.co.uk

Trademark notice: Product or corporate names may be trademarks or registered trademarks and are used only for identification and explanation without intent to infringe.

Publisher's Note
The publisher has gone to great lengths to ensure the quality of this reprint but points out that some imperfections in the original copies may be apparent.

Library of Congress Cataloging-in-Publication Data

Handbook of sustainable apparel production / editor, Subramanian Senthilkannan Muthu.
        pages cm
   Includes bibliographical references and index.
   ISBN 978-1-4822-9937-3 (alk. paper)
   1. Clothing factories--Handbooks, manuals, etc. 2. Clothing trade--Handbooks, manuals, etc. 3. Sustainable engineering--Handbooks, manuals, etc. I. Muthu, Subramanian Senthilkannan, editor.

TT498.H36 2014
338.4'7687--dc23   2014043724

ISBN: 978-1-4822-9937-3 (hbk)
ISBN: 978-1-03-291871-6 (pbk)
ISBN: 978-0-429-18855-8 (ebk)

DOI: 10.1201/b18428

Visit the Taylor & Francis Web site at
http://www.taylorandfrancis.com

and the CRC Press Web site at
http://www.crcpress.com

Contents

Preface ix
Editor xi
Contributors xiii

SECTION I Sustainable Textile Raw Materials and Manufacturing Processes for Apparel Manufacture

Chapter 1 Water-Free Plasma Processing and Finishing of Apparel Textiles 3

Kartick K. Samanta, Santanu Basak, S.K. Chattopadhyay, and T.N. Gayatri

Chapter 2 Sustainability in Clothing Manufacturing and Competitiveness: Is It a New Mind-Set or a Paradox? 39

Wen Ying Claire Shih, and Konstantinos Agrafiotis

Chapter 3 Sustainable Flame-Retardant Finishing of Textiles: Advancement in Technology 51

Kartick K. Samanta, Santanu Basak, and S.K. Chattopadhyay

Chapter 4 Application of Biotechnology in the Processing of Textile Fabrics 77

D. Saravanan, Thilak Vadicherla, K.P. Kannan, and S.N. Sreelakshmi

Chapter 5 Design and Development of Jute-Based Apparels 97

Sanjoy Debnath

Chapter 6 Sustainable UV-Protective Apparel Textile 113

Kartick K. Samanta, Santanu Basak, and S.K. Chattopadhyay

SECTION II Environmental and Social Assessment of Apparel Manufacturing

Chapter 7 Carbon Footprint of Textile and Clothing Products 141

Sohel Rana, Subramani Pichandi, Shabaridharan Karunamoorthy, Amitava Bhattacharyya, Shama Parveen, and Raul Fangueiro

Chapter 8 Environmental Impacts of Apparel Production, Distribution, and
Consumption: An Overview 167

Kim Y. Hiller Connell

Chapter 9 Life Cycle Assessment Studies Pertaining to Textiles and the Clothing Sector 181

Shabaridharan Karunamoorthy, Sohel Rana, Rosina Begam, Shama Parveen,
and Raul Fangueiro

Chapter 10 Social Impacts of the Clothing and Fashion Industry 207

Shanthi Radhakrishnan

SECTION III Sustainability and Consumption Behavior of the Apparel Industry

Chapter 11 Consumer Behavior and Its Importance in the Sustainability of the
Clothing Field 231

Jane McCann

Chapter 12 Consumer Behavior in the Fashion Field 271

Kirsi Niinimäki

Chapter 13 Textile Sustainability: Major Frameworks and Strategic Solutions 289

Arun Pal Aneja and Rudrajeet Pal

SECTION IV Assessment of Sustainable Apparel Production

Chapter 14 Eco-Parameters and Testing of Sustainable Textiles and Apparels 309

Shanthi Radhakrishnan

Chapter 15 Functional Aspects, Ecotesting, and Environmental Impact of Natural Dyes 333

L. Ammayappan and Seiko Jose

Chapter 16 Test Methods Related to Characteristics, Performance, and Ecological and
Safety Parameters of Textiles 351

Luis Almeida

Chapter 17 Environmental Communication and Green Claims of Textile Products 375

Shanthi Radhakrishnan

Contents

Chapter 18 Supplier Assessment in Global Apparel Supply Chains 399

Shams Rahman and Aswini Yadlapalli

Chapter 19 Sustainable Measures Taken by Industry Affiliates, Nonprofit Organizations, and Governmental and Educational Institutions 419

Thilak Vadicherla and D. Saravanan

SECTION V Sustainability and Fashion

Chapter 20 Exploring a Framework for Fashion Design for Sustainability 439

Alison Gwilt

Chapter 21 Fashion Industry and New Approaches for Sustainability 453

Kirsi Niinimäki, Esben Rahbek Gjerdrum Pedersen, Kerli Kant Hvass, and Lisbeth Svengren-Holm

Chapter 22 Eco-Design/Sustainable Design of Textile Products 475

Thilak Vadicherla and D. Saravanan

Chapter 23 Fashion Industry and Sustainability 501

Shanthi Radhakrishnan

Index 531

Preface

Since the last decade, "sustainability" has been one of the frequently heard terms in any industrial segment and has become a hot-button societal issue. The apparel sector, one of the important industrial sectors, cannot exclude sustainability, which is an important part of the agenda in the range of activities on which the apparel sector must focus. Sustainability and its developments are not new to the apparel sector, and it is becoming increasingly popular these days. Perhaps achieving sustainability in the clothing and apparel sector might look like an ambitious aspect, but it is doable. Since the supply chain is rather long, people do doubt its capabilities; however, many approaches are being followed to achieve sustainability. Owing to the longer and massive supply chain in the apparel industry, the assessment and implementation of sustainable measures in this industry are complex, and they cannot be denied. Nevertheless, striving efforts, cooperation, and collaboration from various members involved in the entire apparel supply chain make the complex task practically possible.

Sustainability in apparel production is a vast topic and it has many facets. This handbook divides this wide topic into five sections, and in each section, very important aspects are dealt with in dedicated chapters, written by subject experts. Section I (Chapters 1–6) is dedicated to sustainable raw materials and manufacturing processes and technologies that can pave the way toward sustainability in apparel production. Section II (Chapters 7–10) is earmarked to deal with an important aspect: the environmental and social impacts of apparel production and its assessment.

Chapters 11–13 in Section III revolve around discussions pertaining to the sustainability and consumption behavior of the apparel industry. The assessment of sustainability aspects and parameters is an important subject, and various topics come under this umbrella. Hence, Section IV (Chapters 14–19) is dedicated to this lengthy topic. Section V (Chapters 20–23) is earmarked to detail the design aspects, particularly sustainable design/eco-design, and new approaches to fashion sustainability.

All the important aspects, including even very minute details pertaining to sustainable apparel production, have been collated in this handbook, presented in the 23 chapters. I thank all the chapter contributors for their earnest efforts in successfully bringing out this handbook with the enriched technical content. I am very confident that the readers will benefit from the imperative details presented here pertaining to sustainable apparel production. The handbook will undoubtedly become an important reference for researchers and students, industrialists, and sustainability professionals working in the apparel sector.

Editor

Dr. Subramanian Senthilkannan Muthu is currently working for SGS as a global sustainability consultant, based in Hong Kong. He earned his diploma and bachelor's and master's degrees in textile technology from one of the premier institutes in India. He was awarded a doctorate by the Institute of Textiles and Clothing of Hong Kong Polytechnic University for his dissertation entitled "Eco-Functional Assessment of Grocery Shopping Bags." Dr. Muthu also has more than seven years of industrial experience in textile manufacturing, testing, and sustainability evaluation of textiles and clothing materials. He was an outstanding student throughout his study period and earned numerous awards and medals, including many gold medals. He has more than 75 academic publications in various textiles and environmental journals to his credit. Additionally, he has 2 patents, 2 book chapters, 15 books, and numerous conference publications. He is an acting editor, editorial board member, and reviewer for many international peer-reviewed journals in the textiles and environmental science disciplines. He is the editor-in-chief of the *Textiles and Clothing Sustainability* journal, an open-access journal. Dr. Muthu is also the series editor of two book series, namely, Textile Science and Clothing Technology, and Environmental Footprints and Eco-Design.

Contributors

Konstantinos Agrafiotis
Independent Consultant
Fashion Business Fashion Ventures
London, United Kingdom

Luis Almeida
Department of Textile Engineering
University of Minho
Guimaraes, Portugal

L. Ammayappan
National Institute of Research on Jute and Allied Fibre Technology
Indian Council of Agricultural Research
Kolkata, India

Arun Pal Aneja
Department of Engineering
East Carolina University
Greenville, North Carolina

Santanu Basak
Plasma Nanotech Lab
Chemical and Biochemical Processing Division
Central Institute for Research on Cotton Technology
Mumbai, India

Rosina Begam
PSG Institute of Advanced Studies
Coimbatore, India

Amitava Bhattacharyya
PSG Institute of Advanced Studies
Coimbatore, India

S.K. Chattopadhyay
Plasma Nanotech Lab
Chemical and Biochemical Processing Division
Central Institute for Research on Cotton Technology
Mumbai, India

Kim Y. Hiller Connell
Department of Apparel, Textiles, and Interior Design
Kansas State University
Manhattan, Kansas

Sanjoy Debnath
National Institute of Research on Jute and Allied Fibre Technology
Indian Council of Agricultural Research
Kolkata, India

Raul Fangueiro
School of Engineering
University of Minho
Guimaraes, Portugal

T.N. Gayatri
Plasma Nanotech Lab
Chemical and Biochemical Processing Division
Central Institute for Research on Cotton Technology
Mumbai, India

Alison Gwilt
Art and Design Research Centre
Sheffield Hallam University
Sheffield, United Kingdom

Kerli Kant Hvass
Copenhagen Business School
Centre for Corporate Social Responsibility
Frederiksberg, Denmark

Seiko Jose
National Institute of Research on Jute and Allied Fibre Technology
Indian Council of Agricultural Research
Kolkata, India

K.P. Kannan
Department of Biotechnology
Bannari Amman Institute of Technology
Erode, India

Shabaridharan Karunamoorthy
PSG Institute of Advanced Studies
Coimbatore, India

Jane McCann
University of South Wales
Newport, United Kingdom

Kirsi Niinimäki
School of Arts, Design and Architecture
Aalto University
Helsinki, Finland

Rudrajeet Pal
Academy of Textile, Technology and Management
University of Borås
Borås, Sweden

Shama Parveen
School of Engineering
University of Minho
Guimaraes, Portugal

Esben Rahbek Gjerdrum Pedersen
Copenhagen Business School
Centre for Corporate Social Responsibility
Frederiksberg, Denmark

Subramani Pichandi
School of Engineering
University of Minho
Guimaraes, Portugal

Shanthi Radhakrishnan
Department of Fashion Technology
Kumaraguru College of Technology
Coimbatore, India

Shams Rahman
School of Business IT and Logistics
College of Business
RMIT University
Melbourne, Victoria, Australia

Sohel Rana
School of Engineering
University of Minho
Guimaraes, Portugal

Kartick K. Samanta
Plasma Nanotech Lab
Chemical and Biochemical Processing Division
Central Institute for Research on Cotton Technology
Mumbai, India

D. Saravanan
Department of Textile and Fashion Technology
Bannari Amman Institute of Technology
Erode, India

Wen Ying Claire Shih
Department of Fashion Design
Hsuan Chuang University
Hsinchu, Taiwan, Republic of China

S.N. Sreelakshmi
Department of Biotechnology
Bannari Amman Institute of Technology
Erode, India

Lisbeth Svengren-Holm
The Swedish School of Textiles
University of Borås
Borås, Sweden

Thilak Vadicherla
Department of Textile and Fashion Technology
Bannari Amman Institute of Technology
Erode, India

Aswini Yadlapalli
School of Business IT and Logistics
College of Business
RMIT University
Melbourne, Victoria, Australia

Section I

Sustainable Textile Raw Materials and Manufacturing Processes for Apparel Manufacture

1 Water-Free Plasma Processing and Finishing of Apparel Textiles

*Kartick K. Samanta, Santanu Basak,
S.K. Chattopadhyay, and T.N. Gayatri*

CONTENTS

1.1 Introduction 3
1.2 Generation and Classification of Plasma...5
    1.2.1 Hot and Cold Plasmas..5
    1.2.2 Low- and Atmospheric Pressure Plasma 5
        1.2.2.1 Low-Pressure Plasma..5
        1.2.2.2 Atmospheric Pressure Plasma...6
1.3 Environment-Friendly Plasma Processing...7
1.4 Generation of Plasma and Its Optical Properties 9
1.5 Textile Coloration Using Plasma 11
    1.5.1 Coloration of Cellulosic Textile 11
    1.5.2 Coloration of Wool 12
    1.5.3 Coloration of Silk...13
    1.5.4 Coloration of Other Fibers 15
1.6 Hydrophobic and Superhydrophobic Textile 16
    1.6.1 Hydrophobic Textile Using Fluorocarbon Precursor.....................18
    1.6.2 Hydrophobic Textile Using Hydrocarbon Precursor 21
    1.6.3 Hydrophobic Textile Using Silicone Precursors............................25
    1.6.4 Hydrophobic Finishing of Dyed Textiles and Garments 26
1.7 Other Applications of Plasma in Textile..27
    1.7.1 Desizing..27
    1.7.2 Scouring and Improvement in Water Absorbency.........................28
    1.7.3 Flame-Retardant Finishing...29
    1.7.4 Wrinkle-Resistant Finish..30
1.8 Sustainable Textile Processing Using Plasma 30
1.9 Summary 31
References..32

1.1 INTRODUCTION

Chemical processing of textile is very important, as it removes impurities from the surface of the fibers, thus making it suitable for dyeing and finishing. Besides, it improves aesthetic and functional quality of the textile. However, the traditional wet chemical processing of textile is water and energy intensive (630 toe) due to the involvement of multiple numbers of padding, drying, curing, and post-washing operations.[1–3] They also increase the cost of production. Approximately, 150 L of water is used to process 1 kg of cotton textile starting from its preparation to finishing steps. Of late, due to the stringent environmental effluent regulation that has come into force, the textile industry is slowly gearing up toward the implementation of environment-friendly processing technologies. In this direction, several technological advancements have been taking place in this sector with the aim to reduce consumption of water, chemicals, and energy and to reduce the effluent generation and cost of production. Some of these developments are as follows: (1) low material to liquid processing; (2) spray and foam finishing; (3) enzyme processing; (4) dyeing with natural dyes; (5) digital printing; (6) infrared (IR) dyeing and drying; (7) radiofrequency (RF) drying; (8) ultrasound-assisted dyeing and dispersion; (9) use of natural dyes and other molecules for ultraviolet (UV) protection; (10) applications of aromatic and medicinal plants for antimicrobial, mosquito-repellent, and well-being textiles; and (11) use of aloe vera, neem, and such plant molecules for antimicrobial finishing and skin nourishing. A few more promising technologies that have also been emerged in the laboratory to the industry level are use of (1) nanoparticles, nanofiber, and nanocoating, (2) supercritical carbon dioxide for dyeing, (3) plasma for water-free processing, and (4) UV and laser application in textile. In spite of such developments, textile chemical processes in the real-application scenario have still remained energy and water intensive, besides being associated with air and water pollution due to the use of several non-environment-friendly chemicals and auxiliaries. With the increased global awareness of environmental pollution, climate change, carbon footprint, and health and hygiene in recent years, the need for eco-friendly textiles are increasing exponentially. The natural fiber–based apparel textiles are once again in demand by the eco-conscious customers due to its advantages of biodegradability being produced from renewable sources, good moisture regain, soft feel, adequate to fair strength, and dye friendliness. In this context, the plasma processing of textile in operations like desizing, dyeing, printing, and traditional to high-end finishing is getting considerable attention because of its potentiality in execution of multistep operation in a single step while avoiding the use of water altogether. Plasma is a partially ionized gas composed of many types of species, such as positive and negative ions, electrons, neutrals, excited molecules, photons, and UV light. It has the potential to be commercialized in textile processing for production of value-added home, apparel, and technical textiles at a much lower cost. Unlike in the case of the conventional processing of textiles, the main attraction of plasma for industrial processing is to avoid chemical effluents. Other advantages are low consumption of chemical and energy, lower operating cost, rapid processing, and high efficiency.[4–5] Plasma processing of textiles brings physicochemical changes on the top surface of polymeric substrates without altering the bulk (core) properties of the material. Presently, plasma is used for a number of industrial applications ranging from arc welding, metal hardening, metal coating, nuclear fusion, synthesis of nanomaterials, creation of nanostructures, surface cleaning, and functional polymeric/metal coating. However, all kinds of plasmas are not suitable for surface modification of heat-sensitive polymeric and textile substrates. Only cold plasma, also known as nonthermal plasma with bulk temperature of <25°C–250°C, can be used for nanoscale surface engineering of the textile substrate. The surface modification of textiles using plasma can be achieved with a nonpolymerizing gaseous molecule (small molecule), such as oxygen (O_2), nitrogen (N_2), air, argon (Ar), helium (He), or fluorine (F) for surface activation, cleaning, oxidation, changing surface energy, increasing surface roughness/area, etching, and creating nanostructures. These, in turn, help in improving the textile properties in terms of water absorbency, wetting, wicking, oil absorbency, dye exhaustion, adhesion, and antifelting of wool. On the other hand, plasma reaction with a bigger molecule containing vinyl, hydroxyl, carbonyl, carboxyl, acrylate, fluorocarbon, silicone, and phosphorous leads to the development of water-repellent, oil-repellent, UV-protective, flame-retardant, and crease-resistant textiles. An alternating current (AC) with a high-frequency power supply helps to dissociate various gaseous molecules into a collection of ions, electrons, neutral molecules, and other species. Hence, it is an energetic chemical environment where the generation of plasma species opens up diverse complex reactions resulting in various end applications. In the last one and a half decade, several plasma-enhanced technological advancements have taken place in different domains of textile preparatory processing and finishing. However, still there is a lack of comprehensive information on the effect of plasma treatment, especially atmospheric pressure plasma (APP) on the improvement of surface properties of cellulosic, protein (wool and silk) and other man-made fibers. Similarly, a comprehensive report on the mechanism of plasma reaction of using fluorocarbon, hydrocarbon, and silicone compounds with increasing their molecular weights is not available.

The present chapter discusses the importance of plasma processing of textiles and various plasma–polymer interactions. The coloration of cellulosic, protein, and other thermoplastic textiles pretreated with plasma has also been discussed in terms of improvement in dyeing time, temperature, and shade depth. Hydrophobic finishing of textiles using hydrocarbon, fluorocarbon, and silicone compounds along with their mechanism of plasma reactions has also been discussed in details. The other functional finishing and preparatory processes of textiles, such as flame retardance, crease resistance, hydrophilicity, desizing, and scouring, have also been summarized.

1.2 GENERATION AND CLASSIFICATION OF PLASMA

Among the different methods of plasma generation, an electrical breakdown of a gaseous molecule in the presence of high-frequency AC is the most popular one, commonly used for material processing including textiles. The high frequency of the AC helps to dissociate various gaseous molecules into a collection of ions, electrons, neutral particles, and other species. Plasma is often considered as the fourth state of matter. It was first identified by Sir William Crookes in 1879 and named *plasma* by Irving Langmuir in 1928. The different types of plasmas are discussed in the following text.

1.2.1 Hot and Cold Plasmas

Plasma can be classified into hot/high-temperature/thermal plasma and cold/low-temperature/nonthermal plasma based on the temperature of the plasma zone. Hot plasma occurs when the temperatures of the electrons and atomic and molecular species are extremely high and remain near to the thermal equilibrium state. In that condition, the molecules remain almost fully ionized (100%). The sun and the other stars in the various galaxies of the universe, fusion reactor, and plasma torches are the few examples of the thermal plasma. The temperature of such plasma is around 10^6–10^8 K with a high electron density of $\geq 10^{20}$ m^{-3}. Whereas in cold plasma, while the electrons remain at a higher temperature, the ions and the neutral molecules remain near to the ambient temperature ($T_e \gg T_{ion} \approx T_{gas} = 25°C$–$250°C$, $T_{electrons} \approx 727°C$). In such plasma, the electron density is significantly lower ($n_e \approx 10^{10}$ m^{-3}) and only a small fraction of the gas molecules (about 1%) is ionized.[4] Cold plasma contains reactive chemical species with a high selectivity that has the potential to be utilized for the surface modification of fibrous material.

1.2.2 Low- and Atmospheric Pressure Plasma

1.2.2.1 Low-Pressure Plasma

It is easy to ionize a gaseous molecule by electrical breakdown under a low-pressure condition, and the same has been studied extensively for fluorescent lamp, neon sign, and material processing. The advantages of low-pressure plasma processing are (1) the presence of highly energetic reactive species with the requisite density, (2) uniform glow plasma, (3) temperature below 250°C, (4) lower breakdown voltages, and (5) easy to control. Some of the limitations of low-pressure plasma processing are (1) longer processing time, (2) limited sample size, and (3) batch to semicontinuous operation. Plasma properties and the breakdown voltage of a gas, which are the functions of the plasma pressure and electrode spacing, are governed by Paschen's law, $V_b = f(p, d)$, where V_b is the breakdown voltage; p, the pressure; and d, the interelectrode spacing. To generate a plasma, the applied voltage must be more than the breakdown voltage of a particular gas at a given electrode spacing. The breakdown voltage (V_b) increases rapidly with the increase in pressure of the plasma reactor. Thus, the APP requires more discharge voltage than that required for the low-pressure plasma.[4]

1.2.2.2 Atmospheric Pressure Plasma

Unlike in the case of low-pressure plasma, it is quite difficult and challenging to generate plasma at an atmospheric pressure due to the presence of a high electrical field in between the two electrodes, placed within a narrow spacing. However, if low-temperature stable plasma can be generated at an atmospheric pressure, it can overcome the several limitations associated with low-pressure technology. At the same time, it is possible to integrate the plasma reactor with the existing textile processes for a continuous treatment. Three important categories of atmospheric pressure cold plasmas that are mostly being used in material processing are (1) corona discharge, (2) plasma jet, and (3) dielectric barrier discharge (DBD).

Corona discharge is characterized by an asymmetric electrode pair, which surrounds the nonhomogeneous electrode assembly powered by a continuous or a pulsed DC/AC electrical supply. In a highly nonuniform electric field, such as in a point–plane or wire–cylinder gap, plasma is formed when the high electric field near the point electrode exceeds the breakdown strength of the gas. The plasma discharge gap is kept at about 0.5–2 mm, and it is mostly used for ozone generation from air, electrostatic precipitation in dust collection, and activation of polymeric film in the packaging industry.

The atmospheric pressure plasma jet (AAPJ) consists of two concentric electrodes, through which a flow of mixture of gases is supplied. The inner electrode is coupled with a few kHz to 13.56 MHz RF electrical supply operating at a discharge voltage, somewhere in between a few hundred volts (250 V) and a few kilovolts (kV), with the outer electrode grounded. It operates on a feedstock gas that flows between the outer and inner electrodes, producing highly reactive species with high velocity. Plasma jet is mostly used for the downstream processing of textiles and other substrates for surface etching, activation, and polymeric coating.[5] The properties of different types of plasmas are summarized in Table 1.1.

TABLE 1.1
Comparison of Various Types of Plasmas

Parameters	Corona	DBD	APPJ	Low-Pressure Plasma
Method and type	Sharp-pointed electrode	Dielectric barrier covers the electrodes	RF capacitively coupled	Glow plasma and capacitively/inductively coupled
Excitation	Pulsed DC/AC	AC or RF	RF 13.5 MHz	DC/low RF or high RF (13.56 MHz)
Pressure (bar)	1	1	1	10^{-5} to 10^{-8}
Electron energy (eV)	~5	1–10	1–2	5
Electron density/cm^3	10^9–10^{13}	10^{12}–10^{15}	10^{11}–10^{12}	10^8–10^{13}
Breakdown voltage (kV)	20–50	1–25	0.05–0.2	0.2–0.8
Scalability and flexibility	Yes	Yes	Yes	No
T_{max} Temp (K)	Ambient temperature	Ambient temperature to 600	400	425
Commonly used gas	Air	N_2, O_2, air, helium (He), argon (Ar)	He, Ar	Krypton, argon, xenon, oxygen

Sources: Schutze, A. et al., *IEEE Trans. Plasma Sci.*, 26(6), 1685, 1998; Conrads, H. and Schmidt, M., *Plasma Sources Sci. Technol.*, 9, 441, 2000; Eliasson, B. and Kogelschatz, U., *IEEE Trans. Plasma Sci.*, 19, 1063, 1991; Mohamed, A.H. et al., *Eur. Phys. J. D*, 60, 517, 2010.

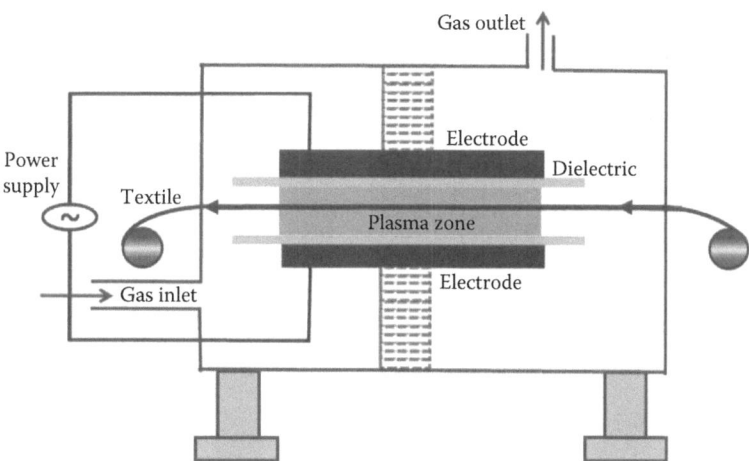

FIGURE 1.1 Schematic of DBD plasma treatment of textile in a continuous manner. (From Samanta, K.K. et al., A status report on surface modification of textile using cold plasma, Central Institute for Research on Cotton Technology, Mumbai, India, 2014, pp. 1–38.)

The DBD consists of two parallel flat electrodes producing a strong thermodynamic, nonequilibrium plasma at an atmospheric pressure and with a moderate gas temperature. At least one of the electrode is covered with a quartz, Pyrex, alumina, glass, silicone, or Teflon dielectric sheet, with a thickness of <1–3 mm as shown in Figure 1.1. Since the dielectric barrier separates the electrodes and the discharge gas, no electrode etching/corrosion happens. An AC voltage with an amplitude of 1–25 kV and RF of 1–100 kHz is applied to ignite the DBD plasma with an electrode spacing of <1 mm to several mm. The DBD plasma is used for plasma-assisted chemical vapor deposition or plasma polymerization for functional coating, surface etching, cleaning, and activation of polymer.

1.3 ENVIRONMENT-FRIENDLY PLASMA PROCESSING

It is well known that during the wet chemical processing of textiles, the industry consumes a large quantity of water (150 L/kg) as a processing medium and generates almost an equal amount of liquid effluents. The cost of the final products also increases due to the multiple drying of the wet textiles. Of late, due to the increased environmental awareness and stringent effluent norms, the textile industry is moving toward the implementation of environmentally friendly low-water- to zero-water-based processing technologies, such as digital printing, spray and foam finishing, and processing with supercritical fluids and solvents for the development of sustainable textiles. From the point of view of eco-friendly textile chemicals and green textile products, some more technological advancement have also taken place in this field, such as application of (1) natural dyes; (2) enzymes for desizing and biopolishing; (3) aromatic and medicinal plants for antimicrobial, UV-protective, and well-being textiles; (4) plant molecules, such as banana pseudostem sap (BPS) and spinach juice (SJ) for flame-retardant finishing; and (5) biomolecules like DNA from herring sperm, salmon fish, whey proteins, and casein for flame-retardant finishing.[11] In this context, low-temperature plasma seems to be a promising environmentally friendly technology for the wet chemical processing of textiles. The plasma treatment of textiles only modifies the surface of the material without altering the bulk properties to increase the uptake of liquid, dye, chemical, and auxiliary. It also enables the processors to accomplish multifunctional textiles in a single step without usage of water or simultaneous dyeing and finishing operation.

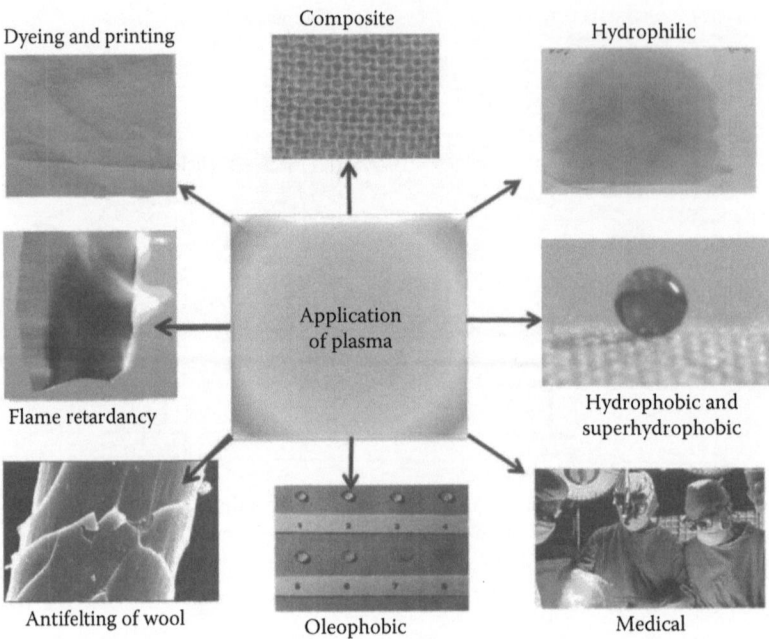

FIGURE 1.2 Application of plasma in textile processing and finishing. (From Samanta, K.K. et al., A status report on surface modification of textile using cold plasma, Central Institute for Research on Cotton Technology, Mumbai, India, 2014, pp. 1–38.)

Imparting of different value-added functionalities—such as water-, stain-, and oil-repellent, hydrophilic, antimicrobial, flame-retardant, UV-protective, and static-resistant properties—and the improvement in desizing, dyeing, printing, biocompatibility, and adhesion processes can be accomplished by modifying the fiber surface at a nanometer level as presented in Figure 1.2.[5,12–24] The main advantages of plasma processing of textiles are as follows: (1) liquid-free processing, (2) single-step operation, (3) requirement of minimal amount of chemicals, (4) cost-effectiveness in terms of processing time and temperature, (5) imparted functionality independent of substrate chemistry, (6) environment friendliness, and (7) surface design preserving the bulk property. Low-pressure plasma has been extensively studied for such applications, but the technology has not been successfully commercialized in the textile arena due to its inherent technoeconomical limitations. On the other hand, atmospheric pressure cold plasma has the potential to overcome the limitations associated with low-pressure plasma and hence is being explored for similar end applications. The APP is becoming popular in textile and allied sectors for academic and industrial applications due to its feasibility integration in line with existing textile processes.

The surface modification of textiles with a desired functionality can be achieved by selecting appropriate plasma processing parameters with a suitable precursor molecule. The fragmentation of the precursor molecule followed by plasma reaction with the textile substrates is the best way of surface engineering to develop value-added apparel, home, technical, and smart textiles. In the plasma processing of textiles, there is no requirement of any pretreatment, such as swelling of the fibers in organic, aqueous, or alkaline solvent. From the physical point of view, the roughening of the fiber surface is responsible for changes in the coefficient of friction, top cohesion, spinnability, yarn strength, and increase in antifelting of wool. From the chemical point of view, fiber surface oxidation or reaction with suitable precursor molecules is the main factor responsible for improving various functional properties of textiles. Plasma chemistry is a complex process that involves a large number of elementary homogeneous and heterogeneous reactions. A homogeneous reaction occurs between the species in the gaseous phases and a heterogeneous reaction occurs between the plasma species and the solid substrate. The different types of plasma–substrate interaction are discussed in the following:

1. *Ion formation*: Reactions due to ion formation that directly lead to a new chemical product, like the formation of NH_3 and NO_2.
2. *Recombination*: When the rate of produced surface radicals is quite high and the air is excluded from the system, a tough cross-linked shell is formed that offers protection against the solvent.
3. *Oxidation*: Treatment with oxygen-containing plasma leads to the formation of various polar groups like ketone, hydroxyl, ether, peroxide, and carboxylic acid to make the surface hydrophilic.
4. *Peroxide formation*: When a polymer is exposed to argon (Ar) or helium (He) plasma followed by exposure to air, a high proportion of reactive sites are converted to peroxide. Since peroxide is known as an initiator for vinyl polymerization, it is used for graft polymerization.
5. *Radical formation*: Carbon-free radicals are formed on the polymer surface, when the energetic ions/photons of plasma break the organic bonds of the polymer.
6. *Polymerization*: Formation of radicals on the polymer surface followed by graft polymerization of a vinyl-based monomer outside the plasma reactor or simultaneous formation of radicals on the polymer surface and in situ reaction with the fragmented species of a precursor leads to the formation of an ultrathin polymeric film (coating) on the textile substrate.
7. *Surface cleaning*: A cleaning process in which argon (Ar), helium (He), or oxygen (O_2) gas is used to ablate organic contaminants, such as oil from the substrate surface.

1.4 GENERATION OF PLASMA AND ITS OPTICAL PROPERTIES

Plasma can be generated either at a low-pressure or an atmospheric condition. The ionization of a gaseous molecule to form plasma can be achieved by applying either sufficient heat energy or a magnetic or an electrical field. Among various methods of plasma generation, the electrical breakdown of a gaseous molecule in the presence of a direct current (DC) or a high-frequency AC is the most popular. The high-frequency AC power supply helps to dissociate various gaseous molecules into a collection of ions, electrons, neutral particles, and other species and is mostly preferred over DC discharge and commonly used for surface modification of fibrous materials. Depending on the ionization pattern of a gaseous molecule and the state of energy of the excited atoms or molecules in the plasma zone, a particular color is produced. Samanta et al. generated plasma at atmospheric pressures in the presence of noble and common gases (with small molecular weight), such as helium (He), argon (Ar), and mixtures of He–oxygen, He–air, and He–nitrogen.[10,25] Plasmas were also generated in the mixture of helium (He) with large-molecular-weight gases and liquid, such as 1,3-butadiene (BD), tetrafluoroethane (TFE), mixture of difluoromethane and pentafluoroethane, and acrylonitrile (liquid). All the aforementioned plasmas were generated with a discharge voltage of 2–6 kV and in a frequency of 16–22 kHz. Somewhat a higher discharge voltage was required for plasma generation in the presence of He–butadiene, He–fluorocarbon, and He–acrylonitrile mixtures due to their higher molecular weight. The pure helium (He) gas on ionization exhibited a light bluish purple color. The 2% (v/v) oxygen along with He produced a milky white color. Similarly, a light milky color was observed in argon (Ar) gas. When air was introduced along with He, it produced a pinkish purple color, but a strong bluish purple color was visible in the presence of He–nitrogen. Helium (He) in the admixture of a small amount of TFE (fluorocarbon) produced a reddish purple color. On the other hand, the BD (hydrocarbon) monomer with helium mixture produced a milkish white color. A similar color was also observed when acrylonitrile (hydrocarbon) vapor was introduced

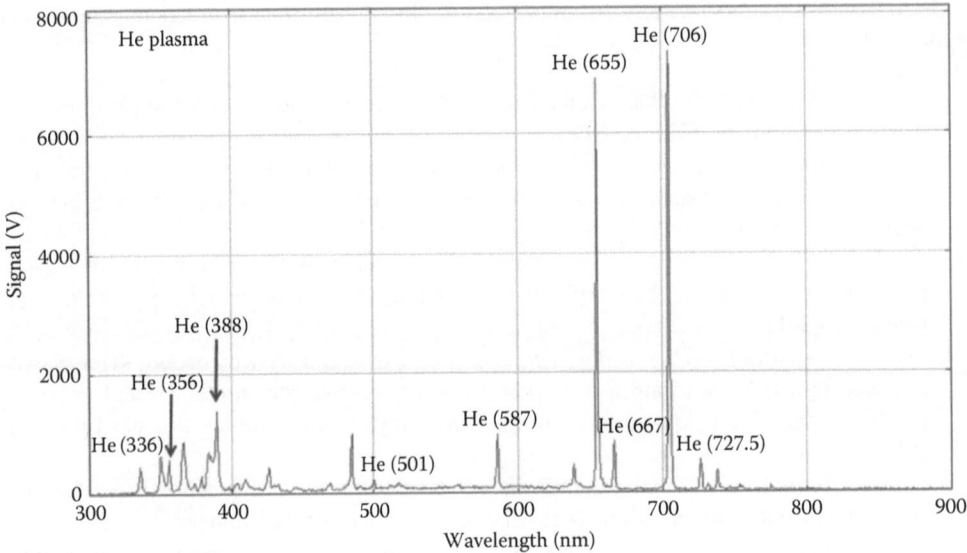

FIGURE 1.3 OES spectrum of He plasma. (From Samanta, K.K. et al., *Int. J. Eng. Res. Technol.*, 3(3), 2467, 2014.)

into the helium plasma. The formation of a distinct color plasma in the presence of different molecules is an indication of the ionization/fragmentation of the precursor molecule. In plasma, the light emitted by the excited atoms and the molecules over the wavelength of 200–1100 nm was collected using an optical emission spectroscopy (OES). Figure 1.3 shows the OES of the He plasma that emits photons at different wavelengths of 706, 655, 667, 587, 727.5, 388, 356, and 336 nm. The other peaks in the spectrum in the wavelength of 300–500 nm are possibly due to the atomic lines of nitrogen (i.e., the trace amount present in He gas).

Figure 1.4 plots the intensity of He 706 nm atomic line with the plasma discharge time.[26] It can be seen that the intensity of He 706 nm atomic line remains constant with the plasma discharge time. Since the emission intensity of a particular species is directly proportional to its concentration, this observation implies the formation of uniform steady state plasma and it is a characteristic of stable (glow) plasma. When oxygen was introduced in He plasma, the two additional atomic lines of oxygen at 776 and 844 nm were observed, in addition to the major helium atomic lines.

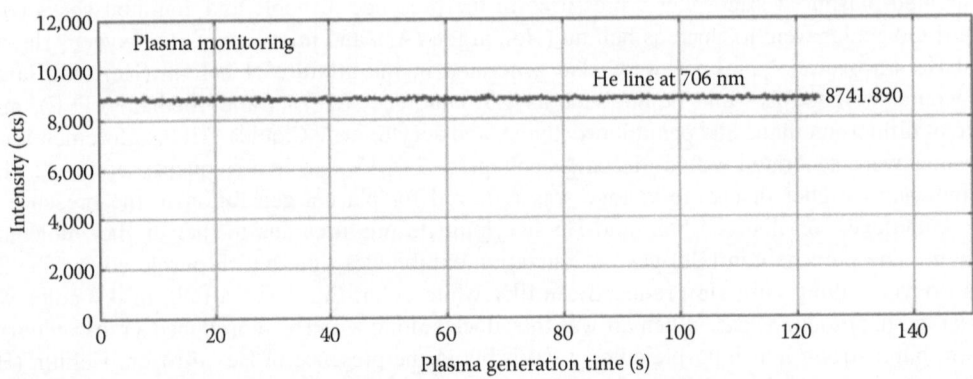

FIGURE 1.4 Intensity of He 706 nm atomic line with plasma discharge time. (From Samanta, K.K. et al., *Int. J. Eng. Res. Technol.*, 3(3), 2467, 2014.)

In the case of He–air and He–nitrogen plasma, several strong atomic lines of nitrogen appeared in the wavelength of 300–450 nm. In all these cases, the plasma current–voltage waveform remained smooth and free from any short or long spikes, which is also an indication of glow plasma. Discharge current, voltage, and waveform were measured using a digital oscilloscope attached with high voltage and current probes. The OES spectrum of helium in the admixture of large-molecular-weight hydrocarbon and fluorocarbon gaseous/liquid molecules has been discussed in Section 1.6.

1.5 TEXTILE COLORATION USING PLASMA

1.5.1 COLORATION OF CELLULOSIC TEXTILE

Plasma treatment has been used for modifying cotton fibers for its improvement in hydrophilic properties and surface roughness. The improvement in hydrophilic property (chemical changes) and increase in surface area/roughness (physical changes) have been utilized for faster dye exhaustion and/or for getting a better color (K/S) value, besides reducing the dyeing time and temperature. DBD plasma treatment on cotton, wool, and polypropylene (PP) was found to change the hydrophobic character of these fibers into hydrophilic. After plasma treatment, the specific surface area increased significantly from 0.1 to 0.35 m^2/g in the cotton fabrics resulting in increased dye uptake.[27] Plasma treatment also brings additional advantages, such as improving the dyeing kinetics, the shade depth, and the dye bath exhaustion. Air–dichlorodifluoromethane (DCFM) plasma treatment on cotton fabric leads to an improvement in dyeability with reactive and natural dyes. However, there was a small decrease in dyeability with the direct dye.[15] Using dichloromethane RF plasma (10 Pa) for 10–45 s on cotton and polyester fabrics, the dyeability with reactive dyes could be enhanced without affecting the other properties.[13] A low-pressure plasma polymerization was carried out for 5–30 min on a cotton fabric in amine ethylenediamine or triethylenetetramine (TETA).[28] The treated fabric was dyed with a reactive Remazol Black B dye and the maximum improvement in color value was found to be 33.9% compared to the untreated sample. This improvement was due to the formation of chemical groups that were suitable to react with the dye molecules. In plasma-modified dyed samples, the rubbing fastness was found to be good for a higher color yield. DBD plasma treatment followed by TETA-treated cotton textile can also be dyed with acid dyes to yield a better color due to the formation of new amine groups, as has been confirmed by Fourier transform infrared (FTIR) spectroscopy analysis.[29] After plasma treatment, an immersion in a cationic solution turned the cotton fiber anionic dyeable. The peak at 1222 cm^{-1} was for the C–N stretching of the primary aliphatic amine (C–NH_2). The increase in the reactive dyeing rate in the plasma-treated sample compared to the untreated cotton sample was visible after 40 min of dyeing time.[30] The exhaustion of the dye was possibly due to the formation of holes during the O_2 plasma treatment that provided a new pathway of the dye penetration. The dyeing of the plasma-pretreated cotton woven fabrics showed a deeper and a brighter shade.[31] Plasma treatment in the presence of air or oxygen increased both the rate of dyeing and the direct dye uptake of chloramine Fast Red K in the absence of an electrolyte.[32] Oxygen plasma treatment was found to be more effective than air plasma treatment. Helium–oxygen plasma treatment for 4 min in an atmospheric pressure could also improve the K/S value of the cotton fabric.[26] The increase in dye uptake of the cotton fabric can be attributed to the cumulative phenomena of (1) oxidative attack on cotton fibers that modifies the surface properties, (2) change of fabric surface area per unit volume caused by surface erosion, (3) etching effect of fiber by plasma and removal of impurities like wax or any remaining size material, (4) chemical changes in cotton fiber leading to formation of carbonyl and carboxyl groups, and (5) formation of free radicals in the cellulosic polymer.[33]

The color yield in the digital inkjet–printed cotton fabric was found to improve significantly by pretreatment of APP.[34] Other properties such as color fastness to crocking, laundering, outline sharpness, and antibacterial properties were also found to improve compared to that noted with the control cotton samples. Plasma treatment and the imparted improved properties were found to be durable to several washing cycles. Man et al. reported that plasma treatment (150 W, O_2 flow 0.4 L/min) of cotton fabric showed an increase in K/S value from 1.22 to 2.26 in the untreated and plasma-treated sample.[35] Plasma treatment illustrates a great deal of improvement in terms of water absorption of cotton fabric due to morphological modification and hydrophilic group formation. In continuous pigment dyeing, the cracks and voids formed by plasma etching, and the hydrophilic groups created by oxygen radicals, cause a high rate of ink absorption in the aqueous pigment. The effective pigment diffusion and its penetration into fiber contribute to a higher color yield under the same padding speed and the pressure. The color levelness data show that the plasma-treated cotton fabric possessed an improved levelness of pigment dyeing, even at a high red pigment concentration. The better fastness to crocking in the red pigment–dyed fabric was due to the better linkage of the polymer binder–pigment system and the formation of oxygen-containing groups in the plasma-treated sample.

1.5.2 Coloration of Wool

Protein fibers, such as wool and silk, were plasma treated in the presence of reactive to nonreactive gaseous plasma, and their effects on the rate of dye exhaustion, final dye uptake, K/S value, and fixation have been studied by several research groups. Wool fiber was plasma treated using various nonpolymerizing gases to improve hydrophilic properties, in terms of water absorbency and surface roughness. The improvement in water absorbency time could enhance the uptake of an anionic dye by 5% at equilibrium compared to the control sample, when dyeing was carried out at 70°C. The improvements in dye uptake in the plasma-treated sample were mainly due to (1) partial destruction of the outer epicuticle layer of the wool fiber, (2) reduction of covalently bonded fatty acid (18-methyl eicosanoic acid) layer in the upper epicuticle, (3) reduction of cystine (S–S) linkage to cysteic acid (–SO_3H), and (4) formation of the additional NH_2 group. X-ray photoelectron spectroscopy (XPS) analysis showed that after 5 min of N_2 plasma treatment, the carbon (C) atomic percentage decreased by 8.5% due to the partial etching and the removal of long-chain carbon hydrophobic fatty acid from the fiber surface. As expected, the atomic percentages of N and O increased in the treated samples by 16% and 39%, respectively. The increase in N_2 percentage promoted in formation of additional NH_2 groups. Surface oxygen percentage increased due to surface oxidation, and the same was confirmed by reduction in sulfur (S) atomic percentage from 2.58 to 2.23 with a shift in sulfur atomic peak from 163 to 168 eV binding energy in the untreated to plasma-treated samples, respectively.[36] Electron spin resonance (ESR) analysis showed an increase in the area under the curve by 18.2 times with G factor 2.007 after plasma treatment, due to the formation of more nitrogen-centered free radicals.[37] Similar results were also observed, when the wool fabric was plasma treated in the presence of O_2 plasma. Two more physicochemical changes were noted: (1) formation of additional –C=O, –OH, and –COOH groups and (2) a little improvement in the crystallinity value. Plasma treatment enhanced surface etching, resulting in an increase of the surface area to 0.35 m^2/g from 0.1 m^2/g in the untreated wool.[27] The improvements in the dye uptake and the shade depth in wool sample have also been reported by Ratnapandian et al.[38] In a similar study, the wool fabric was O_2 plasma treated for 5 min and dyed with acid, chrome, and reactive dyes at 50°C at a pH of 4.5. It was observed that in the case of acid, chrome, and reactive dyes, the dye uptake increased by 72.6%, 6.5%, and 39.4%, respectively, after 1 h of dyeing. The result showed that plasma treatment was useful and helped in faster dye exhaustion. After 2 h of dyeing at equilibrium, the total dye uptake was found to increase by only 2% compared to the untreated fabric for the acid- and the chrome-dyed fabrics. On the other hand, a 25% more dye uptake was found in the reactive dyed sample at the equilibrium. It was possibly due to the formation of oxygen-containing functional groups that could react with the dye molecules. However, plasma etching and new dye site formation were not sufficient for improvement of the acid and chrome dyes at the equilibrium.[39] According to Rombaldoni et al., if O_2 plasma could be used for 5 min at 30 W for the surface modification of wool, followed by dyeing using a combination of

1:2 metal complex dyes at 98°C, 85°C, and 80°C at neutral pH, the final bath exhaustion of the treated sample could be compared to the control sample dyed at 85°C.[40] Plasma pretreatment has also been made feasible for continuous application of natural dyes on wool fibers. Plasma treatment was found to improve the natural dye uptake compared to the control sample, when copper sulfate was used as a mordant.[38] Air plasma treatment was found to improve the adhesion and the penetration of the printing paste on the wool fiber and decrease the surface hairiness. This led to a higher color yield in the wool fabric at a lower steaming treatment in the absence of any wetting agent.[41] An increase in the diffusion coefficient of acid dyes in wool has also been reported in other atmospheric plasma–treated samples.[42]

The effect of atmospheric pressure He plasma treatment on the coloration of wool has been reported using dichlorotriazine-based reactive dye (C.I. Reactive Red 2) at 35°C by the pad batch method.[43] The samples were plasma treated for 6 min in a glow-discharge cold plasma at a discharge voltage of 3 kV and frequency of 20 kHz. The improvement in hydrophilic property after plasma treatment, its stability with the number of washing cycles, and storage time, in addition to the antifelting of wool, were also reported. The dye bath liquor consisted of 5 or 10 g/L reactive dyes, 10 g/L TRO, 0–300 g/L urea, and 0–10 g/L sodium bisulfite at a pH of either 5 or 7. Wool fabric was padded at room temperature (30°C–35°C) and kept for 24 h at 35°C. The dyeing time was varied from 10 min to 30 h, and the K/S values were measured before and after stripping off the unfixed dye. Both the dye pickup and the fixation of dye molecules in the plasma-treated wool fabric were found to increase by almost twice, as expressed by the K/S value. When the total concentration of dye in the treated sample was determined by measuring the optical density, it was found to be 40%–50% higher than that noted in the untreated sample. Even after removal of the unfixed dye, the K/S values were found to increase by almost 100%, that is, 3.7–4.6 in the plasma-treated samples in comparison to 1.5–2.5 in the untreated sample. The percentage fixation was in the range of 80% ± 5% for both the untreated and the plasma-treated samples. The surface chemistry analysis using the secondary ion mass spectrometer (SIMS) showed an increase in HO$^-$ and NH$^-$ intensities with a simultaneous decrease in CH$^-$ intensity after plasma treatment. This is an indication of the formation of additional –NH$_2$ groups resulting in better dye exhaustion and fixation. This difference in the K/S values and the total dye uptake implies the possibility of a large amount of dyes being absorbed and reacted on the plasma-treated fiber surface and a little quantity of it being diffused toward the core of the fiber as shown schematically in Figure 1.5.

1.5.3 Coloration of Silk

Plasma treatment was found to improve the wettability of the degummed silk fabric. Similar to the plasma treatment of wool as discussed earlier, a 5 min atmospheric pressure oxygen plasma treatment of tussar silk showed a wicking height of 100 mm in ~5 min, whereas in the control silk, a maximum height of only 100 mm was observed after a prolonged time. In case of both the eri and the muga silk, plasma treatment improved the wicking rate by nearly twice compared to the untreated silk. After 30 min of low-pressure plasma treatment in the presence of O_2, N_2, and H_2 atmospheres, the coefficient of friction increased to 0.7–0.8 from 0.27 in the untreated sample. It was observed that after 10 and 30 min of O_2 plasma treatments, there were 13% and 22% loss of crystallinity, respectively. The ATR-FTIR spectra showed an increase in β-sheet structure after 15 min of the O_2 plasma treatment, and the silk amide-II random coil band was found to disappear. However, the band was observed, if the characterization could be conducted within a few hours of plasma treatment. It might be due to the formation of β-sheet from amide-II, because of the surface ageing. Further, the tensile properties were also affected by the O_2 plasma treatment.[44] In all the plasma-treated fabrics, the weight loss, surface etching, and roughness were found to increase with the increasing plasma treatment time. A plasma treatment of 30 min duration resulted about 4%–5% loss in the sample weight. Unlike hydrogen (H_2) plasma treatment, the O/C and N/C atomic ratios improved slightly after the O_2 and the N_2 plasma treatments. The effect

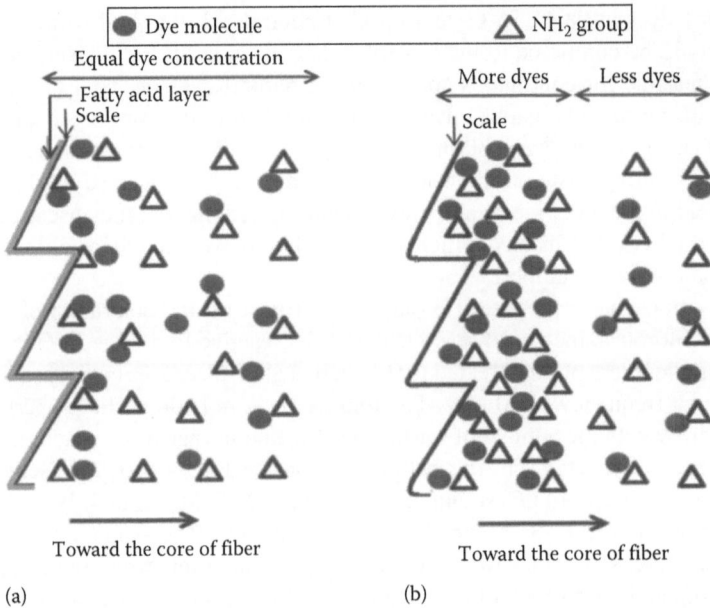

FIGURE 1.5 Schematic representation of dye molecules in (a) untreated and (b) plasma-treated wool. (From Panda, P.K. et al., *J. Appl. Polym. Sci.*, 124(5), 4289, 2012.)

of NH_3 plasma treatment on silk has also been reported considerable improvement in the N atomic percentage of the exposed surface. Plasma treatment was found to improve the dyeability and the color strength of the silk fabric irrespective of the plasma gases used. The degummed silk fabrics were plasma treated in O_2, N_2, and H_2 atmosphere for 5 min and dyed with Remazol reactive dye with M:L = 1:50 at 50°C for 90 min. The K/S value of the treated fabrics improved significantly compared to the control fabric. The 5 min plasma treated and the dyed fabric at 6% shade exhibited an equal color strength to that of the 10% dyed control sample. This might be due to plasma treatment that has helped in the formation of more active dye sites.[45]

In a more recent study, a silk fabric was plasma treated in an atmospheric pressure helium–nitrogen (He–N_2) plasma for 1–10 min, and the nitrogen flow rate varied from 33.3 to 225 mL/min keeping a constant helium (He) flow rate at 450 mL/min.[46] The effects of plasma treatment time and N_2 gas flow rate on water wicking, surface physical and chemical properties, and the dyeing rate were investigated in detail. Energy dispersive x-ray (EDX) analysis showed the presence of 3.2% more nitrogen (N) in the plasma-treated sample, which possibly promoted to the formation of more amine groups resulting in faster exhaustion of anionic acid dye molecules, even at a lower temperature. Three different temperatures, such as 40°C (low), 60°C (medium), and 80°C (high) were used for the dyeing experiment. It was possible to dye the sample without any salt at an ambient temperature of 40°C, in contrast to 90°C required in the conventional dyeing.

Figure 1.6 shows the dye exhaustion percentage against the dyeing time in both the untreated and plasma-treated samples. In all the plasma-treated samples, the dye exhaustion percentages were found to be much higher than their corresponding untreated samples at any particular temperature. The difference in the dyeing rate between the untreated and the treated samples is more prominent when dyeing was carried out at 40°C and 60°C. In the plasma-treated fabrics, 82.7% and 95.1% of the dye get exhausted at 40°C and 60°C, respectively, compared to only 41.8% and 76.2% in their corresponding untreated samples, after 30 min of dyeing. At the end of the dyeing cycle (after 80 min), the dye bath exhaustion was 93.6% in the plasma-treated sample at 40°C, which is similar to the dye exhaustion in the untreated samples at 60°C and 80°C. Hence, in the plasma-treated

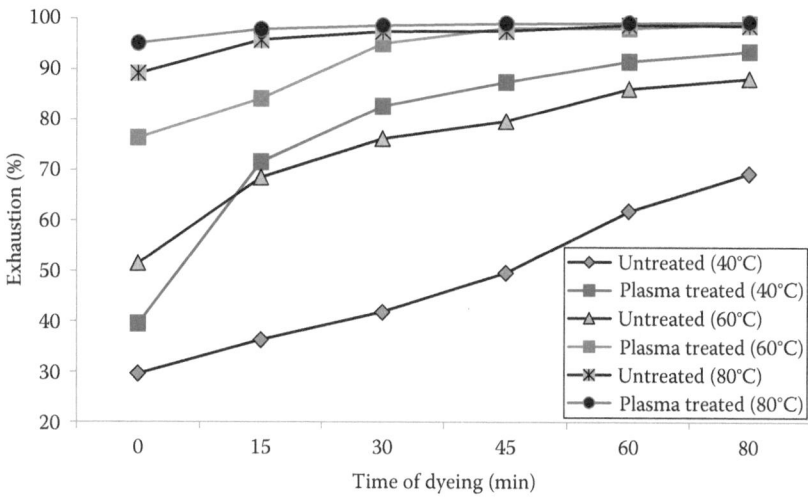

FIGURE 1.6 Dye exhaustion percentage with dyeing time in untreated and plasma-treated samples at different temperatures. (From Pandit, P., Plasma application for performance enhancement of textile fabrics, MTech, thesis, Institute of Chemical Technology, Mumbai, India, 2014.)

sample, it is possible to reduce the dyeing temperature by 30°C–50°C or dyeing time by about 50 min. With the increase in dyeing temperature, the exhaustion was found to increase as shown in the same figure. Plasma treatment time and the N_2 gas flow rate exhibited a marginal effect on the mechanical properties of the silk fabrics.

1.5.4 Coloration of Other Fibers

Polyamide (nylon 6) fabrics were treated with cold plasma using oxygen, argon, and tetrafluoromethane plasma and an improvement in dyeing with basic dyes was noted. This happened mainly due to the etching of the fiber surface and the introduction of polar groups in it. When different monomer coatings were applied to polyester, polyamide, and PP fabrics, it improved the affinity to other classes of dyes.[47] A low-temperature plasma treatment of gray, mercerized cotton, and polyester/cotton blended fabrics before dyeing has been reported to be effective for improvement in dyeing and its quality. Silicon tetrachloride hydrophilic plasma coating on PET fabric was found to increase the surface roughness, resulting in improvement of PET dyeing.[48] The plasma polymerization of acrylic-like coatings on polyester and polyamide fabrics was found to enhance wettability and dyeability toward the basic dyes in addition to the improvement in the soil resistance property.[49] The color depth of the dyed fabric (polyester/cotton blend) was found to increase with an acrylic-like film coating thickness.

Dyeing of polyamide fabrics was carried out after the surface modification with a DBD plasma (power −0.5 to 2.5 kW/min/m).[50] The change in hydrophobic to hydrophilic surface was the key point for the adsorption of aqueous dye solutions to achieve excellent dye uptake, higher rate of dyeing, uniform dyeing, and fastness for the darker shades, using reduced dye concentration, time, and temperature. The physicochemical improvements elicited by the DBD plasma in the dyeing of polyamide 6,6 (PA66) fibers were also investigated.[51] With the plasma dosage energy of 2500 W/min/m², the static contact angle decreased from 133° to 24°. The SEM images of the plasma-treated sample showed the appearance of ripple-like structures in the submicron level formed due to plasma etching. The formation of acidic molecules with lower molecular weights acted as a dye *carrier* and microchannels on the PA66 surface, which probably favored dye diffusion into the fiber structure. Plasma treatment allows a high level of direct dye diffusion and fixation at lower temperatures and shortening of dyeing time compared to the traditional dyeing method. Almost all the dye molecules get exhausted within 60 min of the dyeing cycle in the plasma-treated sample, whereas a significant amount of dye still remained in the dye bath even after 120 min of dyeing in the untreated sample. Dispersive and polar components of the total surface energy of the untreated sample were 10.6 and 1.3 mJ/m^2, respectively, and after 2500 W/min/m^2 of plasma dosing, the dispersive component decreased to 7.0 mJ/m^2 with a simultaneous sharp increase in the polar component up to 73.1 mJ/m^2. The ATR-FTIR spectra of the plasma-treated sample showed a significant increase in the intensity and broadening of the C=O stretching bands, O=C–N bending band, N–H group, and asymmetric and symmetric stretching vibration bands of C–H. XPS analysis indicated that the oxygen atomic percentage increased profoundly from 16.3% to 28.5% in the untreated and plasma-treated samples. A small increase in nitrogen was also noted in the treated sample.

Polyester fiber, owing to its inherent hydrophobic characteristic, suffers from poor water wicking, adhesion, soil deposition, and antistatic charge properties.[52] It was found that the atmospheric pressure air plasma exposure of polyester could increase its wettability, oxygen concentration, roughness, and antistatic properties. A low-temperature DBD plasma treatment (power 30 W, frequency 50 Hz, and voltage 7.5 kV) of the polyester fabric produced at an atmospheric pressure using oxygen gas (3 L/min) has been investigated for dyeing and hydrophilic properties of the fabric.[53] The increase in the K/S values may be correlated with the increase in nanoscale roughness and wettability of the sample that might have increased the water swelling capability and affinity to disperse the dye. The FTIR spectrum of the treated sample depicted a new sharp absorption band at 3449.54 cm^{-1} for –C–OH groups and formation of other oxygen-containing polar groups, such as C=O, –C–OH and –COOH. Two disperse dyes, namely, C.I. Disperse Blue 19 and C.I. Disperse Brown 1, were used for dyeing at 100°C, 110°C, 120°C, and 130°C, and the effects of various gases such as air, argon, nitrogen, and oxygen in the K/S value have been reported. In the oxygen plasma–treated sample, the K/S value increased to 7.00 from 6.64 in the untreated sample, when the dyeing was conducted at 130°C. The oxygen plasma–treated samples showed a higher improvement in the K/S value than that found with the remaining gases. The different gases had a little effect in the color improvement. The improvement in dyeing was more profound, when the dyeing was carried out at 100°C. The percentage improvement in K/S value was more noticeable in the case of C.I. Disperse Brown 1 dye than the C.I. Disperse Blue19.

It is challenging to promote dye, ink, and coating adhesion on PP fiber due to the absence of any reactive chemical group and due to its high crystallinity.[31] The atmospheric pressure glow discharge in air–O$_2$–He plasma was used to promote the adhesion of water-based inks on the PP fabric. The initial surface tension level of 31 dyne/cm in the untreated sample increased to 52 dyne/cm upon plasma treatment, and there was a complete ink surface retention on the treated surface.[54] The PP fabrics were also activated by the DBD plasma in air and argon gases to make it dyeable with leuco and pigment forms of vat dyestuffs. The vat-dyed samples showed a significant increase in color strength, when the PP fabrics were plasma pretreated with either argon or air gases.[55] The atomic force microscope (AFM) study showed an increase in surface roughness value from RSM = 3.1 nm in the untreated sample to RSM = 24 nm in the plasma-treated sample for 5 s. The improved surface energy after plasma treatment in ambient air was found to be 38 mJ/m^2, which is required for the printing with solvent-based inks.[56] The surface modification of PP fabrics by acrylonitrile cold plasma to deposit polyacrylonitrile (PAN)-like layers led to the improvement in water absorption and dyeing.[57] This possibly happened due to the presence of nitrogen- and carbon-based unsaturated linkages and the formation of =C=O groups.

1.6 HYDROPHOBIC AND SUPERHYDROPHOBIC TEXTILE

Hydrophobic or water-repellent textiles are important in many applications to protect the textile as well as its user from unwanted wetting, staining, or chemical contamination in the presence of rainwater, food, beverages, chemical, and pesticides. The hydrophobic textile does not allow the liquid droplet to get absorbed by the fabric, enabling the liquid droplets to roll-off, leaving the underlying material unwetted. There are three important parameters that govern the hydrophobic finishing of the textile: (1) the surface energy of the textile, (2) surface energy of the hydrophobic coating, and (3) surface energy of the liquid. All the solid surfaces have a distinct surface energy, which is a function of surface area and the molecules that are present on the surface. Similarly, each and every liquid has a specific surface tension, which is a measure of the interaction energy between the molecules of the liquid. In the hydrophobic finishing of textiles, the surface energy of the textile material is reduced by coating or grafting with other materials of low surface energy. Hydrophobic or the water-repellent textile is defined by its surface energy or the contact angle value. The contact angle (θ) is the angle at which the liquid–vapor interface meets the solid–liquid interface. The contact angle is determined as the resultant of the adhesive and cohesive forces. The adhesive forces between a liquid and solid materials cause a liquid drop to spread over the surface. On the other hand, the cohesive forces within the liquid cause the droplet to ball up and avoid any contact with the solid surface. In hydrophobic finishing of textiles, the force of adhesion is reduced by attaching a material with low surface energy. As the tendency of a liquid drop to spread out over a flat solid surface increases, the contact angle gets reduced. Thus, the contact angle provides an inverse measurement of a hydrophobic (nonwettable) surface. If the contact angle is below 90° or the surface energy more than 71 dyne/cm (water), the surface is commonly considered as a hydrophilic surface. On the other hand, if the contact angle is between 90° and 140° and surface energy lower than 71 dyne/cm (preferably 15–50 dyne/cm), the surface is called hydrophobic. Nowadays, the concept of a superhydrophobic surface is also getting considerable research attention, so as to produce a surface that self-cleans and is dust-repellent. When the contact angle is beyond 150° with a very low rolling angle, the sample is called a superhydrophobic surface. A superhydrophobic surface shows almost no contact between the liquid droplet and a solid surface, and at that situation, the liquid droplet can easily roll-off from the surface. This phenomenon is also sometimes referred as *lotus effect* on textile.[58] To produce such a surface, a combination of two fundamental properties is required: (1) the surface roughness and (2) modification of solid surface with a low-surface-energy material. The selection of an appropriate method to create surface roughness and lowering the surface energy also depends on the mechanical and physicochemical properties of the material.[59] Hydrocarbon, fluorocarbons, and silicones compounds are mainly used for hydrophobic modification of textiles to increase the contact angle and decrease the surface energy value. Hydrocarbon and silicone compounds help in developing a water-repellent textile. However, the fluorocarbon compound is required to impart both oil and water repellency in textiles. Aluminum salt paraffin dispersions are positively charged due to the presence of a trivalent aluminum salt. This produces a counter polar charge on the fiber surface, which is significant for the adsorption of the product. After drying, the radicals form a so-called brush, perpendicular to the fiber surface that prevents the water drops from penetrating the fiber. Polysiloxanes form a fiber-encircling silicone film with methyl group perpendicular to the surface. The oxygen atoms are facing toward the fiber. The film formation and the direction of the methyl groups are responsible for the hydrophobic characteristic of the finish. Fluorocarbon polymers form a film, where the fluorocarbon radicals remained perpendicular to the fiber axis, thus preventing wetting of the fiber surface. The high hydrophobic and the oleophobic characteristic on the surface are mainly ascribable by the presence of extremely low interfacial tensions of the fluorocarbon chain. When a finish is applied, the surface of the substrates must be covered with molecules in such a way that their hydrophobic radicals are ideally parallel and facing outward.

Hydrophobic finishing of textiles is carried out by the conventional pad-dry-cure method or by spraying of hydrophobic chemicals on to the textile substrate. In the present chapter, the hydrophobic finishing of various textile substrates using hydrocarbon-, fluorocarbon-, and silicone-based precursors has been summarized by the in situ plasma reaction of such compounds or radical formation followed by graft polymerization of a vinyl-based monomer along with their mechanism of plasma reaction.

1.6.1 Hydrophobic Textile Using Fluorocarbon Precursor

Low-pressure SF_6 plasma with an RF power of 25–75 W on PET, silk, cotton, and cotton–silk blended woven textiles produced an abrupt increase in hydrophobicity.[60] Prior to plasma treatment, the water droplets on these fabrics were found to get readily absorbed, and after plasma treatment, the water contact angles were found to increase to 135°–145°. In a similar study, the SF_6 cold plasma was explored to impart a high degree of hydrophobicity in a *Bombyx mori* silk fabric.[61] The IR spectra of the untreated and plasma-treated samples did not show any change in the peak positions of amide I, II, and III bands, which were at 1627, 1513, and 1228 cm^{-1}, respectively. The small changes in the peak intensities in the frequency region of 1000–1050 cm^{-1} suggested the formation of –CF groups. After argon induced the graft polymerization of 5% phosphorous containing phosphate and phosphoramidate monomers followed by SF_6 plasma treatment, it was found that it could impart excellent flame retardancy as well as water-repellent functionalities in the silk fabric.[62] The improvement in the hydrophobic property was attributed to the attachment of –CF, –CF_2, and –CF_3 molecules to the silk polymer, and the plasma reactions were dependent on the pressure and power of the plasma reactor.

The CF_4 plasma produced a Teflon-like structure with a high degree of water repellency to the polyethylene terephthalate (PET) surface, where the contact angle was found to increase from 105° to 120°–155°. However, the O_2 plasma treatment could only increase the surface hydrophilicity of the sample.[63] Plasma treatment with O_2 in the presence of hexafluoroethane was carried out on cotton fabric to impart an effective barrier coating to the aqueous contamination.[64] A mixture of hexafluoroethane (C_2F_6) and hydrogen was used as a barrier coating against the hydrolysis of high-performance aramid fabrics (Nomex©). The deposit of the thin protective layer could maintain the integrity of the structure, when it was immersed in 85% H_2SO_4 solution for 20 h at room temperature, whereas conventional fluorocarbon-finished fibers showed a significant shrinkage and reduction in other properties.[27] Oleophobic and hydrophobic surface engineering was created using fluorine-based molecules.[65] The water-repellent and flame-retardant functionalities were found to improve with a little amount of chemical, without using any water and other auxiliaries. If a fluoroalkane, such as tetrafluoromethane or hexafluoroethane, is used as a process gas, then fluorine will be substituted for the abstracted hydrogen on the substrate surface, causing a reduction in the surface energy. The plasma treatment of a surface with fluorocarbon gas is a complex process, due to the involvement of simultaneous rival processes of deposition and etching.[66] The decomposition of fluorocarbon gases in the plasma first leads to the production of CF_x (1 < x < 3) fragments that can subsequently react with the substrate in the gas phase. The CF_x radicals are the building blocks of the grown fluoropolymer film and their concentration in the gas phase directly affects the polymerization rate.

Samanta et al. generated plasma in the presence of helium and TFE gases at an atmospheric pressure, at a discharge voltage and frequency of 6.0 kV and 16.5 kHz, respectively, to impart hydrophobic functionality in cellulosic textiles.[25,67] The current and voltage waveforms, which were completely free from spikes with fairly low voltage and moderate current of 36 mA, indicated the formation of a glow plasma. After 1 min of He-TFE plasma reaction, the sample became highly hydrophobic and the water absorbency time of the droplet (37 μL) increased to 36 min from <1 s in the untreated sample. In the 2–8 min plasma-treated samples, the water absorbency time was more than 60 min. The contact angle in the untreated sample was ~0°; however, in all plasma-treated samples, it was in the range of 125°–153°. The samples, even after 25 numbers of laundering, could retain most of the imparted hydrophobic functionality. The 8 min treated sample showed a water contact angle as high as 153° and water rolling angle of 5°, exhibiting a superhydrophobic characteristic. It was interesting to note that both the top and bottom surfaces of the fabric showed a similar degree of hydrophobicity in terms of water absorbency time and contact angle. This implies that the fragments of TFE could easily penetrate through the textile structure and modify the fibers present on both surfaces. The OES of He-TFE plasma exhibited eight atomic lines of He at 587, 655.8, 667.5, 706 (strongest), 726.5, 336, 356, and 501 nm and several other atomic lines of F and C (TFE fragments) at 388.5 nm for C and F, 430 nm for C and F, 450 nm for F, 482 nm for C and F, 518 nm for F, 559 nm for C and F, 566 nm for CHF, and 606 nm for C. When the He to TFE gas ratio was increased from 30 to 60 at the total lowest gas flow rate of 305–310 mL/min, the intensity of the aforementioned atomic lines was found to increase profoundly. In the 1 min plasma-treated (soap-washed) sample, the fluorine atomic percentage, as determined by XPS analysis, was as high as 29.0%, and it was 39.6% in the 8 min plasma-treated sample. There was no significant change in the F percentage after washing of samples with soap solution. In the 8 min treated sample, a large amount of fluorine was chemically attached by forming –CF (12%), –CF_2 (1.8%), and –CF_3 (5.5%) bonds, which are responsible for the developed hydrophobicity concurrent with the depletion of the –OH group of the cellulose. This helps in increasing the F^-/OH^- ratio (SIMS analysis) sharply from 0 in the untreated sample to 40.7 and 44.6 in the 1 and 8 min plasma-treated soap-washed samples, respectively. The TFE molecule in the plasma zone was first broken down into various fragments, such as CF_3, CFH_2, CF_2, CFH, and CH_2, those further combined in various proportions to form mainly two types of species. The first one with molecular masses of 73, 93, 184, and 209 amu had a long hydrocarbon backbone with fluorocarbon moieties attached to one end. The other type was principally represented by the molecular species of 151 amu, which was formed by a combination of TFE with a smaller fluorocarbon CF_2 moiety. Clearly, both kinds of molecules were suitable to impart a high degree of hydrophobic functionality. It is possible that fluorine-rich molecules, such as –CF_3, –CFH_2, and CF_3–CHF–CF_2– (151 amu), might have been attached to the cellulose, as the sample showed a high percentage of fluorine rather than carbon in XPS analysis.

A low-temperature APP was also generated in the presence of helium–fluorocarbon (He/FC) gaseous mixture to improve the hydrophobic functionality of cotton textiles.[10,68] After the plasma reaction with a commercial fluorocarbon gas (mixture of difluoromethane and pentafluoroethane), the hydrophilic cotton was found to turn into highly hydrophobic. The FTIR analysis of the treated sample showed the presence of different fluorocarbon species and CF and CF_2 symmetric and asymmetric stretching in between 1160 and 1368 cm^{-1}. After plasma treatment for 1 min, the water absorbency time increased to 2.5 min, from 3 s measured in the untreated sample. The 12 min of plasma reaction could impart a water absorbency of >60 min. The result indicates that the pentafluoroethane and difluoromethane get fragmented inside the plasma zone and reacted with the cotton cellulose. The SIMS negative ion mass spectra of the untreated sample showed the major mass peaks at 1 amu for H^-, 12 amu for C^-, 13 amu for CH^-, 16 amu for O^-, and 17 amu for OH^-, as shown in Figure 1.7, whereas in the helium–fluorocarbon (He–FC)-plasma-treated sample, there was a strong mass peak of F^- at 19 amu. The intensity of F^- peak was much more compared to the

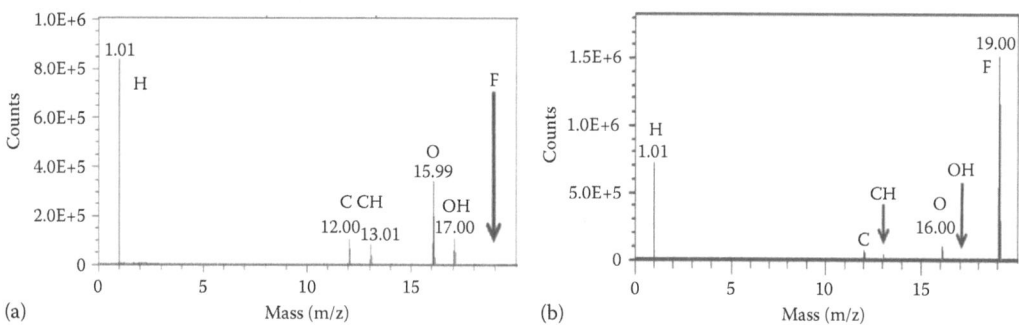

FIGURE 1.7 Negative ion mass spectra of (a) untreated and (b) plasma-treated cotton textiles. (From Samanta, K.K. et al., Plasma processing of textile at CIRCOT. Central Institute for Research on Cotton Technology, Mumbai, India, 2014, pp. 1–17.)

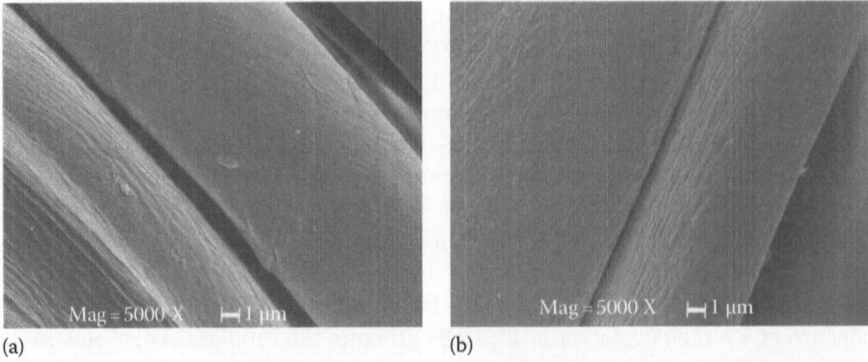

FIGURE 1.8 SEM micrograph of (a) untreated and (b) He–fluorocarbon-plasma-treated samples. (From Samanta, K.K. et al., Plasma processing of textile at CIRCOT. Central Institute for Research on Cotton Technology, Mumbai, India, 2014, pp. 1–17.)

other mass peaks. The appearance of F^- was possibly due to the reaction of $-CF$, $-CF_2$, and $-CF_3$ containing fluorocarbon fragments with cotton in the plasma zone. The presence of such fluorine-rich molecules on the plasma-treated cotton had helped the sample to be hydrophobic.

From the SIMS molecular image, it was seen that C^- (12 amu) and O^- (16 amu) were uniformly distributed over the entire surface area in the untreated sample. Similarly, in the treated sample, C^-, O^-, and F^- molecules were also uniformly distributed over the entire surface area. The plasma-treated sample was mostly bluish in color (blue color for F^-) due to the presence of a large amount of F^- molecule compared to the C^- and OH^- molecules. The scanning electron microscope (SEM) images showed that the 12 min helium–fluorocarbon (He–FC)-plasma-treated sample had a surface morphology similar to the untreated cotton, though the surface became highly hydrophobic (Figure 1.8). This implies that plasma modification has happened only on the fiber surface at nanometer level without blocking the intrafiber/interfiber spacing or features.

Vinogradov and Lunk studied the deposition of a fluorocarbon layer on Si wafers and on the technical textiles in the presence of the DBD plasma, using CF_4, C_2F_6, C_3HF_7, $C_2H_2F_4$, C_3F_8, and C_4F_8 fluorocarbon precursors.[69] The fluorocarbon polymer films composed of superhydrophobic polytetrafluoroethylene (PTFE) could be deposited on the fiber surface only when the ratio of F/C atoms of the starting fluorocarbon molecules was smaller than three (3). The deposition of the fluorocarbon layer was confirmed by FTIR and XPS analyses. The FTIR absorption spectra of the films on Si wafers showed an *amorphous* film structure with the main absorption peak of CF_2 groups at 1200 cm^{-1}. The variation in the F/C ratio of plasma polymerization strongly correlated to the appearance of the *amorphous* PTFE absorption peak at 740 cm^{-1}. The coated surface with a surface tension as low as 11 mN/m could be achieved depending upon the experimental conditions and the gas mixture. McCord et al. studied the effect of CF_4 and C_3F_6 plasmas on the surface properties of cotton fabric.[70] The hydrophobicity was reported to increase after treatment with both plasma gases. XPS analysis showed an increase in surface fluorine content to ~1%–2% and 2.3%–7.8% in CF_4- and C_3F_6-plasma-treated cotton fabrics, respectively. Based on XPS analysis, the possible mechanism of plasma reaction with cellulose has also been proposed.

Unlike the aforementioned plasma reaction with a smaller fluorocarbon gaseous molecule (C1–C4), the long-chain (C6, molecular weight 418.1 g/mol) fluorocarbon monomer, for example, 2-(perfluorohexyl)ethyl acrylate (PFHEA), was also used for plasma-induced graft polymerization on the cotton substrate to achieve a durable water-, alcohol-, and oil-repellent coating.[71] The monomer vapor deposition on the fabrics followed by a glow plasma exposure (equipment size 33 cm × 12.5 cm) was carried out to complete the graft polymerization of C6 fluorocarbon on the fabric surface. A higher water/alcohol repellency rating of 10.5 (which corresponds to 15:85, water/IPA ratio) and an oil repellency rating of 6.5 (in between n-decane and n-octane) were achieved at a high monomer flow rate of 2 mL/min for 30 s. Besides, the water contact angle of 158.8°, that is, the superhydrophobic surface, could also be achieved on the cotton fabric. As expected, the fluorine percentage increased from 0% in the untreated sample to 57.5% in the plasma-treated sample. Similarly, the corresponding carbon and oxygen percentage reduced from 62.51% and 37.49% to 32.26% and 9.94% in the untreated and plasma-treated samples, respectively. The surface morphology of the treated sample looked uniform and smoother than the untreated samples. The plasma-graft-polymerized samples, which showed 85% alcohol repellency, was subjected to home laundering as per the DuPont USA procedure. After five numbers of laundering, the conventionally pad-dry-cured sample (both sides modified) showed 65% alcohol repellency, whereas the plasma-graft-polymerized sample showed 15% repellency. Still a bigger fluorocarbon molecule of C8, 1,1,2,2, tetrahydroperfluorodecyl acrylate (molecular weight of 518.1 g/mol), was used for graft polymerization on PAN fabric in a low-pressure Ar plasma for 10 min in direct contact with the substrate surface. It has been reported that the process required much smaller amounts of fluorinated reactant to achieve both water and oil repellency.[72]

The deposition of a nanoparticulate hydrophobic fluorocarbon film on the cotton substrate by plasma reaction was studied, and the sample showed a superhydrophobic property after 30 s of plasma exposure.[73] A post-high-temperature treatment was also found to be effective to recover the imparted hydrophobic property in the plasma-treated and washed fabric. The contact angles of water and 1-bromonaphthalene on the plasma-treated cotton fabrics were 133° and 124°, respectively. The SEM micrograph of the treated cotton fabric showed a thin film packed with nanoparticles of 10–200 nm tightly wrapped over the fabric.

1.6.2 Hydrophobic Textile Using Hydrocarbon Precursor

Hydrophobic textile using plasma reaction of a hydrocarbon precursor is also an important area of research. Plasma polymerization with a gaseous hydrocarbon monomer, such as methane (CH_4), ethylene (C_2H_4), or acetylene (C_2H_2), can deposit the filmlike coating of cross-linked amorphous hydrocarbon layers, thus imparting hydrophobic properties to the textile.[74] The RF stable-glow plasma was generated over a 1×16 cm^2 area in a cylindrical electrode geometry, for hydrophobic finishing of various substrates including cotton at a speed of 5–10 cm/min. CH_4 plasma polymerization was found to deposit a very smooth hydrocarbon layer composed of CH_2 and CH_3 groups.[75] The inclusion of oxygen-containing species was less than 1%. On the flat substrates, the water contact angle was ~90°, whereas on the rough surface like cotton, the contact angle reached up to 150°. The plasma-modified surface appeared to be quite smoother, and the coating on the cotton fabric did not degrade even after a number of cold water washings (without any mechanical rubbing). The C(1s) peak was the only main peak detected in XPS characterization, wherein the high resolution of the C(1s) peak region showed a single peak at 285 eV binding energies with a small asymmetric tail in the higher binding energy side due to the presence of oxygenated species. The polarization modulation reflection–absorption IR spectroscopy (PM-RAIRS) spectra of the hydrophobic coating deposited on a gold film had showed well-defined C–H vibration peaks at 2958, 2930, 2871, 2840, 1460, 1378, 1708, and 1660 cm^{-1}. These peaks indicate that the atmospheric CH_4 plasma polymerization can deposit a hydrophobic coating composed mostly of CH_2 polymeric backbones with CH_3 side groups.

Similar to the aforementioned plasma reaction using smaller hydrocarbon molecules (C1–C2), such as methane, ethylene, or acetylene, a stable-glow cold plasma was generated in the mixture of gas BD monomer (molecular weight 54 g/mol) and helium (He) at an atmospheric pressure for the hydrophobic finishing of a plain woven cellulosic (viscose rayon) textile.[76] The plasma was generated at a discharge voltage and frequency of 3.96 kV and 21.8 kHz. Helium (He) gas, on ionization, exhibits eight major atomic lines at 388.6, 491.5, 501.2, 587.0, 655.8, 667.5, 706.0, and 726.5 nm wavelengths with a maximum emission intensity at 706 nm. When the gas ratio of He to BD was

TABLE 1.2
Water Absorbency Time and Contact Angle in Untreated and He–BD-Plasma-Treated Cellulosic Textiles

Different Samples (Gas Flow Rate in mL/min)	Water Absorbency Time	Contact Angle (°)
Untreated sample	<1 s	~0
He 2500; BD 500 Plasma time, 1.5 min	28.5 min	<90; immeasurable
He 2500; BD 500 Plasma time, 7 min	≫60 min	143
He 2500; BD 500 Plasma time, 12 min	≫60 min	142
He 500, BD 10 Plasma time, 12 min	≫60 min	138
He 500, BD 40 Plasma time, 12 min	≫60 min	143

Source: Samanta, K.K. et al., *Surf. Coat. Technol.*, 213, 65, 2012.

kept at 12.5 (He 500/BD 40 mL/min), the emission peak of He at 667.5 nm became more intense and the 655.8 nm line was not visible. When increasing the gas ratio to 50 (He 500/BD 10 mL/min), several additional peaks were observed; some of these are C lines at 430.5, 467, 517.7, and 559.7 nm. The result indicated that the fragmentation of BD can be controlled by altering the carrier to precursor gas ratio.

It was observed that after plasma reaction for 1.5 min, the hydrophilic cellulosic textile turned to a highly hydrophobic one, resulting in the improvement of water absorbency time from <1 s in the untreated sample to 28.5 min in the plasma-treated sample as shown in Table 1.2. The treatment time beyond 7 min could impart a water absorbency time >60 min. It was interesting to note that in spite of the reduction in the butadiene (BD) gas flow rate from 500 to 10–40 mL/min and He from 2500 to 500 mL/min, the water absorbency time still remained at >60 min in the 12 min plasma-treated samples. The water contact angle in all the treated samples was as high as 138°–143° compared to ~0° in the untreated sample. The water absorbency time reduced to 250 s, when the sample was washed with soap solution. Gas chromatography mass spectroscopy (GC-MS) analysis revealed the formation of dimer of 1,3 butadiene (110 and 108 amu) and species with seven carbon atoms (96 amu). In the plasma-treated sample, the $-C-C-/-CH_x$ bond percentage increased to 91.2% from 72.7% in the untreated sample, mainly due to the attachment of $-CH$, $-CH_2$, and $-CH_3$ species of butadiene fragments. In the same sample, the $-COH$ bond percentage decreased by 18.1%. The increase in $-C-C-/-CH_x$ bond (hydrophobic species) with a similar depletion in $-C-OH$ bond (hydrophilic species) was responsible for the developed hydrophobicity in the otherwise hydrophilic cellulosic textile. The SIMS showed a decrease in the ratio of hydrophilic to hydrophobic component from 1.87 in the untreated sample to 0.64 in the 12 min plasma-treated and washed samples. Based on the GC-MS, XPS, SIMS, and OES results, it has been proposed that the plasma reaction might have happened because of the breaking off the $-CH-OH$ and $-CH_2-OH$ bonds of cellulose and the simultaneous reaction with various fragments of BD at that location (Figure 1.9). It was possible to control the plasma generation as well as reaction with the substrate by altering the gas flow ratio and the total gas flow. The SEM micrographs did not show any blockage of interfiber or intrafiber spacing or features in the treated sample, and the individual fibers in the yarn structure were also clearly visible. This possibly happened due to the nanoscale surface modification of the fibers in the plasma zone.

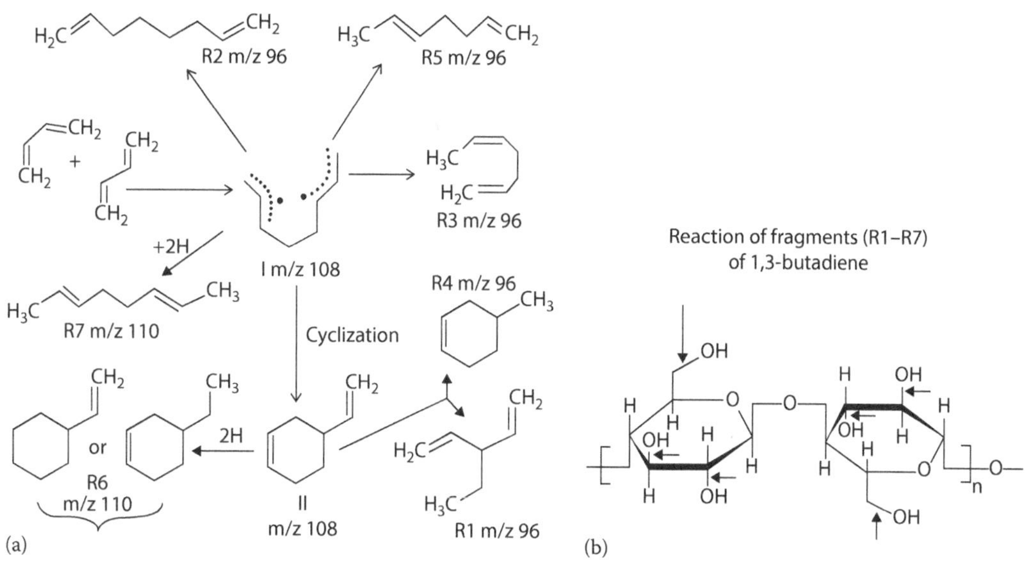

FIGURE 1.9 (a) Fragmentation of BD and (b) plasma reaction with cellulose textile. (From Samanta, K.K. et al., *Surf. Coat. Technol.*, 213, 65, 2012.)

Parida et al. investigated the effect of still bigger molecules, that is, styrene, cyclic C8 hydrocarbon (molecular weight 104.1 g/mol) for the hydrophobic finishing of cotton textiles.[77] The fabric was treated with a styrene–helium atmospheric pressure glow plasma at a discharge voltage of 1.55–6.0 kV and frequency of 17–21 kHz. The helium (He) gas flow rate was kept at 350 mL/min and mixed with the styrene vapor flowed at a rate of 0.026 g/min. After plasma treatment, the water drop disappearance time and the water contact angle were found to improve to 60 min and 133°, respectively, from a few seconds (4–7 s) and ~0° in the untreated sample. The water absorption time was found to increase quickly with the plasma discharge voltage up to 1.73 kV, and later on, it decreased with increasing discharge voltage till 6 kV at all the frequencies level (i.e., 17.9, 18.6, 18.9, 19.8, 20.2, 20.6, and 21.0 kHz). After washing, equivalent to five laundry wash cycles, the water drop disappearance time in the treated samples was found to reduce from 60 min in the as-prepared sample to below 18 min in the washed sample (approximately >50% reduction); the contact angle reduced from 107°–120° to 91°–110°. The reduction in the water drop disappearance time might be due to the removal of physically deposited polymerized/condensed styrene oligomers from the fiber surface during the washing treatment. The residual hydrophobicity after the washing is an indication of plasma reaction (i.e., covalent bonding) of styrene or its fragments with the cotton textile. Raman spectra revealed the appearance of peaks corresponding to the monosubstituted aromatic structure, which is different from poly(styrene). The reduction in –CH_2OH peak was the confirmation of reaction of styrene or its fragments in place of the primary alcohol of the cotton cellulose. OES analysis in real time showed various fragments of styrene, such as CH, CH_2, C_3H_3, C_6H_5, and H_2. It was found that a low-voltage and frequency levels were favorable for radicalization of styrene and/or formations of the aromatic radicals, which were suitable for imparting effective hydrophobic functionality in textile. In contrast, the higher voltage and frequency lead to intense fragmentation of styrene/benzene into low-molecular-weight species that are not much promising for the development of hydrophobic functionality. The GC-MS analysis of styrene–He plasma exhaust revealed the presence of styrene and other saturated versions of styrene. Interestingly, neither benzene radical nor its other fragments could be trapped in *n*-hexane for resulting hydrogenation and polymerization. Based on the Raman spectra, OES, and GC-MS

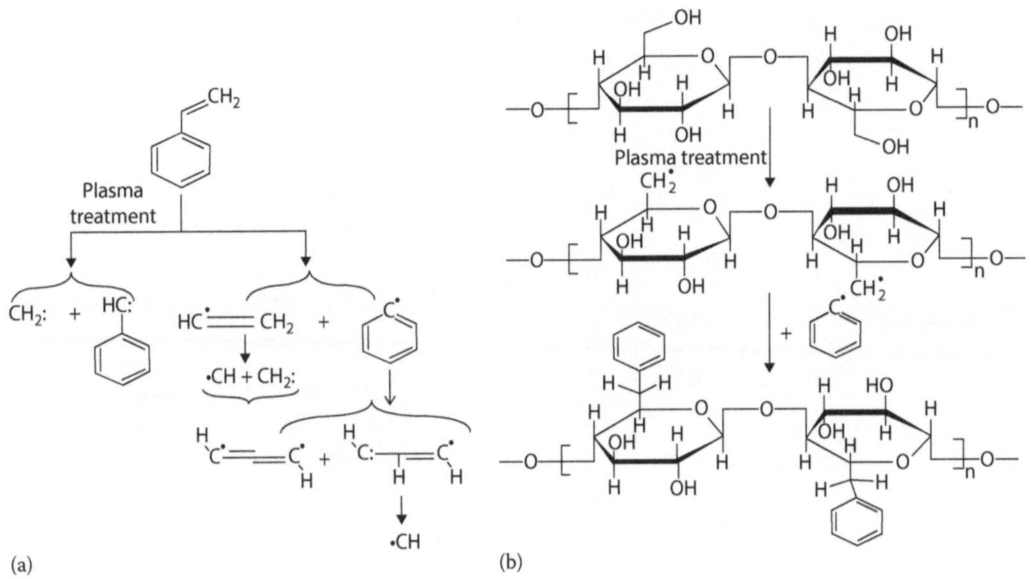

FIGURE 1.10 (a) Fragmentation of styrene followed by (b) plasma reaction with cellulose substrates. (From Parida, D. et al., *Plasma Chem. Plasma Proc.*, 32, 1259, 2012.)

analysis, the mechanism of plasma reaction of styrene with cellulose has been proposed as shown in Figure 1.10.

Further bigger C11 hydrocarbon molecule, such as vinyl laurate (molecular weight 226.3 g/mol), was also used for hydrophobic finishing of cotton textiles. The exposure of cotton fabric in a low-pressure plasma (2.45 GHz) in the presence of argon/nitrogen/oxygen and monomer vinyl laurate initiated a graft copolymerization reaction.[78] The grafted cotton fabric showed a water-repellent property that was durable even to a repeated laundering.

In the similar line, recently, the plasma reaction of still a longer hydrocarbon molecule (C12), dodecyl acrylate (DA) (molecular weight 240.3 g/mol), was attempted for hydrophobic finishing of regenerated cellulose (viscose fabric).[79] The scoured fabric was pretreated with 0.01–0.25 M solution of DA in ethanol by the dip-pad-dry method, followed by a helium plasma treatment at an atmospheric pressure for 0.5–2.5 min. The effects of plasma power density, frequency, gas flow rate, concentration of precursor, and plasma treatment time were also investigated. The plasma-treated samples were thoroughly washed with hexane (solvent for precursor and polymer) and acetone to remove the unreacted and loosely deposited materials before any physical and chemical characterization. The OES spectra of the liquid precursor showed various molecular peaks at 314, 374, 380, and 391 nm wavelengths corresponding to CH, CH$^+$, and CO fragments of the precursor, the new peak at 426.7 nm for CO$^+$/H$_2$/O fragments and at 470 nm to C/H/O fragments of the precursor. When the fabric was plasma treated at different discharge voltages in the range of 1–3 kV only to alter the plasma power density, the water absorbency was found to decrease. It indicates that the higher plasma power is playing a negative role in the hydrophobic finishing due to the two competing effects in the plasma zone. The first one is the fragmentation and activation of DA into the desirable entities that can lead to a desirable hydrophobic reaction with cellulosic substrate. The second one is the intense fragmentation of DA into undesirable smaller entities that are not desirable for suitable hydrophobic finishing of textile incurrence with plasma etching effect. In the treated sample, the water absorbency time was more than 1 h and the contact angle of 143° and it did not change much after the aqueous and solvent washing. In the ATR-FTIR spectra, the new peak at 1731 cm^{-1} for C=O stretching of ester group confirmed the reaction of DA or its fragments with cellulose. In the negative secondary ion mass spectra (SIMS) of the untreated sample, the OH⁻ peak was higher compared to O⁻, likely due to the presence of primary and the secondary hydroxyl groups of cellulose. This ratio, on plasma treatment, changed a little in favor of O⁻ presumably due to the attachment of DA that contains ester groups.

1.6.3 Hydrophobic Textile Using Silicone Precursors

Montarsolo et al. reported that a low-pressure (50 Pa) oxygen plasma pretreatment for 60 s could improve the washing fastness and the abrasion resistance of TEOS and hexadecyltrimethoxysilane-based hydrophobic finishing of cotton textiles.[80] The treated fabric showed a reduction of C–C bond percentage and increase in –C–O bond percentage. Low-pressure plasma polymerizations of HMDS and HMDSO under different power and time were carried out on cotton and polyamide fabrics.[81] The water repellency of the polyamide fabric was found to enhance strongly and the treatments were found to slow down the vertical flame spread in the cotton fabrics for a short plasma treatment time using little amount of chemical. Hexamethyldisiloxane (HMDSO)-derived plasma polymers were used for the hydrophobic finishing of cotton textiles leading to a smooth surface that showed a water contact angle of 130° without changing its water vapor transmission rate.[82] The air permeability of the cotton fabric was reduced after plasma polymerization owing to the deposition of the plasma polymer. The HMDSO plasma polymer had a poor durability to washing. Two concentrations of precursor, namely, 0.95 and 1.25 mL/min, were used and the plasma treatment time was varied from 80 to 320 s at a step of 80 s. It was observed that after plasma deposition, the surface energy of the cotton fabric reduced from 72 dyne/cm in the untreated sample to 43.4 dyne/cm in the treated sample. Cotton being highly hydrophilic in nature showed a water contact angle of 0°, which was found to increase to 130°–140° with the different plasma treatment times and concentrations of the precursor. The plasma treatment time had a profound effect on the contact angle, rather than on the surface energy. The FTIR spectra of all the HMDSO plasma–polymerized samples exhibited the characteristic absorbance peaks at 1256, 840, and 795 cm^{-1} attributed to the band of the CH$_3$ group attached to Si, due to the symmetric bending in Si–CH$_3$, to the Si–C rocking vibrations in the SiCH$_3$ groups, and to the SiCH$_3$ rocking vibrations, respectively. Plasma polymerization of the same monomer was used to lower down the water uptake, increase the contact angle, and create an inorganic surface on the PP. After 800 s of plasma polymerization, the water uptake was found to reduce from 3 to ~0.4 mg/cm². The sample showed a better hydrophobic characteristic with the increase in the plasma treatment time. The structure of the deposited layers exhibited a much higher content of Si–O–Si linkage in comparison to HMDSO. The longer treatment time exhibited a sharp decrease in the FTIR absorption intensities corresponding to CH$_2$ and CH$_3$ vibration. At the same time, it amplified the Si–O–Si and Si–C absorption intensities. The plasma-induced demethylation and dehydrogenation mechanisms were responsible for the formation of insoluble and cross-linked structures.[83]

Hydrophobic films with the desired properties could be designed depending on the input plasma energy and the gas ratio of O$_2$ to HMDSO. The plasma-polymerized SiO$_x$ coating showed an excellent adhesion on the flexible substrates due to the interface formation during the first stage of film growth.[84] The gradient layers can also be deposited in a continuous manner by flowing the gases parallel to the transport direction of the textile and not toward it, which is normally used to obtain a homogeneous deposition. The plasma-polymerized HMDSO coating on the PP fabric was applied to lower down the water uptake and to increase the contact angles. The HMDSO plasma polymer deposited on the polyester textile was used as a selective membrane to separate water from the hydrophobic organic compounds.[85] On the polyester (PET) fabric treated with RF plasma in a mixture of argon gas and gas-phase HMDSO at atmospheric pressure, the water repellency rating was found to improve greatly. The FTIR spectra of (a) HMDSO-coated polyester fiber, (b) plasma-coated fibers of HMDSO for 15 times, and (c) plasma-coated fibers of HMDSO for 20 times showed an increase of Si–O–Si, Si–(CH$_3$)$_2$, and Si–C bonds compared to the untreated PET fiber. As discussed earlier, the content of silicon compounds increased with the

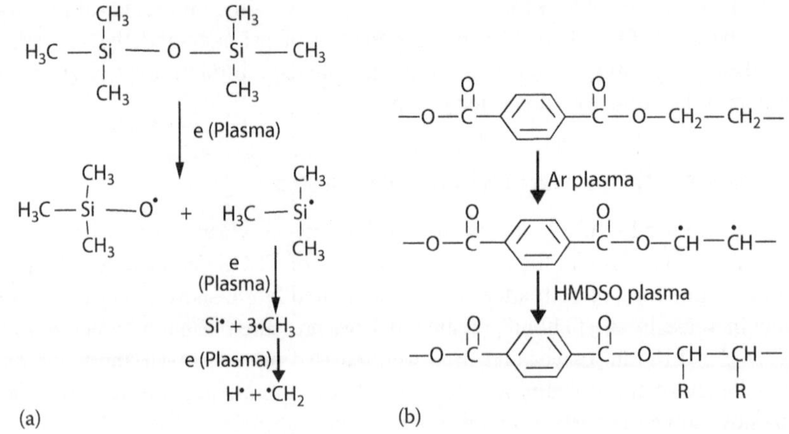

FIGURE 1.11 (a) Possible fragmentation of HMDSO and (b) reaction with polyester in the presence of plasma. (From Ji, Y.-Y. et al., *Surf. Coat. Technol.*, 202(22–23), 5663, 2008.)

plasma treatment time. However, a plasma polymerization time of more than 30 min covered the fiber surface with a white powder.[86] Like in the plasma reaction of fluorocarbon and hydrocarbon precursors with textiles as indicated earlier, the fragmentation of HMDSO in the plasma zone followed by its reaction with polyester textile has also been proposed based on the FTIR analysis as shown in Figure 1.11.

The Si–O (8.3 eV), Si–C (4.5 eV), and C–H (3.5 eV) bonds were broken by high- and low-energy electron collisions. The C–H radical groups were generated on the polyester (PET) fiber surface by the simultaneous interaction of Ar plasma due to the chain scission of C–H (3.5 eV), C–O (3.7 eV), and C=O (7.6 eV) as shown in the same figure. Consequently, the fragmented species of HMDSO reacted with the radical center of the fiber and improved the hydrophobicity of polyester further. The water repellency rating (AATCC 22) in the untreated polyester fiber was 0, and it improved to 90 in the plasma-treated sample. The silicon-compound-coated PET fiber was found to exhibit the water-repellent superhydrophobic characteristic.

The effects of argon (Ar) and argon–air gases on plasma-polymerized HMDSO film on PET were investigated in atmospheric pressure DBD plasma.[87] In the presence of Ar gas, a polymeric film with a structure close to $[(CH_3)_2–Si–O](n)$ was achieved; however, in the argon–air mixture, a silica-like film containing only a few carbon atoms was formed. In the latter case, due to the deposition of the silica-like film, the carbon percentage was much lower (only 3.6%) compared to 34.2% in the Ar plasma–treated sample. The opposite phenomenon was observed for oxygen also, where it was 35.4% and 61.5% in the Ar and Ar–air mixture, respectively. The plasma deposit in Ar resulted in high proportion of $(CH_3)_2SiO_2$ units, quite similar to the linear PDMS polymer, but was only slightly oxidized. In contrast, the polymer film deposited in the argon–air mixtures was highly oxidized, since they have a very high proportion of SiO_4 units and a small amount of CH_3SiO_3 units. The average contact angle on the Ar plasma–enhanced coating was 107° (i.e., hydrophobic), while it was only 24° (hydrophilic) in the argon–air mixture. The deposited coating was quite homogeneous and smooth in the Ar–HMDSO plasma, which was a glow type in nature. In contrast, the microdischarge/filamentary plasma observed in the Ar–air mixture led to the development of a very rough and inhomogeneous deposition of a SiO_2-like coating.

1.6.4 Hydrophobic Finishing of Dyed Textiles and Garments

The effect of plasma treatment on the dyed cotton fabric was carried out to study the feasibility of hydrophobic finishing of dyed textiles, more specifically for its future use in apparels. The cotton

TABLE 1.3
Color Parameters in the Dyed Hydrophilic and Hydrophobic Cotton Textile

Color Parameters	Dyed Untreated (Hydrophilic) Cotton Fabric	Dyed Plasma-Treated (Hydrophobic) Cotton Fabric
L^*	41.6	41.5
a^*	56.5	56.8
b^*	−2.2	−4.3
C^*	56.5	57.0
H^*	357	355
K/S	13.5	14.7

Source: From Samanta, K.K. et al., Plasma processing of textile at CIRCOT. Central Institute for Research on Cotton Technology, Mumbai, India, 2014, pp. 1–17.

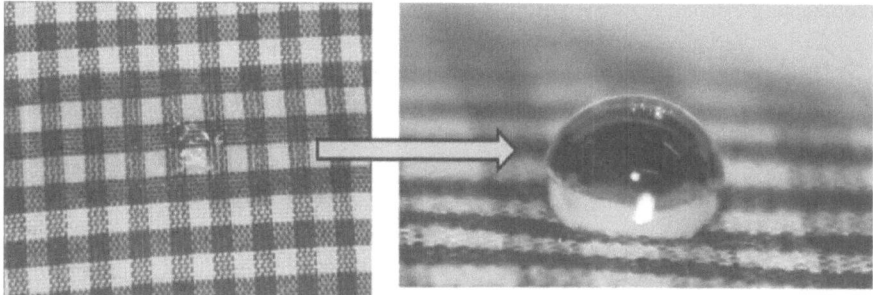

FIGURE 1.12 Hydrophobic finishing of polyester/cotton garment using plasma. (From Samanta, K.K. et al., Plasma processing of textile at CIRCOT. Central Institute for Research on Cotton Technology, Mumbai, India, 2014, pp. 1–17.)

textile was dyed with a reactive dye and then treated with He–fluorocarbon plasma (a mixture of pentafluoroethane and difluoromethane) for different periods of treatment. It was observed that after the plasma reaction, the dyed hydrophilic cotton was converted into a hydrophobic cotton. The water absorbency time increased to 2 min after the plasma reaction for 1 min from 5 s in the untreated sample. With further increase in the plasma treatment time, the water absorbency time was also found to increase linearly. In the 9 and the 12 min plasma-treated samples, the water absorbency time was 30 min and >60 min, respectively. It was interesting to note that the plasma-treated sample appeared to be darker (more K/S value) and there was no significant change in various color parameters as shown in Table 1.3.

Similar to the aforementioned observation, helium–fluorocarbon (He–FC) plasma was also applied in the dyed polyester/cotton blended garment for hydrophobic finishing. It was observed that the blended textile became hydrophobic after the plasma reaction as shown in Figure 1.12. As a result, the water absorbency time was found to increase from 150 s to >3600 s in the untreated and plasma-treated samples, respectively.

1.7 OTHER APPLICATIONS OF PLASMA IN TEXTILE

1.7.1 Desizing

The majority of the natural fibers are not much hydrophilic prior to any chemical treatment due to the presence of oil, fat, wax, and other contaminants on the fiber surface. The different plasma–surface interactions, such as surface activation, surface oxidation, etching, and plasma reaction on the cotton textiles, were utilized to improve desizing, scouring, and bleaching. Polyvinyl alcohol (PVA) is primarily used as a sizing material for synthetic yarns and as a secondary sizing agent to starch for cotton yarns.[54] The complete removal of the PVA size is difficult in conventional processes being an energy- and water-intensive process. The atmospheric plasma has been found to greatly increase the solubility of PVA in cold water without influencing the bulk properties of the cotton fabric.[88] The desizing efficiency of 99% was obtained with both in air–He and air–O_2–He plasma treatment followed by hot and cold washes. The SEM images revealed that both plasma-treated sample surfaces were nearly as clean as the unsized fabric, indicating complete removal of the PVA size. Plasma attacks the PVA molecules and makes them easy water-soluble products. Plasma was also used as a pretreatment prior to the desizing of starch on the cotton fabric.[15] It was found that the rate of enzymatic desizing of a cotton fabric pretreated with air plasma for 2 s in a commercial plasma reactor was higher than that of the control sample in the initial 15 min period. After 20 min of plasma treatment, the desizing loss was found to be 2% higher than the control sample. It has been mentioned that O_2 and N_2 present in the air plasma interact with starch and cellulose, forming new groups like –C=O, –OH, and C–N. This helped in the subsequent removal of starch from the fabric. In a similar study, the desizing efficiency of the indigo-dyed denim cotton fabric was found to increase by 5%, when the fabric was treated with atmospheric pressure O_2 plasma jet (O_2 1%) operated at a speed of 5 mm/s.[89] The formation of cracks and grooves in the plasma-treated fiber along the fiber direction, particularly in the wrinkled portion of it, made the fiber surface rougher, resulting in better desizing. The same research group recently reported that the He–O_2 plasma treatment was found to be effective for the removal of starch-based sizes from the cotton fabric without using any wet treatment.[90] The removal of starch was confirmed by the ATR-FTIR analysis, where the intensity of peak in the regions of 3400–3200 cm^{-1} pertaining to –OH groups of starch get reduced.

1.7.2 Scouring and Improvement in Water Absorbency

Sun and Stylios reported that the scouring time can be reduced by 50% for cotton textiles pretreated with the RF O_2 plasma.[30] It has been reported that the percentage wax removal was nearly 84%, after 1 min of plasma treatment followed by 25 min of scouring process. On the other hand, 40 min scouring time was required in the untreated sample to achieve a similar degree of scouring. Naturally colored cotton is getting attention due to environmental interest, as no bleaching and dyeing process is required except a pretreatment to make the fibers hydrophilic. Atmospheric pressure Ar or air plasma treatment was capable of modifying the surface of the naturally colored cotton fabrics without any loss of any important properties, such as color strength and fastness.[91] To increase the hydrophilic features and decrease the chemical waste from the existing pretreatment, the cotton fabric was treated with corona discharge in air.[65] It was effective to increase the hydrophilic property of the cotton fabric without affecting the integrity of the fiber or yarn. The plasma could incorporate both the chemical and physical changes in the waxy cuticle layer of the cotton fiber without damaging the cellulose backbone. Low-temperature plasma treatment of gray, mercerized cotton, and polyester/cotton blended fabrics before dyeing has been reported to be effective for improvement in dyeing. Both the air and the O_2 plasma treatments had shown that the scouring process could be eliminated.[92] The polyester/cotton (P/C) blended woven textile was treated with DBD plasma to improve the hydrophilic properties using He/O_2 gases.[93] It was found that the plasma process parameters played a critical role in deciding the efficiency of the treatment, and at optimum conditions, the hydrophilicity in terms of vertical wicking was found to be higher than the untreated sample.

The treatment of polyester fibers by the glow discharge in air or oxygen causes a partial degradation of the fiber surface, with an increase in the capillary sorption of iodine or cations in the aqueous solution. Polyester has a hydrophobic surface, as it is made up of ether oxygen (C–O–C) linkages, while the hydrophilic ester oxygen (C=O) linkages are facing toward the core of the fiber.

When it was plasma treated, either the ester oxygen (C=O) came closer to the surface, as a result of etching, or few more C=O groups were formed by the plasma reaction with oxygen making the sample hydrophilic.[1] The glow-discharge He plasma was used to treat PET and nylon 6 fabrics in a continuous manner to improve the surface energy and the water absorbency without any detrimental effect on the mechanical properties of the fabric.[5] It was observed that the hydrophilicity of nylon and polyester fabrics could be improved significantly after 60 s of He plasma treatment. This leads to decrease in the water absorbency time from 540 to 1.1 s and from 700 to 6.7 s in the untreated to treated nylon and the polyester fabrics, respectively.[94] The APP treatment with He–Ar or acetone–argon on wool and PET fabric and on the film produced a wettable surface, whose wettability increased significantly with increasing plasma treatment time. The He–Ar plasma treatment was found to be more effective than by acetone–argon treatment.[95]

The atmospheric air plasma treatment of the PET fiber and polyamide 66 fabrics decreased the wetting time, because of the creation of roughness in the etched surface that caused the wettability. The soil release behavior of the treated fabrics was also improved.[96] The DBD plasma in air at ambient temperature on the polyester fabric produced distinct variations on the textile surface with an increase in hydrophilicity of the fabric.[97] The weight of the absorbed water by the capillary action increased from 12 to 200 mg after application of 300–1000 W dosages of atmospheric pressure air plasma on PET woven fabric. The contact angle of water on the plasma-treated sample decreased from 80° to 40°, indicating an increase in the surface energy of the PET fiber due to the chemical changes.[98] The APP treatment of PP nonwoven in nitrogen and dry air resulted in surface activation and permanent hydrophilic finishing of lightweight nonwoven fabric.[99]

1.7.3 FLAME-RETARDANT FINISHING

Flame-retardant finishing is important as it directly relates to human health and hazards. Cellulosic, lignocellulosic, and protein fibers, such as cotton, flax, ramie, jute, silk, and wool, are mostly used for apparel application. Thermoplastic synthetic fibers, such as polyester, nylon, acrylic, and even to some extent PP, are also used for similar applications. These fibers catch flame readily and burn with a high flame (temperature ≈ 450°C) that makes it difficult to extinguish. The situation is more complex for the thermoplastic fibers, as these fibers during burning also shrink and melt. Single-step multifunctional finishing of cotton textiles to impart water repellency as well as flame retardancy has been reported using argon (Ar) plasma–induced graft copolymerization of acrylate phosphate and phosphonate derivatives in combination with carbon tetrafluoride (CF_4) gas.[100] CF_4 was required to impart a water-repellent property. The fire-retardant chemicals had no adverse effect on the water repellence behavior. The same research group from the national textile center has also developed monomers like cyclotriphosphazene and hexachlorocyclotriphosphazene for deposition on the plasma-preactivated cotton surface followed by polymerization by passing the fabric through the plasma zone to ensure a reaction with the fabric. The different metal and/or metal oxide mixtures, such as aluminum oxide, nanosilver, and titanium dioxide, were applied in the argon plasma–preactivated polyester fabric and the treated fabric showed an improvement in flame retardancy.[101] Cotton fibers were coated with silicone dioxide (SiO_2) layers using an APP. The SiO_2 network armor was obtained through the hydrolysis and condensation of precursor TEOS, and it successively cross-linked on the cotton fiber surface. Due to the protective effects of SiO_2 network armor, the modified cellulosic fibers exhibited enhanced thermal properties and improved flame retardancy. Low-pressure cold plasma was used to deposit a thin film on polyamide 6 (PA6) substrates for the fire-retardant finishing by plasma polymerization of the 1.1.3.3-tetramethyl disiloxane (TMDS) monomer and oxygen gas mixture.[102] The presence of oxygen during the polymerization process of TMDS promoted the formation of more thermally stable coatings. The deposits showed the efficient fire-retardant properties. The coatings formed a thermal barrier that decreased the rate of heat release (RHR) in the coated PA6 by 28%, and the ignition time was found to increase from 105 s in the untreated sample to 219–315 s in the different treated samples. Atmospheric plasma was also used to react predeposited sodium silicate layers on viscose and cotton fabrics.[103] The modified cellulose fabrics were tested for improved flame-retardant properties in a 45° angle flammability tester. The silica network remained attached to the substrate even after the intense ultrasound washes. Lam et al. reported that the combination of plasma pretreatment using a mixture of He (30 L/min) and O_2 (0.2 L/min) gases along with zinc oxide was useful for improving the phosphorus-based flame-retardant finishing of cotton textiles.[104]

1.7.4 WRINKLE-RESISTANT FINISH

The wrinkle resistance treatment of cotton textile is very important as it has inherent high wrinkling tendency. An APPJ (0.1 L/min O_2 flow, 2 mm downstream, 2 mm/s speed) was used for the surface modification of the cotton fiber cross-linked with the BTCA with the TiO_2 catalyst to enhance the wrinkle recovery angle.[105] On plasma treatment with oxygen and/or polycarboxylic acid, the wrinkle-resistant finishing agent on cotton fabric enhanced the effect of postfinishing processes.[104] Chen et al. reported that argon plasma (50 mL/min) treatment on the cotton fabric could increase the bonded cross-linking in terms of nitrogen content, when a mixture of (1:1) DMDHEU and acrylic acid (AA) were used.[106] The process was pad (cross-linking chemical)-dry-plasma-cure. Dry crease recovery angle, wet crease recovery angle (WCRA), and tensile strength retention (TSR) in the pad-dry-plasma-cure-finished fabrics were found to be higher than the pad-dry-cure-finished fabrics. With Ar plasma treatment, more cross-linking agent DMDHEU-AA was deposited on the finished fabrics and reacted with cellulose molecules to improve the crease recovery angle. Indeed, it was found that the number of cross-links per anhydroglucose unit (CL/AGU) and the length of the cross-links of pad-dry-plasma-cure-finished fabrics were higher compared to the traditional pad-dry-cure-finished fabrics at the same resin concentration.[106] The plasma process could help in reducing the formaldehyde release from the finished fabrics. The type of gas used, gas flow rate, treatment time, applied power, jet-to-substrate distance, and pore size or fabric structures are also crucial parameters in determining the efficacy of wrinkle-resistant finish.[107] Titanium dioxide coating together with plasma pretreatment improved the wrinkle recovery property of cotton fabrics. However, the treated fabric had a negative effect on the fabric handle.

1.8 SUSTAINABLE TEXTILE PROCESSING USING PLASMA

The textile industry consumes approximately 150 L of water as a processing medium for its operation starting from desizing to final finishing and also generates almost an equal amount of liquid effluents. The discharge of such processed water as an effluent contaminated with residual dyes, pigments, acid, alkali, salt, starch, and other various chemicals causes environmental pollution. The cost of the final products also increases due to the multiple-step operations and drying of the wet textiles by water evaporation using steam or electrical energy. In spite of several technological developments, still a large number of textile chemicals remain non-environment-friendly along with water-, energy-, chemical-, and time-intensive processes. This makes the traditional wet chemical processing nonsustainable in terms of chemical, cost, water, energy, carbon footprint, etc. Of late, due to the increased environmental awareness and stringent effluent norms, the textile industry is now moving toward the implementation of environmental and sustainable material and processing technologies, such as digital printing; spray and foam finishing; UV excimer irradiation; processing with supercritical fluids; natural dyes; enzymes for desizing and biopolishing; aromatic and medicinal plants for antimicrobial, UV-protective, and well-being textiles; plant molecules such as BPS and SJ for flame-retardant finishing; and biomolecules like DNA from herring sperm, salmon fish, whey proteins, and casein for flame-retardant finishing.[11] In this context, low-temperature plasma seems to be a promising environmentally friendly sustainable technology for the wet chemical processing of textiles. The plasma processing of textiles only modifies the surface of the material without altering the bulk properties to increase the uptake of liquid, dye, chemical, and auxiliary.

Water-Free Plasma Processing and Finishing of Apparel Textiles

TABLE 1.4

Comparison of Textile Chemical Processing by Conventional and Sustainable Plasma Methods

Different Process Parameters	Conventional Wet Chemical Processing of Textiles	Sustainable Plasma Processing of Textiles
Processing medium	Water based.	Ionized gas (no water involved).
Water requirement	Approximately 150 L/kg for cotton processing starting from desizing to finishing.	Zero for a particular process, such as hydrophobic or hydrophilic finishing.
Dyeing time, temperature, and consumption of dye	Majority of the textile dyeing is carried out at boiling temperature for 45–60 min.	Dyeing temperature or time can be reduced. For example, plasma-treated wool and silk can be dyed at <45°C. Similarly, the dyeing time can be reduced up to 30 min to achieve similar dye exhaustion. Plasma-treated sample shows more dye exhaustion resulting in less effluent load.
Dye bath auxiliaries	There is a need of acid, alkali, salt, etc., resulting in more effluent load.	Usage of acid, alkali, or salt may be reduced or avoided resulting in less effluent load.
Time required for a particular textile processing/finishing	Approximately 10–20 min required for hydrophobic finishing due to multistep operations, such as padding, drying, and curing.	Due to the single-step process, only 1–10 min is required for similar hydrophobic finishing.
Energy and chemical consumption	High due to the involvement of drying/curing of wet textile by evaporation and bulk modification.	Less, as only the surface of the fiber is modified, and water-free dry process.
Cost of production	High due to the requirement of more time for water evaporation and more chemicals and energy consumption.	Low due to the water-free, rapid, single-step processing.
Sustainability in terms of water, energy, chemicals, cost, and pollution	Less, as the process uses 150 L of water, time-consuming and chemical-intensive process along with environmental pollution.	More, as dry processing, low chemical consumption, nanoscale surface modification, no liquid effluent, and less time required for a particular process.

Source: Samanta, K.K. et al., Book chapter on environment-friendly textile processing using plasma and UV treatment, in: Muthu, S.S. (ed.), *Roadmap to Sustainable Textiles and Clothing*, Springer Publication, Singapore, 2014, pp. 161–201.

It also enables the process to accomplish single to multifunctional textiles and multiple-step operation in a single step without usage of water or simultaneous dyeing and finishing operation. Some of the advantages of plasma processing over traditional wet processing of textile in terms of sustainability are reported in Table 1.4. Although the plasma processing of textile has several advantages, still it has the drawback that it cannot replace all pretreatment processes; in addition to that, it requires capital investment and skilled manpower. Also the speed of continuous plasma treatment is quite low. Therefore, its processing speed needs to be enhanced for online integration with the man-made fiber spinning (melt, dry, or solution) or with the other textile chemical processing machineries for its successful commercial application.[108]

1.9 SUMMARY

Wet chemical processing of textiles is important for improving its aesthetic and functional values. However, the process requires around 150 L of water as a processing medium, in addition to the large quantity of textile chemicals, auxiliaries, dyes, and pigments. The discharge of the major proportion of such water contaminated with residual dyes, pigments, and other chemicals causes water pollution. In spite of the several technological advancements in the textile sector in the last few decades, such as low material to liquid processing, digital printing, spray and foam finishing, and solvent processing, the textile chemical processes still remain water, energy, cost, and effluent intensive. In this context, cold plasma that does not use water seems to be a promising technology for wet chemical processing of textiles. The emerging plasma technology can be utilized in the different stages of textile processing and finishing, such as in desizing, scouring, dyeing, printing, hydrophobicity, hydrophilicity, oil absorbency, antimicrobial, flame retardancy, UV protection, antistatic, crease resistance, and antifelting finishing. It can also be used for improvement in biocompatibility and adhesion strength. Only a low-temperature plasma (<250°C), either at atmospheric pressure or at low pressure, can be used for such modifications. Though a low-pressure plasma technology has its own several advantages, still the APP, such as corona discharge, DBD, and plasma jet, is becoming more popular in the textile field due to their advantages of rapid processing and feasibility to continuous operation. The surface modification of textiles using plasma can be achieved using small nonpolymerizing gaseous molecules, such as oxygen, nitrogen, air, argon, helium, or fluorine, for surface activation, cleaning, oxidation, changes in surface energy, an increase in surface roughness/area, etching, and creation of nanostructures. These help in the improvement in textile properties in terms of water absorbency, wetting, wicking, oil absorbency, dye exhaustion, chemical penetration, adhesion, and antifelting of wool. On the other hand, plasma reaction with a bigger molecule containing vinyl, hydroxyl, carbonyl, carboxyl, acrylate, fluorocarbon, silicone, and phosphorous leads to the development of water-repellent, oil-repellent, UV-protective, flame-retardant, and crease-resistant functional textiles. In plasma-treated samples, an improvement in coloration has been achieved for cellulosic, protein, and manmade textile in the presence of direct, reactive, vat, natural, acid, basic, and disperse dyes. In the protein fibers, the dyeing time or the temperature can be reduced significantly due to the removal of fatty acid hydrocarbon layers, generation of more NH_2 groups, and improvement in hydrophilicity. Water-repellent hydrophobic to superhydrophobic textiles have been developed using smaller hydrocarbon or fluorocarbon (C1–C4) gaseous molecules, such as methane, ethylene, acetylene, BD, CF_4, C_2F_6, C_3HF_7, $C_2H_2F_4$, and C_3F_8. Large-size hydrocarbon and fluorocarbon molecules (C6–C12), such as PFHEA, 1,1,2,2, tetrahydroperfluorodecyl acrylate, styrene, vinyl laurate, and DA, have also been used for the development of aqueous as well as solvent wash-durable hydrophobic textiles. Also, in the hydrophobic finishing of textiles, a silicone-based HMDSO precursor has mainly been used. Recently, an attempt has also been made for functional finishing of dyed fabric and garment, where it showed that there was no significant change in fabric color parameters, even though the sample became hydrophobic. A single-step flame-retardant and water-repellent finishing of cotton textile has also been made using argon (Ar) plasma–induced graft copolymerization of acrylate phosphate and phosphonate derivatives in combination with CF_4 gas. Similarly, hydrophilic finishing along with improved dyeing properties has also been reported by several research groups. Plasma processing of textiles is carried out in the dry state, and hence, adoption of such emerging technology would help develop superior product quality at lower cost while addressing adverse issues related to the environment.

REFERENCES

1. Karmakar, S. R. 1999. *Chemical Technology in the Pre-treatment Processes of Textiles*. Textile Science and Technology, vol. 12. Elsevier, Amsterdam, the Netherlands, pp. 1243–1249.
2. Parvathi, C., T. Maruthavanan, and C. Prakash. November 2009. Environmental impact of textile industries. *Indian Text. J.* 1–4.
3. Ozturk, H. K. 2005. Energy usage and cost in textile industry: A case study for Turkey. *Energy* 20: 1–23.
4. Nehra, V., A. Kumar, and H. K. Dwivedi. 2008. Atmospheric non-thermal plasma sources. *Int. J. Eng.* 2: 53–67.

5. Samanta, K. K., M. Jassal, and A. K. Agrawal. 2006. Atmospheric pressure glow discharge plasma and its applications in textile. *Indian J. Fibre Text. Res.* 31(1): 83–98.
6. Schutze, A., J. Y. Jeong, S. E. Babayan, J. Park, G. S. Selwyn, and R. F. Hicks. 1998. The atmospheric–pressure plasma jet: A review and comparison to other plasma sources. *IEEE Trans. Plasma Sci.* 26(6): 1685–1694.
7. Conrads, H. and M. Schmidt. 2000. Plasma generation and plasma sources. *Plasma Sources Sci. Technol.* 9: 441–454.
8. Eliasson, B. and U. Kogelschatz. 1991. Non-equilibrium plasma chemical processing. *IEEE Trans. Plasma Sci.* 19: 1063–1077.
9. Mohamed, A. H., J. F. Kolb, and K. H. Schoenbach. 2010. Low temperature atmospheric pressure, direct current microplasma jet operated in air, nitrogen and oxygen. *Eur. Phys. J. D* 60: 517–522.
10. Samanta, K. K., S. Saxena, A. Arputhraj, M. Bhowmick, T. N. Gayatri, A. H. Shaikh, and S. Basak. 2014. A status report on surface modification of textile using cold plasma. Central Institute for Research on Cotton Technology, Mumbai, India, pp. 1–38.
11. Basak, S., K. K. Samanta, S. Saxena, S. K. Chattopadhyay, R. Narkar, R. Mahangade, and G. B. Hadge. 2014. Flame resistant cellulosic substrate using banana pseudostem sap. *Polish J. Chem. Technol.* 17(1).
12. Samanta, K. K., S. Saxena, A. Arputhraj, M. Bhowmick, T. N. Gayatri, A. H. Shaikh, and S. Basak. 2014. Plasma processing of textile at CIRCOT. Central Institute for Research on Cotton Technology, Mumbai, India, pp. 1–17.
13. Jahagirdar, C. J. and L. B. Tiwari. 2004. Study of plasma polymerization of dichloromethane on cotton and polyester fabrics. *J. Appl. Polym. Sci.* 94(5): 2014–2021.
14. Bhat, N. V. and Y. N. Benjamin. 1999. Surface resistivity behavior of plasma treated and plasma grafted cotton and polyester fabrics. *Text. Res. J.* 69(1): 38–42.
15. Bhat, N. V., A. N. Netravali, A. V. Gore, M. P. Sathianarayanan, G. A. Arolkar, and R. R. Deshmukh. 2011. Surface modification of cotton fabrics using plasma technology. *Text. Res. J.* 81(10): 1014–1026.
16. Shin, Y., K. Son, D. Yoo, S. Hudson, M. McCord, S. Matthews, and Y. J. Whang. 2006. Functional finishing of nonwoven fabrics. I. Accessibility of surface modified PET spunbond by atmospheric pressure He/O_2 plasma treatment. *J. Appl. Polym. Sci.* 100(6): 4306–4310.
17. Samanta, K. K., M. Jassal, and A. K. Agrawal. 2010. Antistatic effect of atmospheric pressure glow discharge cold plasma treatment on textile substrates. *Fibre Polym.* 11(3): 431–437.
18. Qiuran, J., L. Ranxing, S. Jie, W. Chunxia, P. Shujing, J. Feng, Y. Lan, and Q. Yiping. 2009. Influence of ethanol pretreatment on effectiveness of atmospheric pressure plasma treatment of polyethylene fibers. *Surf. Coat. Technol.* 203(12): 1604–1608.
19. Krump, H., M. Simor, I. Hudec, M. Jasso, and A. S. Luyt. 2005. Adhesion strength study between plasma treated polyester fibres and a rubber matrix. *J. Appl. Polym. Sci.* 240: 268–274.
20. Nobuyuki, Z., I. Hiroto, and Y. Kazuya. 2008. Plasma-chemical surface functionalization of flexible substrates at atmospheric pressure. *Thin Solid Films.* 516(19): 6683–6687.
21. Park, D. J., M. H. Lee, I. W. Yeon, D.-W. Han, J. B. Choi, J. K. Kim, and S. O. Hyun. 2008. Sterilization of microorganisms in silk fabrics by microwave-induced argon plasma treatment at atmospheric pressure. *Surf. Coat. Technol.* 202: 5773–5778.
22. Zhang, C. and K. Fang. 2009. Surface modification of polyester fabrics for inkjet printing with atmospheric-pressure air/Ar plasma. *Surf. Coat. Technol.* 203: 2058–2063.
23. Naebe, M., P. G. Cookson, J. Rippon, R. P. Brady, X. Wang, N. Brack, and G. Riessen. 2010. Effects of plasma treatment of wool on the uptake of sulfonated dyes with different hydrophobic properties. *Text. Res. J.* 80(4): 312–324.
24. Leroux, F., C. Campagne, A. Perwuelz, and L. Gengembre. 2009. Atmospheric air plasma treatment of polyester textile materials. Textile structure influence on surface oxidation and silicon resin adhesion. *Surf. Coat. Technol.* 203(20–21): 3178–3183.
25. Samanta, K. K. 2010. Surface functionalization of textile substrates using atmospheric pressure glow plasma. PhD thesis, Indian Institute of Technology (IIT), Delhi, India.
26. Samanta, K. K., T. N. Gayatri, S. Saxena, S. Basak, S. K. Chattopadhyay, and A. Arputhraj. 2014. Effect of plasma treatment on physico-chemical properties of cotton. *Int. J. Eng. Res. Technol.* 3(3): 2467–2477.
27. Hocker, H. 2002. Plasma treatment of textile fibers. *Pure Appl. Chem.* 74(3): 423–427.
28. Özdogan, E., R. Saber, H. Ayhan, and N. Seventekin. 2002. A new approach for dyeability of cotton fabrics by different plasma polymerization methods. *Color. Technol.* 118(3): 100–103.
29. Karahan, H. A., E. Ozdogan, A. Demir, and H. Ayhan. 2008. Effects of atmospheric plasma treatment on the dyeability of cotton fabrics by acid dyes. *Color. Technol.* 124: 106–110.

30. Sun, D. and G. K. Stylios. 2004. Effect of low temperature plasma treatment on the scouring and dyeing of natural fabrics. *Text. Res. J.* 74(9): 751–756.
31. Nasadil, P. and P. Benešovsky. 2008. Plasma in textile treatment. II central European symposium on plasma chemistry. *Chemicke Listy* 102: 1486–1489.
32. Spitzl, S. and M. Hildegard. 2003. Plasma pre-treatment of textiles for improvement of dyeing processes. *Int. Dyer* 188(5): 20–25.
33. Guglani, R. 2002. Recent developments in textile dyeing techniques. http://www.fibre2fashion.com/industryarticle/12/1171/recent-developments-in-textile-dyeing-techniques11.asp. (Accessed January 17, 2015.)
34. Kan, C. W., C. W. M. Yuen, and W. Y. Tsoi. 2011. Using atmospheric pressure plasma for enhancing the deposition of printing paste on cotton fabric for digital ink-jet printing. *Cellulose* 18(3): 827–839.
35. Man, W. S., C. W. Kan, and S. P. Ng. 2014. The use of atmospheric pressure plasma treatment on enhancing the pigment application to cotton fabric. *Vacuum* 99: 7–11.
36. Zawhary, M. M., N. A. Ibrahim, and M. A. Eid. 2006. The impact of nitrogen plasma treatment on physical chemical and dyeing properties of wool fabric. *Polym. Plast. Technol. Eng.* 45: 1123–1132.
37. Molina, R., C. Canal, E. Bertran, J. M. D. Tascon, and P. Erra. 2004. Low Temperature plasma modified wool fabrics: Surface study by SEM. In: A. Méndez-Vilas and L. Labajos Broncano (eds.), *Current Issues on Multidisciplinary Microscopy Research and Education*. Formatex, Badajoz, Spain, pp. 242–249.
38. Ratnapandian, S., L. Wang, S. M. Fergusson, and M. Naebe. 2011. Effect of atmospheric pressure plasma treatment on pad dyeing of natural dyes on wool. *J. Fibre Bioeng. Inform.* 4(3): 267–276.
39. Kan, C. W. 2007. Effect of low temperature plasma on different wool dyeing systems. *AUTEX Res. J.* 8(4): 132–139.
40. Rombaldoni, F., A. Montarsolo, R. Mossotti, R. Innocenti, and G. Mazzuchetti. 2010. Oxygen plasma treatment to reduce the dyeing temperature of wool fabrics. *J. Appl. Polym. Sci.* 118(2): 1173–1183.
41. Özdogan, E., A. Demir, H. A. Karahan, H. Ayhan, and N. Seventekin. 2009. Effects of atmospheric plasma on the printability of wool fabrics. *Tekstíl ve Konfeksiyon* 2: 123–127.
42. Holme, I. 2000. Challenge and change in wool dyeing and finishing. In: *10th International Wool Textile Research Conference*, KNL-9, p. 1, Aachen, Germany.
43. Panda, P. K., D. Rastogi, M. Jassal, and A. K. Agrawal. 2012. Effect of atmospheric pressure helium plasma on felting and low temperature dyeing of wool. *J. Appl. Polym. Sci.* 124(5): 4289–4297.
44. Yue, C. Y., L. Hong, R. Yu, H. W. Wang, and L. J. Zhu. 2004. Study of *Bombyx mori* silk treated by oxygen plasma. *J. Zhejiang Univ. Sci.* 5(8): 918–922.
45. Iriyama, Y., T. Mochizuki, M. Watanabe, and M. Utada. 2003. Preparation of silk film and its plasma treatment for better dyeability. *J. Photopolym. Sci. Technol.* 16(1): 75–80.
46. Pandit, P. 2014. Plasma application for performance enhancement of textile fabrics. MTech thesis, Institute of Chemical Technology, Mumbai, India.
47. Vesel, A. and M. Mozetic. 2009. Surface functionalization of organic materials by weakly ionized highly dissociated oxygen plasma. *J. Phys. Conf. Ser.* 162: 012015.
48. Kang, E. T. and K. G. Neoh. 2009. Surface modification of polymers. In: *Encyclopedia of Polymer Science and Technology*, vol. 115. Wiley Inter Science, New York, p. 167.
49. Cireli, A., B. Kutlu, and M. Mutlu. 2007. Surface modification of polyester and polyamide fabrics by low frequency plasma polymerization of acrylic acid. *J. Appl. Polym. Sci.* 104(4): 2318–2322.
50. Souto, A. P., F. R. Oliveira, M. Fernandes, and N. Carneiro. 2012. Influence of DBD plasma modification in the dyeing process of polyamide. *J. Text. Eng.* 19(85): 20–26.
51. Oliveira, F. R., A. Zille, and A. P. Souto. 2014. Dyeing mechanism and optimization of polyamide 6,6 functionalized with double barrier discharge (DBD) plasma in air. *Appl. Surf. Sci.* 293: 177–186.
52. Gotoh, K. and A. Yasukawa. 2011. Atmospheric pressure plasma modification of polyester fabric for improvement of textile-specific properties. *Text. Res. J.* 81(4): 368–378.
53. Kamel, M. M., M. M. El Zawahry, H. Helmy, and M. A. Eid. 2011. Improvements in the dyeability of polyester fabrics by atmospheric pressure oxygen plasma treatment. *J. Text. Inst.* 102(3): 220–231.
54. Sparavigna, A. 2008. Plasma treatment advantages for textiles. arxiv.org/pdf/0801.3727. (Accessed January 17, 2015.)
55. Yaman, N., E. Özdogan, and N. Seventekin. 2011. Atmospheric plasma treatment of polypropylene fabric for improved dyeability with insoluble textile dyestuff. *Fiber Polym.* 12(1): 35–41.
56. Oravcova, A. and I. Hudec. 2010. The influence of atmospheric pressure plasma treatment on surface properties of polypropylene films. *Acta Chimica Slovaca* 3(2): 57–62.
57. Sarmadi, A. M., T. H. Ying, and F. Denes. 1993. Surface modification of polypropylene fabrics by acrylonitrile cold plasma. *Text. Res. J.* 63(12): 697–705.

58. Barthlott, W. and C. Neinhuis. 1997. Purity of the sacred lotus, or escape from contamination in biological surfaces. *Planta* 202(1): 1–8.
59. Balu, B., V. Breedveld, and D. W. Hess. 2008. Fabrication of "Roll-off" and "Sticky" superhydrophobic cellulose surfaces via plasma processing. *Langmuir* 24: 4785–4790.
60. Supasai, T., S. K. Hodak, and B. Paosawatyanyong. 2007. Effect of SF_6 plasma treatment on hydrophobicity improvement of fabrics. *Jurnal Fizik Malaysia* 28(1 and 2): 1–6.
61. Nimmanpipug, P., V. S. Lee, S. Janhom, P. Suanput, D. Boonyawan, and K. Tashino. 2008. Molecular functionalization of cold plasma treated *Bombyx mori* silk. *Macromol. Symp.* 264: 107–112.
62. Kamalangkla, K., S. K. Hodak, and J. L. Grutzmacher. 2011. Multifuctional silk fabrics by means of the plasma induced graft polymerisation process. *Surf. Coat. Technol.* 205: 3755–3762.
63. Wang, H. Z., M. W. Rembold, and J. Q. Wang. 1993. Characterization of surface properties of plasma polymerized fluorinated hydrocarbon layers: Surface stability as a requirement for permanent water repellency. *J. Appl. Polym. Sci.* 49(4): 701–710.
64. Allan, G., A. Fotheringham, and J. Weedall. 2002. The use of plasma and neural modelling to optimise the application of a repellent coating to disposable surgical garments. *AUTEX Res. J.* 2(2): 64–68.
65. Hegemann, D. 2006. Plasma polymerization and its application in textiles. *Indian. J. Fibre Text. Res.* 31(1): 99–115.
66. Jung, H., B. Gweon, D. B. Kim, and W. Chao. 2011. A simple approach to surface modification using PTFE (poly tet-fluoroethylene) with laminar and turbulent flow of microplasma jets at atmospheric pressure. *Plasma Proc. Polym.* 8(6): 535–541.
67. Samanta, K. K.., R. M. Gurjar, S. K. Chattopadhyay, S. Saxena, S. Basak, and T. N. Gayatri. 2013. Water free eco-friendly textile finishing using plasma technology. In: *Conference Proceedings of International Conference on "Environment and Its Impact on Society,"* J. D. Birla Institute, Kolkata, India, August 2013. ISBN 978-93-5126-892-5.
68. Samanta, K. K., T. N. Gayatri, S. Saxena, S. Basak, S. K. Chattopadhyay, A. Arputharaj, and G. B. Hadge. 2014. Hydrophobic finishing of cotton textile using atmospheric pressure plasma. In: *Conference Proceedings of International Conference on Natural Fibres*, Kolkata, India, August 1–3, 2014, p. 163.
69. Vinogradov, I. P. and A. Lunk. 2005. Structure and chemical composition of polymer films deposited in a DBD in Ar/fluorocarbon mixtures. *Surf. Coat. Technol.* 200 (5): 660–663.
70. McCord, M. G., Y. J. Hwang, Y. Qiu, L. K. Hughes, and M. A. Bourham. 2003. Surface analysis of cotton fabrics fluorinated in radio-frequency plasma. *J. Appl. Polym. Sci.* 88(8): 2038–2047.
71. Ramamoorthy, A., A. El-Shafei, and P. Hauser. 2013. Plasma induced graft polymerization of C6 fluorocarbons on cotton fabrics for sustainable finishing applications. *Plasma Proc. Polym.* 10: 430–443.
72. Tsafack, M. J. and J. Levalois-Grützmacher. 2007. Towards multi-functional surfaces using the plasma-induced graft polymerization (PIGP) process: Flame and water-proof cotton textiles. *Surf. Coat. Technol.* 201(12): 5789–5795.
73. Zhang, J., P. France, A. Radomyselskiy, S. Datta, J. Zhao, and W. Van Ooij. 2003. Hydrophobic cotton fabric coated by a thin nanoparticle plasma film. *J. Appl. Polym. Sci.* 88(6): 1473.
74. Girard-Lauriault, P. L., P. Desjardins, W. E. S. Unger, A. Lippitz, and M. R. Wertheimer. 2008. Chemical characterisation of nitrogen-rich plasma-polymer films deposited in dielectric barrier discharges at atmospheric pressure. *Plasma Proc. Polym.* 5(7): 631–644.
75. Kim, J. H., G. Liu, and S. H. Kim. 2006. Deposition of stable hydrophobic coatings with in-line CH_4 atmospheric RF plasma. *J. Mater. Chem.* 16: 977–981.
76. Samanta, K. K., A. G. Joshi, M. Jassal, and A. K. Agrawal. 2012. Study of hydrophobic finishing of cellulosic substrate using He/1,3-butadiene plasma at atmospheric pressure. *Surf. Coat. Technol.* 213: 65–76.
77. Parida, D., M. Jassal, and A. K. Agarwal. 2012. Functionalization of cotton by in-situ reaction of styrene in atmospheric pressure plasma zone. *Plasma Chem. Plasma Proc.* 32: 1259–1274.
78. Abidi, N. and E. Hequet. 2005. Cotton fabric graft copolymerization using microwave plasma. II. Physical properties. *J. Appl. Polym. Sci.* 98(2): 896–902.
79. Panda, P. K., M. Jassal, and A. K. Agrawal. 2013. Functionalization of cellulosic substrate using He/dodecyl acrylate plasma at atmospheric pressure. *Surf. Coat. Technol.* 225: 97–105.
80. Montarsolo, A., M. Periolatto, M. Zerbola, R. Mossotti, and F. Ferrero. 2013. Hydrophobic sol-gel finishing for textiles: Improvement by plasma pre-treatment. *Text. Res. J.* 83(11):1190–1200.
81. Kilic, B., A. Cireli, and M. Mutlu. 2009. Surface modification and characterization of cotton and polyamide fabrics by plasma polymerization of hexamethyldisilane and hexamethyldisiloxane. *Int. J. Cloth. Sci. Technol.* 21(2–3): 137–145.

82. Kale, K. H. and S. Palaskar. 2011. Atmospheric pressure plasma polymerization of hexamethyldisiloxane for imparting water repellency to cotton fabric. *Text. Res. J.* 81(6): 608–620.
83. Sarmadi, A., T. Ying, and F. Denes. 1995. HMDSO-plasma modification of polypropylene fabrics. *Eur. Polym. J.* 31(9): 847–857.
84. Steele, D. A. and R. D. Short. 2011. On the use of SIFT-MS and PTR-MS experiments to explore reaction mechanisms in plasmas of volatile organics: Siloxanes. *Plasma Proc. Polym.* 8(4): 287–294.
85. Ji, Y. Y., H. K. Chang, Y. C. Hong, and S. H. Lee. 2009. Water-repellent improvement of polyester fibre via plasma RF treatment with Ar/HMDSO at atmospheric pressure. *Curr. Appl. Phys.* 9: 253–256.
86. Ji, Y.-Y., Y.-C. Hong, S.-H. Lee, S.-D. Kim, and S.-S. Kim. 2008. Formation of super-hydrophobic and water-repellency surface with hexamethyldisiloxane (HMDSO) coating on polyethyleneteraphtalate fiber by atmospheric pressure plasma polymerization. 6th Asian-European International Conference on Plasma Surface Engineering. *Surf. Coat. Technol.* 202(22–23): 5663–5667.
87. Morent, R., N. De Geyter, S. Van Vlierberghe, P. Dubruel, C. Leys, L. Gengembre, E. Schacht, and E. Payen. 2009. Deposition of HMDSO-based coatings on PET substrates using an atmospheric pressure dielectric barrier discharge. *Prog. Organ. Coat.* 64(2–3): 304–310.
88. Cai, Z., Y. Qui, C. Zhang, Y. J. Hwang, and M. McCord. 2003. Effect of atmospheric plasma treatment on desizing of PVA on cotton. *Text. Res. J.* 73(8): 670–674.
89. Kan, C. W. and C. W. M. Yuen. 2012. Effect of atmospheric pressure plasma treatment on the desizing and subsequent colour fading process of cotton denim fabric. *Color. Technol.* 128: 356–363.
90. Kan, C. W., C. F. Lam, C. K. Chan, and S. P. Ng. 2014. Using atmospheric pressure plasma treatment for treating grey cotton fabric. *Carbohydr. Polym.* 102: 167–173.
91. Demir, A., E. Ozdogan, N. Ozdil, and A. Gurel. 2011. Ecological materials and methods in the textile industry: Atmospheric–pressure plasma treatments of naturally coloured cotton. *J. Appl. Polym. Sci.* 119(3): 1410–1416.
92. Tsriskina, A. L., J. N. Guschchina, B. L. Gorberg, and A. A. Ivanov. July 1991. Referativnyi Zhurnal. *Kotlostroenie*. 42B: 7.
93. Kale, K. and S. Palaskar. 2010. Studies on atmospheric pressure plasma treatment of polyester/cotton blended fabric. Paper presented in *51st Joint Technological Conference of ATIRA, BTRA, SITRA and NITRA*. NITRA, Ghaziabad, India, June 29, 2010.
94. Samanta, K. K., M. Jassal, and A. K. Agrawal. 2009. Improvement in water and oil absorbency of textile substrate by atmospheric pressure cold plasma treatment. *Surf. Coat. Technol.* 203: 1336–1342.
95. Wakida, T., S. Tokino, S. Niu, H. Kawamura, Y. Sato, M. Lee, H. Uchiyama, and H. Inagaki. 1993. Characterization of wool and polyethylene terephthalate fabrics and film treated with low temperature plasma under atmospheric pressure. *Text. Res. J.* 63: 433–438.
96. Nourbakhsh, S. and I. Ebrahimi. 2012. Different surface modification of poly (ethylene terephthalate) and polyamide 66 fibers by atmospheric air plasma discharge and laser treatment: Surface morphology and soil release behavior. *J. Text. Sci. Eng.* 2(2): 109.
97. Píchal, J. and Y. Klenko. 2009. ADBD plasma surface treatment of PES fabric sheets, plasma physics and technology topical issue: 23rd Symposium on Plasma Physics and Technology. *Eur. Phys. J. D* 54(2): 271–280.
98. Guo, L., C. Campagne, A. Perwuelz, and F. Leroux. 2009. Zeta potential and surface physico-chemical properties of atmospheric air-plasma-treated polyester fabrics. *Text. Res. J.* 79(15): 1371–1377.
99. Cernák, M., M. Šimor, J. Ráhel, D. Kovácik, A. Záhoranová, and M. Mazúr. 2002. Atmospheric-pressure plasma treatment of nonwovens using surface dielectric barrier discharges. In: *12th TANDEC International Nonwovens Conference*, Knoxville, TN, pp. 1–15.
100. Tsafack, M. J. and L. Grutzmachor. 2007. Multifunctional surfaces using the plasma induced graft polymerisation (PIGP) process: Flame and waterproof cotton textiles. *Surf. Coat. Technol.* 201(12): 5789–5795.
101. Raslan, W. M., U. S. Rashed, H. E. Sayad, and A. A. E. Halwagy. 2011. Flame retardancy using plasma nano technology. *Mater. Sci. Appl.* 2: 1432–1442.
102. Quede, A., C. Jama, P. Supiot, M. Le Bras, R. Delobel, O. Dessaux, and P. Goudmand. 2002. Elaboration of fire retardant coatings on polyamide-6 using a cold plasma polymerization process. *Surf. Coat. Technol.* 151–152: 424–428.
103. Totolin, V., M. Sarmadi, S. O. Manolache, and F. S. Denes. 2012. Environmentally friendly flame-retardant materials produced by atmospheric pressure plasma modifications. *J. Appl. Polym. Sci.* 124(1): 116–122.
104. Lam, Y. L., C. W. Kan, and C. W. M. Yuen. 2011. Effect of zinc oxide on flame retardant finishing of plasma pre-treated cotton fabric. *Cellulose* 18(1): 151–165.

105. Lam, Y., C. W. Kan, and C. W. M. Yuen. 2012. Effect of plasma pretreatment on enhancing wrinkle resistant property of cotton fibre treated with BTCA and TiO_2. *J. Appl. Polym. Sci.* 124(4): 3341–3347.
106. Chen, C.-C., J.-C. Chen, and W.-H. Yao. 2010. Argon plasma treatment for improving the physical properties of crosslinked cotton fabrics with dimethyloldihydroxyethyleneurea-acrylic acid. *Text. Res. J.* 80(8): 675–682.
107. Wang, Y. 2011. The uniform Si–O coating on cotton fibers by an atmospheric pressure plasma treatment. *J. Macromol. Sci. B Phys.* 50:1739–1746.
108. Moor, A. 2010. The environmental impact of wet processing and how to improve sustainability, Written for MADE-BY. Final thesis, Amsterdam Fashion Institute Hogeschool van Amsterdam, Amsterdam, the Netherlands, pp. 1–100. http://kennisbank.hva.nl/document/220188. (Accessed January 17, 2015.)
109. Samanta, K. K., S. Basak, and S. K. Chattopadhyay. 2014. Environment-friendly textile processing using plasma and UV treatment. In: S. S. Muthu (ed.), *Roadmap to Sustainable Textiles and Clothing*. Springer Publication, Singapore, pp. 161–201.

2 Sustainability in Clothing Manufacturing and Competitiveness

Is It a New Mind-Set or a Paradox?

Wen Ying Claire Shih, and Konstantinos Agrafiotis

CONTENTS

2.1 Introduction 39
2.2 Who Commands the GVCs in the Textile and Clothing Sectors?.................40
2.3 Trade Agreements Alter the Balances of Production Patterns......................41
2.4 Competitiveness and the Theoretical Background of Activities in GVCs ...42
    2.4.1 Transaction Cost Economics 42
    2.4.2 Relational View and Network Resources 42
    2.4.3 Trust in Networks 43
2.5 Key Issues of Sustainability in Production Networks 43
    2.5.1 Environmental Issues..43
    2.5.2 Resource Productivity Issue and Sustainable Materials Strategies ...44
    2.5.3 Labor Conditions Issue 45
2.6 Sustainability Architecture through the Traceability of the Supply Chain...........46
2.7 Concluding Comments and Perspectives..47
References..48

2.1 INTRODUCTION

Humanity may have to learn to detach economic growth from resource consumption, as depleting natural resources can lead us all into an untenable future. More specifically, in the clothing industry, the strategy of time-based competition ushered the fashion world into the fast fashion phenomenon of disposable clothing. Cutthroat competition has led retailers' and manufacturers' brands into an unprecedented race to the bottom for the cheapest price and thus to the lowest standards of production. Additionally, the dispersal of subcontracting makes it difficult to monitor working condition standards and also environmental issues of production.

Against this backdrop of faster deliveries, faster consumption, and cheaper prices based on miserable wages and working conditions, the textile and clothing (T&C) sectors must rethink their manufacturing strategies. Suggestions abound as to what needs to be done, viewed from different perspectives depending on where interests stand. However, to the best of our knowledge, as far as competitiveness is concerned, there is little written on this pressing issue. Simply put, fashion companies cannot exactly figure out their future competitiveness if they source responsibly across the supply chain, taking also sustainability into consideration.

Ethical supply chains can provide some answers, as these are currently exercised by some of the big buyers across fashion markets. After all, in buyer-driven chains, they are the ones who can impose the rules on manufacturing companies. But fashion, by definition, is a fragmented industry, which means that there are a myriad of brands, manufacturers, intermediaries, and so on, who do not act always with good intentions. In fact, guile prevails in most commercial exchanges, as companies strive for value appropriation across supply chains in order to gain a competitive edge. In this chapter, transaction cost economics (TCE) provides a powerful framework of partnering relationships. In these relationships, uncertainty and opportunism are the order of the day, since all the actors involved will seek to capture more value for themselves. TCE prescribes the drafting of detailed contracts that can mitigate opportunistic behavior.

The relational view of the firm espouses a different perspective by substituting contracts with trust, as members of a supply network work collectively sharing valuable resources. These resources also refer to the environment and its protection and to worker conditions within a manufacturing network. Thus, network resources are vital in this form of partnering relationship. Within the ethical sourcing mechanics, transparency of all activities and sharing of information and knowledge are crucial across the supply chain. This is complemented by fair practices toward workers and respect for the natural capital. It is in the interest of all network members to maintain stability in the flow of work in progress as this increases the probabilities of gaining a competitive advantage. In the second scenario of the relational view, there is scope for competitiveness and environmental sustainability. In the following sections, the fundamental issues will be discussed in order to provide our arguments on sustainability.

The first section evaluates the global value chains (GVCs) in the T&C sectors and their role in production networks scattered around the world, together with the influences of trade agreements in the fragmentation of production. The second section examines the competitive forces viewed from TCE and the resource-based view (RBV), and their leverage in network sourcing and resources. The third section investigates the thorny issues of the T&C production and their impacts on the environment, together with the resource productivity (RP) variable and labor conditions in production network. The fourth section reviews the new concept of the traceability of the supply chain and its possible implementation in clothing terms. Finally, the last section presents the concluding comments and possible perspectives arising from the previous discussion of all relevant parameters.

2.2 WHO COMMANDS THE GVCs IN THE TEXTILE AND CLOTHING SECTORS?

The concept of GVCs emerged in the 1990s as a supply system for the T&C sectors. In this supply mechanism, buyers who constitute major fashion brands and retailers in advanced countries command supply chains in a complex system of production dispersed mainly in developing countries around the world (Gereffi, 1999). By understanding the mechanics of GVC together with the trade agreements, we can gain insight into the competitive forces that shape the T&C industry and then proceed to discuss the sustainability of supply in T&C chains in relation to the streams of academic theories presented in this chapter. Viewing sustainability from the supply management perspective can provide us with a possible tenable solution for the T&C industries that can also contribute to their competitiveness.

The governance mechanism of T&C production and distribution relates to the GVC linking numerous suppliers who procure end fashion products to the international markets (Gereffi, 1999; Gereffi and Memedovic, 2003). Continuous shifts of production to cheaper locations caused mainly by bilateral and/or multilateral trade agreements with the developed countries have always led the T&C sectors into a race to the bottom, looking for cost economization at all times (Gereffi and Memedovic, 2003; Martin, 2013).

A typical T&C GVC includes five major component sections: the raw materials, the textile production, the apparel manufacturing, the export trade intermediaries, and the retail marketing section at the interface of brands/retailers and customers (Gereffi and Memedovic, 2003; Martin, 2013). Fundamentally, there are two types of chains in which international companies organize their supply networks: the producer-driven GVC and the buyer-driven GVC. The latter is associated with large fashion retailers and fashion brand owners who *mastermind* the GVC. Their role extends into coordinating production activities and arranging decentralized networks of suppliers.

Gereffi et al. (2005), refining their research further, propose that the captive and the relational value chains are related to clothing manufacturing. Captive value chains refer to lower supplier competences. Suppliers in this case are captive since they face significant switching costs if they wish to break free from the relationship. Buyers, on the other hand, have the ability and the experience to codify product specifications such as design, product development, and the procurement of textile and trim inputs limiting their suppliers into mere assembly operations.

The relational value chains, on the other hand, function differently. In this, the lead company teams up with a supplier who is competent enough to carry out complicated production processes without the buyer having to codify product specifications to the last detail. Mutual dependence together with trust prevails in these relationships, since reputation, territorial proximity, and social coherence come to the fore (Gereffi and Memedovic, 2003; Gereffi et al., 2005). The two types of value chains do not function independently from each other but are rather intertwined. The captive model, according to Humphrey and Schnitz (2001) and Knutsen (2004), refers to *power asymmetries* between manufacturers and buyers. In this asymmetrical arrangement, manufacturers are rather docile in relation to their buying masters. This fact demonstrates who really wields power in the GVC since the lead buyer usually determines in what manner other chain members behave and conduct transactions (Gereffi, 1999; Humphrey and Schmitz, 2001). It must be noted that in any given GVC, textile production is more capital intensive than apparel production, and thus, outsourcing of clothing production to cheaper locations was imperative and occurred much faster than textiles (Abernathy et al., 1999; Cline, 2012).

Jarillo (1988), Gulati et al. (2000), and Berger (2006) discuss the concept of central controllers in production networks. The controllers have the abilities to efficiently organize and coordinate activities and information flows. They usually have the means of obtaining and commanding critical resources that are difficult to imitate, thus configuring a structured hierarchy around docile suppliers (Cox, 1999). In clothing terms, central controllers represent the relational value chain players who are usually large manufacturers who in some cases have achieved the integration of their T&C facilities (Berger, 2006; Shih, 2013). Central controllers are also observed in big trade intermediaries in chains and networks (Gereffi et al., 2005).

Wagner (2006) argues that many manufacturers in the T&C sectors engage more in adversarial relationships in their transactions with buyers than in longer-term partnering arrangements. This demonstrates buyers' reluctance to develop their suppliers. Although longer-term cooperation could be ideal, few manufacturers have the necessary resources and capabilities to contribute to these relationships. Consequently, the synergies that could have been generated rarely materialize (Lam and Postle, 2006). In this framework of transactions, either capable or docile manufacturers can enter and remain in the chain as suppliers to these international buyers (Morris and Barnes, 2008).

Another important issue in GVC is industrial upgrading, where the manufacturers need to form the necessary backward linkages of procuring fabrics and trimming in addition to clothing production. This is so-called full package production where manufacturers are able to offer the complete package of fabrics, trims, patterns, and assembly in order to provide shop-ready merchandise. As a result, the GVC players who can demonstrate these capabilities are usually more competitive (Knutsen, 2004; Morris and Barnes, 2008; Nordas, 2004; Shih, 2013).

2.3 TRADE AGREEMENTS ALTER THE BALANCES OF PRODUCTION PATTERNS

The geographical dispersal of production is also shaped by trade agreements. Given the phasing out of the quota system in 2005, regional sourcing has become common practice dominating trade arrangement in the T&C sectors. This has led to multiple bilateral and multilateral agreements to achieve access to major markets without tariffs (Nordas, 2004). A large number of member countries of the World Trade Organization (WTO) have signed preferential trade agreements in its Generalized System of Preferences. These preferential trade arrangements between groups of developing countries and the developed world have had a profound impact in altering the balances of textile and apparel supplies (Frederick and Gereffi, 2009; Tewari, 2006; World Trade Organization, 2009).

Another important variable in trade agreements is the rules of origin. This requires that the exporting country engages in *double transformation*. According to Audet (2004), for any clothing article sourced from a country under the double transformation agreement, both assembly and procurement of fabric and trim must come from the same country or from neighboring countries that belong to the same free-trade area. In the post quota era, countries and regions strive to put in place all necessary backward linkages of export quality. Governments also sponsor export processing zones, as it is imperative to supply their own clothing manufacture (United Nations, 2008).

2.4 COMPETITIVENESS AND THE THEORETICAL BACKGROUND OF ACTIVITIES IN GVCs

The actors involved in GVCs shaping the T&C industries can be explained by two streams of academic theories. The theories are also instrumental in explaining potential sources of competitive advantage that can be generated through partnering arrangements. The two streams of literature can lead us to construct the argument for sustainability in a more analytical rather than in an empirical manner.

2.4.1 Transaction Cost Economics

TCE explains a company's boundaries in terms of top management's fundamental decisions of *buy or make* (Williamson, 1979). There are two major costs that any company strives to minimize. Production costs refer to costs associated with manufacturing, distributing, managing, and organizing activities; transaction costs consist of the costs a company incurs when negotiating, coordinating, monitoring, and managing an exchange with another company (Das and Teng, 2000; Poppo and Zenger, 2002). Companies decide to make in-house (e.g., vertically integrate) when production costs are low and transaction costs are high. Conversely, companies can decide to outsource activities when production costs are high and transaction costs are low (Das and Teng, 2000). Since transaction costs must be kept low at all times, which operations to internalize and which to contract out become strategic decisions for a company's top managers (Cousins, 2005; Gereffi et al., 2005). This explains fundamentally the coordination and arrangement of T&C production in global terms as well as the strategic decision to outsource production.

Another important variable of TCE is opportunism. A company aiming to appropriate value for itself acts with guile. This gives rise to opportunistic behavior against the other company. In order to safeguard against opportunism, top managers devise detailed contracts that can mitigate the uncertainty of opportunism between transacting companies (Cousins, 2005; Williamson, 1979). In terms of T&C industries, there are cases of manufacturing companies that have managed to upgrade themselves to brand owner status by appropriating value in the form of repeated transactions with other lead companies over a period of time, thus learning trade secrets and proceeding to implementation. Nevertheless, joint ventures are also observed in the T&C sectors, where uncertainty deriving from possible opportunism is dealt with by means of contracts.

2.4.2 Relational View and Network Resources

The RBV takes a different perspective from TCE. Here, it is argued that the company (as the unit of its analysis) comprises a bundle of resources that create tangible and intangible values (Barney, 1991; Grant, 1991; Wernerfelt, 1984). Competitive advantage can be generated by the company as it obtains the ability to pool and apply only its valuable resources (Barney, 1991; Grant, 1991).

An extension of the RBV theory postulates that resources are not only limited to the ones that reside solely within the company, but they can be acquired through purchasing or interfirm relationships. The relational view of the firm thus puts forward the notion that competitive advantage can accrue from a network of companies. The synergies generated by combining and reconfiguring resources form a powerful blend that can lower transaction costs and increase value creation (Barney, 1999; Das and Teng, 2000; Gulati et al., 2000). Duschek (2004) argues that the relational view is indeed a mix of competences, resources, TCE, and social networks. Competitive advantage can be derived from (a) relation asset investments that entail cospecialized resources, (b) knowledge sharing that has the power of increasing product novelty and decreasing opportunistic behavior, and (c) complementary resources that can generate strategic compatibilities of resources (Cox, 1999; Duschek, 2004). The positioning of the company in a network can determine its ability to manage the network strategically as this will affect the appropriate selection of members. Interfirm specialization in the network means that network partners can mobilize and share resources and information. Long-term partnering relationships can accrue from strategic networks in many interfaces of partners' operations (Gulati et al., 2000).

Network resources are closely associated with RBV and the relational view since they can reinforce competitive advantages and provide strategic benefits to the central controlling company (Duschek, 2004; Gulati et al., 2000). Network sourcing and network resources are crucial in forging specialized capabilities in the T&C sectors as no single company can acquire all resources needed in order to compete in global markets. Nevertheless, networks also have a dark side in the clothing sector. Subcontracting production to uncertified garment factories that are hidden from view has been a reality in the sector for decades (Cline, 2012; Harney, 2009).

2.4.3 Trust in Networks

Trust is significant in commercial/production exchanges since there is always the risk of opportunism. Trust can engender confidence in partnerships and avert business risks that contracts alone cannot adequately cover. Under TCE, the use of contracts can police and monitor a partner's behavior, but trust can lower transaction costs and reduce uncertainty. Thus, both trust and contracts can be interchangeably used as they complement each other (Gulati et al., 2000; Poppo and Zenger, 2002). Trust additionally possesses a number of benefits that complement relationships. These include the continuity of partnerships in adverse conditions, dependencies, and mutual commitment of partners, since they all acknowledge that benefits and advantages will be fairly distributed among members of the chain/network (Cullen et al., 2000; Lee and Cavusgil, 2006; McCutcheon and Stuart, 2000; Poppo and Zenger, 2002). Moreover, trust can enhance competitiveness since partners engage in knowledge-sharing relationships (Cheng et al., 2008).

In the next section, the authors attempt to probe into the major sustainability issues that relate to production networks in dispersed geographical locations around the globe.

2.5 KEY ISSUES OF SUSTAINABILITY IN PRODUCTION NETWORKS

There are a number of environmental, productivity, and social issues that arise with regard to sustainability in the GVC. We will address briefly all three starting from the most important, the environment.

2.5.1 Environmental Issues

The term *sustainable development* was coined by Norwegian Prime Minister Brundtland back in 1987 in a landmark report titled *Our Common Future*. The concept addressed social and economic development while simultaneously focusing on the importance of environmental protection and natural resources (Brundtland, 1987). The report led to the 1992 Earth Summit, which produced a global action plan known as *Agenda 21*. In 2012, the United Nations (UN) produced a continuation of Agenda 21 reflecting also on the views of the first sustainable development report and the progress that has been achieved in the 20 years since the Earth Summit. The Sustainable Development in the Twenty-First Century postulates that human activity has exacerbated the global environmental crisis for human enterprise has rapidly depleted resources that humanity will need to depend on for survival. Scientists point out that some of the sustainability thresholds of a *safe operating space* have been violated. The factors that have led to this deterioration of resources are linked to population growth, increasing levels of affluence, and the structure of consumption (United Nations, 2012).

All three factors are related to the T&C sectors. The rising affluence of consumer classes around the world means that they unfortunately follow the same wasteful consumption patterns observed in developed countries (Cline, 2013). If we add speed to the equation of fashion consumption, then the combination becomes *lethal* in sustainability terms (Siegle, 2011). Moreover, research into the environmental impact of components in clothing items throughout their life cycle has found that nearly all had a detrimental effect (Chen and Burns, 2006).

The view of Earth exceeding its *sustainable carrying capacity* is pointed out by a number of researchers (Abdul Hamid and Duraiappah, 2014; Smitha, 2011). They argue convincingly that we are gambling with our future, as corporations around the world plunder natural resources for profit. Smitha (2011) claims that this is an insidious purpose and a pitiful business model as companies focus on convincing us to consume more at the expense of fair practices.

The UN report (2012) also states that although some efficiency gains have been achieved, overall global policies have failed to estimate the scope of actions needed to achieve a truly sustainable future. Consequently, more drastic measures will have to be taken in the medium term. Integrated strategies in clusters of sectors are imperative for sustainability. Integrated strategies refer to systems where system components must relate to each other and be viewed in a holistic mode (Hong et al., 2009). They together with industry standards are indeed the prerequisites in the T&C sectors. This has been the task of the Sustainable Apparel Coalition (SAC), which was formed in 2010 in order to establish a single industry standard (Clifford, 2013).

Moreover, the report also recommends the coordination of efforts at the implementation stages, as multiplicity of ideas and actions or even conflicting interests can derail otherwise sincere endeavors. This holds true for the T&C sectors, as the current state of affairs in both industries is confusing. There is a plurality of views as to what is best for a sustainable future in both sectors from the decelerating of fashion consumption to mending, handcrafting, and clothes leasing, just to name a few (Gardetti and Torres, 2013).

2.5.2 Resource Productivity Issue and Sustainable Materials Strategies

RP refers to the sensible utilization of material resources in a production process by also measuring the environmental impact. The goal is to use less and produce more with given resources. This entails cost efficiencies that contribute toward production optimization processes as economizing on resources saves money for the companies in a production network (Martin, 2013). The Organization for Economic Cooperation and Development (OECD) gives a similar definition in its report for the G8 environment ministers' summit in 2011. The OECD's goal is to establish a resource-efficient economy by enhancing RP built on the three founding principles, reuse–recycle–reduce (3R). Consequently, by improving RP, companies respect the environment by decreasing consumption of virgin materials. Decoupling is one of the keywords in green growth strategies where detaching material consumption from economic growth is the target. The report recommends that the improvement of RP is achievable through sustainable materials management. This requires integrated life cycle–based strategies for waste, materials, and products. Moreover, 3R initiatives are recommended such as circular economy (not linear), integrated supply chain management, and general public encouragement to take "an ethically based responsibility for sustainable growth" (Organisation for Economic Co-operation and Development, 2011). Perhaps one of the best examples comes from a textile and petrochemical company, Far Eastern New Century Corporation, based in Taiwan, which apart from textiles produces the plastic polyethylene terephthalate (PET) used in drinks bottles. This year, the company announced its plans to build two new production lines of recycled PET bottles to be completed by November 2014. The company in its textiles division is one of the principal suppliers of Nike, the sportswear brand, which was the sponsor of ten national teams in the 2014 Football World Cup. Most of the fabrics used for the teams' uniforms came from recycled PET bottles supplied by the Far Eastern Group (Kao, 2014). In smaller-scale companies as well, the same product life cycle analysis can be applied. Patagonia, the climbing equipment and clothing company, has been one of the pioneers in 3R initiatives, where customers return their used garments to the stores for recycling the materials into new clothing uses (Ball, 2010).

RP in manufacturing can come also from upgrading machinery especially in garment factories where in many cases it is outdated. This does not represent an investment on a big scale, although smaller manufacturers may find difficulties borrowing capital, and buyers are sometimes indifferent to the need (Martin, 2013). However, in the event that machinery upgrading occurs, manufacturers calculate the business they can capture to the factory's capacity—because the factory is more specialized—so they may attract buyers of better brands by entering into a more upgraded network. Updated machinery can raise productivity in garment assembly, enhance quality of manufacturing (Abernathy et al., 1999), and, this in turn, can have a positive impact on workers' wages without necessarily triggering the next wave of relocation. Simply put, manufacturing countries have to detach themselves from the low wage and low skills logic (Martin, 2013).

2.5.3 LABOR CONDITIONS ISSUE

It is a well-known fact that the T&C sectors are the first industries of any developing country along the path to industrialization. The clothing industry is especially the most widespread manufacturing sector, and it has long been the epicenter of the globalization process and a testing ground of workers' conditions. Unfortunately, WTO, the official mediator of trade agreements, has failed to include the social dimension in its provisions. This has led to a multiplicity of private initiatives addressing the issue of labor conditions (Bair et al., 2014).

Since the mid-1990s, these initiatives have evolved into corporate responsibility programs, which led to auditing systems with the purpose of monitoring contractors' compliance. These programs are in essence risk management tools that can safeguard against reputational damage for brands and retailers in the event of sweatshop scandals, which unfortunately occur often in garment factories (Bair et al., 2014; Cline, 2013; Martin, 2013).

The issue is far more complicated because on the one hand, developing countries and especially least developed countries (LDCs) rely heavily on clothing exports and also on the employment of large numbers of semi- or unskilled workers; on the other hand, working conditions are often squalid, and wages in garment factories are in many cases pitiful (Cline, 2013; Martin, 2013; Rose, 2014). It is not unusual for the inspections delegated by the buying companies to independent auditors to be fraught with *audit frauds*. In this, auditors either turn a blind eye to breaches of social and/or environmental regulations or recommend minor remedial actions (Siegle, 2011). In other cases, fake certification documents are produced or buyers are shown demonstration factories but the majority of work is subcontracted to uncertified garment factories (Cline, 2013; Martin, 2013). Workers' wages and especially living wages are a thorny issue in clothing production as factory owners may pay legal minimum wages, but these are far from what the World Bank calculates as wages above poverty line. The cost pressure is constant to cut prices and in some cases, drastic cuts are demanded by buyers. It is sad that actual labor costs account for only 1% of the retail prices of fast fashion clothing articles (Cline, 2013), while sales have increased, but retail prices of clothing have dropped 15% in the past decade (Siegle, 2011).

In the next section, the transparency of the supply chain is examined through the lens of traceability of the supply chain. This relates to GVC and production networks.

2.6 SUSTAINABILITY ARCHITECTURE THROUGH THE TRACEABILITY OF THE SUPPLY CHAIN

The transformation of supply chains from the current state to a greener future is certainly not an easy task. Lubin and Esty (2010) argue that sustainability currently evolves into a megatrend. In this, they postulate that companies wishing to achieve a competitive edge have to view sustainability in its strategic context. They propose a road map based on a two-pronged strategy, which must be pursued simultaneously: formulating a vision for value creation and taking actions. Companies need to redefine what they do in order to embrace this new source of value creation and also to transform systems and processes so that they fit sustainability demands.

Transparent supply chains in upstream production networks may provide the answers, and this is the subject of the UN report on the traceability of supply chains, which can serve as a useful guide for the sustainability architecture in the T&C sectors. Traceability is defined as "the ability to identify and trace the history, distribution, location and application of products, parts and materials." In other words, it is a system that monitors the track of inputs as they enter the supply chain and are manufactured into end products. Adding the sustainability parameter into the system, traceability addresses also the sustainable origin of inputs, together with safeguarding good practices, respect for workers, and obviously, the environmental impact of these actions (United Nations Global Compact Office, 2014). Traceability is a similar concept to life cycle analysis used in T&C that helps clothing companies to investigate the origin of their raw materials and also administer improvements in the processes that are related to environmental issues (Seuring et al., 2008). Traceability is slowly becoming an accepted practice. In January 2014, the EU voted a set of directives on certification programs for companies operating within sustainable requirements. Coordination difficulties among stakeholders surface again, but the glue that should bond all stakeholders together is the unified vision of sustainability. Additionally, investments in technology represent a serious challenge as well as remote manufacturing locations, language barriers, and local managerial skills to access technology (United Nations Global Compact Office, 2014).

The traceability report recommends a best practice model based on what it calls the *global collaborative scheme*, which is demonstrated in Figure 2.1.

In this, there must be a coordinating body that guides and works on traceability with all members in the given supply chain. Partners in the scheme participate according to their position in the chain

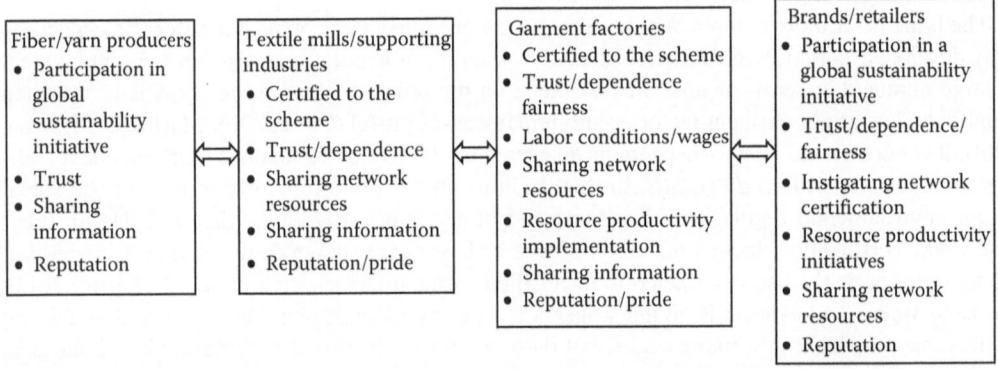

FIGURE 2.1 A global collaborative scheme in the T&C sectors—buyer-driven global value chain. (Adapted from United Nations Global Compact Office, *A Guide to Traceability: A Practical Approach to Advance Sustainability in Global Supply Chains*, United Nations, New York, 2014.)

and also communicate effectively about any problems with their immediate partners in the network (Curwen et al., 2013). Moreover, network members can enjoy reputational benefits since they have demonstrated strong commitment to sustainability principles, once transparency has been achieved. However, traceability is a difficult feat. Companies need to be aware that it is time-consuming and must not be deterred by early disappointments. Partners in the network must establish interim evaluations of the progress and act according to their sustainability objectives (United Nations Global Compact Office, 2014).

2.7 CONCLUDING COMMENTS AND PERSPECTIVES

Lead firms' race to the bottom tactics, combined with the relentless fragmentation of production through waves of relocation, facilitated by trade agreements, has proved to be unethical practices.

Amid the deterioration of natural resources to an alarming state and miserable working conditions, companies in the T&C sectors have initiated some corrective actions. The urgency of sustainability constitutes an imperative priority that needs to build momentum as fast as clothing ranges replenish shelves and racks in retailers' stores.

No one denies that businesses need to survive and thrive in a competitive environment. It has been suggested in the literature that there is scope for competitive practices combined with deference to the environment with sustainability in mind as a vision and actions put in place. The sustainability path is fraught with difficulties. It demands focus, investments, determination, and endeavors for all stakeholders involved, but the prospect of a greener future for all is the biggest reward. We therefore conclude with the following comments.

First, it is imperative that lead firms in buyer-driven GVCs change their course from the current exploitative model to a cooperative one. This will rely on mutual arrangements between the buying firms and central controllers for the betterment of workers' lives and also the *cleaning up* of the supply chains. Easy to say, one might comment, but this needs to be the guiding principle unifying all actions.

The theoretical frameworks presented earlier correspond well with GVCs' competitive principles. In this, the buyer-driven chains in the sectors need to relinquish the captive model of subjugated manufacturers and move into the relational one taking into account the sustainability imperative. GVCs can indeed mutate from the current adversarial conditions to a more collective mode of operation. Opportunism will always be there; after all, guile is a human trait, but the master plan of sustainability is to mitigate these unethical operations.

Second, under the relational model of operations, RP can be improved since network members can work together to raise productivity levels that are not particularly high, especially in the clothing sector in LDCs. Network resources can also be communicated, shared, and improved to the weaker and smaller manufacturing firms since trust prevails in relationships. Companies in the important backward linkages can also mobilize resources, since the collective effort for exports may propel enterprising activities, providing domestic networks with a competitive edge. Moreover, modest investments in upgrading outdated sewing machinery, training staff, and factory repairs for health and safety, among other upgrading actions, are means that the network's central controllers demonstrate genuine interest to provide resources of mutual interfirm specialization and foster new capabilities. Specialization and capabilities, together with trust in partnerships, can lead to sources of competitive advantage in a network. 3R initiatives are also recommended such as wastewater management in laundering facilities (Thorpe, 2011), recycling of clothing, and reuse in new fabrics. This reduces the use of virgin materials but companies still remain competitive since they divert production to reusable resources.

Third, under the relational mind-set, the traceability of supply chain materials is facilitated since all network members contribute meaningfully in the transparency initiative of their network. They also acknowledge that benefits will be fairly distributed among participants. Central controllers become the gatekeepers of traceability and sustainability, by cooperating with other partnering companies in accordance with their positioning in the network. Internal and external communication of the vision and courses of action are paramount and must be provided by the leading firm.

Fourth, the implementation stage demands integrated strategies and not a patchwork of differing views and actions. Networks, as it is understood, welcome integrated strategies as they can provide them with the means to operate in fairness. Sustainability industry standards that are so urgently needed at present can be more easily implemented in partner networks because even sustainable competing networks can acknowledge their usefulness (the SAC is constituted by competing members).

On balance, we believe that relational production networks in buyer-driven chains can be the definitive arrangements to deliver sustainability in upstream operations in the near future. More empirical evidence is required in order to validate findings, as the present chapter was presented in a more analytical manner. It is acknowledged that the path to sustainability may be fraught with difficulties, but humanity is compelled not to ignore the natural capital. It is only a matter of time before the sectors' big global players realize they need to devise sustainability blueprints for the industry and present the sustainability cause to markets, convincing retail customers of *our common future*.

REFERENCES

Abdul Hamid, Z. and Duraiappah, A. 2014. The GDP-well-being gap. *Taipei Times*, May 18, 2014, p. 9.

Abernathy, F. H., Dunlop, J. T., Hammond, J. H., and Weil, D. 1999. *A Stitch in Time-Learning Retailing and the Transformation of Manufacturing-Lessons from the Apparel and Textile Industries.* New York: Oxford University Press, Inc.

Audet, D. 2004. A new world map in textiles and clothing. Organisation for Economic Co-operation and Development.

Bair, J., Dickson, M., and Miller, D. 2014. *Workers' Rights and Labor Compliance in Global Supply Chains, Is a Social Label the Answer?* Oxon, U.K.: Routledge, Taylor & Francis Group.

Ball, J. 2010. *Patagonia Clothing: Making a Profit and Meeting Environmental Challenges. Mother Earth News.* Topeka, KS: Ogden Publications Inc.

Barney, J. B. 1991. Firm resources and sustained competitive advantages. *Journal of Management* 17(1): 99–120.

Barney, J. B. 1999. How a firm's capabilities affect boundary decisions. *Sloan Management Review* Spring: 137–145.

Berger, S. 2006. *How We Compete.* New York: Doubleday of Random House, Inc.

Brundtland, G. H. 1987. *Our Common Future, Chairman's Foreword. Report of the World Commission on Environment and Development.* Oslo, Norway: United Nations.

Chen, H.-L. and Burns, L. D. July 2006. Environmental analysis of textile products *clothing and textiles. Research Journal* 24: 248–261.

Cheng, J., Yeh, C., and Tu, C. 2008. Trust and knowledge sharing in green supply chains. *Supply Chain Management* 13: 283–295.

Clifford, S. 2013. Some retailers say more about their clothing's origin. *The New York Times*, May 8, 2013, p. 1.

Cline, E. L. 2013. *Over-Dressed: The Shockingly High Cost of Cheap Fashion.* New York: Penguin Group.

Cousins, P. D. 2005. The alignment of appropriate firm and supply strategies for competitive advantage. *International Journal of Operations & Production Management* 25(5): 403–428.

Cox, A. 1999. Power, value and supply chain management. *Supply Chain Management: An International Journal* 4(4): 167–175.

Cullen, J. B., Johnson, J. L., and Sakano, T. 2000. Success through commitment and trust: The soft side of strategic alliance management. *Journal of Product Innovation Management* 23: 330–341.

Curwen, L. G., Park, J., and Sarkar, A. A. 2013. Challenges and solutions of sustainable apparel product development. *Clothing & Textiles Research Journal* 31(1): 32–47.

Das, T. K. and Teng, B.-S. 2000. A resource-based theory of strategic alliances. *Journal of Management* 26(1): 31–61.

Duschek, S. 2004. Inter-firm resources and sustained competitive advantage. *Management Revue* 15(1): 53–73.

Frederick, S. and Gereffi, G. 2009. *Review and Analysis of Protectionist Actions in the Textile and Apparel Industries.* London, U.K.: World Bank and the Center for Economic Policy Research.

Gardetti, M. A. and Torres, A. L. 2013. *Sustainability in Fashion and Textiles: Values, Design, Production and Consumption.* Sheffield, U.K.: Greenleaf Publishing.

Gereffi, G. 1999. International trade and industrial upgrading in the apparel commodity chain. *Journal of International Economics* 48(1): 37–70.

Gereffi, G., Humphrey, J., and Sturgeon, T. 2005. The governance of global value chains. *Review of International Political Economy* 12(1): 78–104.

Gereffi, G. and Memedovic, O. 2003. *The Global Apparel Value Chain: What Prospects for Upgrading by Developing Countries.* Vienna, Austria: United Nations Industrial Development Organization.

Grant, R. M. 1991. The resource-based theory of competitive advantage: Implications for strategy formulation. *California Management Review* 33(3): 114–135.

Gulati, R., Nohria, N., and Zaheer, A. 2000. Strategic networks. *Strategic Management Journal* 21: 203–215.

Harney, A. 2009. *The China Price: The Trust Cost of Chinese Competitive Advantage.* New York: Penguin Books.

Hong, P., Kwon, H. B., and Roh, J. 2009. Implementation of strategic green orientation in supply chains: An empirical study of manufacturing firms. *European Journal of Innovation Management* 12(4): 512–532.

Humphrey, J. and Schmitz, H. 2001. Governance in global value chains. *IDS Bulletin* 32(3): 19–29.

Jarillo, J. C. 1988. On strategic networks. *Strategic Management Journal* 9: 31–41.

Kao, C. 2014. Far Eastern to increase capacity of bottle trade. *Taipei Times*, June 27, 2014, p. 13.

Knutsen, H. M. 2004. Industrial development in buyer-driven networks: The garment industry in Vietnam and Sri Lanka. *Journal of Economic Geography* 4: 545–564.

Lam, J. K. C. and Postle, R. 2006. Textile and apparel supply chain management in Hong Kong. *Science and Technology* 18(4): 265–277.

Lee, Y. and Cavusgil, S. T. 2006. Enhancing alliance performance: The effects of contractual-based versus relational-based governance. *Journal of Business Research* 59: 896–905.

Lubin, D. A. and Esty, D. C. May 2010. The sustainability imperative. *Harvard Business Review* 88: 42–50.

Martin, M. 2013. *Creating Sustainable Apparel Value Chains: A Primer on Industry Transformation.* Impact Economy Primer Series. Geneva, Switzerland: Impact Economy.

McCutcheon, D. and Stuart, F. I. 2000. Issues in the choice of supplier alliance partners. *Journal of Operations Management* 18: 279–301.

Morris, M. and Barnes, J. 2008. *Globalization, the Changed Global Dynamics of the Clothing and Textile Value Chains and the Impact on Sub-Saharan Africa.* Vienna, Austria: United Nations Industrial Development Organization.

Nordas, H. K. 2004. The global textile and clothing industry post the agreement on textiles and clothing. Discussion paper no. 5. Geneva, Switzerland: World Trade Organization.

Organisation for Economic Co-operation and Development 2011. Resource productivity in the G8 and the OECD. Paris, France: Paris Organization for Economic Co-operation and Development.

Poppo, L. and Zenger, T. 2002. Do formal contracts and relational governance function as substitutes or complements? *Strategic Management Journal* 23: 707–725.

Rose, M. 2014. Inspections roil garment industry in Bangladesh. *Taipei Times*, June 30, 2014, p. 9.

Seuring, S., Sarkis, J., Muller, M., and Rao, P. 2008. Sustainability and supply chain management. *Journal of Cleaner Production* 16(5): 1545–1710.

Shih, W. Y. 2013. Investigation of the competitiveness of a textile and apparel manufacturer: A case study in Taiwan. PhD thesis, The University of Manchester, Manchester, U.K.

Siegle, L. 2011. Why fast fashion is slow death for the planet. *The Guardian: The Observer*, May 8, 2011.

Smitha, E. 2011. *Screwing Mother Earth for Profit.* London, U.K.: Watkins Publishing.

Tewari, M. 2006. Is price and cost competitiveness enough for apparel firms to gain market share in the world after quotas? A review. *Global Economy Journal* 6(4): 1–46.

Thorpe, L. 2011. Levi Strauss & Co—The Levi style with a lot less water, Best practice exchange. London, U.K.: Guardian News and Media [Online]. Retrieved from: http://www.theguardian.com/sustainable-business/levi-rethinking-traditional-process-water.

United Nations. 2008. *Unveiling Protectionism: Regional Responses to Remaining Barriers in the Textiles and Clothing Trade.* New York: United Nations.

United Nations. 2012. Back to our common future: Sustainable development in the 21st century (Sd21) project. New York: United Nations, Department of Economic and Social Affairs, Division for Sustainable Development.

United Nations Global Compact Office. 2014. A guide to traceability: A practical approach to advance sustainability in global supply chains. New York: United Nations.

Wagner, S. M. 2006. Supplier development practices: An exploratory study. *European Journal of Marketing* 40(5/6): 554–571.

Wernerfelt, B. 1984. A resource based view of the firm. *Strategic Management Journal* 5(2): 171–180.

Williamson, O. E. 1979. Transaction-cost economics: The governance of contractual relations. *Journal of Law and Economics* 22(2): 233–261.

World Trade Organization. 2009. The Global Textile and clothing industry post the agreement on textiles and clothing. Geneva, Switzerland: WTO [Online]. Retrieved from: http://www.wto.org/english/res_e/booksp_e/discussion_papers5_e.pdf, accessed July 2, 2014.

3 Sustainable Flame-Retardant Finishing of Textiles
Advancement in Technology

Kartick K. Samanta, Santanu Basak, and S.K. Chattopadhyay

CONTENTS

3.1 Introduction 51
3.2 Mechanism of Flame-Retardant Finishing of Textiles 53
3.3 Traditional Flame-Retardant Finishing of Textiles 53
    3.3.1 Imparting Flame Retardancy to Cotton 54
        3.3.1.1 Borax and Boric Acid Mixture 54
        3.3.1.2 Nitrogen- and Phosphorous-Based Chemicals 54
        3.3.1.3 Antimony- and Halogen-Based Chemicals 55
    3.3.2 Imparting Flame Retardancy to Jute 56
    3.3.3 Imparting Flame Retardancy to Wool 60
    3.3.4 Imparting Flame Retardancy to Thermoplastic Fibers 60
3.4 Environment-Friendly Flame-Retardant Finishing 61
3.5 Flame-Retardant Finishing Using Nanotechnology 61
3.6 Flame-Retardant Finishing Using Plasma Technology 63
3.7 Flame-Retardant Finishing Using Plant Extract 65
    3.7.1 Thermal Analysis 66
    3.7.2 Chemical Analysis 68
    3.7.3 Mechanism of Flame-Retardant Finishing 71
3.8 Sustainable Flame-Retardant Finishing 71
3.9 Summary 72
References 73

3.1 INTRODUCTION

Among the various functional finishing of textile substrates, flame-retardant finishing is important as it directly relates to human health and hazards. Cellulosic, lignocellulosic, and protein-based natural fibers, such as cotton, flax, ramie, jute, silk, and wool, are mostly used in apparels and home furnishings. Thermoplastic synthetic fibers, such as polyester, nylon, acrylic, and even to some extent polypropylene, are also used for similar applications. Unlike in the case of natural fibers, the situation is more complex for such thermoplastic fibers, as they shrink, melt, and burn in contact with the flame. Although their combustion temperatures are quite high (450°C–550°C), however, they, being thermoplastic in nature, first shrink and then melt in the temperature range of 165°C–265°C. The shrinking of a fabric and dripping of molten polymer can cause different degrees of skin burning. Besides their apparel and home-furnishing applications, both natural and synthetic fibers are also used in hotels, hospitals, automobiles, railways, and airways as tapestries and upholstery. For apparel and home textiles, mostly cotton is preferred due to its advantages of soft feel, good moisture regain, and adequate thermal insulation. But being cellulosic in nature with a low limiting oxygen index (LOI) of 18, cotton catches flame readily and burns vigorously in an open atmosphere, which is difficult to extinguish, and also sometimes causes accidental death. Though cotton and regenerated cellulose (viscose) have combustion temperatures of 350°C and 420°C, respectively, during real-time burning, the temperature can go as high as 400°C–450°C, producing hardly any residual mass. It may be noted that textiles with an LOI ≤ 21 catches flame readily and burns in an open atmosphere rapidly. The samples with an LOI ≥ 21 to ≤ 27 also catch flame and, however, burn slowly in an open atmosphere. On the other hand, the sample with an LOI ≥ 27 is generally considered to be a flame-retardant textile. The major requirement of any textile material to be considered as a flame retardant is that it should have an LOI of more than 27. The situation is slightly better with the lignocellulosic textiles, such as jute with the LOI of 21 or more, making them suitable for packaging of agricultural crops and food products and upholstery and home-furnishing applications. Wool and silk being protein fibers, the situation are still better due to their higher combustion temperature of 600°C (wool) along with higher LOI value of 23.5–25. As the majority of the natural as well as the man-made fibers or fabrics has LOI values <25, there is a need for imparting flame-retardant finishes either during their manufacturing (e.g., polymerization, spinning) or in their final finishing stages making them suitable for various end applications as stated earlier.

The flame-retardant finishing of textiles can be categorized as nondurable, semidurable, and durable based on their efficacy in performance after successive washing cycles. The simple and well-known nondurable flame-retardant chemicals available in the market are inorganic salts, borax and boric acid mixture, diammonium phosphate, and urea.[1,2] Phosphorous-based flame retardants along with nitrogenous compounds due to their synergistic effect provide the most effective treatment for all kinds of textiles as has been reported in the literatures. Chlorine- and bromine-based halogen compounds either alone or with an antimony compound providing synergism have also been introduced in the market in the middle of the nineteenth century. In the last six decades, flame-retardant formulations based on the composition of phosphorous, nitrogen, and halogen, like Tetrakis phosphonium salt and N-alkyl phosphopropionamide derivatives, have been widely used for commercial application.[2,3] However, when such formulations are applied to cotton textiles, the fabric tear and tensile strength get reduced significantly, besides the fabric becoming stiffer. This mainly happened due to the requirement of an acidic condition for application of the aforementioned formulations and involvement of high-temperature drying or curing operations. Although the process is effective for different kinds of fiber, it has the disadvantages of being toxic, hazardous, and costly.[4] To address these drawbacks, quite a few environmentally friendly and cost-effective formulations have been developed in the last few decades, to reduce the amount of formaldehyde release from the treated fabric using butane tetracarboxylic acid (BTCA) as the binding agent.[2,5] Attempts have also been made to develop halogen-free phosphorus–nitrogen-based flame retardants to promote more char formation during burning of the cellulosic substrate.[2] Recently, due to the rapid growth of nanoscience and technology, nano-zinc oxide (ZnO) particles, TiO_2, different clay compounds, and polycarboxylic acid have been reported on cotton and other textiles in an eco-friendly manner.[6,7] Similarly, plasma treatment has been reported either as a pretreatment for increasing the uptake of fire-retardant chemicals, or for graft polymerization of acrylate phosphate and phosphonate derivatives, or as a posttreatment for better reaction. Due to constantly increasing awareness on human health, hygiene, and fashionable textiles with functional values, the demand for cellulosic textiles, processed and finished with natural products, such as natural dyes, enzymes,[8] and aromatic and medicinal plant extracts for antimicrobial, mosquito-repellent, and well-being textiles, is getting importance in academic research and industrial product development for both the traditional and promising upcoming markets.[9,10] Thus, cellulosic textiles have been finished with biomacromolecules, such as DNA from herring sperm and salmon fish, to improve their thermal stability.[10] Whey proteins, casein, and hydrophobins have also been explored for similar such applications.[11,12] In the last few years, efforts have also been made for utilization of agro-residues and other plant molecules for flame-retardant finishing of cellulosic and lignocellulosic textiles. In this regard, applications of banana pseudostem sap (BPS) and spinach extracts for flame-retardant finishing of cotton and jute fabrics due to their advantages of being environmentally friendly, cost effective, and sustainable materials produced from renewable sources are worth mentioning.[13]

The functional flame-retardant finishing of fibrous materials is an emerging area of research and sustainable product development. In this context, this chapter discusses in detail the mechanism of flame-retardant finishing of various textile substrates and the development of various chemical formulations over the years along with their merits and demerits. Further, it describes the application of emerging nano- and plasma technology for imparting flame-retardant finishing of textiles. A special reference to the recent developments on eco-friendly, cost-effective, and sustainable flame-retardant formulations from plant- and animal-based resources has also been made.

3.2 MECHANISM OF FLAME-RETARDANT FINISHING OF TEXTILES

The mechanism of imparting flame-retardant finishing to textiles is related to the combustion process of a fiber or fabric in contact with flame. It makes a difference between the untreated and flame-retardant-treated textiles in terms of LOI, heat required for combustion, burning rate, rate of heat release (RHR), and formation of flammable and nonflammable gases, char, and tar. On application of heat energy to a textile, its temperature goes up and finally reaches to the pyrolysis temperature. At that time, several chemical changes occur in the polymer structure, such as formation of nonflammable gases, like carbon dioxide, water vapor, carbon char, and tar, and flammable gases like carbon monoxide, hydrogen, and many other oxidizable organic volatiles. With further increase in temperature, the fiber reaches to its combustion temperature, when the flammable gases in the presence of oxygen burn rapidly and produce more amount of heat energy. Some portion of this heat energy again accelerates the pyrolysis of the polymer that successively produces more flammable gases, resulting in an intense combustion. Hence, to formulate an effective flame-retardant chemical, the points to be taken care of are either to increase the pyrolysis temperature of the fiber and formation of char and nonflammable gases or to decrease the formation of flammable gases and masking the availability of oxygen. A few approaches followed for the purpose are described in the following:

1. Use those chemicals that thermally decompose through a strong endothermic reaction to ensure less heat production.
2. Alter the pyrolysis temperature of the polymer to ensure formation of more char and less flammable volatile gases.
3. Dilute or replace the surrounding oxygen in the burning microclimate with inert or nonflammable gases to reduce or stop combustion.
4. Apply a suitable chemical that will form a glassy/foamy insulating layer on the textile substrates during combustion to restrict the flow of heat and/or oxygen.

3.3 TRADITIONAL FLAME-RETARDANT FINISHING OF TEXTILES

In the sixteenth century, Sabbatini et al. reported that clay or gypsum pigments could be added in the paint for a theater scenery to impart flame retardancy.[14] Obadiah Wyld et al. published a patent in the seventeenth century on the flame-retardant formulation by a mixture of borax, alum, and ferrous sulfate. In the eighteenth century, Gay Lussac et al. developed fire-retardant jute and cotton fabrics with a mixture of ammonium phosphate, ammonium chloride, and borax.[14,15] Later on, some other research group reported the use of aluminum hydroxide and calcium carbonate as fillers in textiles or polymers. These chemicals thermally decompose on heating through the strong endothermic reactions. Here the pyrolysis temperature of the fiber is not reached and no polymer combustion takes place. Inorganic salts also have been known for a long time for imparting flame retardancy to the cellulosic textiles that will not come in contact with water, rain, or perspiration.

With the advent of synthetic fibers during the period 1935–1980, many flame-retardant formulations and their application mechanism have been developed to impart flame retardancy in polyester, nylon, and polypropylene textiles, either by co-/graft polymerization (modacrylic LOI 29) or by using additives in the fiber-spinning bath. In the middle of the 1960s, many synthetic fibers with high LOI value and thermal resistance have also been developed, for use by the firefighters and in high-performance industrial applications, where the sample has to sustain its existence at an elevated temperature for a longer duration. Such high-performance fibers are basically made of aromatic-structured polymeric chains with a high strength-to-weight ratio and modulus. Some of the examples of these kinds of fibers are poly(para-aramid), better known as Kevlar, and poly(meta-aramid), known as Nomex. These fibers have LOI values of 28–30 and combustion temperature > 500°C. Being their LOI values >27, they are intrinsically considered as flame-retardant fibers/fabrics. A few more fibers in this category are carbon, poly(*p*-phenylene benzobisoxazole) (PBO-Zylon), polybenzimidazole (PBI), poly(tetrafluoroethylene) (PTFE), polyimide, and *Fyrol* (diethyl *N,N*-bis(2-hydroxyethyl aminomethyl phosphonate)). These fibers can sustain very high temperatures in the range of 500°C–700°C and for a longer duration compared to the maximum temperature ≈ 350°C at a shorter duration in the case of most apparel-grade fibers.

3.3.1 Imparting Flame Retardancy to Cotton

3.3.1.1 Borax and Boric Acid Mixture

Borax and boric acid mixture with about 10% add-on are the most well-known chemicals commonly used for the flame-retardant finishing of fibrous material. Borax is used to lower down the melting temperature of boric acid, and on heating, it releases water vapor. Besides, it produces a thick glassy layer of boron trioxide on the substrate surface before reaching to its pyrolysis temperature. The mechanism is to insulate or cover up the textile fully or partially, from heat and surrounding oxygen. However, as these chemicals do not form any chemical bond with the substrates, they lose their efficacy after the laundering operation.

3.3.1.2 Nitrogen- and Phosphorous-Based Chemicals

Diammonium phosphate along with urea and ammonium salts of phosphoric acid can be used as a fire retardant for cellulosic substrates. They act in the condensed phase mechanism. When a nitrogen- and phosphorous-based chemical is applied to improve the thermal stability of the cellulosic substrate, it reduces the pyrolysis temperature and promotes production of more residual char and less flammable volatiles. This happened due to the fact that phosphorous-containing flame retardants produce phosphoric acid at an elevated temperature, which then cross-links with the hydroxyl group of the cellulose, thus causing early dehydration. In such type of formulations, nitrogen acts in synergism with phosphorous and depicts a better efficiency. The interaction between phosphorous and nitrogen also alters the thermal decomposition pathway of cellulose. Possibly, it delays the thermal decomposition of cellulose by depolymerization. As a result, the LOI of the cellulosic fabric is enhanced through accelerated dehydration on heating. In this regard, the synergistic effect of trimethyl melamine (TMM) and dimethyl dihydroxy ethylene urea (DMDHEU) as a nitrogen provider with an organophosphorus compound has been reported in the literature.[15] The TMM serves as a better nitrogen donor compared to DMDHEU, depending on the measurement of improvement in the LOI after the application. A group of researchers reported that probably organic nitrogen helps in controlling the pH during the cross-linking reaction of phosphoric acid. Here, nitrogen gets protonated, thus reducing the amount of available acid required for cross-linking. It also might have been converted into phosphorous acid amide that can catalyze the dehydration and carbonization of cellulose. The water-insoluble ammonium polyphosphate (APP) is also known as an effective flame retardant, when it is applied to binder systems in the coating. However, the majority of the aforementioned formulations are not durable to washing.

The most successful durable flame retardant for cellulosic textiles is based on phosphorous- and nitrogen-containing chemicals that can directly react or cross-link with the fiber. Tetrakis hydroxymethyl phosphonium chloride (THPC) is one of those compounds, which in mixture with urea forms an insoluble structure on the cellulose. The process is commonly known as *Proban* process. THPC is made of phosphine, formaldehyde, and hydrochloric acid. It can be applied to the cotton fabric with urea by pad-dry-cure method. Approximately 25% of THPC along with 15% urea formulation was found to yield 4%–5% phosphorous add-on in the cotton fabric. However, as discussed earlier, the treatment of cotton textile with such chemicals increases the stiffness of the sample with a simultaneous decrease in tensile and tear strengths. Further, it releases formaldehyde, a carcinogenic chemical during the processing. To overcome this drawback, THPC-urea-treated fabric was dried at 15% moisture content and then exposed to the ammonia vapor in an enclosed chamber, followed by oxidation with hydrogen peroxide. The finish achieved by this method resulted in imparting a good fire retardant to the textiles with better retention of fabric physicomechanical properties. However, the process is sensitive to dye molecules and the formulation affects different dye classes, such as reactive, direct, and acid. Therefore, it is recommended that in the fire-retardant finishing of cellulosic textile with THPC-urea, vat dyes should be used. Consequently, in the last five decades, flame retardants based on the composition of phosphorous, nitrogen, and halogen like Tetrakis phosphonium salt and N-alkyl phosphopropionamide derivatives have been widely used for commercial application.[16] In this regard, N-methylol dimethylphophopropionamide (Pyrovatex CP) in combination with trimethylol melamine and phosphoric acid is applied to the cotton fabric by pad-dry-cure process with an add-on of around 20%–25%. After the treatment, a wash with alkali is given to remove the residual phosphoric acid from the treated fabric surface. Unlike in the earlier Proban process, the main advantages of this process are durability to 50–60 wash cycles and low release of formaldehyde (below 40 ppm). It may be noted that melamine formaldehyde has a dual function, namely, as a resin to cross-link Pyrovatex to cellulose and as an additional nitrogen donor to accelerate the dehydration of cellulose. The restriction on choice of dye classes is less in this process compared to the Proban process and the treated fabric remained comparatively softer. Presently, most of the textile industries use N-methylol dimethylphophopropionamide (Pyrovatex) with a melamine resin for fire-retardant finishing of cotton textile with the following recipe:

Sl. No.	Chemicals	Dosage (gpl)
1	Softener	50
2	Wetting agent	30
3	Melamine resin	60
4	Pyrovatex	400
5	Phosphoric acid	25

The fabric is padded with the aforementioned recipe, followed by drying at 100°C for 5 min and curing at 150°C for another 5 min, and then washing in alkali water for neutralization. Though the process is quite advanced and popular in textile industry, but it suffers from higher cost of application and unpleasant odor of formaldehyde emitted during the processing. Besides, the chemicals cited earlier are hazardous, non-eco-friendly, and nonbiodegradable. Due to the release of significant amounts of formaldehyde, the finishing process poses hazard to industrial workers, as it causes mucous and breathing problems. Considering the disadvantages, the Pyrovatex-finished flame-retardant textile is not recommended for home furnishing applications.

3.3.1.3 Antimony- and Halogen-Based Chemicals

Unlike other finishing areas of textiles using nano-, bio-, and plasma technology, very little research has been reported on the flame-retardant finishing of textiles using antimony compounds in the last decade of the twentieth century. Thereafter, the lead was taken by the U.K. home-furnishing industries due to the requirement of fabric to be resistant to cigarette ignition.[2] The same was met by providing halogen-containing back coatings to the upholstered fabrics. This coating prevents the combustion of the substrate by free radical reactions that provide the heat required for the flame propagation. The halogen-containing compounds work in the gas-phase mechanism and, those on heating, release hydrogen halide, which forms long-lived and less reactive free radicals. The formation of free radicals reduces effectively the heat available for fire propagation in concurrent with the dilution of oxygen in the burning microclimate. The main advantage of this process is that it can be applied to both natural as well as synthetic textiles, as it involves radical transfer reactions. Antimony chloride and antimony bromide are the few examples of this category. Linked to it, antimony oxide (Sb_2O_3) with stannous chloride provides a synergistic effect of flame retardancy into textiles. Here, the synergistic effect of antimony comes from the volatility of antimony trihalides and the effectiveness of antimony compounds in scavenging the free radicals. However, when these chemicals are applied to textiles with a binder for better fixation and durability, the fabric drape and handle get affected due to the presence of the binder. Also, there is a possibility of generation of toxic polyhalogenated dioxins and furanes during processing.

Due to the environmental issues associated with halogen compounds and the increasing demand for flame-retardant textiles, one new mechanism of flame retardancy, that is, intumescent technology, has come into the market in 1996. Intumescent chemicals provide a foamlike insulating layer on the fabric surface. This thick coating protects the underlying polymeric material from heat and flame. Successively on heating, the material gets expanded to increase in volume but decrease in density. Consequently, there was a production of porous carbonaceous foamlike char mass that masks the substrates from available oxygen. The reduction in smoke generation and toxic gas formation are the other two main advantages of this process. During exposure of the treated material to flame (i.e., at high temperatures), the temperature of the intumescent material is raised causing melting of the thermoplastic matrix. As the temperature corresponds to the approximate value for the viscosity of the melt, an endothermic gas-producing reaction is triggered. The gas is collected in small bubbles, thus turning the material to be foam. The formation of solid multicellular carbonaceous foamy char provides an insulating layer that slows down the heat and the mass transport required for the flame propagation.[1] The intumescence property of a material depends on the ratio of carbon, nitrogen, and phosphorous atoms.[2] APP is an example of intumescent that is applied as a coating in rigid polyurethane (PU) foams. Melamine is another example of a nitrogen-containing intumescent applied as a fire retardant for varnishes and paints. Few more such examples are ammonium polyphoaphate, ammonium pentaborate, and melamine and its salts for flame retardancy of polyamide fiber (Nylon 6). All these chemicals work in the condensed phase. It was found that APP interacts with nylon 6 and produces alkylpolyphosphoric ester that acts as a precursor of the intumescent char. An APP add-on level of 10%–20% provides an LOI of 23–24, which was found to increase to 41–50, when the add-on was increased to 40%. Though this is a promising technology, it has the limitations of water solubility, brushing, and high cost associated with a higher add-on.

3.3.2 Imparting Flame Retardancy to Jute

In lignocellulosic fibers, cellulose, being the major constituent (70%–95%), also catches flame readily and is quite difficult to extinguish, thus posing risks to human health and damage to textile products.[3] The lignocellulosic textile (jute) has a higher LOI value (≈21) compared to cellulosic textiles (e.g., cotton and viscose) with an LOI of 18–19. In the past, significant efforts have been made to improve the flame-retardant property of jute textiles using various inorganic salts and commercially available synthetic chemicals. Inorganic salt, like sodium potassium tartrate (Rochelle salt), had been used as a fire-retardant agent for jute fabrics during the 1980s. The process could be considered environment friendly and the treated fabric showed good thermal stability while maintaining the mechanical strength of the fabric.[17] However, the main issue with this process is the loss in efficacy after rainwater washing or soap washing. Similar to cotton textiles, the application of common borax and boric acid mixture has also been reported in the literature.[4] Boron-based derivatives provide better stability of finish due to their deeper penetration into the fiber, yarn, and fabric structure. Furthermore, boron compounds have fungicidal and insecticidal properties that are very much beneficial to jute products, as they are used in packaging of agricultural crops and for backing carpets.[18] Similar to the application in cotton as discussed earlier, phosphorous-based flame retardant in combination with nitrogenous compound is also very successful in imparting flame retardancy in jute due to their synergistic effect and formation of more nonflammable products.[19] In this regard, a research group on jute has developed a durable flame-retardant jute fabric with diammonium phosphate, urea, and resin with an LOI of 34, when the add-on percentage was kept at 7%. However, the durability of the imparted finish was not satisfactory and the treatment process had environmental effects. Sulfur along with nitrogen synergism has also been attempted to improve the flame-retardant property of the jute fabric. Sulfur- and nitrogen-based thiourea and thiourea resin has been applied on the jute fabric and it has been found that thiourea-treated fabric showed a fire-retardant behavior.[20] It might have happened due to the fact that thiourea [$SC-(NH_2)_2$] is an organosulfur compound, and upon heating, it releases ammonia, nitrogen oxide, sulfur oxide, and water that protect the substrate from getting O_2 supply, thus improving the LOI value.[19,22] Furthermore, sulfur oxide formed on heating reacts with the atmospheric oxygen (O_2) at elevated temperatures and produces highly reactive sulfur trioxide (SO_3), which further reacts with water vapor or water to produce sulfuric acid.[22] Similarly to phosphoric acid in phosphorous-based flame retardants, the carbonium ion mechanism might have also helped in the dehydration of cellulose in lignocellulosic fibers resulting in increased char formation.

In a similar line, a thiourea resin was applied to the jute fabric to improve flame-retardant and mosquito-repellent functionalities.[23] The treated fabric showed a higher LOI of 45 at 10% add-on compared to an LOI of 21 measured in the untreated jute sample as indicated in Table 3.1. In the treated sample after the flame was stopped, the afterglow remained for a longer duration and continued the fire propagation. In the control sample, no char length was observed as the sample burnt completely with flame, whereas in the urea- and resin-treated samples, 12.5–20 cm char length was observed depending on the concentration of the applications. They have also studied the temperature profile in the untreated and the thiourea-treated samples during the burning of samples in real time using an infrared thermometer in noncontact mode as shown in Figure 3.1.

It can be seen from the figure that the untreated sample burnt completely very fast with flame within 60 s with temperature as high as 400°C–450°C. In contrast, the treated sample burnt only at 100°C for the initial 10 s with an afterglow. Later, after 20 s, the afterglow stopped and the fabric temperature rapidly falls down to the ambient temperature within 20–30 s. In contrast, the control fabric burnt with flame, and a small extent of the afterglow with temperature 250°C–300°C could be seen at the edges of the sample holder. The thiourea-treated sample also showed the mosquito-repellent property, in addition to its flame retardancy. The mosquito repellent was tested quantitatively by the cage test and it has been reported that out of 25 mosquitoes in the cage, 10 mosquitoes were present on the control jute fabric after 1 h, whereas only 3 mosquitoes sat on the treated fabric. Thus, the treated fabric was 70% mosquito repellent. After 90–120 min, some of these mosquitoes were found surrounding the treated fabric sample, but none of them landed on the fabric. In contrast, in the control sample, the number of mosquitoes that landed increased with time. The mosquito-repellent effect in the treated fabric was possibly attributed to the strong odors of thiourea sulfur dioxide and ammonia.[24,25]

The same research group also tried to improve the durability of the thiourea-treated samples with the DMDHEU resin. It was found that with increasing the resin concentration, the LOI of the treated fabrics was also increased due to the presence of more nitrogen, triggered by the DMDHEU resin. However, the thiourea process has the limitation of imparting yellowness to the fabric and reduction in mechanical strength, as the application is done in acidic condition.

To improve the durability of the imparted flame-retardant finish to the water, reactive halogen-based chemicals have been developed. Antimony in combination with various halogen compounds, a well-known formulation for cotton textile, has also been applied to the lignocellulosic substrate.

TABLE 3.1
Flame-Retardant Parameters in Control, SMSN-, and Thiourea-Treated Jute Samples

Flammability Parameters	Control	SMSN Concentration (%)		
		2	4	8
Add-on (%)	0	2	4	8
LOI	21	29	32	43
After flame time (s)	60	10	5	Nil
Afterglow	Completely burnt with flame in 60 s	Completely burnt: initially with flame followed by afterglow for 10 min	Completely burnt: initially with flame followed by afterglow for 30 min	30 s
Char length (mm)	—	—	—	7
Burning rate (mm/min)	250	25	8.3	—

Thiourea-Treated Samples

	Control	Thiourea (80 gpl) + Different Resin Concentration		
		0	40	80
LOI	21	43	45	45
Flashing over the fabric surface	Yes	No	No	No
Flame time (s)	Completely burnt with flame within 60 s	Nil	Nil	Nil
Afterglow (s)		80	60	50
Char length (mm)	—	20	15	12.5
State of fabric	Completely burnt	Fire retardant	Fire retardant	Fire retardant

Sources: Basak, S. et al., *J. Text. Assoc.*, 74(5), 273, 2014; Basak, S. et al., *Man Made Text. India*, 61(11), 386, 2013; Basak, S. et al., *Polish J. Chem. Technol.*, 16(2), 106, 2014.

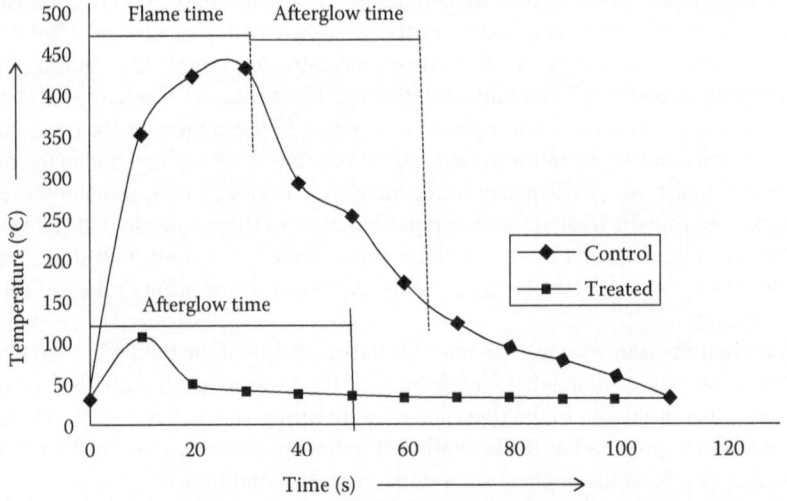

FIGURE 3.1 Temperature profile in control and thiourea-treated jute fabrics during burning. (From Basak, S. et al., *J. Text. Assoc.*, 74(5), 273, 2014.)

Similar to cotton, this particular process was not well accepted commercially due to its anticipated negative impact of halogen compounds on the environment and the requirements of the large quantum (15%–30%) of chemicals to achieve a satisfactory level of flame retardancy.[2] One of the most efficient methods of protecting such lignocellulosic textile from fire is the use of intumescent-based fire-retardant coating, so as to reduce the percentage add-on to achieve a similar degree of fire protection. For example, carbonizing compounds, such as polyhydric alcohols, polyphenols, carbohydrates, and resins; dehydrating compounds like diammonium phosphate, urea, melamine, ammonium sulfate, and foam forming agents like dicyandiamide; and melamine have been reported to enhance more carbonaceous char formation.[16] Similar to this principle to some extent, sodium silicate nonahydrate (SMSN) has been applied on jute fabric and various thermal, mechanical, and other functional properties were evaluated.[27] It has been reported that SMSN not only improves the thermal stability of the samples but also ensures better antimicrobial efficacy. It can be seen (Table 3.1) that the LOI value linearly increased from 21 in the untreated sample to 29 and 32 in the 2% and 4% treated samples, respectively. With an increase in quantity of SMSN, burning with flame time gradually decreased and burning with afterglow time increased from 10 to 30 min in the 2% and 4% treated samples, respectively. In the 8% treated sample due to its high LOI value of 43, it did not burn with flame and also the afterglow was self-extinguishable within 30 s and produced a char length of 7 mm. The total burning time of a similar sample was possible to increase from 60 s in the untreated sample to 610 and 1805 s in the 2% and 4% treated samples, respectively. The SMSN possibly formed a silicate coating on the jute fabric that acted as an intumescent. It also serves as a hydrated plaster and a similar report has been reported in the literature for nontextile substrates.[28] Upon exposure to a high temperature, it swells and increases in volume and forms an insulating carbonaceous foamlike glassy layer on the treated fabric surface.[28,29] The insulating layer contains char-producing polyol that normally burns to produce nonoxidizable gases, like CO, H_2O, and CO_2. These generated gases might have diluted the major flammable gas concentration, such as levoglucosan and pyroglucosan produced during the pyrolysis of the lignocellulosic material in the burning microclimate. Furthermore, the chemically attached nine molecules of water with SMSN absorbed a significant amount of heat during burning, which is also hindered in faster fire propagation. Many of the existing flame-retardant formulations reduce the tensile strength of the fabric up to 30%, as they are applied in acidic condition. As this particular process was applied in alkaline condition, there was only a marginal loss in tensile strength. In the 2% SMSN-treated sample, there was only 3.7% and 7.5% reduction of tensile strength in the warp and weft directions, respectively. In the case of 8% SMSN-treated sample, there was 6% and 15% loss in the tensile strength in warp and weft directions, respectively.

From the thermogravimetric (TG) analysis, it was found that pure SMSN showed 60% mass loss in the temperature range of 50°C–200°C, possibly due to the loss of unbound and bound water molecules in different ways.[33] Beyond 200°C and till 800°C, there was no significant change in the mass. On the contrary, the 4% and 8% SMSN-treated jute fabrics lost more mass below 150°C compared to the control jute fabric due to the loss of nine molecules of water of SMSN along with the water molecules present in the fabric.[22,31] In the second stage, both the 4% and the 8% treated jute fabrics started to lose its mass due to its earlier pyrolysis from 230°C, which was about 35°C lower than the pyrolysis temperature of the control fabric (265°C). It was inferred that the presence of SMSN in the jute fabric had reduced the pyrolysis temperature of the lignocellulose and increased the overall thermal stability. Additionally, dilution of flammable volatile gases by the generation of nonoxidizable CO_2 and H_2O at relatively lower temperatures might have also contributed in providing flame retardancy.[22] In spite of having several advantages, like good fire protection, eco-friendliness, low cost, marginal effect on mechanical strength, ease of application, and stability to weathering, the SMSN-based intumescent technology has not got commercial success, because of its lack of durability in repeated laundering. As a result, Pyrovatex-based technology is widely used in the industry till today due to its main advantages of repeated wash durability and little or no adverse effect on the fabric physical properties, though the processing condition is not so eco-friendly.

Besides the excellent flame-retardant functionality, the SMSN treatment was also found to exhibit good antimicrobial efficacy against both gram-positive and gram-negative bacteria. It was observed that the antibacterial activity was increased gradually from 62% in the 2% SMSN-treated sample to 99.9% in the 8% SMSN-treated sample against *Staphylococcus aureus* (Sa). A slightly better result was obtained against *Klebsiella pneumonia* (Kp) bacteria. Hence, SMSN can be used for dual functional finishing of the lignocellulosic textile. The antibacterial activity in the treated samples may be attributed to the higher pH of SMSN and breakage of the membrane integrity of bacteria and disruption of the cytoplasmic membrane.[32-34]

3.3.3 IMPARTING FLAME RETARDANCY TO WOOL

It is known that protein fiber, like the wool, has a higher LOI of 25. It means that wool fiber is to some extent flame retardant due to the presence of nitrogen-based protein molecule in the wool keratin polymer. Despite this, wool fiber requires an additional flame-retardant treatment in order to meet the flammability standards required for some specific end applications. The well-known process for the same is commonly known as *Zirpo*, in which a mixture of hexafluorozirconate with titanate salt is applied by exhaust or padding method under acidic conditions. The process helps to attain an LOI as high as 35. Flame retardancy takes place in the condensed phase through zirconium ions that enhance or catalyze the carbonaceous char formation. This finish can withstand dry cleaning and aqueous washing at 40°C at neutral pH. Moreover, it has no significant adverse effect on the fabric strength unlike other flame retardants as described earlier. The only bothering issue is that with time the treated fabric slowly turns yellowish, whose degree increases with increasing light exposure. Alternatively, tetrabromophthalic anhydride (TBPA)-based flame retardant with an add-on of 10% in acidic condition also provides effective textile flame retardancy. The imparted finish is durable to mild laundering and dry cleaning, but it generates poly(brominated dioxins) under burning conditions.[1]

3.3.4 IMPARTING FLAME RETARDANCY TO THERMOPLASTIC FIBERS

Thermoplastic fibers, such as polyester, nylon, and polypropylene, are subjected to flame-retardant finishing depending upon their end applications. In case of nylon and polyester, the flame-retardant formulation can be applied, either in the polymerization stage (e.g., during polymerization) or as a spinning bath additive or grafting with a suitable monomer at the fiber stage or even in the later stage similar to an application done in the case of a traditional chemical finishing. Phosphorous and bromine–containing compound is the most effective fire-retardant formulation for the polyester fabric. Bromine-containing phosphate ester composition can be applied to polyester by both the padding and exhaust methods to achieve the desired fire retardancy. The only issue that needs to be tackled is their adverse carcinogenic effect that has resulted in their withdrawal from the market. In the recent time, the most commonly applied fire retardant for polyester is the mixture of cyclic phosphate/phosphonates. With an add-on of 3%–4% by pad-dry-heat steps, it could provide a durable flame retardancy to a wide range of polyester fabrics. Hexabromocyclododecane is another example of a durable flame-retardant agent for the polyester fabric. The material is fully water insoluble and a good flame retardant, when the add-on percentage is kept about 8%.

Similar to polyester, phosphorous and bromine–containing compound is also preferred and commonly used for nylon. Thiourea with the mixture of urea and formaldehyde also exhibited excellent fire-retardant effect on the nylon fiber. The effect of such fire retardant is possibly due to the lowering of the melting point of nylon by 40°C–50°C and also allowing the fiber to drip away from the ignition source. A back coating of nylon fabric with a mixture of antimony trioxide with bromine also provides an excellent fire-retardant effect. Other synthetic fibers, like polypropylene, can also be made flame retardant with the mixture of the aforementioned phosphorous and bromine–containing compound. However, to achieve a satisfactory level of flame retardancy, more quantum of flame-retardant agent needs to be applied.

3.4 ENVIRONMENT-FRIENDLY FLAME-RETARDANT FINISHING

Since year 2000, some of the major issues related to environment friendliness, toxicity, and sustainability of various traditional flame-retardant compounds have become a serious issue with increasing global awareness of sustainable eco-friendly green textile chemicals, auxiliaries, and textile products. Some of the adverse effects of the well-known existing flame-retardant formulations are mentioned in the following:

- Release of formaldehyde during the curing process that is hazardous to the worker and the surrounding environment.
- Toxicity of some of the halogen-based compounds and presence of heavy metals, as they produce toxic dioxins and furanes.
- Phosphorous, halogenated compounds, antimony, and zirconium compounds increase the toxicity of the wastewater and environment.
- A majority of the flame retardants reduce the fabric tensile and tear strengths and increase stiffness, making the fabric harsh in feel, yellow in color, and nondurable. Although industries are mostly using softener and easy-care chemicals to mitigate with such shortcomings in the fabrics, however, it increases the number of processing steps and cost due to the additional requirements of chemical and process, besides the effluent generation.

Thus, there is a need of development of environment-friendly, natural, and biomolecule-based sustainable flame-retardant formulations. As such, containing formaldehyde release, toxicity and reduction in fabric physicomechanical properties are the key challenges associated with many of the existing flame retardants. To address them, significant researches are being carried out both in the academic and the industrial sectors. A research group from the United States has reported a fully formaldehyde-free flame-retardant chemical from the char-forming polycarboxylated species, such as BTCA. The functional groups of BTCA get attached to the hydroxyl group of cotton cellulose and thus develop flame retardancy. Another research group has utilized BTCA as a cross-linker between the hydroxyalkyl organophosphorus oligomer and the cellulose polymer. Researchers have also developed Noflan, a mixture of phosphorous and nitrogen compound, mainly based on alkyl phosphoramidate. It can be applied with salt containing ammonium chloride that may react with the cellulosic substrate through the phosphoramidate NH_2 group. This formulation can be applied with methylated resin for improved durability of cotton and polyester–cotton blended textiles. Similarly, the Swiss-based research group has developed organophosphoramidate-containing flame retardants for cellulosic substrates. Researchers from the USDA Southern Regional Laboratory have synthesized two (2) new monomers, (2-methyl oxiranylmethyl)-phosphonic acid dimethyl ester and [2-(dimethoxy-phosphoryl methyl)-oxyranylmethyl]-phosphonic acid. These two monomers along with dicyanamide and citric acid can provide LOI of more than 28. Similarly, the particular finish is also suitable for cotton and its blended fabrics.[1,2] In our laboratory, eco-friendly flame-retardant formulations have been developed using BPS, an agro-waste plant residue, and spinach juice (SJ), a vegetable extract. After application of these two plant extracts on cellulosic cotton and lignocellulosic jute textiles, the LOI was found to increase over 30 from a value of about 21 in case of their untreated samples. The applications have been discussed in details in Section 3.7.

3.5 FLAME-RETARDANT FINISHING USING NANOTECHNOLOGY

Nanotechnology, an emerging science, has gained profound interest within the research community because of its various industrial applications, such as in the fields of textile, material, medicine, electronics, plastics, energy, aerospace, biotechnology, and agriculture. The novel material properties and low chemical consumption associated with nanoparticles have attracted global interest, across disciplines and industries including the textile industry. Nanotechnology endeavors at manipulating atoms and molecules in a controlled manner in order to build up a new material with novel properties that can be used to produce smart and intelligent textiles, such as water-repellent, oil-repellent, hydrophilic, UV-protective, and flame-retardant properties. Nanomaterials, like nanoparticles, nanoclay, nanofibers, and nanocoating, are applied in textile substrates in order to improve such functionalities.[35]

The antimony nanoparticles have been synthesized using potassium antimony tartrate precursor with particle sizes in the range of 350–550 nm.[36] The x-ray diffraction (XRD) peaks at different 2θ values of 27.7, 32, 35, 45.9, and 57.1 have confirmed the crystal structure of the antimony oxide. Similarly, the wave number in the region of 400–800 cm^{-1} depicted the symmetric and asymmetric vibrations of Sb_2O_3 as observed in the Fourier transform infrared spectroscopy (FTIR) spectrum. After the application of antimony oxide nanoparticles on a cotton fabric along with zinc chloride, the sample was found not to catch the flame, whereas the untreated sample burnt completely within 18 s at a maximum temperature of 450°C. The LOI value was found to increase significantly from 18 in the untreated sample to 24 in the nanoparticle-treated sample. With further increase in the add-on percentage of antimony oxide, the LOI was found to increase. The thermal gravimetric analysis (TGA) curve of the treated sample showed the starting of pyrolysis of cotton at 230°C at a slower rate than the untreated cotton, where a sharp degradation was observed with pyrolysis starting at 330°C. It was further observed that the untreated cotton loses most of its mass before 350°C, whereas it was 500°C in the case of the nanoparticle-treated sample. After washing of the sample with soap solution, the LOI was found to reduce to 21 from 24 noted in the treated sample.

Yet in an another study, the ZnO nanoparticle was used to impart dual flame-retardant and UV-protective functionalities in cotton and 65/35 cotton/polyester blended textiles.[6] The size of the synthesized ZnO particles was 30 nm. The nanoparticles were applied in the textile by the conventional pad-dry-cure method with a mixture of different polycarboxylic acids, such as succinic acid (SA) and BTCAs, and sodium hypophosphite (SHP) catalyst. The effect of catalyst concentration of SHP on flammability was studied in details. It was observed that char length was much more in case of the treated samples (pure and blended fabrics) compared to the untreated sample.

An improvement in the mechanical property of polyamide 6 (PA6) has been reported with nanoclay-added fibers. The incorporation of nanoclay into the PA6 fibers leads to the enhancement in the fire-resistant properties also, thus offering a new technology proposition for imparting highly durable fire-retardant finishing to the PA6 textiles at a lower cost.[37] The presence of nanodispersed montmorillonite (MMT) clay in polymeric matrices showed a significant improvement in the fire-retardant property. This research group has also developed hybrid polymeric materials, including organomodified clays, TiO_2 nanoparticles, silica nanoparticles, and carbon nanotubes (CNTs).[38] These materials exhibited improvement in the fire-retardant property in terms of the heat release rate (HRR). It was found that the peak of HRR was decreased by 50%–70% in a cone calorimeter experiment. In a similar line, the nanoclays and poly(silsesquioxanes) can reduce the peak HRRs in PU-coated cotton and knitted polyester fabrics. However, the presence of these nanoparticles alone reduced the time to ignition and prolonged the time of burning.[38] MMT clay and polyhedral oligomeric silsesquioxanes (POSSs) have been added to PU for coating of cotton and polyester fabrics for improvement in flame retadancy.[39] After application of PU/clay and PU/POSS coating, the polyester or cotton fabrics showed better flame retardancy properties in terms of cone calorimetry and TG analysis. Similarly, the flame-retardant property of the polyester fabric was improved by a novel layer-by-layer nanocoating.[40] Instead of direct application of the silica nanoparticle, the same layer-by-layer technique was applied to impart a multilayer coating. Five bilayers of positively and negatively charged colloidal silica with <10 nm average thickness were found to increase the time to ignition of the PET fabric by 45% and decrease the HRR peak by 20%. In the vertical burning test, this nanocoated polyester sample showed a drastic reduction in the burning time, besides the elimination of more dangerous melt dripping. In another study, the flame-retardant coating of the cotton fabric was carried out with branched polyethylenimine (BPEI) and sodium MMT clay by the layer-by-layer technique.[41] Four different coating recipes were formulated by changing the pH and clay loading (0.2 or 1 wt%). The BPEI at pH 10 produced the thickest films, whereas 1 wt% MMT gave the highest clay loading. The performance was evaluated for the 5 and 20 bilayers. TG analysis showed that the coated fabrics left as much as 13% char after heating to 500°C, which was nearly double in magnitude compared to the uncoated fabric. Additionally, it reduced the afterglow time in the vertical flame tests and could maintain the weave structure of the fabric after burning as observed in SEM images. The calorimeter data showed that for all the coated fabrics, both the total heat release and the heat release capacity reduced. Unlike the traditional finishing of textiles, here the fibers could retain their mechanical strength after coating.

3.6 FLAME-RETARDANT FINISHING USING PLASMA TECHNOLOGY

Wet-chemical processing of textiles is important for its improvement in aesthetic and functional value. However, the process remains water, chemical, and energy intensive due to the involvement of multistep operations, such as padding, drying, curing, and postwashing. Several developments have taken place in the textile sector in the last five decades from a cost and environmental point of view, such as application of enzymes, natural dyes, digital printing, low material to liquid processing, spray and foam finishing, infrared dyeing and drying, ultrasound dyeing and dispersion, and water-free plasma processing.[42–44] Plasma, a partially ionized gas composed of many types of species, such as positive and negative ions, electrons, neutrals, excited molecules, photons, and UV light, can be used for nanoscale surface engineering of textile substrates without altering the major inherent bulk properties. Plasma reaction with a small nonpolymerizing molecule causes surface activation, cleaning, oxidation, change in surface energy, an increase in surface roughness/area, and etching. Similarly, plasma reaction with a bigger molecule leads to plasma polymerization, coating, deposition, and creation of nanostructures. These altogether improve the aesthetics and functional value of the textiles in terms of improvement in water absorbency, water repellency, oil absorbency, oil repellency, UV protection, antimicrobial, flame retardancy, dyeing, desizing, antistatic, adhesion, and antifelting of wool.[44–48] Low-temperature plasma, also known as nonthermal plasma with a bulk temperature of 50°C–250°C, is used for such surface modification of textiles. Plasma is an energetic chemical environment, where the generation of plasma species opens up diverse reactions resulting in various end applications in apparel, home, and technical textiles.

A single-step multifunctional finishing of the cotton textile to impart water repellency as well as flame retardancy has been reported using argon (Ar) plasma.[49] Argon plasma has been used for graft copolymerization of monomers of acrylate phosphate and phosphonate derivatives in combination with carbon tetrafluoride (CF_4) gas. The CF_4 gas was responsible for imparting water repellency, and the fire-retardant chemicals had no opposing effect on it. The same research group from the national textile center has also developed monomers like cyclotriphosphazene and hexachlorocyclotriphosphazene to be deposited on the plasma preactivated cotton surface followed by polymerization by passing the fabric through the plasma zone to ensure formation of a grafted flame-retardant nanolayer on the fabric. They have also studied the low-temperature argon plasma–induced graft copolymerization of monomers like phosphorous diethyl phosphate and diethyl-2-(methacryloyloxyethyl) phosphate (DEMEP) on poly(acrylonitrile) fabrics. Different metal and/or metal oxide mixtures, such as aluminum oxide, nanosilver, and titanium dioxide, have been applied in the argon plasma–preactivated polyester fabric.[50] The treated fabric showed improvement in flame retardancy and thermogravimetry. Besides, the treated fabric showed good antimicrobial property. Similar to plasma treatment, the graft copolymerization of methacrylamide on the UV-irradiated cotton fabric was carried out using benzophenone as a photosensitizer so as to react easily with phosphorous-containing compounds, and the treated fabric showed improvement in thermal stability.[51]

Cotton fibers were coated with silicon dioxide (SiO_2) layers using atmospheric pressure plasma (APP) treatment. The SiO_2 network armor was obtained through hydrolysis and condensation of the precursor TEOS, and it was successively cross-linked on the cotton fiber's surface. Due to the protective effects of SiO_2 network armor, the modified cellulosic fibers exhibited enhanced thermal properties and improved flame retardancy. The scanning electron microscope (SEM) images have confirmed the presence of SiO_2 network on the fabric surfaces, even after intense ultrasound washes. The SiO_2-APP-coated flame-retardant cotton textiles will have numerous applications in the development of upholstery furniture, clothing, and military applications. A dense, thin film was obtained from the tetramethyldisiloxane (TDMS) monomer premixed with oxygen by plasma treatment.[38] The flame-retardant properties of polyamides and polyamide nanocomposites could be improved using such deposit. The LOI of the coated PA6 nanocomposites was greatly improved when the film thickness was equal to 0.6 mm. The mechanism was attributed to the fact that under the burning condition, the nanocomposite structure of the polymer led to the formation of a surface protective layer. The presence of carbonaceous and silica-like layer acts as a barrier, thus preventing heat and mass transfers in the burning zone. Additionally, the polymer slows the toxic gas formation. A recent patented process has demonstrated the use of atmospheric plasma for flash-fire-resistant finishing of pure cotton, flame-retarded cotton, and poly(meta-aramid) fabrics using clay and a silicon-containing monomer such as hexamethylene disiloxane (HMDSO) in various combinations. It was found that at relatively low heat flux of 35 kW/m², the cone calorimetry showed that for pure cotton, both the time-to-ignite (TTI) and the time-to-peak (TTP) values increase for argon plasma–treated fabrics. It was observed that ArClayAr-Cotton showed significantly more TTP value. The effect of the presence of clay was clearly seen in this experiment.[41] Low-pressure cold plasma was used to deposit a thin film on PA6 substrates for fire-retardant finishing by plasma polymerization of the 1.1.3.3-tetramethyl disiloxane (TMDS) monomer and oxygen gas mixture.[52] The presence of oxygen during the polymerization process of TMDS promoted the formation of more thermally stable coatings with efficient fire-retardant properties. The coatings formed a thermal barrier that decreased the RHR of the coated PA6 by 28% and increased ignition time from 105 s in the untreated sample to 219–315 s in the treated samples. Another study on PA6 using low-pressure microwave plasma has described the graft polymerization of fluorinated acrylate monomer 1,1,2,2-tetrahydroperfluorodecyl acrylate (AC8). The deposit showed efficient fire-retardant properties in terms of decreased RHR peak value by 50% compared to the uncoated PA6. This was possibly due to the reaction of CFx radicals in the gas phase with different fragments of polymer generated during the thermal degradation, resulting in dilution of the combustible gases.[53] Joëlle et al. reported the use of some monomer along with the diethyl(acryloyloxyethyl)phosphoramidate (DEAEPN) monomer, photoinitiator, and ethylene glycol diacrylate (EGDA) for multifunctional finishing of cotton textiles, with special emphasis on a simultaneous improvement in dyeing and flame retardancy.[54]

A flame-retardant agent (RF) of an organic phosphorus compound in combination with a melamine resin (cross-linking agent, CL), phosphoric acid (catalyst, PA), and zinc oxide to nano-ZnO (cocatalyst) could impart effective durable flame-retardant properties. In the study, an APP jet was applied as a pretreatment to improve the postfinishing operation (e.g., the flame-retardant finishing) of the cotton fabric. By application of FR-CL-PA in the presence of a ZnO/nano-ZnO cocatalyst to cotton fabrics, it was possible to impart flame-retardant properties. Both the plasma pretreatment–added and the ZnO/nano-ZnO (cocatalyst)–added flame-retardant formulations could improve the cross-linking between the flame-retardant chemical and the cotton fabric, thus minimizing the drawbacks of the existing acid-based flame-retardant formulation. To improve the durability to washing or to harsh weather conditions, the flame-retardant chemicals must be firmly attached to the surface of the textile. It could be most efficiently achieved through the covalent bonding. In this regard, the simultaneous grafting and polymerization of fire-retardant monomers on cotton and polyacrylonitrile (PAN) fabrics induced by argon plasma with four acrylate monomers containing phosphorus, such as diethyl (acryloyloxyethyl) phosphate (DEAEP), DEMEP, diethyl (acryloyloxymethyl) phosphonate (DEAMP), and dimethyl (acryloyloxymethyl) phosphonate (DMAMP), is worth mentioning. The treated samples showed an improved flame-retardant property in terms of the LOI and the thermogrametry.[55] An APP jet was used as a pretreatment to enhance the flame-retardant property of the cotton fabric by either the sputtering or etching effect. The flame-retardant agent (FR) *N*-methylol dimethylphosphonopropionamide, in combination with a melamine resin, cross-linking agent (CL), and a catalyst of phosphoric acid (PA), was used. Titanium dioxide (TiO$_2$)/nano-TiO$_2$ cocatalyst was added to the FR formulation in order to improve the cross-linking of the FR-CL-PA components.[7] Similarly, it has been reported that the plasma containing a phosphorus compound can be used for flame-retardant finishing of acrylic, cotton, and viscose rayon fibers.[56]

3.7 FLAME-RETARDANT FINISHING USING PLANT EXTRACT

In the last few decades, due to the stringent pollution norms and the government legislation that have come into force, various technologies have been developed and implemented in the textile production and processing arena so as to reduce the consumption of water, energy, processing costs, and/or effluent load. Some of these technologies are as follows: (1) low material to liquid processing, (2) spray and foam finishing, (3) use of enzymes, (4) natural dyes, (5) digital printing, (6) infrared dyeing and drying, (7) radiofrequency drying, and (8) ultrasound for dyeing and dispersion. Few more promising and emerging technologies have also been explored in laboratory to industry, such as (1) supercritical carbon dioxide for dyeing, (2) plasma technology for water-free processing, and (3) UV- and laser-based processing and finishing. All these developed technologies are related to pre- or postprocessing of textiles or textile processing machineries. In spite of such developments, textile chemical processes still remain energy and water intensive, besides adding to pollution due to the use of various non-environment-friendly chemicals and auxiliaries. As discussed earlier, many of the present-day flame-retardant formulations, such as those based on phosphorus and nitrogen, antimony–halogen, diammonium phosphate, Pyrovatex, urea, and melamine formaldehyde, have an adverse effect on the environment, human body, and fabric physical, mechanical, and aesthetic properties. In this context, environment-friendly chemicals, such as BTCA, BTCA-hydroxyalkyl organophosphorus compound, Noflan, organophosphoramidates compound, and a few more, have also been demonstrated for commercial applications. Emerging plasma and nanotechnology have recently been explored to impart flame retardancy to the natural as well as synthetic textiles with an aim to partially or fully avoid the usage of water and reduce the number of processing steps and the quantum of chemicals required. However, the same has not met with much commercial success, and the processes still use the traditional types of flame-retardant chemicals. It may also be noted that till date, a very limited research work has been carried out on the development of textile chemicals and auxiliaries utilizing natural resources.

In the last two decades due to the global awareness of environmental pollution, climate change, carbon footprint, health, and hygiene, the demand for organic material is growing exponentially. In this context, organic fruits, vegetable, crops and pulses, and even organic cotton have been quite successful, and they are in high demand in the markets. Due to the advantages of natural fibers, being biodegradable and produced from renewable sources and with good moisture regain, soft feel, adequate to fair strength, and good appearance after chemical treatment, they are commonly used in apparel and home textiles. Buoyed by these advantages, the demand of natural fiber (nonorganic to organic)–based textile is growing profoundly, and the year 2009 was celebrated as the International Year of Natural Fibres. For the sustainable development of textile products, it is required that the textile meant for the specialty apparel and home application should be made of natural fibers and to be processed and finished with eco-friendly chemicals and auxiliaries, preferably with natural products. This will help in adding value to the natural products, while preserving the natural resources. In this regard, a number of plant extract or biomolecules have been successfully used in textile processing and finishing, such as (1) enzyme for desizing, scouring, and biopolishing; (2) natural dye for coloration; (3) natural dyes for UV protection and antimicrobial finishing; (4) aromatic and medicinal plants for antimicrobial, mosquito-repellent, and well-being textiles; (5) chitosan for antimicrobial finishing; and (6) aloe vera and neem for antimicrobial finishing and skin nourishing.[57–61,63]

In this context, there is a need to develop sustainable green-flame-retardant chemicals preferably produced from renewable sources that will be easy to apply; cost effective; environmental friendly; limited to no adverse effect on fabric mechanical properties; versatile to cellulosic, lignocellulosic, and protein textiles; and semidurable to durable. In recent years, a few researches have been reported on flame-retardant finishing of cellulosic cotton textiles using biomacromolecules, such as DNA from herring sperm and salmon fish. It has been reported that the DNA consists of phosphate, carbonaceous deoxyribose units of polysaccharide dehydrate and some essential amino acids, which are responsible for the formation of more carbonaceous char and ammonia release to enhance the thermal stability of the cotton fabric.[10] Efforts were also directed to make cotton fabric flame retardant with whey proteins, casein, and hydrophobins, as these are rich in phosphate, disulfide, and protein that can influence pyrolysis by early char formation.[10] As some of the plants (i.e., the plant molecules) contain phosphorous and other minerals, in our laboratory, two such plant extracts have been utilized for flame-retardant finishing of cellulosic and lignocellulosic textiles. In this regard, BPS, an agro-waste plant extract, and SJ, a vegetable extract, have been utilized, as they are rich in phosphorous, nitrogen, chlorine, silicate, and other many metallic compounds.[11,12] A recent study on BPS reported that potassium chloride, sodium chloride, and metal phosphate are the major composition of salts of the BPS.[13,65] Furthermore, the BPS is abundantly available in many of the countries and could be considered as a green, cost-effective flame-retardant chemical for jute and cotton textiles.

3.7.1 THERMAL ANALYSIS

The BPS was applied on premordanted (5% tannic acid + 10% alum) cellulosic and lignocellulosic textiles by the pad-dry method in an alkaline condition. The flame retardancy of the samples was measured in terms of LOI, burning rate, total heat production, differential scanning calorimeter (DSC), and TGA. As discussed earlier, a textile with LOI ≤ 21 burns rapidly in an open atmosphere and a sample with LOI ≥ 27 is generally considered to be flame retardant. It can be seen from Table 3.2 that the untreated cotton and jute have the LOI values of 18 and 21, respectively. However, when both samples were treated with the BPS, the LOI was found to become more than 27 and, hence, can be considered as flame-retardant textiles. From Figure 3.2, it can also be seen that the BPS-treated fabrics showed a better thermal stability in contact with flame compared to the control fabric.[13]

TABLE 3.2
Flammability Parameters of Control and BPS-Treated Cotton and Jute Fabrics

			Plant Extract–Treated Textiles	
Flammability Parameters	Control Cotton	Control Jute	BPS on Cotton	BPS on Jute
Add-on (%)		—	4.5	3
LOI	18	21	30	33
Horizontal flammability				
Warp way burn rate (mm/min)	75	62	7.5	15
Vertical flammability				
Burning with flame time (s)	60	60	4	5
Burning with afterglow time (s) after flame stopped	0 (as completely burnt with flame)	0 (as completely burnt with flame)	900	600
Total burning time (s)[a]	60+0	60+0	4+900	5+600
Observed burning rate (mm/min)	250	250	16.6	24.8

Sources: Basak, S. et al., *Polish J. Chem. Technol.*, 2014; Smanata, K.K. et al., Eco-friendly coloration and functionalization of textile using plant extracts, in S.S. Muthu (ed.), *Roadmap to Sustainable Textiles and Clothing*, Springer Publication, Singapore, pp. 263–287, 2014; Basak, S. et al., *Int. J. Cloth. Sci. Technol.*, 27(2), 2014.

[a] Total burning time = burning with flame time + burning with afterglow time.

Sustainable Flame-Retardant Finishing of Textiles

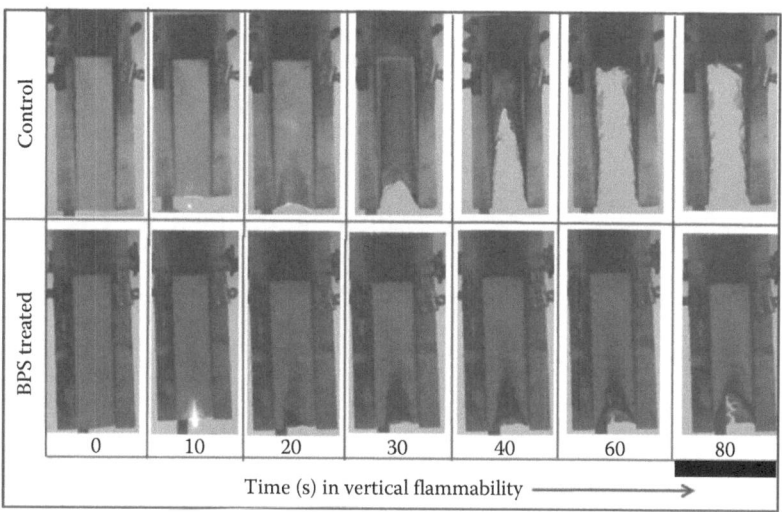

FIGURE 3.2 Burning of control and BPS-treated cotton fabrics at different intervals of time. (From Basak, S. et al., *Polish J. Chem. Technol.*, 2014; Basak, S. et al., *Int. J. Cloth. Sci. Technol.*, 27(2), 2014.)

The BPS-treated cotton fabric burns with flame only for 4 s followed by burning with an afterglow (after the flame is stopped) for 900 s. Hence, the total burning time of a 250 mm × 20 mm sample becomes 904 s, whereas the untreated sample of same size burnt completely within 60 s with the flame as shown in Figure 3.2.

Therefore, it can be expected that the rate of heat production would also be more in case of the control sample compared to the BSP-treated sample. It is important to note that the available time either to escape from the fire zone or to extinguish the fire was much more (904 s) in the treated sample compared to only 60 s in the case of the untreated sample. In the treated sample, the observed burning rate was only 16.6 mm/min, which is significantly lower than the 250 mm/min in the untreated sample. Compared to cellulosic cotton textiles, lignocellulosic jute shows a bit higher LOI value. The treated fabric showed an LOI of 33 and the flame time was only 5 s compared to the untreated sample, where the corresponding values were 21 and 60 s for the complete burning. Though the flame stopped after 5 s in the treated sample, it burnt at a very slow rate of 24.8 mm/min. The advantages of this process are that the flame is self-extinguishable in 4–5 s and the application process is quite simple. It has been reported that the total heat of combustion measured in terms of the gross-calorific value (GCV) measured in the oxygen bomb calorimeter was found to reduce from 16.4 to 13.1 MJ/kg in the untreated and BPS-treated cotton fabrics, respectively. The durability to soap washing of this particular finish was also investigated, and it was found that on washing, the LOI value reduced to 24 from 30 in the as-prepared sample. This implies that the treated fabric could retain some part of the imparted functionality after washing, and hence, it can be considered as a semidurable finish. After application of the BPS finish, there was no significant adverse effect (<5% change) in either the tensile or the tear strength of the treated fabrics, whereas most of the flame-retardant finish causes 10%–30% loss of the tensile strength.[4]

TG analysis was carried out to study the thermal stability of the BPS-treated fabric over a temperature range compared to the control sample.[13] Figure 3.3 shows the TG curves of both the samples in air atmosphere at a heating rate of 10°C/min. Four distinct stages were observed in the BPS-treated sample. In the initial stage at temperatures below 200°C, the mass loss occurred mainly due to the removal of absorbed moisture from the fabric.[30] In the second stage, the main pyrolysis occurred over the temperature range of 200°C–350°C with a sharp first derivative peak at 300°C. Above 380°C, that is, in the third stage, dehydration and char formation happened due to the formation of water and CO_2, and the sample showed one more first derivative peak.

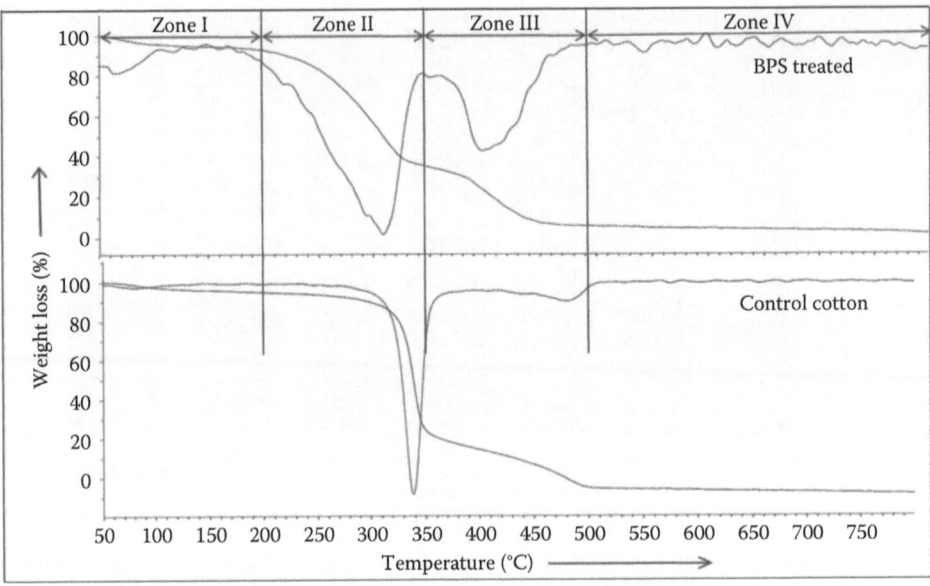

FIGURE 3.3 TG curve of control and BPS-treated cotton fabrics. (From Basak, S. et al., *Polish J. Chem. Technol.*, 2014.)

In contrast, the TG curve of the control sample showed a rapid decomposition in a narrow temperature range of 320°C–340°C with the sharp first derivative peak at 340°C. It lost around 98% of its mass at below 500°C. The BPS-treated samples started to lose mass, that is, the pyrolysis from temperature of 215°C, which is 85°C lower than the degradation temperature of the control sample. In the last stage of the TG curves, both the untreated and the treated fabrics showed char formation and the quantity of char residue remained higher in the treated sample compared to the untreated samples. A similar result was also observed in the BPS-treated jute fabric. The result showed that the BPS could be effectively used for flame-retardant finishing of cellulosic and lignocellulosic textiles.

DSC analysis was carried out to understand the flame-retardant mechanism of the control and BPS-treated lignocellulosic textiles. It was observed that unlike the control sample, the treated sample showed a steep endothermic peak in the initial region, which corresponds to the evaporation of water. The control fabric showed small endothermic peak, corresponding to the depolymerization and degradation of cellulose at 387°C. On the other hand, the treated sample showed a strong exothermic peak at 326°C with onset at 291°C and followed by an endothermic hump at 365°C. This signifies that similar to TG analysis, the degradation of cellulose has shifted toward the lower temperature by 22°C due to the presence of BPS. Further, the large exothermic peak in the treated fabric exhibits an extensive dehydration and char formation.[62]

3.7.2 Chemical Analysis

Figure 3.4 shows the negative and the positive time-of-flight secondary ion mass spectra (ToF-SIMS) of the pure BPS. The different molecules that were detected have been calculated based on their mass/charge (m/z) ratio. The negative ToF-SIMS of BPS showed the presence of major molecules at different mass units, such as H^- (1 amu), C^- (12 amu), CH^- (13 amu), N^- (14 amu), O^- (16 amu), OH^- (17 amu), F^- (19 amu), Cl^- (35,37 amu), Cl_2^- (70,71 amu), PO_2^- (62,63 amu), PO_3^- (79 amu), and KCl^- (74,76 amu). On the other hand, the positive ion mass spectrum of the sample showed

Sustainable Flame-Retardant Finishing of Textiles

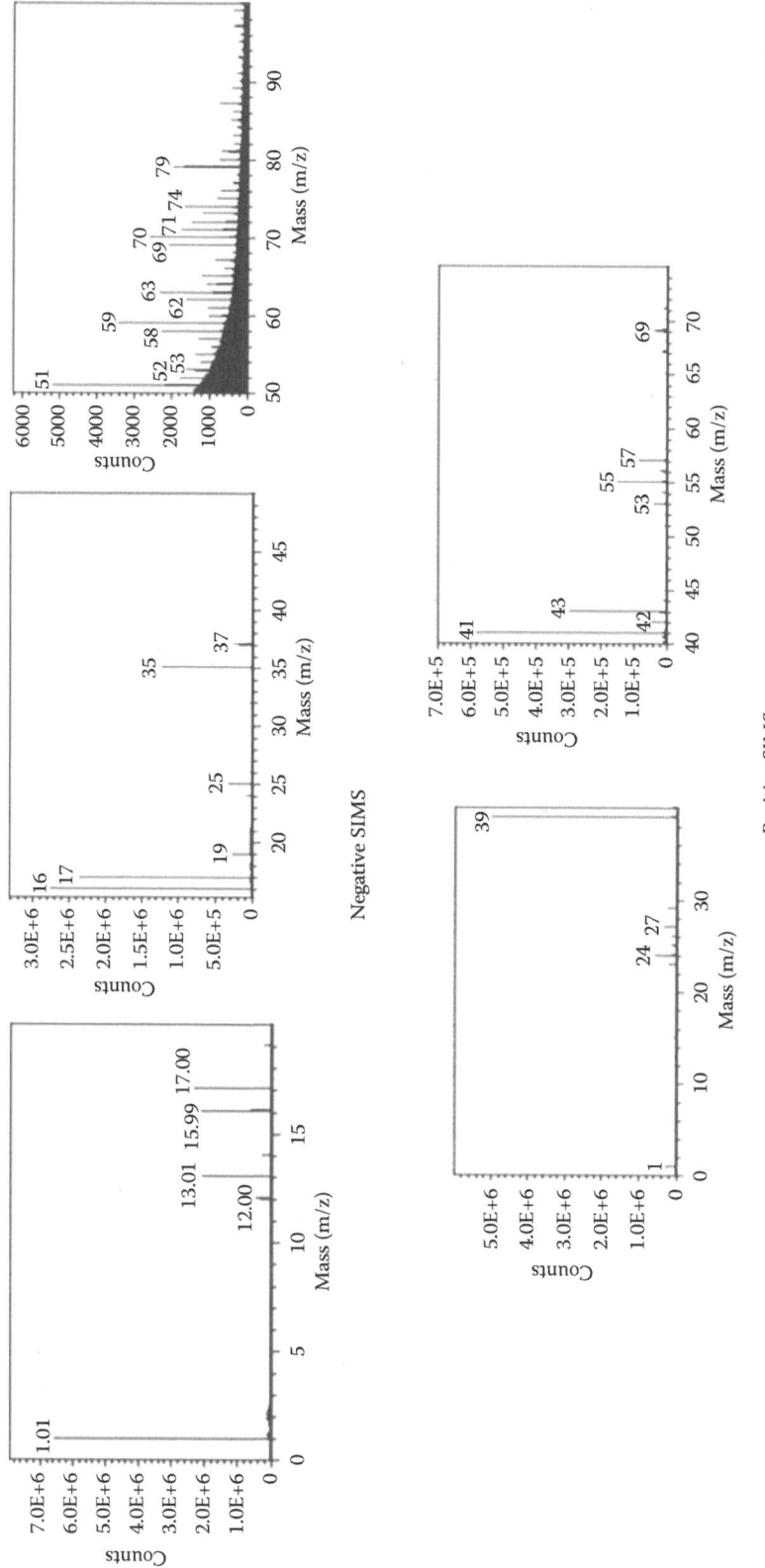

FIGURE 3.4 Negative and positive ion mass spectra of pure BPS. (From Smanata, K.K. et al., Eco-friendly coloration and functionalization of textile using plant extracts, in S.S. Muthu (ed.), *Roadmap to Sustainable Textiles and Clothing*, Springer Publication, Singapore, 2014, pp. 263–287.)

mostly the presence of various metal ions, such as Mg^+ (24,25 amu), Al^+ (27 amu), K^+ (39 amu), Ca^+ (40,41 amu), Mn^+ (55), and Fe^+ (55,56). Similar elemental peaks have also been detected in energy-dispersive x-ray spectroscopy (EDX) analysis of pure BPS.[13] Therefore, the flame-retardant properties in the BPS-treated sample in terms of higher LOI value and more char formation in cotton and jute fabrics were due to the presence of free metal ions and various metal salts like potassium chloride, potassium fluoride, calcium chloride, and phosphate.[13]

Similar to the BPS, the SJ, a vegetable extract, has also been utilized for flame-retardant finishing of the cotton textile.[63] This plant (*Spinacia oleracea*) is available in many countries and it is rich in phosphorous, nitrogen, and other metallic constituents. Unlike the BPS on cotton, the SJ was applied to cotton textile directly without any premordanting. In the 8% add-on sample, the LOI was found to increase to 30 from 18 in the control cotton sample. It was interesting to note that the sample did not catch flame and the afterglow was present only for 400 s, whereas as discussed earlier, the control sample completely burnt within 60 s with a high-temperature range of 400°C–450°C. As the sample burnt slowly, the warp way burning rate decreased from 75 mm/min in the control sample to 10 mm/min in the SJ-treated sample. There was a reduction in LOI from 30 to 22 after washing of sample with soap solution. Hence, like in the BPS treatment, this particular finish can also be considered as a semidurable finishing. There was no significant change either in the fabric tensile or tear strength after the application of this plant extract–based finish. The developed finish is cost effective compared to any other synthetic flame-retardant chemicals, as it is abundantly available and produced from renewable sources.

A single-step multifunctional finishing process for textiles is important to reduce the cost of processing. In this context, the BPS and the SJ, in addition to flame retardancy, can also impart effective UV-protective functionality and natural color. It was observed that the plain-woven bleached cotton fabric showed a poor UPF value of 10. As a result, the majority of the incident UV rays can easily pass through the fabric structure. After mordanting with tannic acid and alum, the UPF value did not improve much. When the mordanted fabric was treated with BPS in alkaline condition, the UPF value was found over 100, that is, a UPF rating of 50+ (excellent category). Furthermore, the UVA and UVB transmittance percentages reduced drastically from 9.5% and 7.2% in the untreated sample to 1.2% and 1.0%, respectively, in the BPS-treated sample. On the other hand, if the bleached fabric was directly treated with an alkaline BPS without any mordant, it showed quite a lower UPF value (40). From these two facts, it can be said that the improved UPF value, that is, the lowering of UV (A and B) transmission percentage, is arising mainly due to the synergistic effect of the BPS and the mordant. The UV protection of the BPS was possibly attributed to the presence of N,N-alkyl benzeneamine, as confirmed by the GC-MS analysis of BPS.[64–66] The UPF value of the treated fabric was 70 after the first ISO-1 washing and 55 after the second washing. Unlike the flame retardancy finishing, the fabric sample could easily retain the imparted UV protection functionality after washing. Similarly, the SJ-treated cotton fabric in alkaline condition showed the UPF value of 125 (rating ≥ 50), which may be due to the presence of organic color and silicate molecules. In the SJ-treated sample, the significant reduction in UVA and UVB percentage has also been reported in comparison with the untreated sample.

It has also been reported that after application of the BPS in the premordanted cotton fabric, the sample color changed from white to khaki. This color was only developed in the premordanted fabric in alkaline condition. The mordant has helped in exhaustion of the BPS and the color formation. In the treated fabric, the K/S value (color depth) improved from 0.04 for the control to 0.80 for the BPS-treated samples. With further increase in the concentration of the BPS, the depth of color increased linearly. Similarly, after the application of SJ on the bleached cotton fabric, the color of the sample changed to white to dark green. Therefore, from various results, it can be said that the BPS and the SJ can be used for imparting multifunctional flame retardance, UV protection, and natural color to textiles in a single step.

3.7.3 Mechanism of Flame-Retardant Finishing

The flame-retardant functionality imparted by the BPS may be due to the presence of phosphate and other mineral salts in the BPS. Different metals that are present in the form of metal chloride and phosphate were detected in positive and negative ion mass spectra. In addition, the elemental peaks of chlorine, phosphorous, magnesium, aluminum, sodium, potassium, calcium, etc., were also detected in EDX analysis. The FTIR analysis of pure BPS sample confirmed the presence of inorganic salts.[67] One recent study on pure BPS reported that 10.5 g inorganic salts was obtained by evaporation of 500 mL aqueous ash extract collected from burning the banana pseudostem. It was found that potassium chloride, sodium chloride, and metal phosphate are the major components of that salt.[67] On the other hand, the flame-retardant property of the SJ-treated cotton fabric was possibly due to the presence of silicate and phosphate, which have also been detected by elemental EDX and FTIR analysis. From the various characterizations, it was presumed that phosphate, silicate, chloride, and other mineral salts present in the BPS and SJ might have increased the thermal stability of the cotton and jute fabrics by forming more char and producing less flammable gases. A similar mechanism has also been reported in the literature wherein the presence of organometallic additives in cellulosic textiles could increase char formation while reducing formation of tar, that is, the volatile gases.[68]

As far as the flame retardancy mechanism is concerned, the BPS mainly acts in the condensed phase. It might be due to the fact that it contains phosphate, phosphite, and other metallic constituents that help influence the pyrolysis of the cellulose polymer resulting in earlier char formation as supported by TGA. From the SEM pictures of the treated fabric, it was also observed that the BPS produced a thick protective coating on the surface of the cotton fiber. When this treated cotton fabric was exposed to flame, it favored the formation of a stable and protective char that limited the exchange of oxygen and combustible volatile substance, resulting in disappearance of flame and reduction in the burning rate. It has also been reported that the BPS coating might have also acted as an intumescent that swells on heating and increase its volume, thus protecting the underlying cellulose polymer from the heat or flame.

3.8 SUSTAINABLE FLAME-RETARDANT FINISHING

As discussed earlier, many of the present-day flame-retardant formulations, such as thiourea, THPC-urea, Pyrovatex–urea–melamine formaldehyde, antimony–halogen, phosphorus–nitrogen, and diammonium phosphate, have adverse effects of formaldehyde release and environmental impact in addition to the reduction in fabric physical, mechanical, and aesthetic properties. For the development of sustainable textiles, it is preferred that the textile should be made of natural fibers and to be processed and finished with eco-friendly chemicals and auxiliaries, preferably with natural products. As far as flame-retardant chemicals and fabrics are concerned, it should not be biologically harmful, and it should not have any detrimental effect on fabric properties. In this regard, formaldehyde- and halogen-free formulations have also been developed, but those are not very successful in the market. Till date, the most popular flame-retardant formulation for commercial application is Pyrovatex, which releases 330 and 550 ppm formaldehyde, when applied either alone or in combination with melamine formaldehyde to cotton fabrics. Even after five home launderings, it releases more than 25 ppm of formaldehyde that is harmful to the end user. To mitigate this problem, BTCA along with citric acid (cross-linker) has also been explored instead of using melamine formaldehyde. This has helped to reduce the formaldehyde release level below 20 ppm after 5 home launderings.[3] Additionally, more substantive intumescent products, nanoparticle additives, and blends with inherently flame-retardant fibers have also been developed to provide environment-friendly durable textiles. Recently, eco-friendly green-flame-retardant chemicals from the plant extract and other biomolecules, such as DNA, whey protein, BPS, and

SJ, have been applied on cotton and other lignocellulosic textiles for the development of sustainable textiles. As these materials are mostly produced from agricultural or other residues, they are renewable, cost effective, easy to apply, and free from any formaldehyde release. Such applications were found not to affect the mechanical properties of fabrics and can also be suitably used for natural coloration of textiles.

3.9 SUMMARY

Wet-chemical processing of textile is important to improve its aesthetic and functional value. Among the various functional finishes of textiles, flame-retardant finishing is important, as it is directly related to the protection of human beings and other valuable products from fire hazards. Natural and synthetic fibers are used for apparel and home-furnishing applications. Due to the several advantages of natural fibers in terms of biodegradability, produced from renewable sources, good moisture regain, soft feel, adequate to fair strength, and good appearance after chemical treatment, natural fibers are preferred over their synthetic counterparts for apparel and home-furnishing applications. Both natural and synthetic apparel-grade fibers have a low LOI value <25; therefore, a flame-retardant finishing is essential for these fibers to make them suitable in apparels, baby products, hospital curtain, tent, automobiles, railways, and for protection of workers who are engaged in oil, gas, and petroleum industries. Although the combustion temperature of such fibers is quite high (450°C–550°C), but they degrade at a much lower temperature through pyrolysis and/or melting. Significant efforts have been made in the past to improve the flame retardancy of cotton and other textiles during their manufacturing or in the finishing stage. The most commonly used chemical for the same is a mixture of borax and boric acid, inorganic salts, diammonium phosphate, and urea. Phosphorous along with nitrogenous compounds and antimony with halogen compounds due to their synergism are also quite popular. Later on, environment-friendly flame retardants, such as urea and ammonium sulfamate, BTCA-hydroxyalkyl organophosphorus, and SMSN, have been developed. The emerging nanotechnology has been explored for similar applications using a number of nanoparticles, such as silica, ZnO, TiO_2, nanoclay, and CNTs. These materials were applied in the textile by layer-by-layer technique or as a polymer coating. Plasma technology has also been used to impart flame retardancy in the textile without usage of water or to accomplish multifunctional finishes in a single step. For example, monomers of acrylate phosphate and phosphonate derivatives in combination with carbon tetrafluoride (CF_4) were used to impart flame-retardant and water-repellent functionalities simultaneously. A majority of present-day flame-retardant formulations have the following disadvantages: (1) formaldehyde release during processing, (2) toxicity from some of the halogen-based compounds, (3) presence of heavy metals, (4) reduction in fabric tensile and tear strength, (5) increase in fabric stiffness and coarseness, (6) yellowing of color, and (7) nondurability. To address some of these issues and in the context of environmentally friendly, sustainable product development in recent years, various biomolecules have been used for flame-retardant finishing of textiles, such as DNA from herring sperm and salmon fish, whey proteins, casein, and hydrophobins. Similarly, BPS, an agricultural residue, and SJ, a vegetable extract, have also been applied on cotton and jute fabrics due to their advantages of low cost and being produced from renewable sources. After application of these plant molecules, the LOI was found to increase to over 30 from 18 and 21 in the untreated cotton and jute samples, respectively. In addition to flame retardancy, they could also impart UV protection and natural color to the textiles. The flame-retardant textiles will provide much longer time either to extinguish the flame or to escape from the fire zone. In spite of several such developments, the majority of flame-retardant formulations still lack the requisite durability to rainwater and/or soap washing, which needs to be addressed on priority. It is felt that the extraction and application of various plant molecules and other biomolecules for flame-retardant finishing of textiles will help to develop eco-friendly green textile products while ensuring much needed value addition to agro-residues.

REFERENCES

1. Schindler, W.D. and Hauser, P.J. (ed.). Flame retardant finishes. In *Chemical Finishing of Textiles*. Woodhead Publishing Limited, Boca Raton, FL, 2004.
2. Horrocks, A.R. Flame retardant challenges for textiles and fibres. *Polymer Degradation and Stability* 96, 3 (2011): 377–392.
3. Katovic, D., Vukusic, S.B., Gragac, S.F. et al. Flame retardancy of paper obtained with environmentally friendly agents. *Fibres & Textiles in Eastern Europe* 17, 3 (2009): 90–94.
4. Banerjee, S.K., Day, A., and Ray, P.K. Fire proofing jute. *Textile Research Journal* 56 (1985): 338–343.
5. Kandola, B.K., Horrocks, A.R., Price, D. et al. Flame retardant treatments of cellulose and their influence on the mechanism of cellulose pyrolysis. *Journal of Macromolecular Science* 36 (1996): 794–796.
6. Hady, A.A.E., Farouk, A., and Sharaf, S. Flame retardancy and UV protection of cotton based fabrics using nano ZnO and polycarboxylic acid. *Carbohydrate Polymers* 92, 1 (2013): 400–406.
7. Kan, C.W., Lam, Y.L., and Yuen, C.W. Fabric handle of plasma-treated cotton fabrics with flame-retardant finishing catalyzed by titanium dioxide. *Green Processing and Synthesis* 1, 2 (2012): 195–204.
8. Sarvanan, D., Lakshmi, S.N.S., Raja, K.S. et al. Biopolishing of cotton fabric with fungal cellulose and its effect on the morphology of cotton fibres. *International Journal of Fibre and Textile Research* 38 (2013): 156–159.
9. Joshi, M., Ali, S.W., and Rajendran, S. Antibacterial finishing of polyester cotton blend fabric using neem: A natural bioactive agent. *Journal of Applied Polymer Science* 106 (2007): 793.
10. Alongi, J., Carletto, R.A., Balsio, A.D. et al. Intrinsic intumescent like flame retardant properties of DNA treated cotton fabrics. *Carbohydrate Polymers* 96, 1 (2013): 296–304.
11. Bosco, F., Carletto, R.A., Alongi, J. et al. Thermal stability of flame resistance of cotton fabrics treated with whey proteins. *Carbohydrate Polymers* 94, 1 (2012): 372–377.
12. Carosio, F., Blasio, A.D., Cuttica, F. et al. Polyester and polyester cotton blend fabrics have been treated with caseins. *Industrial and Engineering Chemistry Research* 53, 10 (2013): 3917–3923.
13. Basak, S., Samanta, K.K., Saxena, S. et al. Flame resistant cellulosic substrate using banana pseudostem sap. *Polish Journal of Chemical Technology* 17, 1 (2014) (in press).
14. http://www.fpl.fs.fed.us/documnts/pdf1984/levan84a.pdf, (Accessed January 18, 2015.)
15. Hull, T.R. and Kandola, K.B. *Fire Retardancy of Polymers: New Strategies and Mechanism*. Royal Society of Chemistry, RSC Publications, London, U.K., 2009, pp. 001–454.
16. Camino, G., Lomakin, S. Intumescent materials. In *Fire Retardant Materials*. A.R. Horrocks and D. Price (ed.), Woodhead Publishing Limited, Boston, MA, 2001, pp. 318–335.
17. Sharma, U. Fire retardancy of jute fabrics with PST (Rochelle salt). *Colourage* 32, 26 (1986): 19–20.
18. Karastergiou, P.S. and Philippou, J.L. Thermogravimetric analysis of fire retardant treated particle boards. In *Wood and Fire Safety*. Aristotelian University of Thessaloniki, Thessaloniki, Greece, 2000, p. 388.
19. Parikh, D.V., Sachinvala, N.D., Sawhney, A.P.S. et al. Flame retardant cotton blends highlofts. *Journal of Fire Sciences* 21 (2003): 385–387.
20. Samanta, A.K., Singhee, D., Basu G. et al. Thermal behaviour and the structural features of chemically and bio-chemically modified jute substrate. *Indian Journal of Fibre and Textile Research* 32, 3 (2007): 355–365.
21. Hirota K., Makela J., and Tokunga O. Reactions of sulphur dioxide with ammonia: Dependence on oxygen and nitric oxide. *Industrial & Engineering Chemistry Research* 35, 10 (1996): 3362–3368.
22. Xing, T.L., Liu, J., Li, S.W. et al. Thermal properties of flame retardant cotton fabric grafted by dimethyl methacryloyloxyethyl phosphate. *Thermal Science* 16, 5 (2012): 1472–1475.
23. Basak, S., Samanta, K.K., Chattopadhyay, S.K. et al. Fire retardant finishing of jute fabric treated with thio-urea. *Journal of Textile Association* 74, 5 (2014): 273–281.
24. http://www.thriftyfun.com/tf596497.tip.html, (Accessed January 18, 2015.)
25. Glemser, E.J., Dowling, L., and Inqlis, D. A novel method for controlling multicoloured Asian lady beetle. *Environmental Entomology* 41, 5 (2012): 1169–1176.
26. Basak, S., Chattopadhyay, S.K., Samanta, K.K. et al. Fire retardant jute fabric using water glass. *Man Made Textiles in India* 61, 11 (2013): 386–390.
27. Basak, S., Samanta, K.K., Chattopadhyay, S.K. et al. Flame retardant and antimicrobial jute textile using sodium metasilicate nonahydrate. *Polish Journal of Chemical Technology* 16, 2 (2014): 106–115.
28. Rowell, R.M. and Dietenberger, M.A. Thermal properties, combustion and fire retardancy of wood. In R.M. Rowell (ed.), *Handbook of Wood Chemistry and Wood Composites*. CRC Press (Taylor & Francis Group), New York, 2013, pp. 127–149.

29. Kashiwagi, T., Gilman, J.W., Butler, K.M. et al. Flame retardant mechanism of silica gel/silica. *Fire and Materials* 24, 6 (2000): 277–289.
30. Motashari, S.M. and Motashari, S.Z. Thermogravimetry of deposited ammonium aluminium sulphate dodecahydrate used as flame retardant for cotton fabrics. *Cellulose Chemistry and Technology* 43, 10 (2009): 455–460.
31. Kyeong, W.P. Ion exchange and application of layered silicate. In A. Kilislioglu (ed.), *Ion Exchange Technologies*. Intech Publishers, Rijeka, Croatia, 2012, pp. 241–259.
32. Huang, H., Williams, S.K., Sims, C.A. et al. Sodium metasilicate affects antimicrobial, sensory, physical, and chemical characteristics of fresh commercial chicken breast meat stored at 4°C for 9 days. *Poultry Science* 90, 5 (2011): 1124–1133.
33. Vatten, D.A., Maitin, V., and Richardsons, C.R. Evaluation of antibacterial and toxicological effects of a novel sodium silicate complex. *Research Journal of Microbiology* 7, 3 (2012): 191–198.
34. Sheng, H.U., Congqin, N., Yue, Z. et al. Antibacterial activity of silicate bioceramics. *Journal of Wuhan University of Technology* 26, 2 (2011): 226–230.
35. Samanta, K.K., Basak, S., and Chowdhury, P. Nanotechnology for advanced textile application. *Man-Made Textiles in India* 62, 2 (2014): 55–59.
36. Samanta, K.K., Saxena S., Vigneshwaran N. et al. Development of flame retardant protective textile using nanoparticles. In *Proceedings of International Conference on Enhancing Health, Wellbeing and Sustainability—Opportunities, Challenges and Future Directions*, Mumbai, India, January 10–12, 2013, p. 42.
37. Joshi, M. and Bhattacharyya, A. Nanotechnology—A new route to high-performance functional textiles. *Textile Progress* 43, 3 (2011): 155–233.
38. Bourbigot, S. and Duquesne, S. Fire retardant polymers: Recent developments and opportunities. *Journal of Materials Chemistry* 17 (2007): 2283–2300.
39. Devaux, E., Rochery, M., and Bourbigot, S. Polyurethane/clay and polyurethane/POSS nanocomposites as flame retarded coating for polyester and cotton fabrics. *Fire and Materials* 26, 4–5 (2002): 149–154.
40. Carosio, F., Laufer, G., Alongi, J. et al. Layer-by-layer assembly of silica-based flame retardant thin film on PET fabric. *Polymer Degradation and Stability* 96, 5 (2011): 745–750.
41. Li, Y.C., Schulz, J., Mannen, S. et al. Flame retardant behavior of polyelectrolyte-clay thin film assemblies on cotton fabric. *ACS Nanotechnology* 4, 6 (2010): 3325–3337.
42. Samanta, K.K., Basak, S., and Chattopadhyay, S.K. Environment-friendly textile processing using plasma and UV treatment. In S.S. Muthu (ed.), *Roadmap to Sustainable Textiles and Clothing*. Springer Publication, Singapore, pp. 161–201, 2014.
43. Muthukumar, M., Sargunamani, D., Selvakumar, N. et al. Statistical analysis of the effect of aromatic, azo and sulphonic acid groups on decolouration of acid dye effluents using advanced oxidation process. *Dyes and Pigments* 63 (2004): 199–304.
44. Samanta, K.K., Jassal, M., and Agrawal, A.K. Improvement in water and oil absorbency of textile substrate by atmospheric pressure cold plasma treatment. *Surface Coating and Technology* 203 (2009): 1336–1342.
45. Samanta, K.K., Jassal, M., and Agrawal, A.K. Antistatic effect of atmospheric pressure glow discharge cold plasma treatment on textile substrates. *Fibers and Polymers* 11, 3 (2010): 431–437.
46. Samanta, K.K., Joshi, A.G., Jassal, M. et al. Study of hydrophobic finishing of cellulosic substrate using He/1,3-butadiene plasma at atmospheric pressure. *Surface and Coating Technology* 213 (2012): 65–76.
47. Wakida, T., Tokino, S., Niu, S. et al. Characterization of wool and polyethylene terephthalate fabrics and film treated with low temperature plasma under atmospheric pressure. *Textile Research Journal* 63 (1993): 433–438.
48. Panda, P.K., Rastogi, D., and Jassal, M. Effect of atmospheric pressure helium plasma on felting and low temperature dyeing of wool. *Journal of Applied Polymer Science* 124, 5 (2012): 4289–4297.
49. Safaek, T. and Grutzmachor, L. Multifunctional surfaces using the plasma induced graft polymerisation (PIGP) process: Flame and waterproof cotton textiles. *Surface and Coating Technology* 201, 12 (2007): 5789–5795.
50. Raslan, W.M., Rashed, U.S., Sayad, H.E. et al. Flame retardancy using plasma nano technology. *Material Science and Application* 2 (2011): 1432–1442.
51. Kaur, I. and Sharma, R.M. Development of flame retardant cotton fabric through grafting and post grafting reaction. *Indian Journal of Fibre and Textile Research* 32 (2007): 312–318.
52. Quede, A., Jama, C., Supiot, P. et al. Elaboration of fire retardant coatings on polyamide-6 using a cold plasma polymerization process. *Surface and Coatings Technology* 67, 5 (2002): 424–428.

53. Errifai, I., Jama, C., Bras, M.L. et al. Elaboration of a fire retardant coating for polyamide-6 using cold plasma polymerization of a fluorinated acrylate. *Surface and Coatings Technology* 180, 8 (2004): 297–301.
54. Grützmacher, J.L., Tsafack, M.J., Kamlangkla, K. et al. Multifunctional coatings on fabrics by application of a low-pressure plasma process. In *13th International Conference on Plasma Surface Engineering*, Garmisch-Partenkirchen, Germany, September 10–14, 2012.
55. Tsafack, M.J. and Grützmacher, J.L. Plasma-induced graft-polymerization of flame retardant monomers onto PAN fabrics. *Surface & Coatings Technology* 200, 11 (2006): 3503–3510.
56. Shah, J.N. and Shah, S.R. Innovative plasma technology in textile processing: A step towards green environment. *Research Journal of Engineering Sciences* 2, 4 (2013): 34–39.
57. Rajendran, R., Radhai, R., and Balakumar, C. Synthesis and characterization of neem chitosan nanocomposite for development of antimicrobial cotton textile. *Journal of Engineering Fibre and Fabrics* 7 (2012): 46–49.
58. Thilagavati, G. and Kannaian, T. Combined antimicrobial and aroma finishing treatment for cotton using microencapsulated geranium leaves extract. *Indian Journal of Natural Products and Resources* 1, 3 (2010): 248–252.
59. Srivastava, A. and Singh, T.G. Utilisation of aloe vera for dyeing natural fabrics. *Asian Journal of Home Science* 6 (2011): 1–4.
60. Varun, S.K. An evaluation of aroma imparted by cotton fabrics dyed by natural colourants. *BMC Dermatology* 4 (2004): 1–15.
61. Vasanth, K.D., Boopathi, N., Karthick, N. et al. Aesthetic finishes for home textile materials. *International Journal of Textile Science* 1, 3 (2012): 5–9.
62. Basak, S., Samanta, K.K., Chattopadhyay, S.K. et al. Self extinguishable fabric using banana pseudostem sap. *Current Science* 108, (2015): (in press).
63. Smanata, K.K., Basak, S., and Chattopadhyay, S.K. Eco-friendly coloration and functionalization of textile using plant extracts. In S.S. Muthu (ed.), *Roadmap to Sustainable Textiles and Clothing*. Springer Publication, Singapore, pp. 263–287, 2014.
64. Sayed, M.E., Mansour, O.Y., Selim, I.Z. et al. Identification and utilisation of banana plant juice and its pulping liquors as anticorrosive materials. *Journal of Scientific and Industrial Research* 60 (2001): 738–747.
65. Basak, S., Samanta, K.K., and Chattopadhyay, S.K. Flame retardant cellulosic fabric using banana pseudostem sap. *International Journal of Clothing Science and Technology* 27, 2 (2015): (in press).
66. Katarzyna, P. and Prezewozna, S. Natural dyeing plants as a source of compounds protecting against UV irradiation. *Herba Policia* 55 (2009): 56–59.
67. Neog, S.R. and Deka, C.D. Salt substitute from banana plant (*Musa balbisiana* Colla). *Journal of Chemical and Pharmaceutical Research* 5, 6 (2013): 155–159.
68. Soares, S., Camino, G., and Levchik, S. Effect of metal carboxylates on the thermal decomposition of cellulose. *Polymer Degradation and Stability* 62 (1998): 25.

4 Application of Biotechnology in the Processing of Textile Fabrics

D. Saravanan, Thilak Vadicherla, K.P. Kannan, and S.N. Sreelakshmi

CONTENTS

4.1 Introduction 77
4.2 Peroxide- and Chlorine-Based Single-Stage Processes......................................78
4.3 Enzyme-Assisted Processes..81
4.4 Combinations of Enzymes..83
4.5 Reusing Amylases in Desizing: A Case Study 84
    4.5.1 Production of Amylase and Desizing 84
    4.5.2 Amylase Inhibition by End Products...85
    4.5.3 Thermal and Mechanical Deactivation of Amylases 86
4.6 One-Step Desizing and Bleaching Using Amylase and Glucose Oxidase: A Case Study 87
    4.6.1 Amylase Desizing..87
    4.6.2 Bleaching of Cotton Using Glucose Oxidase 89
    4.6.3 Combined Desizing and Bleaching 90
4.7 Multienzyme Scouring Process: A Case Study..90
    4.7.1 Multienzyme Treatment..92
4.8 Conclusion 92
References..93

4.1 INTRODUCTION

Cotton fiber has a dominant role in the apparel segment even though many alternative, regenerated cellulosic fibers are available. Besides cellulose, cotton fiber contains many constituents that perform certain designated functions during fiber growth; inside the boll, such constituents are systematically eliminated in the fabric preparation process to facilitate dyeability and finishing and to avoid unwanted reactions in such operations.

    Wet processing of cotton materials consumes ~80% of total energy requirement of all the textile operations, of which ~66% of the energy is consumed in heating and evaporation of water from the fabrics (Harrison, 1986). In spite of improved performances of drying machines and new techniques adopted for drying the textile materials, the increase in energy costs makes every manufacturer to think in terms of conserving the energy required in these areas. Even though intermediate drying is not carried out in many fabric preparation stages, the individual processes also consume more amount of energy since these processes are carried out, generally, at higher temperatures. Attempts have been made, in the past, to optimize unit operations to reduce the energy consumption and to derive the advantages in the respective unit operations (Dickinson, 1987; Trotman, 2000; Gursoy and Hall, 2001). Though continuous bleaching has been made possible with the development of

TABLE 4.1
Sustainability Road Map for Preparatory Process

Processes	Possibilities	Possible Methods	Additional Benefits	Demonstration of Sustainability
Desizing–scouring Bleaching	Combining two or all processes	Conventional chemicals	Shortened process sequence	Labeling of products and process
		Enzyme(s)-assisted processes	Reusing the process liquor and shortened process sequence	

J box, combining desizing, scouring, and bleaching (DSB) was found to be difficult due to difference in process conditions. Desizing can be performed at a certain pH, varying between 1 and 14, while scouring operation using the alkali requires a pH of 14 at boil, and bleaching can be carried out at a pH of either 3–4 or 10–11.5 depending on the bleaching agent used.

Sustainable approaches in the fabric preparation necessarily involve reducing the use of chemicals, energy, and time and enhancing the use of eco-friendly substances. A logical approach to conserve energy and materials in the preparatory processes, considered by both researchers and industry, is to shorten the sequence by combining the operations desizing–scouring–bleaching, popularly known as DSB process. Scouring appeared to be a major hurdle in combining these processes into a single stage. In a combined process, hydrogen peroxide or sodium hypochlorite or sodium chlorite is used with sodium hydroxide or new scourant formulations (Lewin and Epstein, 1962; Anon, 1982, 1992; Gulrajani and Sukumar, 1984; Parikh and Karami, 1989; Ammayappan et al., 2003; Hashem, 2007). Peroxygen boosters were considered to be the key compound in developing a DSB process, which can also interact with starch and PVA sizes and act as desize–scour boosters. Processes without peroxygen boosters were also developed incorporating high strengths of peroxide but were not widely accepted. The role of peroxide activators, namely, trisodium phosphate, tetrasodium pyrophosphate, sodium hexameta phosphate, magnesium sulfate, and sodium salts of EDTA, has been analyzed in combination of sodium silicate in the single-stage preparatory process (Gulrajani and Sukumar, 1984).

Notwithstanding these developments, preparatory processes using enzymes, amylases, pectinases, proteases, lipases, and glucose oxidase have been developed and followed commercially. Combinations of enzymes are used in the processes where multiple substrates are expected to be removed. Sustainable options are available using enzymes and conventional chemicals (Table 4.1) in the preparation of fabrics for dyeing and finishing with relative merits and demerits. Preparation of fabrics using combination of enzymes, one-step desizing–bleaching, multienzyme treatment to remove all the impurities, and reusability of the process liquor have been presented in this chapter.

4.2 PEROXIDE- AND CHLORINE-BASED SINGLE-STAGE PROCESSES

Single-stage preparatory process using hydrogen peroxide has been developed successfully in the past for starch and acrylic-base sized textile materials (Gulrajani and Sukumar, 1984; Ammayappan et al., 2003). Oxidative desizing agents are active against a broad range of sizes; at the same time, many of them can also be used as bleaching agents. In a combined process, hydrogen peroxide is used as the desizing and bleaching agent, and sodium hydroxide can act as a conventional scouring agent.

On activation, hydrogen peroxide degrades the sizing materials at lower temperatures, and bleaching occurs along with completion of desizing at higher temperatures. Higher alkalinity at elevated temperature produces efficient scouring action. Silicate is used as a stabilizer and to obviate the catalytic activation of heavy metals as well as acts as a buffer. Coloring materials are decomposed into soluble products and acidic impurities in cotton fibers. These acidic products lower the pH of the bath, which in turn reduces the rate of decomposition of hydrogen peroxide and the bleaching action. At the end, pH reaches closer to neutral and leads substantial amount of hydrogen peroxide in undecomposed condition on the fabric, that is, up to 21%–28%, and starch removal was found only to the extent of 82%. Buffers like borax (pH 9), sodium bicarbonate (pH 10), sodium hydroxide, and disodium hydrogen phosphate (pH 11) have been tried to counteract the acidic action of the decomposed products on hydrogen peroxide decomposition (Gulrajani, 1989). Buffered systems showed improvement in whiteness index compared to unbuffered systems since pH is maintained throughout the bleaching operation. In unbuffered systems, the loss in tensile strength is often observed after 55%–60% of peroxide decomposition, even though there is no improvement in the whiteness. The strength loss increases steadily when the pH approaches the neutral. Though hydrogen peroxide whitens the fabrics in alkaline conditions (heterolytic cleavage), degradation occurs at neutral pH (homolytic cleavage):

$$H_2O_2 \leftrightarrow H^+ + HO_2^- \quad \text{(heterolytic cleavage)}$$

$$H_2O_2 \leftrightarrow {}^*OH + HO^* \quad \text{(homolytic cleavage)}$$

The perhydroxyl ions under alkaline conditions attack both the cellulose and the colored molecules. Complete decomposition of peroxide is achieved in shorter duration with higher pH, that is, pH 11, than pH 10. Prolonging the treatment beyond complete decomposition also results in reduction in strength to chain breakage. The removal of starch also increases with increasing pH in both buffered and unbuffered conditions.

A self-emulsifiable solvent system of bleaching has been developed to combine the three different processes involved in the preparatory process. The system uses a high proportion of water, very low levels of solvents, and hydrogen peroxide. The presence of hydrogen peroxide helps in both desizing and bleaching, whereas the emulsified solvent is responsible for the scouring of cotton. Since the system involves very low solvent content, there is no need for a solvent recovery plant. Also, the operation can be carried out in plants usually available.

A self-emulsifiable solvent mixture was prepared (Gulrajani and Sukumar, 1985b) consisting of a solvent (10 parts), an emulsifier, and pine oil (50 parts) as a wetting agent (40 parts). Nonionic emulsifier, perchloroethylene and pine oil combination, was found to be more stable than anionic emulsifiers. Hydrophilic–lipophilic balance (HLB) values of the various agents used in the scourant mix were found to be closer to each other (pine oil, 13.5; nonylphenol, 13–14; and perchloroethylene, 13). Due to closer HLB values, the components of the scouring agents are completely miscible and give a stable emulsion with water.

The emulsified solvent mixes with the fats and oils and thus performs the scouring action, while hypochlorite acts as both bleaching agent and oxidative desizing agent. An aftertreatment was given to the fabrics with cold wash, treatment with 1% solution of sodium thiosulfate to remove the residual chlorine at room temperature for 5 min., and finally again cold wash.

Sodium chlorite affects the cotton fibers to the lesser extent and completely destroys the seed coats. The synthetic fibers can be treated with the sodium chlorite without causing damage to the fibers. The main drawback is that, at the temperatures of 80°C–85°C and pH of 2.5–4, the evolution of ClO_2 makes the process highly corrosive and hazardous. So different activators to make sodium chlorite effective at low temperatures like triethanolamine hydrochloride (0.4%), sodium dichloroisocyanate (0.7%), formaldehyde (3 g/L), sodium bisulfite (8 g/L), and formaldehyde adduct have been suggested with varying concentrations at temperatures from 35°C to 55°C. This process can also be carried out with or without buffered conditions, and the recipes need wider range in pH, time, and temperature. In the case of unbuffered conditions, the initial pH has been kept relatively higher (Gulrajani and Sukumar, 1985a; El Sisi et al., 1990) than the buffered conditions.

In the case of sodium chlorite–hydrogen peroxide system, free radical mechanism is responsible for the bleaching action. Various free radicals created during the treatment result in disintegration and destruction of foreign matters in the cotton. The bleaching effect is more pronounced with peroxide than sodium chlorite in spite of the higher concentration. This is mainly due to the difference between the two oxidants with respect to decomposition in alkaline medium. Presence of cooxidants impedes the decomposition of each other, especially at low concentrations. Bleaching in the case of sodium chlorite is taken care mainly by oxygen, while the formation by chain reactions of radicals such as HOO* and HO* is responsible for bleaching action of hydrogen peroxide. Thus, low levels of hydrogen peroxide (≤0.4 g/L) result in poor degree of whiteness. Also, at lower concentrations of hydrogen peroxide, the decomposition products are not sufficient enough to induce activation of sodium chlorite. The mechanism under alkaline medium is

$$HOOH \rightarrow H^+ + HOO^-$$

$$HOO^- + HOOH \rightarrow HOO^* + HO^* + HO^-$$

$$HO^* + H_2O_2 \rightarrow HOO^* + H_2O$$

The HO* and HOO* radicals react with the chlorite ions, and, as the result, a reaction chain is perpetuated as suggested by following reactions:

$$ClO_2^- + HO^* \rightarrow ClO^- + H_2O$$

$$ClO_2^- + HO^* \rightarrow ClO_2^- + HO_2^-$$

$$2ClO_2 \rightarrow ClO^- + ClO_3^-$$

$$ClO_2^* + ClO^- \rightarrow ClO^* + ClO_2^-$$

$$ClO^* + ClO^- + HOO^- \rightarrow Cl_2 + O_2 + HO^*$$

These free radicals enhance the bleaching effect of sodium chlorite by hydrogen peroxide when used at higher concentrations. These free radicals seem to disintegrate the impurities and destroy the coloring matters of the cotton.

In the case of the hypochlorite-solvent-assisted single-stage preparatory process (Whitwell, 1984; Gulrajani and Sukumar, 1985b; Gulrajani, 1987), the whiteness index shows approximately a linear relationship with available chlorine in sodium hypochlorite solution at various treatment times ranging from 45 to 225 min. A whiteness index equivalent to the conventional bleaching is obtained at a concentration around 6 g/L available chlorine. Tensile strength also indicates a linear relationship with the available chlorine concentration. At a concentration of 4–6 g/L of available chlorine, the tensile strength appears to be equivalent to that obtained in the conventional method. But higher degradation in tensile strength is observed at higher concentrations of both available chlorine and scouring agent due to severe oxidation and higher weight loss values. The decrease in the strength also takes place as the temperature of the treatment increases, due to severe action of sodium hypochlorite at a higher temperature. Wetting time decreases with increasing available chlorine concentration; however, no marked difference can be observed beyond 8 g/L. Better absorbency is obtained using the scouring agent concentration of 8%, at a temperature of 50°C–55°C, which is closer to the cloud point of the nonionic emulsifier used in the mix. Yellowing of processed fabrics occurs on storage, similar to that occurring in conventional bleaching. It has been observed that 30%–35% more NaOCl is consumed in the single-stage process in comparison to conventional bleaching.

In the peroxide–alkali process (Ammayappan et al., 2003), increase in weight loss due to scouring action and desizing occurs with increase in concentrations of peroxide and sodium hydroxide, and higher weight loss is achieved at a temperature of 95°C. The whiteness index is found to be directly related to the concentration of sodium hydroxide, which is mainly due to the increased formation of perhydroxyl ion at higher pH that acts as the bleaching species. Increase in whiteness is obtained gradually with increase in temperature as the activation of hydrogen peroxide increases with temperature. Wettability improves with increasing concentration of sodium hydroxide and the lowest wetting time was observed at 95°C.

In peroxide solvent scouring agent–based process, substantial weight loss, strength loss, and low wetting time have obtained a concentration of up to 4% of scouring agent and also at higher levels. For a given temperature and scouring agent concentration, a rapid increase in weight loss and a steep decrease in tensile strength and wetting time are observed with respect to the increase in peroxide concentration of up to 2%. Apparently, a linear relation exists between (1) weight loss and tensile strength, (2) wetting time and weight loss, and (3) strength and wetting time. Significant correlation is observed between wetting time and weight loss. Weight loss increases gradually with increase in concentrations of both peroxide and scouring agent, and similar trend is also observed in the case of wetting time. There is a decrease in tensile strength with respect to the peroxide concentration at all concentrations of scouring agent. However, at lower concentrations, maximum fall in strength occurs below 2% peroxide. Whiteness index shows a linear relationship with the concentration of hydrogen peroxide, and other factors like scouring agent concentration, time, and temperature had a little effect.

In the case of combined process using sodium chlorite–hydrogen peroxide, the whiteness increases with increasing time and pH (8–10) and is efficient for 100% cotton, sized with starch and P/C blends (50;50). Relatively, a lower degree of whiteness is achieved in the case of blended fabrics. The weight loss occurs to the extent of 8.0%–8.5% with the tensile strength and 81%–86% of the grey fabric strength.

In the single-stage preparatory process, an attempt has been made to improve the efficiency by carrying out the process using microwave (Tarakcioglu and Anis, 1996) with the energy of 600 W for a duration of 30 s–3 min. The process involves steaming of the fabric with the simultaneous exposure to the microwave energy. In another novel approach (Lacasse and Baumann, 2004), flash steam has been used to combine all the preparatory processes into a single step. This helps to complete the fabric preparation within 2–4 min. After applying the solution, consisting of alkali, hydrogen peroxide, and an auxiliary to the fabric, it is steamed with saturated steam followed by hot wash. Table 4.2 gives the various recipes developed for the single-stage preparatory process.

4.3 ENZYME-ASSISTED PROCESSES

Enzymes provide sustainable solutions, in operations that require more amounts of energy, at relatively low energy levels with meager amounts of auxiliaries (Underkofler et al., 1958; Palmer, 1981; Kamat, 1995; Etters and Anis, 1998; Tzanov et al., 2000; Carlier, 2001; Gursoy and Hall, 2001; Kirk et al., 2002; Rodriques, 2002; Galante and Formantici, 2003). Enzyme-assisted processes involve diffusion of enzymes from bulk solution to solid surface, adsorption of enzyme onto the surface of substrate, formation of enzyme–substrate complex, catalysis of relevant reaction, and diffusion of soluble degradation products from solid substrate to the reaction medium. Reactions, in these steps, are controlled by diffusion of enzyme molecules, which tends to be slow because of the large size of the molecules. High-twist yarns and heavyweight fabrics slow down quick interchange of solution to interior of the fabrics and also the respective reactions (Naik and Paul, 1997; Yoon, 2005). Reviews and reports are frequently published to demonstrate the advantages of enzyme-based processes in comparison with chemical treatments (Robner, 1993; Naik and Paul, 1997; Karmakar, 1998; Carlier, 2001; Holme, 2001; Paul and Pardeshi, 2002; Rai, 2004; Aiteromem, 2008; Tanapongpipat et al., 2008).

TABLE 4.2
Recommended Recipes for Single-Stage Preparatory Process

No.	Process	Recipe		References
1	Peroxide–alkali	NaOH	10 g/L	Ammayappan et al. (2003)
		H_2O_2	1.0 g/L	
		Na_2SiO_3	5 g/L	
		Temperature	95°C	
		Wetting agent	1 g/L	
		pH	10.5–11.5	
		Time	120 min	
2	Peroxide-solvent-assisted scourant, starch and acrylic size	Scouring agent	–4%	Gulrajani and Sukumar (1984, 1985b)
		Peroxide	2%	
		Sodium silicate	1%	
		Wetting agent	0.1%	
		Temperature	95°C	
		pH	10	
		Time	180 min	
3	Peroxide-solvent-assisted scourant	Scouring agent	4% owf	Gulrajani (1989)
		H_2O_2	1%	
		Na_2SiO_3	2%	
		Na_3PO_4	2%	
		pH	11	
		Time	60 min	
4	Hypochlorite-solvent-assisted scourant	Scouring agent	2%	Gulrajani and Sukumar (1985a), Gulrajani (1987), and Whitwell (1984)
		NaOCl	6 g/L (av. Cl_2)	
		Temperature	40°C	
		Time	180 min	
		M/l	1:20	
		pH	11	
5	Sodium chlorite–peroxide process	$NaClO_2$	3 g/L	El Sisi et al. (1990)
		Na_2HPO_4	10 g/L	
		NI wetting agent	2 g/L	
		pH	10	
		Time	90 min	
		Temperature	95°C	
		M/l	1:20	
6	Sodium chlorite solvent–assisted scourant	Sodium chlorite	0.8%–1.95%	Gulrajani (1989)
		Scouring agent	2%	
		Temperature	30°C–55°C	
		Time	5–14 h	
		pH	4.6 (buffered)	
7	Peroxide–alkali process	Sodium hydroxide	20 g/L	Gulrajani and Sukumar (1985b)
		Hydrogen peroxide	30–40 mL/L	
		Peroxydisulfate	5 g/L	
		Sodium silicate	20 mL/L	
		Surfactant	10 g/L	

Activity of amylases in desizing has been studied extensively using pure starch as a model compound, and attempts have been made to analyze the hydrolytic activities of pancreatic, malt bacterial amylases in desizing. Lipases increase lipid removal from all morphological locations on the cotton fibers including lumen, fiber surfaces, and interfiber capillaries (Munk et al., 2000; Obendorf et al., 2001). Most of the nitrogen-containing compounds of cotton can be removed by a mild alkaline scouring, and a very low residue remains in the scoured and bleached cotton fibers (Wakelyn, 1975; Hartzell and Hsieh, 1998; Najafi et al., 2005). Denaturation of protein substances by aging, heating, and oxidation makes them less accessible to enzymatic degradation (Andrade et al., 2002). Major limitations in the use of pectinases, in scouring, are posed by the difficulties in accessing the substrates, incomplete removal of wax, and little or no effect on mote removal (Kim et al., 2005; Lu, 2005; Li et al., 2007; Halim et al., 2008). Often, efforts are made to improve the efficiency of scouring process with novel attempts (Li and Hardin, 1998a,b; Etters, 1999; Lenting et al., 2002; Semenova et al., 2003; Yachmenev, 2005). Enzyme-assisted processing of textile materials is, conveniently, carried out in equipment such as jigger, jets, and pad-batch and pad-stream ranges, employing different levels of mechanical agitations without necessitating further capital investments. Also, reusing of the process liquor in the subsequent cycles has also been attempted, in desizing, to reduce the water consumption and provide a sustainable solution using enzymes (Sreelakshmi, 2014).

In many situations, combinations of three or four different enzymes including amylase, pectinase, protease, cellulases, glucose oxidase, and various hemicellulases have been attempted since pectinases alone is not effective enough to improve absorbency of the scoured samples and effective removal of cuticle and other impurities from cotton fibers (Csiszar et al., 1998; Diller et al., 1999; Traore and Diller, 2000; Degari et al., 2002; Ledakowics et al., 2005; Sae et al., 2007). Incidentally, the addition of amylases in the scouring also facilitates the desizing process.

4.4 COMBINATIONS OF ENZYMES

Research works have been carried out (Diller et al., 1999; Traore and Diller, 2000; Riegels et al., 2001; Kitchareonseree, 2002; Sae et al., 2007) to study the effect of combined enzymes on the efficiency of scouring, using lipase, pectinase, xylanase, and cellulase, with strong agitation levels, regardless to their compatibility and optimum levels, where the treatment with pectinase alone results in ~7.5%–8.0% weight loss, while the combination of all the enzymes results in the weight loss up to 13.9%, much higher than the sum of natural impurities. Higher wickability is observed in the case of pectinase and xylanase combinations than pectinase and cellulase and pectinase and lipase combinations, with the highest value observed in all enzyme combinations.

Scouring of cotton, polyester, and cotton/polyester blended fabrics using lipase (from porcine pancreas), protease from *Aspergillus oryzae*, and cellulase from *Aspergillus niger* shows a weight loss of 0.2%–1.0% with improvement in the strength and whiteness compared to the treatment when these enzymes are alone (Riegels et al., 2001; Kitchareonseree, 2002; Sae et al., 2007). A two-step scouring of cotton has also been suggested, with lipase and protease in the first step and cellulase in the second step. Pectinase scouring produces 18-fold higher amounts of reducing sugars and galacturonic acid than any of the two-step processes, while lipase/proteases/cellulase scouring produces approximately fivefold higher amounts of amino acids than the pectinase scouring and three major fatty acids, namely, palmitic, stearic, and uric acids (Sae et al., 2007).

In commercial cellulase preparations, the presence of xylanase and pectinase also facilitates removal of seed coats that has often been perceived to be different in DSB process. Cellulase treatment alone destroys approximately up to 70%–85% of motes depending upon its concentrations (Csiszar et al., 1998; Ledakowics et al., 2005). Protease and lipase can also be used together with pectate lyase to remove spinning, coning, and slashing lubricants from the fabrics (Li and Hardin, 1998a; Lange et al., 2001; Miller et al., 2003; Wang et al., 2007). Lipase or protease treatments, alone, do not improve wetting or water retention values or other technical properties of cotton (Buchert et al., 2000).

Combined treatments, using cellulase and pectinase, show higher wettability, while protease with cellulase and pectinase results slightly lower values, followed by proteases alone depending upon the type of fabrics and combination of enzymes (Karapinar and Sariisik, 2004; Opwis et al., 2006). Pectinase–cellulase, pectinase–cellulase with protease, or pectinase–cellulase–xylanase produces better scouring results than those individual enzymes, in terms of dyeability, K/S, and fastness properties. However, after hydrogen peroxide bleach, all the combinations (Wang et al., 2007) show similar whiteness index (68–70). Moreover, the presence of proteases in the same bath, with other enzymes, is expected to cause proteolysis of coenzymes, a fact that has not been looked upon in many of these studies.

4.5 REUSING AMYLASES IN DESIZING: A CASE STUDY

Amylases are a group of enzymes that break down the starch into reducing sugars and are classified into α-amylase, β-amylase, and γ-amylase, which differ in the method of hydrolyzing the polysaccharide bonds. Amylases can be isolated from fungi, yeast, bacteria, and actinomycetes (Pandey et al., 2000). Amylases from microorganisms, especially *Aspergillus* spp., have gained more importance because of their easier availability and high productivity (Saravanan et al., 2014). The main aim of desizing is to eliminate sizing ingredients, which hinder the reactions during subsequent wet processing steps. Nowadays, amylases are commercialized and preferred for desizing due to their high efficiency and substrate specificity, complete removal of the size without harmful effects on the fabrics (Etters and Annis, 1998).

Though there are many advantages of using the enzymes in textile wet processing steps, high costs of hydrolytic enzymes remain a significant obstacle in the application of enzymes in commercial scales (Azevedo et al., 2002). Strategies to reduce enzyme cost consist of increasing enzyme production efficiency, increasing enzyme specific activity, and recycling enzymes for subsequent hydrolysis steps (Cherry, 2003; Tu et al., 2007). Recycling or reusing can considerably reduce the costs associated with enzymatic processing as the enzymes are not consumed during the reactions (Azevedo et al., 2002).

The assumption is that, by recovering the active enzymes after the enzymatic hydrolysis step, it is possible to reduce the amounts of new enzymes required in the next cycle of hydrolysis, and therefore reduce the overall enzyme cost in the process. There are reports on recycling of cellulase enzymes in biopolishing (Azevedo et al., 2002), but surprisingly no attempts to reuse or recycle amylases after desizing have been reported in literature. Washing of cotton fabrics in desizing, using surfactants, needs huge quantities of water due to build up of highly viscous starch removed in the process. On the other hand, enzymatic desizing does not result in the buildup of viscosity due to conversion of starch into low-molecular-weight compounds like glucose and other oligosaccharides.

4.5.1 Production of Amylase and Desizing

The microorganism used in the study *A. niger* was isolated from local soil sample and identified on the basis of morphological characteristics and microscopic identification (Ellis, 1971; Onions et al., 1981). The basal medium (Shukla et al., 2005) was used to produce amylase enzyme containing the following components (g/L): potassium dihydrogen phosphate, 1; magnesium sulfate, 0.1; calcium chloride, 1; disodium hydrogen phosphate, 3; yeast extract powder, 5; Tween 80, 1; and starch, 2. The inoculum was prepared by growing fungal cultures in a 100 mL Erlenmeyer flask containing 10 mL of fermentation medium and incubated at 30°C for 48 h in a rotary shaker. After incubation, 10 mL inoculum was transferred into 90 mL of fermentation medium and incubated at 30°C under shaking conditions. Culture media were collected at intervals for a period of 8 days, filtered using Whatman No. 1 filter paper, and the cell-free filtrate was used as enzyme source.

Application of Biotechnology in the Processing of Textile Fabrics

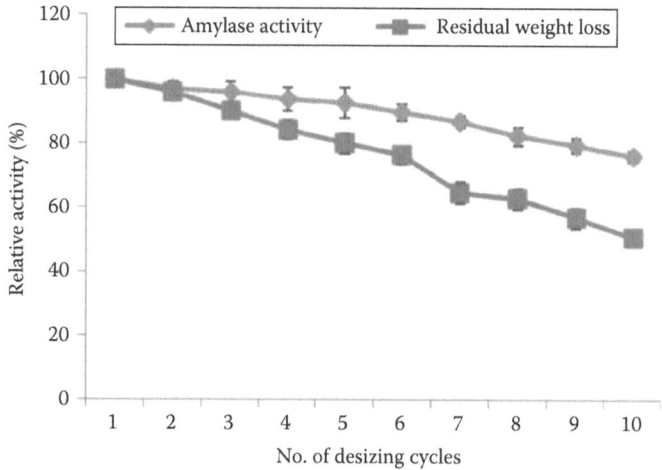

FIGURE 4.1 Effect of desizing cycles on amylase activity.

Amylase activity was assayed by measuring the release of reducing sugars during enzyme substrate reaction by dinitrosalicylic acid method (Miller, 1959). Absorbance was measured at 540 nm in a Lambda 35 model UV–visible spectrophotometer, and one unit of enzyme activity was defined as amount of enzyme required for releasing 1 µg of maltose/mL/min under specified assay conditions. All the experiments were performed independently in triplicates, and the results showed here are the averages of the three.

Prior to biodesizing, fabrics were pretreated by dipping in hot water at 100°C for 2 min to facilitate the swelling of starch layer. Pretreated samples were desized in a shaker bath, at 100 rpm, and incubation was carried out, using the cell-free amylases (8000 U/mL/min) in a temperature-controlled water bath at 40°C for 1 h in 250 mL Erlenmeyer flask using liquor to material ratio of 40:1, in the presence of Triton X-100 as the wetting agent. The pH of the treatment bath was adjusted to 5.0 in order to maintain a constant pH during enzymatic desizing process. After the process, enzyme reaction was stopped by washing the fabrics in distilled water for 10 min at 90°C and dried. After desizing, treatment liquors were collected and centrifuged at 8000 rpm to remove any traces of insoluble material, and the supernatant was analyzed for amylase activity. The reusability of amylase for desizing of cotton fabrics was determined by estimating amylase activity and taking the weight loss of the cotton fabrics, before and after the amylase treatment at regular intervals for 10 cycles. After desizing of cotton fabrics, amylases are expected to be distributed between the liquor and substrate, and hence, the amounts of reusable enzymes present in the liquor and adsorbed to the substrate were determined. Desizing of cotton fabrics was carried out for 10 repeated cycles without addition of fresh enzyme and removal of reaction products. It was observed that the amylases present in the treatment liquor after 10 cycles were still active with activity levels about 50% after the 10th cycle (Figure 4.1). The reduction in the activity toward cotton fabrics could be attributed to the preferential and irreversible adsorptions of amylase enzyme by the cotton fabrics (Manning and Campbell, 1961). It is also suggested that loss of enzyme activity could be due to inhibition activities of end products that are formed during the process and also due to thermal/mechanical deactivation of the enzymes.

4.5.2 Amylase Inhibition by End Products

It is well known that hydrolysis products can inhibit the action of amylase on starch (Robyt, 1984). Hill et al. (1999) reported that starch hydrolysis end products such as maltose and glucose can inhibit amylase enzyme. The amount of maltose, glucose, and dextrins produced during starch hydrolysis

TABLE 4.3
Effect of Reducing Sugar Concentrations on Amylase Activity

Desizing Cycles	Relative Enzyme Activity (%)	Weight Loss (%)	Reducing Sugar		
			Maltose (g/L)	Glucose (g/L)	Dextrose (g/L)
0	100.0	—	—	—	—
1	97.1	5.1 ± 0.09	0.13 ± 0.02	0.03 ± 0.01	0.05 ± 0.03
2	96.0	4.9 ± 0.16	0.25 ± 0.02	0.06 ± 0.01	0.10 ± 0.02
3	93.7	4.6 ± 0.16	1.34 ± 0.02	0.06 ± 0.02	0.15 ± 0.02
4	92.5	4.3 ± 0.24	2.5 ± 0.02	0.15 ± 0.02	0.20 ± 0.03
5	89.7	4.1 ± 0.12	3.5 ± 0.32	1.25 ± 0.02	0.79 ± 0.02
6	86.8	3.9 ± 0.09	4.5 ± 0.28	2.47 ± 0.03	0.94 ± 0.02
7	84.0	3.3 ± 0.20	5.4 ± 0.16	3.20 ± 0.03	1.5 ± 0.16
8	82.2	3.2 ± 0.25	6.7 ± 0.20	4.50 ± 0.05	1.7 ± 0.21
9	79.4	2.9 ± 0.08	8.5 ± 0.12	5.45 ± 0.02	2.5 ± 0.33
10	76.0	2.6 ± 0.12	10.6 ± 0.20	6.60 ± 0.04	2.8 ± 0.21

was determined by DNS assay and compared with amylase activity and weight loss values of the cotton fabrics (Table 4.3). The results of the present investigation indicated that maltose showed a strong inhibition effect on amylase than glucose and dextrins. After 10 cycles, 10.6 g/L of maltose, 6.6 g/L of glucose, and 2.8 g/L of dextrose were present in the desizing liquor, and the effect of amylase toward cotton fabrics reduced to 50.9%. Since maltose inhibition is high, it is important to control its concentration in reaction mixture. Alpha glucosidase (maltase) is the enzyme that is responsible for maltose degradation.

4.5.3 Thermal and Mechanical Deactivation of Amylases

The effect of temperature on amylase activity was investigated by incubating amylase enzyme at different temperatures ranging from 30°C to 80°C for 1 h and 10 min. After the incubation time, the samples were collected and tested for amylase assay.

Thermostability studies recorded a decrease in the stability of the enzymes, as shown by the decrease in enzyme activity, at all the tested temperatures (30°C–90°C) with increase in time. With increase in temperature, reduction in stability was also observed (Figure 4.2) with the enzymes

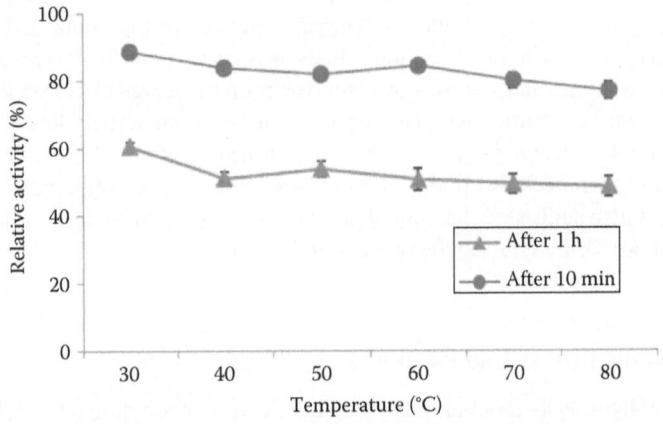

FIGURE 4.2 Effect of temperature on the stability of amylase activity.

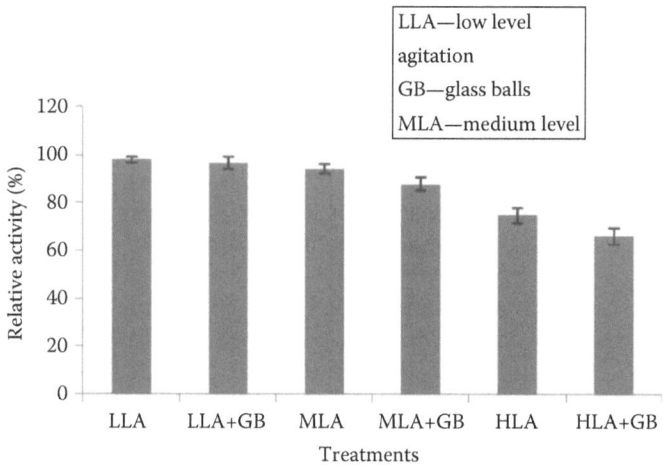

FIGURE 4.3 Effect of mechanical agitation on amylase activity.

showing the least stability at 80°C. This could possibly be due to the fact that at higher temperatures, the rate of thermal inactivation is also faster.

The effects of low, medium, and high levels of agitation on amylase enzymes were analyzed at constant temperature bath. It is evident that mechanical agitation greatly influences the effect of enzymes on those fibers and is directly related to the weight loss of the cotton fabrics. Low agitation and medium agitation levels do not cause significant effects on the enzyme deactivation, but high-level agitation treatments can cause enzyme deactivation as shown in Figure 4.3. Similar works carried out by Azevedo et al. (2002) also reported no significant mechanical deactivation due to the mechanical agitations under medium agitation levels.

4.6 ONE-STEP DESIZING AND BLEACHING USING AMYLASE AND GLUCOSE OXIDASE: A CASE STUDY

α-Amylases cleave the starch into various oligomers and finally to glucose as the end product in the hydrolysis and desizing of the grey cotton fabrics using amylases has been successfully implemented in the commercial processes. Glucose oxidase–catalyzed reaction on glucose, in the presence of oxygen, produces hydrogen peroxide and shows the potential for bleaching of cotton fabrics. Glucose is oxidized to δ galactone by the cofactor of glucose oxidase (flavin adenine dinucleotide [FAD]), which in turn reduces to its hydride form (FADH2). Subsequently, oxygen present in the reaction system is reduced to hydrogen peroxide, while FADH2 is reoxidized to FAD. Attempts have been made to utilize the spent desize bath for bleaching using immobilized glucose oxidase enzyme in the aerated systems. Combinations of pullulanase with amyloglucosidase and glucose oxidase have been attempted to utilize the glucose released in the amylase reaction for bleaching of cotton fabrics. One-step desizing and bleaching of the grey cotton fabrics using α amylase–glucose oxidase combination has been presented in the following text (excerpt from Saravanan et al., 2012).

4.6.1 Amylase Desizing

Bacillus amyloliquefaciens (MTCC 610) cultures were grown in a 100 mL broth containing soluble starch (1.0%, w/v) with peptone (0.5%, w/v), $MgSO_4 \cdot 7H_2O$ (0.2%, w/v), $(NH_4)2HPO_4$ (0.2%, w/v), $CaCl_2 \cdot 2H_2O$ (0.05%, w/v), K_2HPO_4 (1.4%, w/v), and KH_2PO_4 (0.6%, w/v), taken in a 500 mL Erlenmeyer flask. The cultures were incubated in the incubator shaker at a speed of 120 rpm and at a temperature of 37°C for 48 h. Bacterial growth in the inoculums was assessed using absorbance value of the sample drawn from the culture medium, with UV–visible spectrophotometer at regular intervals. Samples drawn from the fermentation broth were centrifuged at 7000×*g* for 10 min at 4°C; supernatant obtained from the centrifuge was collected and used for the measurement of protein content using bovine serum albumin solutions as the standard enzyme activity measurement and characterization studies for pH and temperature activity. A piece of grey cloth was used for desizing using the amylase with a concentration of 6840 U (8 mL/L) at a temperature of 50°C for 50 min at a pH value of 5.5, using the material-to-liquor ratio of 1:30. Duration of desizing and concentration of amylases were selected based on the initial experiments carried out using grey fabric samples at various levels.

Enzyme secretion (protein content) in the bacterial cultures followed a similar trend as that of bacterial growth. Peak enzyme production in terms of protein concentration was about 440 mg/dL, between 28 and 30 h of postinoculum, which translated to the enzyme activity of 871 U/mL/min. Hydrolysis of the amylases increased apparently in a linear fashion from pH 3.0, and maximum pH activity (1106 U/mL/min) was observed in between pH 5.8 and 5.9, followed by a decline phase due to inactivation of amylases at higher pH levels (Figure 4.4a). In the case of

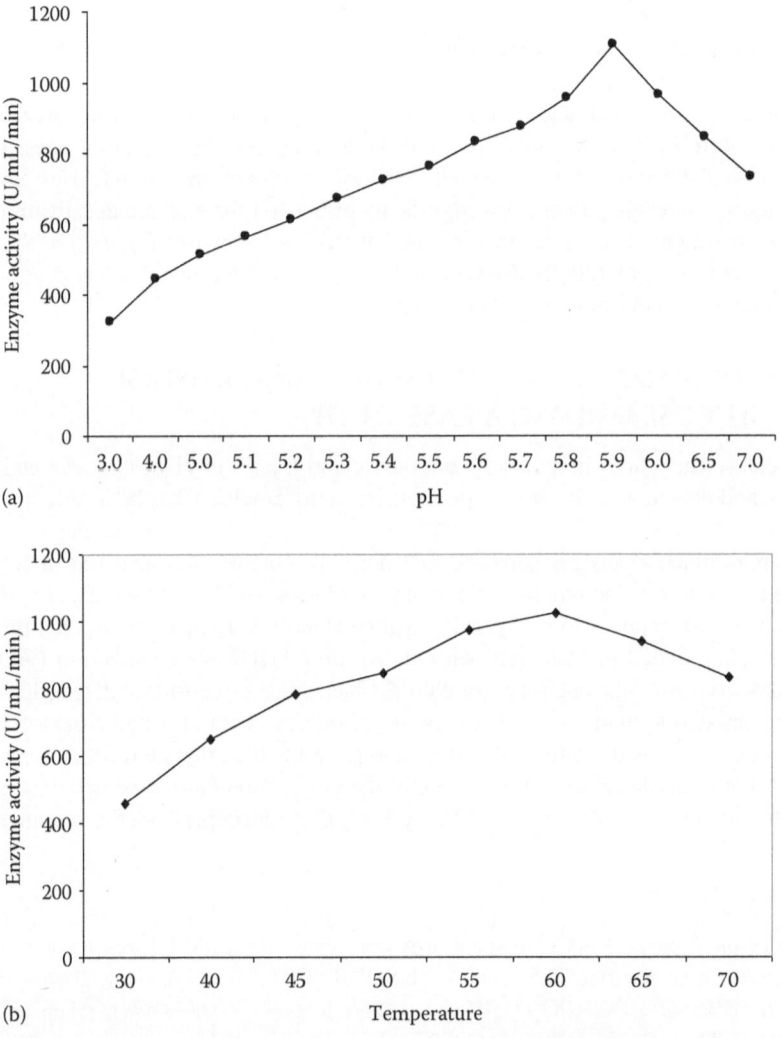

FIGURE 4.4 Characterization of amylases for optimum (a) pH and (b) temperature.

thermal characterization, maximum activity (1026 U/mL/min) was observed, at the temperatures between 55°C and 60°C, followed by a decline in the activities further, obviously due to thermal inactivation (Figure 4.4b).

Desizing of grey cotton fabrics using amylase obtained from *B. amyloliquefaciens* resulted in the weight loss of 5.2%, amounting to 87% size removal from the fabric samples. The extractions obtained from the desized fabric samples did not cause any discoloration of iodine solution, while dark blue color was obtained in the case of the untreated samples. However, the residual impurities from the desized fabrics, assessed using solvent extraction, were found to the extent of 3.53% on account of other natural impurities present in the samples.

4.6.2 Bleaching of Cotton Using Glucose Oxidase

Glucose oxidase of *A. niger* was used in bleaching of cotton fabrics at 40 U/mL (0.04 mg/mL) of glucose solution, and hydrogen peroxide released in the reaction was measured by titration with permanganometric method. The reactions were carried out in a shaker bath, at a speed of 150 rpm, and the mechanical agitations were enhanced by the addition of 10 glass balls, each weighing 1 g. Oxygen from a pressurized cylinder was supplied into the reaction bath to an extent of 5 L/min.

Bleaching of cotton fabrics carried out without the oxygen supply or mechanical agitation did not increase the whiteness values of the fabrics (International Commission on Illumination [CIE] whiteness index of 29.06) (Figure 4.5); however, reduction in the whiteness of the bleached samples was observed due to discoloration of reaction bath to dark brown color, which in turn caused fast tinting effect on the fabric samples. Such discoloration often occurs, in the case of D-glucose, at elevated temperatures in alkaline conditions, a characteristic reaction of D-glucose. This obviously necessitates a higher concentration of glucose oxidase or longer incubation time for increasing the conversion of reducing sugar into hydrogen peroxide. In spite of similar concentrations of glucose and process conditions used in glucose oxidase bleaching process, phenomenon of discoloration has not been reported in the literature. When gaseous oxygen was supplied from a pressurized cylinder to the extent of 5 L/min, whiteness values showed marginal improvement only.

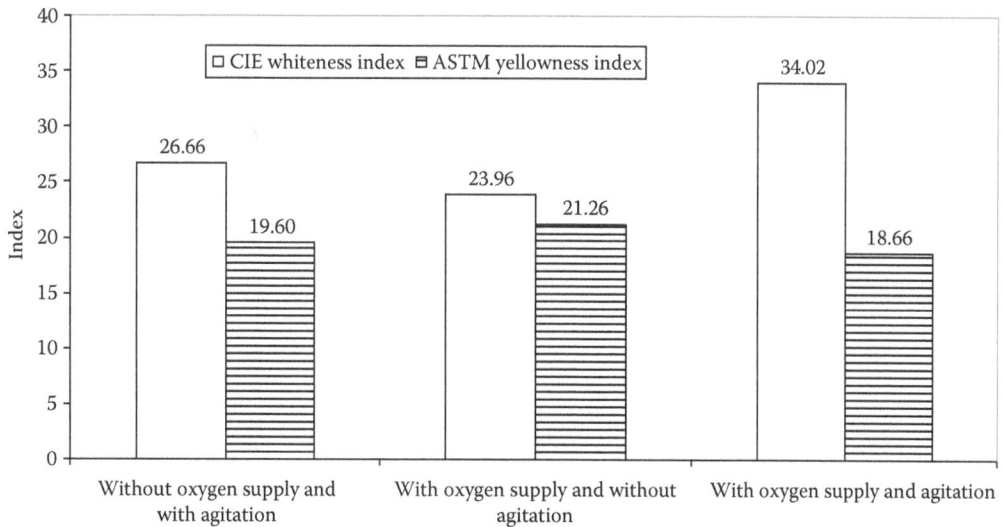

FIGURE 4.5 Effect of mechanical agitations and oxygen supply on bleaching.

4.6.3 COMBINED DESIZING AND BLEACHING

On completion of desizing duration, the reaction bath was cooled down to 37°C; glucose oxidase enzyme was added (quantity as specified in the discussions) to the same bath without draining. Standing bath was maintained for a duration of 90 min for the conversion of glucose into hydrogen peroxide. Bleaching was carried out, using alkaline activation method, at a temperature of 90°C for 60 min at a pH of 11. After bleaching, the samples were washed using hot and cold water thoroughly.

Reducing sugar released in the desizing process was utilized for conversion into hydrogen peroxide, without further addition of glucose, using glucose oxidase enzyme. Glucose oxidase 40 U/mL was added to the cooled desize bath and incubated for the conversion of glucose into hydrogen peroxide, with the introduction of gaseous oxygen into the reaction bath. Though hydrogen peroxide, at acidic pH, releases hydroxyl radicals that result in tendering of cellulose, such reactions require high temperature conditions, which did not prevail in the present study. Hydrogen peroxide released in the process was activated using sodium carbonate and bleaching was carried out for 60 min at 90°C. Higher concentration of glucose oxidase (60 U/mL) resulted in significantly higher whiteness index up to 52.30 (CIE units), with the same level of glucose in the bath, which further increased to 73.6 with the glucose oxidase concentration of 80 U/mL, much closer to the CIE whiteness values (73.0–74.0) obtained in the commercial bleaching processes using hydrogen peroxide.

The presence of hydrophobic impurities on the fiber surface, dried film of starch, and other hydrophobic components (tallow) in the sized yarns contributes to the lower absorbency of the grey fabrics. Treatment with mesophile amylases, carried out at lower temperatures, often results in the incomplete removal of starch and other size ingredients attached to it, resulting in lower absorbency levels. The drop absorbency values of desized fabrics showed the lowest absorbency value time of 24 s; however, when the bleaching treatment is combined with desizing, the residual constituents are expected to be removed and enhance the absorbency of the fabrics. Interestingly, in the case of samples obtained from the one-step desizing–bleaching process, the lowest absorbency value was observed at 2 s and the highest value at 49 s, showing strong influence of the residual impurities (2.13%) and effectiveness of the oxidation treatment in the combined process. The one-step process can be commercially exploited in place of the harsh chemical treatment, which deteriorates the fibers and fabrics in the process.

4.7 MULTIENZYME SCOURING PROCESS: A CASE STUDY

Scouring using cellulase and protease separately and in combination shows no difference in the weight loss, though marginal improvement is realized in the absorbency and residual extractable impurities (Guha and Shah, 2001). Cellulase and pectinase have been combined to improve the removal of pectineus matters in the scouring process, in which cellulases have been reported to open the structure of the primary wall and facilitate the pectinase reaction (Andrade et al., 2002). Combination of pectinase, cellulase, lipase, and protease in two steps results in improved removal of fatty matters with enhanced absorbency. However, such treatment often results in higher weight loss with absorbency equivalent to that of alkali-scoured specimen. Lower absorbency values have been reported using enzyme treatments with amylase, pectinase, protease, cellulase, and lipase (Csiszar et al., 2001; Karapinar and Sariisik, 2004). Scouring of the raw cotton fabrics using combination of cellulase, pectinase, and protease enzymes has been explained in this part (excerpt from Saravanan et al., 2010). Fabric samples were scoured in a single stage using two steps, with pectinase (6.0 g/L) and cellulase (6.5 g/L) in the first step followed by protease (1.5 g/L) treatment in the second step.

Characterization of protease and pectinase enzymes initially shows an increasing weight loss followed by a decrease after reaching a peak, adhering to the characteristic curve of the enzyme activity (Figure 4.6). Extraction of the specimens treated with protease using benzene–alcohol mixture shows the residual extractable content of 2.1%, which obviously indicates the large proportion

Application of Biotechnology in the Processing of Textile Fabrics

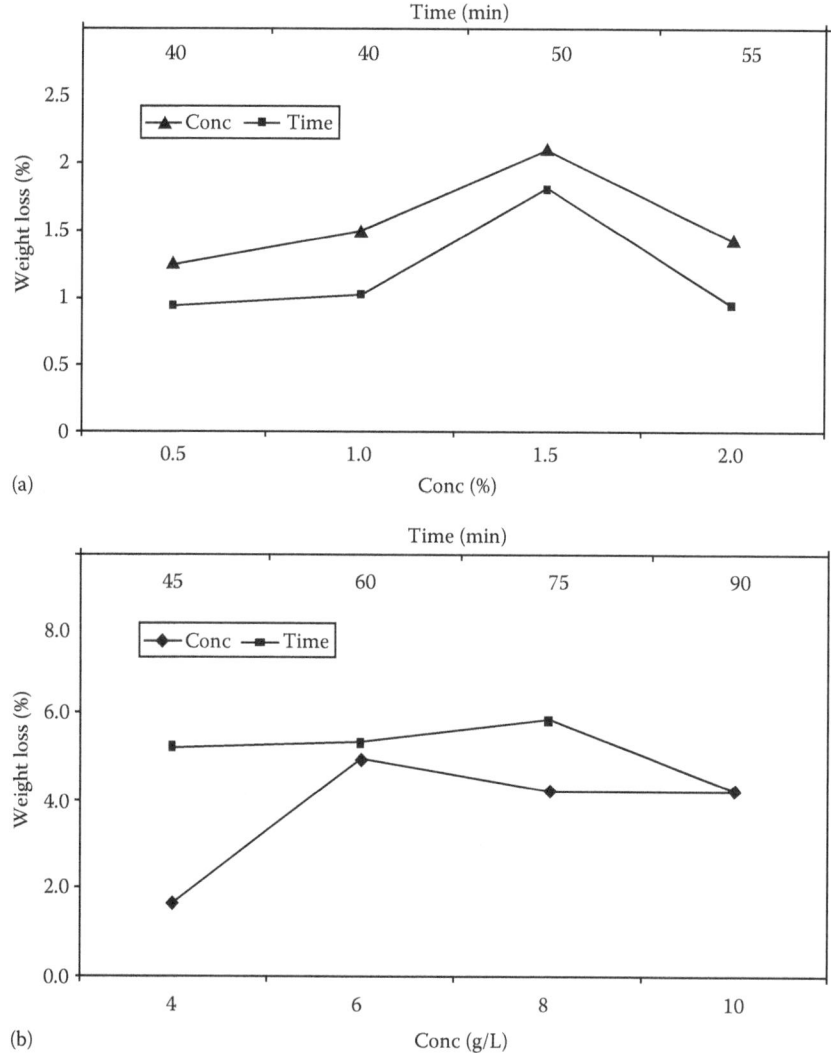

FIGURE 4.6 Effect of (a) protease and (b) pectinase concentrations and scouring time on weight loss.

of residual matters in the protease-treated fabric samples. On the other hand, pectinase treatment results in lower extractable impurity levels of 1.2%. Minimum weight loss of ~1.6% is observed using protease, while the maximum weight loss of ~3.2% is observed in various combinations of experiments conducted in the study. The control sample without enzymes under mechanical agitations shows a weight loss of 0.58%. An average weight loss of ~1.2% obtained with pectinase, for various combinations of control and noise factors, indicates certain hindrances that might have occurred in the hydrolysis process.

Drop absorbency of untreated fabrics shows high values (>100 s), while protease-treated fabric samples show a significant improvement. The lowest absorbency measured in drop test has been observed at 30 s, while the highest value has been observed at 39 s with an average value of 32 s. Wide variations in the absorbency are obtained in the case of pectinase-treated samples, 10 s and upward. Cellulase enzymes are used in the present study with a view to open up the primary wall of the cotton fibers and improve access for protease and pectinase enzymes. Highest weight loss using cellulase was found to be 6.5% and the lowest is 3.3%, while overall average is ~4.8%.

TABLE 4.4
Effect of Enzyme Treatment on Scouring of Cotton

Parameter	Pectinase Treatment	Protease Treatment	Cellulase Treatment	Multienzyme Scouring
Weight loss, %	1.22	2.71	4.83	6.82
Absorbency time, s	9.7	32.0	—	6.0
Tear strength loss, % (warp + weft)	13.98 + 11.79	4.65 + 1.23	—	9.87 + 13.5
Extractable impurities, %	1.2	2.1	1.3	0.6

4.7.1 MULTIENZYME TREATMENT

Among various enzymes used in the process, protease requires alkaline pH, while pectinase and cellulase need acidic pH. In order to overcome this difficulty, prior trials were conducted using two different combinations, namely, treatment of pectinase together with cellulase followed by protease and treatment of all enzymes together at an average pH value. However, when all the enzymes are treated together, a very low weight loss of 2.8% is obtained, while relatively higher weight loss of 3.6% is obtained when they are used in two steps, one after another.

Residual impurities in the multienzyme-treated fabrics are lower by 100% when compared to pectinase treatment and 250% compared to protease treatment (Table 4.4). Extraction of the specimens using benzene–alcohol mixture shows the residual extractable content of 0.6%, obviously indicating the removal of a large proportion of residual matters in the multienzyme-treated fabric samples. Higher removal of impurities could have been facilitated by the incorporation of cellulase in the reaction mixture, which is capable of hydrolyzing the surface layers and the impurities attached to thereon, thus facilitating also the access to the respective substrates.

Weight loss of multienzyme-treated samples is found to be higher than that of individual enzyme-treated samples; the highest weight loss observed is ~6.81%, which is similar or lower to that reported earlier. Weight loss of multienzyme-treated sample is ~150% when compared to protease-treated sample and ~460% when compared to pectinase-treated sample. Higher weight loss obtained in the combined treatment could possibly be due to the cooperative action among these enzymes as reported by others. Lowest absorbency (expressed in drop absorbency), in the samples obtained from multienzyme scouring, is ~6.0 s, while the highest is ~40.0 s, which is about two-third and one-fifth of the values obtained in the case of pectinase- and protease-treated samples, respectively. The lowest value obtained in the study is comparable to those reported in the literature. Scouring of cotton fabrics using a combination of pectinase, protease, and cellulase shows an improved performance in terms of residual extractable impurities, absorbency, and weight loss in scouring as compared to individual enzyme treatment.

4.8 CONCLUSION

Undoubtedly, it has been proved that sodium hypochlorite/hydrogen peroxide/sodium chlorite along with a self-emulsifiable solvent-assisted scouring system can be used for a single-stage preparatory process. Such single-stage process results in 5%–18% savings in energy and proved to be a cost-saving process and up to 30% compared to the conventional process methods. Needless to say that the conventional chemical treatments using acids, alkali, and alkaline bleach solutions for cotton fabric preparation offer better results compared to enzyme-assisted treatments; however, degradation of fibers appears to be inevitable in such processes. Many of the enzyme-based treatments offer advantages compared to conventional treatments without degrading the properties of the fabrics, due to the substrate-specific nature of the enzymes. Lack of compatible bleaching agents in enzyme-assisted treatments and limited efficiency of oxidoreductases necessitate conventional bleaching process. Attempts to address these issues would, definitely, be a viable proposition to counteract the limitations associated with unit operations and single-stage fabric preparation using conventional chemicals. However, reusability of process liquor in the case of enzyme-assisted processing poses an enormous scope to the industry to exercise sustainable solutions in the water-starved locations. Similar efforts also need to be taken in areas of scouring and bleaching.

REFERENCES

Aiteromem A.L. (2008). Sustainable textile manufacturing with new, enzyme based processes, *International Dyer*, 7, 15–19.

Ammayappan L., Muthukrishnan G., and Saravana Prabhakar C. (2003). A single stage preparatory process for woven cotton fabric and its optimisation, *Man Made Textiles in India*, 47(1), 29–35.

Andrade V.S., Sarubho L.A., Fukushima K., Myaji M., Nishimura K., and Takaki G.M.C. (2002). Production of extracellular proteases by *Mucor circinelloides* using D-glucose as carbon source/substrate, *Brazilian Journal of Microbiology*, 33(2), 106–110.

Anon. (1982). Evaluating combined preparation processes for energy and material conservation, *Textile Chemists and Colorists*, 14, 23–35.

Anon. (1992). *Seminar on Energy Conservation in Textile Industry*, The Energy Conservation Center (ECC), Tokyo, Japan, pp. 1–56.

Azevedo H., Bishop D., and Cavaco-Paulo A. (2002). possibilities for recycling cellulases after use in cotton processing, *Applied Biochemistry and Biotechnology*, 101, 61–75.

Buchert J., Peter J., Puolakka A., and Nousiainen P. (2000). Scouring of cotton with pectinases, proteases and lipases, *Textile Chemists and Colorists and American Dyestuff Reporter*, 32(5), 48–52.

Carlier F. (2001). Enzymes, *Industrie Textila*, 54(1334–1335), 121–123.

Cherry J. (2003). Cellulase cost reduction. Enzyme sugar platform project review. Novozymes Biotech Inc., Davis, CA. http://www1.eere.energy.gov/biomass/pdfs/novozymes_esp_review.pdf.

Cherry J. (2003). Cellulase cost reduction. Enzyme sugar platform project review. Novozymes Biotech Inc., Davis, CA. http://www1.eere.energy.gov/biomass/pdfs/novozymes_esp_review.pdf. Accessed October 14, 2003.

Csiszar E., Losonsczi A., Szakacs G., Rusznak I., Bezur L., and Reicher J. (2001). Enzymes and chelating agents in cotton pretreatment, *Journal of Biotechnology*, 89(2–3), 271–279.

Csiszar E., Szakacs G. and Rusznak I. (1998). Combining traditional cotton scouring with cellulase enzymatic treatment, *Textile Research Journal*, 68(3), 163–167.

Degari O., Gepstein, S., and Dosorets C.G. (2002). Potential use of cutinase in enzymatic scouring of cotton fiber cuticles, *Applied Biochemistry and Biotechnology*, 102–103, 277–289.

Dickinson K. (1987). Oxidative desizing, *Review of Progress in Coloration*, 17, 1–6.

Diller G.B. and Traore M.K. (1998). Influence of direct and reactive dyes on the enzymatic hydrolysis of cotton, *Textile Research Journal*, 68(3), 185–192.

Diller G.B., Walsh W.K., and Radhakrishnaiah P. (1999). Effect of enzymatic treatment on dyeing and finishing of cellulose fibres: A study of the basic mechanisms and optimization of the process, National Textile Center Annual Report No. C96-A01, pp. 1–9.

Ellis M.B. (1971). *Dematiaceous Hyphomycetes*. Kew, U.K.: Commonwealth Mycological Institute.

El Sisi F.F., Hafiz S.A.A., Rafie M.H., and Hebeish A. (1990). Development of a one-step process for desizing, scouring, bleaching cotton based textiles, *American Dyestuff Reporter*, 79(10), 39–43.

Etters J.N. and Annis P.A. (1998). Textile enzyme use: A developing technology, *American Dyestuff Reporter*, 77(5), 18–23.

Etters J.N. (1999). Cotton preparation with alkaline pectinase: An environmental advance, *Textile Chemists and Colorists and American Dyestuff Reporter*, 1(3), 33–36.

Galante Y.M. and Formantici C. (2003). Enzyme applications in detergency and in manufacturing industries, *Current Organic Chemistry*, 7(13), 1399–1422.

Guha S.B. and Shah S.R. (2001) Enzymatic scouring of cotton fabrics, *Journal of the Textile Association*, 61(1–2), 215–218.

Gulrajani M.L. (1987). Reply to the comments, *Textile Research Journal*, 57(4), 243–244.

Gulrajani M.L. (1989). Development, optimisation and solarisation of combined preparatory process, *Colourage*, 33(2), 19–28.

Gulrajani M.L. and Sukumar N. (1984). Development, optimisation and economics of a single stage preparatory process, *Journal of Society of Dyers and Colorists*, 100(1), 21–27.

Gulrajani M.L. and Sukumar N. (1985a). Further optimisation of single stage preparatory process—Effect of pH buffers, *Journal of Society of Dyers and Colorists*, 101(10), 330–333.

Gulrajani M.L. and Sukumar N. (1985b). Optimisation of a single-stage preparatory process for cotton using NaOCl, *Textile Research Journal*, 55(10), 614–619.

Gursoy N.C. and Hall M.E. (2001). Optimisation of hydrogen peroxide bleaching, *International Textile Bulletin*, 2001(5), 80–86.

Halim E.S.A., Fahmy H.M. and Moustafa M.G.F. (2008). Bioscouring of linen fabric in comparison with conventional chemical treatment, *Carbohydrate Polymers*, 74(3), 707–711.

Harrison, P.W. (1986). Low-liquor dyeing and finishing, *Textile Progress*, 14, 17–34.

Hartzell M.M. and Hsieh Y.L. (1998). Enzymatic scouring to improve cotton fabric wettability, *Textile Research Journal*, 68(4), 233–241.

Hashem M.M. (2007). An approach towards a single pretreatment recipe for different types of cotton, *Fibres and Textiles in Eastern Europe*, 15, 85–92.

Hill G.A., Macdonald D.G., and Lang X. (1999). Amylase inhibition and inactivation in barley malt during cold starch hydrolysis, *Biotechnology Letters*, 19, 1139–1141.

Holme I. (2001). Biopreparation—The eco-friendly alternative, *International Dyer*, 186(2), 8–9.

Kamat S.Y. (1995). Magic enzymes, *Proceedings of Bilateral Symposium on Eco-friendly Processing*, Indian Institute of Technology, New Delhi, India, November 6–7, pp. 60–62.

Karapinar E. and Sariisik M.O. (2004). Scouring of cotton with cellulases, pectinases and proteases, *Fibres and Textiles in Eastern Europe*, 12(3), 79–82.

Karmakar S.R. (1998). Application of biotechnology in the pretreatment process of textiles, *Colourage Annual*, 45(12), 75–86.

Kim J., Kim S.Y., and Choe E.K. (2005). The beneficial influence of enzymatic scouring on cotton properties, *Journal of Natural Fibres*, 2(4), 39–52.

Kirk O., Borchert T.V., and Fuglsang C.C. (2002). Industrial enzyme applications, *Current Opinion in Biotechnology*, 13, 345–351.

Kitchareonseree P. (2002). Enzymatic scouring of various fabrics, MSc thesis, Chulalongkorn University, Bangkok, Thailand.

Lacasse K. and Baumann W. (2004). Textile Chemicals: Environmental Data and Facts, Springer, New York, p. 609.

Lange N.E.K., Kongsbak L., Schulein M., Bjo M.E., and Husain P.A. (2001). Biopreparation of textiles at high temperatures, US Patent Office, Patent No. 6,258,590.

Ledakowics J.S., Lichawska J. and Pye R. (2005). Enzymatic pre-treatment of cotton fabrics, *Journal of Natural Fibres*, 3(2/3), 199–207.

Lenting H.B.M., Zwier E., and Nierstrasz V.A. (2002). Identifying important parameters for a continuous bioscouring process, *Textile Research Journal*, 72(9), 825–831.

Lewin M. and Epstein J.A. (1962). Functional groups and degradation of cotton oxidised by hypochlorite, *Journal of Polymer Science*, 1962, 1023–1037.

Li T., Wang N., Li S., Zhao Q., Guo M., and Zhang C. (2007). Optimization of covalent immobilization of pectinase on sodium alginate support, *Biotechnology Letters*, 29(9), 1413–1416.

Li Y. and Hardin I.R. (1998a). Enzymatic scouring of cotton surfactants, agitation, and selection of enzymes, *Textile Chemists and Colorists*, 30(9), 23–29.

Li Y. and Hardin I.R. (1998b). Treating cotton with cellulase and pectinases: Effects on cuticle and fibre properties, *Textile Research Journal*, 68(9), 671–679.

Lu H. (2005). Insight into cotton enzymatic pretreatment, *International Dyer*, 120(4), 9–11.

Manning G.B. and Campbell L.L. (1961). Thermostable α amylase of *Bacillus stearothermophilus*, *Journal Biological Chemistry*, 236(11), 2952–2957.

Miller C.A., Jorgensen S.S., Otto E.W., Lange N.E.K., Condon B., and Liu J. (2003). Alkaline enzyme scouring of cotton textiles, US Patent Office, Patent No. 6,551,358.

Miller G.L. (1959). Use of dinitrosalicylic acid reagent for determination of reducing sugar, *Analytical Chemistry*, 31, 426–429.

Munk S., Munch P., Stahnke L., Nissen J.A., and Schieberle P. (2000). Primary odorants of laundry soiled with sweat/sebum: Influence of lipase on odor profile, *Journal of Surfactants and Detergents*, 3(4), 505–515.

Naik S.R. and Paul R. (1997). Application of enzymes in textile processing, *Asian Textile Journal*, 6(2), 48–55.

Najafi M.F., Deabagkar D., and Deepti D. (2005). Potential application of protease isolated from *Pseudomonas aeruginosa* PD 100, *Electronic Journal of Biotechnology*, 8(2), 197–203.

Obendorf S.K., Varanasi A., Mejldal R., and Thellersen M. (2001). Function of lipase in lipid soil removal as studied using fabrics with different chemical accessibility, *Journal of Surfactants and Detergents*, 4(3), 233–245.

Onions A.H.S., Allsopp D., and Eggins H.O.W. (1981). *Smith's Introduction to Industrial Mycology*, 7th edn., Edward Arnold, London, U.K.

Opwis K., Knittel D., Doffler C., and Koppe A. (2006). Combined use of enzymes in the pretreatment of cotton, *Melliand International*, 12(2), 130–136.

Palmer T. (1981). *Understanding Enzymes*, Ellis Horwood Limited, Bristol, United Kingdom, pp. 18–19.

Pandey A., Nigam P., Soccol C.R., Soccol V.T., Singh D., and Mohan R. (2000). Advances in microbial amylases, *Biotechnology and Applied Biochemistry*, 31, 135–152.

Parikh D.V. and Karami J.H. (1989). Use of acid in bleaching cotton textiles, *American Dyestuff Reporter*, 78, 15–23.

Paul R. and Pardeshi P.D. (2002). Enzymes: The marvellous molecular machine, *Asian Textile Journal*, 11(1), 29–35.

Rai I. (2004). *Biotechnology and Textiles, Seminar on Chemical Processing Challenges for Indian Textile Industry*, Mumbai, India, pp. 25–28, 40.

Riegels M., Koch R., Pendersen L.S., and Lund H. (2001). Enzymatic hydrolysis of cyclic oligomers, US Patent Office, Patent No. 6,184,010.

Robner U. (1993). Enzymatic degradation of impurities in cotton, *Melliand English*, 74(2), E63–E65.

Robyt J.F. (1984). Enzymes in the hydrolysis and synthesis of starch. In *Starch: Chemistry and Technology*, R.L. Whistler, J.M. Beniller, E.R. Paschell (eds.), pp. 87–123, Academic Press, New York.

Rodriques K.A. (2002). Textile manufacturing and treating processes comprising a hydrophobically modified polymer, US Patent Office, Patent No. 6,337,313.

Sae B.P., Sangwatanaroj U., and Punnapayak H. (2007). Analysis of the products from enzymatic scouring of cotton, *Biotechnology Journal*, 2(3), 316–325.

Saravanan D., Arun Ramanathan V.A., Karthick P., Vel Murugan S., and Nalankilli G. (2010). Optimisation of multi-enzyme scouring process using Taguchi methods, *Indian Journal of Fibre and Textile Research*, 35(2), 164–171.

Saravanan D., Sivasaravanan S., Sudharshan Prabhu M., Vasanthi N.S., Senthil Raja K., Das A., and Ramachandran T. (2012). One-step process for desizing and bleaching of cotton fabrics using the combination of amylase and glucose oxidase enzymes, *Journal of Applied Polymer Science*, 123, 2445–2450.

Saravanan D., Sreelakshmi S.N., and Vasanthi N.S. (2014). Low-temperature acidic amylases from *Aspergillus* for desizing of cotton fabrics, *Journal of the Textile Institute*, 105(1), 59–66.

Semenova M.A., Grishutin S.G., Gusakov A.V., Okunev O.N., and Sinitsyn A.P. (2003). Isolation and properties of pectinases from the fungus *Aspergillus japonicus*, *Biochemistry*, 68(5), 559–569.

Shukla L., Ananthanarayan L., and Amey J.D. (2005). Production and application of amylase enzyme, *Colourage*, LII(11), 43–48.

Sreelakshmi V.N. (2014). Sustainable practices in cotton fabric preparation, PhD thesis, Anna University, Chennai, India.

Tanapongpipat A., Khamman C., Pruksathorm K., and Hunsom M. (2008). Process modification in the scouring process of textile industry, *Journal of Cleaner Production*, 16(1), 152–158.

Tarakcioglu I. and Anis P. (1996). Microwave processes for the combined desizing, scouring and bleaching of grey cotton fabrics, *Journal of the Textile Institute* 87 Part 1 (3), 602–608.

Tavčer, P.F., Križman, P., and Preša, P. (2005). Combined bioscouring and bleaching of cotton fibres, *Journal of Natural Fibres*, 3, 83–97.

Traore M.K. and Diller G.B. (2000). Environmentally friendly scouring process, *Textile Chemists and Colorists and American Dyestuff Reporter*, 32(12), 40–44.

Trotman E.R. (2000). *Dyeing and Chemical Technology of Textile Fibres*, Charles Griffin and Co., London, U.K., 6th edn., pp. 187–218.

Tu M., Chandra R.P., and Saddler J.N. (2007). Evaluating the distribution of cellulases and the recycling of free cellulases during the hydrolysis of lignocellulosic substrates, *Biotechnology Progress*, 23, 398–406.

Tzanov T., Costa S., Calafell M., Guebitz G.M., and Paulo A.C. (2000). Enzymes for cotton fabrics preparation and recycling of waste water for dyeing, *Colourage Annual*, 47(12), 65–72.

Underkofler L.A., Barton R.R., and Rennert S.S. (1958). Production of microbial enzymes and their applications, *Applied and Environmental Microbiology*, 6(3), 212–221.

Wakelyn P.J. (1975). Amino acid composition of total protein of cotton, *Textile Research Journal*, 45(5), 418–420.
Wang Q., Fan X., Hua Z., Gao W., and Chen J. (2007). Influence of combined enzymatic treatment on one-bath scouring of cotton knitted fabrics, *Biocatalysis and Biotransformation*, 25, 9–15.
Whitwell J.C. (1984). Communication to the editor, *Textile Research Journal*, 36(4), 239–242.
Yachmenev V.G. (2005). The effects of ultrasound on the performance of industrial enzymes used in cotton biopreparation/biofinishing applications, *Journal of Natural Fibres*, 3(2/3), 99–112.
Yoon M.Y. (2005). Denim finishing with enzymes, *International Dyer*, 90(11), 16–19.

5 Design and Development of Jute-Based Apparels

Sanjoy Debnath

CONTENTS

5.1 Introduction	97
5.2 Categories of the Apparels and Their Development Methodologies	98
5.2.1 Categories of the Apparels	98
5.2.2 Apparel Development Methodologies	99
5.3 Chemical Modification of Jute Fibrous Structure	99
5.4 Characteristics of Alkali-Treated Jute Fiber Structure	99
5.5 Woollenized Jute–Wool Blended Yarn	100
5.6 Woollenized Jute–Polypropylene Blended Yarn	100
5.7 Woollenized Jute–Polyester, Jute–Hollow Polyester, and Jute–Acrylic Blended Knitting Yarn	100
5.8 Jute and Hollow Polyester Blended Fine Yarn for Winter Apparels	101
5.9 Development of Jute-Based Fabric for Jacket	102
5.10 Finishing Treatment on Jute–Hollow Polyester Fabrics for Warm Garment	103
5.11 Properties of Developed Jute-Based Warm Fabrics for Jacket	107
5.12 Development of Warm Garments from Jute–Hollow Polyester and Cotton Blended Fabrics	107
5.13 Evaluation of Jute-Based Apparels	108
5.14 Effect of Washing Jute-Based Apparels	109
5.15 Conclusions	110
References	110

5.1 INTRODUCTION

Jute is one of the natural fibers grown abundantly in India and Bangladesh. This fiber is very popular due to its successful use as a packaging and carpet backing material. Though this fiber is grown much earlier in India, the British introduced the machinery and processing mill for jute yarn and fabrics. Jute-based products are still used in different applications. With the advancement of science and technology, the partial share of the jute market has been taken by the synthetic industry. Hence today, many of the jute-based conventional products are finding stiff competition in the global market. To overcome this, researches in different countries, mostly in jute-growing countries, have concentrated for its diversified applications (Gupta and Bhattacharyya, 1984; Singh et al., 2004; Debnath et al., 2008, 2009). Out of the different diversified applications, jute in fashion application has been covered in this chapter. The road map for jute-based apparel development, starting from raw jute fiber to yarn, sustainable chemical modifications of fiber and yarn, fabric development, garment development, and finishing of such jute-based fabrics, has also been focused. The properties of the materials in the different processes involved in jute-based apparel manufacturing have also been covered. More precisely, details of design and development of jute-based apparels for sustainable development have as well been emphasized.

Jute, an annually renewable and biodegradable bast fiber, is sustainable due to the sequestration of a huge amount of carbon dioxide during its cultivation. Hence, due to its natural degradability, apparels produced from this eco-friendly fiber can be utilized as manure after the end of its life cycle. As a whole, the process is sustainable because of its multibenefits. Jute is also well known due to some of its specific characteristics like color, surface roughness, strength, coarseness, and extensibility. However, some of the important properties like coarseness, surface roughness, and extensibility hinder its direct applicability in the apparel industry. Furthermore, the applications in apparels are also restricted for use as outerwear/secondary application. This is mainly due to harsh as well as pricking feelings of jute fibers. Under these circumstances, jute without treatment is avoided, but with suitable surface modifications of jute fiber/yarn/fabric, to some extent, it can be used as direct skin contact materials. Jute has its natural golden color, which has an esthetic appeal in fashion garments. Presently, most of the documented researches in the area of jute-based apparel development are mostly confined in fiber and yarn modifications and to some extent in fabric development. Hence, there is a dearth of knowledge in the whole process for apparel design and development from jute-based materials. This chapter will open up the new avenues and possibilities for sustainable apparel development from jute-based textiles.

Apart from these, today, with successive developments in agricultural sciences, fine jute variety has been developed, and cultivation of such fibers produces very fine fibers, like JRC-321 (Majumdar, 2002), sustainably. In spite of this fine fiber, the yarn/fabric structures made out of this are not suitable directly for apparel applications due to the heavier weight of the final material. Mostly, the apparels developed out of this jute or jute-based materials are heavier, thicker, and coarser in nature. There are different means by which jute fiber materials can be suitably applied in the development of sustainable winter apparels.

5.2 CATEGORIES OF THE APPARELS AND THEIR DEVELOPMENT METHODOLOGIES

5.2.1 CATEGORIES OF THE APPARELS

Apparels can be categorized based on the manufacturing process, materials used, and chemical/physical treatments applied thereon. However, in most of the cases, apparels are classified as winter or summer wear. Generally, the fabrics used to develop these two types of apparels are the same, but the other accessories like the lining material and laminating material (if any) are added to differentiate the summer from the winter apparels. The manufacturing process is different for producing the two categories of seasonal apparels. Winter apparels are also developed from bulked yarn from natural materials like wool or blended with synthetic materials like polyester, polypropylene (PP), and acrylic fibers (Debnath, 2014). Sometimes, 100% acrylic yarns are also used to manufacture the winter apparels. Hence, the basic raw material plays an important role to decide its category as summer or winter clothes. Some of the winter apparels are developed from bulked yarn of wool/wool–polyester/wool–acrylic or 100% bulked acrylic yarns, wherein the yarn is treated with heat energy, popularly known in the industry as the steaming process (Sinha and Basu, 2001). This steaming process generates a bulk, due to the lateral shrinkage of the yarn structure. Jute and jute–synthetic fiber blended yarns are also developed to increase the bulk in the structure when treated under favorable chemical treatments. The bulked yarn contains more voids in the structure and holds more static air, which acts as a thermal insulation material. As a result, the thickness of the fabric makes it more suitable for manufacturing winter apparels. Selected fiber materials generated additional bulk in the structure when subjected to physical or chemical treatment. This process is normally adopted for making of winter apparels (Anonymous, 2009).

On the other hand, summer apparels are mostly lighter in weight, are lesser in thickness, and have a more open structure that provides better breathability. Natural fibers mostly used are cotton or linen and ramie fibers blended with cotton or blended with man-made fibers like polyester and

Design and Development of Jute-Based Apparels

viscose. Among the synthetic fibers, mostly, polyester fiber as a whole is also used for summer apparels, which is a very lightweight thin fabric that is sometimes used as a party wear. Thus, the differences in raw material, chemical/thermal processing, and manufacturing process, solely or in combination, together differentiate the apparel usages.

5.2.2 Apparel Development Methodologies

Since jute and jute-based materials are well known for their use as traditional packaging materials, the process know-how is well documented. In case of apparel made from jute-based material, the whole process is entirely different. In this chapter, the raw material and jute fiber and yarn modification are covered step by step. To obtain sustainable apparel from this fiber, certain man-made fibers are blended in a small proportion and modifications incorporated in fiber/yarn structure to suit the requirements of the apparel. Also, constructional design modifications have been covered in weaving as well as in apparel manufacturing stages. The sustainable properties of the yarn, fabric, and apparels out of such materials have also been discussed in detail.

5.3 CHEMICAL MODIFICATION OF JUTE FIBROUS STRUCTURE

When we talk about fiber modification, the first thing that comes into picture is woollenization or alkali treatment (Saha et al., 1961). The raw jute does not have any natural crimp, and hence to introduce crimp in the structure of the jute, it should be treated with a strong alkali like NaOH. Wool fiber, being one of the animal fibers, has limitations in production, and it is very expensive to be used extensively in warm clothes as compared to plant based jute fiber. Wool has natural crimps and scales on its surface. On the contrary, jute fiber does not contain any such crimps or scales in the raw stage, but this can be generated using chemical treatment (Sao and Jain, 1995). The scales as well as crimps that are present naturally on the wool restrict the fibers from coming close to each other. As a fact, the yarn/fabric structures made out of this treatment are sustainable and are bulkier in nature (Anonymous, 2008b). The static air present in the pores of the structure of wool acts as thermal insulation, and thus it is suitable for development of winter garment materials (Anonymous, 2011). In case of jute fiber, woollenization process produces crimp similar to wool as explained by Chakravarty (1962, 1963), Ganguly and Sao (1985), and Ganguly et al. (1985). This introduces more voids in the structure. It has been noticed that after this treatment, the elongation at break has been increased considerably due to formation of crimps/expansion in diameter of the yarn (Sao and Jain, 1995). Woollenization is a process where jute is treated with NaOH solution and the process has been optimized so that the treatment achieved at 18% NaOH solution gives the best result (Saha et al., 1961). This alkali treatment removes a significant amount of the hemicellulose and a small quantity of soluble lignin present in the jute structure, and as a result, a reduction in the strength of the jute has also been noticed (Sinha and Gupta, 1986). Several attempts were also made to recover NaOH after this delignification process, from the spent liquor.

5.4 CHARACTERISTICS OF ALKALI-TREATED JUTE FIBER STRUCTURE

This woollenization process introduces several changes in properties of the jute fiber. Hence, yarn made out of jute fiber differs significantly before and after woollenization treatment (Ganguly and Sao, 1985). After this process, the yarn surface becomes rougher due to further bleaching process. Hence, due to increase in surface roughness, 100% woollenized jute yarn is not a suitable material for development of wrappers or shawls as winter apparels. However, to overcome this problem, different softening agents are suggested for alkali-treated yarn such as Velan PF and Sopamine OC. Further, to improve the sustained water repellence property of the treated yarn, a different silicone treatment is recommended.

5.5 WOOLLENIZED JUTE–WOOL BLENDED YARN

Woollenized jute can be well mixed with wool at different ratios like 50:50 and 70:30 (Singh et al., 2004). The objective behind blending such dissimilar fibers is not only to improve the sustainable thermal insulation property but also to reduce substantially the fiber coarseness of the blended yarn. We can also produce coarse knitting yarn for apparels in woollen/worsted systems as recommended by Debnath et al. (2009) and Singh et al. (2004). Apart from this wool fiber, other natural or synthetic fibers such as PP can also be blended with woollenized jute to produce different knitted warm fabrics as well as blanket materials (Anonymous, 1981). Bleaching of blended woollenized jute and wool can be achieved with 0.75 vol. of H_2O_2 solution treated for 2 h at 80°C (Anonymous, 1981). Twelve different shades were obtained by dyeing woollenized jute blended yarn; both bleached and unbleached woollenized jutes were used and they were mixed with wool at a ratio of 35:65 (Chakravarty, 1962; Sao et al., 1983; Sao and Jain, 1984).

5.6 WOOLLENIZED JUTE–POLYPROPYLENE BLENDED YARN

In several studies, woollenized jute fibers are blended with synthetic fibers or blended jute–synthetic yarn subjected to woollenized treatment (Gupta et al., 1982) to obtain sustained bulk properties. PP is one of the inert fibers that can be blended with jute to achieve jute–PP blended bulked yarn. In this blended yarn, the PP component introduces additional bulk in the woollenized treated jute–PP blended yarn (Sinha et al., 1988). To be specific, higher bulk and comparable stretchable yarn is possible due to preferential migration to the yarn surface that occurred during woollenization treatment of jute and PP blended yarn. Due to this preferential fiber migration in the yarn, this bulked yarn structure is highly sustainable than whole jute yarn. This results in higher bulk and tenacity in comparison to the yarn produced from an Indian variety, that is, Chokla wool (Singh et al., 2004). However, the wool yarn shows comparatively higher breaking elongation and uniformity in diameter. Tenacity of textured jute–PP blended yarn is also higher than the wool yarn (Indian Chokla wool), though at the time of woollenization, the tenacity of the jute component drops significantly (Singh et al., 2004). It has been reported that the ratio of tenacity (wet/dry) is higher in case of jute–PP blended textured yarns than the wool yarn (Gupta et al., 1982, 1984; Sinha et al., 1988).

5.7 WOOLLENIZED JUTE–POLYESTER, JUTE–HOLLOW POLYESTER, AND JUTE–ACRYLIC BLENDED KNITTING YARN

Jute as such has no crimp as mentioned earlier; the crimp/bulk can be introduced through woollenized treatment in the jute yarn structure and is temporary in nature. Since the bulk is only temporary in full jute yarn, efforts have been taken to blend this yarn with synthetic fibers. Ghosh and Samanta (1997) studied the comparative properties of texturized jute–polyester staple yarn spun in conventional rotor and jute apron draft spinning systems. They found that a 70:30 blend ratio of jute and polyester fibers spun in a rotor spinning system showed the optimum results compared to its counterpart. Later, Sinha and Basu (2001) demonstrate the use of jute and shrinkable acrylic fiber to develop bulked knitted yarn. They used a simple technique to develop bulk in the raw jute–acrylic blended yarn. The bulking in the raw yarn can be obtained by boiling the yarn in water for a duration of 15–30 min or steam treatment for the same duration. This bulked structure is also quite sustainable in nature. Chaudhury and Basu (1998) studied the properties of woollenized jute and acrylic blended friction spun yarn at different concentration of alkali treatments. It has been proved that the bulking effect of jute and acrylic blended yarn can be made by two successful methods, one is chemical/alkali treatment and the other is heat treatment. The jute–synthetic fiber blend trend continues with the blend of jute–hollow polyester fiber, wherein a jute and hollow polyester fiber (80:20) blended fine yarn (130 tex) has been developed. This single yarn was further plied into 2-ply and 3-ply yarns (Debnath et al., 2007a). To develop a bulked yarn,

TABLE 5.1
Properties of Yarn Shrinkage and Weight Loss Behavior of 80:20 Jute–Polyester (Hollow) Blended Bulked Yarn

Chemical Process Stage	2-Ply Blended Yarn		3-Ply Blended Yarn	
	Shrinkage (%)	Weight Loss (%)	Shrinkage (%)	Weight Loss (%)
Bleaching	10	15.37	12.5	14.27
Dyeing	10	18.01	12.5	16.74

Source: Debnath, S. et al., *J. Inst. Eng. (India), Text. Eng.*, 87(2), 11, 2007a.

woollenization treatment was done for both 2-ply and 3-ply blended yarns. Later, bleaching followed by dyeing were also done. Yarn shrinkage and weight loss properties were also studied at each step of the process and the results were compared with the untreated yarns (Table 5.1). Other physical properties, namely, tenacity, breaking strain, diameter, coefficient of friction, specific work of rupture, and bulk density, have also been studied (Debnath et al., 2007b). They also compared these properties with similar commercial woollen and acrylic yarns. This study (Debnath et al., 2007a) established the fact that jute–hollow polyester blended yarn has higher bulk over similar commercial yarns due to its low yarn packing fraction. The 3-ply jute–hollow polyester blended bulked yarn has sustainable higher bulk, regularity (cv%), extensibility, pliability, and work of rupture than those of 2-ply yarn. They also investigated that the decrease in tenacity value of 3-ply blended bulked yarn occurs after 18% (w/w) NaOH chemical treatment, while breaking extension increases significantly. However, the coefficient of variation% of tenacity and breaking extension values decreases after bleaching and dyeing processes. This yarn shows almost similar tenacity with lower extension compared to wool yarn. The lowest value of specific work of rupture has been observed in the case of 3-ply jute–hollow polyester blended bulked yarn. The 3-ply blended bulked yarn shows lower coefficient of friction than wool yarn (Table 5.2). This shows a new avenue in application in the area of handspun knitted garments. The specific flexural rigidity of jute–hollow polyester blended yarns is also lower than woollen yarns but comparatively higher than acrylic yarns (Debnath et al., 2007a).

5.8 JUTE AND HOLLOW POLYESTER BLENDED FINE YARN FOR WINTER APPARELS

The main hindrance for development of apparel from jute-based material is the heaviness of the final garment. This is because heavier material is not comfortable to wear for long periods of time. The main reason is that the fabric made out of the jute-based yarn is heavier in nature due to the use of heavier yarn linear density in fabric development. This can be overcome easily by using lightweight fabric. There are many ways to reduce the fabric weight and one is reduction in the warp and weft density; the outcome is a lighter but open-structured fabric. In another process, increasing the fineness of the yarn and maintaining the warp–weft yarn spacing can also reduce fabric weight. In this second process, the fabric will be lighter without altering much the cloth cover factor.

In this study, to reduce the final weight of the fabric, very fine yarn of 122 tex has been spun using jute spinning machinery (Anonymous, 2006). The polyester of hollow cross-sectioned fiber (6 denier × 110 mm) was used for lightweight jute–polyester blended yarn. The reason behind the use of hollow polyester fiber is due to its inherent bulkiness and ease of availability in local market. Blending this synthetic fiber component will add the lightness and bulkiness of the final yarn. Hence, jute–hollow fiber blended yarn has higher bulk density compared to other blended yarns as reported by Singh et al. (2004) and Debnath et al. (2007b). This blended yarn was developed in a

TABLE 5.2
Comparative Properties of 80:20 Jute–Polyester (Hollow) Bulk Yarn with Commercial Acrylic and Wool Yarns

Yarn Properties	Commercial Yarns			3-Ply Dyed Jute–Hollow Polyester Yarn
	3-Ply Acrylic	4-Ply Acrylic	4-Ply Wool	
Linear density, tex	170	290	300	370
Diameter, mm	1.19	1.71	1.35	1.85
Packing factor	0.128	0.106	0.161	0.094
Packing factor, C.V.%	15.46	11.56	11.61	16.14
Bulk density, g/cm^3	0.153	0.126	0.210	0.138
Tenacity, cN/tex	7.44	9.10	4.00	3.84
Tenacity, C.V.%	7.01	6.55	10.12	9.09
Breaking strain, %	37.35	45.42	18.01	8.6
Breaking strain, C.V.%	9.83	11.57	16.73	7.83
Specific work of rupture, mJ/tex m	16.85	14.00	4.47	1.27
Coefficient of friction of yarn, μm	0.81	0.80	0.90	0.84
Specific flexural rigidity, mN mm^2/tex$^2 \cdot 10^{-4}$	9.57	5.20	23.20	11.98

Source: Debnath, S. et al., *J. Inst. Eng. (India), Text. Eng.*, 87(2), 11, 2007a.

jute spinning system. The jute fiber reeds were processed through a conventional softener machine, wherein jute batching oil is used (2% on the weight of jute), and then placed in a tight chamber for 48 h for piling. Later, the reeds were taken out and fed to the breaker card followed by the finisher card. This card-processed jute sliver was then taken for blending with preprocessed polyester sliver obtained from a flax card/woollen card/synthetic card machine. The blending is made in the first jute gill-drawing machine twice followed by the second and third jute gill-drawing machines. Finally, the yarn is spun in a jute apron draft spinning machine. It has been optimized that 80:20 jute/hollow polyester gives better result in terms of strength. On the contrary, a 70:30 ratio jute/hollow polyester fiber blended yarn produces a more bulkier and softer yarn while sacrificing a reduction in strength of 1.5% (Debnath and Sengupta, 2009). The blended yarn sample produced from spinning was of 122 tex with 6 t.p.i. (twist per inch), Z twist. It has been suggested that better yarns can be obtained from a jute apron draft spinning machine over a slip draft spinning machine (Anonymous, 2011).

5.9 DEVELOPMENT OF JUTE-BASED FABRIC FOR JACKET

Normally, the lighter weight fabric can give comfort to the wearer without affecting the thermal property. To develop the lightweight jute-based fabric, union fabric construction has been selected. In this, the developed jute-based (jute–polyester) fine yarn in weft direction and very fine cotton yarn of 5.9 tex (100s Ne) in warp direction have been taken into consideration (Anonymous, 2008a). Weaving has been done in a traditional handloom with a 400-hook jacquard attachment to generate ornamentation in the fabric weaving stages. The target is to produce lighter fabric similar to *khadi* or woollen warm garment cloths. Further, to reduce the weight of the fabric, weft mixing has been done. Cotton and jute–polyester yarns have been used in weft direction (four jute-based and two cotton yarns, four jute-based and four cotton yarns, six jute-based and two cotton yarns) alternatively. This process not only reduces the overall weight of the fabric but also introduces some rib effect in the fabric (Anonymous, 2006). This undulated rib surface increases the thermal insulation and esthetic look of the fabric with reduction in the fabric weight. In most of the cases, the 72s

Design and Development of Jute-Based Apparels

FIGURE 5.1 Ornamented woven jute–polyester and cotton blended thermal insulating fabric for apparel.

Stockport reed (72 dents per 2 in. or 5 cm) has been used. In warp direction, alternative use of two different colored cotton yarns in groups was used. These produced a check pattern in the fabric. Two other different types of jacket fabric have been developed, wherein all cotton yarns were used in the warp direction and all jute–polyester yarns were used in the weft direction (Figure 5.1). The ornamentation in the fabric has been incorporated using extra warp or extra weft or both extra warp and extra weft designs using a handloom jacquard (Debnath, 2014).

5.10 FINISHING TREATMENT ON JUTE–HOLLOW POLYESTER FABRICS FOR WARM GARMENT

The raw handloom fabrics are used without any pretreatment for chemical finishing treatments. The finishing chemicals used were Leomin HBN (*cationic softener*), Ceraperm 3P plus (*macropolysiloxane emulsion*), Ceraperm MW (*cationic micropolysiloxane emulsion*), Ceraperm TOWI (*nanopolysiloxane emulsion*), Ceraperm UP (*nonionic macropolysiloxane emulsion*), Ceraperm OEW (*nonionic softener*), Ceraperm UP (*macropolysiloxane emulsion*), and Sandoclean PCJ (*nonionic detergent*) and were supplied by M/s Clariant Chemicals (India) Ltd., Tiruppur, Tamil Nadu. All other chemicals used elsewhere were graded as analytical reagents.

For each finishing formulation, 0.25 gpl Sandoclean PCJ as nonionic detergent solution was added. The pH was adjusted to 5.0 ± 0.2 by adding 0.5% acetic acid. The fabric sample was taken and impregnated with the finishing solution for 5 min at 25°C. It was further padded to 80% ± 5% expression under 1.5 kg/cm^2 using a laboratory padding mangle (RB Engineering Ltd., Gujarat, India). After padding, the fabric was dried and cured in a high-temperature steamer (RB Engineering Ltd., Gujarat, India) as per conditions mentioned in Table 5.3. After curing, it was conditioned in ambient condition (Anonymous, 2011).

The performance properties of finished and unfinished samples were evaluated by measuring different properties such as finish add-on, bending length, flexural rigidity, dry crease recovery angle (DCRA), air permeability, and thermal insulation values (TIVs), which were evaluated as per

TABLE 5.3
Finishing Formulations for the Fabric

Sl. No.	Recipe	Conditions
1	Leomin HBN = 80 gpl	Pad → dry (100°C/5 min)
2	Leomin HBN = 40 gpl Ceraperm UP = 20 gpl Ceraperm MW = 20 gpl	Pad → dry → cure (130°C/5 min)
3	Leomin HBN = 40 gpl Ceraperm UP = 20 gpl Ceraperm TOWI = 20 gpl	
4	Ceraperm 3P plus = 40 gpl Ceraperm OEW = 40 gpl	
5	Ceraperm TOWI = 80 gpl	
6	Ceraperm TOWI = 40 gpl Ceraperm MW = 40 gpl	
7	Ceraperm TOWI = 40 gpl Ceraperm UP = 40 gpl	
8	Ceraperm TOWI = 40 gpl Leomin HBN = 40 gpl	

standard procedure. For surface study, control and finished jute fiber samples were magnified in JEOL scanning electron microscope using a JSM 6360 model.

This study on the effect of finishing treatment of jute–polyester–cotton blended fabric reveals that the finish add-on is higher in nanopolysiloxane-based finishing samples (>2.8%) than other non-nanopolysiloxane-based finishing (<2.73%). However, pretreatment was not done for this fabric; the amount of cationic softener added on the fabric is less than 2.3% (Anonymous, 2011).

The DCRA of finished and unfinished jute blended fabrics in both the directions shows that the improvement is significantly witnessed only when they were finished with resin finishing, that is, chemical that masked free hydroxyl groups of jute and cotton fibers. Micropolysiloxane emulsion–based finishing could only improve the crease recovery of this fabric both in warp (10%) and weft (4%–7%) directions. The DCRA of both finished and unfinished fabric in warp direction is higher than weft direction due to the presence of jute component in weft yarn (Anonymous, 2011).

This study also reveals the bending length of finished and unfinished jute blended fabric in both directions; the control fabric is higher in weft direction than warp direction due to a wide variation in linear density between warp and weft yarns, so the stiffness of jute fiber used in weft direction is higher than warp direction. The warp yarn linear density is 5.9 tex, while the weft yarn linear density is 122 tex. Except for the finishing formulations 1 and 2, other finishing formulations show reduction in bending length when compared to control fabric both in warp and weft directions. In warp direction, macropolysiloxane emulsion–based finishing showed more reduction in bending length (10%) than nanopolysiloxane-based finishing (5%), while in weft direction the reduction in bending length is comparatively less than in warp direction irrespective of finishing treatment. This is because in warp cotton yarn, more amount of OH group is present compared to the jute component. In comparison with the control sample, the nanopolysiloxane-based finishing presents more reduction in bending length (9%–14%) than other finishings (0%–8%) (Anonymous, 2011).

Surface morphology was studied using scanning electron microscopy (SEM) of control (Figure 5.2) and finished fabric samples. It revealed nanopolysiloxane finished, the combination of nanopolysiloxane- + micropolysiloxane-finished and nanopolysiloxane + macropolysiloxane-finished fabric samples as shown in Figures 5.3 through 5.5, respectively.

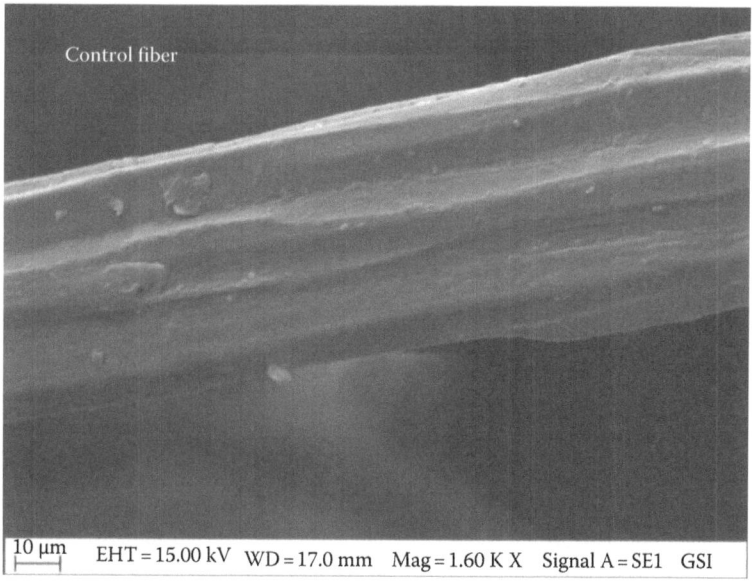

FIGURE 5.2 SEM photograph of controlled jute–polyester and cotton blended fabric.

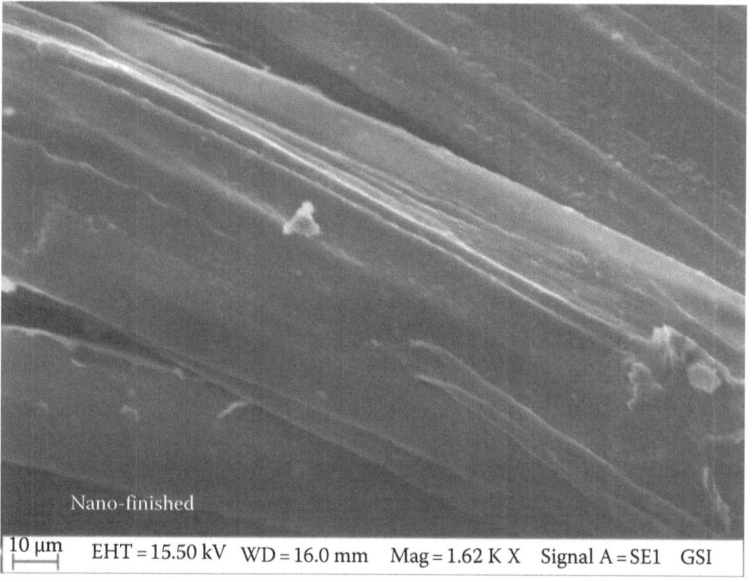

FIGURE 5.3 SEM photograph of nanopolysiloxane-treated jute–polyester and cotton blended fabric.

It is inferred from the SEM pictures that untreated jute has irregular grooves on the surface of the fiber, and after finishing treatment, finishing chemicals form a polymer film on the surface of the fiber. This coating that covers the fiber grooves is better in nanopolysiloxane-finished samples compared to fabric samples finished with other finishing combinations. Since the add-on is higher in nanopolysiloxane finishing, the coverage is more and better than others.

The air permeability and TIVs of the chemically finished and unfinished fabric samples were evaluated by using Shirley air permeability tester and NIRJAFT TIV tester using

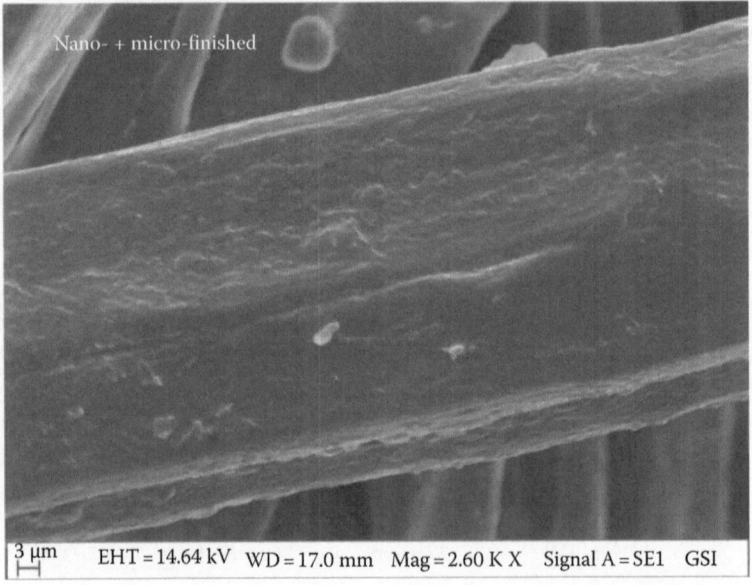

FIGURE 5.4 SEM photograph of nano- + micropolysiloxane-treated jute–polyester and cotton blended fabric.

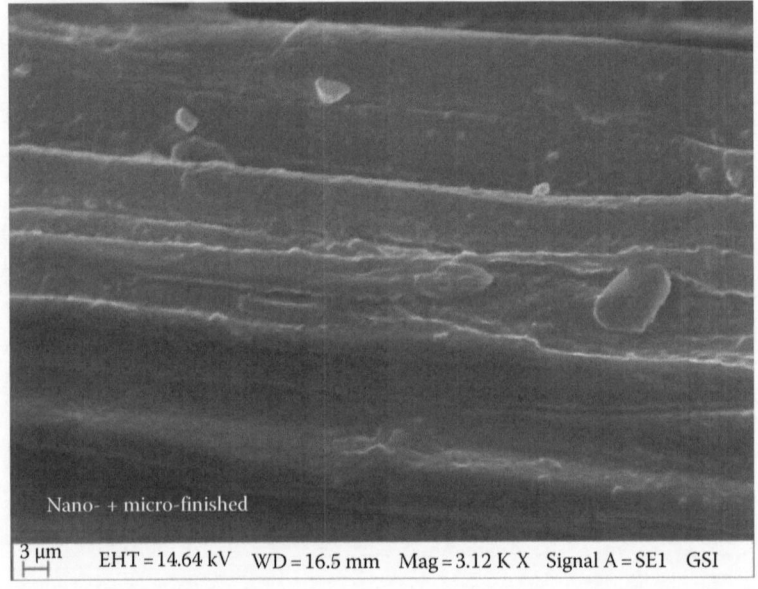

FIGURE 5.5 SEM photograph of nano- + macropolysiloxane-treated jute–polyester and cotton blended fabric.

standard procedures. All these tests were carried out in standard atmospheric condition maintained at 65% ± 2% RH and 20°C ± 2°C as fiber properties change with varying temperature and relative humidity as suggested by Booth (1976). The fabrics were conditioned for 72 h in the aforementioned atmospheric conditions before testing. No significant difference in air permeability as well as TIVs has been observed between treated and untreated, chemically finished jute–hollow polyester fabric samples.

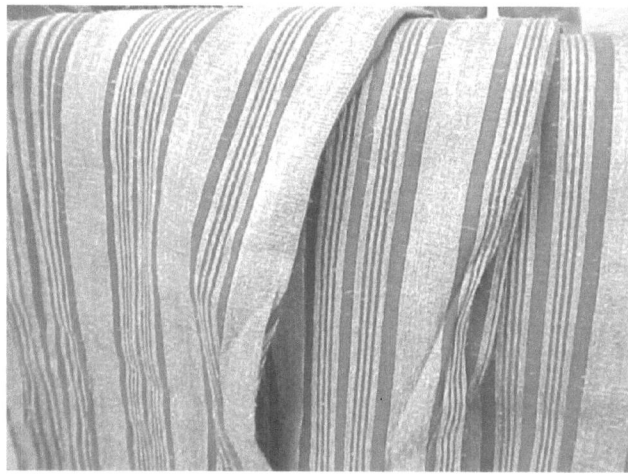

FIGURE 5.6 Jute–polyester and cotton blended rib structured fabric for winter apparel.

5.11 PROPERTIES OF DEVELOPED JUTE-BASED WARM FABRICS FOR JACKET

The fabric weight and TIVs were measured for the developed jacket fabrics and compared with the commercial (*khadi*) cotton and acrylic jacket fabrics. The developed jacket fabric shows 30% and 62% higher TIVs compared to commercial acrylic and cotton jacket fabrics, respectively (Anonymous, 2008a). This developed fabric is also 8% and 17% lighter in weight compared to commercial acrylic and cotton jacket fabrics, respectively. The fabric thickness of the developed jute-based fabric is 47% and 19% lower compared to commercial acrylic and cotton jacket fabrics, respectively (Debnath et al., 2008). Jute–polyester and cotton yarns, used alternately in weft direction, have been utilized to improve the esthetic and physical properties of developed fabrics (Figure 5.6).

5.12 DEVELOPMENT OF WARM GARMENTS FROM JUTE–HOLLOW POLYESTER AND COTTON BLENDED FABRICS

Different types of winter jackets have been developed using the jute–polyester and cotton blended fabrics woven in handloom. The weight of most of the developed jute-based fabrics is around 136 g/m^2. The weft yarn density and combination of warp and weft yarns used in weft direction is designed in such a way that final fabric weight is close to 136 g/m^2. This is close to the commercial cotton/woollen *khadi* jacket fabrics. This parameter is necessary to maintain, as the fabric weight is one of the parameters that decide the final weight of the jacket.

Different lining materials have been tried, namely, cotton lightweight fabric of 75 g/m^2, acrylic fabric in case of reversible jacket (Figure 5.7) of 90 g/m^2, and polyester lining fabric of 60 g/m^2 for normal lightweight jacket.

Normally, these consist of three layers. Mostly, the outer layer is the jute–hollow polyester and cotton blended union fabric and the inner layer is the lining material. To avoid shrinkage of the developed jute–hollow polyester and cotton blended fabric, the inner layer of the fabric has been further laminated with commercial polyester laminating material. This lamination is made with hot ironing process. This laminated composite material structure will also improve the crease recovery property of the final jacket. Further, to improve the thermal insulation of the jacket, apart from the three-layered material (jacket fabric, lining fabric, and laminating fabric), a bulked jute-based nonwoven material has also been utilized. In this very lightweight jute and polyester (80:20), blended needle-punched nonwoven material of 80 g/m^2 has been used (Debnath and Madhusoothanan, 2011).

FIGURE 5.7 Reversible jacket made from jute-based jacket fabric.

Since the very lightweight needle-punched material has a very poor strength, a cotton gauge cloth of 5 g/m² has been used as reinforcing material so that the total weight of the nonwoven material becomes 85 g/m²; this material has been stitched with normal stitching (lockstitch) method, that is, face to face with the lining material with a stitch density of 2 stitches/cm. Between two stitchings, a spacing of 5 cm was maintained in both width- and lengthwise directions of the fabric to anchor the nonwoven with the lining material.

5.13 EVALUATION OF JUTE-BASED APPARELS

The complete developed jackets were evaluated to compare their property performance with commercial jacket materials. Jacket weight, thermal insulation, and thickness properties of the developed jute blend jackets have been evaluated and compared with commercial jacket. Commercial Commercial polyester jacket (Oswal make) has been considered wherein, jacket fabric is used in the outer layer, hollow polyester fiber is used in the inner layer, and another polyester fabric is used as the lining material. Commercial windcheater jacket has also been taken into account to compare the performance of the developed jute-based jackets (Table 5.4). The thermal insulation of the jackets was measured in tog using NIRJAFT digital thermal insulation tester (Debnath and Madhusoothanan, 2010, 2011). This method of measurement of thermal insulation is based on hot plate method (Roy et al., 2009). In this thermal insulation test, the specimen test area considered was 706.85 cm² ($\varphi = 30$ cm). An average of three different readings was considered. The thicknesses of the jackets were measured in mm using a thickness gauge, where the pressure foot area was 5.067 cm² ($\varphi = 2.54$ cm). The dial gauge attached with the thickness tester has a least count of 0.01 mm and maximum displacement of 10.5 mm. An average of twenty readings at ten different places was taken into consideration. Shirley air permeability tester (SDL-21) has been used for measuring the air permeability through the jacket (Debnath and Madhusoothanan, 2011). The air permeability values have been expressed as unit volume of air in cm³ that passed per second

TABLE 5.4
Some Important Properties of Jute–Polyester–Cotton Jacket of XL Size

Types of Jackets	Jacket Weight (g)	Thermal Insulation (tog)	Thickness (mm)	Air Permeability (cm³/cm²/s)
Warp all cotton, weft jute–polyester and cotton (gents)	386	0.63	0.8575	3.78
Warp all cotton, weft all jute–polyester (gents)	642	0.551	0.8624	3.81
Reversible jute–polyester–cotton–acrylic jacket (gents)	652	0.91	0.8946	2.2
Jute–polyester–cotton ladies jacket Warp all jute and weft all jute–polyester and extra weft	452	0.61	0.8551	2.84
Jute nonwoven as added inner lining material	707	2.31	1.2	2.37
Commercial polyester jacket (*Oswal* make)	426	0.52	0.8755	0.86
Commercial windcheater	200	0.14	0.0873	0.14

through a 10 cm water head pressure following ASTM standards (ASTM, 1978). The test area of the instrument was 5.07 cm².

This study reveals that the thermal insulation of the jute-based jackets increases with the construction and the materials used to develop such apparels. Higher TIVs of the developed jackets are observed than commercial jackets (Anonymous, 2008a, b). The thermal insulation of jackets can be lower or higher depending on the jacket fabric constructional design and lining material. The weight of the jackets can be also lighter without affecting the thermal insulation, depending on the design of the jacket fabric and the jacket. Except for reversible jacket and jacket with nonwoven lining material, the thickness values of other developed jackets are lower than commercial (Oswal make) jackets (Table 5.4). The needle-punched nonwoven material as an additional lining material increases the thickness of the jacket, generating more voids in the jacket structure. This improves the thermal insulation of the final jacket. The air permeability values of the jute-based jackets are much higher than the commercial polyester jackets made with windcheater materials. This higher air permeability of the jute-based jacket allows better breathability in the jacket material compared to commercial ones without affecting the thermal insulation property.

5.14 EFFECT OF WASHING JUTE-BASED APPARELS

The effect of repeated washing on thermal insulation, thickness, and shrinkage of jute-based garments has been studied. In this, a normal detergent has been used for soaking for 30 min followed by washing and drying under shade. This sequential treatment cycle has been repeated for three consecutive times, and after completion of each treatment cycle, different properties were evaluated as mentioned earlier. After this repeated three washing treatments, no significant change in dimensional property has been observed in the jackets made out of jute–polyester and cotton blended union fabrics. Similarly, TIVs in tog remain unchanged after three repeated washes. This same jacket fabric has also been used for making a blazer (Figure 5.8). After one dry wash (dry cleaning), no change in thermal insulation has been observed. Based on these, it can be recommended that the developed jute–polyester blended fabrics can also be used for preparation of other warm garments like shawl, cap, and other protective winter fashionable apparels.

FIGURE 5.8 Blazer made from jute–polyester and cotton blended fabrics.

5.15 CONCLUSIONS

It has been concluded that judicious application of jute fibrous material for apparel application can lead to sustainable growth in the apparel sector. Jute as being an annually renewable natural fiber needs to be modified, and admixture with little quantity of man-made fiber is sufficient to use in the warm garment sector. There are huge possibilities to use jute fibrous material in the area of warm apparels for sustainable development. The jute-based materials for apparels seem to have favorable properties as far as the winter apparel is concerned. Proper application of chemical finishing will further improve the handle properties of the jute-based apparel fabrics. The constructional design parameters of the fabric as well as the apparel also influenced the overall properties of the jute-based apparels. The jute-based apparel properties are comparable to that with some commercial ones. In some cases, it is superior than apparels made out of synthetic materials available commercially. Hence, sustainable development of jute-based materials can share a significant role in winter apparel production.

REFERENCES

Anonymous. Annual report. Jute Technological Research Laboratories, ICAR, 12, Regent Park, Kolkata, India (1981).

Anonymous. Annual report 2005–2006. National Institute of Research on Jute & Allied Fibre Technology, ICAR, 12, Regent Park, Kolkata, India (2006): pp. 69–70.

Anonymous. Annual report 2007–2008. National Institute of Research on Jute & Allied Fibre Technology, ICAR, 12, Regent Park, Kolkata, India (2008a): p. 33.

Anonymous. Annual report 2007–2008. National Institute of Research on Jute & Applied Fibre Technology, ICAR, 12, Regent Park, Kolkata, India (2008b): pp. 76–83.

Anonymous. Annual report 2008–2009. National Institute of Research on Jute & Applied Fibre Technology, ICAR, 12, Regent Park, Kolkata, India (2009): pp. 20–21.

Anonymous. Annual report 2010–2011. National Institute of Research on Jute & Applied Fibre Technology, ICAR, 12, Regent Park, Kolkata, India (2011): pp. 25–27.

Anonymous. Annual report 2011–2012. National Institute of Research on Jute & Applied Fibre Technology, ICAR, 12, Regent Park, Kolkata, India (2012): pp. 25–27.
ASTM. *Annual Book of ASTM Standards*. Part-32. ASTM, West Conshohocken, PA (1978): pp. 358–365.
Booth, J. E. *Principles of Textile Testing*. 3rd edn., Newnes-Butterworth, London, U.K. (1976): pp. 100–101.
Chakravarty, A. C. Crimp produce in jute fibre by treatment with solution of sodium hydroxide. *Textile Research Journal* 32(6) (1962): 525–526.
Chakravarty, A. C. Woolenization of jute: Some physical aspect. *Jute Bulletin* 26(1) (1963): 16–17.
Chaudhury, A. and Basu, G. Studies on the properties of DREF spun acrylic yarn. *Indian Journal of Fibre and Textile Research* 23 (1998): 8–12.
Debnath, S. Chapter 20: Designing of jute–based thermal insulating materials and their properties, in: *Textiles History, Properties and Performance and Applications*. M. I. H. Mondal (ed.), Nova Publishers, New York (2014): pp. 499–518.
Debnath, S., Bhattacharya, G. K., and Singh, U. S. A blanket from jute-hollow polyester blended bulk yarn and method of preparing the same. Indian Patent Application No. 1102/KOL/2009, August 28, 2009.
Debnath, S. and Madhusoothanan, M. Thermal insulation, compression and air permeability of polyester needle-punched nonwoven. *Indian Journal of Fibre & Textile Research* 35(1) (2010): 38–44.
Debnath, S. and Madhusoothanan, M. Thermal resistance and air permeability of jute-polypropylene blended needle-punched nonwoven. *Indian Journal of Fibre & Textile Research* 36(2) (2011): 122–131.
Debnath, S. and Sengupta, S. Effect of linear density, twist and blend proportion on some physical properties of jute and hollow polyester blended yarn. *Indian Journal of Fibre & Textile Research* 34(1) (2009): 11–19.
Debnath, S., Sengupta, S., and Singh, U. S. Properties of jute and hollow-polyester blended bulked yarn. *Journal of The Institution of Engineers (India), Textile Engineering*, 87(2) (2007a): 11–15.
Debnath, S., Sengupta, S., and Singh, U. S. Comparative study on the physical properties of jute, jute-viscose and jute-polyester (hollow) blended yarns. *Journal of The Institution of Engineers (India), Textile Engineering* 88(1) (2007b): 5–9.
Debnath, S., Sengupta, S., and Singh, U. S. A method for producing jute-hollow polyester blended yarn, union fabric of said yarn and method of preparing said union fabric and shawl from the said yarn. Indian Patent Application No. 1187/KOL/2008, July 09, 2008.
Ganguly, P. K. and Sao, K. P. Effect of treatment time on swelling of jute. *Textile Research Journal* 55(6) (1985): 376–377.
Ganguly, P. K., Sao, K. P., and Ambally, C. Crimp in alkaline treated jute: Effect of treatment time. *Textile Research Journal* 55(4) (1985): 253–254.
Ghosh, P. and Samanta, A. K. Chemical texturing or bulking of rotor-spun jute/polyester fibre blended yarns. *Journal of Textile Institute* 88(Part 1)(3) (1997): 209–231.
Gupta, N. P. and Bhattacharyya, G. K. Performance of yarns and blankets from chemically treated jute-polypropylene blends. *Indian Journal of Textile Research* 9(4) (1984): 160–163.
Gupta, N. P., Mazumdar, A., Bhattacharyya, G. K., Sur, D., and Roy, D. Chemically texturizing jute and jute-polypropylene bleached yarns. *Textile Research Journal* 52(11) (1982): 694–702.
Majumdar, S. Prediction of fibre quality from anatomical studies of jute stem: Part I-prediction of fineness. *Indian Journal of Fibre & Textile Research* 27(3) (2002): 248–253.
Roy, G., Naskar, M., and Ghosh, S. K. Development of digital thermal insulation value tester for jute products. *Indian Journal of Textile Research* 34(1) (2009): 36–40.
Saha, P. K., Chatterjee, K. K., and Sarkar, P. B. *Monogram: Woollenised Jute as Wool Substitute*. Indian Central Jute Committee, Calcutta, India (1961).
Sao, K. P. and Jain, A. K. On crimp measurement in alkali treated jute. *Journal of Textile Association (India)* 46 September (1984): 155–159.
Sao, K. P. and Jain, A. K. Mercerization and crimp formation in jute. *Indian Journal of Fibre & Textile Research* 20(4) (1995): 185–191.
Sao, K. P., Jain, A. K., and Anantha Krishanan, S. R. A comparative study of the woollenised jute fibers of several strains. *Textile Trends* 26(7) (1983): 47–49.
Singh, U. S., Bhattacharya, G. K., Bagchi, N. N., and Debnath, S. Comparative study of woollen blanket and jute-wool blanket. *Textile Trends* 47(9) (2004): 27–28.
Sinha, A. K. and Basu, G. Studies on physical property of jute-acrylic bulked yarn. *Indian Journal of Fibre and Textile Research* 26 (2001): 268.
Sinha, A. K. and Gupta, N. P. Performance of texturized jute blended carpet. *Indian journal of Textile Research* 11(1) (1986): 35–37.
Sinha, A. K., Mathew, M. D., and Roy, D. Properties of jute polypropylene blended yarns texturised by sodium hydroxide solution. *Indian Journal of Textile Research* 13(1) (1988): 26–30.

6 Sustainable UV-Protective Apparel Textile

Kartick K. Samanta, Santanu Basak, and S.K. Chattopadhyay

CONTENTS

6.1 Introduction 113
6.2 UV Radiation and Its Impact 115
6.3 Evaluation of UV-Protective Performance 116
6.4 UV and the Sun Protective Textile 118
    6.4.1 Effect of Fiber and Fabric Properties 119
    6.4.2 Effect of Color 119
6.5 Traditional UV-Protective Finishing of Textile 120
6.6 UV-Protective Finish Using Nanotechnology 122
    6.6.1 UV Protection Using ZnO Nanoparticles 123
    6.6.2 UV Protection Using TiO_2 Nanoparticles 125
6.7 UV-Protective Finishing Using Plasma 128
6.8 Sustainable UV-Protective Finishing of Textile 129
    6.8.1 Sustainable UV-Protective Textile Using Plant Extract 129
    6.8.2 Sustainable UV-Protective Textile Using Nanoparticles of Natural Polymers 132
    6.8.3 Sustainable UV-Protective Textile Using Natural Polymer 133
6.9 Summary 134
References 134

6.1 INTRODUCTION

In recent years, consumers are showing more concern on the need of protection from sun rays and ultraviolet (UV) radiation that may induce skin damage from excessive exposure. In this context, UV-protective textiles are important in our daily life, as about 6.1% of the total solar emission reaching the Earth is composed of UV radiation. The remaining percentages in the emission spectra are a mixture of visible light (51.8%) and infrared (42.1%) radiation. The average photon energies in the infrared, the visible light, and the UV light of the solar emission are 63, 200, and 315–400 kJ/mol, respectively. Due to the lower wavelength and higher photon energy of UV rays (UVRs) compared to the two other components, its penetration ability into the top layer of the skin is also high, which may cause acute and chronic diseases, such as acceleration of skin ageing, sunburn, blotches, wrinkles, weak immune system, suntanning, photocarcinogenesis, skin cancer, eye damage, and DNA damage. UV radiation is categorized into three zones based on their wavelength, namely, (1) UVC (200–280 nm), getting mostly absorbed in the ozone layer of the upper atmosphere; (2) UVB (280–320 nm), which penetrates the top layer of the skin and may cause skin ageing, sunburn, etc.; and (3) UVA (320–400 nm), which penetrates through the skin and may cause skin ageing.[1] Therefore, for an effective UV protection, the textile has to be designed and finished in such a way so as to block the transmission of both UVA and UVB, as they are capable of penetrating the human skin. The UV-protective performance of a textile is defined by the ultraviolet protection factor (UPF), and a UPF of ≤10 for a textile is considered as providing no protection from UV radiation. On the other hand, if the fabric has a UPF of ≥50, it is considered to be an excellent UV-protective textile. The fabric's ability to protect from UVRs is dependent on several parameters, such as fiber type and additives, constructional characteristics (e.g., weave, ends and picks/unit length, thread count, cover factor [CF]), the nature and depth of the color applied, the presence of an optical brightening agent (OBA), pigments or finishing agents, launderings and washing conditions, stretch, and moisture.

As natural fibers like cotton, linen, silk, and wool have a lower capacity of UV absorption than synthetic fibers, they require to be finished with UV-absorbing chemicals or agents to equip them to perform as UV-protecting textiles. In the past, various chemicals, such as silicate, triazine class–hindered amine, derivatives of *o*-hydroxyphenyl diphenyl triazine, OBA, and metal oxide particles of TiO_2 and ZnO, have been explored for improving the UV-protective performance of natural as well as synthetic textiles.[2-5] Later on, due to the rapid growth of nanoscience and technology, nano-sized metal particles play a promising role, with their successful commercial application for the multifunctional finishing of textiles. They offer major advantages of being nontoxic, environmentally friendly, and stable even in processing conditions besides the requirement in a minimal quantity that does not alter any fabric properties.[6] Similarly, plasma nanotechnology has been explored either to synthesize nanomaterial, nanocoating, or surface activation prior to chemical finishing due to their advantage of being a water-free process avoiding the generation of liquid effluent altogether. The plasma polymerization of hexamethyldisiloxane (HMDSO) on the textile has demonstrated a successful blocking of the UVRs in the wavelength of <425 to >350 nm.[7]

Due to constantly increasing global awareness on green/ecotextile products and their market demand, research has been intensified in the area of development of sustainable UV-protective textiles using plant extracts and other natural polymeric materials or biomolecules. The textile industry is also slowly exploring the possibility of processing and finishing of natural and synthetic textiles using natural chemicals, auxiliaries, and polymers to mitigate with the adverse impact of environmental pollution and to improve carbon footprint and sustainability of producing value-added products suitable for such niche markets. In this regard, plant extracts (e.g., the plant molecules) of banana pseudostem sap (BPS) and spinach juice (SJ) have been found suitable for UV protection and flame-retardant finishing of cellulosic and lignocellulosic textiles. Similarly, various natural dyes extracted from the root, bark, flower, leaf, and fruit of a certain plant have been found to be very successful for natural coloration, UV protection, and antimicrobial finishing of textiles. It has been reported that the natural polymer, such as lignin in nanoform, acts as a good UV absorber and can provide a UPF value of 50 in the nonwoven linen textile. Other natural polymers, such as sericin and aloe vera, have recently been established as promising industrial materials for high-end biomedical applications, offering excellent oxygen permeability, cell protection, antioxidant action, moisture regulation, UV protection, and skin nourishment.[8] After application of sericin, the UPF value was found to improve to 125 for both natural and synthetic textiles. As these materials, such as a natural dye, agro-residue, lignin, sericin, and aloe vera, are produced from renewable sources, they are cost-effective, have almost no adverse effect on the environment and the human body, and hence can be best used to develop sustainable UV-protective apparel textiles.

The functional UV-protective finishing of textiles is an emerging area of research and sustainable product development. After getting considerable attention at present, there is a need to collate the information for the benefit of all concerned with the subject. Accordingly, this chapter discusses in brief UV radiation and its hazards on the human body, followed by the traditional mechanism of UV-protective finishing of natural and synthetic textiles. It also discusses the recent technological advancements in this area, such as UV-protective finishing using nano- and plasma technology, with a special reference to the latest developments on eco-friendly, cost-effective, and sustainable UV-protective formulations and their applications on cotton, silk, wool, linen, jute, and other textiles using various plant extracts, natural polymers, and biomaterials along with descriptions on the simultaneous improvement of other functional properties, such as flame retardancy, antimicrobial property, natural coloring, hydrophilicity, and skin nourishment.

6.2 UV RADIATION AND ITS IMPACT

During sunbathing, UV exposure can be considered appropriate, as it promotes the circulation of blood, invigorates the metabolism, and improves the body's resistance to various pathogens. During exposure to sunlight, epidermal 7-dehydrocholesterol (7-DHC) causes photolysis of previtamin D_3 in a reaction activated by UVB radiation.[1] Later on, previtamin D_3 forms vitamin D_3 in a thermal process. Whole body exposure at suberythemal doses produces a significant elevation in serum vitamin D_3 in an unclothed subject. This could be controlled by the seasonal clothing. However, some concerns were expressed that an extensive coverage of the human body by the fabric could result in a serious deficiency in vitamin D. Therefore, a reasonable care is to be taken, while dealing with individuals, so as to optimize the minimum or the maximum exposure requirement of the UV radiation.

In spite of its inherent advantage, however, UV radiation penetrates into the top layer of the skin during excessive exposure, causing damage to the lower layer of the skin and producing premature ageing and other effects, like roughening, blotches, sagging, wrinkles, and onset of squamous cell and basal cell cancer.[9] Many people love sunbathing for a prolonged duration that may in the long term pose risks to their health. Persons working in an open atmosphere, particularly agricultural workers, are more prone to keratose, a precursor of skin cancer. Similarly, due to the increased undertaking of leisure and sport activities by human beings worldwide and with the simultaneous increase in prolonged sunlight exposure, it has become the major cause of onset of malignant cutaneous melanoma.[6] These effects were recognized even in the early 1990s. Australia has the highest levels of solar UV radiation, mainly because of its geographical location. Similarly, New Zealand, the United States, Switzerland, Norway, Scotland, Britain, and Scandinavian countries also have high melanoma rates.[9] In addition, the intensity or dosage of UV radiation also depends on the cloud cover, the sun's altitude, hole-free ozone layer, its scattering in the upper atmosphere, and environmental condition. With the increasing awareness on human health and skin care, many researches have been reported on the impact of UVRs on various living organisms, including the human being, and a relationship between the skin disease and UV dosage has also been developed.[9–11]

In the solar spectrum, approximately 6%–7% UV radiation is present that causes several skin-related diseases.[1] The energy of UV radiation reaching the Earth is quite similar to the bond energy of a polymeric material (organic), resulting in the photoinduced degradation of the polymer in an open atmosphere.[9] The total UV spectrum of the solar ray can be classified into three zones based on their wavelength, namely, (1) UVA in the region of 320–400 nm, (2) UVB in the region of 280–320 nm, and (3) UVC in the region of 200–280 nm. As the energy or the adverse effect of UV radiation is inversely proportional to its wavelength, UVC is the most harmful among the three zones. Fortunately, it gets absorbed by the ozone layer in the upper atmosphere. The photon energy of the different solar radiation along with their relative intensity is given in Table 6.1. If the wavelength

TABLE 6.1
Relative Intensity and Photon Energy of Different Solar Radiations

Sl. No.	Wavelength (nm)	Radiation Type	Relative Intensity (%)	Photon Energy (kJ/mol)
1	280–320	UVB rays	0.50	400
2	320–400	UVA rays	5.60	315–350
3	400–700	Visible rays	51.8	200
4	700–3000	Infrared rays	42.1	63

Sources: Menter, J.M. and Hatch, K.L., Clothing as solar radiation protection, in: Elsner, P. and Hatch, K., eds., *Textiles and the Skin, Current Problems in Dermatology*, Vol. 31, Karger, Basel, Switzerland, 2003, pp. 50–63; Schinder, W.D. and Hauser, P.J., Ultraviolet protection finishes, in: *Chemical Finishing of Textiles*, Woodhead Publishing in Textiles, pp. 157–164.

of UV radiation is less, that is, the energy is more, then the radiation has the most capability to penetrate through the skin, and depending upon the duration of exposure, it can cause different levels of skin diseases. An excessive UV radiation can also lead to cell damage, inflammation of human skin, and erythema or sunburn.[9] These kinds of diseases were found to occur more at the wavelength of 308 nm. Also, the total UVRs reaching the skin are important factors in the occurrence of both erythema and skin cancer, although there is not much established scientific evidence available behind these two. In terms of sensitivity to light and tendency to pigmentation, there are six basic types of skins that require different levels of UV protection: (1) white-I, (2) white-II, (3) brownish, (4) brown-I, (5) brown-II, and (6) dark brown.[9]

Similar to the harmful effects of UVRs on human beings, as the photon energies of UVA and UVB are more than the carbon–carbon single bond energy (335 kJ/mol) of a polymer, polymer degradation, cross-linking, and other reactions get initiated. Therefore, in a UV-protective finishing, the textile has to be designed and finished in such a way that it should effectively block the (300–320 nm) radiation part. As the UV photon has more energy than visible light as stated earlier, the polymer exposed to UV light causes photoinduced degradation, resulting in gradual loss of fiber integrity and mechanical properties.[12] Textile grade fibers with a large surface-area-to-volume ratio make them more susceptible to UV attack, compared to polymers in other forms. The degree of degradation also depends on the chemistry of the fiber-forming polymer and the presence/absence of a UV blocker. There are reported data on the reduction in tensile strength of 100%, 23%, 34%, and 44% in the nylon, wool, cotton, and polyester (PET) samples, respectively, after 30 days of exposure. Additionally, decrease in elasticity and little increase in crystallinity have also been reported for the nylon sample.[13,14]

6.3 EVALUATION OF UV-PROTECTIVE PERFORMANCE

The most important function to be performed by a garment is to protect the wearer from high or low temperature and from the harmful rays of sunlight. The solar radiation in the wavelength region of 200–400 nm is known as UV radiations. The UV-blocking property of a virgin fabric is enhanced when either a dye or a pigment or a delustrant or a UV absorber is added for absorbing the incident UV radiation while restricting its transmission to the skin through the fabric. The effect of UV radiation on biological organisms has been extensively studied, and various numerical formulae have been developed, such as UV index, UPF, and sun protection factor (SPF), to provide a single measurable index and to create general awareness among people. UV radiation being an electromagnetic ray, when it falls on a material, some part of it gets reflected, and the other parts get either absorbed or transmitted. In this regard, the UV index provides a numerical indication of the maximum potential solar UVR level during the day, and the higher its number, the higher is the solar UVR hazard. The SPF is the ratio of the potential erythemal effect to the actual erythemal effect transmitted through the fabric by the radiation and can be calculated from spectroscopic measurements.[12] The dose that results in a minimal erythema extending to the borders of the irradiation is used to determine the SPF of the fabric, which can be calculated as[1]

$$\text{SPF (in vivo)} = \frac{\text{Minimum erythemal dose through the test fabric}}{\text{Minimum erythemal dose in absence of the test fabric}}$$

The fabric treated with a UV absorber ensures that the clothes effectively reflect the harmful UVRs of the sun, reducing a person's UVR exposure and taking due care of the skin. The extent of skin protection required by different people depends on the UV intensity and its distribution in a particular geographical location, daytime period, and the season. The UPF is also sometimes referred to as the SPF; the higher the SPF value, the better is the protection from UV radiation.[15]

In Europe and Australia, the SPF is referred to as the UPF.[12] The SPF value is also commonly used with a *sun-blocking* skin cream, as a relative indication of how long a person can be exposed to sunlight before skin damage is visible. Typically, an SPF value >40 is considered to have an excellent protection against UV light as per the Australian/New Zealand standard (AS/NZS) 4399.[12] In the summer season, as people get exposed to more sunlight than in the winter season, they prefer lightweight and light color woven and knitted garments. Therefore, the UV-protective finishing of shirts, blouses, T-shirts, swimwear, beachwear, and sportswear is very important. The UV-protective finishing of textiles/polymers used for making an automobile's body or textile is also important, as they are parked in an open atmosphere for a significantly longer duration that leads to fading of the color.

The UPF indicates how long a person wearing the textile can stay in the sun before skin reddening starts, compared to a person without the said textile as a cover. The UPF value is meant to create an awareness of the negative impact of UV radiation among end users. The UPF is determined by the following equation[1,9,16]:

$$UPF = \frac{Risk_{unprotected}}{Risk_{protected}}$$

$$UPF = \frac{\int_{\lambda=280nm}^{400nm} E_\lambda S_\lambda \Delta_\lambda}{\int_{\lambda=280nm}^{400nm} E_\lambda S_\lambda T_\lambda \Delta_\lambda}$$

where

E_λ is the solar irradiance (W/m²/nm)
S_λ is the erythema action spectrum describing the harmfulness of different wavelengths
T_λ is the spectral transmittance through the specimen at wavelength λ
Δ_λ is the wavelength interval of the measurements (nm)

The UPF value of a textile is calculated using this formula based on the transmission value of the UVRs through the substrate, and it depends on (1) the specific fiber material used; (2) structural characteristics of the fabric; (3) moisture content; (4) depth of the color; (5) presence of OBAs; (6) specific finishing chemical, for example, UV absorbers; and (7) laundering conditions of the garments.[17] The UPF and SPF values for any textile material should be the same in a similar incident spectral distribution of the fabric specimen. However, this has not been proven to be the case.[1] Gies et al. reported a very good agreement between the in vitro UPF and in vivo SPF for the 16 fabrics under study, with protection factors ranging between 10 and 200 (SPF>50 is reported as "50+").[18] On the other hand, Menzies et al. reported that five out of six cases of in vivo SPF values were less than the in vitro UPF value, if human testing was done *on skin*.[1] A better agreement between the SPF and the UPF values was achieved, when human testing was performed at 2 mm *off skin* (8 mm in one case). In another study by Greenoak and Pailthorpe, in 21 cases out of the 22 fabrics studied, the obtained ratio of SPF to UPF was significantly lower than unity.[1] Several other studies have also yielded insights into the reason for the observed disparities. In this regard, Menzies et al. observed that in contrast to his results on the six tested fabrics, the UPF values obtained from the standardized neutral density thin film meshes were in a good agreement with the SPF values.[1,18,19] From the result, it was also concluded that the lack of similarity between the fabric UPF and the SPF values was possibly due to the nonuniformity of fabric transmission (i.e., the *hole effect*). In the majority of cases, the UPF value has been calculated mainly to characterize the fabric, and based on the degree of protection (i.e., the UPF value) of a textile against the UVRs, it has been categorized into *no* to *excellent* protection as given in Table 6.2.

TABLE 6.2
Classification of UV-Protective Textile Based on the UPF Rating

UPF Value	UPF Rating	UV Radiation Transmittance (%)	UV-Protective Performance
<15	5, 10	>6.7	No protection
15–24	15, 20	6.7–4.2	Good protection
25–39	25, 30, 35	4.1–2.6	Very good protection
40–50, >50	40, 45, 50, 50+	≤2.5	Excellent protection

Sources: Pailthorpe, M., *Mutat. Res.*, 422, 175, 1998; Seungsin, L., *Fibers Polym.*, 10(3), 295, 2009; Erdem, N. et al., *J. Appl. Polym. Sci.*, 115, 152, 2010.

TABLE 6.3
Different Standards and Test Methods of UV-Protective Textiles

Organization	Standard/Test Method	Title
Australia/New Zealand	AS/NZS 4399	Sun protective clothing—evaluation and classification
American Society for Testing and Materials	Proposed ASTM	A standard guide for labeling of UV-protective textiles
International Test Association for Applying UV Protection	UV standard 801	UV standard 801
British Standard Institution	BS 7949	Children's clothing; requirement for protection against erythemally weighted solar UV radiation
British Standard Institution	BS 7914	Method of test for penetration of erythemally weighted solar UV radiation through clothing fabrics
Comite Européen de Normalisation	Proposed CEN	Classification and marking of UV-protective apparel
Comite Européen de Normalisation	CEN/TC 248/WG 14	Apparel fabric; solar UV-protective properties, method of test
American Association of Textile Chemists and Colorists	AATCC TM 183	Transmittance or blocking of erythemally weighted UV radiation through fabrics

Sources: Menter, J.M. and Hatch, K.L., Clothing as solar radiation protection, in: Elsner, P. and Hatch, K., eds., *Textiles and the Skin, Current Problems in Dermatology*, Vol. 31, Karger, Basel, Switzerland, 2003, pp. 50–63; Schinder, W.D. and Hauser, P.J., Ultraviolet protection finishes, in: *Chemical Finishing of Textiles*, Woodhead Publishing in Textiles, pp. 157–164; Pailthorpe, M., *Mutat. Res.*, 422, 175, 1998.

Due to the general awareness of the adverse effect of UVRs and for export/import/trade purposes, several standards have been developed and proposed worldwide by different organizations. Some of these important standards are given in Table 6.3. There is a significant difference between the various methods of evaluation/characterization of UV-protective fabrics, and as such, in the newly developed test methods, there is no need of sunburning of the test participant.[12] In all these methods, the UV transmittance through the fabrics is measured and UPF/SPF is calculated using the standard chart of the solar spectrum and the erythemal effect.

6.4 UV AND THE SUN PROTECTIVE TEXTILE

In order to mitigate the adverse effect of UV radiation, individuals are encouraged to use sunscreens, sunglasses, and UV-protective clothing.[16] As discussed earlier, the UV-protective performance of a textile depends on the quantum of UV light absorbed, scattered, or transmitted through the fabric structure. This is governed by the chemistry of the fiber-forming polymer, physicochemical properties of the fiber, the presence of UV absorbers, construction of the fabric (woven, knitted, or nonwoven), thickness, porosity, CF, extension, moisture content, color, imparted finish, and previous history of the fiber/fabric.[16,23] It also depends on the wear condition of the textile, moisture, and additives present in the fiber and incorporated during processing as discussed in the following.[16]

6.4.1 Effect of Fiber and Fabric Properties

As the UPF is calculated based on the transmittance value of the UVRs through the fabric, it has an inverse relationship with the fabric porosity, and a linear relationship with the fabric CF, and is defined as UPF = (100 − CF). It has been further reported that the UPF values of 200, 40, 20, and 10 can be achieved with the fabric CFs of 99.5%, 97.5%, 95%, and 90%, respectively.[20] A fabric should have more than 93% CF to achieve a minimum UPF rating of 15, and a small increase in fabric CF (>95% CF) can lead to substantial improvements in the UPF of the textile.[16] Natural fibers, such as jute, cotton, silk, wool, and ramie, show more UV stability due to the presence of natural coloring pigments and other impurities in them in small quantities.[20] As a result, the gray cotton fabric exhibits more UPF value than the bleached one that has undergone improvement in optical transparency after the removal of wax and pectin. The other lignocellulosic fibers, such as linen and hemp, reported to have UPF values of 20 and 10–15, in spite of the presence of lignin in their structure. Thus, these fibers are not very much promising in terms of offering UV protection. On the other hand, the protein fibers have mixed effect on UV protection; wool absorbs UVRs strongly in the UV region of 280–400 nm and even beyond 400 nm, while silk's color, strength, and resiliency in both dry and wet conditions get reduced.[24] Wool tricot possesses a UPF value as high as 45. Mulberry silk gets affected more than the muga silk in the presence of UV light. The bleached silk and PAN fabrics showed UPF values as low as 9.4 and 3.9, respectively. PET fibers absorb more UVA and UVB radiation than the aliphatic polyamide fibers, possibly due to the presence of benzene ring in the structure, thus exhibiting 26 UPF value.[12]

The high relative humidity or moisture content reduces the fabric interstices due to the swelling of fibers, thus ensuring a higher UPF value. On the other hand, the presence of water molecule ensures more transmittance of UV light due to the less scattering of light, resulting in a lower UPF[9,11] and there was an increase in the UV transmittance percentage from 15%–20% to 50% in the case of dry to wet cotton garments. The UVA and the UVB transmittance percentages of below 6% and 2.5%, respectively, are preferred for a good UV-protective textile.[25]

6.4.2 Effect of Color

As the majority of dyes and pigments can restrict the transmittance of UV light through the fabric, the dyed and printed fabrics generally exhibit good UPF ratings than the fabric with higher gram/meter^2 (GSM).[9,20] This helps to produce lightweight summer apparel. If the fabrics are identically woven, then the darker shades, like black, navy, and dark red, can absorb more UV light compared to the light pastel colors.[16,23] Some of the direct, reactive, and vat dyes are also capable of providing a UPF of 50+. The direct dyes also have substantial capability in improving the UPF of a bleached fabric. In naturally dyed cellulosic and lignocellulosic fabrics, the enhancement in UPF was in the range of 15 to 50+, depending upon the color, its concentration, and type of mordant used.[12,26] It has been discussed in detail in Section 6.8. Similarly, naturally colored cotton, because of the presence of light green to tan and brown pigments in them, showed excellent UV protection with UPF of 64 and 47, in comparison to normal cotton with a UPF value in the range of 4–10. The naturally green pigmented cotton provided UPF as high as 64 in comparison to only 4 in the bleached cotton fabric. Furthermore, the fabric could retain the same UPF value, even after repeated laundering, with and without the bleaching process. Though both the samples had the UPF rating of 50+, the unbleached sample showed a better UPF (i.e., with lower UVA and UVB transmittance) compared to the bleached fabric. Even after a light exposure of 80 AFUs in the American Association of

Textile Chemists and Colorists (AATCC) fading unit, the sample was found to possess a UPF in the range of 32–43.[27] Thus, the naturally colored cotton can demonstrate an excellent UV-protective functionality, much superior to conventional, bleached, or unbleached cotton fabrics. Green-, tan-, and brown-colored samples showed UPF values of 30 to 50+, 20, and 40 to 50+, respectively, compared to 4 and 8, respectively, in the conventional bleached and unbleached fabric.[27] Much similar to the adverse effect of UV light on the human body, a negative impact has also been noticed in the plant body, like in the cotton ball formation.[12] The digital printing of textiles has also been explored as a unique method of textile coloration and UV protection.[23] Cotton fabrics were ink-jet printed with cyan, magenta, yellow, and black colors for the structural changes in them to reduce fabric porosity while simultaneously improving the UV-blocking properties. The control sample showed the UPF in the range of 0–5, and it was found to increase to 50+, when dyed with reactive black. The reactive printed samples with cyan, magenta, yellow, and black colors, with and without washing, showed a very good to excellent UV protection. The UPF was a bit lower in the case of only pigment printed sample.

6.5 TRADITIONAL UV-PROTECTIVE FINISHING OF TEXTILE

The requirement of a chemical to be considered as an effective UV-protecting agent is that it should efficiently absorb the UV radiation in the region of 300–320 nm.[13] Sun protection must involve a combination of sun avoidance and use of suitable textiles, hats, sunglasses, sun creams, etc.[20] A UV-protective textile can be developed by altering the fabric construction, dyeing, and finishing. In addition to these, several UV-protective chemicals are also available in the market. Among those, the most common and simple one is the OBA or the fabric whitening agent that is used during washing or finishing operations of textiles. The OBAs are commonly used to enhance the whiteness of textiles due to the emission of visible blue light by absorbing UV light. Most of the optical brighteners have a maximum excitation in the range of 340–400 nm of the UVRs. They can improve the UPF of cotton and blended textiles; however, the effect is not so promising in the case of 100% thermoplastic PET or nylon textiles. The presence of 0.5% OBA in the PET/cotton (67:33 P/C) blended fabric was sufficient to improve the UPF from 16.3 to 32.2, which is similar to the improvement obtained using 0.2% UV absorber. However, the fabric practically looses this improved functionality after 10 washes only. Another limitation of the OBA-based UV-protective finishes is that they mostly work in the UVA part of the daylight but has a weak absorption in the UVB region that causes more skin damage.

Several organic or inorganic colorless compounds are also considered as UV absorbers, as they have a strong absorption in the UV range of 290–360 nm.[13] The presence of such UV absorbers in the fibers should convert the energy of the UV radiation into vibration energy of the UV-blocking molecules and, subsequently, to heat energy without causing any photoinduced degradation of the polymer. The high energy, that is, the short wavelength of the UVR, excites the UV absorber to a still higher energy level, and the absorbed energy is dissipated at a long wavelength. The majority of the UV absorbers are applied as dye bath additives during dyeing. The important requirements are convenience in textile applications and lack of added color to the treated fibers. In this regard, sunscreen lotions containing the UV absorbers are considered as a method of physical blocking. The most widely used UVB screen matching to the skin is 2-ethylhexyl-4-methoxycinnamate with a high refractive index. It is preferred that an effective UV absorber must absorb the UVRs throughout the UV spectrum to avoid skin disease and degradation of the polymer altogether.[28] The UV absorber is incorporated into the spinning dope prior to fiber extrusion for the production of a synthetic filament, or it is added as a dye bath auxiliary to improve the light fastness of certain dyes with pastel shades and the weatherability of the spun-dyed fibers. The application of 0.6%–2.5% UV absorber is sufficient to enhance the UV protection of textiles. Concurrent with UV protection, it also reduces the photoinduced degradation of PET, nylon, silk, and wool fibers or photoinduced yellowness of wool fiber. Some of the UV absorbers are not very stable to laundering and light fastness.

The UV-protective finishing of polypropylene (PP) is more important compared to other synthetic fibers, as it is susceptible to photodegradation due to the presence of tertiary hydrogen.[22] For the same, the antioxidants and UV stabilizers are frequently applied to mitigate such type of polymer degradation. Triazine class–hindered amine light stabilizers are used in PP to improve its UV stability. Even pigmented PP or mass-colored PP requires UV stabilizers, when the textile is meant for an outdoor application. The high-energy UV absorbers suitable for PET textiles are the derivatives of *o*-hydroxyphenyl diphenyl triazine. They are suitable for application in dye bath, pad liquor, or print paste. Some of the other UV absorbers are the derivatives of *o*-hydroxybenzophenone and *o*-hydroxyphenylbenzotriazole.[2] The chemical structures of such traditional UV absorbers suitable for the natural and synthetic fibers are given in Figure 6.1.

Unlike in the organic UV absorber, the presence of inorganic pigments in the fabric results in more diffused reflection of light, thus providing an extra protection. For example, the TiO_2 particles added in the melt-spinning dope as a delustering agent also performs as a UV absorber. Similarly, the ceramic materials and zinc oxide (ZnO) particles also exhibited good UV-protective performance in the region of 280–400 nm of wavelength. Compared to micron- to submicron-sized

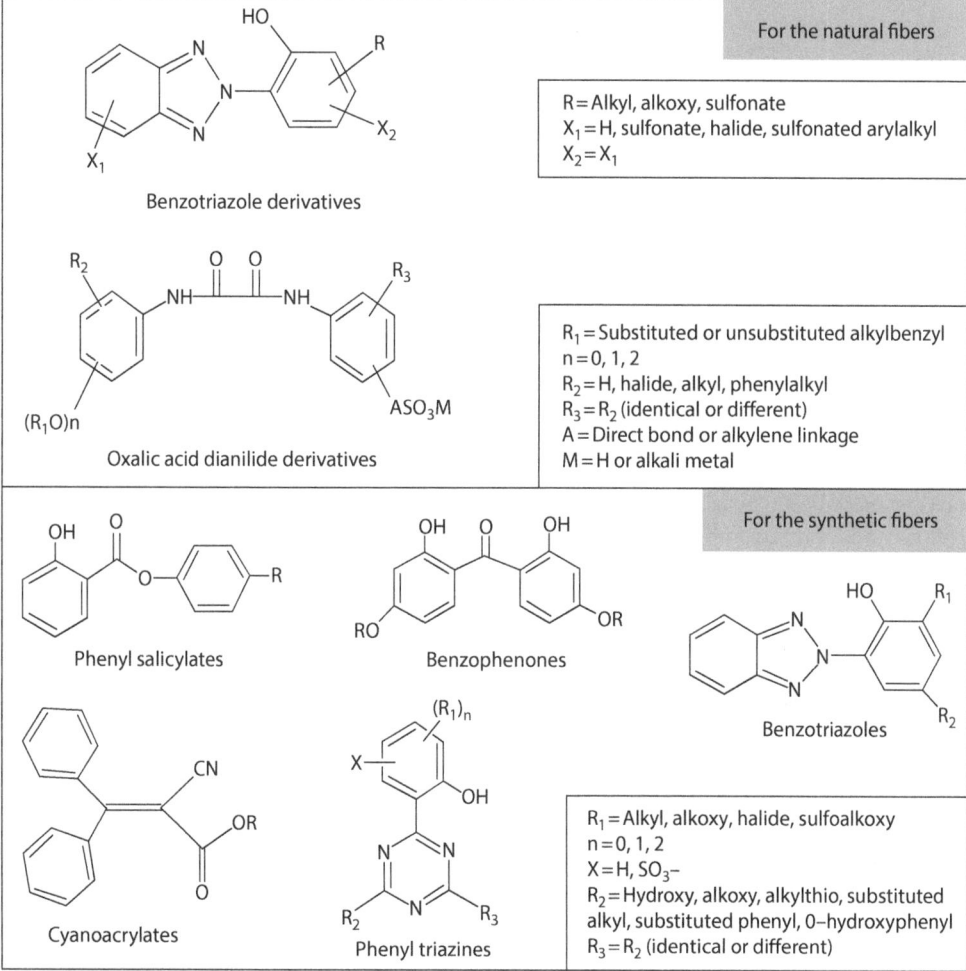

FIGURE 6.1 Structure of UV absorbers suitable for natural and synthetic fibers. (From Schinder, W.D. and Hauser, P.J., Ultraviolet protection finishes, in: *Chemical Finishing of Textiles*, Woodhead Publishing in Textiles, pp. 157–164.)

particles of TiO_2, ZnO, Al_2O_3, and silicate, their nanoform, when applied, exhibited much better efficacy as discussed in Section 6.6.[22] A novel formulation to impart UV protection to cotton, PET, and cotton/PET fabrics has been reported while using gamma irradiation for surface curing.[29] The naturally occurring aluminum potassium sulfate (alum) was applied alone and in combination with zinc oxide (ZnO). It was observed that approximately 0.3 g/mL alum was sufficient for an abrupt increase in the UV protection compared to the uncoated fabrics. On the other hand, alum along with ZnO could increase the UV protection performance by two- to threefold compared to what could be achieved only by the alum coating, especially for the PET fabrics. There was a little decrease in water absorbance and moisture regain in the ZnO- and alum/ZnO-coated fabrics over the blank samples. In contrast, the alum-treated samples showed a hydrophilic character. In spite of the fact that TiO_2 has a stronger UV scattering/absorbing characteristic with a higher refractive index of 2.6, ZnO particles are preferred over titanium dioxide (TiO_2) so far because it is relatively more safe to human beings. This is also because of the fact that ZnO can cover a wider range of UVA and UVB, and it is less reactive compared to titanium oxide. As revealed through the cell-culture study, TiO_2 forms radicals that were found to be harmful, if it is absorbed. It has been concluded that formulations with alum and ZnO are sound for the UV-protective finishing of all kinds of fabrics. The inorganic UV blockers are more preferred than the organic UV blockers, as they are nontoxic and chemically stable under elevated application temperatures. In recent years, as research in natural polymeric-based UV absorbers (organic material) progressed, the nanoform polymers and the agro-residues have shown promises from the viewpoint of sustainable UV textile development, as discussed in Section 6.8 subsequently.

6.6 UV-PROTECTIVE FINISH USING NANOTECHNOLOGY

Nanotechnology, an interdisciplinary science, is gaining much attention both in the academic research and in the industrial product development for applications in textile, material, medicine, electronics, optics, plastics, energy, aerospace, biotechnology, and agriculture.[6] It is no wonder that the textiles, having the long tradition and history of evolution based on numerous innovations, have quickly accepted and adopted nanotechnology for novel product development. The nanotechnology is forecasted to be the second industrial revolution in the world. Novel material properties requiring lesser chemicals (nanoparticles/nanomaterials) have attracted global interest across research disciplines and industries, including the textile sector.[30] This has to develop fashionable textiles, yet has to incorporate consumer-demanded functionalities, either in vogue or bizarre, for both the traditional and the upcoming niche markets. As stated by the *European Technological Platform for Textiles and Fashion*, the textile industry must improve the quality of the product and reduce the costs of production; at the same time, it should offer innovative products in traditional markets and new to luxury products for the upcoming markets. Nanotechnology aims at manipulating atoms, molecules, and development of nanosized material in a precise and controlled manner in order to build up new materials, which can be effectively used to produce functional, smart, and intelligent textiles, such as water-/stain- and oil-repellent, hydrophilic, hydrophobic, UV-protective, and flame-retardant textiles. Mainly, nanoparticles, nanowires, nanorods, nanocapsules, nanofibers, and nanocoating are applied in such application.[6] Nanomaterials with size of a few nanometers to around 500 nm are used for improving the aforementioned textile attributes. Nanoparticles, due to their large surface-area-to-volume ratio, high surface energy, and better affinity toward textile substrates and to the requirement of less chemicals, can increase the durability of the nanofinish while preserving the fabric feel and breathability. Additionally, the van der Waals force acting between the nanoparticles and the textile is also responsible for better durability of the imparted finish. To further increase the washing fastness, the nanoparticles are applied along with a binder by the padding or dipping method. Inorganic nanomaterials, such as metal to metal oxides, have attracted lots of attention in the last two decades. Mainly metal oxides such as TiO_2, ZnO, MgO, and CaO and also silver (Ag) nanoparticles have been explored, as they are stable under harsh process conditions

TABLE 6.4
Name of Common Nanoparticles and Their End Application

Nanoparticles	End Application
Titanium dioxide (TiO_2) and organic or inorganic modified TiO_2	Antibacterial, photocatalyst, self-cleaning, UV protection, dye degradation, air and water purifier, solar cell, gas sensor, cocatalyst for cross-linking of cotton
Silver (Ag)	Antimicrobial, disinfectant, antifungal, UV protection, electrically conductive
Zinc oxide (ZnO)	Antibacterial, UV protection, superhydrophobic, photocatalytic, solar cells, sensors, sunscreens
Copper (Cu)	Antibacterial, UV protection, electrically conductive
Clay	Antibacterial, UV protection, flame retardant
Gold (Au)	Antibacterial, antifungal, electrically conductive
Carbon nanotube (CNT)	Antimicrobial, flame retardant, electrically conductive, antistatic, chemical absorber
Nanolignin	UV absorber, antimicrobial, and antistatic
Nanocellulose	Composite, paint, gas permeability regulation, hydrophilic
Antimony oxide (Sb_2O_3)	Flame retardant
Silica (SiO_2)	Biomedical, additive for rubber and plastics, filler for composite

Source: Dastjerdi, R. and Montazer, M., *Colloids Surf. B Biointer.*, 79(1), 5, 2010.

and quite safer to human beings and animals.[31–33] The application of small amounts of such metallic nanomaterial helps to achieve a better UV-protective performance, making them a preferred choice over the organic UV blocker, as discussed earlier.[32] Also, inorganic UV blockers are stable under high temperature and UV irradiation compared to organic blockers.

In the last two decades, TiO_2, ZnO, and silver nanoparticles have been extensively studied for their textile applications, like photocatalysis of organic color, UV protection, antimicrobial and hydrophilic finishing, and water and air purification.[3] Some of the examples of common nanoparticles and their end applications are reported in Table 6.4. Inorganic UV blockers are mostly semiconductor metal oxides, such as TiO_2, ZnO, SiO_2, and Al_2O_3. It has been established that the nanosized titanium dioxide and zinc oxide are the two most efficient compounds in UV absorbing and scattering, thus providing a better UV protection compared to the normal-sized material.[22] Therefore, the UV-protective finishing of textiles using these two nanoparticles has been discussed in detail in the following text.

6.6.1 UV Protection Using ZnO Nanoparticles

Application of nanoparticles to textiles is aimed at producing new or multifunctional textiles using least quantum of chemical. For example, ZnO nanoparticles can be used for antibacterial and UV-protective finishing. It has also been found highly suitable for alternative applications in solar cells, sensors, electroacoustic transducers, photodiodes and UV light-emitting devices, sunscreens, gas sensors, UV absorbers, antireflection coatings, photocatalysis, and catalysts.[35] The large surface-area-to-volume ratio is the key point in their effectiveness in blocking incident UV radiation, when compared to their bulk materials. Due to their higher surface energy and more affinity toward the fabrics in nanoform, there is improved durability of finish. Nanosized ZnO particles have been characterized and applied on cotton and PET/cotton blended textiles for UV-protective finishing (Kathirvelu et al).[32] It has been reported that the performance of ZnO nanoparticles as a UV absorber can be efficiently transferred to the fabric with the application of ZnO nanoparticles on the surface of both the pure cotton and the blended textiles. The UV protection test indicated a significant improvement in the UV-absorbing activity of the ZnO-coated textile. A fabric finished with ZnO nanoparticles demonstrated an excellent UV-blocking property even after 55 home launderings.[36–38] The ZnO nanoparticles were applied in different concentrations on the cotton fabrics by the pad-dry-cure method, and the UV protection property of both the untreated and the treated samples was evaluated using a UV–visible spectrophotometer. The treated fabric showed a higher UV absorption, resulting in a lower transmission of it through the fabric. The 2% nano-ZnO-coated cotton fabric was found to have about 75% blockage of the incident UVRs due to the presence of coating.[4] The physical properties of the fabric, such as the strength and whiteness, did not get affected due to the application of such finish.[39]

Zinc oxide nanoparticles of the size 24–71 nm were applied to PP nonwoven fabrics produced by the electrospinning method for development of UV-protective materials.[21] The ZnO nanoparticles were added to 10%–15% solution of polyurethane and it was electrospun. It was laid over the PP nonwoven fabric of 26 g/m^2. The ZnO nanoparticles were applied in the layered fabric systems composed of nanocomposite fiber webs with various concentrations of zinc oxide. The UPF value was found to increase from 2 in the untreated sample to 50+ in the 2% ZnO-treated samples with 5 g/m^2 web density. It was possible to block 97% of the UVA and 99.5% of the UVB compared to the untreated sample, where the blockages were only 38.9% and 39.4%, respectively. The effect of zinc oxide concentration and the web area density on the UV-protective properties of the composite fabric has also been reported. A very thin layer of nanocomposite fiber mat showed a significant improvement in the UV blocking with a UPF value of >40 (i.e., the excellent protection). UV protection was found to increase with increasing zinc oxide concentrations in the nanocomposite fiber web. Similarly, the synthesized ZnO nanoparticles were applied in four types of cotton and PET/cotton (45/55) woven and knitted fabrics.[31] The UV-protective performance of the different samples was evaluated in the UV–visible spectrophotometer. The Fourier transform infrared (FTIR) spectroscopy absorption peak at 430 cm^{-1} depicted the presence of nano-ZnO. It was observed that the different untreated cotton and the PET/cotton blended textiles had UVA and UVB protection values in the range of 1–2.3 and 1–5.5, which increased to 4–11.8 and 4.7–16.2, respectively, in the treated samples.

The ZnO nanorods were synthesized on the SiO$_2$-coated cotton fabric by a hydrothermal method. Successively, a hot water treatment at 100°C or above was given for transformation of ZnO nanoparticles present on the cotton fabric from sphere to rod or needle shape, with a much smaller diameter of 24 nm.[3] The UV-blocking property of both the untreated and the treated fabrics (in terms of UPF) was measured by a UV transmittance analyzer (UV-1000F, Labsphere, USA) according to AS/NZS 4399:1996 (Sun Protective Clothing: Evaluation and Classification). The x-ray diffraction (XRD) peaks at 2θ value of 31.8°, 34.4°, and 36.3° indicated the formation of ZnO crystallites on the cotton fabric before and after the boiling water treatment. It was observed that the UPF value gets reduced from 105 (50+) in the unwashed sample to 47.7 (45) after 5 washing cycles as per the AATCC 124-2005 standard. After 20 washing cycles, it further reduced to <40. The hot water treatment was found to be effective in improving the UPF value, rather than the boiling water treatment.[3] The ZnO/carboxymethyl chitosan bionanocomposite was prepared at different temperatures to impart the UV-protective property in the cotton textile.[35] In the 2%–6% bionanocomposite concentrated suspensions, the cotton fabrics were immersed at a wet pickup of 100%, followed by drying and curing at 100°C and 160°C for 5 and 3 min, respectively. The samples after thoroughly washing were then taken for UV-protective characterization using a UV-Shimadzu 3101 PC spectrophotometer. The UV–visible spectra peaks at 310 and 300 nm showed the bionanocomposite, and the peak at 380 nm was for the bulk ZnO.[35] It was observed that the improvement in the UPF value was quite poor, that is, from 5 to 7.6 in the control and the treated samples.

In order to impart functional UV protection to cotton fabric of 75 g/m^2, the ZnO nanoparticles were synthesized using zinc nitrate and sodium hydroxide as precursors, and soluble starch as the stabilizing agent.[4] Acrylic binder of 1% was used for anchoring the 2% ZnO nanoparticles on the surface of the cotton fabric. The UV–visible peak at 361 nm was for the synthesized ZnO nanoparticles with an average size of 40 nm in the presence of 0.5% soluble starch. In the nano-ZnO-treated sample, there was a loss in mechanical strength. In contrast, there was a significant improvement in air permeability of the sample coated with nanoparticles and a binder formulation. As expected, the coefficient of friction between the fabric to fabric and metal was more in the treated fabrics. It was found that about 80% of UV light can easily pass through the untreated fabric, whereas only 50% and 25% could pass through the bulk ZnO and nano-ZnO particle–treated samples. This shows that the nanoparticles are more effective in UV blocking compared to their micro-sized bulk material. In another study by the same research group, the size of the particles was estimated to be 38 ± 3 nm with the transmission electron microscope (TEM). The absorption at 361 nm in the UV–visible spectra was due to quantum confinement.[40] The starch content in the nano-ZnO–starch composite was 37.7%. The wurtzite lattice parameters (a, c) were calculated from the XRD spectra and were found to be 3.250 and 5.208 Å for the bulk ZnO and 3.259 and 5.223 Å for the nano-ZnO. All the peaks could be indexed to the hexagonal ZnO and were very close to the other reported data. Also, the wurtzite structure of ZnO is the thermodynamically stable crystallographic state. It was observed that both the bulk ZnO and nano-ZnO showed a similar antimicrobial efficacy of >99.5% against *Klebsiella pneumoniae* (gram-negative) bacteria. On the other hand, the nano-ZnO-incorporated sample showed a better efficacy of 99.9% against *Staphylococcus aureus* (gram-positive) bacteria compared to only 91.8% in the case of the bulk ZnO-treated sample. This was in spite of the fact that the latter sample contained more ZnO particles (2.2%) on the fabric compared to only 1.6% in the case of nano-ZnO. The UV–visible transmission spectra of the fabrics showing the percent blockage of UVA (315–400 nm) and UVB (280–315 nm) were 19% and 32% in the control sample, and those values were 42% and 59%, respectively, for the bulk ZnO (2.0%)-treated sample and 65% and 68%, respectively, for the nano-ZnO (1.0%)-treated sample. The UPF of the control cotton fabric was calculated to be 1.54, while it was 2.48 and 3.71 for the bulk ZnO- and nano-ZnO-coated fabrics, respectively.[40]

6.6.2 UV Protection Using TiO$_2$ Nanoparticles

Similar to ZnO nanoparticles, titanium dioxide (TiO$_2$) nanoparticles also possess a good UV-blocking property, and it has been well explored in practice due to their advantages of being nontoxic and chemically stable at elevated temperatures and under UV exposure. Similar to ZnO nanoparticles, they have also been explored for imparting properties like antimicrobial, self-cleaning, antifading, and stain-resistant pigmentation. The rapid growth of nanoscience and technology has provided a new pathway for making better UV-protective films, fibers, and fabrics using TiO$_2$ nanoparticles.[41] Erdem et al. investigated the effects of nano-TiO$_2$ on UV protection and the structural properties of PP filaments.[22] The PP/TiO$_2$ nanocomposite was prepared by melt compounding. The filaments were loaded with 0.3%, 1%, and 3% TiO$_2$ nanoparticles, and the UV-protective properties of composite filaments were found to improve. The nano-sol of titania was produced from titanium isopropoxide by the solgel technique using water (T$_1$) or ethanol (T$_2$) as a medium and then applied in the fabric sample by the pad-dry-cure method. The treated fabrics were tested for antimicrobial activity, UPF, and self-cleaning activity (photodegradation) as per the AATCC standard procedures.[42] The durability of the imparted finishes was also investigated after a repeated laundering.[33] The UV reflectance spectrum of nano-TiO$_2$-treated PET fabrics reveals that the presence of nano-TiO$_2$ causes a profound decrease in the reflectance percentage of the irradiated UVR compared to the unmodified sample, implying higher blockage of the incident UVRs leading to better UV protection.[43]

Apparel grade cotton textiles hardly show any protection from the UV light. Hence, TiO$_2$ nanoparticles have been synthesized by an in situ method in the cotton textile. TiO$_2$ nano-sol, which is an intermediate state of nanoparticles and its precursor, was first synthesized using the titanium tetra-isopropoxide (TTIP) precursor, alcohol, acid, and deionized water.[44] The nanoparticle prepared by the solgel technique has advantages of in situ synthesis and uniform application in textiles.[5] The size of the particles was in the range of 10–300 nm as measured by TEM, atomic force microscopy (AFM), and the particle size analyzer as shown in Figure 6.2. XRD analysis confirmed the formation of the anatase grade of TiO$_2$ particles on the cotton fabric. After application of the nanoparticles, the UPF was found to increase significantly from 10 in the untreated sample to 50+

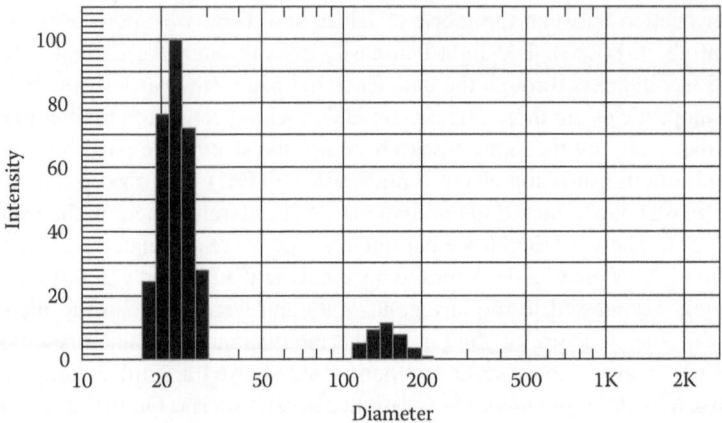

FIGURE 6.2 Particle size of TiO_2 nanoparticles measured by a particle size analyzer. (From Samanta, K.K. et al., Cotton textile incorporating titanium dioxide nanoparticles and method to manufacture the same, Indian Patent Number: 3468/MUM/2012, 2012.)

in the nanoparticle-treated sample, as measured according to the AATCC 183:2000 method on a Labsphere UV transmittance analyzer.[44] It was found that the UVA (315–400 nm) transmittance percentage reduced from 15% in the untreated sample to 6.2% in the TiO_2-treated sample, and similarly, the UVB (290–315 nm) transmission percentage reduced from 8% to 0.1% as shown in Figure 6.3.

The result indicates that in the TiO_2 nanofinished textile, most of the incident UV light gets absorbed by the fabric without passing through it. The UV-protective textile was subjected to standard washing cycles as per the AATCC-61-2A method. The sample, after 30 washing cycles, could maintain a similar UPF rating, and UVA and UVB transmittance percentages, as it was before the washing.[44] This indicates that the TiO_2 nanoparticles are firmly present in the structure of the cotton textile and do not come out during the washing of sample with soap solution. This novel technology

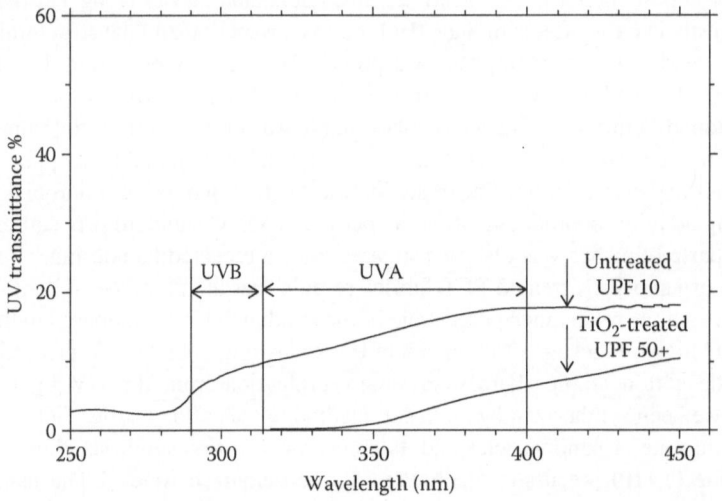

FIGURE 6.3 UVA and UVB transmittance percentage in untreated and TiO_2 nanoparticle–treated cotton textiles. (From Samanta, K.K. et al., UV protective finishing of cotton textile using TiO_2 nanoparticles, in: *Oral Talk in APA International Congress on Advances in Human Healthcare Systems*, Healthcare India, New Delhi, India, 2012.)

helped to achieve a durable UV-protective textile without usage of any binder/cross-linking agent. In the sample after 30 washings, 3% of the Ti atom was present in the form of TiO_2 particles, as detected by energy-dispersive x-ray (EDX) analysis. After application of TiO_2 nanoparticles, there was no significant change in fabric comfort property in terms of air permeability and also in tensile strength.[45] It was estimated that a minimum of 0.24% TiO_2 loading was required in the cotton fabric to achieve a UPF rating of 50+. The secondary ion mass spectrometer (SIMS) was used to analyze the presence of TiO_2 molecules in the washed sample using gallium ion source. The sample was subjected to 30 numbers of home launderings before the test, and the presence of Ti for TiO_2 was detected at a molecular mass of 48 amu as shown in Figure 6.4. This also substantiates the fact that the imparted UV-protective finish was durable to soap washing, although no binder was used. Similar to cotton, in the lignocellulosic jute textile, the UPF value was found to improve from 25 in the control sample to 50+ in the TiO_2 nanoparticle–treated sample. As a result, both UVA and UVB transmittance were reduced from 5.7%–3.0% and 3.5%–1.8% in the control to TiO_2-treated samples, respectively.

A bleached PET fabric (115 g/m²) was made UV protective using TiO_2 nanoparticles of 6 nm size, as measured in TEM analysis.[46] The UV-protective performance of the fabrics was measured in a UV–visible spectrophotometer Cary 100 Scan (Varian), and the UPF values were automatically calculated from the measured data according to AS/NZS 4399:1996 using a StarTek UV fabric protection application software version. One gram of PET fabric was immersed in 20 mL of 0.1 M TiO_2 colloidal solution for 5 min and dried at room temperature, followed by curing at 100°C for 30 min. Thereafter, the samples were washed in deionized water and dried before any physical and chemical characterization. The UPF value was found to enhance from 40 in the untreated sample to 50+ in the TiO_2-treated sample. There was a little decrease in the UPF value, when the sample was washed. However, when the sample was pretreated with alginate, followed by TiO_2 loading, the sample could maintain a 50+ rating even after washing. In addition to UV protection, the sample also exhibited an excellent antibacterial property against gram-negative *Escherichia coli* bacteria, despite five washing cycles. TiO_2 nanoparticles of 100 nm sizes, mixture of anatase, and rutile grade were added at three different levels (0.3%, 1%, and 3%) in the master batch PP, and a composite filament was developed in a melt-spinning machine to enhance its UV-blocking performance.[22] The physical and structural properties of the samples were characterized using scanning electron microscope (SEM), XRD, differential scanning calorimeter (DSC), tensile strength, and UPF value. The UV transmittance property was measured in SDL Atlas Camspec M350 UV–visible spectrophotometer after

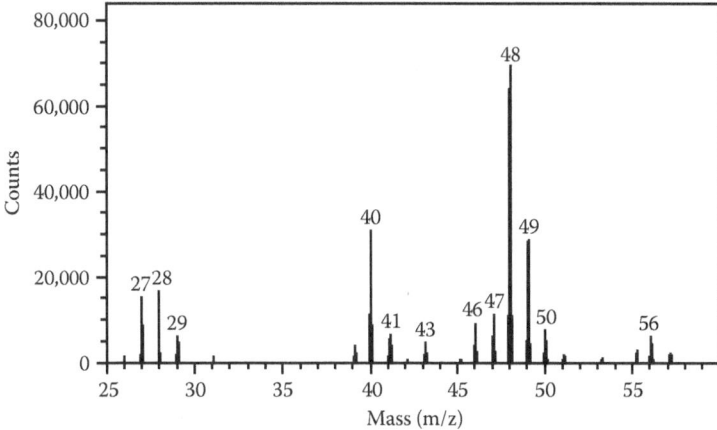

FIGURE 6.4 SIMS positive mass spectrum of TiO_2 nanoparticle–treated cotton textile. (From Samanta, K.K. et al., Cotton textile incorporating titanium dioxide nanoparticles and method to manufacture the same, Indian Patent Number: 3468/MUM/2012, 2012.)

scanning the sample over the wavelength of 200–400 nm. It was observed that after incorporation of the TiO_2 particles, there was no significant change in the fiber tensile strength and crystallinity, as the typical α form of PP crystals were observed in all the samples. In all the 3 samples with different loadings, the UPF value was 50+, whereas it was as low as 5 in the untreated sample. The UVA and UVB transmittances were almost zero in the 1% and 3% TiO_2-loaded samples compared to 18% and 13.9% in the untreated sample, respectively.[22]

In another study, TiO_2 nanoparticles were synthesized by the solgel method using a titanium isopropoxide precursor and applied by the pad-dry-cure method for a UV-protective, antimicrobial, and self-cleaning finishing of a cotton textile of 118 g/m².[32] When water was used for the synthesis of TiO_2 nanoparticles, it led to the formation of smaller particles of 7 nm compared to 12 nm that were observed in the case of an ethanol-based preparation. The synthesized nanoparticles were confirmed by the FTIR absorption band at 432 cm^{-1} for TiO_2. A similar result was also observed from the XRD peaks at 29.9, 47.9, 56.9, and 74.7 for TiO_2. The fabrics were finished with 1%, 1.5%, and 2% TiO_2 nanoparticles in the presence of 1% acrylic binder. There was only a marginal effect on antimicrobial activity with the increasing percentage loading of TiO_2. The antimicrobial efficacy was found to reduce slowly from 77.6% in the unwashed sample to 34.7% after 20 washes, according AATCC Method 61 (1996) test no. 2A, possibly due to the removal of some nanoparticles from the substrate. Similar result was also observed, where the UPF value get reduced from 44.8 to 18.8 in the unwashed to washed sample with 18 washing cycles.[32] As expected, the smaller-sized particles exhibited better antimicrobial and self-cleaning properties than the bigger ones. Thus, the TiO_2 finish could be effectively used with only a small amount for achieving a multifunctional textile.

The UV-blocking treatment of the cotton fabrics was developed by the solgel method. A thin layer of titanium dioxide was formed on the surface of the cotton fabric, which provided an excellent UV protection property. The effect was found to be durable, even after 50 home launderings. Apart from titanium dioxide, zinc oxide nanorods of 10–50 nm in length were also applied to the cotton fabric for a similar end use. The fabric treated with zinc oxide nanorods demonstrated an excellent UV-protective rating.[47] Silver (Ag) nanoparticles along with octyltriethoxysilane (OTES) were used for UV protective, antimicrobial, and superhydrophobic finishing of cotton textiles.[48] The Ag-coated fabric reacted with OTES to form a low surface energy layer on the fabric surface. It was observed that the untreated sample has a UPF value of 13, and it increased to 228 in the silver nanoparticle–treated sample and further to 266 when OTES coating was applied.

6.7 UV-PROTECTIVE FINISHING USING PLASMA

The wet-chemical processing of textiles is important for its improvement in aesthetic and functional value. However, such processes are water, chemical, and energy intensive due to the involvement of multistep operations, such as padding, drying, curing, and postwashing. In the last six decades, several technological advancements have taken place in the textile sector in order to economize the cost, improve quality, and take care of the environment. They constitute an application of enzymes, natural dyes, digital printing, low material-to-liquid processing, spray and foam finishing, infrared dyeing and drying, RF drying, ultrasound dyeing and dispersion, and water-free plasma processing.[49–51] Plasma, a partially ionized gas composed of both positive and negative ions, electrons, neutrals, excited molecules, photons, and UV light, can be used for nanoscale surface engineering of textile substrates, without altering the inherent bulk properties of the material. Plasma reaction with a small nonpolymerizing molecule causes surface activation, cleaning, oxidation, change in surface energy, increase in surface roughness/area, and etching. Similarly, plasma reaction with a bigger molecule leads to plasma polymerization, coating, deposition, and creation of nanostructures. These altogether improve the aesthetics and functional values of the textiles such as improvement in water absorbency, water repellency, oil absorbency, oil repellency, UV protection, antimicrobial, flame retardant, dyeing, desizing, antistatic, adhesion, and antifelting of wool.[51–55] The low-temperature plasma, also known as nonthermal plasma with the bulk temperature of 50°C–250°C, is used for such surface modification of textiles. Plasma is an energetic chemical environment, where the generation of plasma species opens up the diverse reactions resulting in various end applications in apparel, home, and technical textiles.

The UV-protective finishing of textiles is important to avoid skin ageing, sunburns, weak immune system, suntanning, photocarcinogenesis, and even skin cancer. In the past, various chemicals, such as silicate, micron- to submicron-sized metal or metal oxide particles, OBA, and natural dyes, have been explored to improve the UV-protective functionality of apparel grade textiles. However, in these processes, there is an involvement of water as a processing medium, followed by drying/curing operation. This makes the process water and energy intensive, in addition to the generation of a large quantity of liquid effluent. Therefore, the main attraction of plasma processing of textile is to accomplish the multistep operation in a single step while avoiding the usage of water. In this regard, HMDSO–ethanol pretreated cotton fabrics were plasma treated at an atmospheric pressure to obtain a thin, uniform Si–O coating on the fabric at a plasma power of 80 W for 60 s. The Si–O functional groups bonded to the cellulose by a Si–O–cellulose network in the plasma zone, resulting in better cohesion between the fiber and the chemical. The applied coating was found to be continuous and uniform; however, the fiber surfaces became much rougher than the uncoated fibers as observed under the SEM. The UV transmission in the plasma-treated cotton fabric was reduced by 10%–15% compared to the untreated sample due to the presence of such coating.[7] The UV blocking was prominent in the treated samples in the range of <425 to >350 nm wavelength. The finished sample was also durable to 20 numbers of washing cycles. The deposition increased the interfacial reflection and, hence, reduced UV transmission. Similarly, it has been reported that HMDSO plasma can also be used for UV-protective finishing of PET/cotton blended textiles.[56,57] It has also been reported that the UPF of the finished multifunctional textile was nearly as high as 2000 after plasma treatment, thus making it highly protective against UV radiation.[58]

The TiO_2 films can be deposited on the cotton knit wear using titanium tetrachloride ($TiCl_4$) as a precursor in an oxygen atmosphere. This deposition could enhance the antimicrobial activity and UV blockage in synthetic fibers also.[59] The dielectric barrier discharge (DBD) air plasma treatment for 2–10 s with plasma power of 1.3 W was used for the surface activation of the PET fabric to facilitate the deposition of aluminum oxide (Al_2O_3), nano-silver (Ag), and nano-titanium dioxide (TiO_2) for producing the multifunctional textiles.[60] Aluminum oxide with particle size of 100–200 nm and the titanium dioxide and silver with <100 nm were used in the study. Both SEM and TEM results showed the presence of Al_2O_3 and nanoparticles of TiO_2 and Ag on the fiber surface, even after several washing cycles. All the fabrics treated with plasma Al_2O_3, by spreading the powder either on the sample or on the electrode disk, did not wet the fabric. It was observed that the untreated sample had a UPF value of 8, and it was found to increase to 16 and 48.3 in the air plasma Al_2O_3- and the TiO_2-treated samples, respectively.

6.8 SUSTAINABLE UV-PROTECTIVE FINISHING OF TEXTILE

6.8.1 SUSTAINABLE UV-PROTECTIVE TEXTILE USING PLANT EXTRACT

The UV-protective finishing of textile has become an important area of research for developing, particularly, health-care textiles. In the past, UV-protective textiles have been developed by altering the fabric structural parameters, using natural color and pigments, UV absorber, nanoparticles, and plasma-induced nanocoating, as discussed earlier. In recent years, due to the global awareness of eco-/green textile products, research has been intensified for the development of UV-protective textile using a number of plant molecules (extract) that are applied to the natural fiber–based textile substrate. A few recent studies have also reported the presence of active molecules in some of the natural dyes that can absorb UV light, which do not allow UVR to pass through the fabric. Earlier, a group of European researchers has reported that the dyes extracted from the madder woods, knotgrass, fenugreek, and marigold possess a very good UV-protective property. They have dyed linen, hemp, and silk fabrics with these natural plant molecules. The linen fabric was dyed with India madder, which showed a UPF rating of more than 50, considered to be an excellent.[61] Similarly, in a few recent studies from India, it has been reported that grapefruit oil enhanced the UV-protective property of bamboo, Tencel fiber, and the regenerated blended fabrics. The honeysuckle-extracted molecules also provide a good UV-protective attribute to the wool fabric. Another type of textile, such as pomegranate and alum-mordanted lignocellulosic-bleached jute fabrics, was made UV protective after dyeing with babool, manjistha, annatto, and ratanjot extracts.[62] The UV-protective property of these textiles is reported in Table 6.5.

From Table 6.5, it can also be observed that the plain woven bleached cotton fabric showed a poor UPF value of 10, as the majority of the incident UVRs can easily pass through the fabric. After mordanting with tannic acid and alum, the UPF value does not improve much. However, when the mordanted fabric was treated with an alkaline BPS, an agricultural waste material available in India in large quantity, the UPF value was found to improve more than 100 (i.e., a 50+ rating). Furthermore, the UVA and the UVB transmittance percentages get reduced drastically from 9.5% to 7.2% in the untreated samples to 1.2% and 1% in the BPS-treated samples, respectively. On the other hand, if the bleached fabric is directly treated with the alkaline BPS solution without any mordant, it showed a lower UPF value of 40. From these findings, it can be said that the improvement in UPF and the corresponding decrease in UV (A and B) transmission percentages are arising mainly due to the synergistic effect of the BPS and the mordant. The UV protection of the BPS-treated sample is arising possibly due to the presence of N,N-alkylbenzeneamine as confirmed by GC–MS analysis of the BPS.[65,66] The durability of the imparted finish was investigated after washing the sample as per the ISO 1 standard. It was found that the UPF value was 70 and 55 after 1st and 2nd washes, respectively. This indicates that there was no significant decrease in the UPF in the successive washing cycles. Similarly, the SJ treated on the bleached cotton fabric in alkaline condition has a UPF value of 50+ (125), and the UVA and UVB transmission percentages reduced to 2% and 1.8%, respectively. The improvement in the UV protection in that sample was possibly due to the presence of organic color and silicate molecules in the SJ. It was interesting

TABLE 6.5
UV-Protective Performance of Different Plant Molecule–Treated Fabrics

Samples	UPF Mean	UPF SD	UPF Rating	UVA Transmittance (%)	UVB Transmittance (%)
Bleached jute fabric	16	2.5	10	10.7	7.1
Mordant-treated jute fabric	17	3.9	15	7.2	5.7
Babool-treated jute fabric	43	5.2	40	2.7	2.4
Manjistha + mordant-treated jute fabric	44	4.8	40	2.9	2.4
Annatto + mordant-treated jute fabric	36	3.9	35	3.8	3.0
Ratanjot + mordant-treated jute fabric	21	5.8	20	5.6	4.8
Grapes oil–treated cotton fabric	—	—	50+	—	—
Bleached cotton fabric	11	3.2	10	9.5	7.2
Mordant-treated cotton fabric	14	4.2	10	7.5	6.2
BPS-treated cotton fabric	43	5.2	40	2.7	2.4
BPS + mordant-treated cotton fabric	145	8.2	50+	1.2	1.0
SJ-treated cotton fabric	125	9.0	50+	2.0	1.8

Sources: Chattopadhyay, S.N. and Pan, N.C., *J. Text. Inst.*, 104(8), 808, 2013; Basak, S. et al., *Pol. J. Chem. Technol.*, accepted, 2014; Smanata, K.K. et al., Eco-friendly coloration and functionalization of textile using plant extracts, in: Muthu, S.S., ed., *Roadmap to Sustainable Textiles and Clothing*, Springer, 2014, pp. 263–287.

to note that these two green materials could also improve the thermal stability of the textile. BPS and SJ were applied on cotton and jute woven fabrics, and the limiting oxygen index (LOI) was found to increase to over 30 from the value of 18 and 21 in the untreated cotton and jute samples, respectively. The BPS-treated cotton fabric burns with flame only for 4 s, followed by burning with afterglow (after flame stopped) for 900 s. Hence, the total burning time of a 250 mm × 20 mm test specimen becomes 904 s, whereas the untreated sample of the same size burnt completely within 60 s with the flame.[63] The SJ, a vegetable extract, has also been utilized for a similar flame-retardant finishing of the cotton textile without any mordanting.[64] This plant (*Spinacia oleracea*) is available in many of the countries and it is rich in phosphorous, nitrogen, and other metallic constituents. In the 8% add-on treated sample, the LOI was found to increase to 30 from 18 in the control sample. It was interesting to note that the sample did not catch flame and the afterglow was present only for 400 s, whereas, as discussed earlier, the control sample completely burnt within 60 s with a high temperature of 400°C–450°C. As the treated sample burnt slowly, the warp way burning rate decreased from 75 mm/min in the control sample to 10 mm/min in the SJ-treated sample. Unlike the UPF result, there was a decrease in the LOI from 30 to 22 after washing of the sample with the soap solution. From the various characterizations, it was postulated that the presence of phosphate, silicate, chloride, and other mineral salts in the BPS and SJ might have increased the thermal stability of the cellulosic and lignocellulosic textiles by the formation of more char while ensuring production of less flammable volatiles.

Salah reported the utilization of banana peel saps, an agro-waste for improving the UV-protective property of the cotton fabric.[67] The sap-treated fabric also exhibited excellent antimicrobial properties and a natural color to the textile, in addition to UV protection. They have reported that the cotton fabric made of Egyptian fiber showed a UPF of around 19.8, whereas the same fabrics after mercerization and sap dyed exhibited a UPF of around 60. The reduction in UV transmittance through the fabric might be due to the combined effect of reduction in fabric porosity due to the alkaline swelling of the fiber after mercerization, mordant–fabric complex formation, and the presence of luteolin in banana peel saps. The antimicrobial efficacies in the untreated fabric against both gram-positive and gram-negative bacteria were only 65% and 32%, respectively, due to the absence of any antimicrobial agent. In contrast, the bleached, mordanted, and banana peel sap–treated cotton fabric showed more than 90% bacterial reduction against both these bacteria. The ferrous sulfate was found to be the best mordant for improving the color of the cotton fabric using the banana peel sap.

Sarkar has reported that madder (*Rubia tinctorum*) and indigo (*Indigofera tinctoria*) extracts have good UV-protective property.[68] To understand the mechanism of action of UV protection, they have applied these extracts on the alum-mordanted plain and twill woven cotton textiles with an initial UPF of 3.8 and 19.2, respectively. It was observed that 6% madder extract and indigo-dyed plain woven cotton fabric showed a UPF of 16.6 and >50, respectively. It might be due to the fact that the dark blue color of indigo dye has helped in improving the UPF value of the treated fabric. When correlated, they observed that the UPF value increased with the increasing color strength of the treated fabric. In contrast, the twill woven cotton fabric showed a UPF of more than 50, when it was dyed with both the extracts, possibly due to higher thickness, more fabric cover, and initial high UPF value of the fabric. Similarly, Subramaniyan et al. have reported that the anthocyanin extracted from the mulberry fruit could improve the UV protection property of the cotton fabric.[69] As expected, the higher concentration of anthocyanin at lower pH of 4.5 helped to provide a better UPF. They have also observed that the crude mulberry extract could provide the same degree of UV protection, when applied at an acidic pH with the same concentrations. Recently, Vijayalakshmi and Ramachandran reported the use of sweet spring citrus oil for its application in antimicrobial, antifungal, and UV-protective finishing of lightweight denim fabrics, such as 100% cotton, 100% Tencel, cotton/PET blend, and Tencel/PET blend.[70] The fabrics were initially treated with the enzyme, followed by citrus oil application by the exhaust method for 30 min at room temperature. The pH and the material-to-liquid ratio of the solution of citrus oil were kept at around 6 and 1:5, respectively. Both the Tencel/PET and cotton/PET blends exhibited better antimicrobial efficacy compared to the 100% cotton and the 100% Tencel fabrics, when tested against both *S. aureus* (gram-positive) and *K. pneumoniae* (gram-negative) bacteria. As far as the UPF is concerned, 100% cotton and Tencel/PET blended fabrics showed higher UPF values of 40–50, compared to the 100% Tencel and the cotton/PET blended fabrics, with UPF of 20–30. As the denim garments are worn for a sufficient period of time, such multifunctional antimicrobial and UV-protective finishes will be beneficial for producing eco-friendly textile products.

Rungruangkitkrai et al. studied the effect of the eucalyptus leaf extract on the UV protection of the protein fabric (wool). The dye was extracted from the eucalyptus leaves and applied to a premordanted wool fabric by the pad dry method.[71] It was found that by increasing the concentration of the dye, the UPF was also improved. The initial fabric showed a UPF of 10, which, after treatment with 20 g/L dye solution, but without any mordanting, increased to 48.5. However, if the fabric was premordanted, the UPF was found to improve further. The ferrous sulfate mordant was found to provide an excellent UPF value in the treated fabric compared to the nonmordanted sample. The treated fabric provides good wash durability due to the presence of tannin in the eucalyptus leaf extract, whose phenolic structure might have helped to form metal chelate in the presence of the mordant prior to dyeing. In this regard, Mongkholrattanasit et al. reported that besides tannin, the eucalyptus leaf extract also contains rutin and quercetin as the coloring matter that might have also helped in improving the color strength, simultaneously with the improvement in the UV-protective performance.[72] Similar to wool, the dyeing and UV-protective finishing of silk with vegetable extract of flos sophorae have also been investigated.[73] Alum premordanted silk fabric showed a UPF of 69, when dyeing was carried out at 70°C for 45 min. It was observed that the treated fabric showed a higher UPF value, when the dyeing was carried out at a higher temperature of 100°C for a shorter duration of 30 min while maintaining the same material-to-liquid ratio of 1:10. As expected, the UPF was found to increase with the increasing concentration of color of the vegetable extract. In addition to natural color and UV protection, this plant-extracted solution also exhibited very good thermal stability as an additional functionality. In the treated sample, the color fastness to washing, perspiration, and hot-pressing was also found to be satisfactory. Przewozna and Kowalinki have reported an interesting research on the application of plant extract to improve the UV protection of linen, hemp, and knitted silk fabrics.[74] The research group has used different colors of the natural dyes, like the yellow color of turmeric (*Curcuma longa*), blue color of indigo (*Isatis tinctoria*), and red color of madder (*R. tinctorum*) and also lac (*Kerria lacca*), henna (*Lawsonia inermis*), and so on. Different types of the mordants, such as alum, citric acid, and washing soda, were used. After application of such natural colors, the UPF value of the linen fabric was reported to improve to 30, when it was mordanted with washing soda and dyed with lac. On the other hand, the unmordanted lac-dyed fabric showed a UPF of around 10. This signifies that mordant plays a vital role in UV-protective finishing. More or less the same result has also been observed in the BPS-treated sample. When the same fabric was mordanted with alum, followed by dyeing with madder, the sample showed a UPF of 25. A similar improvement was also observed when the sample was mordanted with copper sulfate, followed by dyeing with yellow color turmeric dye. It was interesting to note that the linen fabric without mordanting but dyed with henna showed a very good UPF value of 35. As far as hemp fabric is concerned, the nonmordanted turmeric-dyed sample showed a low UPF value of 15, and it improved to 30 upon mordanting with copper sulfate prior to dyeing.

6.8.2 Sustainable UV-Protective Textile Using Nanoparticles of Natural Polymers

Similar to synthesis and application of metal oxide nanoparticles in textile, the organic polymeric nanomaterial, that is, nanolignin, has also been attempted for UV protection, antimicrobial, and antistatic finishing of cellulosic textiles.[75] Natural fibers like hemp and flax contain lignin as one of their fiber-forming chemical. Natural pigments and lignin in the natural fibers act as a UV absorber and ensure a good protection. Nanolignin was applied by the padding technique for several times with silicone emulsion (5–25 g/L) at a temperature of 18°C–20°C for 2–5 min, followed by drying at 40°C–60°C. The coated fabric could maintain a very good UV-protective characteristic even after washing. There was no significant change in the fabric stiffness and air permeability due to such application. Nanolignin was obtained from kraft lignin by ultrasound treatment, and the particle size was measured to be in the range of 5–170 nm. However, the majority of the particles were below 40 nm in size.[2] The highest achievable UPF was 25 in the treated linen sample compared to 5 in the untreated sample. The presence of silicone emulsion with lignin helped in better UV protection (UPF > 45), by the improved fixation of particles in the fabric structure. However, the improvement was not so promising in the case of fabric made from hemp. It was observed that a plasma treatment with a power of 2000 W for 5 min in an oxygen mixture was useful for surface activation, prior to application of nanolignin in the linen fabric.[76,77] In the case of a flax nonwoven fabric, the lignin-treated sample showed almost double the UPF value (50) compared to the unmodified sample.[77,78] An antibacterial test was conducted in the nanolignin-treated linen fabrics against eight bacteria cultures, which are mostly found in the environment. To improve the fixation of particles in the fabrics, different polymeric binder dispersions, such as acrylic, polyurethane, vinyl acetate, and dimethylol dihydroxy ethylene urea, have also been used.[77] As lignin is produced from a renewable source, it is quite cost-effective and can be used for sustainable UV-protective finishing of textiles.

6.8.3 Sustainable UV-Protective Textile Using Natural Polymer

Similar to the UV-protective finishing of textiles using nanolignin and other plant extracts, the silk sericin has also been explored for the UV-protective finishing of various fibrous materials. Silk derived from the silkworm *Bombyx mori* is composed of two major proteins: fibroin and sericin. Sericin or silk glue is a globular protein, which has around 25%–30% of silk proteins. It consists of 18 amino acids, and the majority of these have strong polar side chains consisting of hydroxyl, carboxyl, and amino groups. Its high hydrophilic characteristic originates from the high content of serine and aspartic acid, approximately in the proportions of 33.4% and 16.7% of the sericin present in silk.[79] The silk sericin has the potential for biomedical applications, such as an excellent oxygen permeability, cell protection, antioxidant action, moisture-regulating ability, protection from UV radiation and microbes, wound healing, and anticancer and anticoagulant properties.[8,79,80] The sericin generated in the degumming process of silk and discharged as an effluent is a by-product that can be tailored for application in improving the moisture content, UV absorption, and antimicrobial properties of textiles.[8] The PET fabric, after washing with Lissapol N, was treated in both sides under an excimer lamp for 1, 3, 5, 7, 10, 12, and 15 min at a distance of 5 mm in atmospheric conditions and then was functionalized with sericin. The UPF value of the treated fabric was measured as per the U.S. AATCC Test Method 183-2000, and it was found to increase to 125 ± 6 in the treated sample from 55 ± 5 in the untreated sample.[80] The sericin acts as a UV absorber, thus converting the electronic excitation energy into the thermal energy. The high-energy, short-wavelength UVRs excite the UV absorber to a higher energy state, and then, the absorbed energy possibly gets dissipated in the longer wavelength, thus avoiding skin damage. In addition to UV protection, the sericin-treated sample also exhibited very good hydrophilic property, where the moisture regain value was found to improve from 0.6% to 2.3% in the PET sample. When the antioxidant property was evaluated, the radical scavenging activity was found to increase by 56% over the control sample. The antistatic property was measured qualitatively, in which less ash was found to form in the treated sample. Aloe vera (*Aloe barbadensis* Miller), belonging to the family Liliaceae, has been utilized for its traditional medicinal properties including cosmetic application. It has an excellent skin care property that includes anti-inflammatory and antiageing attributes. Kimberly-Clark Inc., Ltd., has patented an aloe vera application as an antiageing and moisturizing agent. Similarly, DyStar Auxiliaries GmbH has developed a textile product containing a mixture of vitamin E, aloe vera, and jojoba oil in a silicon matrix for moisturizing and UV-protective finishing of different textile substrates.[8] As described earlier, the natural protein polymers, sericin, and aloe vera are the important functional materials produced from natural resources and hence can be effectively used for eco-friendly and sustainable application in textiles and cosmetics.

6.9 SUMMARY

The UV-protective finishing of textiles is an emerging area of research and product development to mitigate with the harmful effects of UV radiation. Approximately 6%–7% of the total solar emission reaching to the Earth is composed of UV radiation, which is composed of UVB (280–320 nm) and UVA (320–400 nm). Hence, a UV-protective finishing of textile needs to be designed in such a way as to absorb or restrict the transmission of 280–400 nm wavelengths of UV radiation. Transmittance of such radiation to the skin through the fabric causes accelerated skin ageing, sunburn, blotches, wrinkles, weak immune system, suntanning, photocarcinogenesis, skin cancer, and eye damage. Today's young generation is more concerned and caring about their skin and does not like any patchy and black/brown spot on their skin. A little amount of UV exposure from sunlight/sunbath has been taken into consideration as an appropriate action, which promotes the circulation of blood, invigorates the metabolism, improves resistance to various pathogens, and generates vitamin D. On the other hand, excessive UV exposure causes the aforementioned health hazards. To get rid of the adverse effect of UVRs, people frequently use sun cream, sun lotion, and UV-protective goggle, umbrella, hat, and fabric. The UV-protective performance of a textile is governed by the chemistry of the fiber-forming polymer, physicochemical properties of the fiber, the presence of UV absorbers, construction of the fabric (woven, knitted, nonwoven, thickness, porosity, CF), extension, moisture content, color, imparted finishes, and previous history of the fabric. Though heavier (high thickness) fabric and/or dark color fabric can provide good protection against UVRs, people longing for some exposure to sunlight/sunbath in the summer season prefer to have lightweight and light-colored apparels. As the majority of the apparel grade light color textile show a UPF value < 10 (i.e., no protection), the UV-protective finishing of such textiles is getting considerable attention. Traditionally, both natural and synthetic textiles were treated with 0.6%–2.5% OBA, 2-ethylhexyl-4-methoxycinnamate, and a derivative of o-hydroxybenzophenone for UV protection. However, the rapid growth of nanoscience and its application in textiles in recent years has proved a boon, and it has been found that the application of only 0.5%–3% metal or metal oxide nanoparticles, such as ZnO, TiO_2, Ag, and Cu, can help to achieve a UPF value of 50+ (i.e., excellent rating) with simultaneous improvement in the antimicrobial, self-cleaning, and antifungal functionalities. Plasma-enhanced nanocoating, nanoparticle deposition, and surface activation have also demonstrated the production of UV-protective textiles without using water as a processing medium. Further, due to the global awareness of eco-friendly and sustainable textile processes and products more recently, various natural materials including biomaterials, such as agro-waste, lignin in nanoform, silk sericin, aloe vera, and natural dyes, have been used in textile processing and finishing. As these materials are produced from renewable sources, easy to apply, cost-effective, and safer to human beings and the environment, they can be used for the development of sustainable UV-protective textiles while ensuring additional improvement in antibacterial, antistatic, and antioxidant properties.

REFERENCES

1. Menter, J.M., Hatch, K.L. 2003. Clothing as solar radiation protection. In: *Textiles and the Skin. Current Problems in Dermatology*, Vol. 31. Elsner, P., Hatch, K., eds. Karger, Basel, Switzerland, pp. 50–63.
2. Zimniewska, M., Batog, J. 2012. Ultraviolet-blocking properties of natural fibres. In: *Handbook of Natural Fibres. Processing and Application*, Vol. 2. Kozlowski, R., ed. Woodhead Publishing Limited with the Textile Institute, Cambridge, pp. 141–167.

3. Mao, Z., Shi, Q., Zhang, L. 2009. The formation and UV blocking property of needle-shaped ZnO nanorod on cotton fabric. *Thin Solid Films* 517: 2681–2686.
4. Yadav, A. et al. 2006. Functional finishing in cotton fabrics using zinc oxide nanoparticles. *Bulletin of Material Science* 29(6): 641–645.
5. Samanta, K.K. et al. 2012. Cotton textile incorporating titanium dioxide nanoparticles and method to manufacture the same. Indian Patent Number: 3468/MUM/2012.
6. Samanta, K.K., Basak, S., Chowdhury, P. 2014. Nano-technology for advanced textile application. *Man Made Textiles in India* XLII: 55–59.
7. Wang, Y. 2011. The uniform Si-O coating on cotton fibers by an atmospheric pressure plasma treatment, *Journal of Macromolecular Science Part B: Physics* 50(9): 1739–1746.
8. Gulrajani, M.L. 2008. Bio- and nanotechnology in the processing of silk. http://www.fibre2fashion.com/industry-article/pdffiles/16/1517.pdf.
9. Saravanan, D. 2007. UV protection textile materials. *AUTEX Research Journal* 7(1): 53–62.
10. Reinert, G., Fuso, F., Hilfiker, R., Schmidt, E. 1997. UV protecting properties of textile fabrics and their improvement. *AATCC Review* 29(12): 31–43.
11. Bajaj, P., Kothari, V.K., Ghosh, S.B. 2000. Some innovations in UV protective clothing. *Indian Journal of Fibres and Textile Research* 35(4): 315–329.
12. Schinder, W.D., Hauser, P.J. (ed.) Ultraviolet protection finishes. In: *Chemical Finishing of Textiles*. Woodhead Publishing in Textiles, pp. 157–164.
13. El Zaher, N.A., Kishk, S.S. 1996. Study of the effect of UVR on the chemical structure, mechanical properties and crystallinity of Nylon-6 films. *Colourage* 11: 25–30.
14. Hunt, R. 2003. Opportunities in UV protection. *Knitting International* 2: 51–53.
15. Kayavirasu, R. 2010. Application of nano technology. fibre2fashion.com, Retrieved from www.fibre2fashion.com/industry…/application-of-nano-technology1.asp. Accessed January 21, 2015.
16. Kursun, S., Ozcan, G. 2010. An investigation of UV protection of swimwear fabrics. *Textile Research Journal* 80(17): 1811–1818.
17. Saleh, A.F.A. 2011. Using nanotechnology in the finishing of cellulosic fabrics, Thesis. Chemistry Department, University of Duisburg-Essen, Essen, Germany.
18. Gies, H.P., Roy, C.R., Holmes, G. 2000. Ultraviolet radiation protection by clothing: Comparison of in vivo and in vitro measurements. *Radiation Protection Dosimetry* 91: 247–250.
19. Menzies, S.W. et al. 1991. A comparative study of fabric protection against ultraviolet-induced erythema determined by spectrophotometric and human skin measurements. *Photodermatology Photoimmunology* 8: 157–163.
20. Pailthorpe, M. 1998. Apparel textiles and sun protection: A marketing opportunity or a quality control nightmare. *Mutation Research* 422: 175–183.
21. Seungsin, L. 2009. Developing UV-protective textiles based on electrospun zinc oxide nanocomposite fibers. *Fibers and Polymers* 10(3): 295–301.
22. Erdem, N. et al. 2010. Structural and ultraviolet-protective properties of nano-TiO_2-doped polypropylene filaments. *Journal of Applied Polymer Science* 115(1): 152–157.
23. Grace, W. et al. 2010. Improving UV protection of cotton fabrics through digital textile printing. In: *Beltwide Cotton Conferences*, New Orleans, LA.
24. Gogoi, S., Baruah, B., Sarkar, C.R. 1999. Effect of ultraviolet light on silk fabric. *Colourage* 46(2): 23–29.
25. Holme, I. 2003. UV absorbers for protection and performance. *International Dyer* 188(4): 9–10.
26. Gupta, D., Jain, A., Panwar, S. 2005. Anti UV and antimicrobial properties of some natural dyes on cotton. *Indian Journal of Fibre and Textile Research* 30(2): 190–195.
27. Hustvedt, G., Crews, P.C. 2005. The ultraviolet protection factor of naturally-pigmented cotton. *The Journal of Cotton Science* 9: 47–55.
28. Gantz, G.M., Sumner, W.G. 1957. Stable ultraviolet light absorbers. *Textile Research Journal* 27(3): 244–251.
29. Maged, H. et al. 2009. Novel UV-protective formulations for cotton, PET fabrics and their blend utilizing irradiation technique. *European Polymer Journal* 45(10): 2926–2934.
30. Samanta, K.K., Ray, D.P. 2014. Synthesis and application of nanomaterial for processing and finishing of textile. In: *Jute and Allied Fibres—Processing and Value Addition*, Nag, D. and Ray, D.P., eds. New Delhi Publisher, New Delhi, India, pp. 225–230.
31. Kathirvelu, S., D'Souza, L., Dhurai, B. 2009. UV protective finishing of textile using ZnO nanoparticles. *Indian Journal of Fibre and Textile Research* 34: 267–273.

32. Sundaresan, K. et al. 2012. Influence of nano titanium dioxide finish, prepared by sol-gel technique, on the ultraviolet protection, antimicrobial, and self-cleaning characteristics of cotton fabrics. *Journal of Industrial Textiles* 41(3): 259–277.
33. Chowdhury, P., Samanta, K.K., Basak, S. 2014. Recent development in textile for sportswear application. *International Journal of Engineering Research and Technology* 3(5): 1905–1910.
34. Dastjerdi, R., Montazer, M. 2010. A review on the application of inorganic nano-structured materials in the modification of textiles: Focus on anti-microbial properties. *Colloids and Surfaces B: Biointerfaces* 79(1): 5–18.
35. Shafei, A.E., Okeil, A.A. 2010. ZnO/carboxymethyl chitosan bionano-composite to impart antibacterial and UV protection for cotton fabric. *Carbohydrate Polymer* 83(2): 920–925.
36. Becheri, A., Dürr, M., Baglioni, P. 2008. *Journal of Nanoparticle Research* 10: 679–689.
37. Riva, A., Algaba, I., Pepio, M. 2006. Modelisation of the effect. *Cellulose* 13: 697–704.
38. Gopalkrishnan, D., Maithili, K.G. Functional nano-finish for textile, Retrieved from http://www.fibre2fashion.com/industry-article/technology-industry-article/functional-nano-finishes-for-textiles/functional-nano-finishes-for-textiles1.asp
39. Deshpande, R.H. 2011. Coating of cotton fabrics with acrylate nano ZnO composite for imparting UV Protection Finish. In: *Proceedings of the Seventh International Conference on ATNT*.
40. Vigneshwaran, N. et al. 2006. Functional finishing of cotton fabrics using zinc oxide–soluble starch nanocomposites. *Nanotechnology* 17: 5087–5095.
41. Hongying, Y. et al. 2004. Studying the mechanisms of titanium dioxide as ultraviolet-blocking additive for films and fabrics by an improved scheme. *Journal of Applied Polymer Science* 92: 3201–3210.
42. Wang, H., Wu, Y., Xu, B.Q. 2005. Preparation and characterization of nano sized TiO_2 for photo catalysis. *Applied Catalysis B* 59: 139–145.
43. Hashemiza, S. et al. 2012. Nano TiO_2 on alkali modified polyester fabric with improved antibacterial properties. In: *Proceedings of the Fourth International Conference on Nanostructures (ICNS4)*, Kish Island, Iran.
44. Samanta, K.K. et al. 2012. UV protective finishing of cotton textile using TiO_2 nanoparticles. In: Proceeding of *in APA International Congress on Advances in Human Healthcare Systems*, Healthcare India, New Delhi, India.
45. Samanta, K.K. et al. 2012. Development of UV protective healthcare textile using TiO_2 nanoparticles. In: *Oral Talk in National Seminar on Value Added Apparel and Home Textile from Natural Fibres*, Mumbai, India.
46. Mihailovic, D. et al. 2010. Functionalization of polyester fabrics with alginates and TiO_2 nanoparticles. *Carbohydrate Polymers* 79: 526–532.
47. Wong, Y.W.H. 2006. Selected applications of nanotechnology in textiles. *AUTEX Research Journal* 6(1): 1–8.
48. Khalilabad, M.S., Yazdanshenas, M.E. 2013. Fabrication of superhydrophobic, antibacterial and ultraviolet-blocking cotton fabric. *The Journal of the Textile Institute* 104(8): 861–869.
49. Samanta, K.K., Basak, S., Chattopadhyay, S.K. 2014. Environment-friendly textile processing using plasma and UV treatment. In: *Roadmap to Sustainable Textiles and Clothing*, Muthu, S.S., ed., Springer, Singapore, pp. 161–201.
50. Muthukumar, M. et al. 2004. Statistical analysis of the effect of aromatic, azo and sulphonic acid groups on decolouration of acid dye effluents using advanced oxidation process. *Dyes and Pigments* 63: 199–304.
51. Samanta, K.K., Jassal, M., Agrawal, A.K. 2009. Improvement in water and oil absorbency of textile substrate by atmospheric pressure cold plasma treatment. *Surface Coating and Technology* 203: 1336–1342.
52. Samanta, K.K., Jassal, M., Agrawal, A.K. 2010. Antistatic effect of atmospheric pressure glow discharge cold plasma treatment on textile substrates. *Fibers and Polymers* 11(3): 431–437.
53. Samanta, K.K. et al. 2012. Study of hydrophobic finishing of cellulosic substrate using He/1,3-butadiene plasma at atmospheric pressure. *Surface and Coating Technology* 213: 65–76.
54. Wakida, T. et al. 1993. Characterization of wool and polyethylene terephthalate fabrics and film treated with low temperature plasma under atmospheric pressure. *Textile Research Journal* 63: 433–438.
55. Panda, P.K., Rastogi, D., Jassal, M. 2012. Effect of atmospheric pressure helium plasma on felting and low temperature dyeing of wool. *Journal of Applied Polymer Science* 124(5): 4289–4297.
56. Shah, J.N., Shah, S.R. 2013. Innovative plasma technology in textile processing: A step towards green environment. *Research Journal of Engineering Sciences* 2(4): 34–39.
57. Desai, A.A. January 2008. Plasma technology: A review. *The Indian Textile Journal*.
58. Shahidi, S., Ghoranneviss, M. Method of preparation of multifunctional technical textile by plasma-treatment, US 8632860 B2, Publication date: January 21, 2014, Application Number US 13/287,447.

59. Hegemann, D. 2006. Plasma polymerization and its application in textiles. *Indian Journal of Fibre and Textile Research* 31: 99–115.
60. Raslan, W.M. et al. 2011. Ultraviolet protection, flame retardancy and antibacterial properties of treated polyester fabric using plasma-nano technology. *Materials Sciences and Applications* 2: 1432–1442.
61. Sun, S.S., Tang, R.C. 2011. Adsorption and UV protection properties of the extent form honeysuckle into wool. *Industrial Engineering and Chemistry Research Journal* 50: 4217–4219.
62. Chattopadhyay, S.N., Pan, N.C. 2013. Development of natural dyed jute fabric with improved colour yield and UV protection characteristics. *The Journal of the Textile Institute* 104(8): 808–818.
63. Basak, S. et al. 2015. Flame resistant cellulosic substrate using banana pseudostem sap. *Polish Journal of Chemical Technology*, 17(1) (in press).
64. Smanata, K.K., Basak, S., Chattopadhyay, S.K. 2014. Eco-friendly coloration and functionalization of textile using plant extracts. In: *Roadmap to Sustainable Textiles and Clothing*, Muthu, S.S., ed., Springer, Singapore, pp. 263–287.
65. Sayed, M.E. et al. 2001. Identification and utilisation of banana plant juice and its pulping liquors as anticorrosive materials. *Journal of Scientific and Industrial Research* 60: 738–747.
66. Katarzyna, P., Prezewozna, S. 2009. Natural dyeing plants as a source of compounds protecting against UV irradiation. *Herba Policia* 55: 56–59.
67. Salah, S.M. 2011. Antibacterial activity and ultraviolet (UV) protection property of some Egyptian cotton fabrics treated with aqueous extraction from banana peel. *African Journal of Agricultural Research* 6(20): 4746–4752.
68. Sarkar, A.K. 2004. An evaluation of UV protection imparted by cotton fabrics dyed with natural colorants. *BMC Dermatology* 4(15): 1–8.
69. Subramaniyan, G., Sundarmoorthy, S., Andiappan, M. 2013. Ultraviolet protection property of mulberry fruit extract on cotton fabrics. *Indian Journal of Fibre and Textile Research* 38: 420–423.
70. Vijayalakshmi, D., Ramachandran, T. 2013. Application of natural oil on light weight denim garments and analysis of its multifunctional properties. *Indian Journal of Fibre and Textile Research* 38: 309–312.
71. Rungruangkitkrai, N. et al. 2012. UV protective property of the wool fibre dyed with eucalyptus extract by padding technology. In: *RMUTP Conference: Textile and Fashion*, Bangkok, Thailand.
72. Mongkholrattanasit, R., Punrattanasin, M. 2012. Properties of wool fabric dyed with eucalyptus, rutin, quercetin and tannin by padding technique. In: *RMUTP Conference: Textile and Fashion*, Bangkok, Thailand.
73. Wang, N., Jia, S., Zhou, Q. 2009. Research on dyeing and ultraviolet protection of silk fabric using vegetable dyes extracted from Flos Sophorae. *Textile Research Journal* 79(15): 1402–1409.
74. Przewozna, K.S., Kowalinki, J. 2008. Light fastness property and the UV protective factor of naturally dyed linen, hemp and silk fabric. In: *International Conference of Flax and Other Bast Plants*, Poznan, Poland.
75. Kozlowski, R., Zimniewska, M., Batog, J. 2008. Cellulose fibre textiles containing nanolignins, a method of applying nanolignins onto textiles and the use of nanolignins in textile production. WO 2008140337 A1, Application Number PCT/PL2007/000025.
76. Zimniewska, M., Kozlowski, R., Batog, J. 2008. Nanolignin modified linen fabric as a multifunctional product. *Molecular Crystal Liquid Crystal* 484: 409–416.
77. Zimniewska, M. et al. 2012. Functionalization of natural fibres textiles by improvement of nanoparticles fixation on their surface. *Journal of Fiber Bioengineering & Informatics* 5(3): 321–339.
78. Kozłowski, R. et al. 2012. Nanolignin as an effective UV blocker for fabrics, nonwoven and polymers foils. *Scientific Israel—Technological Advantages* 14(1): 69–73.
79. Aramwit, P., Siritientong, T., Srichana, T. 2012. Potential applications of silk sericin, a natural protein from textile industry by-products. *Waste Management & Research* 30(3): 217–224.
80. Gupta, D., Chaudhary, H., Gupta, C. 2014. Sericin-based polyester textile for medical applications. *The Journal of the Textile Institute* 105(5): 1–11.

Section II

Environmental and Social Assessment of Apparel Manufacturing

7 Carbon Footprint of Textile and Clothing Products

*Sohel Rana, Subramani Pichandi,
Shabaridharan Karunamoorthy, Amitava Bhattacharyya,
Shama Parveen, and Raul Fangueiro*

CONTENTS

7.1 Introduction	141
7.2 Greenhouse Gases	142
7.2.1 Types of Greenhouse Gases	143
7.2.2 Sources of Greenhouse Gases	144
7.3 Carbon Footprint	145
7.3.1 Definition of Carbon Footprint and Other Related Parameters	145
7.3.1.1 Global Warming Potential	145
7.3.1.2 Carbon Footprint	145
7.3.1.3 Energy Intensity	146
7.3.2 Global Carbon Footprint and Its Effects	146
7.4 Carbon Footprint of Various Textile Processes	147
7.5 Carbon Footprint of Natural Fibers and Their Products	149
7.5.1 Carbon Footprint of Cotton Fiber Products	152
7.5.1.1 Carbon Footprint of White Long Shirt	153
7.5.1.2 Comparison of Carbon Footprint of Different Cotton Products	155
7.5.2 Carbon Footprint of Wool Fiber and Products	155
7.5.3 Carbon Footprint of Jute Fiber and Products	156
7.5.4 Carbon Footprint of Linen Fiber Products	157
7.6 Carbon Footprint of Synthetic Fibers and Their Products	158
7.6.1 Carbon Footprint of PP Shopping Bags	158
7.6.2 Carbon Footprint of Products Produced from Regenerated Fibers	160
7.7 Modern Strategies to Reduce Carbon Footprint of Textile Processing	161
7.8 Conclusions	163
References	164

7.1 INTRODUCTION

There exists an enormous pressure on the earth to protect its natural environment. Human activities, increase in human needs, and sophistication lead to deterioration of the natural environmental system. Consumption of electricity, food, clothing, etc., is steadily raising, leading to a continuous growth of related industries. So the amount of greenhouse gas (GHG) emission is also continuously increasing. Excessive emission of GHG elicits pollution and worsens the condition of nature.[1]

There are two types of carbon footprints:

1. Primary carbon footprint
2. Secondary carbon footprint

Primary carbon footprint is the result of direct emission of GHG due to combustion of fossil fuels, and this type of carbon footprint is in our direct control. Primary carbon footprint is the outcome of transportation, domestic energy consumption, etc. On the other hand, secondary carbon footprint is the result of indirect emission of GHG during the entire life cycle of different products. This may occur due to use of clothing, recreation and leisure goods, etc.[1,2]

Many organizations have been formed to keep an update and control over the carbon footprint. The following are some of the organizations working internationally to reduce GHG emissions and that developed some of the standards or standard methods to evaluate GHG emissions:

- International Standards Organization (ISO)
- United Nations Framework Convention on Climate Change (UNFCCC)
- Intergovernmental Panel on Climate Change (IPCC)
- World Resources Institute (WRI) and World Business Council for Sustainable Development (WBCSD)
- Organisation for Economic Co-operation and Development (OECD)–International Energy Agency (IEA)
- U.S. Department of Energy (DOE)
- Lawrence Berkeley National Laboratory (LBNL)
- California Climate Change Registry

Some of the agencies such as IEA and U.S. DOE maintain and publish statistics on the energy consumption and emission levels of CO_2 and thereby help to control the emission of GHG by various countries. The former works for ensuring reliable, affordable, and clean energy for its member counties and the rest of the world, and the latter works for the United States. UNFCC was established in 1992 and helps to stabilize the concentration of GHG in the atmosphere at a certain level, preventing dangerous human interface with the climate system. IPCC is a scientific body that gives a clear view on climate change all over the world. Currently, more than 195 countries are members of the IPCC that actively participate in the assessment of climate change. LBNL is associated with and managed by the University of California. It is conducting research on a wide range of topics including environmental assessment and emission of GHGs. California Climate Change Registry collects the verified reports on GHG emission in the region of California and the rest of the regions too. It helps to stabilize the emission of GHGs in the recorded regions of California.

These organizations have developed standards to study the presence of GHGs and quantify GHGs; equivalent factors for GHG; GHG protocol; guidelines for monitoring, evaluation, reporting, verification, and certification of energy-related projects; and saving of water.[3] Even after the effective control of GHG emission, many regions such as Latin America, Asia, and China are still above and continuously increasing than the global average CO_2 emission of 1.3 gigaton in 2009–2010.[4] In this chapter, various aspects, types, and sources of GHG, carbon footprint and its importance, carbon footprint of various textile processes, and natural as well as synthetic fiber products will be discussed in detail.

7.2 GREENHOUSE GASES

GHGs are the basis of assessment of carbon footprint of various processes, products, and entities. The sun produces radiation that reaches the earth mainly in three wavelength regions, namely, ultraviolet, visible, and infrared. These radiations coming to the earth are partly reflected and partly absorbed. The absorbed radiation increases the temperature of the earth and radiates some of the energy. When equilibrium is achieved between absorbed and radiated energy, almost constant average temperature is achieved. Such emitted radiations may not be reflected completely from the earth and may be partially absorbed and trapped by the gases present in troposphere. The absorbed gases are reflected in all directions and some of the radiations are returned back to the earth. This leads to increase in temperature of the earth resulting in global warming. This effect is called as a *greenhouse effect*. The gases responsible for this effect are GHGs.

7.2.1 Types of Greenhouse Gases

There are six different types of gases present in the atmosphere giving significant impact on the greenhouse effect:

- Carbon dioxide (CO_2)
- Methane (CH_4)
- Nitrous oxide (N_2O)
- Hydrofluorocarbons (HFCs)
- Perfluorocarbons (PFCs)
- Sulfur hexafluoride (SF_6)

These gases can be divided into two broad categories based on their presence in the atmosphere. Some of these gases are naturally present in the atmosphere, and concentration of these gases increases continuously due to human activities. CO_2, CH_4, and N_2O are examples of such gases. The other type of gases is not naturally present in the atmosphere and is created only due to human activities. Chlorofluorocarbons are examples of such gases.

Figure 7.1 shows the volume of GHGs emitted in the collective regions of Australia, Europe, the United Kingdom, and the United States, as reported by UNFCCC for the year 2012. From the figure, it can be commented that CO_2 stands for the highest position in the emission of radiation responsible for global warming, accounting for approximately 81%, followed by CH_4 (≈10%), N_2O (≈6%), and fluorides (≈2%).[5]

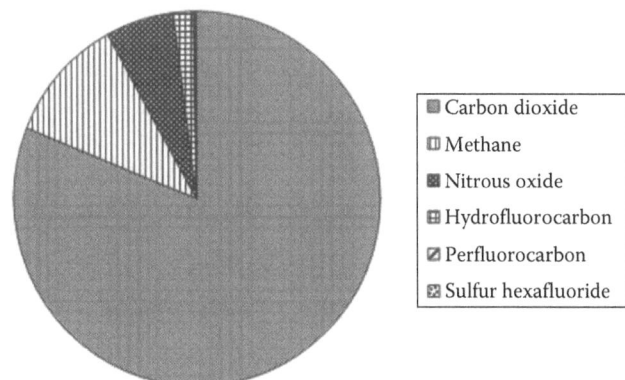

FIGURE 7.1 Emission of GHGs in 2012. (From Hertwich, E.G. and Peters, G.P., *Environ. Sci. Technol.*, 43, 6414, 2009.)

7.2.2 Sources of Greenhouse Gases

Many sectors emit GHGs that affect the natural environment. The following list gives some of these sectors that play a major role in the emission of GHG:

- Residential sector
- Industrial sector
- Commercial sector
- Transportation
- Agricultural sources
- Waste management

Transportation accounts for 12% of total emission of GHG in Australia. It has been reported that the emission of GHGs due to domestic transport increased by 27% between 1990 and 2006. Studies reported that public transport can minimize energy per passenger per kilometer by 65% as compared to a motor vehicle.[6] According to the regulation imposed by the IPCC, the United Kingdom should reduce GHG by at least 80% by 2050 in various sectors, including industries like aviation

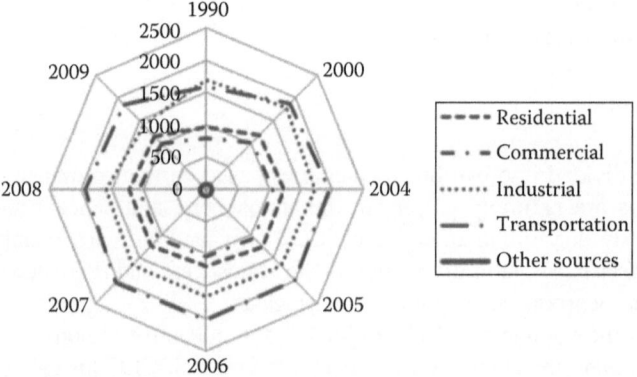

FIGURE 7.2 Emission of carbon dioxide by different sectors. (From U.S. Census Bureau, Statistical abstract of the United States: 2012, pp. 221–242.)

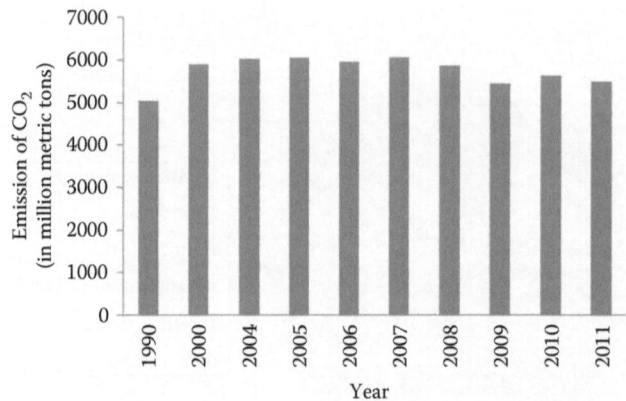

FIGURE 7.3 Year-wise emission of carbon dioxide in the United States. (From U.S. Census Bureau, Statistical abstract of the United States: 2012, pp. 221–242.)

Carbon Footprint of Textile and Clothing Products

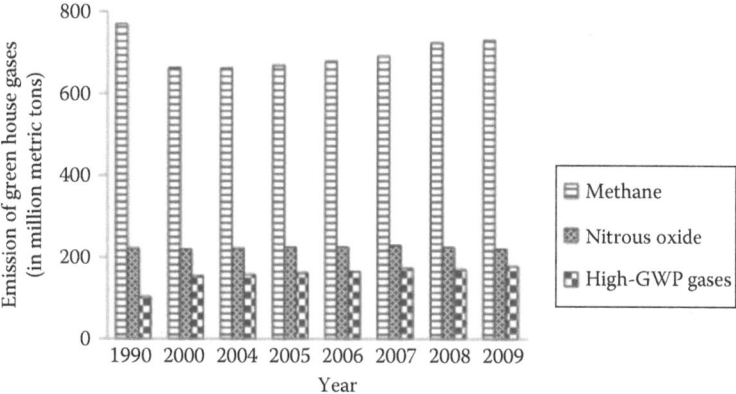

FIGURE 7.4 Year-wise emission of GHG in the United States. (From U.S. Census Bureau, Statistical abstract of the United States: 2012, pp. 221–242.)

and shipping. From 2011 onward, all flights arriving and departing from EU airports have been covered under these regulations.[7]

In India, the total emission of GHG in 2007 was around 1727.71 million tons, among which 57.8% GHG emission resulted from the energy sector, which includes transportation, electricity, residential, and others.[13] Figure 7.2 shows the emission of CO_2 by different sectors. It is clear that the transportation sector plays a vital role in the emission of CO_2, followed by industrial, residential, commercial, and other sources.

Figures 7.3 and 7.4 show the total emission of CO_2 and other GHGs from 1990 to 2009 in the United States. It can be noticed from these figures that the emission of CO_2 was much higher in volume than the emission of other gases.

7.3 CARBON FOOTPRINT

Carbon footprint is the term used to evaluate the total emission of GHGs by human activities. Few other terms are also used to quantify the emission of GHG and slightly vary in their method of measurement and system of representation. The following section gives the definition of different terms that are regularly used in the field of carbon footprint.

7.3.1 Definition of Carbon Footprint and Other Related Parameters

7.3.1.1 Global Warming Potential

Global warming potential (GWP) is a relative measure of how much heat a GHG traps in the atmosphere to contribute toward global warming and compares the amount of heat trapped by a certain mass of the gas under study to the amount of heat trapped by a similar mass of CO_2. This system of representation may be useful when we evaluate the condition of nature over a long period of time. GWP is the relative effect of climate change over a certain period of time. For example, for 20 years of duration, it can be represented as GWP_{20} and for 100 years, GWP_{100}.[9]

7.3.1.2 Carbon Footprint

Carbon footprint is the measurement of the amount of GHG produced through burning of fossil fuels for electricity, heating, transportation, etc., and it is expressed in terms of tons or kilogram or gram CO_2 equivalent.[9] It can be calculated in terms of GWP using the following equation:[9]

$$\text{Climate change} = \sum_i \text{GWP}_{a,i} \times m_i \qquad (7.1)$$

where

GWP$_{a,i}$ is the GWP for the substance i integrated over a specified number of years
m_i (kg) is the quantity of substance i emitted

The result is expressed in kilogram or tons of the reference substance, CO_2.

7.3.1.3 Energy Intensity

It is defined as the ratio of total energy consumption to gross domestic product (GDP).[10]

7.3.2 GLOBAL CARBON FOOTPRINT AND ITS EFFECTS

The world's average emission of GHG per capita is around 5.8 tons CO_2.[13] The carbon footprint of countries like UAE, Luxembourg, the United States, and Australia is much higher than the world's average value when emission of GHG per capita is considered. In Singapore, the emission of CO_2 had reduced by 53% in 2008 as compared to the highest emission in 1997.[11] The decrease in emission of CO_2 with respect to per capita was up to 64% during this period. Even though the per capita actual rate was reduced to 6.8 tons/year, the emission of CO_2 was above Asia's average emission of CO_2 (3.3 tons/year/capita) and much higher than the world's average.[11] By 2030, New York has targeted to reduce its citywide CO_2 emissions by 30% than the emission level in 2005.[12] On the other hand, per capita emission of developing countries like India is far lower than some of the other developing and developed countries. In India, per capita emission of CO_2 in the year 2007 was around 1.5 tons/year, which is much lower than the world's average. All over the world, 1.5 billion people are in short supply of electricity and around 27% of these people (404.5 million) are living in India.[13] Use of less electricity is responsible for lower per capita CO_2 emissions in India. In 2007, combustion of fossil fuel accounted for 93% of energy consumption in China.[10]

Agriculture accounts for around 14% of total GHG emission contributing to 52% of CH_4 emitted all over the world and around 84% of world's N_2O emission. N_2O is capable of trapping 310 times higher heat than the heat trapped by CO_2 and CH_4 is able to trap 21 times more heat than CO_2.[14]

IPCC identified that combustion of fossil fuel is one of the major sources of CO_2 emission leading to global warming. In 1997, the Kyoto Protocol was signed, and according to this, 37 industrialized countries are supposed to reduce their GHG emission in the period of 2008–2012 by 5.2% lower than the GHG emission in 1990.[14]

Figures 7.5 and 7.6 show the carbon footprint of different countries in the world in 2001.[5] In this study, GHG emissions associated with the final consumption of goods and services in eight different sectors such as construction, shelter, food, clothing, mobility, manufactured products, services, and trade were quantified. From these figures, it can be seen that Europe records the highest carbon footprint and Africa gives the lowest carbon footprint when the volume of GHG emission is considered (Figure 7.5). But when the emission of GHG per capita is considered, Australia stands for the highest, followed by Europe, North America, Africa, South America, and Asia.[5]

Some of the effects of increase in the emission of GHGs are as follows[1]:

- Effects of heat waves and other extreme events (cyclones, floods, storms, wildfires)
- Changes in patterns of infectious disease
- Effects on food yields

Carbon Footprint of Textile and Clothing Products

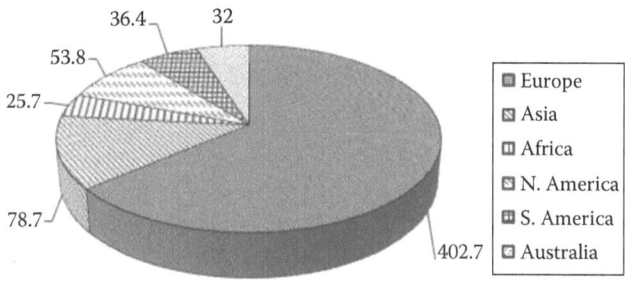

FIGURE 7.5 Carbon footprint of various parts of the world (ton/year). (From Hertwich, E.G. and Peters, G.P., *Environ. Sci. Technol.*, 43, 6414, 2009.)

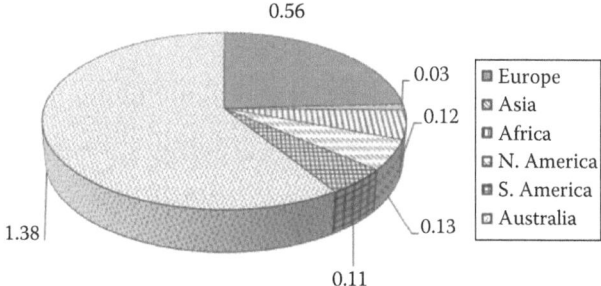

FIGURE 7.6 Carbon footprint per capita of various parts of the world (ton/year). (From Hertwich, E.G. and Peters, G.P., *Environ. Sci. Technol.*, 43, 6414, 2009.)

- Effects on freshwater supplies
- Impaired functioning of ecosystems (e.g., wetlands as water filters)
- Displacement of vulnerable populations (e.g., low-lying island and coastal populations)
- Loss of livelihoods[1]

7.4 CARBON FOOTPRINT OF VARIOUS TEXTILE PROCESSES

Textile industry is identified as one of the largest producers of GHG all over the world.[15] Kissinger et al. reported that textiles and aluminum generate the highest GHG emission per unit of material.[15] The textile industry is indicated as the fifth largest contributor of CO_2 emission.[16,17] In 2008, the global production of textiles reached 60 billion kg of fabric. It had consumed around 1074 billion kW h of electricity (which is equivalent to 132 million tons of coal) and approximately 9 trillion liters of water. Among the total consumption of electricity for textiles, only 15%–20% was consumed in the production of textiles and most of the remaining electricity was consumed in laundering processes. Electrical energy was reported to be one of the major energy consumption sectors in the textile industry.[18] Electrical energy is primarily spent for the following processes:

- Driving machinery
- Cooling
- Temperature control
- Lighting and office equipments

The electrical energy breakdown for a composite plant is shown in Table 7.1. It can be seen from the table that the spinning industry takes the major share of electricity with 41%, followed by weaving and wet processing units.[18,19]

TABLE 7.1
Typical Breakdown of Electricity Use in Composite Textile Industry

Sector	Electrical Energy Consumption (%)
Spinning	41
Weaving preparatory	5
Weaving	13
Humidification	19
Wet processing	10
Lighting	4
Others	8

Source: Choudhury, A.K.R., *Text. Prog.*, 45, 3, 2013.

TABLE 7.2
Machine-Wise Breakdown of Electrical Energy in Spinning Industry

Machine	Electrical Energy Consumption (%)
Blow room	11
Carding	12
Drawing machine	5
Combing machine	1
Simplex	7
Ring spinning machines	37
Open end machines	20
Winding machine	7

Source: Choudhury, A.K.R., *Text. Prog.*, 45, 3, 2013.

Table 7.2 shows the machine-wise electrical energy breakdown in the spinning sector. From the table, it can be seen that the spinning machines, namely, ring and open end spinning, jointly consume more than 50% of the energy consumed in the spinning industry.

Table 7.3 shows the percent share of global consumption of different types of textile fibers all over the world in 2008. From the table, it is clear that polyester and cotton stand for more than 75% of global consumption.[16] Therefore, the carbon footprint of textile industries mostly comes from the production, processing, and use of the products made from these fibers.

Recycling of textile materials is an important process in the textile industry that has a strong impact on the carbon footprint. The recycling industry diverts approximately 10 lb/capita or 2.5 billion lb of postconsumer waste from landfill.[20] In the United States, 70 million lb of scrap is deposited as landfills annually.[21] In Japan, around 2 million tons/year of textiles are sent to landfills. In the United Kingdom, around 3.3 million tons/year of textiles are recovered in which around 2 million tons are exported to other developing countries and 1.2 million tons of textiles are recycled. Around 70% of the world population mainly uses secondhand clothes.[22] Wool is comparatively easy to recycle than other fibers. In the United Kingdom, 40% of the wool garments are recycled, 7% of wool garments are incinerated, and 53% are disposed as landfill.[22] It has been reported that the energy required for the reuse or recycling process of polyester is only 1.8% of the total energy consumed by the virgin fiber. Also, reuse of 1 ton of cotton fiber needs only 2.6% of the energy required for the virgin material.[23] Therefore, recycling and reuse are important processes to reduce

TABLE 7.3
Global Consumption of Textile Substrates

Fiber	Consumption (%)
Polyester	39
Cotton	36
Polyamide	6
Other cellulosic fibers	5
Acrylics	3.5
Wool	2
Silk	0.2
Other fibers	8.3

Source: Athalye, A., Carbon footprint in textile processing, 2012, http://www.fibre2fashion.com, Accessed March 18, 2014.

carbon footprint of textiles. Some examples of the applications of recycled materials are as follows: T-shirts can be used as wipes and polishing clothes, fibers recovered from carpet waste can be produced as nonwovens and mats, and polymers from carpet melt can be used for automotive and other consumer products and also for matrix of composite materials.[24]

7.5 CARBON FOOTPRINT OF NATURAL FIBERS AND THEIR PRODUCTS

The emission of CO_2 in case of natural fibers occurs during preparation, planting, and field operations (weed control, mechanical irrigation, pest control, and fertilizers), harvesting, and yields. During production of natural fibers, normally two types of fertilizers are used such as manure and synthetic chemicals. The use of synthetic fertilizers is a main component of conventional agriculture leading to significant carbon footprint. The production of 1 ton of nitrogen fertilizer emits approximately 7 tons of CO_2 equivalent GHG.[25,26]

According to studies carried out by the Stockholm Environment Institute on behalf of the Bio Regional Development Group,[25] the energy used and CO_2 emitted to manufacture 1 ton of fiber is much higher for synthetic than natural fibers (cotton and hemp). The details of CO_2 emission of various natural fibers as well as polyester fiber are provided in Table 7.4.

TABLE 7.4
Kilogram of CO_2 Emissions per Ton of Spun Fiber

	Crop Cultivation	Fiber Production	Total
Polyester (USA)	0.00	9.52	9.52
Cotton, conventional (USA)	4.20	1.70	5.90
Cotton, organic (USA)	0.90	1.45	2.35
Cotton, organic (India)	2.00	1.80	3.80
Hemp, conventional	1.90	2.15	4.05

Sources: http://www.sei-international.org/mediamanager/documents/Publications/SEI-Report-Ecological FootprintAndWaterAnalysisOfCottonHempAndPolyester-2005.pdf, accessed on March 10, 2014; http://oecotextiles.wordpress.com/2011/01/19/estimating-the-carbon-footprint-of-a-fabric/, accessed on March 10, 2014.

Natural fibers, in addition to having lesser carbon footprint[25] in the production of spun fiber,[25,26] have several additional advantages:

- Fibers are capable of being degraded by microorganisms (biodegradation) and composted (improving soil structure); in this way the fixed CO_2 in the fiber will be released and the cycle will be closed.
- Sequestering carbon: sequestering carbon is the process through which CO_2 from the atmosphere is absorbed by plants through photosynthesis and stored as carbon in biomass such as leaves, stems, branches, roots, and soils. For instance, 1 ton of dry jute fiber leads to absorption of 2.4 tons of carbon.[26]

Producing cotton fibers through organic way provides lot of advantages over conventional process such as less GHG emission, use of less energy for production, and environmental benefits.[25,26] According to a study published in Innovations Agronomiques (2009), organic agriculture emits 43% lesser GHG than conventional agriculture. The research carried out by Cornell University revealed that organic farming required just 63% of energy required for conventional farming. In addition, it is found that organic farming adds 100–400 kg of carbon per hectare to the soil each year and when this stored carbon is included in the carbon footprint, it reduces the total GHG even further.[25,26]

In case of the life cycle of cotton textiles, it is seen that about 50% of CO_2 emissions occur during fiber production, manufacturing of goods, trade, and transport and the remaining 50% are caused by daily usage. Figure 7.7 shows the key CO_2 sources during cotton textile manufacturing process from fiber to garment.[27]

In Europe, light oil and gas are the primary energy sources, but in China the preferred energy source is usually the coal. CO_2 emissions from natural gas are only around 50% of those produced when coal is used as the energy source (see Figure 7.8). In China, around 80% of electricity is produced by thermal power plants. As a result, textiles made in China have a carbon footprint that is around 40% greater than in Turkey, Europe, or South America, simply on the basis of the selected energy source.[27]

Within the full supply chain cycle of cotton textiles/garments (apart from the consumer use phase, i.e., washing and drying), the fiber manufacturing phase emits the most GHG. Cotton incorporated (2009) assessed that GHG emissions were around 1.8 kg CO_2e/kg of fiber. In a parallel study performed on Australian cotton, GHG emissions were assessed around 2.5 kg CO_2/kg of fiber,

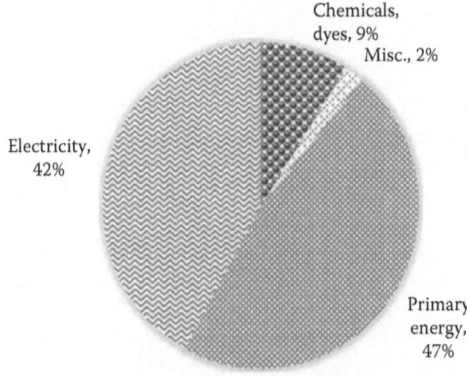

FIGURE 7.7 CO_2 sources within the textile value-added chain for a pair of trousers made of 100% cotton manufactured in China in 2012. (From Strohle, J., Textile achieve ecological footprint new opportunities for China, http://www.benningergroup.com/uploads/media/Carbon_Footprint_China_EN.pdf, accessed on March 10, 2014.)

Carbon Footprint of Textile and Clothing Products

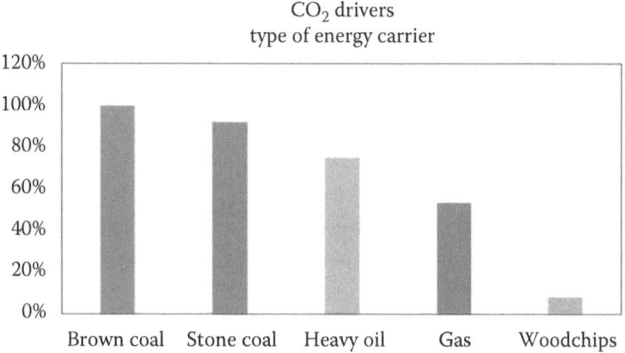

FIGURE 7.8 CO_2 emissions for different energy sources. (From Strohle, J., Textile achieve ecological footprint new opportunities for China, http://www.benningergroup.com/uploads/media/Carbon_Footprint_China_EN.pdf, accessed on March 10, 2014.)

including emissions from fertilizers, chemicals, fuel, and electricity. The GHG emissions in three cotton farming systems, among which one was an irrigated system and the other two were dryland farming systems in Queensland, Australia were assessed. Estimation considering emissions from transportation of farm inputs (farm machinery, agrochemicals, and fertilizers), production, packing and storage, production, extraction, and use of electricity for irrigation, and N_2O emissions from soils due to N-fertilizer usage, revealed that GHG emission was lowest from dryland double skip (1376 kg CO_2e/ha), slightly higher from dryland solid plant (1376 kg CO_2e/ha), and the highest from irrigated cotton farming (4841 kg CO_2e/ha).[28]

Textile finishing is an important process for cotton textiles leading to significant amount of carbon footprint. CO_2 emissions are caused directly by the energy consumers and indirectly by the consumable such as lubricants and chemicals. The dissemination of CO_2 emissions in a fully continuous textile finishing process for cotton textiles shows that about 40% comes from washing and steaming, 50% comes from drying, and 10% from the use of chemicals. In knitwear finishing using the exhaust process, the largest part of emissions, that is, 60%, is caused by heating of water.[27] Table 7.5 provides details of energy consumption in cotton or cotton blend finishing process.

TABLE 7.5
Energy Consumption in Cotton or Cotton Blend Finishing Process

Process/Consumer	Primary Source of Energy Used	CO_2 Emissions
Singeing	Gas	Low
Washing/heating energy	Steam	Very high
Steaming/reaction processes	Steam	Moderate
Drying	Gas/coal/steam	Very high
Fabric transport	Electricity	Low
Air conditioning technology/exhaust air	Electricity	Low
Chemicals	No date	Low

Source: Strohle, J., Textile achieve ecological footprint new opportunities for China, http://www.benningergroup.com/uploads/media/Carbon_Footprint_China_EN.pdf, accessed on March 10, 2014.

7.5.1 CARBON FOOTPRINT OF COTTON FIBER PRODUCTS

Studies reported that the highest CO_2 emissions in case of cotton clothing (e.g., T-shirt) occur during the usage phase of the garment. Also, significant GHG emissions occur during the production of raw materials.[29] As shown in Figure 7.9, the least GHG-intensive processes in case of cotton T-shirt life cycle are fabric and garment production processes besides product disposal stage. This figure also reveals the rough equivalence between GHG emissions and energy use in the garment life cycle. According to another study, CO_2 emissions in different stages of cotton T-shirt life cycle are presented in Figure 7.10. It can be noticed that a major improvement in terms of GHG emissions

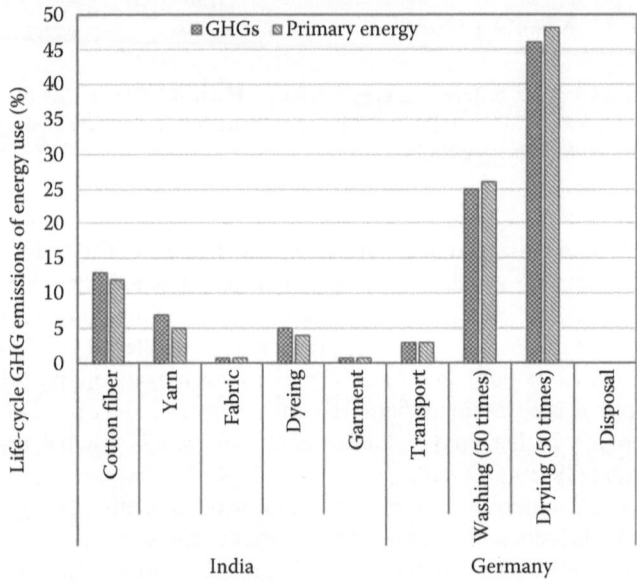

FIGURE 7.9 Garment life cycle GHG emissions of cotton T-shirt. (From Strohle, J., Textile achieve ecological footprint new opportunities for China, http://www.benningergroup.com/uploads/media/Carbon_Footprint_China_EN.pdf, accessed on March 10, 2014.)

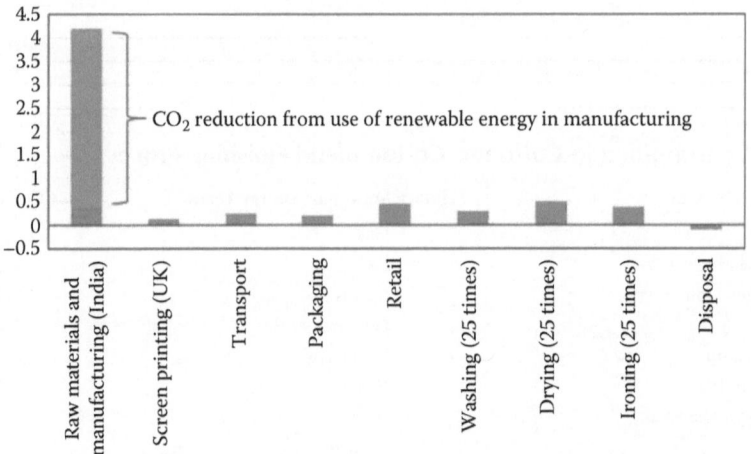

FIGURE 7.10 Garment life cycle GHG emissions of cotton T-shirt, showing amount of CO_2 per stage. (From Strohle, J., Textile achieve ecological footprint new opportunities for China, http://www.benningergroup.com/uploads/media/Carbon_Footprint_China_EN.pdf, accessed on March 10, 2014.)

Carbon Footprint of Textile and Clothing Products

can be achieved through the use of renewable energy in production processes. Besides production of raw materials and fabric, various processes responsible for GHG emissions are printing, transport, packaging, retail, washing, drying, and ironing.[30] According to this study also, use phase is the major cause of GHG emissions, causing nearly 50% of the actual total.[29]

7.5.1.1 Carbon Footprint of White Long Shirt

Figure 7.11 shows the carbon footprint of white long shirt made of cotton. To produce these long shirts, raw materials were cultivated in the United States and garments were made in Bangladesh for German customers. The estimated carbon footprint of this cotton product during its life cycle was 10.75 kg CO_2e.[30]

Carbon footprint due to cotton growing including ginning is up to 1.27 kg CO_2e. As shown in Figure 7.12, almost half of the emissions are caused by direct and indirect N_2O, which has a GWP of 298 relative to CO_2. Direct emissions of N_2O depend on soil structure, the use of fertilizer, water, temperature, etc. The manufacturing phase leads to an emission level of 3.0 kg CO_2e per functional unit during shirt production. CO_2 emissions in the production stage of the shirt are shown in Figure 7.13. Approximately 1/3 of the carbon emissions are caused by heating processes and 2/3 by electricity. CO_2 emissions in the distribution processes are shown in Figure 7.14. During the distribution phase, CO_2e emission is 87 kg, which is more than half resulting from returns by customers.

It was observed in this study that consumers can contribute significantly to reduce the carbon footprint of the products. Use of energy-efficient devices can significantly reduce the carbon footprint during the use phase. Household devices with a better level of energy efficiency may decrease the carbon footprint in the use phase by one-third as compared to the household stock. Also, the carbon footprint in the use phase is influenced by the washing temperature and actual loading of

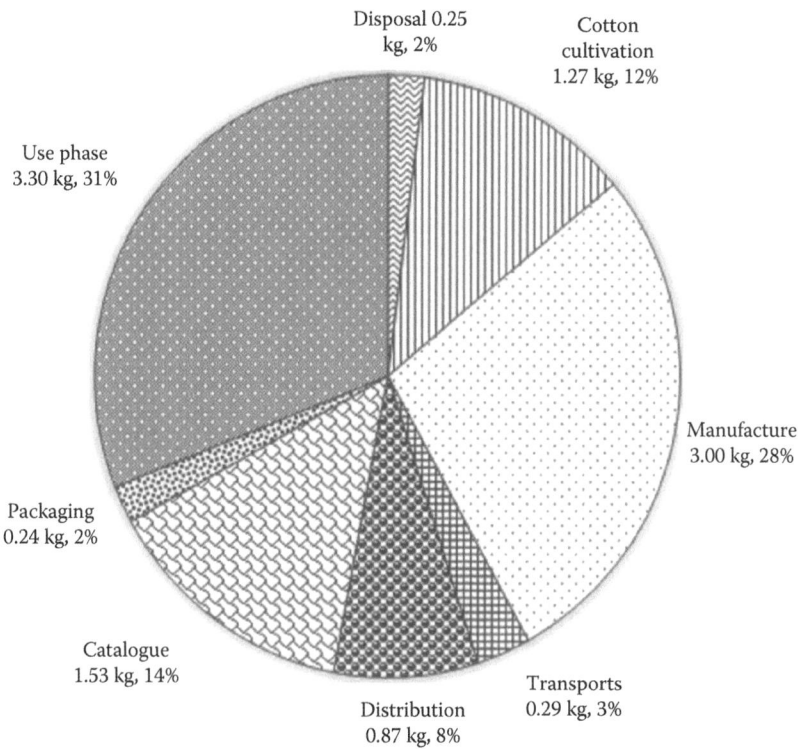

FIGURE 7.11 Carbon footprint of cotton white long shirt during its life cycle. (From Jungmichel, N., *The Carbon Footprint of Textiles*, Systain Consulting, Berlin, Germany, 2010.)

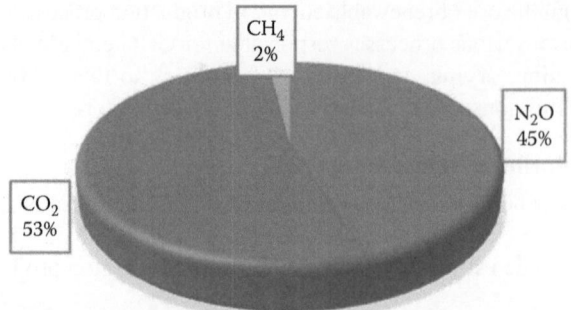

FIGURE 7.12 Different GHG emissions during cotton growing. (From Jungmichel, N., *The Carbon Footprint of Textiles*, Systain Consulting, Berlin, Germany, 2010.)

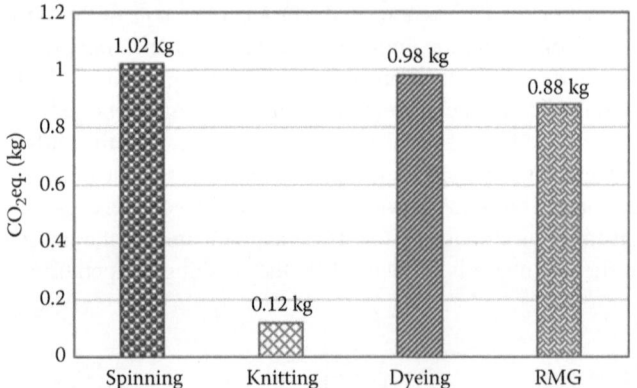

FIGURE 7.13 CO_2 emission in the production process of a shirt. (From Jungmichel, N., *The Carbon Footprint of Textiles*, Systain Consulting, Berlin, Germany, 2010.)

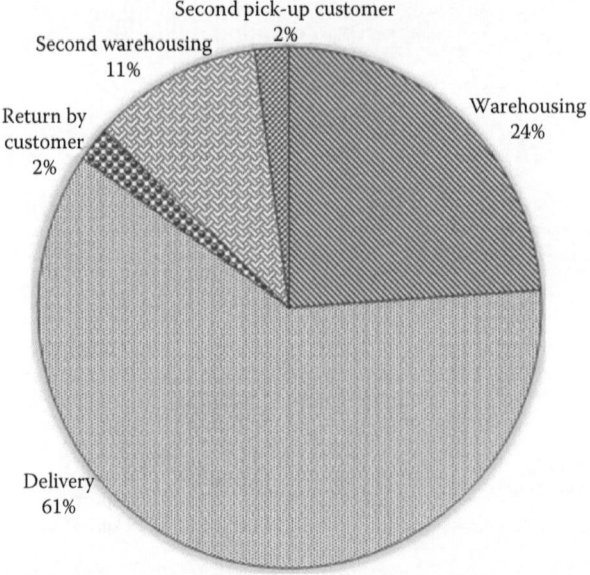

FIGURE 7.14 CO_2eq. emissions during the distribution process. (From Jungmichel, N., *The Carbon Footprint of Textiles*, Systain Consulting, Berlin, Germany, 2010.)

appliances. It was observed that a washing temperature of 40°C instead of 60°C could cut the carbon footprint of the use phase by 45% and 30°C instead of 40°C by 40%.[30]

7.5.1.2 Comparison of Carbon Footprint of Different Cotton Products

Studies also compared the carbon footprint of different cotton products and also with products made with synthetic fibers, as shown in Figure 7.15. The details of these products are provided in Table 7.6. The carbon footprint of the three products during their life cycle is presented in Figure 7.16.

It can be noticed from Figure 7.16 that the acrylic jacket has lower carbon footprint in the manufacturing and use phases as compared to the cotton products. However, disposal of acrylic good leads to significantly higher CO_2 emissions than the cotton products. Among the cotton long shirt and sweat jacket, the long shirt gives a lower carbon footprint in raw material, manufacturing, and disposal phases, but a higher carbon footprint in the distribution phase.

7.5.2 CARBON FOOTPRINT OF WOOL FIBER AND PRODUCTS

According to studies, the energy required for wool production is 38 MJ/kg. New Zealand Merino study estimated an energy usage of 46 MJ/kg to produce wool top, half of which is used in the farm and CO_2 emission for production of wool staples is 2.2 kg CO_2/kg (considering 50 g CO_2/MJ of energy).[31] The energy consumption and CO_2 emission of wool fiber are compared with other

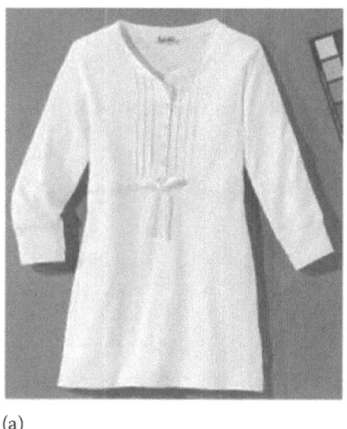  

(a) (b) (c)

FIGURE 7.15 Three different textile products: (a) cotton long shirt, (b) sweat jacket with hood, and (c) jacket for kids. (From Jute Eco-label: Life cycle assessment of jute products, 2006, http://www.jute.com:8080/c/document_library/get_file?uuid=e39c1527-75ed-47e9-9c88-c415ac11cf09&groupId=22165, accessed on November 25, 2013.)

TABLE 7.6
Details of Three Different Products

White Long Shirt	Sweat Jacket	Jacket for Kids
100% cotton (USA)	100% cotton (Africa)	100% acrylic
Net weight—222 g	Net weight—446 g	Net weight—266 g
Cotton from the United States, production in Bangladesh, offered by OTTO	Cotton from Benin, production in Turkey, offered by BAUR	Acrylic from China, production in Bangladesh, offered by OTTO

Source: Jungmichel, N., *The Carbon Footprint of Textiles*, Systain Consulting, Berlin, Germany, 2010.

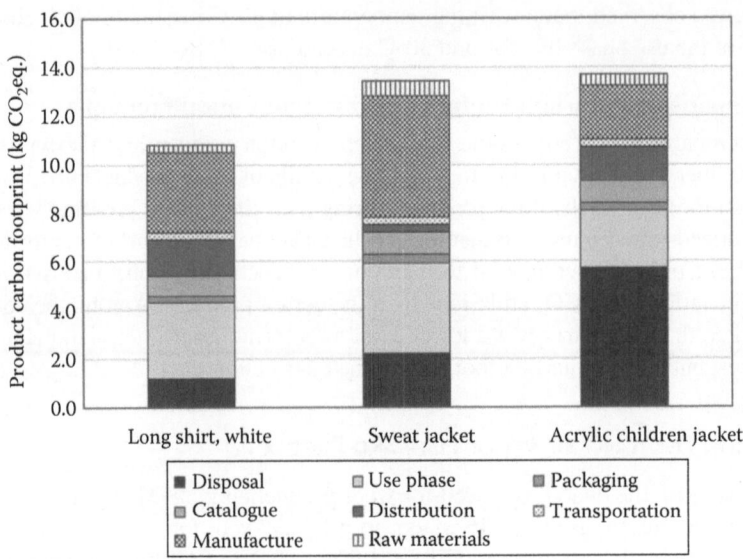

FIGURE 7.16 Carbon footprint of three different textile products during its life cycle. (From Jungmichel, N., *The Carbon Footprint of Textiles*, Systain Consulting, Berlin, Germany, 2010.)

TABLE 7.7
CO_2 Emissions in kg/kg of Different Textile Fibers Based on Energy Consumption (kW h/kg Fiber)

Fiber Type	Energy Consumption kW h/kg Fiber	CO_2 Emissions in kg/kg Fiber
Nylon	69	37
Acrylic	49	26
Polyester	35	19
Polypropylene	32	17
Viscose	28	15
Cotton	15	8
Wool	13	7
Hemp	5	3

Source: http://www.metrocon.info/images/uploads/SWhittaker-METROCON12.pdf, accessed on March 11, 2014.

textile fibers in Table 7.7. It can be noticed that wool fiber consumes lower energy and also leads to lower carbon footprint than the other listed fibers, except hemp fibers, which have the lowest carbon footprint.

7.5.3 Carbon Footprint of Jute Fiber and Products

The emission of GHG of a jute fiber yarn in different phases such as cultivation and retting phase, manufacturing phase, and product disposal phase has been studied and listed in Table 7.8 for 684 tons of jute yarn. This study considered the credits of jute product incineration for energy production to replace fossil fuel utilization and took into account only 50% of CH_4 emission considering capture of the remaining 50% during jute product disposal through landfill.[33]

It can be concluded from these results that the overall GHG emission effect in the cultivation and retting phase is negative. This implies that the jute plantation process acts as a carbon absorber.

TABLE 7.8
Impact of GHG Effect of Jute Fiber at a Different Phase of the Life Cycle

Phase	IPCC–GHG Effect (Direct 100 Years)	Value	Unit
Cultivation and retting phase of final raw jute	CO_2, CO_2 equivalent CH_4	−4,502,370	g·eq. CO_2
Manufacturing phase	CO_2, CO_2 equivalent CH_4	485.71	g·eq. CO_2
Disposal of product through incineration	CO_2, CO_2 equivalent CH_4	−6.895	g·eq. CO_2
Disposal of product through landfill	CO_2, CO_2 equivalent CH_4	14.124	g·eq. CO_2

Source: Jute Eco-label: Life cycle assessment of jute products, 2006, http://www.jute.com:8080/c/document_library/get_file?uuid=e39c1527-75ed-47e9-9c88-c415ac11cf09&groupId=22165, accessed on November 25, 2013.

Even though the emission of CH_4 during the retting process contributes to the GHG impact, this effect is balanced by the carbon sequestration of jute plants during their farming. The manufacturing phase contributes to the GHG effect due to CO_2 emissions resulting from the use of fossil fuel, electricity, and transportation. It can be also noted that the disposal of jute products into an unmanaged landfill leads to the GHG effect due to CH_4 emission. Conversely, this impact reduced considerably when the disposal of the jute product is done through incineration to produce energy for replacement of fossil fuel–based energy.[33]

7.5.4 CARBON FOOTPRINT OF LINEN FIBER PRODUCTS

The linen shirt studied in this research was manufactured in China for use in France. It can be noticed from Figure 7.17 that primary energy is not always a good representation for GHG emissions depending on the type of energy source used. Figure 7.18 illustrates the relative GHG

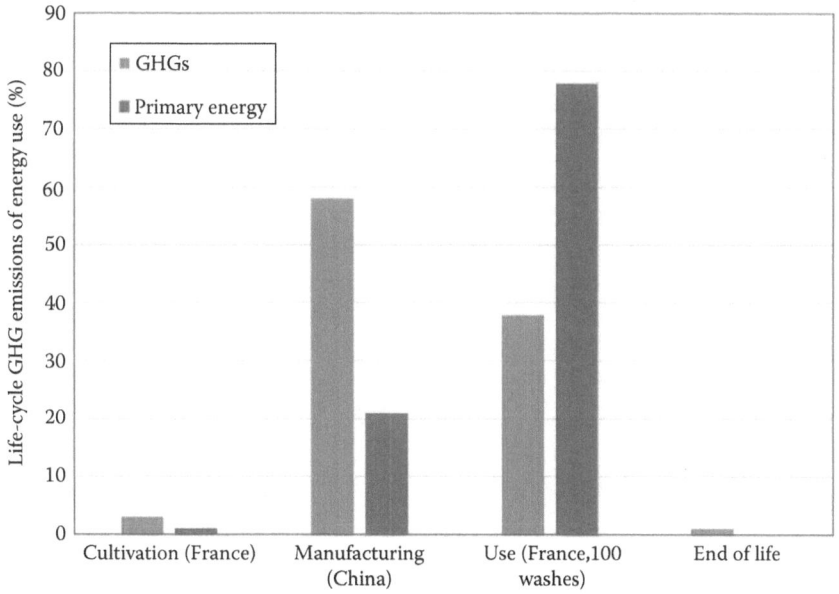

FIGURE 7.17 Garment life cycle GHG emissions of linen shirt. (From Strohle, J., Textile achieve ecological footprint new opportunities for China, http://www.benningergroup.com/uploads/media/Carbon_Footprint_China_EN.pdf, accessed on March 10, 2014.)

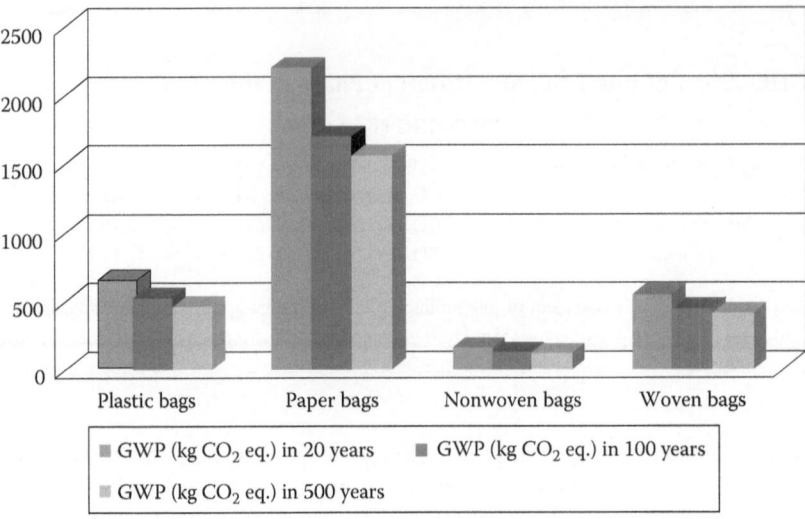

FIGURE 7.18 Carbon footprint results without usage and disposal option in China and Hong Kong. (From Muthu, S.S. et al., *Atmos. Environ.*, 45, 469, 2011.)

emissions produced by a linen shirt during its manufacturing process. Since the linen shirt was manufactured in China, where coal-fired power plants produce most of the used electricity, the amount of GHG emissions is high. On the contrary, during its use in France, a low amount of GHG is emitted due to the use of nuclear power as the energy source.[29] Higher energy is necessary in the use phase of the linen shirt due to the high-energy requirement to iron the linen shirt. It can also be noticed that cultivation of raw material (flax) leads to low amount of energy use and GHG emissions.[29]

7.6 CARBON FOOTPRINT OF SYNTHETIC FIBERS AND THEIR PRODUCTS

In the case of synthetic fiber, the key factor related to carbon footprint is that the fibers are produced from fossil fuels. The extraction of oil from the earth and the production of synthetic polymers require a high amount of energy and therefore emit a much higher amount of CO_2 as compared to natural fibers (refer to Table 7.4).

Acrylic fiber requires 30% more energy during its production than polyester and for nylon it is even higher. Not only the quantity of GHG emissions is of prime concern for synthetic fibers, but also the type of GHGs produced is important. Nylon, for example, emits N_2O, which is 300 times more damaging than CO_2 and because of its long life, it can reach and diminish the layer of stratospheric ozone.[25,26] Moreover, synthetic fibers do not decompose and in landfills they release heavy metals and other additives into the soil and groundwater. Recycling needs expensive separation, while burning produces pollutants. In the case of HDPE, 3 tons of CO_2 is emitted due to burning of 1 ton of material.[25,26]

7.6.1 CARBON FOOTPRINT OF PP SHOPPING BAGS

Nonwoven shopping bags, mostly made of polypropylene fiber, are popular and are commonly used as a reusable item. The process of manufacturing polypropylene (PP) bags starts from fiber production followed by spun bonding and different steps of bag manufacturing such as cutting, screen printing, sewing, and packaging. Shopping bags are made using two different methods, namely, the sewing process (A), that is, joining two sides of the bags by stitching, and thermal bonding (B), that is, joining two sides of the bags through heat application.[34] The results of the carbon footprint evaluated for these two types of shopping bags through the IPCC 2007 GWP V 1.1 method for 100- and 20-year time periods are presented in Table 7.9. It was observed that the shopping bag produced through the sewing process had a lower carbon footprint as compared to the one produced through the thermal bonding process.

Studies also compared the carbon footprint of different types of shopping bags used in China, Hong Kong, and India such as plastic, paper, nonwoven (PP), and woven cotton bags. GHG emissions of these shopping bags are listed in Tables 7.10 and 7.11.[1]

The results of life cycle impact assessment (LCIA) performed using the IPCC 2007 method to evaluate the carbon footprint of these shopping bags (without considering usage and disposal) in

TABLE 7.9
GWP Potentials

Impact Category, Unit	A	B
IPCC GWP 100 a, kg CO_2 eq.	60.7	86.3
IPCC GWP 20 a, kg CO_2 eq.	62.5	88.6

Source: Muthu, S.S. et al., *Fibers Text. Eur.*, 3(92), 12, 2012.
100 a, 100 years; 20 a, 20 years.

TABLE 7.10
Life Cycle Inventory Data of Plastic, Paper, Nonwoven, and Woven Bags in China and Hong Kong

Alternative	Weight/Bag (g)	Bags/Year	Material Consumption	GHG Emissions (CO_2 eq.) (kg)	Primary Energy (MJ)
Plastic bag	6	1095	6.57 kg	12.8	442.2
Paper bag	42.6	1095	46.65 kg	24.8	1518.3
PP fiber nonwoven bag	65.6	10.95	718. 32 g	5.17	122.2
Woven cotton bag	125.4	21.9	2.75 kg	6.06	385

Source: Muthu, S.S. et al., *Atmos. Environ.*, 45, 469, 2011.

TABLE 7.11
Life Cycle Inventory Data of Plastic, Paper, Nonwoven, and Woven Bags in India

Alternative	Weight/Bag (g)	Bags/Year	Material Consumption	GHG Emissions (CO_2 eq.)	Primary Energy (MJ)
Plastic bag	6	150	900 g	1.74 kg	60
Paper bag	42.6	150	6.39 kg	3.41 kg	210
PP fiber nonwoven bag	65.6	1.5	98.4 g	708 g	16.73
Woven cotton bag	125.4	1.3	376.2 g	831 g	52.74

Source: Muthu, S.S. et al., *Atmos. Environ.*, 45, 469, 2011.

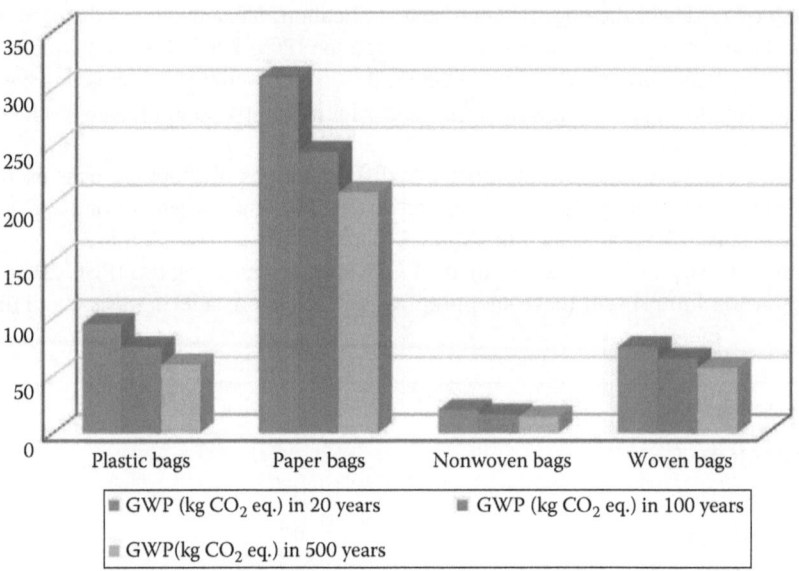

FIGURE 7.19 Carbon footprint results without usage and disposal option in India. (From Muthu, S.S. et al., *Atmos. Environ.*, 45, 469, 2011.)

terms of GWP of 20 years, 100, and 500 years are presented in Figures 7.18 and 7.19. The results show that nonwoven bags made of PP performed better than other bags, followed by woven cotton bags. Paper and plastic bags have very high GWP for 20, 100, and 500 years compared to nonwoven (PP) and cotton woven bags. It can be seen that nonwoven bags consume lesser energy and materials and also emit a lower amount of GHG as compared to other shopping bags used in China, Hong Kong, and India.[1]

7.6.2 Carbon Footprint of Products Produced from Regenerated Fibers

The carbon footprint of products produced from the most commonly regenerated cellulosic fiber, that is, viscose rayon, has also been reported in the literature. One of such studies reported GHG emissions due to the production of Pashmina shawl in India, including the staple fiber production stage up to the manufacturing of this product.[35] Both direct and indirect GHG emissions were considered for carbon footprint evaluation. The direct GHG emissions included the emissions from the production site such as from electricity consumption and fuel combustion. On the other hand, indirect GHG emissions included the emissions from fiber production, that is, the embodied energy of fiber produced, transportation of raw materials and fuel to the production place, and transportation of Pashmina shawl from the production place to the warehouse, that is, the transportation of carbon footprint.[35] Potential sources of CO_2 for the manufacturing of this product, as identified in this study, were rayon staple fiber production (embodied carbon footprint), transportation of raw material (rayon staple fiber) and fuel (diesel/petrol) to the production site, transportation of produced Pashmina shawl from the production site to the warehouse (transportation carbon footprint), etc. Among these, manufacturing of rayon staple fiber (embodied carbon footprint) and production of Pashmina shawl were found to be the main sources of carbon footprint. Approximately 25% of carbon emissions in the total production process of one Pashmina shawl come from the dyeing and packing unit, whereas 57% come from the spinning and weaving units. Emissions from transportation and packing of Pashmina shawl were lesser. Table 7.12 provides the emission inventory data for shawl production.[35]

TABLE 7.12
Emission Inventory for Pashmina Shawl Production

Emission Inventory	kg CO_2/Pashmina Shawl
Fiber	1.6
Transportation	0.063
Manufacturing	0.8164
Chemicals	0.06
Packing	0.02
Total	2.5594

Source: Shawl, P., Product carbon footprint, http://www.ecotechenergy.in/pdf/carbon-footprint-pashmina-shawl.pdf, accessed on March 11, 2014.

7.7 MODERN STRATEGIES TO REDUCE CARBON FOOTPRINT OF TEXTILE PROCESSING

Carbon footprint reduction is an important issue for all human beings in order to protect our planet for our future generations. Our commitments toward future compel us to create a green and sustainable environment. With the growth of civilization, textiles become much more than a primary need for individuals. The per capita consumption of clothing increases exponentially with the improved living standard and fashion consciousness of the people.[20] As discussed in the previous sections, textile products create carbon footprint in each phase of their life cycle. The complicated supply chain of textile industry leads to significant amount of carbon footprint in its each segment. With a huge production volume, it is found to be one of the major sources of emissions of GHG globally.[36,37] In 2008, global textile production was estimated at 60 billion kg of fabric. The energy and water needed to produce that amount of fabric are 1074 billion kW h of electricity or 132 million metric tons of coal and 6–9 trillion liters of water.[37]

Looking into this huge consumption of energy and water, strategies have been initiated to reduce carbon footprints in textile processing. More and more energy-efficient processes are being used. The intermediate products and processes are improved with new innovations for lower carbon footprints. All the processing steps starting from harvesting to packing, transporting, usage, and disposal add carbon footprint to textiles.[1]

It is reported that other synthetic fiber production is more energy intensive.[38] Thus, the first corrective measure taken to reduce the carbon footprint is the use of natural fibers. Natural fibers have lower carbon footprint than the synthetic ones. Use of organic fibers further reduces the carbon footprint. The less energy-intensive processes create less carbon footprint. Natural fibers have other advantages like biodegradability and sequestering carbon from atmosphere.[39] DuPont came up with a fiber called Sorona having much lower carbon emission as compared to other synthetic fibers. Unlike most of the synthetic fibers, it is an agricultural product and not petrochemical derived. Sorona has high renewable ingredients content.

Energy- and water-intensive textile processing is the major contributor as far as carbon footprint is concerned. Strategies have been proposed to reduce the carbon footprint of textile processing. Low energy- and water-intensive processes and products are now commercially available in the market. Some of the initiatives are as follows: reduction of water consumption during pretreatment, dyeing, washing, and finishing is achieved through the use of low and ultralow liquor ratio machines. Less water leads to less energy for heating at different processing steps and also less effluent treatment load. A study carried out by Benninger shows that the reduction and reuse of water and energy is the key parameter to reduce carbon footprint in the textile industries.[27] The knitwear industries use a soft flow machine during processing that consumes high water and energy. Continuous dyeing and finishing machines for knitwear are the recent developments that reduce the water and energy usage to a significant level. A number of eco-efficient solutions have been introduced into the market that are environmentally friendly and contribute to saving of resources. The processing time and water consumption can be reduced with these chemicals as compared to the conventional systems. A one-step process of textile dyeing and finishing, combining the dyeing, washing, and finishing steps into one step, can reduce the processing time and energy cost. Consumption of nonrenewable energy can be reduced by preheating of water through solar energy or heat exchanger in waste water line. The heat loss is minimized by adequate insulation of processing machines and appropriate heat recovery systems. The caustic and water recovery plants reuse the water and alkali to a great extent and thus reduce the carbon load in processing.

The combined approach for processing helps in significant reduction of carbon footprint. Combined singing–desizing, desizing–scouring–bleaching, one bath dyeing of polyester/cotton blends, etc., reduce the number of textile processing stages and thereby reduce consumption of water and energy. Since the drying process takes a high amount of energy, all preparatory processes are carried out without drying, except the last stage where fabrics ready for dyeing or finishing are prepared. Continuous dyeing processes like cold pad batch (CPB) and thermosol require less water for dyeing. After dyeing, the washing process also has been improved with less number of wash chambers and wash liquor. The whole process saves water and energy. Continuous processing of knits is still under development. New processing techniques like waterless dyeing, CO_2 dyeing, foam dyeing, finishing, and coating are gradually gaining their acceptance in industries. A combination of water/air is tried successfully for dyeing that reduces the water consumption during dyeing. The cost associated with these processes is also reducing in recent years. Gaston Systems, USA, has developed a foam finishing machine that saves lots of water. To reduce carbon footprint, entire reprocessing of textile materials, which not only burdens the processing cost but also increases GHG emissions, needs to be avoided. Hence, right first time (RFT) and right every time (RET) dyeing performances are essential for reduction in carbon footprint of textile processing. Thus, advanced software to improve the lab to bulk conversion ratio are utilized by many industries.

Enzyme-based processing like desizing, scouring, bleach neutralizing, bio-softening, and post-dyeing wash-off is in the market. Enzyme suppliers are offering specialized products for combined processes to reduce the number of processing steps. These enzymes replace harsh chemicals used to remove impurities from the fiber or fabric. Their use reduces energy costs and water consumption and also improves the feel of the fabric. Cationic cotton is successfully utilized for salt-free dyeing with reactive and direct dyes. High fixation reactive dyes are useful for less carbon footprints. These dyes require reduced salt for high exhaustion. In the printing area, digital inkjet printings and low-temperature curing pigment printings are commercially available.[17] Huntsman has developed inks from the dyes to use in a digital printer directly for printing on fabrics. This digital printing process significantly reduces the environmental footprint as no water, salt, or other chemicals are required.

Apart from these technological advances, two significant strategies have been adopted for reduction in carbon footprints in the textile sector, namely, reuse and recycle of textiles. The recycle is effective only when the carbon loading to recycle is less than the disposal. Reuse of textile waste can be done in different ways. Some common practices are releasing clothes into the marketplace through secondhand shops and donating to some charity organizations or informally among family members. Reports indicate that large amounts of secondhand clothing are dispatched abroad for selling on the global market in Eastern Europe or Africa.[40] According to statistics, 26,000 tons of used garments and shoes were collected by charity organizations in Sweden during 2008 as donations to Africa and Eastern Europe.[41]

There are many ways to reduce the carbon footprint throughout the entire life cycle of textile products, and one of the promising ways to minimize the carbon footprint is to recycle the process waste instead of disposing at landfill and also to recycle the textile products at the end of life. Reuse of textile products has environmental benefits. However, the amount of energy saved and avoided

Carbon Footprint of Textile and Clothing Products   163

emissions by applying reuse of discarded textiles, the amount of energy usage, and GHG emissions during collection, sorting, and reselling of the used clothes should be assessed and compared with the energy demand and emissions of manufacturing new products.[42] In the present decade, due to the alarming environmental impacts and other reasons such as rewards in terms of monetary benefits given to people when they return the product for recycling, governmental policies, etc., people have started supporting recycling activities as compared to the last decade.[23] Recycling is one of the proven and promising ways of reducing carbon footprint.[23] There are many strategies to recycle textile products, which can reduce carbon footprint. The use of life cycle analysis software is helpful for estimating the recycling potential of textile products.[9]

It is evident from the studies that the textile sector is highly energy intensive. The bulk of the carbon footprint of the textile industry is actually due to the usage of energy, and hence all strategies are directed toward using less energy, reuse of waste energy, and use of renewable sources of energy as far as possible.[43] During fiber production stage, more focus is on the use of organic natural fibers for their less carbon footprint as compared to synthetic ones.[26] Though the crop cultivation stage involves CO_2 emission for natural fibers, the production of synthetic fibers is a higher energy-intensive process and results in high CO_2 emission. Considering the carbon footprint addition due to transportation of raw materials and finished products,[1] composite textile mills and organic farming may be a good option for the future.

7.8 CONCLUSIONS

Carbon footprint is the measurement of the amount of GHG produced through burning of fossil fuels for electricity, heating, transportation, etc., and it is expressed in terms of tons or kg CO_2 equivalent. Various GHGs are CO_2, CH_4, N_2O, and CFC, and their emissions may result in serious problems such as heat waves and other extreme events (cyclones floods, storms, wildfires), changes in patterns of infectious disease, reduced food yields and freshwater supplies, impaired functioning of ecosystems, displacement of vulnerable populations, and loss of livelihoods. The textile industry is identified as one of the largest producers of GHG all over the world and has been reported to generate the highest GHG emission per unit of material. Electrical energy is one of the major energy consumption sectors in the textile industry, and electrical energy is spent for driving machinery, cooling, temperature control, lighting, and office equipment. Among the various textile industries, the spinning industry takes the major share of electricity with 41%, followed by weaving and wet processing units. The emission of CO_2 in case of natural fiber production occurs during preparation, planting, field operations (weed control, mechanical irrigation, pest control, and fertilizers), harvesting, and yields. However, energy used and CO_2 emitted to manufacture 1 ton of natural fiber are much lower as compared to synthetics fibers. In case of cotton textiles, 50% of CO_2 emissions occur during fiber production, manufacture, trade, and transport and the remaining 50% are caused by daily usage, that is, washing and drying. Textile finishing is an important process for cotton textiles leading to significant amount of carbon footprint resulted from CO_2 emissions of 40% from washing and steaming, 50% from drying, and 10% from the use of chemicals. Production of wool fiber uses lower energy and has lower carbon footprint as compared to cotton fibers. The plantation process of jute fibers acts as a carbon absorber, and although emission of CH_4 during the retting process contributes to the GHG impact, this effect is balanced by the carbon sequestration of jute plants during their farming. Cultivation of linen fibers also results in very less GHG emissions. However, higher energy is necessary in the use phase of linen products due to high energy requirements to wash, dry, and iron the linen materials. In case of synthetic fibers, extraction of oil from earth as well as production of synthetic polymers require high amount of energy and therefore emit much higher amount of CO_2 as compared to natural fibers. Among the synthetic fibers, nylon and acrylic leads to higher carbon footprint than polyester fibers. Potential sources of CO_2 emissions for the production of viscose rayon fiber products are rayon staple fiber production (embodied carbon footprint), transportation of raw material (rayon staple fiber) and fuel (diesel/petrol) to the production site, transportation of product from the production site to the warehouse (transportation carbon footprint), etc., and among these, manufacturing of rayon staple fiber (embodied carbon footprint) and products is main source of carbon footprint. In recent times, various modern strategies and processes are being practiced to reduce the carbon footprint of textile industries such as promoting more use of natural fibers; reduction of water consumption during pretreatment, dyeing, washing, and finishing through the use of low and ultralow liquor ratio machines; combining dyeing, washing, and finishing steps into one step; use of continuous dyeing processes like CPB and thermosol; new processing techniques like waterless dyeing, CO_2 dyeing, foam dyeing, finishing and coating; enzyme-based processing like desizing, scouring, bleach neutralizing, bio-softening, and postdyeing wash-off; and reuse and recycle of textile goods.

REFERENCES

1. Muthu SS, Li Y, Hu JY, Mok PY. Carbon footprint of shopping (grocery) bags in China, Hong Kong and India. *Atmos Environ* 45 (2011): 469–475.
2. Kennedy C, Steinberger J, Gasson B et al. Greenhouse gas emissions from global cities. *Environ Sci Technol* 43 (2009): 7279–7302.
3. Hammons TJ, Mcconnach JS. Proposed standard for the quantification of CO_2 emission credits. *Electr Power Compon Syst* 33 (2005): 39–57.
4. IEA Statistics (2012), CO_2 Emissions and Fuel Combustion Highlights, International Energy Agency, France. http://www.columbia.edu. Accessed on March 18, 2014.
5. Hertwich EG, Peters GP. Carbon footprint of nations: A global, trade-linked analysis. *Environ Sci Technol* 43 (2009): 6414–6420.
6. Quirk M. Transport, land use, the built environment and greenhouse emissions: An overview. *Aust Planner* 48 (2011): 37–45.
7. Randles S, Bows A. Aviation, emissions and the climate change debate. *Technol Anal Strat Manage* 21 (2009): 1–16.
8. U.S. Census Bureau. Statistical abstract of the United States: 2012, pp. 221–242.
9. Muthu SS, Li Y, Hu JY et al. Carbon footprint reduction in the textile process chain: Recycling of textile materials. *Fibre Polym* 13 (2012): 1065–1070.
10. Kuby M, He C, Trapido LB et al. The changing structure of energy supply, demand, and CO_2 emissions in China. *Ann Assoc Am Geogr* 101 (2011): 795–805.
11. Velasco E, Roth M. Review of Singapore's air quality and greenhouse gas emissions: Current situation and opportunities. *J Air Waste Manage Assoc* 62 (2012): 625–641.
12. Dickinson J and Desai R, Inventory of New York city greenhouse gas emissions, 2007. http://www.nyc.gov/planyc2030. Accessed on March 18, 2014.
13. Upadhyaya P. Is emission trading a possible policy option for India?. *Clim Policy* 10 (2010): 560–574.
14. Agarwal P, Kumar A, Hooda SS et al. Anthropogenic carbon emissions in India: An econometric analysis. *J Bus Perspect* 14 (2010): 79.
15. Kissinger M, Sussmann C, Moore J et al. Accounting for greenhouse gas emissions of materials at the urban scale-relating existing process life cycle assessment studies to urban material and waste composition. *Low Carbon Econ* 4 (2013): 36–44.
16. Athalye A. Carbon footprint in textile processing. 2012. http://www.fibre2fashion.com. Accessed March 18, 2014.
17. Athalye A. Carbon footprint in textile processing. *Ind Text J* 122 (August 2012): 20.
18. Choudhury AKR. Green chemistry and the textile industry. *Text Prog* 45 (2013): 3–143.
19. Hasanbeigi A. Energy-efficiency improvement opportunities for the textile industry, LBNL-3970E. Ernest Orlando Lawrence Berkeley National Laboratory, Berkeley, CA, 2010.
20. Wang Y. *Recycling in Textile*. Woodhead Publishing Ltd., Cambridge, U.K., 2006.
21. Hawley JM. Digging for diamonds: A conceptual framework for understanding reclaimed textile products. *Cloth Text Res J* 24 (2006): 262–275.
22. Russell M. Sustainable wool production and processing. In Blackburn RS (ed.), *Sustainable Textiles*. Woodhead Publishing, Cambridge, U.K., 2009, pp. 63–87.
23. Michaud JC, Farrant L, Jan O, Kjær B, Bakas I. Environmental benefits of recycling—2010 update, 2010.
24. Wang Y. Fiber and textile waste utilization. *Waste Biomass Valor* 1 (2010): 135–143.

25. Cherrett N, Barrett J, Clemett A, Chadwick M, Chadwick MJ. 2005. http://www.sei-international.org/mediamanager/documents/Publications/SEI-Report-EcologicalFootprintAndWaterAnalysisOfCotton-HempAndPolyester-2005.pdf. Accessed on March 10, 2014.
26. http://oecotextiles.wordpress.com/2011/01/19/estimating-the-carbon-footprint-of-a-fabric/. Accessed on March 10, 2014.
27. Strohle J. Textile achieve ecological footprint new opportunities for China. http://www.benningergroup.com/uploads/media/Carbon_Footprint_China_EN.pdf. Accessed on March 10, 2014.
28. Agarwal B, Jeffries B. Cutting cotton carbon emissions: Finding from Warangal, India. Report, 2013.
29. Apparel Industry life cycle carbon mapping, 2009. Business for Social Responsibility. https://www.bsr.org/reports/BSR_Apparel_Supply_Chain_Carbon_Report.pdf. Accessed on March 10, 2014.
30. Jungmichel N. *The Carbon Footprint of Textiles*. Systain Consulting, Berlin, Germany, 2010.
31. Russel IM. Sustainable wool production and processing. In Blackburn RS (ed.), *Sustainable Textiles: Life Cycle and Environmental Impact*, 1st edn. Woodhead Publishing Ltd., Cambridge, U.K., 2009, pp. 63–87.
32. Burn I, Bennotti p. http://www.metrocon.info/images/uploads/SWhittaker-METROCON12.pdf. Accessed on March 11, 2014.
33. Jute Eco-label: Life cycle assessment of jute products, 2006. http://www.jute.com:8080/c/document_library/get_file?uuid=e39c1527-75ed-47e9-9c88-c415ac11cf09&groupId=22165. Accessed on November 25, 2013.
34. Muthu SS, Li Y, Hu JY et al. Carbon footprint of production processes of polypropylene nonwoven shopping bags. *Fibers Text Eur.* 3(92) (2012): 12–15.
35. Shawl P. Product carbon footprint. http://www.ecotechenergy.in/pdf/carbon-footprint-pashmina-shawl.pdf. Accessed on March 11, 2014.
36. Documentation for emissions of greenhouse gases in the United States, 2003. http://www.eia.doe.gov/emeu/aer/txt/ptb1204.html. Accessed on March 21, 2014.
37. Vivek D. Carbon footprint of textiles, April 3, 2009. http://www.domain-b.com/environment/20090403_carbon_footprint.html. Accessed on March 21, 2014.
38. http://oecotextiles.wordpress.com/2013/10/03/fabric-and-your-carbon-footprint/.
39. Why natural fibers. FAO, Rome, Italy, 2009. http://www.naturalfibers2009.org/en/iynf/sustainable.html.
40. Fletcher K. *Sustainable Fashion and Textiles: Design Journeys*. Earth Scan, London, U.K., 2008.
41. Palm D. *Improved Waste Management of Textiles*. IVL Swedish Environmental Research Institute Ltd., Stockholm, Sweden [online], 2011. http://www.ivl.se/download/18.7df4c4e812d2da6a416800080103/B1976.pdf. Accessed on March 21, 2014.
42. Muthu SS, Li Y, Hu JY, Mok PY. Carbon footprint of shopping (grocery) bags in China, Hong Kong and India, *Ecol Indic* 18 (2012): 58.
43. http://www.esmap.org/sites/esmap.org/files/Tech%20Report%20-%20Manufacturing%20Nov%2009.pdf.

8 Environmental Impacts of Apparel Production, Distribution, and Consumption
An Overview

Kim Y. Hiller Connell

CONTENTS

8.1 Introduction 167
8.2 Overarching Environmental Impacts 168
8.3 Environmental Impacts of Fiber and Yarn Production... 169
    8.3.1 Environmental Impacts of Cotton Fiber and Yarn Production............................ 169
        8.3.1.1 Environmental Impacts of Agrochemicals 169
        8.3.1.2 Environmental Impacts of Water Demands... 170
        8.3.1.3 Environmental Impacts of Cotton Yarn Production 171
    8.3.2 Environmental Impacts of Polyester Fiber and Yarn Production 172
8.4 Environmental Impacts of Textile Production.. 172
    8.4.1 Textile Production and Water Consumption 173
    8.4.2 Textile Production and Water Quality 173
8.5 Environmental Impacts of Apparel Manufacturing and Distribution 174
8.6 Environmental Impacts of Consumer Care 175
    8.6.1 Clothing Care and Water 175
    8.6.2 Clothing Care, Energy Consumption, and Air Pollution..................................... 176
    8.6.3 Environmental Impacts of Dry Cleaning 176
8.7 Environmental Impacts of End of Life/Disposal.. 176
    8.7.1 Environmental Impacts of Solid Waste 177
8.8 Conclusions... 177
References.. 178

8.1 INTRODUCTION

On an annual basis, global demand for apparel and textile products is increasing. According to Marketline (2012), in 2011 the global apparel and textile market had revenue totals just over $3 trillion USD, with an annual growth rate of 3.7% between the years 2007 and 2011. Unfortunately, from the raw materials needed to make textiles to the garment end-of-life impacts, textiles and apparel damage the natural environment and are associated with significant environmental impacts. Through manufacturing outputs such as the release of toxic chemicals in wastewater, the emission of greenhouse gases into the air, and solid waste accumulation, apparel and textile products contribute toward environmental change (Slater 2003) and pose remarkable threats to the sustainability of the Earth's ecosystems. A report by the University of Cambridge Institute for Manufacturing found some of the most significant ways in which the apparel and textile industry negatively alters systems within the natural environment through decreasing air, water, and soil quality; decreasing biodiversity; creating dangerous greenhouse gases; depleting water and other renewable resources; and reducing nonrenewable resources (Allwood et al. 2006). Furthermore, the apparel and textile industry is one of the most significant contributors to greenhouse gas emissions and accounts for nearly 10% of total global carbon emissions (Zaffalon 2010). Current levels of demand for apparel and textile products and the expected significant increase in demand worldwide due to growing prosperity in developing countries have the potential to expand the apparel and textile industry impact on ecological systems.

This chapter takes a supply chain approach to detail the environmental impacts of apparel and textiles. The chapter starts with an examination of overarching environmental impacts that crosscut all aspects of the apparel and textile supply chain—high energy demand and high emissions of atmospheric pollutants such as greenhouse gases. Then the chapter proceeds to specify environmental impacts specific to each stage of the supply chain: fiber, yarn, and textile production, apparel manufacturing and distribution, consumer care, and finally garment disposal.

8.2 OVERARCHING ENVIRONMENTAL IMPACTS

This chapter's overview of the environmental impacts of apparel and textile production, distribution, and consumption begins with a discussion of two environmental issues that dominate throughout the supply chain: the high energy demands and the high emissions of greenhouse gases and other air pollutants.

Similar to many industrial processes, manufacturing of both textiles and apparel depends heavily on large and complicated machinery, typically powered through the burning of fossil fuels. Therefore, textile and apparel factories carry high energy demands. As Zaffalon (2010) states, "The estimated consumption for an annual global production of 60 billion kilograms of fabric boggles the mind: 1 trillion kilowatt hours of electricity" (para. 10). Furthermore, the energy demands continue throughout the apparel and textile supply chain. For example, the production and consumption of a t-shirt that is 100% cotton consumes 109 MJ of energy. This includes the energy required to grow and process the cotton fibers, manufacture and dye the yarns, knit the textile, construct the t-shirt, launder it 25 times, and incinerate the shirt after consumer disposal (Allwood et al. 2006).

The high energy demand of the apparel and textile industry is concerning because of our global dependence on fossil fuels for energy generation. According to projections in the International Energy Outlook, world energy consumption will increase by 56% over the next 30 years (US Energy Information Administration 2013a). Considering that fossil fuels (coal, natural gas, and petroleum) supply 87% of global energy (BP 2013), current systems of energy consumption are unsustainable. Furthermore, although oil makes up 33.1% of worldwide energy consumption, with coal at 29.9% and natural gas at 23.9%, in many countries where apparel and textile manufacturers dominate, coal is the primary source of energy. For example, in China and India, two of the top producers of apparel and textiles, coal supplies approximately 70% of each country's energy (US Energy Information Administration n.d.); and China and India combined are responsible for 54% of global consumption of coal (US Energy Information Administration 2013a).

Electricity generating power plants, particularly when coal-fired, release pollutants into the air, resulting in serious consequences to both human and environmental health. For instance, by-products of burning coal include sulfur dioxide (a cause of acid rain), nitrogen oxides (results in smog and susceptibility to respiratory diseases), and toxic heavy metals such as mercury, lead, and cadmium (known neurotoxins and carcinogens) (Center for Hazardous Substance Research 2009). The dangers to human health are such that "an estimated 3 million people die each year, more than 8000 a day, from breathing polluted air—much of it from burning coal" (Brown 2009: 74). However, more significantly, electricity generation is the largest contributor to global greenhouse gas emissions, the primary driver of climate change (Stocker et al. 2013). Furthermore, due to its carbon-intensive nature, electricity generated from coal-fired power plants produces up to two times more carbon dioxide emissions per unit of energy produced compared to natural gas and petroleum and accounts for 63% of recent increases in global temperatures (Brown 2009). With electricity generation contributing 33% of greenhouse gas emissions, the connection between energy demands and climate change cannot be understated (US Energy Information Administration 2013b).

Due to the apparel and textile industry's heavy reliance on fossil fuels for energy, and even more specifically, coal, it is one of the largest industrial contributors to greenhouse gas emissions and climate change, accounting for nearly 10% of total global emissions (Zaffalon 2010). In 2010, the electricity demands of the apparel and textile industry consumed 132 million tons of coal (Siegle 2011). Additionally, the Business for Social Responsibility (2009) calculated aggregate clothing life cycle greenhouse gas emissions and determined that fiber production accounted for 18% of total clothing greenhouse gas emissions, spinning for 16%, and consumer case for 39%. The other stages along the supply chain ranged between 1% and 7%. Therefore, the apparel and textile industry is a major player in global climate change.

We know that the earth is getting warmer. In fact, it has increased by 0.8°C since 1880, with two-thirds of that increase occurring over the last 40 years (Stocker et al. 2013). We also know that climate change is one of the most serious global threats to sustainability. Climate change impacts the environment and society in numerous ways, including (but not limited to) increasing or decreasing rainfall, decreasing crop yields, biodiversity loss, acidifying oceans, shifting habitats, melting glaciers, and rising oceans (Environmental Defense Fund 2014). As stated by Brown (2009: 59),

> The effects of rising temperature are pervasive. Higher temperatures diminish crop yields, melt the mountain glaciers that feed rivers, generate more destructive storms, increase the severity of clouding, intensify drought, cause more frequent and destructive wildfires, and alter ecosystems everywhere.

Throughout the apparel and textile industry, the two environmental issues of high energy demands and emissions of greenhouse gases and other pollutants dominate and are intrinsically correlated. Therefore, given the urgent global necessity to stabilize climate change and reduce dependence on nonrenewable energy sources, it is critical for the apparel and textile industry to immediately confront and work to find sustainable solutions to these issues.

8.3 ENVIRONMENTAL IMPACTS OF FIBER AND YARN PRODUCTION

The majority of the world's clothing is made from a range of natural plant and animal fibers or synthetic chemicals, for which petroleum is the primary material input. However, in the apparel and textile industry, cotton and polyester fibers dominate. In fact, more than 80% of global textiles are either cotton or polyester (Baugh 2008). Therefore, this section focuses on discussing the environmental impacts of fiber and yarn production for cotton and polyester.

8.3.1 Environmental Impacts of Cotton Fiber and Yarn Production

At around 35% of total world fiber consumption (United States Department of Agriculture 2013), cotton is one of the main fibers in the world, with an estimated annual global production of 24 million tons (Draper et al. 2007). Due to the high demand for the fiber, typical cotton production occurs through large-scale agricultural practices and results in environmental impacts from both a heavy reliance on agrochemicals and high water demands.

8.3.1.1 Environmental Impacts of Agrochemicals

The negative environmental impacts of conventional cotton fiber production on soil, air, water, and living organisms are well documented (Arthington 1996; Bedford 1996; Chouinard and Brown 1997; Myers and Stolton 1999; Walsh and Brown 1995). A primary reason why conventional cotton is a significant cause of environmental degradation is that cotton is one of the most chemically intensive crops, requiring some of the world's most toxic agrichemicals (Draper et al. 2007). Synthetic agrochemicals, including herbicides, pesticides, insecticides, fertilizers, and defoliants, are applied to cotton crops throughout its cultivation. The four main classes of chemicals most commonly used in cotton production are organophosphates, carbamates, pyrethroids, and organochlorides (Lee and Hamnett 2007), and the World Health Organization has classified many of these chemicals as highly hazardous (Draper et al. 2007). The cotton plants absorb and utilize only a small percentage of those chemicals—leaving large amounts of harmful, sometimes toxic, chemicals to disperse through the air, leech into soil and groundwater, and run off into the surface water.

The ecological impacts of these chemicals are numerous for both human and ecosystem health. Agricultural workers, because of their direct contact with the agrochemicals, suffer from a range of pesticide-related illnesses, including cancers and respiratory diseases (Reeves et al. 2002). And a study by Scarborough et al. (1989) found that communities near cotton fields in California had 60%–100% higher complaints of human health ailments such as nausea, diarrhea, and eye, nose, and throat irritation, when compared to communities near other types of agriculture.

In terms of ecosystem health, the outcomes of the intensive chemical applications include impacts to both soil and water. In the context of soil, many of the most common agrochemicals applied to cotton result in both decreased soil biodiversity and fertility. As already stated, the cotton plants only absorb a small percentage of the chemicals applied throughout the cultivation process; and as a result, large amounts of the chemicals leach into the soil. These chemicals, designed to eradicate cotton pests and weeds, also interrupt other biological processes and destroy much of the beneficial biodiversity of the soil, including microorganisms, plant, and insect life (Pesticide Action Network n.d.). A second, soil-related, negative impact of the agrochemical inputs is the result of fertilizer applications and nitrogen residuals left in the soil. Increased nitrogen leads to soil mineralization and increased decomposition of organic matter. With less organic matter in soil, it becomes more difficult for the roots of the cotton plants to break through the ground and absorb any natural nutrients present in the soil—resulting in crops even more dependent on synthetic fertilizers (Kramer et al. 2006).

The second significant environmental implication of agrochemical applications to cotton crops relates to ground and surface water quality. First, residues of pesticides and other chemicals applied to cotton plants leach into and contaminate ground and surface waters. Second, fertilizer residues, high in nitrates and phosphates, seep into aquatic systems and contribute to the phenomenon of eutrophication—the result of accelerated growth of algae in lakes and other bodies of water. Algae blooms in an aquatic system disrupt the natural photosynthesis process. The thick layer of algae on top of the water prevents sunlight, necessary for aquatic organisms to produce energy, from penetrating the water. Furthermore, because of the interruption in photosynthesis, levels of dissolved oxygen in the system decrease and aquatic animals suffocate—leading to dead zones within lakes and other bodies of water (Schindler and Vallentyne 2008).

8.3.1.2 Environmental Impacts of Water Demands

Another environmental concern associated with cotton fiber production is because cotton crops demand large volumes of water inputs. Although cotton crops in some regions of the world, such as West Africa and Brazil, are predominately rain fed and require minimal water inputs through irrigation (Grose n.d.), in other, more arid regions of the world (such as Uzbekistan, Egypt, Turkey, and India), cotton fiber production requires large inputs of irrigated water. In fact, a study by Soth and de Man (2000) estimates that up to 73% of global cotton is irrigated. Additionally, inefficient agricultural practices compound the water requirements, leading to needing 1400–3400 gallons of irrigated water to grow 1 lb of cotton (Baugh 2008).

Irrigation of cotton crops results in several signification ecological consequences, including salinization of soil, sedimentation of aquatic systems, and depletion of water resources. Salinization is the increase of salt content in soil. Fresh water always contains some level of dissolved salts. However, when soil is over- or inefficiently irrigated with poor drainage, croplands become waterlogged and salinized. As cotton plants absorb the water, the salts remain and, over time, accumulate in the soil. The amassed salts are detrimental to plant growth as they limit the ability of roots to draw up the water. Ecologically, the effects of salinization include decreased soil fertility and soil erosion (Ghassemi et al. 1995). Furthermore, sedimentation of aquatic systems (the deposition of particles of soil into a water system), with very similar ecological consequences as eutrophication, is a direct effect of soil erosion.

Finally, the depletion of fresh water resources resulting from the need to irrigate cotton in order to meet water demands is contributing to conditions of water stress throughout the world. The amount of water used in cultivating cotton is environmentally consequential because worldwide water usage is increasing and the expanding global population has placed fresh water supplies in many regions of the world under stress. Water use is actually already outpacing population growth and projections are that global water demands will increase by 40% over the next decade (Gleick 2014). This means that by 2025, two-thirds of the world's people will live in regions classified as water-stressed (Malik 2013).

The Aral Sea Basin in Eastern Europe is an exemplary case of the environmental risk associated with cotton production. This body of water touches the northwestern border of Uzbekistan, a country that is one of the world's top producers of cotton (National Cotton Council of America n.d.). Uzbekistan and the Aral Sea Basin are geographically located in a very arid region of the world that is not naturally well suited for the cultivation of cotton. Therefore, the cotton industry in this country is highly dependent on irrigation to meet agricultural needs; and since the 1960s, the Aral Sea Basin's tributaries have had vast amounts of water diverted for this purpose. The outcome of these diversions is that over the last 40 years the volume of water in the Aral Sea has decreased to just 1/10th of its original volume (Walters 2010), resulting in severe environmental consequences including loss of wetlands, decreased biodiversity, and local climate changes (Allwood et al. 2006). The water that remains in the basin is highly contaminated from excessive saline levels, fish cannot survive, and the sea basin's ecosystem is essentially decimated (Bedford 1996; Myers and Stolton 1999; The Aral Sea Crisis 2008). Also greatly affected by the region's environmental disaster is human health with increased rates of various diseases, birth defects, and miscarriages, with infant mortality rates in the region the highest of the country (The Aral Sea Crisis 2008).

8.3.1.3 Environmental Impacts of Cotton Yarn Production

After cultivation and harvest, cotton fibers go through a series of mechanical processes necessary to convert the raw cotton fiber into yarns that are suitable for knitting and weaving of textiles and other end uses. These processes include blending, mixing, cleaning, carding, drawing, and roving, and serve the purposes of cleaning, lengthening, and aligning the fibers for the spinning process. The fibers are then spun and transformed into cotton yarns (Kadolph 2010). Assuming the yarns are not dyed at this particular stage in the supply chain, the primary inputs needed for yarn manufacturing include the energy needed to operate the machines (which is considerable) and the water and chemicals required to prepare the fibers for spinning. For example, present on raw cotton is a waxy layer that has to be removed during the fiber preparation process and the most common practice for doing so is scouring the fibers in a sodium hydroxide solution (Karthik and Gopalakrishnan 2014). Sodium hydroxide, classified by the United States Environmental Protection Agency as a hazardous waste, is a highly corrosive substance and when discharged with a manufacturing facility's wastewater, can have significant environmental impacts such as increased pH levels and toxic effects on aquatic systems. Considering the other required inputs (energy and water), the environmental impacts of cotton yarn production are primarily the same as the previously discussed consequences of energy production, as well as water and energy consumption and water pollution.

8.3.2 ENVIRONMENTAL IMPACTS OF POLYESTER FIBER AND YARN PRODUCTION

Besides cotton, the synthetic fiber, polyester, is very important throughout the apparel and textile industry. Additionally, while global demand for cotton has remained relatively constant in recent years, since 1990 the demand for polyester has nearly doubled (Allwood et al. 2006). The primary raw material used in manufacturing the majority of synthetic fibers, including polyester, is petroleum. As a result, manufacturing of polyester contributes to the depletion of nonrenewable fossil fuels and the many ecological risks associated with their extraction and processing (Slater 2003). In addition to the utilization of petrochemicals, the production of synthetic fibers also requires environmentally dangerous inputs and creates unsafe emissions as outputs. For example, polyester manufacturing commonly uses antimony (a known carcinogen) as a catalyst during production and emits into water and air dangerous substances such as heavy metals, sodium bromide, and antimony oxide. Therefore, if manufacturing facilities do not have adequate wastewater and air filtrations systems, these dangerous substances can be released into both water and air (Natural Resources Defense Council 2011).

Additionally, both polyester fiber and yarn production are energy intensive, requiring large energy inputs and outputting high greenhouse gas emissions—considerably more compared to the production of cotton (Cherret et al. 2005; Natural Resources Defense Council 2011). The total energy required to product 1 ton of spun polyester fiber ranges between approximately 104,500 and 127,000 MJ. On the other hand, the same amount of conventional cotton requires approximately 25,600 MJ of energy (Cherrett et al. 2005). Logically, because CO_2 and other greenhouse gas emissions strongly correlate to energy demands, polyester fiber production emits considerable more emissions compared to cotton. While conventional cotton production emits between 2.35 and 5.89 kg of CO_2/tons of cotton fiber, CO_2 emissions from polyester fiber production range from 7.2 to 9.5 kg/tons of fiber (Cherrett et al. 2005).

8.4 ENVIRONMENTAL IMPACTS OF TEXTILE PRODUCTION

Following fiber and yarn production is textile production (through either weaving or knitting), a stage that requires further energy, water, and chemical inputs. For example, cotton textile production utilizes environmentally hazardous polyvinyl or polyacrylic compounds (referred to as sizing) to reduce yarn breakage during weaving (Slater 2003) and knit textile production involves the application of oils and other compounds.

It is generally after textiles are woven or knit that the dyeing, printing, and finishing occurs.* Prior to these stages, some textiles must be further prepared through desizing (cotton), scouring, bleaching, mercerizing, and drying, all processes that utilize energy and water, chlorine and hydrogen peroxide bleaches, and other chemicals such as sodium hydroxide and sulfuric acid (Ren 2000; Slater 2003). As a whole, the cleaning processes result in a wide range of chemicals that end up as toxic effluent in water systems. Furthermore, many of these substances are slow to biodegrade and difficult to treat.

Overall, in addition to the overarching environmental issues of carrying a large energy load and producing high levels of greenhouse gases, two significant environmental issues are present at the stage of textile production: water depletion and water pollution. A discussion of both of these issues follows.

---

* Dyeing is a process that may occur at multiple stages within the supply chain, including at the fiber, yarn, textile, or garment manufacturing stages. The environmental impacts will vary depending on the stage. However, for the sake of simplicity, this chapter presents the environmental impacts of dyeing at the textile stage.

8.4.1 TEXTILE PRODUCTION AND WATER CONSUMPTION

Textile production requires large amounts of water, first to prepare fibers and yarns for weaving and knitting and then to color and finish the textiles. The textile industry is the second heaviest industrial consumer of water (behind only agriculture) (Huntsman n.d.). Textiles, frequently resistant to dye absorption, can sometimes require as many as eight dye baths before the desired level of color saturation is achieved, with each bath entailing additional fresh water inputs (Hessel et al. 2007). The Natural Resources Defense Council (Greer 2010) claims that 200 tons of fresh water are needed per ton of manufactured textiles; the daily water consumption of an average-sized textile manufacturing facility is 1.6 million liters (Khan and Malik 2014); and, in 2010, textile production consumed 7 trillion liters of water in order to meet global textile demands (Siegel 2011). Considering predictions that by the year 2050, over 1 billion people will not have adequate access to water (Huntsman n.d.), the quantity of water consumed in textile production is of critical concern.

8.4.2 TEXTILE PRODUCTION AND WATER QUALITY

The wet finishing processes of textile production (dyeing, printing, and finishing) also significantly contribute to the pollution of water and decreased water quality. These processes depend on the application of additional, often toxic and/or carcinogenic, synthetic chemicals including acids, alkalis (sodium hydroxide, potassium hydroxide, sodium carbonate), salts (sodium chloride), stabilizers (sodium silicate, sodium nitrate), surfactants, and of course dyes (Environmental Technology Best Practice Programme 1997). In fact, the textile dyeing and finishing component of the supply chain is one of the most chemically intensive industrial processes, with over 8000 different chemicals used throughout various aspects of dyeing, printing, and finishing. These processes are only second to agricultural in terms of industrial contributions toward water pollution, and according to World Health Organization estimates, 17%–20% of all industrial water pollution is from textile dyeing and finishing (Simlai 2013).

Some of the most commonly used dyes are azo and triphenylmethane compounds, both of which contain carcinogens and endocrine disruptors. Furthermore, many dyes contain dioxins (carcinogens and hormone disruptors), heavy metals such as chromium, cadmium, cobalt, and zinc (carcinogens), and formaldehyde (also a carcinogen) (Lewis and Gertsakis 2001). Textiles also frequently have finishes applied to them to alter the properties of the fibers and, for example, to prevent staining or make the product water repellant or flame retardant. Like dyes, many of the chemical finishes are environmentally harmful, biologically persistent, and slow to biodegrade. Some of the common finishes include hazardous halogen compounds, antimony oxides, organocompounds, and chlorophenylides (Lewis and Gertsakis 2001).

Water effluent from textile dye factories is highly toxic and the chemicals are both persistent in waste water and resistant to treatment processes (Hessel et al. 2007; Ren 2000). This is because not all of the dyes and other chemicals fix to the fibers during wet processing. Reactive dyes dominate the industry because they are used for cotton and, unfortunately, they also have the lowest rate of fixation. Estimates are that 20%–50% of unfixed dye remains in the wastewater effluent postfinishing (Environmental Technology Best Practice Programme 1997). In general, the wastewater created during the textile coloration process has unfixed dye effluence that contains excessive levels of biological oxygen demand (BOD) and chemical oxygen demand (COD). The effluent is also high in toxic substances, salts, metals, colorants, suspended solids, and volatile organic compounds (VOCs), many of which are not easily removed from the effluent by traditional waste water treatments (Hessel et al. 2007; Ren 2000).

The environmental consequences of textile production are acutely evident in cities where textile manufacturing dominates and where governments frequently do not mandate wastewater treatment. For instance, the city of Xintang, in the province of Guangdong, China, makes a third of all jeans sold in the world (Guang et al. 2013). However, the wastewater resulting from bleaching, dyeing, and finishing denim in the city's factories are flushed directly into nearby rivers and streams, frequently without treatment. Reports indicate obvious signs of indigo dye in the rivers and high levels of toxic heavy metals such as lead, copper, and cadmium (Greenpeace 2011). Not only are the substances in the wastewater poisoning aquatic systems and depleting fish stocks, the wastewater is being discharged into water utilized by millions of people.

The seriousness of the ecological consequences of the water demands of the textile industry and its significant contributions toward water contamination cannot be overstated. Global water quantities and quality is commonly cited as one of the top environmental concerns facing society (Circle of Blue 2009). Adequate access to clean water is fundamental to both human and ecosystem health. Depletion of fresh water resources is accelerating and has the potential to lead to devastating consequences such as food scarcity, land and water conflicts, and displacement of millions of people (Brown 2009).

8.5 ENVIRONMENTAL IMPACTS OF APPAREL MANUFACTURING AND DISTRIBUTION

After textile manufacturing, the next stages in the apparel and textile supply chain are apparel manufacturing and distribution; and environment impacts continue in these stages. Similar to processes just described, garment manufacturing is dependent on energy-intensive machinery. This stage also requires additional material inputs such as thread, buttons, zippers, and trims and produces considerable solid waste through unused textiles—which can amount to as much as 15% of the fabric (Abernathy et al. 1999). After the garment is constructed, the application of more chemical finishes sometimes occurs; and the products may be washed again before leaving the manufacturing facility (Slater 2003). Therefore, this stage in the supply chain further depletes natural resources and creates even more solid waste, along with continued air and water pollution.

In assessing the total environmental impacts of the apparel and textile industry, it is also important to consider the environmental costs of transportation. Through globalization, fiber, textile, and apparel production has spread throughout many different nations; and a single garment could potentially travel the globe before purchase by a consumer. For example, cotton fiber that is grown in California may be processed and spun into yarns in South Carolina, knit into a textile in China, and sewn into a garment in South America. Moreover, because of the globalization of the apparel and textile industry, many countries import the vast majority of clothing. The United Kingdom imports 95% of all clothing sold in that country (Draper et al. 2007) and the United States imports 97% (American Apparel and Footwear Association 2012).

Global transportation of apparel and textiles occurs through a combination of truck, rail, ship, and airplane, all of which rely on fuels that deplete nonrenewable resources, create greenhouse gases and other pollutants, and contribute to climate change. The form of transportation plays a large part in determining the overall environmental impact. Airplanes emit the most pollution and clothing retailers can drastically reduce ecological consequences by selecting to transport cargo overseas via ship. In fact, the Natural Resources Defense Council (2012) states that up transporting clothing by ship versus plane emits up to 99% fewer greenhouse gases. Similarly, rail transportation emits fewer greenhouse gases and other air pollutants compared to trucks (Natural Resources Defense Council 2012).

The distribution stage of the apparel and textile supply chain also includes the retailing component of the industry and all of the activities involved in moving product from distribution centers to retail outlets to consumers. In addition to the further transportation required for these activities, there are consequences of operating distribution and retail facilities, including the environmental impacts associated with generating the energy required to heat, cool, and light buildings. Additionally, there is considerable solid waste produced at the distribution and retailing stage of the apparel and textile supply chain. The retailing industry generates large volumes of paper, cardboard, plastic, and other nonhazardous solid waste through excessive packaging and marketing and promotional materials—much of which is never recycled. Therefore, in considering the overall environmental impacts of the textiles and clothing industry, it is unwise to overlook the activities involved in distribution and retailing.

8.6 ENVIRONMENTAL IMPACTS OF CONSUMER CARE

As consumers use and care for their clothing, environmental degradation continues. Throughout much of the Western world, typical practices for washing and drying clothes are among the most energy- and water-intensive activities occurring within the household (EPA 2012a). In fact, for some garments, dry cleaning and home laundry may result in more environmental harm than any other stage of garment's life cycle (Allwood et al. 2006; Business for Social Responsibility 2009; Chouinard and Brown 1997; Franklin-Associates 1993; Madsen et al. 2007) and up to 80% of a garment's carbon footprint occurs during consumer care (Business for Social Responsibility 2009; Draper et al. 2007). It is because of this that scholars suggest, "The biggest gains in environmental performance for many fashion and textile pieces can be made by tackling the impact arising from their washing and drying" (Fletcher 2013: 76).

The environmental consequences of clothing care vary according to both fiber content and individual habits. For example, compared to polyester and other synthetic fibers, cotton fibers absorb a greater degree of water while being washed. Therefore, cotton garments require higher temperatures and more time in the dryer, and will have a greater environmental impact (Franklin-Associates 1993). Additionally, individuals vary significantly in their laundry habits related to behaviors such as water temperatures, propensity to use a dryer, and size of loads, all of which change environmental impacts (Laitala and Boks 2012). Despite these variables, home laundry practices consume water and energy and create water and air pollution, resulting in a heavy environmental burden.

8.6.1 CLOTHING CARE AND WATER

In the United States, the average washing machine uses about 41 gallons of water per load (EPA 2012a). Furthermore, a typical U.S. household does 400 loads of laundry each year. Therefore, in the United States each year, home laundry practices consume approximately 16,400 gallons of water per household (EPA 2012a). The amount of water used in washing clothing is environmentally consequential for the same reasons detailed previously related to water consumption in other stages of the supply chain. Therefore, water conservation strategies during clothing care are essential to global sustainability.

In addition to the water demands of laundry practices, clothing care also contributes to water pollution, with the primary source being the chemical ingredients used in laundry detergents. Detergents typically include surfactants and other components such as builders, bleaches, colorants, optical brighteners, and solvents. At the conclusion of the wash cycle, these chemicals remain in the effluent, resulting in billions of tons of laundry chemicals ending up in wastewater on an annual basis (Bajpa and Tyagi 2007). These chemicals are frequently resistant to water treatment and have the potential to harm both aquatic and human health. For example, the group of surfactants, nonylphenol ethoxylates (NPEs), is highly toxic, a known endocrine inhibitor, and harmful to aquatic reproductive systems (Hogue 2006). Additionally, linear alkyl benzene sulfonates (LAS), also a group of surfactants, have been linked to health effects such as nausea, vomiting, and diarrhea (Bajpa and Tyagi 2007). Finally, many solvents included in detergents are both toxic to aquatic organisms and slow to biodegrade, builders (such as phosphates) have the potential to cause excessive algae growth and eutrophication in fresh water, and sodium hypochlorite bleach is also toxic (EPA 2013a).

An additional water pollution concern associated with laundering relates to the fact that, during the wash cycle, polyester garments shed tiny polyester fibers called microplastics into the water effluent. Research indicates one polyester garment can produce as many as 1900 fibers during one wash cycle. Unfortunately, due to the design of the facilities, wastewater treatment plants do not filter the microplastics out of the water. Instead, the microplastics are flushed back into water systems where they do not biodegrade and contaminate aquatic systems. Of particular concern is that microplastics are likely ingested by and harmful to numerous fish and other aquatic organisms (Browne et al. 2011).

8.6.2 Clothing Care, Energy Consumption, and Air Pollution

In home laundry processes, the heating of the water and operation of both the washing and drying machines requires energy inputs. According to the EPA (2012a), 90% of a washing machine's energy demands goes to heating the water and machine drying is the largest consumer of energy within the home clothing care process (Business for Social Responsibility 2009; Stamminger 2011). As previously detailed, globally we rely on nonrenewable energy sources and fossil fuels to generate a majority of our energy. For example, in the United States, nonrenewable energy sources and fossil fuels (coal, natural gas, and petroleum) generate 68% of the country's electricity, with coal generating the highest portion at 37% (US Energy Information Administration 2014). Therefore, a significant environmental impact of home laundry practices is the depletion of nonrenewable resources.

Due to the high energy demands of washing and drying clothes, our reliance on energy generated by fossil fuels, and the serious environmental consequence of burning fossil fuels, consumer laundry practices considerably contribute to general air pollution and greenhouse gas emissions. In fact, for garments regularly washed and dried, "laundering accounts for 40%–80% of the total life cycle GHG emissions" (Business for Social Responsibility 2009: 3). Therefore, we must not overlook the contributions of clothing care practices to air pollution and global climate change.

8.6.3 Environmental Impacts of Dry Cleaning

Finally, related to the environmental impacts of clothing care, many of the chemical solvents used in dry cleaning processes, such as tetrachloroethylene/perchlorethylene (PERC), are toxic to humans and hazardous to the natural environment (EPA 2012b). Long-term health effects include adverse effects to organs and both immune and reproduction systems, and the EPA (2012b) classifies the solvent as a likely carcinogen. Tetrachloroethylene is also attributable to environmental contamination of soil, water, and air (Greer 2011). So much so that, in the United States, three-quarters of dry cleaning facilities using PERC have contaminated nearby soil and water resources (State Coalition for Remediation of Drycleaners 2010).

8.7 ENVIRONMENTAL IMPACTS OF END OF LIFE/DISPOSAL

Finally, as clothing products reach their end of life, the contribution toward environmental degradation continues through the creation of solid waste. The textile industry is highly efficient in capturing preconsumer fiber and textile waste, and it recycles 75% of the solid waste created through industrial processes (Zimring and Rathje 2012). Common uses for preconsumer waste fibers and textiles include raw materials for automotive seating, furniture, mattress, coarse yarn, home furnishings, paper, and other industries. However, postconsumer recycling of clothing, textiles, and related products is significantly less than of preconsumer products. According to the Council for Textile Recycling (2014), on an annual basis U.S. consumers prevent approximately 3.8 billion lb of textile products (including clothing, footwear, accessories, towels, bedding, and draperies) from entering the municipal solid waste stream through donation and recycling programs. However, that represents only 15% of the country's textile waste. A much larger portion (85%) end up in landfills. On a yearly basis, U.S. consumers throw away approximately 70 lb of textiles per person (Council for Textile Recycling 2014). Furthermore, the volume of postconsumer textile waste going to U.S. landfills is increasing—between 1999 and 2009 it grew by 40% (EPA 2010).

8.7.1 Environmental Impacts of Solid Waste

Currently, 5.2% of municipal solid waste is postconsumer textile waste (EPA 2014), and this textile waste, in combination with other solid waste, has serious environmental costs. Of course there are the obvious irritants of landfills attracting pest such as rats and other animals as well as having an odor and being noisy. However, more considerable are the impacts of landfills on local environments in terms of air and soil contamination and the production of methane, a potent greenhouse gas.

In landfills, the decomposition of organic materials produces methane as an off-gas. As a greenhouse gas, methane is the second largest contributor to climate change and compared to carbon dioxide, it is actually 20 times more powerful in terms of impact on climate change (EPA 2013b). Therefore, the climate change implication of methane generation in landfills is not to be underestimated. Landfills also release other air pollutants and toxins, many of which include carcinogens.

Although most modern landfills, through careful design and strict regulations, control the release of pollutants, municipal solid waste landfills do contribute toward soil and water pollution through the generation of both gases and leachate—a liquid that forms as waste decomposes. Leachate, highly toxic, filters through the landfill and into the surrounding soils as well as ground and surface water systems, posing serious environmental and health risks and interrupting ecosystems (El-Fadel et al. 1997).

8.8 CONCLUSIONS

It is difficult to produce, distribute, and consume clothing without depleting natural resources and creating solid and sometimes hazardous waste and pollution. As our society continues to value materialism and consumption, we also place more stress on the natural environment, deplete natural resources, and strain the earth's carrying capacity. From the acquisition of raw materials to end of life, apparel production and consumption contributes toward global environmental change and degradation. Therefore, the environmentally unsustainable consumption of apparel and textile products is an increasingly important phenomenon.

Fortunately, current activity responding to the challenge of environmental concerns related to the apparel and textile industry suggests a growing commitment to sustainable practices. For example, developments within the agricultural industry such as integrated pest management and the introduction of genetically modified cotton seeds with insecticidal activity has successfully decreased the number of insecticide applications to U.S. cotton crops by almost 50% (International Cotton Advisory Committee 2010). From an apparel manufacturing perspective, there are also numerous examples of progress toward a more sustainable industry. Levi Strauss developed a process termed Water > Less, a finishing process that eliminates the use of water for its top-selling 501 jeans (Joule n.d.); and Nike is partnering with a Netherland firm who has developed the first waterless textile dyeing machine using recaptured carbon dioxide gas (Nike 2012). Similarly, Dow Corning is exploring the potential use of environmentally preferably silicone textile printing inks for textile printing (Dow Corning 2009).

Collaborative industry initiatives have also evolved to develop a number of indices to assist the apparel and textile supply chain in improved decision making. The outdoor clothing industry led the development of the Eco Index, a metric used to evaluate the environmental criteria of garments and footwear (Binkley 2010). Even more promising, apparel and textile companies such as Nike, Gap, Walmart, H&M, Marks and Spencer, Patagonia, and many others launched the Sustainable Apparel Coalition, a nonprofit organization focused on both environmental and social impacts in the industry (Sustainable Apparel Coalition, 2014). The primary focus of this organization, with member companies representing over a third of the apparel and footwear market, is the Higg Index—an assessment tool for measuring the environmental impacts of garments and footwear across the product life cycle and throughout the supply chain. So while the environmental consequences associated with the production, distribution, and consumption of apparel and textiles are serious, there are promising signs that a significant paradigm shift toward sustainability is beginning.

REFERENCES

Abernathy, F.H., J.T. Dunlap, J.H. Hammond, and D. Weil. 1999. *A Stitch in Time: Lean Retailing and the Transformation of Manufacturing—Lessons from the Textile and Apparel Industries.* Oxford, U.K.: Oxford University Press.

Allwood, J.M., S.E. Laursen, C.M. de Rodriguez, and N.M.P. Bocken. 2006. *Well Dressed? The Present and Future Sustainability of Clothing and Textiles in the United Kingdom.* Cambridge, U.K.: University of Cambridge, Institute for Manufacturing.

American Apparel and Footwear Association. 2012. AAFA releases apparel stats 2012 report. https://www.wewear.org/aafa-releases-apparelstats-2012-report/. (Accessed January 13, 2015.)

Arthington, A.H. 1996. The effects of agricultural land use and cotton production on tributaries of the Darling River, Australia. *GeoJournal* 40(1–2): 115–125.

Bajpai, D. and V.K. Tyagi. 2007. Laundry detergents: An overview. *Journal of Oleo Science* 56(7): 327–340.

Baugh, G. 2008. Polyester vs. cotton—Which is better for the environment? *FIBER2.* http://www.udel.edu/fiber/issue2/responsibility/. (Accessed January 13, 2015.)

Bedford, D.P. 1996. International water management in the Aral Sea basin. *Water International* 21(2): 63–69.

Binkley, C. 2010. How green is my sneaker? *Wall Street Journal.* http://online.wsj.com/news/articles/SB10001424052748703724104575379621448311224. (Accessed January 13, 2015.)

BP. 2013. BP statistical review of world energy 2013. http://www.bp.com/content/dam/bp/pdf/statistical-review/statistical_review_of_world_energy_2013.pdf.

Brown, L. 2009. *Plan B 4.0: Mobilizing to Save Civilization.* New York: W.W. Norton & Company.

Browne, M.A., P. Crump, S.J. Niven et al. 2011. Accumulation of microplastic on shorelines worldwide: Sources and sinks. *Environmental Science & Technology* 45(21): 9175–9179.

Business for Social Responsibility. 2009. Apparel industry life cycle carbon mapping. http://www.bsr.org/reports/BSR_Apparel_Supply_Chain_Carbon_Report.pdf. (Accessed January 13, 2015.)

Center for Hazardous Substance Research. 2009. Human health effects of heavy metals. https://www.engg.ksu.edu/CHSR/outreach/resources/docs/15HumanHealthEffectsofheavyMetals.pdf. (Accessed January 13, 2015.)

Cherrett, N., J. Barrett, A. Clemett, M. Chadwick, and M.J. Chadwick. 2005. *Ecological Footprint and Water Analysis of Cotton, Hemp, and Polyester.* Stockholm, Sweden: Stockholm Environment Institute.

Chouinard, Y. and M.S. Brown. 1997. Going organic: Converting Patagonia's cotton product line. *Journal of Industrial Ecology* 1(1): 117–129.

Circle of Blue. 2009. Water issues research. http://www.circleofblue.org/waternews/wp-content/uploads/2009/08/circle_of_blue_globescan.pdf. (Accessed January 13, 2015.)

Council for Textile Recycling. 2014. About CTR. http://weardonaterecycle.org/about/index.html. (Accessed January 13, 2015.)

Dow Corning. 2009. Dow Corning silicone textile printing inks. http://www.dowcorning.com/content/publishedlit/26-1446.pdf. (Accessed January 13, 2015.)

Draper, S., V. Murray, and I. Weissbrod. 2007. Fashioning sustainability: A review of sustainability impacts of the clothing industry. http://www.forumforthefuture.org/sites/default/files/project/downloads/fashionsustain.pdf. (Accessed January 13, 2015.)

El-Fadel, M., A.N. Findikakis, and J.O. Leckie. 1997. Environmental impacts of solid waste landfilling. *Journal of Environmental Management* 50(1): 1–25.

Environmental Defense Fund. 2014. Climate change impacts: The effects of warming on our world can be seen today. http://www.edf.org/climate/climate-change-impacts. (Accessed January 13, 2015.)

Environmental Technology Best Practice Programme. 1997. Water and chemical use in the textile dyeing and finishing industry. http://www.wrap.org.uk/sites/files/wrap/GG062.pdf. (Accessed January 13, 2015.)

EPA. 2010. *Municipal Solid Waste in the United State: Facts and Figures.* Office of Solid Waste. Washington, DC: United States Environmental Protection Agency.

EPA. 2012a. Laundry room & basement. http://www.epa.gov/greenhomes/Basement.htm. (Accessed January 13, 2015.)

EPA. 2012b. Tetrachloroethylene (perchloroethylene). http://www.epa.gov/ttn/atw/hlthef/tet-ethy.html. (Accessed January 13, 2015.)

EPA. 2013a. Key characteristics of laundry detergent ingredients. http://www.epa.gov/dfe/pubs/laundry/techfact/keychar.htm. (Accessed January 13, 2015.)

EPA. 2013b. Overview of greenhouse gases: Methane emissions. http://epa.gov/climatechange/ghgemissions/gases/ch4.html. (Accessed January 13, 2015.)

EPA. 2014. Municipal solid waste. http://www.epa.gov/epawaste/nonhaz/municipal/index.htm. (Accessed January 13, 2015.)

Fletcher, K. 2013. *Sustainable Fashion and Textiles: Design Journeys*. London, U.K.: Earthscan.

Franklin-Associates. 1993. *Resource and Environmental Profile Analysis of a Manufactured Apparel Product: Women's Knit Polyester Blouse*. Prairie Village, KS: Franklin Associates.

Ghassemi, F., A.J. Jakeman, and H.A. Nix. 1995. *Salinisation of Land and Water Resources: Human Causes, Extent, Management and Case Studies*. Wallingford, U.K.: CAB International.

Gleick, P.H., ed. 2014. *The World's Water Volume 8: The Biennial Report on Freshwater Resources*. Vol. 8. Washington, DC: Island Press.

Greenpeace. 2011. Dirty laundry: Unravelling the corporate connections to toxic water pollution in China. http://www.greenpeace.org/international/en/publications/reports/Dirty-Laundry/. (Accessed January 13, 2015.)

Greer, L. 2010. Revolutionizing the global textile industry. https://www.nrdc.org/international/files/revolutionizing.pdf. (Accessed January 13, 2015.)

Greer, L. 2011. How to care for the planet and your health while also caring for your clothes. http://www.nrdc.org/living/stuff/files/CBD-DryCleaning-FS.pdf. (Accessed January 13, 2015.)

Grose, L. (n.d.). Sustainable cotton production and processing. http://www.sustainablecotton.org/images/media/Sustainable_Cotton_Production_&_Processing.pdf. (Accessed January 13, 2015.)

Guang, L., J. Mingzhuo, and L. Guang. 2013. The denim capital of the world: So polluted you can't give the houses away. https://www.chinadialogue.net/article/show/single/en/6283-The-denim-capital-of-the-world-so-polluted-you-can-t-give-the-houses-away. (Accessed January 13, 2015.)

Hessel, C., C. Allegre, M. Maisseu, F. Charbit, and P. Moulin. 2007. Guidelines and legislation for dye house effluents. *Journal of Environmental Management* 83(2): 171–180.

Hogue, C. 2006. Surfactant scrutiny: EPA considers requiring texts on nonlyphenol ethoxylates for long-term effects on aquatic life. *Chemical Engineering News* 85(42): 33–35.

Huntsman. n.d. DyeCoo delivers sustainable textiles with waterless dyeing. http://www.huntsman.com/corporate/a/Innovation/DyeCoo%20delivers%20sustainable%20textiles. (Accessed January 13, 2015.)

International Cotton Advisory Committee. 2010. Factors influencing the use of pesticides in cotton in the U.S. https://www.icac.org/seep/documents/reports/2010_usa.pdf. (Accessed January 13, 2015.)

Joule, E. n.d. The Levi's® brand introduces waterless jeans. http://www.levistrauss.com/news/press-releases/levis-brand-introduces-waterless-jeans. (Accessed January 13, 2015.)

Kadolph, S. 2010. *Textiles*. Upper Saddle River, NJ: Prentice Hall.

Karthik, T. and D. Gopalakrishnan. 2014. Environmental analysis of textile value chain: An overview. In *Roadmap to Sustainable Textiles and Clothing*, S.S. Muthu (ed.), pp. 153–188. Kowloon Bay, Hong Kong: Springer.

Khan, S. and A. Malik. 2014. Environmental and health effects of textile industry wastewater. In *Environmental Deterioration and Human Health*, A. Malik, E. Grohmann, and R. Akhtar (eds.), pp. 55–71. Dordrecht, the Netherlands: Springer.

Kramer, S.B., J.P. Reganold, J.D. Glover, B.J.M. Bohannan, and H.A. Mooney. 2006. Reduced nitrate leaching and enhanced denitrifier activity and efficiency in organically fertilized soils. *Proceedings of the National Academy of Sciences of the United States of America* 103(12): 4522–4527.

Laitala, K. and C. Boks. 2012. Sustainable clothing design: Use matters. *Journal of Design Research* 10(1): 121–139.

Lee, M. and K. Hamnett. 2007. *Eco Chic: The Savvy Shopper's Guide to Ethical Fashion*. London, U.K.: Gaia.

Lewis, H. and J. Gertsakis. 2001. *Design + Environment: A Global Guide to Designing Greener Goods*. Sheffield, U.K.: Greenleaf Publishing.

Madsen, J., B. Hartlin, S. Perumalpillai, S. Selby, and S. Aumônier. 2007. *Mapping of Evidence on Sustainable Development Impacts That Occur in Life Cycles of Clothing: A Report to the Department for Environment, Food and Rural Affairs*. London, U.K.: Environmental Resources Management.

Malik, K. 2013. *Human Development Report 2013. The Rise of the South: Human Progress in a Diverse World*. New York: United Nations Development Programme.

Marketline. 2012. Global textiles, apparel and luxury goods. http://www.reportlinker.com/p016087-summary/Global-Textiles-Apparel-Luxury-Goods.html. (Accessed January 13, 2015.)

Myers, D. and S. Stolton. 1999. *Organic Cotton: From Field to Final Product*. London, U.K.: Intermediate Technology Publications.

National Cotton Council of America. n.d. Production ranking MY 2013. http://www.cotton.org/econ/cropinfo/cropdata/rankings.cfm. (Accessed January 13, 2015.)

Natural Resources Defense Council. 2011. Polyester. http://www.nrdc.org/international/cleanbydesign/files/CBD_FiberFacts_Polyester.pdf. (Accessed January 13, 2015.)

Natural Resources Defense Council. 2012. Clean by design: Transportation. http://www.nrdc.org/international/cleanbydesign/transportation.asp. (Accessed January 13, 2015.)

Nike. 2012. Nike, Inc. announces strategic partnership to scale waterless dyeing technology. http://nikeinc.com/news/nike-inc-announces-strategic-partnership-to-scale-waterless-dyeing-technology. (Accessed January 13, 2015.)

Pesticide Action Network. n.d. Cotton. http://www.panna.org/resources/cotton. (Accessed January 13, 2015.)

Reeves, M., A. Katten, and M. Guzman. 2002. Fields of poison 2002: California farm workers and pesticides. http://www.panna.org/sites/default/files/FieldsofPoison2002Eng.pdf. (Accessed January 13, 2015.)

Ren, X. 2000. Development of environmental performance indicators for textile process and product. *Journal of Cleaner Production* 8(6): 473–481.

Scarborough, M.E., R.G. Ames, M.J. Lipsett, and R.J. Jackson. 1989. Acute health effects of community exposure to cotton defoliants. *Archives of Environmental Health: An International Journal* 44(6): 355–360.

Schindler, D.W. and J.R. Vallentyne. 2008. *The Algae Bowl: Over Fertilization of the World's Freshwaters and Estuaries*. Edmonton, Alberta, Canada: University of Alberta Press.

Siegle, L. 2011. *To Die For: Is Fashion Wearing Out the World?* London, U.K.: Fourth Estate.

Simlai, N. 2013. *Water Stewardship for Industries: The Need for a Paradigm Shift in India*. Hyderabad, India: WWF-India.

Slater, K. 2003. *Environmental Impacts of Textiles: Production, Processes, and Protection*. Cambridge, U.K.: Woodbridge Publishing Limited.

Soth, J. and R. de Man. 2000. *Cotton and Freshwater in West Turkey: A Feasibility Study for a Conservation and Monitoring Project*. Gland, Switzerland: WWF International.

Stamminger, R. 2011. Modelling resource consumption for laundry and dish treatment in individual households for various consumer segments. *Energy Efficiency* 4(4): 559–569.

State Coalition for Remediation of Drycleaners. 2010. December Newsletter. http://www.drycleancoalition.org/download/news1210.pdf. (Accessed January 13, 2015.)

Stocker, T.F., D. Qin, G.-K. Plattner et al. 2013. *Climate Change 2013. The Physical Science Basis. Working Group I Contribution to the Fifth Assessment Report of the Intergovernmental Panel on Climate Change-Abstract for Decision-Makers*. Geneva, Switzerland: Intergovernmental Panel on Climate Change.

Sustainable Apparel Coalition. 2014. Home. http://www.apparelcoalition.org/. (Accessed January 13, 2015.)

The Aral Sea Crisis. 2008. Impacts to life in the region. http://www.columbia.edu/~tmt2120/impacts%20to%20life%20in%20the%20region.htm. (Accessed January 13, 2015.)

United States Department of Agriculture. 2013. Cotton & wool. http://www.ers.usda.gov/topics/crops/cotton-wool.aspx#.UtB5NrQeanI. (Accessed January 13, 2015.)

US Energy Information Administration. 2013a. International energy outlook 2013. http://www.eia.gov/pressroom/presentations/sieminski_07252013.pdf. (Accessed January 13, 2015.)

US Energy Information Administration. 2013b. Electricity explained. http://www.eia.gov/energyexplained/index.cfm?page=electricity_in_the_united_states. (Accessed January 13, 2015.)

US Energy Information Administration. 2014. What is US electricity generation by energy source? http://www.eia.gov/tools/faqs/faq.cfm?id=427&t=3. (Accessed January 13, 2015.)

US Energy Information Administration. n.d. Countries. http://www.eia.gov/countries/. (Accessed January 13, 2015.)

Walsh, J.A.H. and M.S. Brown. 1995. Pricing environmental impacts: A tale of two t-shirts. *Illahee* 11(3): 175–182.

Walters, P. 2010. Aral Sea recovery? http://news.nationalgeographic.com/news/2010/04/100402-aral-sea-story/. (Accessed January 13, 2015.)

Zaffalon, V. 2010. Climate change, carbon mitigation and textiles. http://www.textileworld.com/Issues/2010/July-August/Dyeing_Printing_and_Finishing/Climate_Change-Carbon_Mitigation_And_Textiles. (Accessed January 13, 2015.)

Zimring, C.A. and W.L. Rathje, eds. 2012. *Encyclopedia of Consumption and Waste: The Social Science of Garbage*. Vol. 1. Thousand Oaks, CA: Sage.

9 Life Cycle Assessment Studies Pertaining to Textiles and the Clothing Sector

Shabaridharan Karunamoorthy, Sohel Rana, Rosina Begam, Shama Parveen, and Raul Fangueiro

CONTENTS

9.1 Introduction 181
    9.1.1 Definition of LCA 181
    9.1.2 Phases of LCA 182
    9.1.3 Organizations Working on LCA and Standards 184
9.2 LCA Model for Textiles 184
9.3 LCA Studies on Natural Fibers and Textiles 187
9.4 LCA Studies on Regenerated Fibers and Textiles 196
9.5 LCA Studies on Synthetic Fibers and Textiles 197
9.6 LCA of Textile Recycling Process 201
9.7 Life Cycle Impact Assessment of Other Textile Processes 203
9.8 Limitations of LCA 203
9.9 Conclusion 204
References 204

9.1 INTRODUCTION

The Society of Environmental Toxicology and Chemistry (SETAC) is the first international body that stood as the umbrella for the concept of life cycle assessment (LCA). The efforts made by the organization started from 1989, in which year the first workshop on LCA was held in Smugglers Notch, Vermont. This workshop was followed by the next one in Leuven, Belgium, and acted as the basis for opening two different schools for LCA: one in North America and the other one in Europe. SETAC has also developed the "Code of Practice" for LCA, which was the first initially accepted technical framework for the modern LCA concept. In this chapter, an introduction on LCA, definitions, different phases of LCA, organization involved in LCA studies, and the standards developed to monitor LCA that are provided for LCA model, which can be used for textile materials, have been described. Various studies performed on LCA of natural fibers such as cotton and wool, regenerated fibers such as viscose and its blends, and synthetic fibers such as polyester and its blends have been discussed in detail. LCA of recycled textile materials has also been discussed. Finally, LCA of other textile processes and limitations of LCA have been presented in this chapter.

9.1.1 DEFINITION OF LCA

The International Organization for Standardization (ISO) defines LCA as "the compilation and evaluation of the inputs, outputs and potential environmental impacts of a product system throughout its life cycle." In general, two different approaches are followed in the LCA studies of products. The first and the default approach is the cradle-to-grave approach. The second approach is the cradle-to-factory gate or cradle-to-gate approach. In the first approach, the energy spent in each and every stage of the product, covering raw material, conversion stage or manufacturing process, packaging, distribution or transportation, consumer use and recycling, and disposal or end of life, is considered. In the second approach, the energy spent on the product until the product reaches the factory gate is considered. The transportation, consumer use, and disposal stages are excluded from the LCA study.[1]

9.1.2 Phases of LCA

The U.S. Environmental Protection Agency (EPA) defined LCA to be composed of four separate but interrelated components[2]:

1. Goal and scope
2. Life cycle inventory
3. Life cycle impact analysis
4. Interpretation

The diagrammatic representation of the four different phases of LCA and their interaction with each other are provided in Figure 9.1, and the details of LCA model are presented in Figure 9.2.

The cradle-to-grave approach is the *holistic* approach in which the final consumption or end product is the driving force for the economy. It indirectly helps to manage the environment. It also reduces the *problem shifting* (i.e., to solve one problem, shifting the problem to the next or successive levels of the product's life cycle), which may not give a good solution. When the entire life cycle, that is, cradle to grave of the product, is considered, such problem shifting can be eliminated.

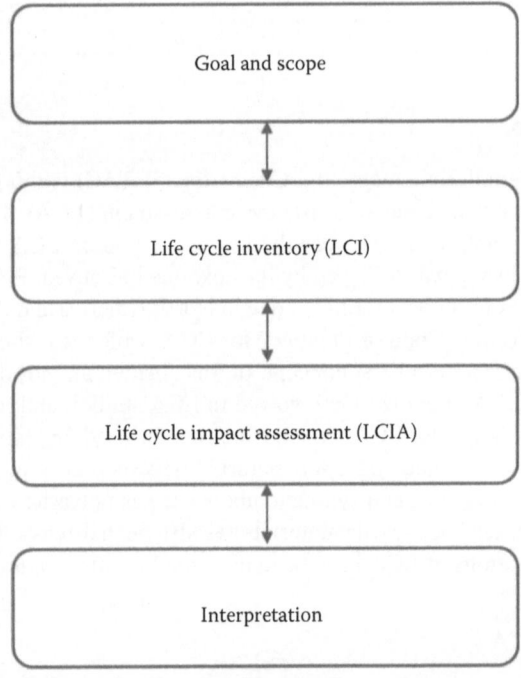

FIGURE 9.1 Representation of LCA model. (From Connell, D., The environmental impact of the textiles industry, in: Carr, C.M. (ed.), *Chemistry of the Textiles Industry*, Springer, Dordrecht, the Netherlands, 1995, pp. 333–354.)

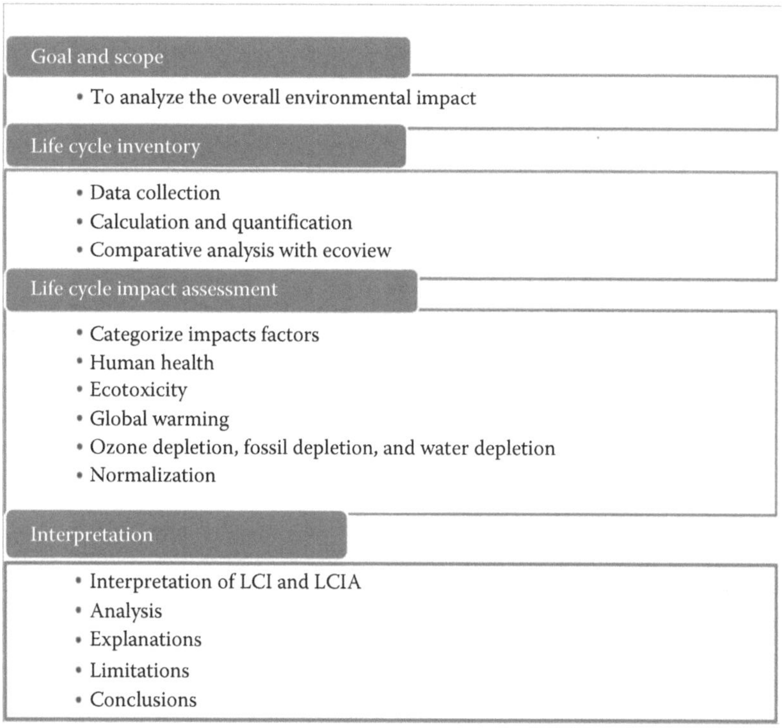

Life Cycle Assessment Studies Pertaining to Textiles and the Clothing Sector

Goal and scope
- To analyze the overall environmental impact

Life cycle inventory
- Data collection
- Calculation and quantification
- Comparative analysis with ecoview

Life cycle impact assessment
- Categorize impacts factors
- Human health
- Ecotoxicity
- Global warming
- Ozone depletion, fossil depletion, and water depletion
- Normalization

Interpretation
- Interpretation of LCI and LCIA
- Analysis
- Explanations
- Limitations
- Conclusions

FIGURE 9.2 Details of LCA model. (From Connell, D., The environmental impact of the textiles industry, in: Carr, C.M. (ed.), *Chemistry of the Textiles Industry*, Springer, Dordrecht, the Netherlands, 1995, pp. 333–354.)

In LCA, a detailed analysis of environmental impacts should be done. So, it is necessary to understand the details of the environment. The term *environment* can be defined as the circumstances surrounding an organism or group of organisms. Connell[3] has described environment into four different stages[4]:

1. Immediate environment
2. Local environment
3. Regional environment
4. Global environment

The immediate environment can be defined as the environment that interacts between an individual and his or her surrounding, which is clothing. The comfort and fit are the primary attributes of clothing that make it comfortable to the wearer. The local environment is the one that affects a group of people and is related with their locality. The regional environment is the environment that produces effects from another location far from the residing or current location. A typical example of regional environmental impact is the acid rain in Norway and Sweden as a result of fossil fuel burning in other Western European countries. The global environment is the one that changes and affects each and every human directly (or) indirectly. Examples of global environmental impact are ozone layer depletion and greenhouse effect. The following are the applications of LCA study of any product:

The first application of LCA is the analysis of origin of a problem pertaining to a product. For example, if the problem is identified in the manufacturing section of the product, usual approach will be to handle the problem in a localized manner. The technologist may not be able to solve the problem beyond the boundary conditions of the production unit. But, with the use of LCA approach, the origin of the problem can be identified and solutions can be found out. The second application of the LCA is to improve the performance of the product through comparison. For example, alternative raw materials or energy sources for the product can be identified to improve the life or performance of the product and to increase the environmental benefits. The third application of LCA is to develop a new product. The fourth application is to choose the best one among the comparable products. The selection will be based on the life and environmental impact of different products.

9.1.3 Organizations Working on LCA and Standards

Many international organizations are working for LCA. Some primary organizations and the milestones achieved in the process of LCA are discussed in following paragraphs.

SETAC is the first international organization that explored the concept of LCA. ISO is a private organization in which many interested institutes and organizations from different parts of the world take participation to improve their work practices and quality of products. ISO has made significant contribution in the field of LCA by developing a series of standard methods[5-7]:

- ISO 14001:2004 Environmental management systems—Requirements with guidance for use
- ISO 14004:2004 Environmental management systems—General guidelines on principles, systems, and support techniques
- ISO 14006:2011 Environmental management systems—Guidelines for incorporating ecodesign
- ISO 14064-1:2006 Greenhouse gases—Part 1: Specification with guidance at the organization level for quantification and reporting of greenhouse gas emissions and removals
- ISO in the 14040 series (environmental management—life cycle assessment)
- ISO 14040:2006 Environmental management—Life cycle assessment: Principles and framework
- ISO 14044:2006 Environmental management—Life cycle assessment: Requirements and guideline
- ISO/TR 14047:2012 Environmental management—Life cycle assessment: Illustrative examples on how to apply ISO 14044 to impact assessment situation
- ISO/TS 14048:2002 Environmental management—Life cycle assessment: Data documentation format
- ISO/TR 14049:2012 Environmental management—Life cycle assessment: Illustrative examples on how to apply ISO 14044 to goal and scope definition and inventory analysis
- ISO/TS 14071:2014 Environmental management—Life cycle assessment: Critical review processes and reviewer competencies; additional requirements and guidelines to ISO 14044:2006
- ISO 14020 series: Environmental labels and declarations
- Publically available specification (PAS 2050:2011)—Specifications for the assessment of life cycle greenhouse gas emissions of goods and services
- Product life cycle accounting and reporting standard (WRI and WBCSD)

The United Nations Environmental Programme, which is represented by the Department of Technology, Industry and Economics, Paris, has organized a series of workshop on LCA in association with the U.S. EPA.

9.2 LCA MODEL FOR TEXTILES

Textile materials can be broadly classified into two categories, namely, natural and man-made fibers. Textile materials that are used by the consumers can be divided into four different phases. The following flow chart (Figure 9.3) represents the four different phases of life cycle of a textile material.

Life Cycle Assessment Studies Pertaining to Textiles and the Clothing Sector

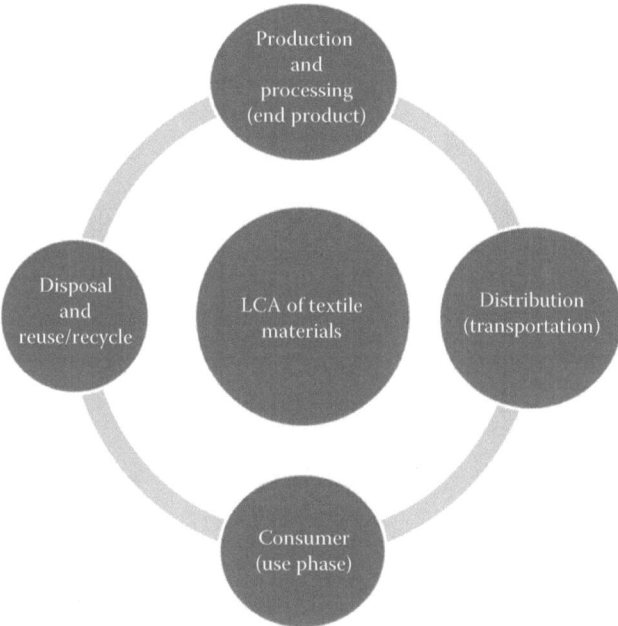

FIGURE 9.3 Different phases in LCA of textile materials.

The production and processing of natural and synthetic fibers vary widely. The flow chart presented in Figure 9.4 shows the different steps in the life cycle inventory of plant-based natural fibers such as cotton, flax, jute, coir, and hemp. In the case of animal fiber, the process of extraction of fiber is eliminated, and instead of that, the removal of fiber from animals, scouring, top making, carbonizing, and bleaching processes are included before the process of yarn manufacturing. All other processes remain the same. In the case of man-made fibers, fiber production process using raw materials and other chemicals needs to be added for life cycle inventory instead of cultivation and extraction of fibers.

The following are some of the important life cycle impacts of textile materials to be considered:

- Climate change
- Ozone layer depletion
- Photochemical oxidant formation
- Terrestrial acidification
- Human toxicity
- Terrestrial ecotoxicity
- Freshwater ecotoxicity
- Fossil depletion
- Water depletion
- Freshwater eutrophication
- Agricultural land occupation
- Urban land occupation
- Natural land transformation

In Europe, environmental impact due to consumption of textiles and clothing contributes around 2%–10% of the total environmental impact. Around 70%–80% accounts for food, drink, transport, private housing, etc. Apparently, annual consumption was estimated as 9,547,000 tons of textile

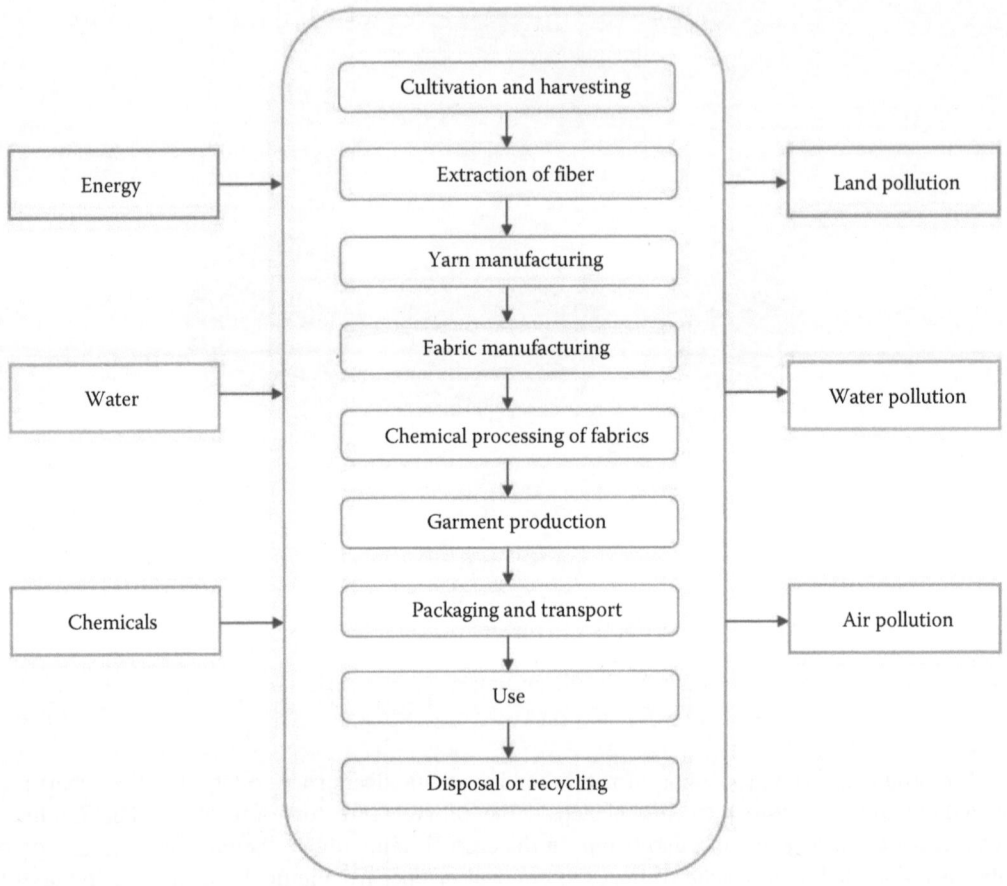

FIGURE 9.4 Simplified flowchart showing different steps used in LCA of natural fiber-based textile products.

products that gave an approximate consumption of 19.1 kg/citizen and year.[7] Among the total textile consumption, 6,754,000 tons were clothes and 2,793,000 tons were household textiles. In clothing, cotton was reported to dominate, sharing 43%, followed by polyester (16%) and the rest of the fibers, namely, viscose, wool, and acrylic, each sharing 10%.[8] Table 9.1 gives the share of each end product used in Europe.

Figure 9.5 shows the share of household textile by end product in Europe.[7]

Transportation and distribution of textile materials can be carried out by four major means:

1. By air
2. By water
3. By rail
4. By road

It has been reported that textile materials from Europe are distributed to the rest of the world from the manufacturing units by sea freight and air freight of 92% and 8%, respectively. Air transportation causes the major impact on climate change, which is about 100 times higher than sea transportation. In distribution phase, air freight contributes approximately 90% of the environmental impact among 8% of the transported textiles.[7]

The chart presented in Figure 9.6 shows the systematic approach toward the recycle and disposal of clothing waste collected in Europe (Figure 9.6). An amount of 20% clothing waste is subjected

TABLE 9.1
Share of Textiles by End Product in Europe

Product	Share of the Product (%)
T-shirt, jersey, jumpers, and pullovers	36.7
Underwear, nightwear hosiery	24.2
Bottoms	20.4
Jacket	7.7
Dress	5.3
Suits and ensembles	2.8
Gloves	1.0
Sportswear	0.9
Swimwear	0.6
Scarves, shawls, ties, etc.	0.4

Source: Beton A, Dias D et al. (2011) Environmental improvement potential of textiles (IMPRO-Textiles). In: Wolf O, Cordella M (eds.), *JRC Scientific and Technical Reports*. European Commission United States Department of Agriculture (USDA) Official USDA Estimates. Retrieved from http://www.fas.usda.gov/psdonline/psdQuery.aspx.

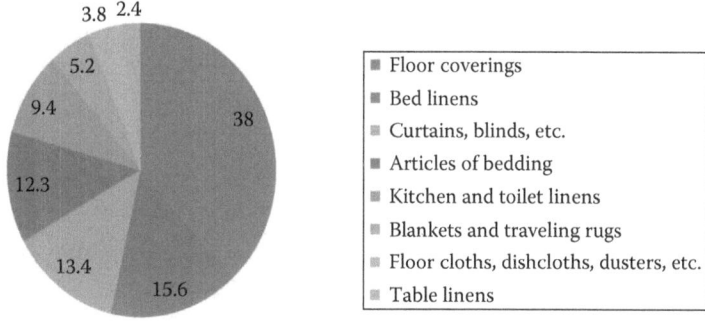

FIGURE 9.5 Share of household textiles in Europe. (From Beton A, Dias D et al. (2011) Environmental improvement potential of textiles (IMPRO-Textiles). In: Wolf O, Cordella M (eds.), *JRC Scientific and Technical Reports*. European Commission United States Department of Agriculture (USDA) Official USDA Estimates. Retrieved from http://www.fas.usda.gov/psdonline/psdQuery.aspx.)

to collection and sorting. From the total waste subjected to collection and sorting, 10% of waste is again sent to the ultimate disposal after sorting is carried out. Table 9.2 shows the final share of each category of waste in Europe.

9.3 LCA STUDIES ON NATURAL FIBERS AND TEXTILES

Among all natural fibers, cotton has the major share consumed by the users. Three countries in the world, namely, the United States, China, and India, together account for 67% of the world's cotton production in 2010.[8] A study was conducted on LCA of cotton-based clothing by selecting four regions in the world, namely, Turkey, China, India, and Latin America. All the four regions together represented 66% of knit wear and 51% of woven fabric of the world's total fabric manufacturing. Figures 9.7 through 9.11 show the different environmental impacts of knit shirts and woven pants produced from cotton fiber. From the figures, it can be observed that in all environmental impacts (particularly in consumer use), the impact due to woven pants is slightly higher than the knit shirts.

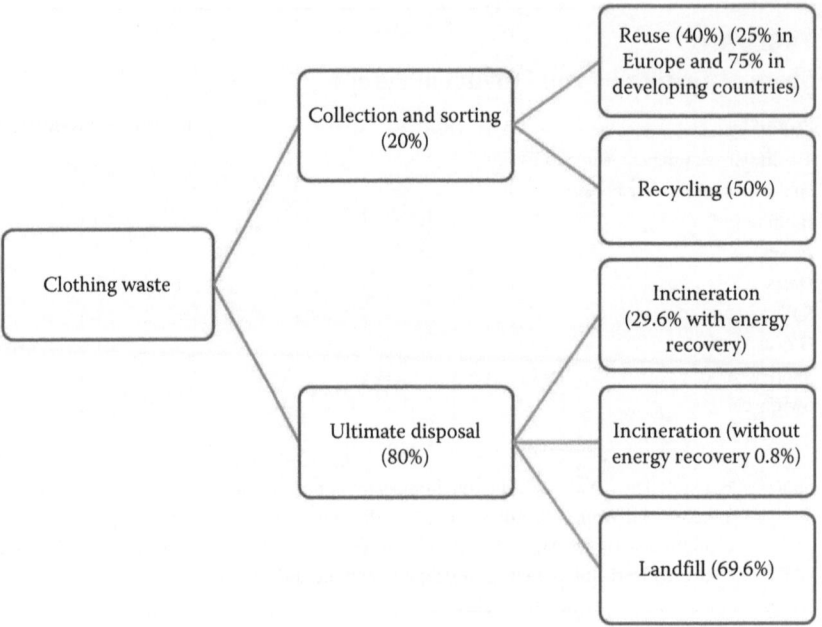

FIGURE 9.6 End of life cycle of clothing waste in Europe. (From Beton A, Dias D et al. (2011) Environmental improvement potential of textiles (IMPRO-Textiles).In: Wolf O, Cordella M (eds.), JRC Scientific and Technical Reports. European Commission United States Department of Agriculture (USDA) Official USDA Estimates.)

TABLE 9.2
Final Share of Clothing Waste

Type of Clothing Waste	Final Share of Waste (%)
Reuse	8
Recycling	10
Incineration (with energy recovery)	24.3
Incineration (without energy recovery)	0.6
Landfill	57.1

Source: Beton A, Dias D et al. (2011) Environmental improvement potential of textiles (IMPRO-Textiles). In: Wolf O, Cordella M (eds.), *JRC Scientific and Technical Reports*. European Commission United States Department of Agriculture (USDA) Official USDA Estimates. Retrieved from http://www.fas.usda.gov/psdonline/psdQuery.aspx.

This is attributed to the higher number of washing cycles used in case of woven pants (72) as compared to the knitted shirts (56). Among all three phases, agricultural phase has shown lesser impact on the environment except water consumption. Agricultural phase has relatively higher impact in water consumption primarily due to irrigation (Figure 9.11). Consumer phase occupies higher share in impacts related to global warming potential and energy demand (Figures 9.9 and 9.10). It is mainly due to the laundering of clothes that consumes higher energy as compared to other phases.[8]

A similar study on cotton has been conducted by Steinberger et al.[9] In their study, they have studied LCA of a cotton T-shirt that was subjected to 50 washes and drying after each wearing and after that it was disposed. Figures 9.12 through 9.14 showed the details of energy used and green house gas (GHG) emissions due to cotton T-shirt throughout its life cycle.[9] The primary energy for

Life Cycle Assessment Studies Pertaining to Textiles and the Clothing Sector 189

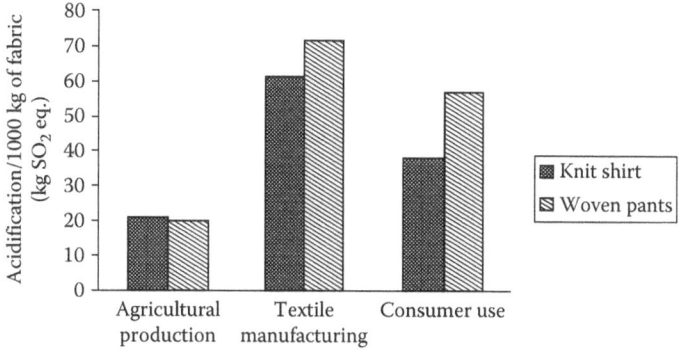

FIGURE 9.7 Acidification potential for 1000 kg of cotton fabric. (From The life cycle inventory & life cycle assessment of cotton figure & fabric—LCA executive summary. Cotton Incorporated and PE International, 2012.)

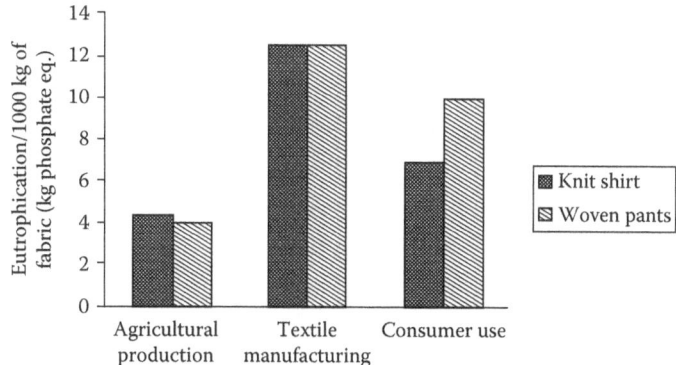

FIGURE 9.8 Eutrophication potential per 1000 kg of cotton fabric. (From The life cycle inventory & life cycle assessment of cotton figure & fabric—LCA executive summary. Cotton Incorporated and PE International, 2012.)

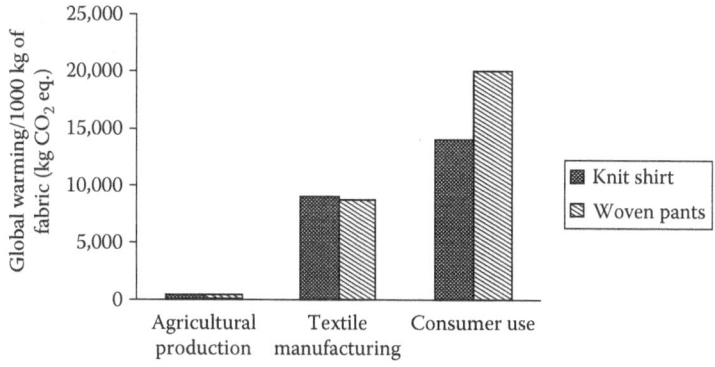

FIGURE 9.9 Global warming potential per 1000 kg of cotton fabric. (From The life cycle inventory & life cycle assessment of cotton figure & fabric—LCA executive summary. Cotton Incorporated and PE International, 2012.)

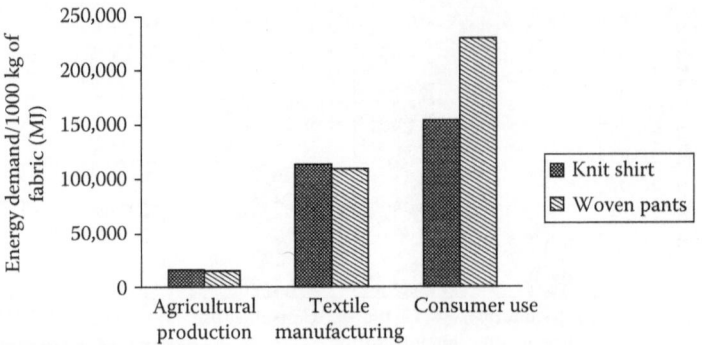

FIGURE 9.10 Energy demand by consumer for 1000 kg of cotton fabric. (From The life cycle inventory & life cycle assessment of cotton figure & fabric—LCA executive summary. Cotton Incorporated and PE International, 2012.)

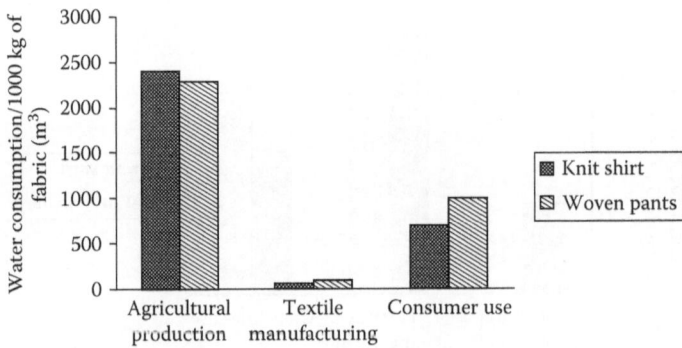

FIGURE 9.11 Water consumption per 1000 kg of fabric. (From The life cycle inventory & life cycle assessment of cotton figure & fabric—LCA executive summary. Cotton Incorporated and PE International, 2012.)

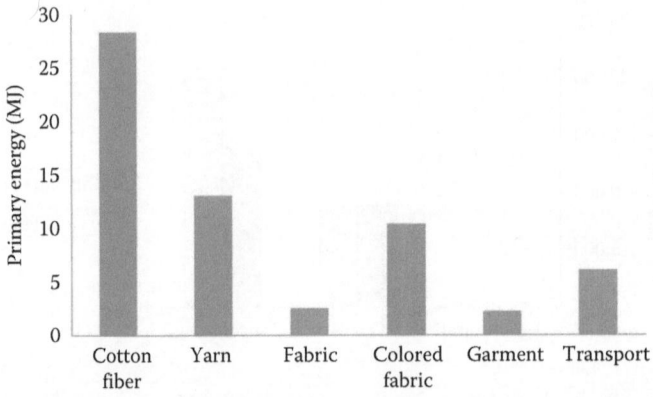

FIGURE 9.12 Primary energy spent on cotton T-shirt till factory gate. (From Steinberger, J.K., et al. *International Journal of Life Cycle Assessment* 14, 443, 2009.)

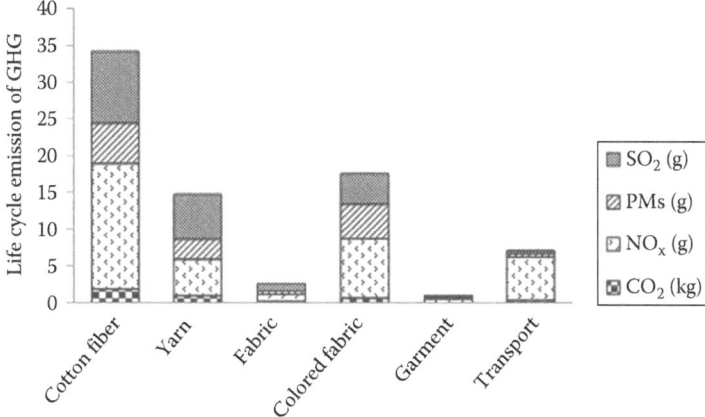

FIGURE 9.13 Emission of GHG by cotton T-shirt till factory gate. (From Steinberger, J.K., et al. *International Journal of Life Cycle Assessment* 14, 443, 2009.)

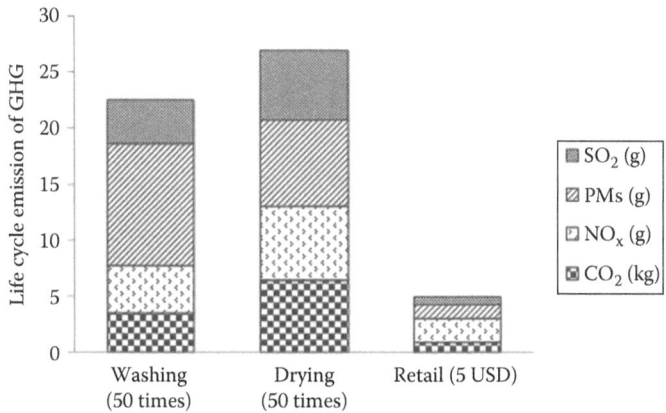

FIGURE 9.14 Life cycle emission of GHG by cotton T-shirt during consumer use. (From Steinberger, J.K., et al. *International Journal of Life Cycle Assessment* 14, 443, 2009.)

cotton T-shirt till the factory gate (Figure 9.12) accounts for only 26% of the total primary energy consumed by cotton T-shirt. The remaining 74% of the primary energy is spent during its use by consumer for washing and drying, accounting for 62.16 and 114.23 MJ, respectively.

Cotton is one of the most water-and-pesticide-intensive crops.[10,11] The United States is the second largest producer of cotton in the world. The use of pesticides in farming of cotton affects both human and wildlife. In the United States, the damage to the society due to the use of pesticides was found to be 9.6 billion USD annually.[11]

A comparison of different types of cotton based on their cultivation techniques, namely, conventional cotton, organic cotton, and genetically modified cotton, has been reported.[7] In Figure 9.15, the ratio of yield, pesticide use, and fertilizer use with respect to the values obtained for conventional cotton is shown. From the figure, it can be observed that the yield of organic cotton is lesser than that of conventional cotton. On the other hand, the yield of genetically modified cotton is higher than that of conventional cotton. Lower yield is the main drawback of organic cotton as compared to cotton grown with pesticides (conventional cotton). But organic cotton is much favorable for *eutrophication* impact. It reduces around 12% and 15% of freshwater and marine eutrophications.

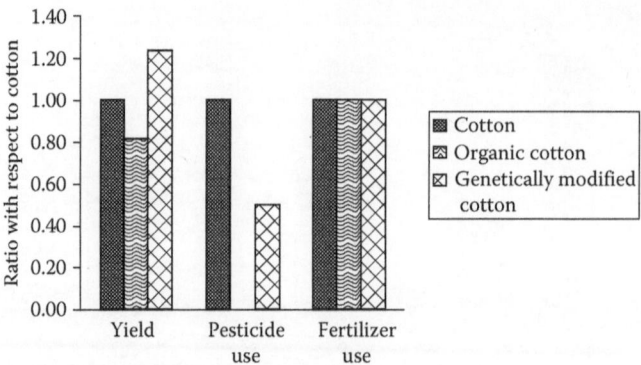

FIGURE 9.15 Comparison of yield, pesticide, and fertilizer use between types of cotton. (From Beton A, Dias D et al. (2011) Environmental improvement potential of textiles (IMPRO-Textiles). In: Wolf O, Cordella M (eds.), *JRC Scientific and Technical Reports*. European Commission United States Department of Agriculture (USDA) Official USDA Estimates. Retrieved from http://www.fas.usda.gov/psdonline/psdQuery.aspx).

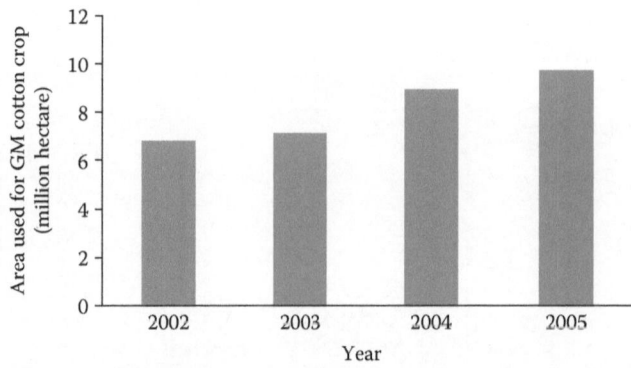

FIGURE 9.16 Area used for cultivation of genetically modified cotton. (From Anderson K, Valenzuela E et al. (2006) Recent and prospective adoption of genetically modified cotton: A global CGE analysis of economic impacts. World Bank Policy Research Working Paper. No. 3917.)

Also, it reduces 75% of terrestrial ecotoxicity, which is mainly due to the absence of pesticides in agricultural processes.

Even though the genetically modified cotton uses half the amount of pesticides used for conventional cotton, it gives higher yield and therefore reduces the total toxicity and life cycle impact on mass basis. Figure 9.16 shows the production of genetically modified cotton over the period of time. It can be seen that the production of genetically modified cotton shows an increasing trend from 2002 to 2005. The benefit through farming of genetically modified cotton was around 2.3 billion USD.[12] India, Pakistan, sub-Sahara Africa, Zambia, and Turkey are the top five regions in the world that gain benefits from farming of genetically modified cotton.

The production of wool has gone down over the years. The share of wool fiber in the global market in 1977 was 9% from which it has declined to 6.5% in 2007.[13] Australia and China are almost equally contributing to the world's wool supply, each contributing approximately 19%, followed by New Zealand (8%). These three countries together contribute around 46% of global wool production. Wool fiber gives higher GHG (methane) emission during farming of sheep as compared to the processing and other stages. On the contrary, plant fibers result in lower GHG emission in the production stage. The energy use and GHG emission for 1 ton of textile fibers were studied using cradle-to-factory gate approach[14, 15] and are listed in Tables 9.3 through 9.5. Emission of methane per sheep varies from 5 to 19 kg/head/year.

Table 9.4 lists the category of impact and extent of impact on the environment due to wool carpets.[14] Table 9.5 shows the relative GHG emission of different fibers by assuming the emission of wool fiber to be the maximum of 100%. It can be observed that linen is having substantially lower GHG emission due to lesser pesticide, fertilizer, and irrigation.[15]

TABLE 9.3
Energy Use and GHG Emission of Wool Fiber

Parameter	On Farm	Processing	Transport
Total energy (MJ/t wool top)	22,550	21,700	1490
GWP (tCO$_2$eq/t wool top)	1,655	471	103

Source: Henry B (2011) Understanding the environmental impacts of wool: A review of life cycle assessment studies. Report prepared for AWI: IWTO Oct 2011.

TABLE 9.4
Environmental Impact of Wool Carpet for 10 Years

Impact Category	Impact Per Square Meters of Carpet for 10 Years
Energy use (MJ)	157
GWP (kg CO$_2$ eq.)	64.3
Eutrophication (kg PO$_4$ eq.)	1.55
Acidification (kg SO$_2$ eq.)	0.17
Ozone (kg ethylene eq.)	0.018

Source: Henry B. Understanding the environmental impacts of wool: A review of life cycle assessment studies. Report prepared for AWI: IWTO Oct 2011.

TABLE 9.5
Relative GHG Emission of Different Fibers

Fiber	Relative CO$_2$ eq. Emission
Cotton	15–20
Linen	<10
Viscose	10–15
Polyester	20–25
Acrylic	35–40
Nylon	30–35
Wool	100

Source: Apparel industry life cycle carbon mapping, BSR, 2009.

A study on LCA of sericulture and reeling of silk was conducted by Vollrath et al.,[16] mainly focusing on cumulative energy demand (CED). Table 9.6 shows the details of energy distribution in various stages of the production of silk. From the table, it can be understood that among 100% of energy required for raw silk cocoon process, 47% energy is consumed for the production of cocoon, and heat for the cooking process consumes the major amount of energy that is around 51%.[16]

Figure 9.17 shows the distribution of energy used in the entire production of silk filament.

Among all silk production processes, the silk reeling is the most energy-intensive part of raw silk production. For example, 2 kW energy is used only for 10 min time period in order to heat up to the temperature of 60°C–90°C. Table 9.7 provides the comparative values of approximate energy consumption in the production of different fibers. Table 9.8 shows a comparative evaluation of global warming potential for 100 years with cradle-to-factory gate approach.[17]

It has been reported that more than 80% of freshwater aquatic ecotoxicity is caused by the release of one type of insecticide, namely, aldicarb, to the soil. The USDA's survey revealed that an amount of 0.67 lb/acre of aldicarb was applied to the 19% of the U.S. cotton fields in 2005.[18]

TABLE 9.6
CED for the Production of Raw Silk

Name of the Process	Energy Consumption
Cocoon production (%)	47
Heat for the cocoon cooking process (%)	51
Fertilizer used (kg/ha/year)	375
Nitrogen consumption (kg/ha/year)	520
Pesticides (kg/kg of raw silk)	0.04
Irrigation system (kWh/ha/year)	3130
Bedding system in rearing sheds (kg/kg of raw silk)	1.7

Source: Vollrath, F., et al. Life cycle analysis of cumulative energy demand on sericulture in Karnataka, India. Paper presented at the *Sixth BACSA International Conference on the Building Value Chains in Sericulture*, Padua, Italy, 2013.

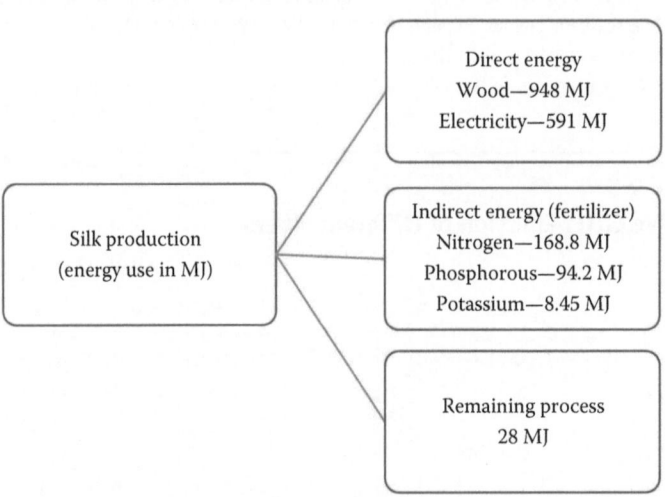

FIGURE 9.17 Energy use in silk production. (From Vollrath, F., et al. Life cycle analysis of cumulative energy demand on sericulture in Karnataka, India. Paper presented at the Sixth BACSA International Conference on the Building Value Chains in Sericulture, Padua, Italy, 2013.)

TABLE 9.7
Comparative Energy Use in the Production of Different Fibers

Fiber	Energy Use (MJ/kg of Fiber)
Cotton	50
Linen	25
Wool	40–45
Viscose (raw material production)	<5
Viscose (fiber production)	35
Polyester (raw material production)	50
Polyester (fiber production)	10–15
Acrylic (raw material production)	50–55
Acrylic (fiber production)	50
Nylon (raw material + fiber production)	85–90

Source: Henry, B. Understanding the environmental impacts of wool: A review of life cycle assessment studies. Report prepared for AWI: IWTO Oct 2011.

TABLE 9.8
Global Warming Potential of Different Fibers

Fiber	GWP_{100a} (tCO_2 eq./t of Fiber)
Cotton (the United States and China)	2.0
PET (Western Europe)	4.1
Polypropylene (Western Europe)	2.8
PLA fiber (without wind energy)	2.6
PLA fiber (with wind energy)	0.9

Source: Shen, L., et al. *Lenzinger Berichte* 88, 1, 2010.

TABLE 9.9
Comparative Environmental Impacts of Cotton and Linen

Parameters	Cotton	Linen
Primary energy (MJ)	5	6
Consumption of water (L)	26	6.4
GHG emission (CO_2 eq. in g)	128	130
Water eutrophication (mg of phosphate eq.)	125	105
Toxic risk for aquatic ecosystem (eq. in g of 1,4 dichlorobenzene)	90	11

Source: Apparel industry life cycle carbon mapping, BSR, 2009.

A comparison has been made on LCA of linen fiber with cotton. Linen fiber was grown in France and cotton was grown in China. Shirts were made from both the fibers and worn for 60 times. The functional unit was defined as wearing a shirt for 1 day. The shirts were washed after each wearing. Five important environmental impacts of the shirts were analyzed and are presented in Table 9.9.

Washing, drying, and possible ironing require the largest amount of energy, about 40%–80% of the total GHG emission for garments.[15] The use of washing machines accounts for significant amount of energy and water use and carbon dioxide emissions.[19] Top-loaded washing machines increase the CO_2 emission per kg of cotton fabric to almost twice as that of side-loaded washing machines. Similarly, the emission of CO_2 for cold washing is less than half of the emission caused by warm washing (11 g for cold water, whereas 64 g CO_2 for top-loaded washing machine with warm water).[15] Machine drying leads to the emission of approximately 175 g of CO_2/kg of cotton clothing. Ironing of clothes leads to the emission of 50–60 g of CO_2/kg of cotton clothing.

9.4 LCA STUDIES ON REGENERATED FIBERS AND TEXTILES

Shen and Patil[17] have studied LCA of regenerated cellulosic fibers. The consumption of viscose fiber according to its end use is given in Table 9.10.

In this study, cradle-to-factory gate approach was followed to find the energy spent on the fibers. CED has been calculated from the sum of nonrenewable energy use (NREU) and renewable energy use (REU) and is listed in Table 9.11.

It was observed that viscose fibers that are produced in Asia require higher nonrenewable energy than viscose fibers produced in Austria and Modal fibers. This was attributed to the relatively inefficient coal-based heat and power production in Asia region. Process energy from natural gas has been reported as the biggest contributor for NREU for the production of Lyocell fiber. It accounts for more than 70% of NREU in 2009, whereas in 2012 it was reduced by 50% (from 42 to 21 GJ/ton). The reduction was achieved through the supply of process energy from municipal solid waste incineration. Table 9.12 shows the LCIA of different regenerated cellulosic fibers.

TABLE 9.10
Viscose Fiber Consumption by End Use in 2005

End Use	Percentage of Consumption (%)
Viscose staple in textile use	70
Viscose staple in nonwovens	15
Viscose filament in textile use	12
Viscose filament in industrial use	3

Source: Shen, L., et al. Lenzinger Berichte 88, 1, 2010.

TABLE 9.11
NREU and REU for 1 ton of Regenerated Cellulosic Fibers

	Cradle-to-Factory Gate CED (GJ/t)	
Fibers	NREU	REU
Viscose—Lenzing (Asia)	61	45
Tencel (Austria)	42	59
Modal (Lenzing)	25	52
Tencel (Austria) in 2012	21	44
Viscose—Lenzing (Austria)	19	51

Source: Shen, L., et al. Lenzinger Berichte 88, 1, 2010.

TABLE 9.12
LCA of Different Regenerated Fibers

Fibers	Forest Land for Production (ha/a/t of Fiber)	Water Use to produce 1 ton of Fiber		GWP_{100a} of 1 ton of Fiber (CO_2 eq./ton of Fiber)
		Process Water (m³)	Cooling Water (m³)	
Viscose—Lenzing (Asia)	0.33	11	308	3.8
Tencel Austria	0.24	20	243	1.1
Modal—Lenzing	0.70	43	429	−0.03
Tencel (Austria) in 2012	0.22	20	243	0.05
Viscose—Lenzing (Austria)	0.69	42	403	−0.25

Source: Shen, L., et al. *Lenzinger Berichte* 88, 1, 2010.

TABLE 9.13
CED of Different Fibers for 1 ton of Staple Fibers

Fiber	Cradle-to-Factory Gate CED (GJ/t)
Cotton (the United States and China)	55
Polyester (Western Europe)	96
Polypropylene (Western Europe)	89
PLA fiber	96
Viscose—Lenzing (Asia)	106
Tencel (Austria)	101
Modal—Lenzing	78
Tencel (Austria) in 2012	65
Viscose—Lenzing (Austria)	70

Source: Shen, L., et al. *Lenzinger Berichte* 88, 1, 2010.

More than 90%–95% of water was used for the cooling purpose. In irrigation of forest, much water was not required both for beech trees in Europe and eucalyptus trees in Asia region. However, in this study, rainfall was not considered.[17]

The use of forest land for the production of viscose fiber in Asia was lower than that of viscose fiber produced in Austria and Modal fiber. This was attributed to the higher yield of trees grown in Asia as compared to those in Europe. The average yield of beech tree (both hardwood and softwood) in Europe was approximately 3.4 odt/ha/year, whereas the yield of eucalyptus tree available in Asia region was around 12 odt/ha/year.

The global warming potential for 100 years of viscose fiber produced in Asia was comparatively higher. The fiber produced in Asia uses coal- and oil-based energy system with an emission factor of 87 kg fossil CO_2 eq./GJ NREU. Table 9.13 provides the CED of different fibers for 1 ton of staple fiber according to cradle-to-factory gate approach.

9.5 LCA STUDIES ON SYNTHETIC FIBERS AND TEXTILES

The electricity required to produce polyester fiber may be generated from coal-fired power plants in China emitting 1 kg/CO_2/kWh. On the other hand, hydroelectric power plants in Brazil emit

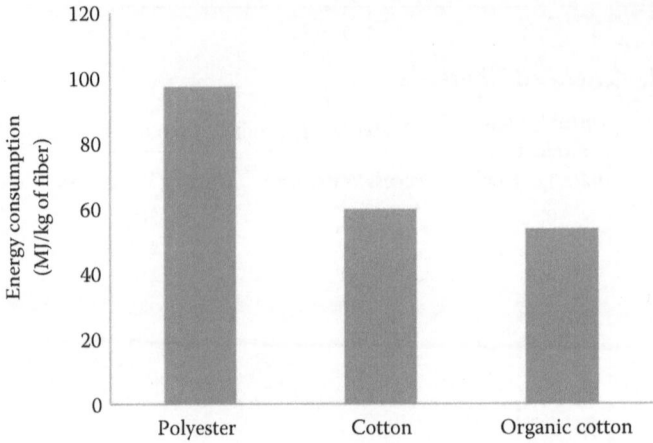

FIGURE 9.18 Energy consumption of three different fibers during its production. (From Kalliala, M.E. and Nousiainen, P. *AUTEX Research Journal* 1(1), 1999.)

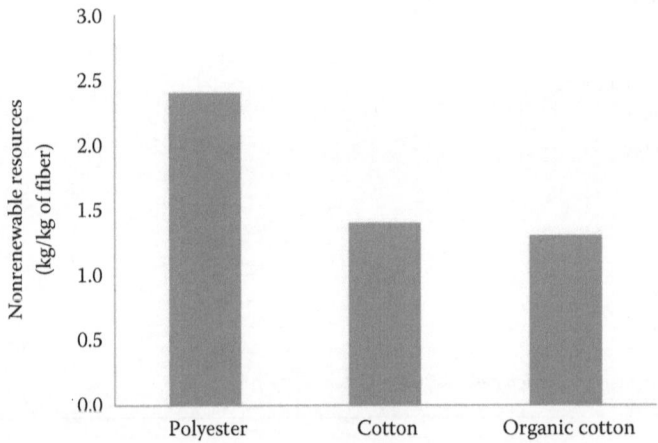

FIGURE 9.19 Nonrenewable energy resources of polyester, conventional cotton, and organic cotton fibers. (From Kalliala, M.E. and Nousiainen, P. *AUTEX Research* Journal 1(1), 1999.)

virtually nothing.[2] Polyester fiber manufacturing requires six times average energy of that required for cotton production. It leads to particulate pollution and CO_2, NO_x, So_x, and CO emissions.

A study has been carried out by Kalliala and Nousiainen (1999), to compare different types of fibers used for hotel textiles. They have compared three different fibers, namely, polyester, cotton, and organic cotton. They have made the life cycle inventory analysis on three different products till the fiber production stage. Figures 9.18 through 9.21 show the energy consumption and GHG emissions of three different types of fibers during its production.[20] From the figures, it can be seen that polyester requires higher primary energy and nonrenewable resources as compared to cotton and organic cotton (Figures 9.18 and 9.19). This is due to the use of raw materials that is obtained from fossil fuels. But the emission of CO_2 is much lower than the emissions caused by the both cotton varieties (Figure 9.20). This is due to the higher amount of fertilizers used in the cultivation of cotton. Even though pesticides are not used in the cultivation of organic cotton, required amount of fertilizers is used for efficient crop production. So, the emission of CO_2 is higher in case of organic cotton fibers.

Life Cycle Assessment Studies Pertaining to Textiles and the Clothing Sector

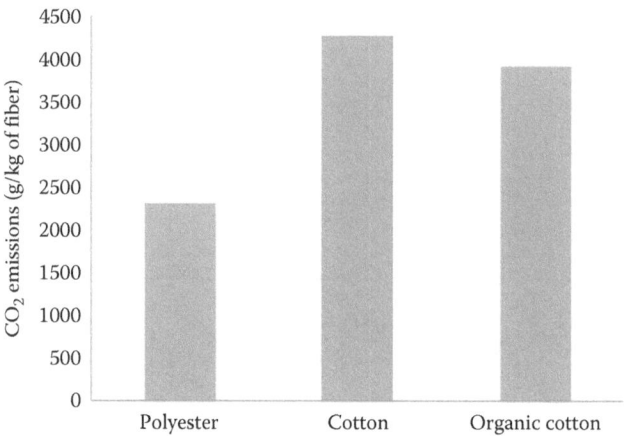

FIGURE 9.20 CO_2 emission for polyester, conventional cotton, and organic cotton fibers during production stage. (From Kalliala, M.E. and Nousiainen, P. *AUTEX Research Journal* 1(1), 1999.)

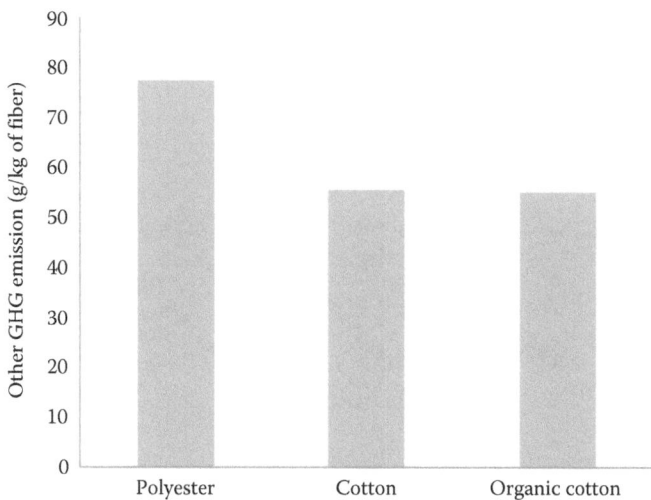

FIGURE 9.21 GHG emissions for polyester, conventional cotton, and organic cotton fibers during production stage. (From Kalliala, M.E. and Nousiainen, P. *AUTEX Research Journal* 1(1), 1999.)

Energy requirements and environmental emissions of a short-sleeve pullover of women's blouse were studied or assessed by Smith and Barker.[2] They have collected data from resin manufacturer, fiber manufacturing, fabric manufacturing, dye manufacturing, apparel manufacturing, consumer use, and garment disposal. The average energy required for industrial processes at each stage was quantified. In environmental emission, atmospheric emission, waterborne wastes, and solid wastes were considered.[2] Figures 9.22 through 9.24 show the energy requirements of a polyester blouse. From these figures, it can be noticed that the energy requirement was highest for fabric manufacturing, followed by the resin manufacturing excluding the consumer use phase (Figure 9.22). In the weaving industry, the energy required for picking and beat up is very high. In loom shed the consumption of power is high. In addition, the energy to generate the required amount of power is also very high. So, fabric manufacturing plays a vital role in energy requirement. Similarly, raw material

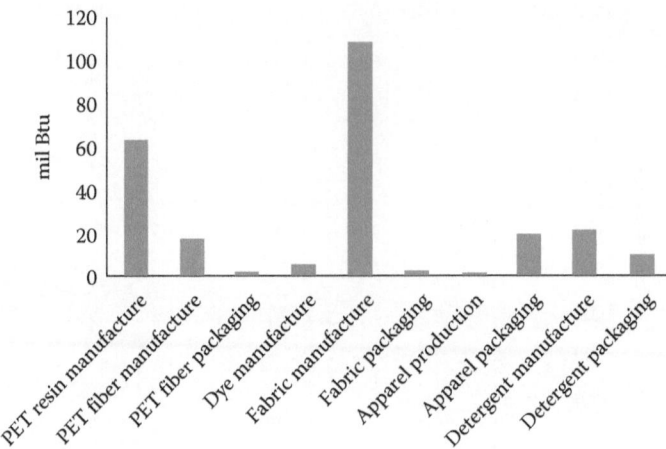

FIGURE 9.22 Energy requirement for 1,000,000 uses of polyester blouses. (From Smith, G.G. and Barker, R.H., *Resour. Conserv. Recy.*, 14, 233, 1995.)

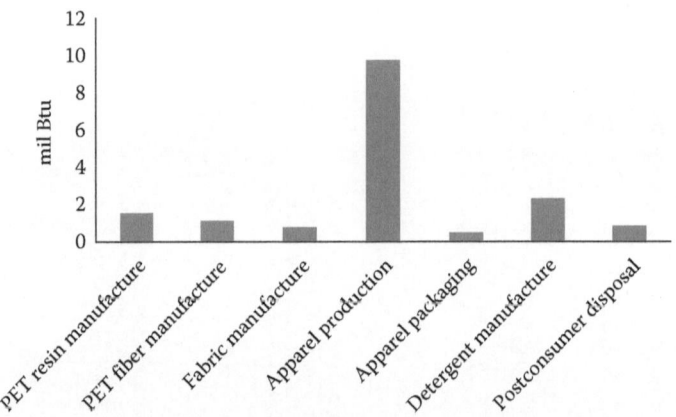

FIGURE 9.23 Transport requirement for 1,000,000 uses of polyester blouses. (From Smith, G.G. and Barker, R.H., *Resour. Conserv. Recy.*, 14, 233, 1995.)

is the major source of energy consumption in the entire life cycle for polyester fiber. So the energy required for raw material is comparatively higher (Figure 9.24).

Cartwright et al.[21] conducted an LCA study on cotton/polyester shirts (35/65) for industrial wear. The shirt has been subjected to 52 times laundering and discarded at a landfill in California. The shirt totally uses an energy of 102 MJ in which 37 MJ is consumed for the production and 65 MJ during its usage. Similarly, the global warming potentials for 100 years are 1.5 and 4.1 kg CO_2 equivalent for production and usage, respectively. Among the total water usage of 2729 L, only 645 L was used for production, and the remaining 2084 L of water was used for washing by the consumers.[21] From the preceding facts, it is clear that similar to the other products, polyester-blended products also involve the maximum usage of energy in the consumer phase.

Kalliala and Nousiainen[20] have made an attempt to perform comparative LCA studies on textiles used in hotels. Two different types of fabrics, namely, 100% cotton and 50/50 polyester/cotton blend used for bed spreads, have been compared for their life cycle impact. Table 9.14 shows the results obtained from the study. For 100 laundering cycles, 100% cotton sheets have consumed 72%

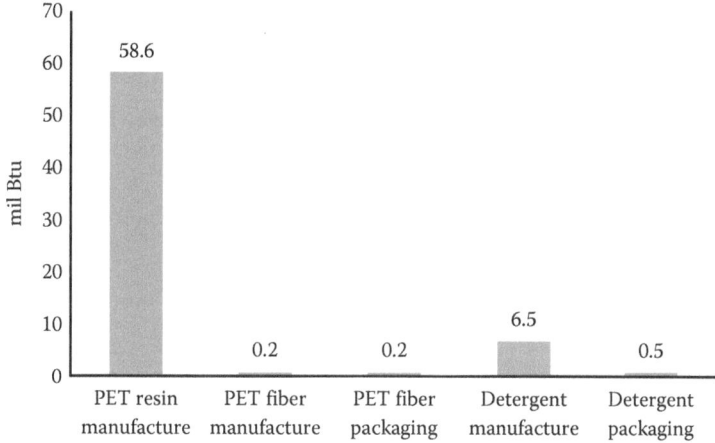

FIGURE 9.24 Energy of material resource requirement for 1,000,000 uses of polyester blouses. (From Smith, G.G. and Barker, R.H., *Resour. Conserv. Recy.*, 14, 233, 1995.)

TABLE 9.14
Life Cycle Impact Assessment of Bed Sheets Used in Hotels

Impact Category	100% Cotton	50/50 Polyester/Cotton
Energy consumption (MJ)	98	115
Nonrenewable resources (kg)	2.2	2.7
Water consumption (kg)	25,600	12,400
GWP (kg CO_2 eq.)	6.4	5.1
Methane (kg)	0.7	0.5
Pesticides (g)	19	9
Fertilizers (kg)	0.526	0.254

Source: Kalliala, M.E. and Nousiainen, P. *AUTEX Research Journal* 1(1), 1999.

more energy than the 50%/50% polyester/cotton sheets. Similarly, 100% cotton bed sheet consumed 300% more water in laundering than 50/50 polyester/cotton fabrics. Due to the aforementioned facts, global warming potential and acidification potential of 50/50 polyester/cotton bed sheets were observed to be 38% lower than that of 100% cotton sheets.[20]

9.6 LCA OF TEXTILE RECYCLING PROCESS

Textile recycling industry prevents 2.5 billion pounds of consumer products from entering into solid waste. A study was conducted to analyze the life cycle energy inventories of garments produced newly and recycled from the used clothing. A cotton shirt and a viscose blouse were produced and compared for their energy inventories. The T-shirt was made up of a single jersey combed yarn. The use phase included 25 washes at 60°C, followed by tumble drying and ironing. In case of the viscose blouse, the use phase included 25 washes at 40°C, followed by hang drying and no ironing.[22] Table 9.15 shows the life cycle inventories for the textiles produced from unused (new) fibers.

Figures 9.25 and 9.26 show the life cycle inventory of a cotton T-shirt and a viscose blouse. From these figures, it can be noticed that the energy required for cotton during its usage is very high

TABLE 9.15
Life Cycle Inventories of Textiles Produced from New Fibers

Parameter	Cotton T-Shirt (MJ)	Viscose Blouse (MJ)
Material	16	33
Production	24	11
Transportation	7	3
Use	65	7

Source: Sahni, S. et al. Textile remanufacturing and energy savings. MITEI-1-g-2010.

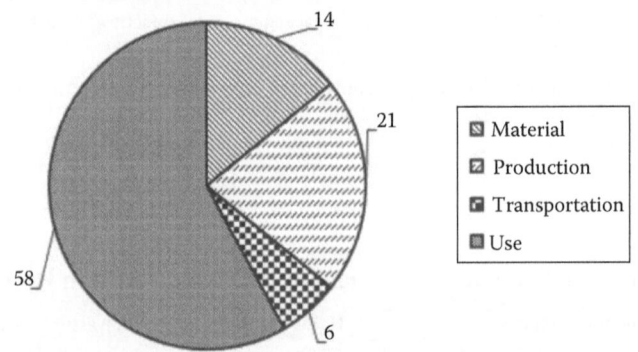

FIGURE 9.25 Life cycle energy inventory for a cotton T-shirt. (From Sahni, S. et al. Textile remanufacturing and energy savings. MITEI-1-g-2010.)

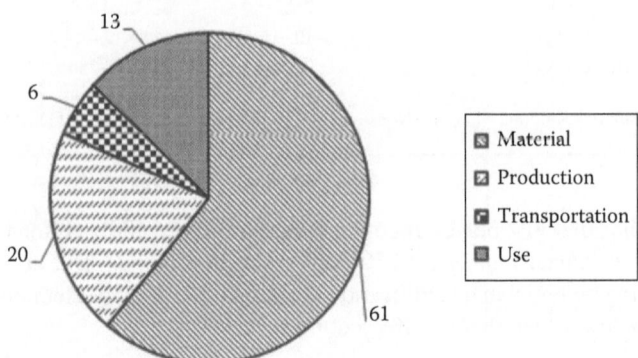

FIGURE 9.26 Life cycle energy inventory for a viscose blouse. (From Sahni, S. et al. Textile remanufacturing and energy savings. MITEI-1-g-2010.)

compared to the viscose blouse. This is due to the higher moisture retention capacity of cotton than viscose, resulting in higher amount of energy consumption for drying of cotton clothes (Figures 9.25 and 9.26).

It has been reported that approximately 18 MJ energy per kg of clothing is spent in the recycling of textiles. According to the previous example, considering the weight of the T-shirt and blouse, the energy needed to reuse the products was estimated as around 4.5 and 3.6 MJ for the T-shirt and blouse, respectively, until the resale of the products. When the products were sold, they were considered as new products and all life cycle energy inventories were calculated based on T-shirt and blouse data. It was found that around 40% and 85% energy saving can be achieved by recycling T-shirt and blouse, respectively.[22] This equation was used for calculating energy saving by recycling the textile products[22]:

$$\text{Energy saving}(\%) = \frac{(\text{LCI of new textiles} - \text{LCI of reused textiles})}{\text{LCI of new textiles}} \times 100 \quad (9.1)$$

Carpets and rug stand rank three in home construction next to brick and ready-mix concrete in the United States causing high environmental impacts. The freshwater aquatic ecotoxicity potential and terrestrial ecotoxicity potential factors are 48 and 39, respectively, contributing above the mean level. It has also been emphasized that among the building materials, recycling of carpet can account for most of the potential avoided impacts, including 9%–13% reduction in life cycle stratospheric ozone depletion potential.[23] It has been estimated that around 16.9% yield rate can be achieved by recycling nylon salvaged carpet.

9.7 LIFE CYCLE IMPACT ASSESSMENT OF OTHER TEXTILE PROCESSES

The process of sizing involves significant amount of starch. Avoiding starch in sizing reduces around 8% of eutrophication. The main reason for the decrease in eutrophication is the use of fertilizers in the farming of potato.[7]

The use of enzymes instead of chemical treatment, modern knitting such as fully fashioned knitting and integral knitting, and reduction in dye liquor in dyeing process reduces the final life cycle impact by only less than 1% in each case. For example, in case of fully fashioned and integral knitting techniques, even though they reduce the cutting costs completely, the energy consumed for knitting is very high with these modern techniques. The conventional flat bed machine requires only 120 Wh, whereas fully fashioned and integral knitting requires 1200 and 3250 Wh, respectively, therefore not allowing to reduce the life cycle impact to a greater extent.[7]

Similarly, modification of dyeing machine by incorporating controllers results in water and chemical use by 69.6% and 59.4%, respectively, resulting in total environmental benefit of less than 1%. This is due to a lesser share of water in dyeing process as compared to the total water consumption in the entire life cycle.[7]

Water recycle is a very important process in a textile mill. There are two types of processes widely used to recycle water, namely, reverse osmosis and ion-exchange technique. Using reverse osmosis, approximately, 81% of water can be recovered for reuse. With the use of ion-exchange method, approximately, 95% of water can be recovered for reuse.[7] The environmental benefit due to wastewater recycling is significantly high. The water depletion indicator is reduced by 22% and 25% by recycling wastewater using reverse osmosis and ion-exchange techniques, respectively.

9.8 LIMITATIONS OF LCA

There are certain limitations of LCA:

- LCA cannot address the localized impact: For example, with the use of LCA, it is possible to identify the region of emission, but may not be possible to find out the functioning of facility due to which the emission has occurred in that specific locality.
- Time aspect: LCA is performed in steady state. Future technological developments may not be considered in the study made in present condition.
- LCA study does not include market mechanism and technological developments. It is a scientific tool working based on linear modeling.
- LCA describes only about the environmental aspects and does not elucidate about their economic and social characteristics.

- It involves a number of technical assumptions and value choices.
- Availability of data: Many counties are still developing their database on various resources.
- It provides information for *decision support* and cannot change the decision-making process itself.

9.9 CONCLUSION

In this chapter, different aspects of LCA of textile products, inventories, and impacts and LCA studies on natural, regenerated, and synthetic fibers as well as recycled textiles have been discussed. Finally, a short note on the limitation of LCA has been included. The concept of LCA has been evolved in the past two decades. The assessment of life cycle impact of different textile products will lead to a reduction in environmental impacts in many aspects. But the awareness about LCA is not to a greater extent in the developing countries. From various studies included in the chapter, it can be concluded that the consumer phase is responsible for the highest energy consumption and environmental impacts than the manufacturing and processing phases. The increase in awareness among the consumers will help to reduce the environmental impacts. Organizing conferences, workshops, and seminars on LCA will facilitate the people to recognize the effect of small changes in their approach toward the benefits of the environment.

REFERENCES

1. Tukker A (ed.) (2004) *Handbook on Life Cycle Assessment*, Vol. 7. Kluwer Academic Publishers, Dordrecht, the Netherlands.
2. Smith GG, Barker RH (1995) Life cycle analysis of a polyester garment. *Resources, Conservation and Recycling* 14:233–249.
3. Connell D (1995) The environmental impact of the textiles industry. In: Carr CM (ed.), *Chemistry of the Textiles Industry*. Springer, Dordrecht, the Netherlands, pp. 333–354. (Life cycle assessment: Principles and practice, EPA/600/R-06/060, May 2006).
4. Tukker A, Huppes G et al. (2006) DG Environment and DG Joint Research Centre (2006) Environmental impact of products (EIPRO), summary of the final report, May 2006.
5. Tukker A, Jansen B (2006) Environmental impacts of products: A detailed review of studies. *Journal of Industrial Ecology* 10:159–182.
6. Tuncer B, Schroeder P A key solution to climate change: Sustainable consumption and production. SWITCH—Asia Network Facility, Wuppertal, Germany.
7. Beton A, Dias D et al. (2011) Environmental improvement potential of textiles (IMPRO-Textiles). In: Wolf O, Cordella M (eds.), *JRC Scientific and Technical Reports*. European Commission United States Department of Agriculture (USDA) Official USDA Estimates, Spain. Retrieved from http://www.fas.usda.gov/psdonline/psdQuery.aspx.
8. The life cycle inventory & life cycle assessment of cotton fiber & fabric—LCA executive summary. Cotton Incorporated and PE International, 2012.
9. Steinberger JK, Friot D et al. (2009) A spatially explicit life cycle inventory of the global textile chain. *International Journal of Life Cycle Assessment* 14:443–455.
10. Chapagain AK, Hoekstra AY et al. (2005) The water footprint of cotton consumption. Value of Water: Research Report Series, Vol. 18. UNESCO-IHE Institute for Water Education, the Netherlands.
11. Pimentel D (2005) Environmental and economic costs of the application of pesticides primarily in the United States. *Environment, Development and Sustainability* 7:2.
12. Anderson K, Valenzuela E et al. (2006) Recent and prospective adoption of genetically modified cotton: A global CGE analysis of economic impacts. World Bank Policy Research Working Paper. No. 3917.
13. Barber A, Pellow G (2006) Life cycle assessment: New Zealand merino industry, Merino wool total energy use and carbon dioxide emissions. Report by the Agribusiness Group, Auckland, New Zealand.
14. Henry B (2011) Understanding the environmental impacts of wool: A review of life cycle assessment studies. Report prepared for AWI: IWTO Oct 2011.
15. BSR (Business for Social Responsibility) (2009) Apparel Industry Life Cycle Carbon Mapping, www.bsr.org/reports/BSR_Apparel_Supply_Chain_Carbon_Report.pdf, accessed March 25, 2012.

16. Vollrath F, Carter R et al. (2013) Life cycle analysis of cumulative energy demand on sericulture in Karnataka, India. Paper presented at the *Sixth BACSA International Conference on the Building Value Chains in Sericulture*, Padua, Italy.
17. Shen L, Patel KM et al. (2010) Life cycle assessment of man-made cellulose fibres. *Lenzinger Berichte* 88:1–59.
18. USDA (May 2006) Agricultural chemical usage 2005 field crops summary. United States Department of Agriculture, National Agricultural Statistics Service (NASS). http://usda.mannlib.cornell.edu/usda/nass/AgriChemUsFC/2000s/2006/AgriChemUsFC-05-17-2006.pdf. accessed March 25, 2014.
19. Bole R (2006) Life-cycle optimization of residential clothes washer replacement. University of Michigan, Ann Arbor, MI, April 21, 2006.
20. Kalliala M E, Nousiainen P (1999) Life cycle assessment environmental profile of cotton and polyester-cotton fabrics. *AUTEX Research Journal* 1(1), 8–20.
21. Cartwright J, Cheng J et al. (2011) Assessing the environmental impacts of industrial laundering: Life cycle assessment of polyester/cotton shirts. http://fiesta.bren.ucsb.edu/~missionlinen, accessed March 25, 2012.
22. Sahni S, Boustani A et al. (2010) Textile remanufacturing and energy savings. MITEI-1-g-2010.
23. Janjic K, (2013) Analysis of the life cycle impacts and potential for avoided impacts associated with single-family homes. EPA 530-R-13-004, July 2013.

10 Social Impacts of the Clothing and Fashion Industry

Shanthi Radhakrishnan

CONTENTS

10.1 Introduction 207
10.2 Key Social Issues in the Textile and Clothing Industry 209
    10.2.1 Globalization..209
    10.2.2 Worker's Rights 210
    10.2.3 Health and Safety... 210
    10.2.4 Raw Materials for Textile Production.. 211
    10.2.5 Wages.. 211
    10.2.6 Work Quality Issues in the Textile Sector 211
10.3 Corporate Social Responsibility 213
    10.3.1 Core Elements of CSR... 213
    10.3.2 Implications of CSR... 214
    10.3.3 Implementation of CSR 215
10.4 Social Impact Measurement 215
    10.4.1 Key Elements of Impact Measurement.. 216
    10.4.2 Dimensions of Social Impact Measurement Methods....................................... 216
10.5 Tools for Social Impact Measurement 217
10.6 Organizations for Social Change 220
10.7 Future Directions 223
References..225

10.1 INTRODUCTION

The United Nations (UN) Environmental Programme has envisaged the fashion industry, which constitutes textile and apparel manufacture and production, as the second largest industry on a global scale that is valued at $1.44 trillion (Boone 2012). This diverse and complex industry incorporates high volume and extensive technological advances along with handicraft and cottage industries. The manufacture and supply chain of the fashion industry spreads far and wide in search of new concepts for design development and the production and conversion of textile materials to apparels or end products for consumer use (Boone 2012).

According to *Business Dictionary* (2014), social impact is the effect of an activity on the social environment of the community and well-being of the individuals and families. Social environment implies the demographics of a said area that may include the ethnic composition, wealth, educational levels, employment rate, and regional values. The effect of the functioning of the clothing and fashion industry in terms of welfare, attitudes and values, employment, economic status, health, safety, and security of the community or society is of utmost importance.

The textile and clothing industry is considered significant in economic and social terms as it provides income and jobs to the community that results in sustained economic development and contributes to the long-term growth and development. A study on the textile industry in Madagascar revealed that a job or employment in the textile and apparel industry would increase the purchasing power of the individual by 24% on an average and this could bring them out of poverty (Nicita and Razzaz 2004). When the social aspects are analyzed, there are two major aspects that are to be considered. As there are no concrete policies and institutions in developing countries, workers do not have the skill sets to take up high-value activities like design development and marketing in the clothing value chain. The workers in the developing countries are hired for low-value jobs like garment assembly and therefore get lower wages than the employees of their headquarter firms in developed countries. Further, this industry is able to fetch better wages for the community especially women, when compared to the other activities like agriculture and other domestic industries that is dominated by gender inequalities. It has been studied that the textile and clothing industry provide better opportunities, employment, and financial remuneration to women employees when compared to agriculture and other domestic industries in developing countries like Pakistan, Bangladesh, Sri Lanka, the Philippines, Thailand, and Zambia. In short, the low economic condition of the developing countries lead to the dependency of the opportunities provided by this industry though the returns are lower than those compared with the developed countries (Razzaue and Eusuf 2007).

The trends in the textile and clothing industry after the expiry of the WTO Agreement on Textiles and Clothing in 2004 have led to liberalization and the basis for economic crisis. There has been a shift of the production centers of this industry to the developing countries and a growth of numerous multinational import enterprises to expand their market shares. The global outsourcing has led to a decline of jobs to thousands in the industrialized countries to open avenues of work for many workers in the developing nations, of which the majority to benefit are the female population. It has been estimated that 80,000 jobs have been lost in Germany alone in the textile and clothing industry over the past 40 years (Wick 2009). Clothing production in the industrialized countries like Germany, United States, Great Britain, and Italy has become hubs for a high-tech textile branch Technical Textiles, of which Germany is a world leader. Most of the headquarters of the multinational trading enterprises and brand-name companies that control the global procurement systems comprising direct and indirect suppliers are located in industrialized nations. Further, the prime areas of design and marketing, special parts of production, and highly priced products belong to the industrialized countries.

The liberalization policies have paved the way for stiff international competition among the producing countries, and it has been estimated by the International Labour Organization (ILO, 2009) that the number of unemployed will increase from 39 to 59 million by 2007–2009, while the number of poor people will increase by 200 million. The wage ratio in approximately two-thirds of the countries has declined, and the goal of reducing the number of people living in extreme poverty by 50% by 2015 will become a distant dream (Wick 2009). Women may have been privileged to get jobs in par with men, but the results of their position in the labor markets and society have kept them in a disadvantaged position in both the developing and industrialized countries. However, the multinational corporations have the economic power to govern the workers at different parts of their global links of production and supply chains, and they gain profit by encouraging the fierce competition among the producers (Wick 2009).

The economic crisis has many dimensions. The multinational enterprises have shifted their focus to the expansion of their market shares, leading to deterioration in social standards. Most developing countries have neglected their domestic markets that are undersupplied when compared to the saturation of the major consuming regions of the United States, the EU, and Japan. Due to the emergence of numerous producers who offer products at lower prices in the liberalized world market, the export prices dropped, causing revaluation and devaluations in the global currency system. The leading industrialized countries have taken to their advantage many instruments of trade and policy like tariff escalation, rules of origin, antidumping, and safeguard procedures to put pressure on the developing countries to lower the tariff on textile and clothing products. In order to face the competitive markets, the labor-intensive textile and clothing industries in the developing countries lower product costs by reducing their labor costs. This has resulted in worsening of social standards that have affected the people and economies of the developing countries over the last few years.

This industry has a complex structure of production that results in fluctuations in employment statistics on a global scale. It has been found that approximately 160 production countries (Wick 2009) are located in the informal economy and in free-export zones (FEZs) and the jobs of employees in these countries are meagerly or not protected by labor and social rights. It has been reported that the textile and clothing industry directs production in the majority of the 3500 FEZs in 130 countries with 66 million workers (Boyenge 2007, Wick 2009). The quality of employment of these workers is nowhere near the prescribed social standards and often covered with practices of overtime, low wages, repression of trade unions, discrimination against women, and poor working conditions and labor relations. The rising prices and mounting inflation have subjected workers of the textile and clothing industries in developing countries to huge social and human hardship (Wick 2009).

The concept of fast fashion has led to many practices that are considered unethical when compared to the norms specified by various organizations who work for a social cause and for the welfare of workers and employees of business organizations. Leaders in the apparel field have brought down the cost of apparel sold in the market due to the shift of manufacturing sites to developing countries where there is an abundant supply of cheap labor. Though an improvement has been noticed in the lives of the people who work for the apparel industries, many social aspects have been ignored by the authorities in the factories. This has led to appalling unsafe conditions that are coming to the knowledge of the consumers who purchase the products in the developed European and U.S. markets. People are more concerned about a sustainable future and hope to buy products that have environmental and social certifications. Despite these growing attitudes, there are many gray areas that have to be addressed, and the social issues related to the textile and clothing industry are being taken up to find the right solutions for future welfare and brighter prospects.

10.2 KEY SOCIAL ISSUES IN THE TEXTILE AND CLOTHING INDUSTRY

The shift of the global clothing industry to developing nations, to curtail prices of products, has resulted in low-cost economies where labor cost is low. Poor treatment of labor, bad working conditions, and human rights have given rise to *sweatshops*. *Sweatshop* is a term for a workplace or industry that has unacceptable working conditions. Deplorable working conditions were found to exist in sweatshops that included unsafe working conditions like exposed electrical wiring, blocked aisles, unguarded machinery, and unsanitary bathrooms and poor lighting and temperature control and ventilation.

Experts believe that the intensive nature of low pricing in the competitive garment industry encouraged manufacturers and contractors to break labor laws. An influx of a large population of undocumented immigrant workers who work for long hours for a low pay regardless of labor laws for garment manufacturing is another reason for the availability of cheap labor. Analysts have also identified that the female workers are higher in number, leading to gender discrimination. In many cases, child labor laws are also violated (General Accounting Office, 1994).

The clothing industry involves numerous production processes that affect people in many ways right from labor involved in raw material procurement and factory workers to those who are involved in disposal and death of garments. Some of the issues that affect the personnel involved in the textile and clothing industry are discussed as follows.

10.2.1 GLOBALIZATION

Garments that are sold in developed countries are produced by the developing countries of the world. Outsourcing is being carried out on a global scale and competitive pricing is the vision of many multinational companies. This target has been achieved by global trade and tie-ups with developing countries where labor is cheap and available in plenty. The poverty-stricken people work hard for their livelihood, and many contractors take advantage of the situation by maintaining poor worker's rights, low pay, and violation of all labor laws.

10.2.2 Worker's Rights

Most manufacturers and contractors prefer to employ low-skilled or unskilled labor as they lack knowledge of worker's rights and can be easily prevented from uniting together to form trade unions (Occupational Safety and Health Administration [OSHA]).

The workers of the Indonesian garment industry had faced systemic human rights violation, and the jury of the Peoples' Tribunal was appointed to assess the human rights abuses existing in the garment industries (Clean Clothes Campaign 2014a). A living wage offers decent living conditions, and if the garment workers of the industry, especially women, still remain in poverty, then the brand cannot claim that they are a truly sustainable company. Great concern was expressed for the lack of urgency and transparency with regard to the provision of a living wage and the right to freedom of association, among leading brands. It was described that making way for a living wage was an important human right that should be an integral part of any sustainability corporate accountability framework. The charges were suspension of minimum wage payments, compulsory and illegal overtime, inhumane productivity measures, denial of social security payments, gender discrimination, and suppression of the right to form associations. Global brands like Nike, GAP, Walmart, H & M, and Adidas mentioned in the testimonies of the workers were invited for the public hearing (Clean Clothes Campaign 2014b). When leaders in apparel business are called for a public hearing, it shows that the sustainability claims by these leaders are not really true in real practice. It is also overwhelming to understand that the evidence of violation of basic human rights prevail and action must be taken by all the stakeholders involved. The judges stressed that the global brands must admit their involvement in the human rights violations and their responsibility for the conditions in those factories. Recommendations were given for action to be taken by the Indonesian government, labor organizations, and the global players to enforce the basic human rights in the said factories.

The UN Global Compact has identified 10 principles and application areas to promote corporate social responsibility (CSR), human rights, labor standards, environmental ethics, and anticorruption. These principles are based on the Universal Declaration of Human Rights, Fundamental Principles and Rights at Work formulated by the ILO, the Rio Declaration on Environment and Development, and the UN Convention against Corruption. The forum declares that business organizations must support and respect the international human rights and should not be involved in human rights abuses. Further, it claims that all business organizations should preserve the freedom of association and right to collective bargaining among the working community and eliminate all types of forced or compulsory labor, child labor, and discrimination in employment and occupation (Gardetti and Torres 2013).

10.2.3 Health and Safety

The regulations for labor and working conditions in garment units in developing countries have lesser restrictions and regulations when compared to the labor force of the developed nations. The poor and appalling working conditions lead to health problems and deterioration of workers' health, causing backache, eyestrain, burns, and other injuries. Restrictions to use the rest rooms to combat time loss and low production culminate in severe kidney problems among workers. In some of the countries, long hours of work like 16–18 h per day cause fatigue that can be the reason for accidents (Morey 2000, Women and Global Human Rights [WGHR]).

10.2.4 Raw Materials for Textile Production

Farming and nurturing of animals form an integral part for accumulating raw materials for producing yarns and fabrics. In the case of cotton cultivation, pesticides have been used in plenty toward the cotton weevil. These pesticides are toxic in nature and cause risk to worker's life and health. It has been reported that the pesticides used for cotton farming range from extremely hazardous and highly hazardous to moderately and slightly hazardous chemicals as classified by WHO toxicity classes (OSHA). It has been reported that pesticides nearly $2.6 billion of worth are sprayed on cotton fields each year in the United States, which accounts for more than 10% of total pesticide used and 25% of insecticides used worldwide. Collateral damage to animal, aquatic, and bird life was noted. In Alabama, illegal use of pesticides containing endosulfan and methyl parathion to cotton fields led to the loss of 240,000 fish due to pesticide-contaminated rainwater runoff reaching the rivers. Further, Australian beef was found to be contaminated with the pesticide chlorfluazuron as the cattle were fed with the contaminated cotton straw, leading to the ban of beef imports from Australia. The spraying of methyl parathion to cotton fields three miles away from the breeding colony of laughing gulls near Corpus Christi, Texas, had killed innumerable adult birds and 25% of the chicks. These chemicals have been banned as they have been listed in the toxicity classes by WHO (Pesticide Action Network [PAN]).

In the case of animal rearing, poor practices could result in neglect and mistreatment of animals, malnutrition, infections, and symptoms for potential illness. Regulations for the protection of animal welfare are less restrictive in many countries especially with regard to slaughter of animals, poor transportation, and maintenance processes like tooth grinding and mulesing. Tooth grinding is a practice of filing down the teeth of female sheep with a rotary stone to prevent forced cull from the broken teeth. Mulesing is the slicing of skin of animal away from the tail and breech area to reduce blowfly attack. These inhumane practices are prohibited in the United Kingdom (Centre for Remanufacturing and Reuse). Ecological wool is sourced from nonmulesed sheep that are raised on holistically managed farms, improving animal welfare and care for the environment. Animal welfare is an important measure to be adopted to maintain sustainability in raw material procurement, and the term *cruelty-free products* is used by many brands while displaying their products to consumers. Sustainability in apparel production starts with sustainable raw materials.

Although a ban on fur farming prevails in many countries around the world, fur fashion is still carried on in catwalks and fashion line presentations. Leather used in the fashion industry is not always a by-product of the meat industry; exotic leather is obtained from animals like ostriches and reptiles that are totally forbidden. The conditions in which animals are reared, transported, and slaughtered are a major issue of concern. Ahimsa or Peace Silk is an alternative to the traditional method of boiling the silk cocoons for reeling (Forum for the Future 2007).

10.2.5 Wages

Unreasonably low wages, excessive working hours or overtime, and dangerous working environment are part of poor treatment of workers. A recent report claims that the wage per hour in Pakistan is $0.23/h, $0.86/h in China, and $0.09/h in Bangladesh, while the wage per hour is $11.16/h in the United States. The apparel production in Bangladesh is low in cost and the U.K. companies like TESCO, ASDA, and Primark procure clothes from the factories in Bangladesh (Forum for the Future 2007).

10.2.6 Work Quality Issues in the Textile Sector

Working Hours and Work fatigue: The Occupational Safety and Health Administration (OSHA) explains that a normal work shift is a work period not more than eight consecutive hours per day, five days a week with a minimum of eight hour rest. Any shift which takes up more continuous hours and or days is known as extended work shift and can be used during emergencies till normalcy is maintained. Extended and unusual work shifts are a cause for increased fatigue, stress and lack of concentration (OSHA).

The Social Accountability 8000 (SA8000) is a standard that specifies the requirements that an establishment or industry should provide for improving worker's rights, workplace conditions, and effective implementation of these conditions through proper management. This standard deals with nine criteria like child labor, forced or compulsory labor, health and safety, freedom of association and right to collective bargaining, discrimination, disciplinary practices, working hours, remuneration, and management system.

The SA8000 (2014) forbids the employment of a person under 15 years of age by any industrial organization since it is termed as *child labor*. Further, it states that all work or service performed by a person under the threat of punishment or retaliation or as a means of repayment of debt is classified as forced or compulsory labor. The standard advocates that all personnel in the organization have the right to join and form trade unions to collectively bargain with the organization where they are employed for necessary benefits or improvement. The organization should respect this cooperative venture and work in unison with the trade unions and their representatives.

This standard propagates the provision of a safe and healthy workplace environment by eliminating all causes of hazards and by preventing occupational injury and illness. This is strengthened by onsite job-specific training to safeguard the health and safety of the employees of the organization. The standard forbids discrimination in any form as well as situations that give rise to discrimination with regard to hiring, remuneration, training, promotion, termination, or retirement. The standard promotes the treatment of all personnel with dignity and respect and should not indulge in harsh or inhumane treatment of the employees of the organization.

The remuneration should meet the basic needs of the personnel and make way for the provision of some discretionary income. The wages should be calculated for a normal work week excluding overtime and must meet the minimum standards of the industry. Finally, the organization should develop procedures and a system to implement and monitor the policies of the SA8000 and also make provision to publicly announce and display its commitment to abide to all the requirements of the SA8000 standard (Social Accountability Standard 2014).

It must be noticed that there is a transformation in the actual nature of work due to the introduction of new technologies and increased commercial constraints where the worker is in contact with the client resulting in time pressures. The physical environment and work station design remain traditional and lead to many problems. The most common trend in recent times is the intensification of work. This may be due to many factors like changes in work organization, reduction in work force by means of restructuring, and budget implications or cost cutting. Intensification of work is always accompanied with a greater incidence of stress, musculoskeletal disorders, harassment at work, and increased time pressure (European Foundation for the Improvement of Living and Working Conditions 2002).

- *Mistreatment of women*: Apparel industries in the developing countries have a majority of women employees, and about 90% of them are young and uneducated. They are usually undocumented immigrants who are not aware of their legal rights and are subjected to mistreatment in miserable working conditions. The common injustice includes low wages that do not meet the basic costs of living, unsafe and unhealthy living conditions, long hours of overtime without compensation, and sexual harassment. Many women tend to pay recruitment and contract fees with a hope of getting well paid in foreign lands and get locked into contractual obligations that can last for many years (Greenhouse 2001). Most factories and women living quarters are located behind barbed wire fences and monitored by armed guards. Visitors are prohibited and they cannot mention their grievances to anyone as they are always under the threat of corporal punishment (Greenhouse 2001).

Saipan, a part of the U.S. Commonwealth of the North Marianas Islands, imports women from Asia to work in the garment factories that are similar to labor camps with extremely poor working conditions, which encompasses physical abuse, debt bondage, and sexual harassment. The women work under a system of fear and domination and are deceived into enslavement. Many U.S.-based companies like The Gap, Walmart, J C Penney, and Sears, Roebuck & Company imported goods from Saipan. Many retailers in the United States faced litigation for using false advertising and unfair business practices regarding the sale of goods manufactured under poor conditions in the sweatshops in Saipan (Collier 1999). Another case study reports the violation in human rights at the Daewoosa factory in American Samoa, a Pacific territory of the United States. The employees were from Vietnam and fed with poor and undernourished diet. The women workers were subjected to physical abuse and sexual harassment, punishment, and nonpayment of wages. In 2001, this factory was closed due to a lawsuit filed against the Daewoosa plant for nonpayment of wages and severe contract violations. There are many such cases all around the world where many countries fail to protect the rights of the workers despite the human rights agreement on both the domestic and international scales, for example, the trafficking of Thai women in Japan, Bolivian girls in Buenos Aires, and workers in Mexico and China (WGHR).

10.3 CORPORATE SOCIAL RESPONSIBILITY

As global sourcing has become an important part of manufacturing and marketing of apparel goods, the attention of many large buyers has focused on the developing countries to make use of the advantage of the subsidies and cheap labor available in plenty. Although many labor laws exist in the third world countries, the workers producing garments for global systems are susceptible to exploitation, due to the poor enforcement of the laws. This calls for the multinationals and international supply chains to recognize their social obligations and meet the ethical requirements for apparel production. In the 1990s, there were several factors that persuaded and pressurized many companies to adopt and implement CSR practices on a voluntary basis rather than a mandatory origin.

As economic activity has shifted to a global scale, there is a huge competitive market where any edge is used by the manufacturer or supply chain institutions, as an initiative to stay in the business trend. The factory contractors were reluctant to disclose their supplier list for the reason that retailers and labels may cut out the contracting firm from the supply chain due to improper practices. CSR was slowly penetrating into the large-scale apparel businesses, and the smaller- and medium-size enterprises started realizing the need to be termed as *no sweatshop* manufacturer. A major trend was the decreasing role of the state in regulating business behavior and the change to international dissemination of information about working conditions. There was an increasing number of nongovernmental labor rights campaigns, and significant brands and corporates were vulnerable to bad publicity in case of any violation of social and environmental practices in the apparel and fashion business. It was reported that there were 65,000 transnational companies in the world of which only 4,000 companies produced reports about their social and environmental practices and performance (Diviney and Lillywhite 2007).

The concept of voluntarily integrating social and environmental practices into business operations is CSR. The aim of CSR is to produce an overall positive impact on society by the judicious management of business processes with regard to social and environmental ethics. UNIDO defines CSR as a management concept, whereby companies integrate social and environmental concerns in their business operations and interactions with the stakeholders. It is a tool by which a company achieves a balance of economic, environmental, and social imperatives coupled with the effort to address the expectations of shareholders and stakeholders.

10.3.1 CORE ELEMENTS OF CSR

It is very essential that every business organization should form a CSR policy to enable strategic planning putting forth its CSR initiatives. This should be framed in consultation with various levels of personnel in the organization and should be a vital part of the overall business policy toward the achievement of business goals. The CSR policy should be approved by the board. The most important core elements of CSR are care for all stakeholders, ethical functioning, respect for workers' rights and welfare, respect for human rights, respect for environment, and activities for social and inclusive development.

- *Care of all stakeholders*: Stakeholders are those who affect or those who are affected by the functioning of the business organization, strategy, or project. These members may be internal or external or at junior or senior levels, having the capacity to react, negotiate, or change the structure of the business enterprise. CSR calls for the respect and responsiveness of the company toward all stakeholders that include shareholders, employees, customers, suppliers, society, and people affected by the project. The company should create value for all stakeholders, actively interact with them, and alleviate any inherent risks that may occur.
- *Ethical functioning*: Ethics, transparency, and accountability should be the motto for the ethical functioning of an organization. To achieve ethical functioning of a business enterprise, it should not engage in abusive, unfair, or corrupt practices.
- *Respect for worker's rights and welfare*: The dignity of employees should be maintained by providing a safe, hygienic, and humane workplace environment, training and skill development for career advancement, recognition of freedom of association and collective bargaining, effective grievance redressal system, provision of equal opportunities without discrimination, and abolition of child labor.
- *Respect for human rights*: In any organization, human rights should be respected and all human rights abuses should be dealt without complicity by the organization or any third party.
- *Respect for environment*: Pollution should be curtailed by undertaking proper checks and adopting clean technologies; optimal use of resources like land, water, and energy to be ensured; and recycling and reduction of waste to be monitored. Adoption of energy-efficient and environment-friendly technologies will help industries to face the challenges of climate change and its impact on the livelihood of human beings.
- *Activities for social and inclusive development*: Business organizations should aim at serving the community and disadvantaged sections of the society for economic and social development by undertaking activities like skill building, education, health, and cultural and social welfare (Ministry of Corporate Affairs 2009).

10.3.2 IMPLICATIONS OF CSR

Indirect benefits result by adopting CSR principles within the company framework and by orienting the mission and vision of the organization toward these goals. *Triple bottom line* is a term used for indicating people, planet, and profit and the impact of the business on these three aspects. The fair and beneficial business practices undertaken by the business enterprise with regard to labor, community, and region are related to the term *people*. Planet relates to sustainable environmental practices that ensure that harmful and dangerous by-products or waste is not released or dumped into the environment or neighborhood. Economic value created after deducting cost of all inputs and capital invested is termed as *profit*. These measures adopted by business organizations would make the companies realize their social and moral responsibilities and also create a positive image in the society (Kanji and Chopra 2010).

A CSR program aids in human resource recruitment and retention, and many students favor an organization with a sound CSR policy. The staff of these companies are very loyal and help to improve the image of the organization among its personnel who are involved in many activities involving community interaction. A unique CSR program can place an organization in a special place that helps in building customer loyalty based on the ethical values adopted by the business enterprise. This serves as an advertisement for the company and helps to separate them in the competitive marketplace where many companies strive to capture the minds of the consumers. CSR helps in maintaining the reputation of the organization by avoiding the evidence of unwanted incidents like corruption scandals and environment-related accidents, which would tarnish the image of the organization. Building an honest authentic culture within the business setup would help to avoid many risks and pave the way for image building and good reputation among all stakeholders (Confederation of Indian Industry 2013). CSR serves as a source for value creation by innovation and by promoting a sustainable business model that deals with values shared with institutions and communities. It also develops corporate philanthropy and sponsorships on a limited scale with strategic operational impact.

CSR initiatives have given rise to ethical consumerism as consumers today are becoming more aware of their social and environmental implications of their day-to-day consumer decisions with regard to the purchase and use of products. The pursuit of corporations toward globalization makes them encounter challenges that impose limits on their growth and progress, for example, government regulations, environmental restrictions, and labor problems. CSR monitors the labor practices not only within the organization but also along the entire supply chain, leading to a positive brand building image. Ethical investments or socially responsible investments have become a popular agenda in business organizations, and nongovernmental organizations are taking a major role in the implementation of such thoughts and ideas. The growth of ethics training inside corporations has reduced corrupt practices and promoted employee loyalty and pride in the organization. Though issues and situations vary from place to place, the adoption of CSR as a tool for the social, political, and cultural structures within the countries, has developed unity and a sense of oneness among different parts of the world (Mehrdost 2012).

10.3.3 IMPLEMENTATION OF CSR

The implementation of the CSR policy should develop a strategy that defines identification of activities, setting of targets and time frames, responsibilities and mechanism of organization, and monitoring techniques with schedules. Assessment must be on the need and impact of the CSR activities, and independent evaluation can also be undertaken. Allocation of budget and networking of companies to share their experiences are essential. Information about the activities and progress on the CSR policy should be disseminated in a structured manner to all the stakeholders of the company. All these efforts would help in building the company's image of being socially responsible and contribute to the long-term sustainability of the business (Ministry of Corporate Affairs, Government of India 2009).

10.4 SOCIAL IMPACT MEASUREMENT

Any business organization has an impact on the lives of the people or organizations with whom they work. The consequence of one's action is known as impact, and the term *social impact* deals with the actions that are taken to address the social needs of the community or the change that occurs in the social setup of the people and organizations involved in the said business. Every organization can measure its social impact by analyzing the extent of change by means of a number of indicators. The organization can start measuring one indicator, and this could lead to measurement of many more indicators. The measurement of social impact has many benefits apart from giving a numerical value to the impact caused. It helps to improve the integrity of the organization and help people to lay faith on the organization; inspires the personnel and volunteers of the organization to continuously improve the services offered; indicates the commitment, honesty, and sincerity of the organization to its stakeholders; forms a powerful medium for publicity and for funding proposals; and makes people conscious of the impact made and in turn strive harder for greater levels of achievement (Meldrum et al.).

10.4.1 KEY ELEMENTS OF IMPACT MEASUREMENT

Social impact measurement deals with gathering important information about the activities of the organization under concern and measuring the overall change that has occurred in the lives of the people and environment over a particular period of time. However, it offers value on four basic elements:

1. *Organization information*: This serves as a basis for measuring social impact, and the measurements taken should be targeted toward the outcomes that is of great importance to the social auditor. Information about the outcomes helps to estimate the elements that help in causing the change and in the identification of areas of improvement. The results of social measurement also help in decision making for the future.
2. *Stakeholders and community*: The results of social impact measurement help to inform others the role and achievement of the organization in the social front. The confidence level of funders and investors increase, and contracts can be gained by displaying the outcomes and impact measurement results. The process of informing the public about the outcomes and achievements of the organization using the results of social impact measurement will raise the public awareness of what has been accomplished and also what can be done in the near future.
3. *Impact on beneficiaries and staff*: Social impact measurement will enable beneficiaries of the organization to understand the services and outcomes offered and the benefits that accompany them. The extent of the social impact can be assessed and the organization can give rewards for achievements and also inspire new beneficiaries to involve themselves with the organization for progress. This elevates the morale of the staff as they will be able to sense the change they have made in the lives of the people around them and also see the results of their efforts.
4. *The sector*: Social impact measurement results and reports will attract different organizations to interact and share knowledge and information leading to greater cooperation for effective approaches and techniques leading to improvements across the sector. The body of information obtained from social impact measurement helps representatives of the sector to negotiate with the government with matters related to planning and policy (Investing for Good 2012).

10.4.2 DIMENSIONS OF SOCIAL IMPACT MEASUREMENT METHODS

A wide range of tools are available for social impact measurement, and there is a need to modify or tailor the methods to meet the requirements of the different types of business organizations. It must also be noted that no single tool would enable to capture a whole range of impacts, and hence, from the various methods available for measuring social impact, one must be able to select the most appropriate method for usage. There are certain dimensions that are to be considered before choosing suitable methods for social impact measurement, namely, purpose, time frame, orientation, length of time frame, perspective, and approach as given in Table 10.1.

- *Purpose*: Social impact measurement tools are developed for different purposes like screening, monitoring, reporting, and evaluation. Evaluation of investment openings and their performance in terms of investor's social and financial objectives are facilitated by tools for screening. Operational decision making by management of the organization, provision of data for investor understanding, and identification of business model modifications and market opportunities by the entrepreneurs are conceivable by the tools aimed at monitoring. Reporting tools enable preparation of reports for external stakeholders or other entities on a regular basis. Tools developed for the purpose of evaluation are used for impact assessment of previous performance, achievements, and organizational learning.

TABLE 10.1
Dimensions of Social Impact Measurement Methods

Sl. No.	Dimensions	Types
1.	Purpose	Screening/monitoring/reporting/evaluating
2.	Time frame	Prospective/ongoing/retrospective
3.	Orientation	Input/output
4.	Length of time frame	Short term/long term
5.	Perspective	Micro (individual), meso (corporation), macro (society)
6.	Approach	Process methods/impact methods/monetarization

Source: Data from Maas, K. and Liket, K., Social impact measurement: Classification of methods, 2011. https://www.erim.eur.nl/.../social%20impact%20measurement.doc.

- *Time frame*: Assessment of social impact may be done in different time frames. The social impact of future plans and programs (prospective) will help to choose different options, design suitable modifications in cases of problems, and assist decision makers to choose the best option. Methods that focus on ongoing events are useful for current testing assumptions, while retrospective methods aid in the evaluation of past activities.
- *Orientation*: Social impact measurement tools have an orientation as assessment of input methods or output methods. Input-oriented methods evaluate the differences in input due to a social activity, while output-oriented methods examine the differences in output due to the outcome of a particular social activity.
- *Length of time frame*: This dimension can be evaluation for a short or long time frame. The focus of measurement usually is short term, but it is always found that long-term assessment is important as social impacts may occur after a long time.
- *Perspective*: Social impact measurement tools may use different perspectives that call for different indicators to draw conclusions. The indicators that are taken into account for a microlevel business perspective will be very different from the indicators that are used for a macrosocioeconomic perspective. Thus, the perspective used for the measurement is very important in deciding the results of the social impact.
- *Approach*: Process methods, impact methods, and monetarization methods are the three broad approaches that can be used for measuring social impact. Process methods monitor the cost-effectiveness and efficiency of ongoing operations. Though they do not give absolute results of social impact, the extent to which the output has caused the desired social outcome can be evaluated. Impact methods measure the operational outputs and their impact on stakeholders and society by using one or more modes of measurement, for example, corporate social performance and corporate financial performance assessment methods may be linked for impact measurements. The third method quantifies social and environmental indicators into monetary values to enable comparison with the traditional financial data (Clark et al.; Maas and Liket).

10.5 TOOLS FOR SOCIAL IMPACT MEASUREMENT

Social impact measurement methods aim at capturing, measuring, and assessing the impact of an activity, action, policy, or program that has been implemented to bring about a change in the target group. Social impact is also considered as a return on investment following investment and allocation of funds and resources and decisions on finance. In the past, numerous methods and approaches were introduced under the banner of social impact measurement, but currently, the outcome or impact has been developed into measurable terms by various tools of social impact measurement.

Many tools are available, but the selection of tools and the right combination of different tools will help to draw conclusions as to the impact of the social change. Before selecting the tools for assessment, the aim and objectives are converted to indicators of change, the samples are to be selected to study the change before and after a process has been initiated, and the tools are used for the data collection and evaluation of the impact on short-term and long-term basis. This will help to draw conclusions about the resultant impact or social change. Though innumerable tools are available, a few important techniques or methods are discussed here:

- *Social return on investment*: Business enterprises that work today use lesser resources for more returns, which include social and economic outcomes along with economic costs and benefits by using economic modeling techniques. Monetary values are given to social and environmental outcomes, thereby demonstrating their impact. This information helps in efficient decision making, produces better outcomes for clients and communities, and maximizes the impact of their work (Maas and Liket).
- *Multicriteria appraisal*: It is an appraisal technique that addresses unavoidable conflicts and competing interests in a nonfinancial path and accounts for both social and environmental factors.
- *Outcome evaluation*: This is the measurement of the actual change that has resulted from a specific activity on a range of stakeholders and based on the principle of measuring what is of importance from the stakeholder's point of view. This methodology is based on certain principles that include the measurement of the positive and negative changes in the lives of people, communities, and environment as a result of a policy; assigning a value to these impacts on the same terms as traditional costs and benefits; providing evidence for planning and decision making; avoiding overclaiming of activities, transparency, and accountability; and conducting a SWOT analysis for identifying reasons for failure and focusing on success.
- *Cost-effective analysis* (CEA): This determines the unit cost of an activity compared to a single predetermined outcome, for example, cost per infection averted due to the prevention measure and cost per extra year. It is useful to assess the value of money for activities that have one major outcome.
- *Social cost–benefit analysis*: It includes social outcomes into a cost–benefit framework and is useful for appraising activities or policies that have strong social outcomes.
- *Social value appraisal:* This assesses the creation of social, economic, and environmental values as part of public investments to effectively abide by the Social Value Act 2013 (The New Economic Foundation 2014).
- *AA1000 accountability principles for sustainable development* is based on the principles of inclusivity, materiality, and responsiveness. Inclusivity principle involves the acceptance of accountability to the stakeholders and involving them for achieving a strategic response to sustainability. Materiality is a principle that determines the influence of an issue on the organization and its stakeholders with regard to the decisions, actions, and performance. The third principle is concerned with the response of the organization to stakeholders issues (AA1000 APS 2008). Toolkits like Local Multiplier 3 and Prove and Improve are available for easy calculation of social impact of an organization.
- *The Global Reporting Initiative* is a leading organization in the sustainability field that has its secretariat in Amsterdam, the Netherlands, with regional offices in Australia, Brazil, China, India, South Africa, and the United States. This organization holds partnerships with the UN Environment Programme, the UN Global Compact, the Organisation for Economic Co-operation and Development, the Internal Organization for Standardization, and many others. Global reporting initiative (GRI) has developed a comprehensive reporting tool Sustainability Reporting Framework that includes reporting guidelines, sector

guidance, and other resources that facilitate organizational transparency and accountability. The framework has been developed by the collaborative efforts of professionals and organizations from various sectors, constituencies and regions, international working groups, stakeholders, and public comment periods that assure its credibility and reliability, making it suitable for all organizations. Many organizations all around the world use this framework to understand and communicate their sustainability performance to the world. The GRI has released its fourth-generation guidelines G4 in May 2013 (The Global Reporting Initiative).

- *The Economy for Common Good* is a tool for economic, political, and social change that works for human betterment all around the world. It is supported by local groups, stakeholder network, companies and organizations, communities and regions, and different associations. The key tool, the Common Good Balance Sheet, has its core center on the welfare of human beings and all living things and proclaims standards for human relationship in the economic context. There are twenty principles that form the foundational values of the economy of common good framework. The common good matrix builds a system around human dignity, global fairness and solidarity, ecological sustainability, social justice, and democratic participation. This organization is working for the emergence of a legislation that requires companies and business enterprises to produce such balance sheets in the near future and would also encourage their initiatives by offering certain tax advantages and preferential treatment in public procurement. On completion of a balance sheet, an external audit can be requested or it can be subjected to a process of peer evaluation (Economy for the Common Good 2010–2013).
- *The GoodCorporation Standard* is based on principles that define a framework for responsible management in an organization. GoodCorporation uses an individual assessment process and has a five-point scale assessment. This standard takes into account employees, customers, suppliers and contractors, community, environment, shareholders, and management commitment (The GoodCorporation Standard 2010).
- *Synergy CodEthic 26000* is a social responsibility and sustainability commitment management systems for organizations. This standard is built around ISO 26000 and is composed of normative documents CodEthic 26000, CodEthic 26004, CodEthic 26011, and CodEthic 26021, which makes this standard a certifiable ethical commitment management system. The particulars of this standard are given in Table 10.2 (Synergy CodEthic 26000).

TABLE 10.2
Synergy CodEthic 26000—Normative Documents

Sl. No.	References	Title
1.	CodEthic 26000	Social responsibility and sustainability commitment management systems for organizations—ethical business good practice requirements
2.	CodEthic 26004	Social responsibility and sustainability commitment management systems for organizations—guidelines, self-assessment criteria, and method of the organization
3.	CodEthic 26011	Social responsibility and sustainability commitment management systems for organizations
4.	CodEthic 26021	Social responsibility and sustainability commitment management systems for organizations—requirements for bodies providing audit and certification of ethical commitment management systems of organizations according to CodEthic 26000

Source: Data from Synergy CodEthic 26000, Social responsibility and sustainability commitment management systems of the organisations—Ethical management and social responsibility, http://www.synergy-gss.com/SynergyStandards/Codethic26000.php.

- *EarthCheck* is an organization that recognizes sustainable practices and increased profits with an aim of conserving our planet. It provides solutions like EarthCheck Assessed, EarthCheck Certified Community, EarthCheck Certified Company, EarthCheck Schools, EarthCheck Sustainable Design, EarthCheck Parks and Ski Fields, EarthCheck Responsible Meetings and Events, and EarthCheck Risk App. EarthCheck tools are very stringent and provide relevant benchmarking for improvement (EarthCheck).
- *Well-being valuation* is based on data obtained from large national survey data that include responses of people to questions related to well-being, aspects, and circumstances of their lives. These data are used to estimate the impact of the good, services, and income on the well-being of the people, and these estimates are converted to monetary values that produce the equivalent impact on well-being (Fujiwara; Trotter 2013).
- *Social LCA* is a technique that assesses the social and economic aspects of products and their impacts along the life cycle of products with the aim of achieving human and societal well-being. Data are collected from stakeholders, company, and partners in both qualitative and quantitative forms, social impact categories are identified, and their impacts are analyzed. Conclusions are used for three different purposes—management social life cycle analysis (SLCA) to identify the social hotspots, consequential SLCA to understand the impact of products, and educative SLCA to use the data for bringing about changes in the existing system. Some of the issues in SLCA are the complexity of social issues, no common unit for assessment, no standards for assessment, lack of availability of data, cost of data collection, and interpretation of data. However, this technique is used to evaluate the social aspects and impacts of products for many buyers and manufacturers (Dasmohapatra 2012).

10.6 ORGANIZATIONS FOR SOCIAL CHANGE

Critical issues face textile and clothing industries, and many multinational enterprises in this sector have violated many norms in order to procure low-cost products from developing and underdeveloped countries. The deplorable conditions and the work culture of these production centers have paved the way for organizations to put a check to these developments by specifying that organizations have to be certified with regard to the social and environmental impact of their business activities. There are many organizations who work for social change for the transformation of culture and attitude in many business organizations. Laws are enforced to safeguard and protect the personnel involved in the trade as well as to produce positive impacts on the environment, community, society, and world as a whole.

The salient features of the most important organizations who are involved in working for the welfare of the working population are mentioned here:

- *The ILO* was founded in 1919 to pursue a vision that lasting peace can be established by having a strong foundation on social justice. The main aims of ILO are promotion of rights at work, decent employment opportunities, enhancement of social protection, and encouragement of dialogue on work-related issues. The key issues addressed by the organization are youth employment, creation of jobs and livelihoods, and social protection. The youth unemployment crisis that looms over the world is very large with 73 million youth all around the world facing unemployment. The networking of ILO with different countries of the world helps to contribute to the quality and quantity of employment and the labor demand and supply. Social security coverage involves access to health care and income security during old age, sickness, unemployment, invalidity, work injury, maternity, and loss of main income earner. The ILO extends policies and assistance to many countries to provide adequate social protection to all members of the society (International Labour Organization).

Social Impacts of the Clothing and Fashion Industry

- *The UN Development Fund for Women (UNIFEM)* works for the improvement of women in developing countries with focus on women empowerment. It works for reducing poverty among women, prohibiting violence against women, preventing the spread of AIDS among women and girls, and achieving gender equality in democratic issues on national, regional, and global scales. The work of the organization is focused on six priorities, which include involvement of women in decision making at all levels, empowering poor and excluded women economically to attain benefits, bestowing women and girls with a violence-free life, and nurturing peace, security, and humanitarian action by women leadership and participation, to promote national planning with accountability for gender equality and to establish a dynamic set of global norms to be implemented in all countries through governmental and stakeholder action at all levels. UNIFEM works with NGOs and takes effort to strengthen its ties with the civil society, collaborates with the private sector, and supports the realization of indigenous human rights for women welfare and upliftment (UN Women).

- *The Clean Clothes Campaign (CCC)* is committed to improve the working conditions and to provide support for the empowerment of workers of global garment and sportswear industries. This organization works for the networking of organizations and unions in garment producing countries to support workers in achieving their goals. This organization is based on certain principles where all workers have the right to have a good and safe working environment and associate freely, bargain collectively to earn a living wage to allow them to live in dignity, and be involved with the public to address any issues regarding workers' rights. This organization works for the implementation of binding legislations that meet the standards specified in the ILO conventions and also monitors if the brand-name garment and shoe companies follow good labor practices and norms specified in the CCC Model Code. The Code of Labour Practices for garment and sportswear industry has listed the ILO Declaration of Fundamental Principles and Rights at Work and the Universal Declaration on Human Rights. The organization calls for transparency, signing of international framework agreements for social dialogue, and regional, national, and global cooperation for open and active communication, consensus building, and constructive criticism (Clean Clothes Campaign).

- *Global Exchange* is an organization that values the rights of workers, health of the planet, international collaboration, green economy, and international human rights and is dedicated to the promotion of social, economic, and environmental justice around the world. The main aims are people power and not corporate power, community rights, economic activism, and end of dirty energy and building positive alternatives like fair trade, green festivals, local clean energy and green economy, peace, democracy, and human rights (Global Exchange Inc. 2013).

- *The International Labor Rights Forum* is dedicated to achieve dignity and justice to workers all around the globe with a noble vision of providing workers the power to speak and defend their rights, form unions for collective bargaining, and acquire jobs free from discrimination, forced labor, and child labor. The threefold core work is to hold global corporations accountable for violations of labor rights along their supply chains and propose laws and policies to protect workers and to advocate worker's rights. This organization believes that labor rights, which are the central pillar of social justice and economic development, are universal. It is considered that child labor and forced labor are the results of social injustice and labor rights can be maintained with global support. It has also been felt that transparency and accountability would help in uniting consumer and worker interests for just and dignified jobs. Some of the issues handled by the organization include Bangladesh factory safety, China program, cocoa campaign, cotton campaign, forced labor in Vietnam, sweat-free communities, and tobacco campaign (International Labor Rights Forum).

- *Jobs with Justice* is an organization that unites workers, community, students, and local- and national-level movements that are formulated for worker's rights and an economy that benefits everyone. This organization works to achieve a base network of coalitions and leaders, undertake research and policy analysis, engage in legislative and regulatory initiatives, and form tactical alliances by advocacy and organizing skills to promote leadership and a good economy for all. Campaigns of strategic importance include Change Walmart, Change the Economy, Caring across Generations, and Debt-Free Future (Jobs with Justice).
- Workers involved in difficult jobs like sewing garments, mining, and building dams and prisons are susceptible to abuse and exploitation. *CorpWatch* is dedicated to investigate and expose cases of violations of human rights, environmental crimes, fraud, and corruptions at corporate level around the globe. There always seem to be a war between employees and top-level management. Some examples of these differences are the murder of union officials in Columbia and Walmart fighting in local courts against laws that insist that the business industries have to make provision of minimum amount of health insurance benefits to its workers. Labor unions, health insurance, sweatshops, and pensions are some of the issues handled by the organization (CorpWatch).
- *Rugmark Foundation, India*, was formed in 1994 by the Indian carpet manufacturers, exporters, nongovernmental organizations, UNICEF, Indo-German Export Promotion Project, and the German Agency for Technical Cooperation (GTZ) against the use of child labor in the carpet industry. This private, voluntary, nonprofit body certifies and assures importers and carpet buyers with a label that the said product has been manufactured without the employment of child labor and has been subjected to a strict surveillance procedure. Initially, many campaigns against the use of child labor in the carpet industry led to a total boycott of the import of carpets from India. The adverse effect of this move on workers, exporters, and carpet manufacturing regions made carpet manufacturers realize the abolition of child labor that resulted in Rugmark certification. Rugmark Foundation provides free education up to the primary level and free books and uniforms and also spreads awareness among parents to understand their responsibility toward their children. Many schemes like provision of schools and rehabilitation centers, mobile clinics for health, adult education programs, vocational training programs for children, and self-help groups for women have helped to avert the child labor problems to a certain extent.

The inspection and monitoring system is very stringent, and the looms of all licenses are regularly monitored by experienced and proficient inspectors specially trained by the foundation. The inspectors do not know the location of the inspection site until the inspection day morning to avoid corruption. The inspection covers all areas like weaving, washing, dyeing, and clipping. If any manufacturing unit is detected for employment of children, withdrawal of license will result. There are more than 350 licenses/carpet exporters until December 2009 (Rugmark India).

- *Sweatshop Watch* was founded in 1995 as a partner of different sectors like labor, community, civil rights, immigrant rights, women, religious and student organizations, and committed individuals. The major aim of this organization was to eliminate the exploitation that occurs in sweatshops with a focus on garment workers in California and national and international agencies. The strong belief is that workers should earn a living wage in a safe and decent working environment. Accomplishments of this organization is the provision of justice and release of workers from the El Monte, California, sweatshop, raising the minimum wage in California, hosting a living wage summit, winning lawsuits with several major retailers for the atrocities and worker abuses in the factories of Saipan, bringing into existence the sweatshop reform law for California, and conducting programs for educating the consumers (Sweatshop Watch 2013).

Currently, many organizations all around the globe are working for a social change to happen that will protect and safeguard the dignity and safety of the garment industry employees. Many laws and regulations exist in many countries, but still it has become mandatory that organizations are necessary to monitor the apparel industries and their working environment to imbibe the importance of social responsibility. Clothes offered to the consumers should be safe and sourced and manufactured in such a way that it protects people and planet with the optimum profit. Thus, these organizations play a vital supervisory role and also highlight the atrocities that happen in many parts of the developing nations. The workers and the affected employees can turn to these organizations for judiciary support and also get their claims resolved. It is necessary that organizations aiming at social change fight for human rights and labor rights by facing and withstanding the pressures of the industrial giants and multinational corporates to maintain the social aspects of clothing manufacture and distribution.

10.7 FUTURE DIRECTIONS

The textile and clothing industry has been marked with increasing competition since the removal of quota system in 2005. There are a few key factors that will have a bearing on the survival of manufacturers and retailers to continue in this field of production. They include international trade relations, the organization and structure of the industry, new and emerging technologies, and human resources and enforcement of international rules and conventions. The least developed nations enjoy the benefits of free entry into the EU when compared with the other nations. This trend has caused a shift in production centers to developing and underdeveloped countries, leading to many social problems.

Skill development and training is an important concern as it can procure decent wages. There are severe shortages of skilled personnel, and the training levels are almost nonexistent in many industries, leading to low wages. Further, in industrial production systems, attention should be given to plant layout and alignment of the departments with special emphasis on ergonomy, ambient working conditions, stress, and workload. Unfavorable working conditions exist in sweatshops, which leads to violation of human rights issues, safety, and health (Martinuzzi et al. 2011).

Organizations today should take keen interest in the career life of employees and balance the workload by sharing the duties and responsibilities among the different employees in the industry. This will enable the employees to balance their work life and personal life to improve their performance coupled with job satisfaction. Work life balance is a broad term that aims at creating an equilibrium between career and ambition (work) on one hand and pleasure, leisure, family, and spiritual development (life) on the other hand.

All business organizations must consider every employee as exclusive and valuable and give time for their own personal needs. The company should initiate good distribution of workload to balance the work life of the individuals; otherwise, it leads to poor performance, absenteeism, and sickness. The work environment that has appropriate provisions in place, the right combination of participation defined by hours and working conditions, effective organizational systems, and supportive management will be able to make provision for good work life balance. The organization should analyze work and family problems faced by employees and the overall satisfaction level of the employees and plan work life balance programs to reduce family work tensions and conflicts. Increments given to workers for balancing their work life would serve as a source of encouragement (Meenakshisundaram and Panchanatham 2012).

Huge variations exist across industries and sectors with regard to working time, duration and organization of work, work management practices, paid training, employee representation, and exposure to physical and psychosocial risks. Job quality is analyzed by four indicators, which are earnings, prospects, intrinsic job quality, and working time quality. The textile and clothing sector scores relatively poor in job quality, and the workers are faced with multiple disadvantages like low pay, relatively high levels of exposure to risks, irregular working time arrangements with minimum or no control over working hours, and very few prospects for career improvement. These conditions promote poor mental and physical well-being among workers.

Policy makers should address this situation and improve job quality by promoting good working environments as working conditions associated with positive employee's health and well-being are associated with high worker motivation, creativity, and commitment, leading to good levels of sustainability in work and improved productivity of organizations (Houten et al. 2013).

Implementation of social rights strengthens democracy in the working environment and provides economic competitiveness with social progress. Social partners include all stakeholders in the organizational level, and they should be involved in discussion, contribution, cooperation, and partnership on company-level issues and in the delivery of their opinion on policy initiatives. This includes the right for information, consultation, and participation of workers in company issues and their involvement in policy matters. The organization should also visualize how these contributions can secure and improve the difficult path toward participatory democracy in the company (ETUI 2011).

CSR demonstrates the company's responsibility toward community and environment. This is usually achieved by expressing the organization's environmental policy toward waste and pollution reduction processes and contribution to educational and social programs and by obtaining adequate returns on employed resources. However, many challenges are to be addressed and the European Commission in its Social Responsibility Policy suggests an action plan for meeting the challenges facing CSR. They include enhancement of the visibility of CSR and disseminating good practices by way of awards to deserving organizations; improvement and tracking levels of trust in business by conducting surveys in this regard; improving self- and coregulation processes by developing code of good practices to work with other business organizations; market reward for CSR-based organizations; improving company disclosure of social and environmental information; integration of CSR into education, training, and research; emphasizing the importance of national and subnational CSR policies; and better alignment of global approaches to CSR (European Commission 2014).

To direct and strengthen the social responsibilities of organizations and business enterprises, many lessons have to be learnt from the series of cases undertaken by different social organizations to safeguard the rights of garment industry workers. To overcome the barriers that inhibit the adoption of social policies, certain directives are suggested (Allwood et al. 2006):

- Consumer education is vital and the consumer must be aware of the fact-based information on the specific impacts of products before he or she goes in for the purchase.
- Reduced material consumption will result when product designing is done with *durability* as the main component of fashion. This will reduce the impacts on the environment and consumers will also be willing to pay more for a durable product.
- Work life balance and improvement in working conditions and environment would increase the benefits of better intrinsic job quality and job satisfaction.
- Technology development will lead to many refined methodologies and innovations that will require less effort and lower impact on environment and society.
- Social dialogue and worker participation and representation in the issues of the organization will benefit both the employer and employees.
- With inflation and rising prices, a living wage, which enables workers to meet their basic needs (nutritious food, clean water, shelter, clothes, education, health care, and transport) as well as provision for a discretionary income, is to be developed.
- Assessment of social impacts along with environmental impacts should become mandatory for all business enterprises.
- The governmental policies should incorporate changes to promote reduction of total and embedded impact of products with due considerations to those arising not only in their own country but all around the world as manufacturing and distribution has become global.
- The involvement of the government in negotiating international agreement on trade could be used to promote environmental and social responsibility.

The social impacts of the clothing industry have come to the forefront and have gained importance in this era of sustainable apparel development and use. Retailers and traders promote their products with green concern, and the consumer wants to know all the relevant information from the sourcing of raw material to the end product presentation, to enable him to make a sustainable choice. *Consciousness in action* is the motto of the day and the attitude of the people has turned toward wise decisions that lead to sustainability. Sustainability and commercial longevity go hand in hand, and it is important that there should be keen understanding of the social and environmental implications of apparel products to compete in the global market.

REFERENCES

AA1000 APS. 2008. AA1000 accountability principles standard 2008. http://www.accountability.org/images/content/0/7/074/AA1000APS%202008.pdf. (accessed July 17, 2014.)

Allwood, J.M., Laursen, S.E., de Rodríguez, C.M., and Bocken, N.M.P. 2006. Well dressed? The present and future sustainability of clothing and textiles in the United Kingdom. www.ifm.eng.cam.ac.uk/uploads/Resources/Other…/UK_textiles.pdf. (accessed June 1, 2014.)

Boone, T. 2012. Fashion Industry: A story of consumption and waste. Rediff.com Business, July 17. http://www.rediff.com/business/slide-show/slide-show-1-special-fashion-industry-a-story-of-consumption-and-waste/20120717.htm. (accessed June 1, 2014.)

Business Dictionary. 2014. Social Impact. http://www.businessdictionary.com/definition/social-impact.html. (accessed June 1, 2014.)

Boyenge, J.P.S. 2007. ILO database on export processing zones (revised), Working Paper 251. April. http://www.ilo.org/public/libdoc/ilo/2007/107B09_80_engl.pdf. (accessed June 29, 2014.)

Centre for Remanufacturing and Reuse (CRR). Key issues, barriers and opportunities. http://www.uniformreuse.co.uk/corporatewear-issues.html. (accessed June 9, 2014.)

Clark, C., Rosenzweig, W., Long, D., and Olsen, S. Double bottom line project: Assessing social impact in double bottom line ventures—Methods catalog. www.riseproject.org/DBL_Methods_Catalog.pdf. (accessed July 17, 2014.)

Clean Clothes Campaign. Who we are, what we believe in. http://www.cleanclothes.org/about/principles. (accessed July 22, 2014.)

Clean Clothes Campaign. 2014a. Indonesian garment industry receives human rights trial. June 20. http://www.cleanclothes.org/news/press-releases/2014/06/20/indonesian-garment-industry-receives-human-rights-trial. (accessed June 20, 2014.)

Clean Clothes Campaign. 2014b. Indonesian wage trial: Human rights violations 'systemic'. June 24. http://www.cleanclothes.org/news/2014/06/24/indonesian-wage-trial-human-rights-violations-systemic. (accessed June 24, 2014.)

Collier, R. 1999. Saipan workers describe slavery of sweatshops/they say American Dream turned into nightmare. *Chronicle*, January 22. http://www.sfgate.com/news/article/Saipan-Workers-Describe-Slavery-of-Sweatshops-2950970.php.

Confederation of Indian Industry. 2013. *Handbook on Corporate Social Responsibility in India*. Confederation of Indian Industry, Gurgaon, India. http://www.pwc.in/assets/pdfs/publications/2013/handbook-on-corporate-social-responsibility-in-india.pdf. (accessed July 8, 2014.)

CorpWatch. About CorpWatch. http://www.corpwatch.org/section.php?id=169.

Dasmohapatra, S. 2012. Social Life Cycle Analysis (SLCA). September 25. http://www4.ncsu.edu/~richardv/documents/Lecture14SocialLCA_final.pdf.

Diviney, E. and Lillywhite, S. 2007. Corporate social responsibility in the Australian garment industry. Brotherhood of St Laurence, Melbourne, Victoria, Australia. http://www.bsl.org.au/pdfs/DivineyLillywhite_ethical_threads.pdf. (accessed June 24, 2014.)

EarthCheck. EarthCheck-Solutions. http://www.earthcheck.org/solutions.aspx. (accessed July 21, 2014.)

Economy for the Common Good. 2010–2013. Economy for the common good-vision and mission. http://economia-del-bien-comun.org/en/content/vision-and-mission. (accessed July 20, 2014.)

ETUI. 2011. Social dialogue and worker representation in EU2020: Underappreciated and underplayed. www.etui.org/content/download/2121/23522/file/Chap+8.pdf. (accessed June 7, 2014.)

European Commission. 2014. EU Corporate Social Responsibility policy: The Commission seeks stakeholders' views on achievements and future challenges. April 29. http://europa.eu/rapid/press-release_IP-14-491_en.htm.

European Foundation for the Improvement of Living and Working Conditions. 2002. Quality of work and employment in Europe—Issues and challenges. February. www.eurofound.europa.eu/pubdocs/2002/12/en/1/ef0212en.pdf. (accessed July 4, 2014.)

Forum for the Future. 2007. A review of the sustainability impacts of the clothing industry. March. http://www.forumforthefuture.org/sites/default/files/project/downloads/fashionsustain.pdf. (accessed June 2, 2014.)

Fujiwara, D. Measuring the social impact of community investment: The methodology paper. https://www.google.co.in/search?num=20&newwindow. (accessed July 17, 2014.)

Gardetti, M.A. and Torres, A.L. 2013. Sustainability. Responsibility. Accountability. March. www.greenleaf-publishing.com/content/pdfs/sifs1intro.pdf. (accessed July 7, 2014.)

General Accounting Office (GAO). 1994. Garment industry efforts to address the prevalence and conditions of sweatshops. November 2. http://www.gao.gov/archive/1995/he95029.pdf. (accessed June 30, 2014.)

Global Exchange Inc. 2013. 25 years of resisting injustice envisioning alternatives taking action—Annual report 2012–2013. http://www.globalexchange.org/sites/default/files/GX.AnnualReport2013.pdf. (accessed July 10, 2014.)

Greenhouse, S. 2001. Beatings and other abuses cited at Samoan plant that supplied US retailers. *The New York Times*, February 6. http://www.nytimes.com/2001/02/06/us/beatings-other-abuses-cited-samoan-apparel-plant-that-supplied-us-retailers.html. (accessed July 6, 2014.)

Houten, G.v., Cabrita, J. and Vargas, O. 2013. Working conditions and job quality: Comparing sectors in Europe. www.eurofound.europa.eu/pubdocs/2013/84/en/1/EF1384EN.pdf. (accessed July 6, 2014.)

International Labour Office (ILO). 2009. Global employment update. May. http://www.ilo.org/wcmsp5/groups/public/---ed_emp/---emp_elm/---trends/documents/publication/wcms_114102.pdf. (accessed July 22, 2014.)

International Labor Rights Forum. International Labor Rights Forum is a human rights organization that advocates for workers globally. http://www.laborrights.org/about. (accessed July 21, 2014.)

International Labour Organization. About the ILO. http://www.ilo.org/global/about-the-ilo/lang--en/index.htm.

Investing for Good. 2012. Guidelines for how to measure and report social impact. www.goodanalyst.com/fileadmin/ifg_users/pdf/IFG_guidelines_01.pdf.

Jobs with Justice. About us. http://www.jwj.org/about-us. (accessed July 22, 2014.)

Kanji, G.K. and Chopra, P.K. 2010. Corporate social responsibility in a global economy. *Total Quality Management & Business Excellence* 21 (2): 119–143. http://dx.doi.org/10.1080/14783360903549808.

Maas, K. and Liket, K. Social impact measurement: Classification of methods. https://www.erim.eur.nl/.../social%20impact%20measurement.doc. (accessed July 14, 2014.)

Martinuzzi, A., Kudlak, R., Faber, C., and Wiman, A. 2011. CSR activities and impacts of the textile sector. RMIS Working Papers No. 2, pp. 1–28. http://www.sustainability.eu/pdf/csr/impact/IMPACT_Sector_Profile_TEXTILE.pdf. (accessed June 1, 2014.)

Meenakshisundaram, M. and Panchanatham, N. 2012. A study of work life balance of employees with reference to a garment industry—Unit. *AMET International Journal of Management*. July–December, 52–58. http://www.ametjournal.com/attachment/ametjournal-4/Dev-Article-7-MeenakshiSundaram.pdf. (accessed July 6, 2014.)

Mehrdost, H. 2012. Strategies to promote social responsibility in cultural organizations (Case study: Social and Cultural Department of Tehran Municipality, District 8). *International Journal of Business and Social Science*. 3 (6) 231–246. http://ijbssnet.com/journals/Vol_3_No_6_Special_Issue_March_2012/29.pdf. (accessed July 6, 2014.)

Meldrum, B., Read, P., and Harrison, C. A guide to measuring social impact. http://www.illuminateict.org.uk/sites/www.illuminateict.org.uk/files/a_guide_to_measuring_social_impact_v2.pdf.

Ministry of Corporate Affairs, Government of India. 2009. Corporate social responsibility voluntary guidelines 2009. *India Corporate Week*, December 14–21. http://www.mca.gov.in/Ministry/latestnews/CSR_Voluntary_Guidelines_24dec2009.pdf. (accessed July 6, 2014.)

Morey, M. 2000. Nike slammed over Indonesian Factories. September 7. http://workers.labor.net.au/70/news1_nike.html. (accessed July 7, 2014.)

Nicita, A. and Razzaz, S. 2004. Who benefits and how much? How gender affects welfare impacts of a booming textile industry. World Bank Policy Research Working Paper (3029), April 15. http://www.launch.org/sites/default/files/Material%20Patterns_042013.pdf. (accessed June 1, 2014.)

OSHA, United States Department of Labour. Extended Unusual Work shifts. https://www.osha.gov/OshDoc/data_Hurricane_Facts/faq_longhours.html. (accessed July 3, 2014.)

Pesticide Action Network (PAN). Cotton. http://www.panna.org/resources/cotton. (accessed July 23, 2014.)

Razzaue, A. and Eusuf, A. 2007. Trade development and poverty linkage: A case study of ready made garment industry in Bangladesh. April. http://www.cuts-citee.org/tdp/pdf/case_study-ready_made_garment_industry_in_bangladesh.pdf. (accessed June 1, 2014.)

Rugmark India. Rugmark India—About us. http://www.rugmarkindia.org/Rugmark/rugmark_india.html. (accessed July 22, 2014.)

Social Accountability Standard (SAI). (2014). Social accountability 8000—International standard. June. http://www.saintl.org/_data/n_0001/resources/live/SA8000Standard2014.pdf. (accessed June 1, 2014.)

Sweatshop Watch. 2013. Sweatshop Watch. December 20. http://www.change.org/organizations/sweatshop_watch.

Synergy CodEthic 26000. Social responsibility and sustainability commitment management systems of the Organisations—Ethical management and social responsibility. http://www.synergy-gss.com/SynergyStandards/Codethic26000.php. (accessed July 20, 2014.)

The Global Reporting Initiative. The sustainability reporting framework overview. https://www.globalreporting.org/information/about-gri/what-is-GRI/Pages/default.aspx. (accessed July 20, 2014.)

The GoodCorporation Standard. 2010. The GoodCorporation Standard. July. http://www.goodcorporation.com/documents/Standard_000.pdf. (accessed July 20, 2014.)

The New Economic Foundation. 2014. Evaluations and impact assessment. http://www.nef-consulting.co.uk/our-services/evaluation-impact-assessment/. (accessed July 14, 2014.)

Trotter, L. 2013. Understanding wellbeing valuation. October 29. http://www.hact.org.uk/blog/2013/10/29/understanding-wellbeing-valuation#sthash.K6AJplDE.dpuf. (accessed July 20, 2014.)

UN Women. United Nations entity for gender equality and the empowerment of women (UN women). http://www.un-ngls.org/spip.php?page=article_s&id_article=829. (accessed July 21, 2014.)

Wick, I. 2009. The social impact of the liberalized world market for textiles and clothing – Strategies of trade unions and women's organisations. OBS—Workbook, 62. https://www.otto-brenner-shop.de/uploads/tx_mplightshop/AH62_en_01.pdf. (accessed June 24, 2014.)

Women and Global Human Rights. Women and sweatshops http://www2.webster.edu/~woolflm/sweatshops.html. (accessed July 6, 2014.)

Section III

Sustainability and Consumption Behavior of the Apparel Industry

11 Consumer Behavior and Its Importance in the Sustainability of the Clothing Field

Jane McCann

CONTENTS

11.1 Introduction 232
11.2 Factors That Impact on the Sustainability of the Clothing Field 233
    11.2.1 Global Structure of the Textile and Clothing Industry 233
    11.2.2 Changing Retail Practice: Going Online 235
    11.2.3 Perception of Sustainability 236
    11.2.4 Relative Low Cost of Clothing 237
11.3 Consumer Decision Making 237
    11.3.1 Consumer Confusion with Size and Fit 238
    11.3.2 Confusing Information: Eco-Labeling 240
11.4 Neglected Market 242
    11.4.1 New Consumer Majority: The Ageing Demographic 242
    11.4.2 Understanding Older Consumers 242
    11.4.3 Ageing Consumer and Clothing 244
        11.4.3.1 Introducing a Finnish Study: An Exploration of How Mature Women Buy Clothing—Empirical Insights and a Model 244
        11.4.3.2 Findings from the Finnish Study 245
    11.4.4 Outdoor Industry Market Research 247
11.5 Case Study: Functional Clothing for Active Ageing Men and Women 248
    11.5.1 Adopting Co-design Methodology: New to Clothing Design 248
    11.5.2 Initial "Show and Tell": What Does the Consumer Understand? 249
    11.5.3 Developing Shared Language: Key to Effective Communication with Users 250
    11.5.4 Capturing Size and Shape: A Key Consideration for Future User Customization 251
    11.5.5 Gathering User Feedback to Identify Real Needs 251
    11.5.6 Size, Shape, and Fit: A Key Challenge in the Global Clothing Chain 252
    11.5.7 Prototype Co-design Development: In Collaboration with Users 252
    11.5.8 Integration of Technologies: Selected and Tested with Users 252
    11.5.9 Garment Specification: Collaboration between a Product Designer and Architect 253
11.6 Summary of Key Findings 254
    11.6.1 Size and Shape 255
    11.6.2 Styling 255
    11.6.3 Color 255
    11.6.4 Fabrication 256

        11.6.5 Retail Experience..256
        11.6.6 Price..257
        11.6.7 Branding 257
        11.6.8 Merging Research Findings..257
11.7 Way Forward..261
        11.7.1 Benefits of Adopting Co-design Process 261
        11.7.2 How to Bring New Products to a New Consumer 261
        11.7.3 Moving toward a Circular Economy 262
        11.7.4 Guidance to Producers and Consumers: Good Practice in the Outdoor Industry 263
        11.7.5 Bringing Garment Production (Back) Nearer to Market...........................264
11.8 Conclusion 266
References...267
Further Information..269

11.1 INTRODUCTION

Due to globalization, design and sourcing decisions, in relation to offshore clothing manufacture, have to be made well in advance of garments being delivered to retail. This results in lay members of the public, when comparing outdoor garment ranges, asking "Why does everything look the same?" Meanwhile, natural resources are being depleted with clothing disposal contributing to landfill. In referencing a Japanese trade report on fiber production, Teijin notes "653,740 tons of clothes disposed and landfilled each year" (www.teijin.com). The complex nature of the textile and garment industry, with its specialist terminology and the vast range of brands with conflicting claims, is highly confusing to the consumer. The limited flexibility of the supply chain, with timelines involved in the international sourcing of materials and products, is unknown to the general public. As a result, consumers have become accustomed to the erosion of the value of clothing, with the relatively cheap cost of mass-produced garments taken for granted, and now even expected. The United Kingdom alone discards approximately 1 million tons of clothing a year (www.gov.uk, 2011).

Both the EU and the Department for Environment, Food and Rural Affairs (DEFRA) have highlighted that clothing is considered *high impact,* accounting for some 5%–10% of the EU's total environmental impact (Bray, 2009). Bray states that U.K. clothing sales represent a fast-growing retail sector, currently accounting for 15% of total consumer expenditure (ibid). In 2013, the U.K. Waste and Resources Action Programme (WRAP) reported on the economic and environmental impacts of the textile recycling industry finding that more than two-thirds of textile waste in Britain is sent to landfill, equating to around 1.4 million tons of material, despite there being spare capacity at recycling plants (www.wrap.org.uk). In addition to environmental concerns, there are a range of ethical issues to be addressed associated with the challenges of, for example, child labor, poor working conditions, low wages, and health and safety risks as well as animal welfare contraventions and inequitable trading practices. "As 90% of UK clothing is imported, our activities have a significant overseas 'footprint', particularly in India, China and other developing countries" (www.gov.uk, 2011).

Until now, when the clothing industry is being forced to question current practices, lay consumers have been largely unaware of the damage to the environment caused by fiber and fabric production. Recently, the Greenpeace *Detox* campaign has drawn consumer attention to the use of hazardous chemicals in technical textiles: "Outdoor clothing is in close contact with a whole range of chemicals before use. Yarns, fabrics and even ready-made garments are treated with chemical substances to enhance functionality and make items easy to care for" (Santen and Kallee, 2012). The CEO of Patagonia (U.S. outdoor clothing brand) predicts that "if the population climbs from 7 to 9 billion people by 2050 and, even more importantly, our growing and increasingly global high-consumption economy continues to draw down on natural resources, we will exceed the planet's capacity by 300–500 per cent, putting us into ecological bankruptcy" (Chouinard and Stanley, 2012).

… Consumer Behavior and Its Importance in the Sustainability of the Clothing Field

The designer and creative advisor to brand leaders Michael Wolff suggests, "Not many people yet feel viscerally the danger that bees are facing around the world, or equate that danger to their own lives and the lives of their children and grandchildren." But, as designers, we must realize that we are "an accessory to the slow and relentless killing of our planet" causing us to think about what we should be doing and what should be changed in our own lives (Wolff, 2014). Given the size of the clothing market, and the ethical issues that surround clothing manufacture, there is a need for research to explore the role that ethical considerations may play in consumers' assessment and selection of clothing lines (Bray, 2009). The *Ecotextile* journalist John Mowbray questions, "But is there such a thing as the sustainable consumer anyway? And who decides what is 'sustainable' behavior and what is not?" (Mowbray, 2013) He predicts that "in societies where freedom of consumer choice is highly valued, changing consumer behavior and buying patterns will be a tough challenge. Yet it is widely agreed that with population growth, and limited natural resources, this is exactly what manufacturing industries need to do if they want to lower their environmental impact and at the same time, attract a new generation of so-called 'sustainable' consumers" (Mowbray, 2013).

This chapter shows that the breadth of choice is a particular requirement of the ageing population with respect to the demands of the changing body in terms of size, fit, and cut for older figure types, appropriate color for ageing complexions, protection for a range of environments, and style and image to suit different lifestyles. It will introduce recent practical research, with a focus on collaborative design process, involving academic and industry stakeholders working with active ageing consumers, in adopting the co-design process to address clothing product development that is stylish and fit for purpose. It will discuss the value of consultation with end users as a more responsible approach in identifying how functional textiles can address everyday clothing requirements for participation in healthy exercise. It will draw attention to some of the ethical and environmental concerns that the outdoor industry is facing ahead of enforced regulation. It will conclude that co-design methodology has the potential to cut down waste, throughout the iterative stages of design development, and how it has contributed to user understanding of the sophistication of textiles and clothing and helped to reintroduce the forgotten *real* value of quality products.

11.2 FACTORS THAT IMPACT ON THE SUSTAINABILITY OF THE CLOTHING FIELD

11.2.1 Global Structure of the Textile and Clothing Industry

The globalization of the textile and clothing industry has resulted in the relocation of garment manufacturing facilities to less developed countries with low labor costs and *tax haven* incentives. Hines recalls the transactional approaches of the 1990s that gave way to "relationships that were often referred to as akin to partnerships" as suppliers reengineered their factory systems to be more efficient, fast, lean, and flexible, to match the demand for quick response, shortening the time in the processes of design, manufacture, warehousing, and delivery to retail outlets. This constituted a complex network known as the *global supply chain*. Since 2000, this intensified to become *fast fashion* that may now be viewed as wasteful *disposable fashion*. Hines compares the 1990s as the age of barcode technology with the 2000s as the age of radio-frequency identification tags (RFIDs), now used by nearly all high-street retailers to manage their supply chains. "We no longer know the people who make the goods we buy" (Hines, 2010).

A wide geographical gap now exists between the markets for the finished products, in the West, and offshore production sites, initially in the East, and increasingly in the global south (www.stringtogether.com). Koszewska reflects that the largest brands and retail chains dominate and control 75% of the clothing market. "Retailers are in effect agents of the consumer society looking after the interests of the consuming buyer, not the paid supplier" (Koszewska, 2011). With the escalation of fast fashion, Hines notes that retailers today refer to months representing "12 seasons in the year" rather than four seasons (Hines, 2010). As a result of the extended supply chain, in garment sourcing, Koszewska reports that "transactions are carried out through a complex network of agents, subcontractors and suppliers. The manufacturing end of the clothing industry is so scattered that even the companies awarding production contracts do not always know where garments are made and what production conditions are" (ibid).

Despite the growing concern about the environmental impact of the global textile production chain and the traceability of materials, components, and finishes, there appears to be less attention to control over inaccuracies in clothing specification that often results with garments in landfill. A major challenge, from a technical design perspective, is that designers in the West have become detached from *hands-on* involvement in the technical design process. In particular, designers are no longer involved in the generation pattern blocks and style patterns. With the production base overseas, designers are primarily involved in the generation of design concepts and design selection based on 2D computer-generated *flat*-line drawings. In the author's experience, flat drawings, with measurements, cannot communicate the subtlety of skilled pattern cutting especially for sophisticated sportswear garments with tailored fit and intricate articulation to achieve arm lift elbow and knee movement and hood adjustment.

This gap, with its communication and language barriers, results in the requirement for high numbers of sample iterations, often as a direct consequence of inadequate product design specification. Specification methods are inconsistent from company to company, and even within the same company. Rick Fowler, of YoungOne Corporation, supplier to brand leaders in the sports sector, receives varying standards in design specification from brands across Europe and the United States, with inconsistent information that may range from "twenty page detailed specifications to sketches on the back of a napkin." In many cases, there is no 3D pattern development to back up the specification. Fowler comments: "In 50 countries and well over 500 factories, where I have worked, the problem remains the same" (Fowler in conversation with Lewis, 2012). Substantial waste is generated through inaccurate interpretations of the design, with time involved in queries fed back through the chain. Fowler's aim is to reduce the "average sample iterations done by the factories from five samples to two" (Fowler, 2012).

Due to this detachment, Western designers seldom become involved in designing *in the round* or in 3D pattern development. An experienced designer, in the outdoor industry, has described the breakdown in communications as comparable to *Chinese whispers* as garment designs, and specifications are passed down the chain from one sourcing office to the next with no surprises as *a curved line becomes straight in the process*. One brand's strategy has been to introduce common templates for designers to work within (McCann, 2011). Sizing across different countries adds to the confusion along with scant information on the degree of fit, or ease on top of body measurements, for a given style. It has been stated that some factories in Asia prefer not to work with blocks or patterns generated by Western designers, using their own initiative to interpret pattern development, with the potential for different versions of a product produced in different batches from different factories.

The pressure involved in *fast fashion*, which has driven ever-shorter production times, often promotes exploitation of the workforce and depletion of the natural environment. In terms of exploitation of the workforce, Koszewska comments that "the social problems that the textile industry has to cope with are particularly acute in the developing countries, where child employment, forced and slave labor, workers' exposure to physical and mental harassment, very long working hours, pay below the minimum rates, dangerous working conditions and discriminatory practices have reached record-high levels" (Koszewska, 2011). In terms of the depletion of natural resources, a key challenge for textile supply chains in the twenty-first century is sustainability in terms of "using energy that is renewable, raw materials that are renewable and where waste products in the manufacturing process have limited negative impact on the workforce and environment" (Hines, 2010).

Ecological and social problems occur with varying intensity throughout the life cycle of textile and clothing products. The growth of some plant-based raw materials used in the textile industry often involves the use of fertilizers and pesticides and finishing processes, with dying, printing, and washing consuming huge amounts of chemical substances (Koszewska, 2011). However, recent ethical challenges exposed in relation to working conditions in Bangladesh, and the Greenpeace Detox campaign on environmentally hazardous substances, have begun to raise public awareness. The "growing community of conscious and demanding consumers will insist more and more strongly that manufacturers respect the principles of ethical conduct, people and the natural environment" (ibid). This has already led to more stringent technical and ethical auditing of offshore production where "factories are almost audited to death" (Taylor and Goodwin, 2014).

11.2.2 CHANGING RETAIL PRACTICE: GOING ONLINE

Lorna Fitzsimons, Director of the U.K. Alliance Project, proposes that the consumer is leading demand through increasing online capacity with, for example, Marks & Spencer (M&S) taking more online sales in 2013 than from the whole of the in-store footprint (Conference Presentation, *ASBCI*, February 15, 2014). Online retailing was predicted to reach £90 million turnover in 2013, representing a 12% rise on the previous year. However, many online shop owners are not having the same degree of success as the giants of the Internet, for example, Amazon, a company that can do more business than an entire nation (Jones, 2013). Kantar Worldpanel reports that online fashion accounts for the highest value sales, now over £1.6 billion, with department store and supermarket websites accounting for the highest growth over the last 5 years of 358% and 840%, respectively (Mitchell, 2013).

For smaller or independent retailers, what really matters is psychology: "When a shopper enters a fashion store they are surrounded by a range of engaging elements that connect with them psychologically. There is the lighting, the colors, the temperature and even the smell" (Jones, 2013). In addition, the sales staff may help customers find what interests them. "This makes the shopper feel positive because there are plenty of subconscious activities going on, such as body language, tone of voice and so on. Altogether a fashion store provides deep psychological engagement." However, when *online almost none of this exists* with little subconscious connection; "all a store has online is a two-dimensional image on a screen and a bit of text. As a result, the average time people spend on websites is measured in seconds. No longer does a fashion store have minutes to engage with customers; online they have seconds, a blink of an eye" (Jones, 2013).

Jones identifies five factors that may enhance online customer engagement: sites should be convenient, *likeable, informative, customizable,* and *knowledgeable.* Convenience is a powerful psychological motivator. If the site is intuitive and can be used quickly and on the customer's own terms, they will be encouraged to stay longer and engage more. Shoppers should also be made to feel that they have a relationship with the store, with the opportunity to ask questions, as being informative is another factor in gaining engagement. Customers also want customization with the site personalized to their needs. Jones suggests that "few sites do this anywhere near to the degree of Amazon." Finally, people want the site to demonstrate that the company is knowledgeable and trustworthy (Jones, 2014).

Mitchell, of Kantar Worldpanel, reports that instant engagement with the target consumer in store or online, locally or globally, is becoming vital in securing sales with customers who have so much fashion choice, less money, and so little time for shopping around. Survey findings indicate that one in three garments is bought in a sales promotion. Consumers are said to increase their intention to purchase (up to 300%) if the brand's body image message is relevant to them (Mitchell, 2013). Franklin, of All Walks Beyond the Catwalk, asks fashion suppliers to look carefully at the messages and images that they are using to sell fashion and to question whether these are what most consumers want. In fact, much of the fashion promotion may be detrimental to consumer's self-esteem. "The body ideal of young, thin, and Caucasian is not representative of a global consumer base and indeed alienates consumers which makes bad business practice" (Franklin, 2013).

Mitchell reports that 46% of Kantar's 15,000 strong consumer panels have said price is the most important factor followed by 32% quality, 15% look, 4% brand, and 3% trend (Mitchell, 2013). Inclusive, ethical, and environmentally sound practices also make for good business practices, with consumers feeling better about their purchase if it is green. Mark Sumner, when sustainability raw materials expert at M&S, stated that "consumers are looking for businesses to help simplify their actions and choices to make it easier for us all to behave responsibly and sustainably" (Sumner, 2013). M&S was one of the first high-street retailers to end providing consumers with free carrier bags and also to introduce *wash at 30* labels. "Consumer behavior changes when it is made easy for them to do so and when they benefit financially" (ibid).

Graham Jones, Internet psychologist and web customer behavior specialist, describes ASOS as one of the "half dozen businesses who are dominating what happens on the Internet. With the choice overwhelming, it engages customers with daily edits and HOT off the press features. It has extended the size range to increase relevance and global appeal and has invested in 'shop on the go' mobile-based applications to serve customers on any device in any language by any payment method and anywhere in the world" (Batty, 2013).

Mitchell suggests, "As 5.5 m consumers already use smartphones for shopping they will increasingly use them for 'show rooming' in finding the best deal. Retailers will get better at personalized targeting opportunities that these technologies can deliver" (Mitchell, 2013).

11.2.3 Perception of Sustainability

Wasaty suggests that "those ominous, towering stacks of discarded clothing we have all seen. When the average consumer visits their favorite store or shops online, they do not want to be reminded of the world because fashion is an innately personal, self-interested indulgence. Fashion is of the now. It is not like the organic food movement, although it could take a few pages from its book" (Wasaty, 2013). She continues, "Consumers, at the moment, will not exercise the same purchasing decisions, such as selecting organic perishables or even beauty products, because the outcome does not create an immediate sense of wellbeing." This sentiment is echoed in the Kurt Salmon retail survey (2014), in respect of *supply chain transparency* where the chief executive of a fashion chain is quoted: "What I have noticed is that any scandal involving what you put in your mouth, customers will vote with their feet but not necessarily if the scandal involves something you wear" (Kurt Salmon, 2014).

Wasaty's view is that sustainable fashion has been presented as an antithesis. "From the onset it has been distinguished not as fashion but as something entirely different. It is given special indicators on e-commerce sites, special sections at fashion weeks and trade shows, consolidated retail spaces, special launches and events as if to say, 'we do not belong with everyone else'" (Wasaty, 2013). However, the clothing industry now has the increasing scrutiny of the media, as well as social media, as consumers become more aware of the ethical and environmental impact of the textile and clothing industries (Bray, 2009). Negative publicity of issues surrounding *sweatshop*-type manufacturing conditions and campaigns by NGOs such as Greenpeace on the use of hazardous chemicals and Four Paws (with regard to animal rights) has grown significantly with resulting partial boycotts of affected brands.

Michael Wolff comments on design sustainability in that "when I think about what's unsustainable, so many challenges crowd into my mind that I'm left feeling impotent. For me, and for many designers, sustainability usually means figuring out how we design everything as well as we possibly can to reduce waste, to be recyclable and to avoid massive and dangerous landfill sites. This is undeniably important" (Wolff, 2014). In discussing the design needs of the ageing population, he continues, "Almost everything in the shopping experience is designed for young people without disabilities. Some ATMs are still unreachable for people in wheelchairs. Too many packages in our supermarkets are impossible for old or unwell hands to open. Clothes made for seniors are resolutely unattractive." Wolff asks, "Why aren't there more ideas, innovations and designs that make life far more comfortable without looking embarrassing?" (ibid).

11.2.4 Relative Low Cost of Clothing

This fast fashion churn has led to the erosion of the value of textiles and clothing. "Textiles are unbelievably cheap by comparison with years ago – almost disposable" (Taylor and Goodwin, February 25, 2014). Cline recalls that "well into the 20th Century, clothes were pricey and precious enough that they were mended and cared for and reimagined countless times" (Cline, 2012). For hundreds of years, American style was handmade or made by a tailor until after World War II when "the average American gained wealth and expenditure on clothing grew alongside paychecks as middle-class life and consumer society had arrived" (ibid). By 1950, average incomes were $4237 of which $437 was spent on clothes as "Americans started to accumulate far more clothing than they could regularly wear and to follow fashion in the ways that hinted at what was to come" (ibid). In 1955, Cline finds that the Sears catalogue promoted a 100% nylon ballerina gown for $15.89, which would convert to $130 today, and *lowest price* dresses, at $2.49, would be comparable with today's budget fashion chains at approximately $20, when adjusted to inflation. Cline makes reference to *Time* magazine in recording that, in 1963, for the middle-market brand, Jonathan Logan, *junior size prices* were $14.98, which represents more than $100 today (ibid).

At a point when 2% of clothing is made in the United States, down from 50% in 1990, Cline states that "Clothing has seen such dramatic declines in price that it's gone from budget-buster and a defining purchase for the American household to discretionary spending" (ibid). The BBC program, *Addicted to Cheap Clothing: How Clothes Get Cheaper*, in recording the opinions of shoppers in a mall, comments that in the U.K. value sector, the cost of clothing has fallen by 35% in the last decade (www.bbc.co.uk). Cline quotes an article in a recent *Vogue*, in respect of a dress priced at $4.95: "Do I get a coffee? A snack? Or something to wear?" (Cline, 2012) Alarmed by the erosion of the value of clothing, Cline proposes that "Clothes could have more meaning and longevity if we think less about owning the latest and cheapest thing and develop more relationship with the things we wear" (ibid).

This situation demands a return to a more considered, respectful, and engaged approach to product design and development. As Walker states, "what is clear is that many of our conventions and practices are no longer valid for the context in which we find ourselves" (Walker, 2006).

11.3 CONSUMER DECISION MAKING

The consumption of textiles and clothing, by both individuals and groups, involves a complex set of behaviors, some intentional and some habitual, which are usually constrained by factors such as time, economic resources, fashion, culture, social networks, and even infrastructures (roads, energy, etc.) (Mowbray, 2013).

The consumer decision model proposed by Blackwell et al. (2001) consists of three main components: the *decision-making* process, the *information* process, and the *judgment* process (Langen, 2013). The decision-making process starts with the awareness of a problem that is activated by different stimuli affecting the individual, such as marketing stimuli and peer influence. When recognizing a problem or need, or in responding to a marketing stimulus, a stimulated customer must decide how much information is necessary in order to arrive at a decision. "If the need is strong and there is a product or service that meets the need close to hand, then a purchase decision is likely to be made there and then" (ibid).

The information search succeeds the first phase if no direct solution for the problem is available. The shopper may obtain information from several sources such as peers, advertising, salespeople, retailers, packaging and point-of-sale (POS) displays, the media and consumer organizations, and his or her own experiences with the product. The usefulness and influence of the information will vary by product and by customer. An evaluation is then made between the products, brands, and services according to the shopper's personal preferences, related to attitudes, personality, lifestyle, etc., with interpersonal variables affected by norms and values. "An important determinant of the degree of evaluation is whether the consumer feels 'involved' in the product. In this context, involvement refers to the degree of perceived relevance and personal importance that accompanies the choice" (Langen, 2013). Where a purchase is highly involving (with high expenditure), the customer is likely to carry out an extensive assessment. In contrast, low involvement purchases (such as low-cost supermarket items) are presumed to have very simple evaluation processes.

> The final phase is the post-purchase evaluation of the decision. It is common for consumers to experience concerns after making a purchase decision. This arises from a concept that is known as 'cognitive dissonance'. The customer, having bought a product, may feel that an alternative would have been preferable. In this case the shopper will not repurchase, but is likely to switch brands next time.
>
> **Langen (2013)**

With relevance to this chapter, an exploration of how clothing is bought suggests that the process may be modeled to consist of three subprocesses: building up need and awareness, searching for and fitting clothing, and evaluating and purchasing the clothing (Holmlund et al., 2011). Bray suggests that there are a number of inherent challenges with studies into clothing purchase due to the diversity of purchase motive possible and the variety of roles that clothing can perform. Studies have shown, for example, that the attributes of choice differ between casual clothing and smart clothing, that body shape influences preferences, that significant differences exist when looking at a product in store or observing it in a catalogue, and that demographic variables alter the key attributes assessed. The variety of product attributes considered can, however, be broadly categorized as functional or symbolic (Bray, 2009).

11.3.1 Consumer Confusion with Size and Fit

Fit affects every stage in the clothing product development life cycle, from design and manufacturing through to in-store retail and e-commerce. Traditionally, most systems for sizing ready-to-wear garments have been based on very limited information. A major sizing survey was conducted in the United Kingdom between 2002 and 2004. SizeUK represented a collaboration of the U.K. government, 17 major U.K. retailers, leading academics, and technology companies. It was the first national survey of the U.K. population since the 1950s and the first time that the shape of the population had been captured and analyzed. Three-dimensional laser body scanning was implemented resulting in a point cloud for each subject from which to generate customized basic pattern blocks. With a target consumer group in mind, this has enabled the SizeUK partner clothing retailers to update and amend their size charts and garment specifications (www.size.org). Subsequent national sizing surveys have been carried out in, for example, Germany and the United States. However, both male and female consumers are confused by the lack of a universal sizing system. In the United Kingdom, this confusion contributes to 70% of all returned garments bought online being fit related (Drapers Etail Report). "Apart from the actual cost of the lost sale and processing these returns, clothing brands also lose their customer loyalty. 52% will buy less from the online store and 23% will buy less from any store!" (Haldre, 2014).

In a recent consumer survey into what women really want from fashion, a questionnaire was sent to more than 40 women with ages that ranged from 18 to 84, who represented different shapes, sizes, and nationalities, living across the westernized world, and with a wide range of incomes. The aim was to hear opinions from those outside the fashion *bubble*. Above all, what was found was that they would like clothes to fit. "We want more petite and tall sizes; we want half sizes, in clothes and in shoes; we want legs and sleeves in several lengths' and, ideally in some clothing sectors we want in-house alteration" (Willis, 2013). The concept of *vanity sizing*, which is calculated to make customers feel better about themselves and become more inclined to buy because they can fit into a smaller-than-usual size, is frustrating with a 49-year-old respondent commenting "I used to be a size 12 and now I'm a size 10, but I haven't got any thinner." Willis reported that women want clothes "to flatter their bodies and not their minds." The market needs "a middle way between couture and off-the-peg, a bit of pin and tuck" with many consumers willing to pay to have their clothes altered to fit properly (Willis, 2013).

In a study at Cambridge University, it was found that

> Women increased their purchase intentions by more than 200% when the models in the mock ads were their size. In the subgroup, over size 6, women increased their purchase intentions by a dramatic 300% when they saw curvier models. Conversely, when women saw models who didn't reflect their size, they decreased their purchase intentions by 60% and women over size 6 dropped their purchase intentions by 76%.
>
> **Barry (2014)**

However, the U.K. industry continues to be reluctant to abandon a core size 12 as the standard for women's wear as "this looks better on the hanger." Julia Mercer, technical manager, M&S, Lingerie, believes that retailers have a key *assisted shopping* role to play in helping women to understand the real shape of their body so that they may "buy the best fit regardless of the size on the label." For its new *assisted shopping* shapewear finder service, M&S has rejected "the traditional apple and pear body shape descriptions in favor of less emotive V, A, O, X and H body shape descriptives" (Mercer, 2013). The service asks customers to respond to six questions about their body shape from which it makes five or six product purchase recommendations. This is said to have resulted in high conversion rates from 5% or 6% to 20% (ibid).

At Matalan, the U.K. clothing and homeware chain, Jane Pye (technical service manager) attributes customer dissatisfaction with fit that is costing the company millions of pounds. One-third of garments returned is due to fit problems factoring in not only the value of the garment but also the additional protracted, hidden costs of processing returns, in handling and rehanging garments and returning them to the store or to the online distribution center. Pye suggests that the actual cost could be more than double and, directly related to fit, states that "It's unbelievable that designers and buyers don't understand the importance of (pattern) blocks that can be shared across global supply chains." Matalan has installed development blocks and fit forms throughout the supply base and insists that every supplier has video conferencing facilities so that they can share visuals. Pye proposes that retailers and brands should "work together to define what is size 12" and "to review a core size strategy that is based on the average size of real UK women which is a size 14/16" (Pye, 2013).

Heikki Haldre (CEO of Fits.me) also believes in virtual fitting technologies to reduce the risks of purchasing online. It is claimed that, when consumers enter their age, height, body weight, the "Fits.me" online virtual fitting room helps them create a personalized model so that they can virtually try on different sizes that fit their sense of style. In addition, a robot mannequin can quickly and easily adjust its body measurements according to requirements (Haldre, 2014). In 2013, the U.K. chain store Warehouse has made virtual *try-on* technology available to consumers on its website in the belief that consumers would be more likely to buy and less likely to return their purchases. However, this service has had mixed response: "It's so depressing. I'm 5ft 5 and 10 stone so don't feel fat. But when I compare my Me Model with the 5ft 10 size 10 model the difference is so stark. No wonder I always feel frumpy next to my taller friends even if the clothes look great on the website. I'd have to weigh about 8 stone to get the proportions the same. Should I just accept the hefty rugby player look?" (http://www.mumsnet.com) (April 13, 27, Saturday 08:54:07) "The horrible thing is I think it's probably quite accurate. I've often gone out thinking I look pretty good (flattering mirrors at home) then feel a bit gutted when I see photos but I've been putting it down to bad angles. Now I see its reality – arrgghh!" (http://www.mumsnet.com) (April 13, 27, Saturday 10:04:46).

Alan Cannon-Jones (London College of Fashion) attributes the disparity over men's sizing charts to decimalization with, in 1971, the confusion beginning with menswear suppliers "tweaking (the tables) according to their own size and fit models as it was too difficult and expensive to access data from sizing surveys." Putting this into a global supply chain added to the confusion. Alan proposes that the solution is to determine the largest and smallest sizes for an identified target consumer group, then decide on the number of sizes needed, and then apply the appropriate grading increments. This logical approach partly explains why the consumer is confused to find a breadth of different sizing fit and nomenclature across commercial ranges (Cannon-Jones, 2013).

When implemented successfully, customized production has the potential to cut down waste to a minimum with very little discounting to shift unsold stock. Ed. Gribbon (President of the U.S. Size and Fit Company, Alvanon) believes that mass customization is viable but that essential elements are needed for it to flourish such as "enabling technologies (including innovations in online, e-mobile and self scanning technologies or reverse engineering algorithms); willing buyers (consumers who are prepared to use the technologies and engage in the customized clothing concept); retailers and brands who are willing to invest in the mass customization proposition; and clothing manufacturers who are willing to make 'one off' garments." Mike Fralix (of [TC]2) has a "vision of the future where mass customization would be accessible through home-based 3D technical printing and knitting technologies. Consumers would produce their own garments to fit their own bodies with no waste" (ASBCI, 2014).

11.3.2 CONFUSING INFORMATION: ECO-LABELING

Growing awareness of the ethical and environmental impact of textile and clothing production has encouraged the development of *ethical* garment ranges along with the establishment of eco-labeling. There is growing evidence to suggest that shoppers are predisposed to being *sustainable* and *want* more of these types of products, but retailers find that these buying signals don't necessarily translate into hard cash at the checkout. So if the sustainable consumer does exist, then how can retailers get them through their doors? (Mowbray, 2013) Wasaty suggests that consumers "are not equipped with the right information to make an informed decision, so in the meantime skepticism will prevail" (Wasaty, 2013). In terms of consumer choice, "To make good decisions, decision makers must have information that is available, accurate, and timely, but they also have to be able to comprehend that information and its meaning. They need to be able to determine meaningful differences between options and weight factors to match their needs and values. Finally, they must be able to make trade-offs and ultimately to choose" (ibid).

Koszewska finds that, by comparison with other sectors, the communication of independently audited corporate social responsibility (CSR) reporting, on working conditions in global supply chains, is relatively common between textile and clothing manufacturers and their consumers: "Manufacturers apparently wish to respond to the growing expectations of the public and understand very well how important communication and business transparency are in the globalizing economy" (Koszewska, 2011). In the outdoor industry, Patagonia is considered a pioneer that has led the way, in consultation with both internal staff and external stakeholders, in campaigns for more responsible practices (Chouinard and Stanley, 2012).

Koszewska reports that more and more textile and clothing manufacturers have applied for process certification to label their products. She comments that, by comparison with social labeling, the certification of eco-labels has been established much earlier being created by international organizations such as the European Community, World Trade Organization (WTO), United Nations Environmental Programme (UNEP), and the International Organization for Standardization (ISO). Social labeling has its roots in the trade-union movement although most of the social labels that are applied today were created in the 1990s. She explains that

Eco-labeling, and increasingly more often social labeling, is a method of differentiating products that better meet social (ethical) and ecological standards with respect to traditional products. Eco- and social labels (special quality marks) are awarded by public or private organizations that aim to popularize and promote products that are kinder to humans or to the environment while having comparable usability and functional characteristics.

Koszewska (2011)

Koszewska reports that most consumers prefer to make reference to information affixed directly to the product, that is, simple and visible, rather than taking account of regular social reporting. Labeling systems can be divided according to different criteria such as territorial coverage or thematic scope. In 2005, a survey on the type of information of interest to the British public, to facilitate buying decisions, found that respondents pointed to *no child labor* 68%, *the fabric composition* 58%, *not tested on animals* 53%, *fair pay for workers* 53%, *the producers' country* 46%, *environmentally friendly* 42%, and *good labor conditions* 36%. However, Koszewska has found that the influence of labeling on the consumer continues to be relatively weak due to the confusing diversity of marks and labels, with lack of transparency, eroding credibility.

Textile industry consultant Katharina Schaus comments on the fact that there are currently well over 100 different quality marks used in the textile and apparel industry suggesting that "the plethora of labels is one of the reasons why the proportion of eco-textiles in the market as whole remains relatively small and suggests fewer standards might lead to higher demand" (*Ecotextile News*, Issue No: 57 October/November 2013, p. 11). It has proved difficult to award labels when the entire textile and garment making production chain is so complex and challenging to audit. Koszewska's research has found that consumers are frequently skeptical about the labeling messages, some of which are frequently abused (Koszewska, 2011). In addition, the burden of certification and licensing costs can be prohibitive for producers in developing countries. In addition, the declarations of companies are coming under the scrutiny of NGOs, such as Greenpeace, with some reports found to misrepresent the facts.

Mowbray feels that it is debatable whether in-house labels contribute massively to more sustainable consumption: "Surely sustainability and consumption are at opposite ends of the spectrum?" "Sustainable consumption is an oxymoron?" Despite the associated problems, Koszewska's review concludes, "Certification and labeling systems belong to the most effective instruments that can induce positive changes in Consumer behavior." However, for greater benefit, the harmonization and standardization of existing systems would enhance transparency supported by new approaches to consumer education. The notion of a single, unified label for textiles covering environmental and social standards is on the political agenda in Germany after H&M's CEO Karl-Johan Persson envisioned "a global generic label for the (Textile) industry – similar to the fair-trade label on coffee" (*Ecotextile News*, Issue No: 57 October/November 2013, p. 11).

A strong message coming from the Association of Suppliers to the British Clothing Industry (ASBCI) trade events points to the need for retailers to help inform consumers in their decision making. The leading retailer M&S launched Plan A (in January 2007) "with the ultimate goal of becoming the world's most sustainable major retailer" in working with customers and suppliers "to combat climate change, reduce waste, use sustainable raw materials, trade ethically, and help our customers to lead healthier lifestyles." However, M&S has been criticized for not displaying products well enough and not making the best use of links through their website between environmental campaigns and products on sale. Information on "Doing the Right Thing," in terms of donating unwanted garments to Oxfam, or "Loving the Planet" (where more sustainable fibers are explained) is not prominent in store. As M&S staff in store have pointed out, greater attention is given behind the scenes to cutting down waste in terms of recycling packaging, while, in store, attention to do with better practice in food production is relatively prominent.

The message is that manufacturers have got to convince consumers. The U.K. ASBCI has staged recent trade conferences on reshoring. If, however, garment production is to be brought *home*, the questions arise: How can consumers identify British product? Are they looking for it? How can the provenance of products be verified? Currently, M&S clientele does not expect to find *Made in Britain* products in their ranges. Buyers and retailers must also be encouraged to locate local production. This requires them to consider the full cost of a product. In making a comparison between offshore and near-to-market production, there seems to be a lack of understanding of the real gross cost versus net cost of products. And if *localization* is really to be the new *globalization*, there is an urgent need for a U.K. sourcing directory of companies. The existing and revived network of small U.K. manufacturers is fragmented. Can a *one place* directory be aggregated from current databases? (ASBCI Conference, 2014).

11.4 NEGLECTED MARKET

11.4.1 NEW CONSUMER MAJORITY: THE AGEING DEMOGRAPHIC

The inclusive design pioneer Michael Wolff reflects that "When I think about what I can do in my own work, what resonates most with me is the type of design we offer people as they get older, as well as people who are differently abled – those of us who see less well, hear less well, move less well, think less well and feel less well than many. On the whole, it's a grim and dismal picture" (Wolff, 2014). In the United Kingdom, 10 million people are over 65 years old with latest projections indicating for 5½ million more elderly people in 20 years time and the number nearly doubling to around 19 million by 2050 (www.parliament.uk). "In researching and targeting the over-50 age group for product development and marketing strategy, it is a big mistake for marketers to consider it as a single consumer group. As the over-50 market accounts for 38% of the population, it is invariably going to include consumers from widely varying spectrums of life, who have similarly differing needs and outlook" (Hiscock, 2000). In no other market would such a disparate group be treated as a homogeneous whole. The active ageing consumers, also known as the *baby boomers*, have been the center of attention as teenagers, leading a fashion revolution. "Why are older and differently abled people denied glamorous, amusing, extraordinary, sexy and fashionable things? And what does this denial say about us?" (Wolff, 2014).

Sometimes referred to as the *gray market*, the over 50s are often viewed as passive, but the statistics suggest these are consumers who are very much active and with greater disposable income than younger cohorts, and with time to spend it. "Post 50s know their own strength – but it still seems like brands are having a hard time recognizing it when it comes to marketing to boomers. Leaving this 80-million-strong group out in the cold in favor of the coveted 18-to-49 demo is a huge mistake" (Brady, 2012). Despite this, many marketers continue to ignore them as consumers. "This group is often perceived as a 'dying' audience" while "marketers believe they have to aim their products at the youth market in order to have a renewable reservoir of consumers with brand awareness and loyalty" (Hiscock, 2000). "To a lot of companies, the idea of addressing themselves to an older market is a death-wish." However, "With boomers poised to make up half the U.S. population in 2017, marketing to boomers just makes good sense," says Beth Brady, global head of Nielsen Marketing (2012). She continues, "This group is redefining aging. They grew up in the age of consumerism. Their sheer size helped define brands; they're rebellious… this is not your grandma."

11.4.2 UNDERSTANDING OLDER CONSUMERS

The World Health Organization has suggested that "although there are commonly used definitions of old age, there is no general agreement on the age at which a person becomes old. The common use of a calendar age to mark the threshold of old age assumes equivalence with biological age, yet at the same time, it is generally accepted that these two are not necessarily synonymous" (www.who.int). A cohort may represent a group of individuals who have some characteristic in common or who have experienced a particular event during a specified period of time. The cohort most often studied by social scientists is the human birth cohort. Age is a condition—not an event (Glenn, 2005). Age is a variable that captures many socioeconomic (e.g., income and generation) and individual differences (e.g., cognitive ability and emotionality) characteristics (Cole et al., 2008). Age ordering sometimes refers to the *active ageing* and *the older old* or *eightysomethings*. This chapter looks at active agers; "Active ageing is the process of optimizing opportunities for health, participation and security in order to enhance quality of life as people age. It applies to both individuals and population groups" (www.who.int).

Given changes in goal selectivity with ageing, we would expect that older consumers would be more prevention focused and less promotion focused compared to younger consumers, especially in domains in which older consumers expect or experience losses, such as some cognitive abilities, health, or fitness (Cole et al., 2008). Biological and physical factors, which change with age, include poorer eyesight or hearing ability, reduced dexterity, a loss of ability to reach and stretch, and more difficulty walking or driving. These physical changes could lead older consumers to simplify choice processes or to restrict the number of stores they visit. However, it is also possible that due to improved medical treatments and differences across countries, there will be an increase in variance of cognitive abilities, goals, and motivations across people of the same age (Cole et al., 2008; Hupert, 2003).

Most consumer behaviors are performed on a routine basis, and most routine consumer behaviors are driven by habit. Habits begin as associations in memory. However, there do appear to be differences between age groups in terms of the kinds of habit behaviors reported. For example, compared to young adults, the elderly emphasize habits related to interpersonal relationships, such as friend behaviors like giving and helping. Specifically, the elderly tend to focus more on personal experiences and emotion (Cole et al., 2008; Metz and Underwood, 2005). The ageing consumer is mostly "experiencing ebbing of materialistic influences and rising influence of experiential aspirations on their behavior" and "consumer research and marketing have failed to realize the relationship between changes in the leading values, views, and behaviors of the marketplace and the New Customer Majority." It seems that "thus far marketing remains rooted primarily in the materialistic values that generally hold the most sway over people in the pre-middle-age years of adulthood. Because of this, many members of the new Customer Majority feel marginalized by companies and their marketers" (Wolfe and Snyder, 2003).

In terms of brand choice, "designer labels are about making social statements, and as people move into and through the second half of life, interest in designer labels falls off because they are not as compelled to make social statements by their brand choices" (Wolfe and Snyder, 2003). The active ageing consumers want less *stuff* as they seek more balance in reprioritizing their lives to become more self-reliant. They tend to be more concerned with value and the usability of the product (fitness for purpose) so consequently they do not usually figure among the early adopters of a product unless there is considerable advantage to them. It follows that new products are not designed initially for older people: they are expected to follow in the early or late majority, when the technology is becoming more mature and the selling points no longer stress the novelty of technology, rather other aspects such as ease of use (Metz and Underwood, 2005). There is a tendency for older consumers to prefer long-established brands and tend to be consistently more brand loyal than younger consumers (Cole et al., 2008). Different complementary mechanisms may lead to such results: "The nostalgia mechanism suggests consumers develop preferences during a 'critical period,' say between 15 and 30, and keep them for life. Another explanation relies on a possible age-associated absence or decrease in innovativeness accompanied by a possible inverse relationship between age and willingness to explore new options" (ibid).

The active ageing community, from diverse backgrounds, with a lifetime of experiences, have developed strong opinions that are not easy for marketers to categorize and address. Hiscock gathered a breadth of opinions on *separating the shades of gray* reported in 2000 that still ring true. The head of market research at Saga considered that "the best way to approach researching the over-50s age group is with qualitative methods such as focus groups, customer workshops, and 'depth interviewing' in couples. You need to take on different research methods for this group. They need longer discussion periods as they have more experience and more to say, so it takes longer," while other comments and advice included the fact that "the over-50s have time to fill in forms, they understand the motivations behind what they are doing and they can often participate in focus groups during the day" (Hiscock, 2000). Another observation was that marketers and agency staff are rarely over 40 and so cannot empathize with the mature audience!

11.4.3 Ageing Consumer and Clothing

In order to inform design guidance to offer a way forward for more sustainable clothing design development and cut down waste associated with clothing consumption, the author has looked at two comparative studies on older consumer design requirements that have been carried out almost concurrently with Design for Ageing Well. A Finnish study provides findings on the everyday clothing design needs of older women, while a German marketing study offers findings specific to the outdoor sports sector. The author has found that both studies correlate well in underpinning findings from *Design for Ageing Well* case study. The author provides an overview of the Finnish study with a view to making subsequent comparisons.

11.4.3.1 Introducing a Finnish Study: An Exploration of How Mature Women Buy Clothing—Empirical Insights and a Model

The Finnish study examines how mature women go about buying clothes with a focus on working women between 50 and 63 years of age buying clothing with their own money for personal use. Ten mature Finnish women were interviewed in August 2007, prior to observations made during 2 days in July 2007, during three hours per day in one major department store in Helsinki (Homlund et al., 2011).

The Finnish informants admitted that, in the decade preceding their involvement in the research, importance given to fashion clothing had changed and that their present identity differed from that of the past. They still wanted to be fashionable but to avoid the latest fashion trends, which are considered to be for the young and therefore less suitable for mature women. Colors were considered to be the most important attribute, in the choice and purchase of clothes, having a more pronounced role for them than when they were younger. They preferred their personal favorites by comparison with fashion trend colors. Due to changes in ageing body shape, good fit was found to be one of the most important requirements to influence their clothing and comfort preferences, equal to the importance of good quality. Finding the right size was said to be complicated, not only because of weight changes but also due to different size criteria.

Although ageing, in terms of general health and changing (mainly increased) body weight, appeared to cause changes in the women's design requirements, they shared quite detailed preferences. They preferred loose and comfortable fit with elasticated waistlines to suit their body shape and sizes ranging from petite to large. They preferred good quality fabric, in natural fibers, with good finishing, and soft handle, along with breathability and stretch for comfort. Regardless of price or age, they wanted garments to be flattering, making them look slimmer or taller, and they memorized brands that seemed to suit them. Garment cut and seam lines were important details that were examined and evaluated very carefully. Low-waist trousers, or tight trousers, T-shirts and sweaters, and T-shirts that would reveal their waists were unacceptable. Some informants avoided high necklines, while two appreciated wider necklines because of hypothyroidism.

Typically, the informants wanted more information on materials and appreciated detailed care instructions on labels as well as information offered by shop assistants. During the observations, it did not seem that the women were looking for any particular brands. They were discussing styles and avoided those that were revealing or too trendy, but also wanted to purchased clothes to match their current wardrobe. The informants found brand recall difficult, only able to recall surprisingly few brand names, between 5 and 13 clothing brands, and did not consider themselves to be particularly brand loyal. More important to them was the comfort and the quality of the clothes and how garments coordinated. However, brands were seen as insurance for good quality in terms of keeping their shape after washing, as well as a guarantee of familiar sizing.

The Finnish informants patronized a limited number of stores and emphasized the importance of being familiar with the store and its clothing ranges. The study showed that older women seem to appreciate actions that make shopping easier. Many reported long relationships developed with favorite shops and mutual interactive relationships with small shopkeepers. The women valued the ease of dealing with familiar shop assistants where they appreciated personal service, expecting to get new ideas and clothing tips, and help in finding correct sizes, so that they did not have to have to return clothes after fittings. They felt annoyed when fitting rooms were inconvenient or narrow, when having to queue to use fitting rooms, or if not able to get help and assistance when trying on clothes.

Even though they planned their shopping carefully and bought less frequently compared to previous buying habits, they bought more at a time than earlier. They also preferred more expensive clothes and planned to wear them for a long time. It seemed that decision making was made on rational grounds, with the utilitarian and functional attributes of the garments evaluated in relation to their needs. The informants assigned more importance to factors such as well-fitting design, with appropriate color, design details, and good quality, plus a good fitting room and shopping convenience, more than to price. However, even when not searching for the lowest price, they saw it as an important criterion. The women stressed the importance of predilection over price and said that they also took into consideration how long they intended to wear the garment. They looked for the best price-to-quality ratio, suggesting that good quality and good fit implies a higher price.

During the interviews, it was reported that the women talked a great deal about the expected time of wear and price matters, sometimes commenting on the price being too low, and sometimes items considered too expensive were bought. They glanced at the price tags before or after they had examined the care labels. The following quote was reported from one of the informants:

> Price matters but it should be evaluated together with the purpose and need. Many times I evaluate the time of wear of the garment; how long I can use it. Price is also an indicator of the quality; high quality and low prices don't go together.
>
> **Raija**

The empirical data revealed four different sources of influence: environment, peer groups, consumer characteristics and preferences, and fashionability (Figure 11.1). In reality, the influencing elements affect the buying in a complex and intertwined manner (Homlund et al., 2011).

11.4.3.2 Findings from the Finnish Study

Homlund et al. state that "designers that see the mature woman as a modern representative of one consumer group without any stereotypical notions of someone boring and conservative can fulfill the clothing expectations and requirements of the modern mature woman" (Homlund et al., 2011).

In abstracting the findings from this study, the key points are constituted to advise brands on how to move forward both from the point of technical design development and also in terms of preparing the route to retail. They suggest that the informed designer is in an appropriate position to advise on POS communication and sales staff training to ensure optimal range coordination and messages on key themes.

- *Brands*: Clothing brands must develop appropriate strategies and concepts for mature women with the targeting of a specific clearly identified consumer group essential.
- *Size, shape, and fit*: Continuity of shape and fit to be based on pattern blocks that address the physiological changes that affect the specified target group within the age group. In terms of shape and posture, it is important to note that the same body measurements do not necessarily represent similar body shapes. Mature women expect brands to follow stable clothing sizing and fit despite style changes.

	Buying process of apparel			
Wearing and combining apparel	Building need and awareness	Searching and fitting	Evaluating and actual purchasing	Wearing and combining apparel
Consumer's style and clothing habits that precede the buying	Development or emergence of need for clothing and how this steers consumer's attention and behavior toward buying (situations, media, weight fluctuation, information processing, strolling, planning and forecasting	Active, concrete searching for clothing and different activities that enable buying of preferred apparel (store selection and visits, trying on clothes, fitting room convenience, sensing the fabrics, assistance, emotions such as fun/enjoyment)	Mental processing prior to purchase (quality, pricing, price sensitivity, brands, design, comfort, fit, lifetime, decision-making style and speed) and rejecting/accepting apparel	Consumer's style and clothing habits that follow the buying

Above the table (model boxes):

Environment:
Cultural, social, and business settings of the consumer and the fashion market that affect the buying (societal values, available options, promotion, advertisements, cycles trends)

Peer groups:
People closest to the consumer (friends, family, colleagues, work position and environment, and age group)

Individual characteristics and preferences:
Consumer's background and preferences of shopping style, retailers, clothing fit (motivation, personality, previous shopping, economic status, clothing expenditure, shopping frequency and volume, preferred brands, preferred retailers, desired service level, shopping convenience change over time)
–Shopping style
–Retailer preferences
–Clothing fit preferences

Individual fashionability
Attitude toward fashion clothing and fashion involvement (fashion significance, fashion knowledge, total well-being, fashion functions, targeted impression, high fashion, color and fabric preferences, relative to fashion style, role models, age consciousness, body type fit)

FIGURE 11.1 The author has modified "a model of mature women's buying of apparel." (From Holmund, M. et al., *J. Fashion Mark. Manage.*, 15(1), 116, 2011, Emerald Group Publishing Limited, pp. 1361–2026, doi: 10.1108/13612021111112377.)

- *Flattering styles*: Mature women know their own body and wish to emphasize their best features with flattering styles with details and cuts that do not make them look older and fatter. Again, it is essential that brands follow continuing fit despite style changes.
- *Good quality*: Longevity of use
- Consumers will purchase long-lasting designs with good quality a greater consideration than price.
- *Point of purchase communication and sales training*: Traditionally, designers have not had responsibility for input to retail point-of-purchase information or delivery. However, findings suggest that designers could make a useful contribution to brand placement in store and to retailer and store personnel training. Guidance on range coordination and themes would underpin the collections, with the potential to improve retailer communication with customers.
- *Choice of media for marketing*: The choice of media is essential in all kinds of marketing as mature women have their most-favored media. Newspapers and brand catalogues reach the mature women better than fashion magazines.
- *Mature models*: The use of older models is a growing trend that has permeated most sectors, except fashion. Fashion houses and magazines as well as high-street stores have failed to utilize mature models to any great degree.
- *Socioeconomic factors*: Retailers should understand the biological, social, economic, and psychological changes that affect consumers as they age in order to better meet the product, service, and delivery needs of the mature consumer. However, the diversity of mature consumers and their needs may vary by region, income, and, particularly, lifestyle.
- *Longevity and value for money*: As discussed, the mature consumer values good fit and quality more than the latest trends and prefer familiar retailers and brands. However, longevity, value for money, and hardwearing attributes are considered to be anathema to fashion, which depends on transience.
- *Age-neutral marketing*: If marketers aim for mature consumers, it seems better to reject age segmentation in favor of age-neutral marketing. Any campaign that is too inclusive or too exclusive, or that patronizes certain age groups, involves a high risk of failing because mature women define themselves along attitudinal lines. Vocabulary in particular is one of the most challenging matters when addressing mature consumer. "'Mature', 'elderly' or 'senior citizens' or any other age allusive word may cause offence" (Homlund et al., 2011).

11.4.4 Outdoor Industry Market Research

Findings from *Best Ager* market research carried out in Germany, led by Claudia Boesl of Generation Sport (2008–2011), confirm that the active ageing community is frustrated with what is on offer in the retail environment. *Best Ager* market information has been collected and elaborated over recent seasons with a view to advising the outdoor market of the potential size and value of the ageing consumer. Findings indicated that many older users engage in active outdoor pursuits but are often unable to find suitable clothing products, in terms of age-appropriate style, shape, and fit. They were also unhappy with the nature of the retail environment, oriented to the youth market, where it is difficult to try garments on in store, and the sales service provided is not in tune with older user needs. Many older potential consumers are looking for quality and age-appropriate design. Since 2012, Generation Sport has continued to have a presence at the leading sportswear trade fair, ISPO (Munich), where it promotes sporting goods associated with health and wellness linked to the adoption of assistive wearable technologies (www.generation-sport).

Best Ager market findings underpin those found in the following case study.

11.5 CASE STUDY: FUNCTIONAL CLOTHING FOR ACTIVE AGEING MEN AND WOMEN

The performance sportswear sector, as an early adopter of textile and technology innovation, is targeted primarily at athletes and the youth market. To date, advances in smart clothes and wearable technology, with assistive technologies and monitoring devices, for older users, have normally been developed for *ill people* with little aesthetic appeal and with technology user interfaces often difficult for the wearer to read, understand, and use. Wearable technology, for everyday use, has not been readily available to older users as little has been done to address older wearers changing physical and cognitive limitations, and style requirements, when developing products and services. However, the attributes of modern textile innovations and novel garment construction techniques, in an adaptable sports-type layering system, in age-appropriate size and fit, may be utilized to promote health and well-being.

Recent cross-disciplinary collaborative research has begun to address the gap in design guidance for functional clothing, in respect of the aesthetic and practical design needs, of the growing active ageing demographic. In this study a co-design approach has been adopted, involving all stakeholders, including an older user reference group (URG), throughout the research and development of functional outdoor clothing for men and women in the 60–75-year age group. Industry collaboration has enabled access to practitioner expertise and to the value in kind of novel materials and processes for co-design planning and subsequent practical development of near market prototypes. The ultimate aim has been the provision of new design direction, in plain language (both in text and clear visual format), to guide the responsible development of functional everyday clothing that is attractive and fit for purpose, for the identified target consumer, for engagement in healthy exercise.

Design for Ageing Well: Improving the Quality of Life for the Ageing Population Using a Technology Enabled Garment System has been carried out within the RCUK-funded New Dynamics of Ageing (NDA) program. The aim of this practical research has been to bring smart clothing and wearable technologies to the active neglected community (60–75 years) through the development of a sports-type clothing *layering system* to encourage healthy exercise and social engagement with a focus on walking. The functional clothing layering system comprises a close-fitting base layer, mid-layer insulation layer, and outer protective wear, with wearable electronics, and their technology user interfaces, integrated within the system.

This design-led collaborative research embraced a representative network of cross-disciplinary academic competencies from across the partner universities. Specialist work packages (WPs) were established in three areas, which were interdisciplinary in their own right. WP1, *behavior*, linked the psychology of behavior (University of Westminster) with sociogerontology (University of Salford); WP2, *clothing*, led design and technical development in the application of technical and smart textiles in functional clothing (University of Wales, Newport—now merged into the new University of South Wales); and WP3, *technology*, addressed wearable electronics (University of Ulster). The process began with the recruitment of the URG of older research participants, within a 60–75-year age range, with an interest in walking.

This design-led, cross-disciplinary, collaborative research project has looked at how to make an intervention to *business as usual* where range development is driven predominantly by last season's sales in combination with trend. This pattern leaves no record of the lost sales to dissatisfied and frustrated older shoppers and no *upbeat* trend information to guide designers.

11.5.1 ADOPTING CO-DESIGN METHODOLOGY: NEW TO CLOTHING DESIGN

The co-design methodology adopted was new to clothing design. Fuad-Luke reminds us that "the people who ultimately will use a designed artifact are entitled to have a voice in determining how the artifact is designed." Adopting a co-design process enables "design for the future *with, for* and *by* society" where "the design of any project where the multi-stakeholders are engaged by the designers in a co-design process has the potentiality to create fresh solutions" (Fuad-Luke, 2007). It is important to study older people in real (or realistic) situations in which they can use the expertise they have developed over the years (Cole et al., 2008).

Once the academic network was in place, the recruitment of older walkers began from within a 30 mile radius of the lead university. Respondents were asked to provide information on their height and weight to determine their BMI classification for potential body scanning. Once the URG was established, a series of preparatory workshops was planned for the development of a shared cross-disciplinary language between all stakeholders. This was necessary for the explanation of complex terminologies and practices across the range of disciplines embracing technologists in electronics, clothing and textiles, and gerontology and to aid communication with the URG.

11.5.2 INITIAL "SHOW AND TELL": WHAT DOES THE CONSUMER UNDERSTAND?

An initial focus group was staged by the clothing WP, with a facilitator from the *behavior* WP, as an expert in user engagement, to ensure that all participants were able to make a contribution, to sum up, and to indicate the way forward. This event was planned to gather initial views from 12 users (men and women) on what type of walking do active agers participate in, what do they currently wear, and what they might currently use in terms of wearable technology. Older users were asked to bring personal garments of interest to the discussion. The event was recorded by video and participants contributed to flip chart notes taken by researchers. "Show and Tell" uncovered preferences that demonstrated differing levels of user understanding in relation to available products. This provided initial insights into the low level of understanding of the technical clothing layering system; modern fibers, fabrics, and finishes; and novel manufacturing techniques. It showed confusion with textile and clothing terminology and quality differences, mistrust of the claims of brands, and the diverse sourcing of garments from a breadth of retail outlets.

A perceived preference for natural fibers (cotton and wool) considered *good* and a mistrust of synthetics, considered *bad*, was attributed to this generation's early introduction to the poor image of the sweaty nylon shirts and crimplene dresses of their youth. As older users, they have become accustomed to cotton base layers, often wool insulation, and a variety of waterproof and other outer layer garments including corduroy trousers, a heavy tweed hunting jacket, and an army poncho! Examples of their personal walking clothing, with opinions and stories related to the pros and cons of the disparate items, included several references to *clothing as an old friend*.

The concept of wearable technology was new to the group, although, when briefly explained, some of the user group acknowledged the use of mobile phones, cameras, and pedometers. Health monitoring, in clothing, was virtually unknown to them. Taylor discusses user confusion with the profusion of complex POS material, prevalent in the retail environment, in more depth (Taylor, 2010). The cost of clothing reoccurred throughout the discussion, with reference made to bargains found at a range of nonspecialist outlets. It was generally expected that clothing should be cheap, with the participants retaining little memory of the relative high cost of clothing when they were young.

Predictably, the users had a lack of appreciation of the sophisticated nature of fibers and fabric developments, and finishes adopted within performance clothing. They were also unaware of the restrictions in choice associated with mass-produced transient fashion with limitations on short production runs of specialist colors. As the project progressed, it became increasingly evident that other WP researchers were equally unaware of the languages and cultures of the textile, clothing, and electronic trades. However, there was little apparent awareness of the impact of the outdoor textiles and clothing on the environment or of ethical concerns, given that meetings took place prior to publicity on factory conditions in Bangladesh. It became evident that all stakeholders would benefit form joint training. Valuable insights had been gained in helping the academic team to plan the detail of the subsequent co-design process.

11.5.3 DEVELOPING SHARED LANGUAGE: KEY TO EFFECTIVE COMMUNICATION WITH USERS

The author introduced her *requirements to consider in designing a brief* with relevance to design process for the development of functional clothing that is fit for purpose. A hierarchical *tree* of design requirements, central to her previous work, relates to the user needs–driven application of technical and smart textile attributes within the performance sportswear clothing layering system. The original tree is presented in the author's thesis, *Establishing the Requirements for the Design Development of Performance Sportswear* (McCann, 1999), and further adapted in *End-User Based Design of Innovative Smart Clothing* (McCann, 2009). This design process covers aspects of the *form* (aesthetic design appropriate to the culture of the user/peer group) and *function* (technical design that is deemed suitable for the intended end use), as guidance, to be further amended and elaborated with respect to the needs of the active ageing community. The philosophy is that end-user needs–driven design is the first step in a more responsible process to cut down waste (Figure 11.2).

Then followed intense user-centered training workshops to enable knowledge transfer through product demonstrations, with specialist input, presented by industry practitioners. Co-design workshops introduced users to the concept of the clothing layering system and the scope for the application of smart materials, with cross-reference to the demands of different levels of participation in walking. The semistructured process involved a *hands-on* review of current products, working systematically from base layer to mid- and outer layers, including legwear, in separate sessions, with either direct industry stakeholder involvement or products and catalogues sent to the events. Co-design engagement was led by designers acting as facilitators, observers, translators, and interpreters to enhance understanding, between academic and industry contributors, and to enable older participants to have the confidence to express their views. Industry stakeholders welcomed users to company showrooms and lead users to trade fairs, thus increasing user familiarity with products while breaking down practitioner's preconceptions of user preferences and attitudes.

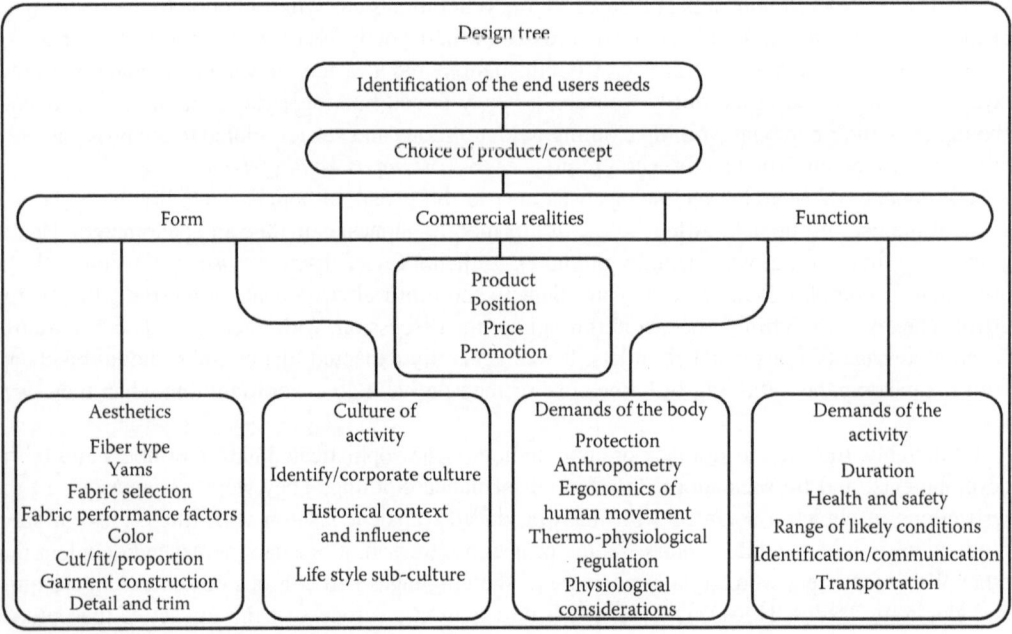

FIGURE 11.2 Requirements tree: From the author's MPhil thesis. (From McCann, J., Establishing the requirements for the design development of performance sportswear, MPhil thesis, University of Derby, Derby, U.K., 1999.)

11.5.4 Capturing Size and Shape: A Key Consideration for Future User Customization

Selected members of the user group, based at the lead university, were body scanned to capture their individual size and shape. From this group, size and shape models were chosen to represent BMI average measurements for the age group (based on SizeUK data) ranging from small to extra large men and women. User engagement was key to the development of individual basic blocks developed from the body scans that, in turn, were adapted to articulated blocks that were cut for enhanced movement. These were further developed into style patterns to accommodate the major design requirements informed through co-design workshop findings. Garment layers were cut to be compatible in terms of styling and ease to avoid impeding movement, with lengths and proportions verified, across the layering system. Design detail was then transferred to flat patterns. In outerwear garments, ergonomic design lines were to be used to accommodate the integration of the textile-based electronic architecture and the interconnections between devices and their controls and powering devices. The final patterns were stored in computer-aided design (CAD) format.

11.5.5 Gathering User Feedback to Identify Real Needs

A wealth of user design requirements came from the co-design engagement as the process gradually uncovered user attitudes and preferences, as well as their perceived limitations, both with clothing and the usability of wearable technologies. Data were gathered by a variety of methods as follows:

Video recordings of stakeholder co-design workshops with audio transcripts
Still photography to record the key design details of user interest or concern
Flip chart notes from workshops to capture key design-related issues
"Hands-on" user materials selection to indicate user design preferences
Comparison of garments with opportunity to try on and evaluate styles
Observation by designers while acting as translators, interpreters, and visualizers
Report sheets to capture user written feedback on presented clothing products
Collages of images and drawings of users to communicate user style preferences
Observation in workshops by technology team to capture user needs re functionality
In-workshop visual questionnaire re preferences for positioning of technology
Still photography against a measurement grid to record posture re movement
Dialogue between experts and users to note technology usability and limitations
Comparison of smartphones to identify user preferences and/or limitations
Color selection workshop with trade explanation of timelines and minimums
Development of collaged "design boards" to represent range plan overview
Written feedback on project from users both requested and volunteered

The interdisciplinary design team observed the interaction with older research participants as they tried on, evaluated, and commented on the design and usability of the variety of garments, and wearable technologies, provided by industry stakeholders. Designers made notes and sketches and took photographs of key aspects of interest, whether observed (and with tacit understandings) or with additional user comment. For example, users might not notice problems with garment balance, or restricted arm lift, while discussing a closure or pocket detail. These problems could be noted and addressed, with or without comment, by the designs team while the technology team captured feedback on specific aspects of functionality and usability.

Visiting speakers from industry explained color forecasting, on a national and international scale, and how buyers and retailers may interpret trend information. It was noted that the fashion industry has subjected consumers to more than a decade of black as the predominant color on the high street leaving older users confused as to where to find interesting and flattering age-appropriate colors. Research participants had been asked to bring in examples of garments and accessories in colors and textures that they liked to wear. The co-design process benefited from longevity of user involvement throughout the project from body scanning to garment cutting, materials sourcing, and selection and positioning of wearable electronics, to development and evaluation with users.

11.5.6 SIZE, SHAPE, AND FIT: A KEY CHALLENGE IN THE GLOBAL CLOTHING CHAIN

Basic garment blocks, generated from the body scans, were made up as toiles (calico prototypes) and fitted on the selected models. The developed blocks represented the starting point for the development of base, mid, soft-shell, and outer layer garments. (Flexibility and adequate range of movement are essential for many activities with loss of range, especially in the shoulder joint, common in old age.) The customized blocks, developed from the scans, were further amended, in terms of sleeve pitch and articulated cut to enhance movement, through the layering system. Users became aware of the technical demands of performance clothing cutting that caters for extended *arm lift*, elbow articulation, and sophisticated hood design, normally found in products exclusive to extreme sports practitioners, due to limited size and shape. Design lines, related to the placement of appropriate materials around the body, incorporated ease of movement in tandem with other textile attributes such as moisture management, thermal regulation, and protection. Parallel work was carried out by an expert intimate apparel designer in the development of user-specific body layers. Sports-type bras were designed and cut for individual levels of support, and close fitting vests were developed for men.

11.5.7 PROTOTYPE CO-DESIGN DEVELOPMENT: IN COLLABORATION WITH USERS

The co-design process demanded experts with a range of competencies in size and shape, pattern cutting for movement, knitwear for larger figure types, and intimate apparel design. In addition, the clothing-oriented designers were supported by product designer and an architect, skilled in the development of innovative working drawings and specifications. The team had external support in the CAD plotting and storage of pattern development, essential to effective garment specification. The co-design development process involved close collaboration with the URG size and shape models for regular fittings with further design adjustments normally made in discussion with the wider group of participants. Industry stakeholders provided input at key stages to support materials sourcing and selection and access to a range of specialist manufacturing processes. Fabric selection represented a realistic compromise between end-user and designer aspirations and commercial fabrics, available within the time span and scope of the project. The overall shape, fit, and proportion of the coordinating layering system was developed and the positioning of the design detail confirmed, including the incorporation of the electronic components. The practical prototype garment development was carried out in collaboration with a U.K. outdoor brand (with production in China); a U.K. functional clothing manufacturer, with additional offshore production; and a regional small to medium-sized enterprise (SME) with an interest in the integration of electronics into performance clothing.

11.5.8 INTEGRATION OF TECHNOLOGIES: SELECTED AND TESTED WITH USERS

Instead of the prevalent *technology push*, the co-design process identified the functionality desired by older users in terms of the selection and usability of the wearable electronic devices and their user interfaces within the garment layering system. The breadth of interest generated through co-design, involving stakeholders from industrial fiber, fabric, and trims suppliers, with the input of electronic providers, in consulting with older users, enabled practical research and development to move nearer to market. In terms of textile-based wearable electronics, knitted sensors were incorporated in body layer garments (as sports bras for women and vests for men), to connect a wireless device for vital signs monitoring. Flexible heated panels were integrated into mid-layer gilets, and potentially jackets. Both soft shells and outer protective garments incorporated textile-based cabling and soft switch controls to connect and operate a smartphone, which acted as the hub, for location and communications. Practical design issues involved in specifying the positioning of the electronic devices and user interfaces within the clothing layering system are highly complex tasks that are still to be resolved on an industrial scale.

11.5.9 Garment Specification: Collaboration between a Product Designer and Architect

The accuracy of garment specification, as discussed, is one of the key issues to address in terms of waste within the clothing chain. The complexity of developing a clothing layering system as a *platform* for textile-based electronics relied on transferring cross-disciplinary knowledge and skills between designers and both technology academics and industry technology providers. Soft elements, such as textile-based conductive fibers and ribbons, knitted sensors, and electronic components, had not previously been within the traditional realm of the fashion designer. Designers from product and architectural disciplines (new to clothing design) engaged in the development of shared technical language, beyond that of traditional fashion design, for the specification of smart clothing with wearable technologies. These designers had previously enjoyed the benefits of CAD with software programs such as SolidWorks (product design) and AutoCad (architecture). In these programs, decision making is aided by a knowledge of industrial processes with constraints, programmed into the software, that reflect limitations that occur in production, in order to verify realistic and accurate stages in the design specification.

These practitioners found a lack of equivalent, currently available, CAD options related to garment specification. They found that clothing designers often use Illustrator software to produce line drawings, but this is unrelated to pattern-digitizing programs. Flat drawings make reference to body scale measurements, often in limited detail, with patterns seldom created by designers in the West. This lack of clarity in specification would never be the case in product design development or in architectural practice where key dimensions are drawn in plan, elevation, and section and sent directly to be measured and made to scale. The capturing of body size and shape, through laser scanning and the generation of customized 3D patterns and mannequins, is being linked to 3D pattern generation but not widely in the global supply chain (http://www.human-solutions.com).

Without patterns generated, and a digitized record made, there is no accurate reference from which garments may be made. The product designer and the architect worked together in bringing their methodologies and skills to merge with the experienced clothing designer/pattern cutter, to help resolve the description, both visual and text, of the novel material assemblies. They generated 2D computer illustrations, showing both full garments and enlarged technical details, with cross-references linked to the digitized style patterns development. The mix of electronic components, to be integrated into textiles and clothing, demanded clothing specialists to work in liaison with electronics experts for the optimal positioning of both hard devices and soft technologies (textile-based sensors, cables, and switches) within the garment layering system. The smartphone acted as the wearable computing hub within the system, requiring a choice of locations within the various layers. Electronic cables and soft switches were to be encapsulated, to be waterproof as necessary, while hard devices were to be removable for washing and/or upgrading. An ongoing concern was that some components, such as power for the heated panels, demanded relatively heavy batteries.

The product designer and architect were proactive in resolving issues to do with integrating the interconnections and switches for the various devices. This involved visits to industry partners to verify the processes involved in, for example, the heat bonding of seams to join the various fabric assemblies, with their specialist coatings and finishes. The positioning to enable usability of the technology user interface was of particular concern in relation to garment assembly. Hybrid knowledge and skills from product design and architecture helped to describe the placement of components, interfaces, and powering devices within the clothing layers. Once the positioning of the electronic architecture was confirmed and prototypes assembled, they refined the drawings to describe the assembly of the garments with the intention that others might replicate the process.

Imagery included photography, diagrams, and *exploded* drawings to clarify the detail of garment construction with references to toiles and working prototypes as relevant.

The product designer liaised with industrial collaborators to verify the development of a generic garment specification sheet. The recommendation from the trade was to communicate with limited text, to avoid language translation problems, but with clear visual references. The provision of accurate digitized garment patterns, which showed the positioning and integration of the technology components, with electronic architecture, and cross-referenced to flat specification drawings, was supported by garment samples. In theory, this design *package* would correlate with circuit diagrams to be produced by the technology partners. Predictably, the specification sheet was designed to provide information such as

> *Gender,*
> *Sample size,*
> *Garment/Style reference number*
> *Material selection and trims*: Including source and references re quality, weight, etc.
> *Flat garment drawings*: Front, back, side view, etc., as necessary
> *Enlarged garment detail*: With photographic and other visual references
> *Exploded drawings*: To show electronic architecture and components
> *Small-scale record of digitized pattern pieces*: With positioning of electronics
> *Photograph(s)*: Sample garment, toile and/or final working prototype

The industrial stakeholder network embraced fiber and fabric producers and clothing companies, from a U.K. outdoor brand, with garment sourcing in Asia, to another U.K. brand with both national and offshore manufacturing plant, to a regional SME, with an affinity to wearable electronics. All provided significant levels of support in textile sourcing, prototype development, and project dissemination. This collaboration enabled the academic team to test and verify the clarity of the prototype specification processes with feedback from industry collaborators. Industry feedback included face-to-face liaison with the regional SME, showroom meetings at the U.K. base of an outdoor brand, and email/electronic communication direct to offshore manufacturers. The SME was proactive in testing the wearable electronics in collaboration with the project's designers, while the brands progressed with greater access to materials but with longer lead times due to offshore production.

Regular consultation between the older research participant fit models and the academic team ensured that the electronic devices and their components were considered comfortable, in terms of their placement within the layers and the positioning and usability of the technology user interface. In the university setting, co-design evaluation, with designers and the technology team, enabled older participants to test and verify the placement of the electronic system such as the smartphone, with Bluetooth hub, PTT Module, the heat pad and its battery, and soft switch controls, within the garment system. This project has succeeded in raising awareness of the importance of the active ageing market with the various industry stakeholders, engaged with different product categories within the range plan, collaborating on the production of prototypes. These were tried and tested by the user group culminating in a final winter walk with both users and researchers in January 2013.

11.6 SUMMARY OF KEY FINDINGS

Collaboration with end users has been effective, throughout the iterative co-design process, in raising the industry's awareness of needing to address the real design requirements of this potential (and predictable) growing consumer market. User engagement in co-design process enabled the identification of older user real life needs and aspirations, in terms of functional clothing and wearable technology. As participants, they became familiar with new products available, and their functionality, becoming better informed in decision-making judgments. The findings uncovered are prioritized to create a hierarchy of design requirements to inform subsequent practical design development.

11.6.1 Size and Shape

Garment fit is one of the major considerations by older wearers in product selection, alongside color choice. There is unanimous confusion around the topic of the commercial sizing of garments, with little standardization or synchronization of sizes between different brands and even between different styles within one brand. Users complain that current clothing ranges often look very similar and with size and fit that is unpredictable. The lack of clarity is both in terms of measurements and in the nomenclature of sizes, exacerbated by the degrees of ease (tolerance) added to body measurements (that vary in the size charts from brand to brand) dictated by the style and mood of the range. Users request a choice of garment lengths to better cater for the variation in height within this age group and in body length, sleeve, and hemlines and the need for more inclusive sizing to extend from extra small to extra large. In general, the demand is for fewer and less transitory styles to be available in a wider size range. Women, in particular, request that waists be designed with adjustability for changing older figure types, possibly through partial elastication, but not *baggy*!

11.6.2 Styling

Having a *personal style* becomes more and more important for the active ageing community who want to look good, and maybe smart, and be comfortable. Participants were clear in that they would welcome fewer styles that are better fitting, with a greater choice of lengths, in age-appropriate colors and textures. They are more concerned with subtle, nontransient, age-appropriate style, rather than seasonal trend, with design lines to flatter the changing body as opposed to following dramatic swings in fashion. Some value multifunctional styling, suitable for a range of everyday life activities, to encompass both countryside and urban environments, in garments cut with sufficient ease to accommodate varied inner layers. Users indicated preferences for subtle discrete design detail to preserve clean lines that offer the opportunity to coordinate differing layering combinations, to cater for individual taste, with personal accessories to suit the wearer.

In general, there is an agreement from male and female participants for subtle shoulder emphasis, such as softened saddle-shoulder lines, to take attention away from their less-defined ageing waistlines. They request style lines that profile or *skim* the body, being not too tight or too *loose*. There is a preference for softer organic shapes, as opposed to angular seaming, with fluid contours to accommodate the ergonomics of movement such as arm lift and elbow and knee articulation. They request optional fit at waistlines through full or partial draw cords in garments of *a decent length* (covering the seat) for the natural outdoors and with longer lines for urban walking. Side panels are proposed as flattering to both men and women to accommodate varied shape requirements (from bust/chest to waist to hip). Side panels enable the elimination of side seams and any associated friction in walking. Emphasis is put on ease of fastenings, especially in outdoor conditions. The research participants are impressed with the slick appearance of new manufacturing methods as yet not commonplace in everyday clothing for the older market: the use of heat bonding for zips and seaming (possible on fabrics with synthetic content) and seam-free knit construction for base layer garments.

11.6.3 Color

Color, in combination with shape and fit, is a key motivation in garment selection within the female members of the active ageing user group. They look for colors that are sympathetic to ageing complexions and hair color. There is a preference for marl effects and textures, to introduce a natural *feel* to man-made fibers, with reference to colors of the countryside, of fruits, flowers, berries, and lichens and natural stone tones. These are not transient choices but levels of color that the older participants were familiar with and enjoyed wearing. They confirm that they do not want relentless, transient, swings in trend-driven color. If they do happen to find an attractive, enduring style, women admit to often buying two or more garments and, if appropriate, in several colors. Male feedback indicates that initial decisions are made on color, and *what am I going to wear it for*, prior to checking sizing and fit for comfort and ease of movement.

11.6.4 FABRICATION

Outdoor clothing design is driven by modern fiber and fabric innovation that occurred during the lifetime of the baby boomer generation. However, co-design engagement has revealed that the positive attributes of these materials are not initially well understood by this community. Some have had a suspicion of synthetics, since the early poor image and handle of bri nylon and crimplene, and have retained the misconception that natural fibers, such as conventional cotton and wool, are considered *good*, while synthetic and man-made fibers are *bad*. Co-design workshops enabled an introduction to the positive attributes of modern fibers and fabrics with the potential to enhance overall comfort within the garment layering system. Users gained awareness of wicking base layer materials, nonabsorbent mid layers, and the increased breathability of waterproof protective fabrics. Users had little understanding of how to differentiate protective outer layers.

They were introduced to fabric innovation that has enabled *hard-shell* outer layer garments to become increasingly lightweight, fitted, and stylish, with enhanced comfort, through moisture management, in waterproof *breathable* membranes with stretch properties. The concept of *body mapping* was explained for the seam-free engineering of textile stitch structures, involving the selection of areas or *zones* with appropriate fiber/yarn selection, in relation to comfort factors such as breathability, protection, ease of movement, and articulation. Soft-shell fabrics were introduced that are water repellent rather than waterproof and, as they do not require seam taping, are softer and less noisy than more protective *hard-shell* fabrics and often with a degree of stretch to enhance comfort and fit. As extreme protection is not a daily requirement, soft-shell fabrics offer attractive, adaptable, and comfortable protection for everyday use.

As the co-design process progressed, with industrial partners presenting fabric innovations to the research participants, with trial demonstrator garments, the participants became increasingly enthusiastic about synthetic and man-made fiber types. The comfort-enhancing properties of modern textiles became readily accepted especially in those materials, and fiber blends, that have a *natural* handle. For base layers, users like TENCEL, wool, and COOLMAX materials and their blends. For mid-layer insulation, they are amazed to learn of the breadth of textures and finishes available in polyester fleece pile fabrics. They are particularly interested in *soft-shell* materials that incorporate the protective elements of outer shells combined with mid-layer comfort and insulation.

There was little prior user awareness of the bad environmental impact of textile production such as traditional cotton growing and processing, other than that related to organic cotton, or of the current wider debates with regard to ethical concerns within the textile and clothing industries. All stakeholders were made aware of initiatives within the outdoor trade for the introduction of more sustainable practices throughout the textile and garment chain, highlighted by the commitment of the European Outdoor Group's Sustainability Working Group (EOG, SWG), since superseded by the launch of the Higg Index 2.

11.6.5 RETAIL EXPERIENCE

What type of shop should consumers go to: specialist, high street, warehouse store, or department store? Are the garments displayed with hanger appeal? This is a particular problem for sports apparel, especially base layers such as Santoni knits. A general comment indicated that knowledge of the outdoor trade and its products is not easily accessible by lay members of the public in terms of knowing what is available. The importance of performance features will vary depending on the walking type with some users participating in range of different walking types. Is the sales staff trained to help older customers and do the retail outlets have suitable changing facilities? Customers must have the confidence to try on garments in relative comfort. Some female participants go into a retail outlet, look around, and come out again if nothing *catches the eye*, while others go in and immediately consult the advice of a sales assistant, if they are sure of their intended end use.

To enable older wearers to be more confident in product selection and promote uptake, clear information, in plain language, is needed on the concept and attributes of the layering system in order to avoid suspicion of the marketing hype of brands. Even if older consumers have the confidence to go into the specialist retail environment, POS labels have print that is too small with confusing technical information promoted with a plethora of complex diagrams and icons. Older consumers are suspicious of the claims of brands and the profusion of marketing hype. Another frustration, highlighted by Willis in quoting survey respondents, is consumer resentment in being fed seasonal ranges that are "way out of kilter with the weather" and she questions: "Why do shop assistants look so wounded when you ask for jerseys when it is snowing and they have their summer range installed already?" (Willis, 2013).

It was found that the active ageing group, who had become more aware of the attributes of the functional layering system through the co-design process, had gradually changed their attitudes as they had become more conversant with the design development process and the specialist terminology involved. Due to the confusion with different claims of the brands, they had developed strong views about needing clear label indicators, possibly with a color code, or *traffic light* system, in terms of a measuring scale for degrees of functionality and related benefits: What are the technical features of a product, and, for example, will it keep you dry?

11.6.6 PRICE

When making a purchase in the retail environment, participants were clear that, if they had compared available products, with a helpful sales assistant, and made a decision, they would invest slightly more than going online to search for a lower price. Most stated that "If spending a lot, and you have found what you want, and are convinced, you will buy in the store despite the potential for a cheaper price elsewhere. If there is a problem you can go back to get it right!"

11.6.7 BRANDING

Strong feelings in feedback from the older research, within the case study participants, indicated that there should be *no* branding on garments! This community had grown up with the garment maker's label on the inside of the garment. They resent promoting stitched, embossed, and printed brand logos, however discrete, on the exterior of garments. The message was clear: No external logos at all for women and, from choice, no logos for many men and, for others, if more tolerant, any branding must be very discrete. A brand stakeholder asked about tonal branding and this was dismissed. The active ageing wish for the brand label to be on the inside and on the POS material. They consider that the style and *look* of the product should provide credible and distinctive brand recognition. Their attitude is, "Why pay for a garment to advertise the brand?" Women, in particular, resent branding as it prevents multifunctional garment end use, especially in the crossover of functional wear into everyday lifestyle, as the bold image of logos conflicts with the wearing of jewelry and other, sometimes, subtle accessories.

11.6.8 MERGING RESEARCH FINDINGS

A comparison of the research findings discussed provides valuable insights into the design requirements of the neglected active ageing consumer with strong messages to the industry in how to address their needs. The author has merged key findings in Table 11.1.

TABLE 11.1
Mature Consumer Clothing Design Requirements: Merging Key Findings from Studies Carried Out in Finland (2007) and the United Kingdom (2009–2012)

Topics/Attributes	How Mature Women Buy Clothing (Finland: 2007)	Design for Aging Well (UK: 2009–2012)
Informants	10 working women aged between 50 and 63 recruited in the Helsinki area to carry out a study (observation) in a major departmental store with subsequent face-to-face interviews in their homes.	20 men and women aged between 60 and 75 recruited from walking groups within a 30 m radius from the University of Wales, Newport to engage in a co-design process, plus a control group in Manchester area.
Clothing need	For personal use	For walking and everyday lifestyle
Fashion/trend/style	Fashion—part of informant's identity but avoids latest youth-oriented trends. Wants to look stylish and unique; choose simple, flattering clothes and avoid revealing or very trendy styles. Buy clothes to match their current clothes.	Personal style more important than fashion. Smart but comfortable multifunctional garments to mix and match with "skimming fit"/flattering fit garments—not too loose or too tight and covering the seat. Fewer styles in a wider choice of colors and fabric types. Softened lines in organic V angular shapes.
Color	Most important attributes: – More pronounced role than when young. – Prefer own favorite colors rather than fashion trends. – Consider black and brown as appropriate for mature rather than young consumers.	Very important: – Age-appropriate color to flatter changing complexion and hair color. – Natural, cheerful colors of the countryside, like fruits, berries, lichens. – Sometimes buy more than one color – They ask, "Why is fashion always black?"
Shape/fit and size	Good fit and comfort found to be most important—equal to quality Finding the right size is difficult. Conscious of changes in body shape Prefer loose and comfortable fit with elasticated waistlines to suit their body shape. Want sizes ranging from petite to large.	Most important—equal to color. Size and fit confusing and unpredictable with little standardization/synchronization between different brands. Will rather have fewer styles with choice of garment lengths: for arms, body, and leg. Inclusive sizing from X small to X large. Skimming fit and garments coordinated in cutting to accommodate movement. Elasticated waist for comfort.
Brand preferences/branding	Not looking for any particular brands Brand recall difficult—not brand loyal However, they see brand names as quality indicators.	Many outdoor ranges seem to "all look the same." All consumers know of Gore-Tex! Unaware of how to differentiate between brands available. Want brand labels to be on the inside of garments: No logos!

(Continued)

TABLE 11.1 (Continued)
Mature Consumer Clothing Design Requirements: Merging Key Findings from Studies Carried Out in Finland (2007) and the United Kingdom (2009–2012)

Topics/Attributes	How Mature Women Buy Clothing (Finland: 2007)	Design for Aging Well (UK: 2009–2012)
Fabrication	Informants prefer good-quality, soft-to-touch fabric, natural fibers, good finishing, with breathability and stretch for comfort. Fabrics to be practical and easy-care. They wanted more information on the material and appreciated detailed care instructions on labels.	Initially unaware of the principle of the sports layering system—later appreciated. Assumption that natural means "good" and synthetic "bad." When informed, they welcome the comfort of man-made fibers and blends in fabrics with natural handle, for example, Tencel, Wool; and Coolmax. Pleased to discover soft shell technology. Want more varied colors and textures, in simple, good fitting styles.
Design detail	Cut and seamlines are important details examined and evaluated very carefully. Low-waist trousers, or tight trousers, T-shirts and sweaters that reveal their waists are unacceptable. Some avoid high necklines while two appreciate wider necklines because of hypothyroidism.	Ease of fastenings, especially in outdoors. Single-direction zips easier. Storage of hoods in collars difficult. Like slick new manufacturing details, for example, laser cut openings with bonded seams and zips. Seamlessly (Santoni) knit, but not too tight. "Cozy" convertible collar design. Velcro considered noisy and damaging to textiles.
Retail preferences	Like to be familiar with store. Value help from sales assistants. Store location important. Comfort of fitting rooms important. Mature consumers visit fewer clothing outlets than in the past.	Where to go? Specialist, high street, warehouse store, or department store? Display of garments important and facility to try on garments in relative comfort. Need helpful, knowledgeable sales staff plus POS in plain language and clear text.
Sources of information/influence	Source of info: Women's magazines, newspapers, customer magazines, and brand catalogues. Advertising seen as for the young. Only one woman uses the Internet. They often shop in pairs. Peer group influence and friends.	Most users not very knowledgeable about brands and products available. Knowledge on outdoor products seems difficult to access—too confusing. Skeptical about claims of brands. Peer group experience from walking groups; mostly couples plus some individuals.

(Continued)

TABLE 11.1 (Continued)
Mature Consumer Clothing Design Requirements: Merging Key Findings from Studies Carried Out in Finland (2007) and the United Kingdom (2009–2012)

Topics/Attributes	How Mature Women Buy Clothing (Finland: 2007)	Design for Aging Well (UK: 2009–2012)
Shopping style	Buy less frequently compared to previous buying habits, but buy more at a time than earlier. Prefer more expensive clothes and plan to wear them for a long time. Impulse buying once or twice a month typically focused on cheaper garments such as T-shirts and sweaters. Memorize and learn from failures—do not make as many mistakes as earlier.	No obvious patterns for outdoor purchases. Sometimes buy waterproof clothes for rainy holidays. If the consumer (possibly a couple) spends time in a store, and the shop assistant is knowledgeable and helpful, users will invest slightly more on the selected product despite the potential for going online to search a cheaper price. If there is a problem it can be taken back! Purchases often replace an old existing garment or are to buy an alternative color.
Price/quality	Give more importance to pleasant and well-fitting design, color, quality, good fitting, and shopping convenience than to price. Importance of predilection over price, taking into consideration how long they intend to wear the garment. Look for best price-to-quality ratio. Good fit tends to imply a higher price. The need for garment influences relative importance of price. Women willing to pay much more for business clothes than for leisure wear. They sometimes comment on price being too low.	Initially expected outdoor products to be relatively cheap being unaware of the complexity of clothing production. Initially did not expect to pay as much for sportswear as other garments. Greater clothing value recognized as a result of co-design engagement when the sophistication and multifunctional attributes of stylish (well fitting) outdoor clothing are appreciated as well as the potential for cross over to "lifewear." NB: Older users can remember higher value of clothing in the past.
Environment and ethics	Consider quality and longevity of use. Fit-for-purpose: decision-making seems to be made on rational grounds; informants evaluate the utilitarian and functional attributes of the garments in relation to their needs. "Price matters but it should be evaluated together with the purpose and need. Many times I evaluate the time of wear of the garment; how long I can use it. Price is also an indicator of the quality; high quality and low prices don't go together" (as commented by a Finnish informant).	Users do not appreciate transient swings in "fast fashion." They consider quality and longevity of use; clothing is often considered "an old friend." Find labeling confusing. Confusion over sizing leads to dissatisfaction and waste. Poor garment specification leads to waste. Users initially unaware of environmental and ethical risks associated with functional clothing and textile chain. (NB: Research was undertaken prior to Bangladesh and Detox publicity.)

11.7 WAY FORWARD

11.7.1 Benefits of Adopting Co-design Process

Co-design engagement has resulted in the development of new hybrid processes and new *shared language*. All stakeholders have realized the value of adopting a co-design approach to gain new insights into the real design needs and aspirations of this growing and, potentially valuable, market. A co-design approach has begun to break down barriers between the textiles and clothing cultures and that of electronics, in both academia and industry. This has wider implications for design guidance to inform, for example, sector skills, standards, and education, and on how to address older age. Findings from this case study may inform the development of a design manual, in clear language (both text and visual) to guide the design development process, supported by the examples of working prototypes as proof of concept. Co-design relies on the importance of designers but gives them a new role in enabling consumers to express their clothing needs. It has promoted a *slower*, value-added, design process with the potential for less wasteful product that is fit for purpose. This concept would work well if production was to be brought *back* closer to the consumer. The surprisingly good response from the industry engagement and the value of findings from this process should now be transferred to address design challenges in bringing products to market. Michael Wolff predicts that "Those designers who can't find a way to tap into the power of inclusive design should fear for the future of their enterprises – and those who do will have a bigger, brighter, satisfying and enriching future ahead" (Wolff, 2014).

11.7.2 How to Bring New Products to a New Consumer

The success of bringing new products to a new consumer will depend on reassessing all the links in the chain contributing to the product cycle. Developing product to meet those needs will involve rethinking the whole design process and the methods of manufacture and supply logistics, and with pricing, advertising, and POS communication. To cut down waste, designers should become closer to prototype development with the potential for production to return to being nearer to the market. A collaborative design approach, more accepted in the disciplines of product and industrial design, would enable a meaningful engagement with consumers in identifying and addressing their real needs. For the potential adoption of smart textiles and wearable technologies, with their supporting powering and communication devices and their services, designers and the product development team must adopt a new hybrid design methodology to inform and guide the design specification and guide the design development process. Co-design process, involving end users throughout the iterative product development process, needs to be extended to embrace the creation of age-appropriate POS material and retail staff training. In particular, this research has looked at design for the active ageing where this research has confirmed "Older people are the experts in making mature choices" (Metz and Underwood, 2005).

This research has synergy with current industry discussions on bringing back textiles and clothing production nearer to *home* that has been the focus of two industry conferences in recent months: *Made in the UK* (ASBCI, Manchester, September 15, 2012) and *A New Dawn: Rebuilding UK Textile Manufacturing* (the Clothworkers' Company, the Dyers' Company, and the Weavers' Company, City of London, November 2, 2012). Addressing the changing demography, which embraces the rapidly growing active ageing community, could be an ideal market to address by U.K. manufacturers. This consumer is already worried by transient, disposable fashion and anxious that, if they find something that suits them, they may not be able to replace it. Slower, value-added design would be appropriate for both the consumer and for more considered production. Fewer styles in better-fitting product, with a choice of lengths, and in colors and textures that are sympathetic to baby boomers could be personalized with meaningful accessories to suit the individual wearer. Operators could build up skills in shorter runs, of known styles, in manufacture with a focus on quality, fit, and appropriate styling, in alluring but more enduring colors and in fabrics that will not date easily.

The clothing needs of the active ageing group could represent a predictable market where value-added, more considered, and sustainable production could be developed to cater for price points ranging from elite to mid market level. This market is not going away and is growing and could be a less frenetic sector to be addressed by U.K. manufacturers. Design for Ageing Well findings offer guidance for the realization of design fit for purpose from functional, aesthetic, and cultural perspectives. Cross-disciplinary methodology has enabled the realization of the smart clothing layering system as an integrated *wearable technology platform* that enhances comfort, independence, and enjoyment for varying levels of participation in walking. Greater understanding between academic researchers and design practitioners has extended to beneficiaries within the international textile, clothing, and related *soft* electronics supply chain, with the potential to promote slower, more considered, responsible, and, ultimately, more sustainable approaches to product development.

11.7.3 Moving toward a Circular Economy

Ravasio, now CSR and sustainability manager of the EOG, states that

> Statistically, it is reckoned that by 2020 the planet will have an extra 1.5 billion middle-class consumers – roughly doubling the consumption potential that currently exists. By 2030 this number is set to treble, with supply levels required to increase accordingly, and further pressure on commodity availability- and prices. With this background in mind, the term Circular Economy (CE) has over the last two years frequently been introduced as a possible solution to the challenges at hand.
>
> **Ravasio (2014)**

The circular economy (CE) is an economic model that is based on an entirely different paradigm, from raw material acquisition through to product design and product's end of life. The Ellen MacArthur Foundation describes CE as a generic term for an economy that is regenerative by design. Material flows are categorized into two types: *biological materials*, where products do not quickly become waste but are reused to extract their maximum value before safely and productively returning to the biosphere, and *technical materials*, designed to recirculate with minimal loss of quality, in turn entraining the shift toward and economy ultimately powered by renewable energy. For business leaders, such an economy can deliver growth with innovative product designers and business leaders already venturing into this space (www.ellenmacarthurfoundation.org). Conceptually, the CE consists of three complementary components:

System level: The redesign of entire industrial systems that is particularly relevant for industries like textiles and fashion that rely on far-off production locations.
Production level: Product design processes and procedures that fully incorporate cradle-to-cradle (circular) material flows and production.
Consumption level: To initiate and benefit from new patterns of consumer behavior in order to determine resource pathways.

When looking at the challenges to address in the context of fashion and textiles, most efforts revolve around the production level with brands and manufacturers geared toward what has been described, by the Ellen MacArthur Foundation, as *the first mile* in a CE that involves resource recovery, usually of postconsumer textile waste. An example of a pioneering brand in the outdoor market is the U.S. company Patagonia with the founder Yvon Chouinard described as *a hero for the planet* (*Time* Magazine, 1999) and as *value-centered entrepreneur* having led the way in using his business "to inspire and implement solutions to the environmental crisis" (Choi and Gray, 2011). The company's *Common Threads Recycling Partnership*, initiated in 2005, directly encourages consumers to *reduce*, *reuse*, *repair*, *recycle*, and *reimagine* their consumption of garments, with one approach inviting them to return their worn-out polyester garments to be recycled by Patagonia's textile supplier in Japan (www.patagonia.com). Patagonia's campaign has brought consumers into discussion through events at stores and through social media as well as via presentations at conferences and other media opportunities. Catalogue and web essays represent two major veins of thought: an increase in resource efficiency through industrial-scale change and a revival of local self-reliance, farm-to-table food, and decentralized alternative energy.

Manufacturers that represent what is described as *the second mile* offer fiber and fabric ranges that rely on a steady stream of input materials from *the first mile*. For example, the Japanese fiber and fabric producer Teijin, with particular relevance to the outdoor industry, has taken the initiative to establish the *closed-loop recycling system: Eco Circle*. As a material supplier to Patagonia, Teijin takes back garments, melts them down, and reextrudes *new* fiber for reuse in knitted base layer and fleece fabrics and woven outerwear qualities (www.teijin.com). As part of M&S's "Plan A" clothes recycling program, a *Shwopping campaign* collected and sent used and unwanted apparel worth £450,000 to Oxfam (http://plana.marksandspencer.com). In addition to the recycled products, alternative biofibers are being produced that include Lenzing's lyocell (branded as TENCEL) and plant-based PET, already prevalent in outdoor clothing products. During the co-design process, older consumers became educated in the existence and use of these products.

It is said that *pure* CE is still considered a hypothetical concept but with aspects that will progressively find their way into mainstream manufacturing and retail models. The recovery and material aspects are intrinsically linked to large brands that can initiate the take-back schemes that represent the foundation stone as to what will guarantee them reliable access to an affordable, continued source of raw material in the long term (www.ellenmacarthurfoundation.org). However, Rick Ridgeway (vice president of environmental affairs, Patagonia) is conscious of smart people talking about "impressive corporate initiatives to create a more circular economy and to achieve resource efficiencies through technological and process improvements." He reminds us that "the elephant in the room is that no one seems to acknowledge is growth – and the existential need for companies and nations to run after growth in the form of ever more feverish consumption, toward the edge of the cliff and over" (Ridgeway, 2013).

Worn Again CEO Cyndi Rhoades suggests that "the challenge is communicating the value of clothing to the wearer … an efficient closed loop depends on this." She calls for industry to come together to engage with consumers, with the idea of one, universal *return to sender* symbol to be placed on tags and other clothing labeling and encourage consumers to value their clothes and recycle as a matter of course, by means of global, integrated collection systems. "The key driver is the perceived value (of the clothing item) … not just of the end of life of garment itself but of the resources in it." Rhodes recognizes that the question "Should we just consume less?" represents "a huge challenge because we have had generations that who have been told to buy more and use more. Maybe a closed loop model could enable the types of consumerism people have become used to" (Rhodes, 2012).

11.7.4 Guidance to Producers and Consumers: Good Practice in the Outdoor Industry

Designers and product developers within the outdoor sector are now obliged to address the environmental and ethical concerns affecting the industry due to the campaigns of NGOs such as Greenpeace and Four Paws. Over the past decade, awareness has grown due to, in Europe, the EOG's SWG's contribution to the Eco Index that has since contributed to wider industry-led initiatives led by the Sustainable Apparel Coalition (SAC). Most recently (December 2013), the self-assessment tool Higg Index 2.0 was launched to further improve the measurement of the sustainability impacts of apparel and footwear products and encourage supply chain transparency, contributing to a more sustainable textile sector (www.apparelcoalition.org). The Higg Index is described as "primarily an indicator based assessment tool for apparel and footwear products. The Index asks practice-based, qualitative questions to gauge environmental sustainability performance and drive behavior for improvement." Version 2 has been "significantly enhanced through a pilot testing period and over 14 months of organizations using the Higg Index 1.0" (www.apparelcoalition.org). The content of this self-assessment tool is based largely on the Eco Index, Nike's Apparel Environmental Design Tool, Global Social Compliance Programme reference tools, and Social/Labor Best Practice Tools.

The Higg Index 2.0 represents a starting point of engagement, education, and collaboration among stakeholders in advance of more rigorous assessment efforts. It is reported that 35 U.K. companies and retailers have now signed the Sustainable Clothing Action Plan (SCAP) 2020 (www.ecotextile.com, March 2014). This commitment, led by WRAP, sees brands pledge to reduce the environmental footprint of clothing through; fiber and fabric selection, the measuring and tracking changes in their footprint, the development of messaging to influence consumer behaviors, the increase of reuse and recycling to recover maximum value from used clothing, the development of actions to help keep clothing out of landfill and, over the long term, to work with supply chain partners to reduce the environmental footprint of their processes. Supporters and signatories include Whistles, the Arcadia Group, M&S Tesco, ASOS, New Look, Sainsbury's, the Centre for Sustainable Fashion, GOTS, and Made-By (www.ecotextile.com). In response to the question "When will there be a consumer-facing score or label based on the Higg Index?", the SAC has stated that it sees the need and value of a consumer-facing product declaration but appreciates the complexity involved in arriving at a single numeric score or other product representation. Thus, to date, our focus has been on making the Higg Index as robust and holistic as possible and to gain confidence in the scoring. No timetable has been set for development of a consumer-facing label" (http://www.apparelcoalition.org).

The U.K. retailer M&S, with their focus on Plan A, has aimed to be one of the world's most sustainable retailers by 2015. In aiming for more responsible and transparent supply chain management within the clothing and textile sector, M&S works in collaboration with selected suppliers. For example, the Hirdaramani Group's Mihila *green factory* (Sri Lanka) demonstrates excellent practice in eco-friendly, carbon-neutral production. This factory, which claims to have zero waste, also manufactures for Patagonia (visited by the author, October 2013).

11.7.5 BRINGING GARMENT PRODUCTION (BACK) NEARER TO MARKET

Bringing garment manufacture back to the United Kingdom has been the subject of high-profile debate at recent trade conferences in the United Kingdom. In 2012, *A New Dawn: Rebuilding UK Textile Manufacturing* was cosponsored by three textile-related City of London livery companies: the Clothworkers', Dyers', and Weavers', constituting the first form of trade associations, many of which are hundreds of years old. Today, the organizations continue to be keen supporters of the industry, in particular skills and education. At this conference, leading representatives of the fashion and textile community supported the concept of reshoring. The wider trade community, and academia, was introduced to *The Alliance Project* established to examine the potential for repatriating textile manufacturing in the United Kingdom. The work was commissioned by Lord David Alliance and the Greater Manchester Combined Authority (GMCA) with the support of the government through the Department for Business, Innovation, and Skills (BIS-GOV.UK).

Based at New Economy, work undertaken to date has identified the following pillars of activity to focus on: skills (to address the current shortage), investment (to grow the capacity of the U.K. textile manufacturing base), innovation (to ensure the revitalized industry does not stagnate), and reconnecting supply and demand (the development of trade links between manufacturers and retailers) (http://neweconomymanchester.com). This project is currently researching the supply chain in the Greater Manchester area, with 120 suppliers, in order to grow data for retailers and buyers to use. Lorna Fitzsimons, director of the U.K. Alliance Project, stresses the finding that "having the right design represents 25% of value" and that "price will be hiked if design is right" (Fitzsimons, February 25, 2014). Other complementary themes have been covered at conferences staged by the

ASBCI: Made in the UK (September 25, 2012) and *Making It in the UK—Ready or Not? The Reality of Manufacturing Fashion in the UK* (February 25, 2014).

Issues that are driving these initiatives are to do with the increasing labor costs in offshore production, fuel costs increasing, the demand for ever-faster fashion, and the increase in attention to ethical trading as retailers and manufacturers, and all involved in the supply chain are under the scrutiny of the media and the consumer power base of social media. The concept of *Made in the U.K.* has continued in the background for high-cost traditional brands but the question being asked is, "Are we ready to build capacity again for mass production?" The issues to be addressed are to do with revitalizing the labor force and investment in resources. Will people want to join the industry? How do we make it attractive? Do we have/can we get the necessary skills? How do we attract young people to the industry? A major challenge involves the prospect of getting industry-related subjects on the curriculum of schools and colleges. The industry must recognize that it needs to be prepared to offer careers and not just *jobs*.

Reshoring is gaining momentum with existing U.K. manufacturers in a leading position such as Burberry, Mulberry, John Lewis, Barbour, John Smedley, and David Nieper. An extensive Kurt Salmon retail study (2014) has shown that closer decisions to sale improve margins considerably, favoring local supply (Adamski, February 25, 2014).

In coining the phrase "Localization is now Globalization," Adamski showed the sequence of the key decisions with regard to fabric, color, silhouette, and quality. The study considers the complexity of supply chains demanding the ability to ship direct from suppliers both at home and abroad. "To make this work, closer more collaborative relationships with suppliers are vital as they are increasingly being asked to act as the medium/agent who delivers to the customer without any direct intervention from the retailer. This is a symbiotic relationship, at a single item level, which has to engender much greater levels of trust, data and execution than ever before" (ibid). The issue of single customization was discussed but also the need for customers to touch and feel clothing products prior to purchase. Adamski was the only speaker to mention *slow fashion* and the notion that attitudes are beginning to change where "consumers want to keep products longer and cherish them."

Kate Hills, founder of *Make it British*, provides insight into the perceptions of the British consumer through survey findings that measure the impact of the *Made in Britain* label that answer the question: "Does the customer care?" Responses were as follows:

- 43% would definitely pay more for a product made in the United Kingdom even if it cost more.
- Men are twice as likely as women to pay more for British-made goods.
- Age plays a big part: 70% of the over 60s versus 16% of the under 30s would be willing to pay more for a product made in the United Kingdom.

In terms of valuing better quality, 60% of the respondents agreed with the statement, "If I know that a product is made in Britain I believe it to be of good quality," and 6 out of 10 said that if a company promotes its products as *Made in Britain*, it gives them more confidence to buy products.

However, Hills finds that some suppliers currently making in the United Kingdom are "not shouting about it." There is also the question to address in terms of what defines the country of origin. She explains that labeling is, generally speaking, not compulsory in the EU. "However, it is necessary to have an origin label if, without such a label, the consumer could be misled as to the true origin of the garment. For instance, if a garment carried the British flag on it but it was made in Hong Kong, then the garment should include a label to that effect. Fraudulent origin labeling is illegal" (http://www.ukft.org). Hills suggests that the origin may be considered as the last country where a substantial change has been made. For example, a garment may be made in the United Kingdom from a foreign fabric, whereas a T-shirt made elsewhere and then printed in the United Kingdom would still be considered a product made outside the United Kingdom. She points out that clarification is clearly needed, as the *rules* have not been updated for many years.

Hills reports that many small manufacturers find it frustrating that buyers focus on price rather than quality and often see British manufacture as somewhere for *overspill* or reprocessing from the Far East. Often specifications from retailers are not quite right, as they have been accustomed to free sampling from offshore, having become used to just waiting for the next, revised, sample to arrive. Retailers come "with no idea of what they want and expect us to make it happen." In addition, pricing needs to be better understood. In the comparison between offshore sourcing and *Made in Britain*, companies often make the mistake of applying net margins rather than gross to offshore production costs. Hills has found herself *hand-holding* companies that want to begin to work with the United Kingdom. "We try and work with the ones who genuinely want British-made products rather than those that have come unstuck with Far Eastern suppliers." Hills suggests *steps for success*:

- Play with the U.K. manufacturer's strengths as labor intensive work is bound to cost more than offshore.
- Visit the factory and meet them face-to-face.
- Do not over sample with no intention of putting down a bulk order.
- Accept that you are going to have to pay for samples.
- You are unlikely to make the same margin as elsewhere but offset that against other benefits.
- Staff is in short supply so do not expect them to be at your beck and call.
- Make a commitment to your U.K. supply base and you will be rewarded in the long run.

The successful family-run company David Nieper was established in Derbyshire in 1961. Christopher Nieper explains how, in the 1970s and 1980s, when boutiques were replaced by high-street chains and "all looked the same," the business model changed to mail order, with the young Joanna Lumley, Marie Helvin, and Yasmine Le Bon modeling "all the elements of women's wardrobe." The aim has been "to make the right size and color – just in time." Bringing more and more elements in-house, including the printing of the catalogues, has resulted in a 3-week lead time with no discounts needed for stock lines. "Customers will buy on value and not on price." To prove this point, a silk and lace slip, originally priced at £49, was put up to £80 with a resulting 40% increase in sales! Nieper considers the "magic ingredients" to be brand, quality, service, and people. Building a brand engenders loyalty that must be earned with customers, staff, and suppliers. Quality garments are made to last and to fit. Service, with a personal touch, is vital in dealing directly with customers. As with other well-established U.K. manufacturers, this company has an ageing workforce. In valuing people, they have committed to training future staff both in establishing a school (Academy), which provides work experience, and offering university scholarships. Nieper recognizes that future staff deserves a *career*, not just a *job*.

11.8 CONCLUSION

"None of us should just carry on as we do. Our current ways are becoming too dangerous, a threat to our children and grandchildren. And as we become older and less physically able, we're condemning ourselves to a desert of gloomy products and places" (Wolff, 2014). From the perspective of a designer, Wolf believes that "we're failing to use our creativity to bring delight to billions of ageing people, as well as a billion people with particular needs. At the same time, we thwart our ability to export innovations and designs to this expanding sector of the global community." He asks, "Why don't our design colleges, designers and businesses wake up to this opportunity?" and proposed that "It's time for designers like me to put an end to product lines that limit the choices and self-expression of billions of people we aren't reaching today, and use design to reach and serve this ignored and growing constituency" (ibid).

If the answer is bring production nearer to *home*, it is important to understand the marketplace and know what customers want. There is general acknowledgment among U.K. retailers in that the importance of design and innovation is "Key to setting yourself apart" (Mitchell, 2013). U.K. manufacture can never be the cheapest, but good design can demand a higher price. Creating points of difference can involve innovation both in terms of product and process. Added value means "something you want to buy! Versus something cheaper" (Mitchell). In terms of the value of clothing, there is the view that "Now that 'Quality' is considered a given, perhaps 'Fitness-for-purpose' is a better definition?" (Taylor, February 25, 2014). Retailers are increasingly wanting to buy British for a variety of reasons including the *feel good factor*, selecting a unique product or brand, addressing customer demand, acknowledging social responsibility, as well as speed of response and price and flexibility (margin) as a result of being nearer to market (ASBCI Conference, 2014).

Michael Wolff advocates that "Sustainability in every aspect of the way we live is essential, but that idea has yet to reach a sufficient level of corporate awareness" (Wolff, 2014). However, Bray acknowledges that the consumption environment and influences on the consumer decision-making process are continuing to change and evolve, concluding that

> Consumer behavior research highlights a number of factors which are currently influencing changes in consumption choices and practices. These include: Increasing concern for environmentalism, Increasing politicization of the consumer whereby the consumers' assessment of companies' ethical standards proves influential in their consumption choices, Growing awareness of global issues such as resource depletion, and the working practices in developing nations. Due to the moral dimensions of these factors, it is commonly stated that 'ethical consumerism' is growing.
>
> **Bray (2009)**

REFERENCES

Adamski, A. The reality of manufacturing fashion in the UK today. Paper presented at *Making it in the UK—Ready or Not? The Reality of Manufacturing Fashion in the UK, ASBCI Conference*, Hinckley, U.K., February 25, 2014.

Arnold, C. *Ethical Marketing and the New Consumer*. Chichester, U.K.: John Wiley & Sons, 2009, ISBN: 978-0-470-74302-7.

Barry, B. Does my bottom line look big in this? A study by University of Cambridge, Judge Business School. *ASBCI Clothing Industry Handbook*, Association of Suppliers to the British Industry, 2014, p. 33.

Batty, M. Right product, right time, right place—meeting the needs of global consumers, Presentation, Fashion Impossible Conference, (ASBCI) Marriott Hotel, Leeds, January 10, 2013.

Bray, J.P. Ethical dimensions in clothing purchase. Paper presented at the *First Annual Ethics in Everyday Life Conference*, Salzburg, Austria, March 17–19, 2009.

Choi, D.Y. and Gray, E.R. *Values-Centred Entrepreneurs and their Companies*. New York: Routledge, 2011.

Chouinard, Y. and Stanley, V. *The Responsible Company: What we've Learned from Patagonia's First 40 Years*. Ventura, CA: Patagonia Books, 2012.

Cline, E.L. *Overdressed: The Shockingly High Cost of Cheap Fashion*. New York: Penguin Portfolio, 2012.

Coates, J.F. From my perspective: The future of clothing. *Technological Forecasting & Social Change* 72(1) (2005): 101–110.

Cole, C., Laurent, G., Drolet, A., Eber, J., Gutchess, A., Lambert-Pandraud, R., Mullet, E., Norton, M.I., and Peters, E. Decision making and brand choice by older consumers. *Marketing Letters* 19 (2008): 355–365 (Springer Science+Business Media, LLC, doi: 10.1007/s11002-008-9058-x).

Fitzsimons, L. The alliance project—Remit and reality. Paper presented at *Making it in the UK—Ready or Not? The Reality of Manufacturing Fashion in the UK, ASBCI Conference*, Hinckley, U.K., February 25, 2014.

Fletcher, K. *Sustainable Fashion and Textiles: Design Journeys*. London, U.K.: Earthscan, 2008.

Franklin, C. Fashion for everybody—sizing up the market, Presentation, Fashion Impossible Conference, (ASBCI) Marriott Hotel, Leeds, January 10, 2013.

Fuad-Luke, A. Re-defining the purpose of (sustainable) design: Enter the design enablers, catalysts in co-design. In: Chapman, J. and Gant, N., eds., *Designers, Visionaries + Other Stories: A Collection of Sustainable Design Essays*. London, U.K.: Earthscan, 2007, pp. 19–46.

Glenn, N.D. *Cohort Analysis*. Thousand Oaks, CA: Sage, 2005.

Haldre, H. Shopper behaviour insights for online fashion retailers: New report, Published in Author Archives for Heikki Haldre, Founder and VP, business development, February 27, 2014. http://www.Fits.me/author/heikki/, accessed September 6, 2014.

Hills, K. What are the next steps for UK manufacturers? Paper presented at *Making it in the UK—Ready or Not? The Reality of Manufacturing Fashion in the UK, ASBCI Conference*, Hinckley, U.K., February 25, 2014.

Hines, T. Trend in global textile supply chains. *Textiles* 2 (2010): 18–20.

Hiscock, J. How to separate all the shades of grey—To target the grey market, you must first understand its structure? *Marketing*, November 2000, p. 3. Available via DIALOG. http://www.marketingmagazine.co.uk/article/72339/market-research-separate-shades-grey-target-grey-market-first-understand-its-structure, accessed February 28, 2004.

Holmund, M., Hagman, A., and Polsa, P. An exploration of how mature women buy clothing: Empirical insights and a model. *Journal of Fashion Marketing and Management* 15(1) (2011): 108–122 (Emerald Group Publishing Limited, pp. 1361–2026, doi: 10.1108/13612021111112377).

Hupert, F.A. Designing for older users. In: Coleman, R. and Clarkson, J.P., eds., *Inclusive Design: Design for the Whole Population*. London, U.K.: Springer, 2003, pp. 30–49.

Jones, G. How online fashion retailers can engage with more customers. Feature presentation by G. Jones, *Internet Psychologist. ASBCI Clothing Industry Handbook*, 2014.

Koszewska, M. Social and eco-labeling of textile and clothing goods as means of communication and product differentiation. *Fibres and Textiles in Eastern Europe* 19(4) (2011): 87, 20–26.

Langen, N. *Ethics in Consumer Choice*. Wiesbaden, Germany: Springer Fachmedien, 2013. doi: 10.1007/978-3-658-00759-1_2.

Maclean, I. Going global with a British brand. Paper presented at *Making it in the UK—Ready or Not? The Reality of Manufacturing Fashion in the UK, ASBCI Conference*, Hinckley, U.K., February 25, 2014.

McCann, J. Establishing the requirements for the design development of performance sportswear, MPhil thesis. Derby, U.K.: University of Derby, 1999.

McCann, J. End-user based design of innovative smart clothing. In: McCann, J. and Bryson, D., eds., *Smart Clothes and Wearable Technology*. Cambridge, U.K.: Woodhead, 2009.

McCann, J. Translating the hybrid methodologies and practical outputs of smart textile-oriented research, in clothing for the growing ageing market, for the benefit of all stakeholders. Paper presented at *Textile Institute World Conference*, Kuala Lumpur, Malaysia, May 15–17, 2012a.

McCann, J. Smart protective textiles for older people. In: Chapman, R., ed., *Smart Textiles for Protection*. Cambridge, U.K.: Woodhead, 2012b.

Metz, D. and Underwood, M. *Older Richer Fitter: Identifying the Customer Needs of Britain's Ageing Population*. London, U.K.: Age Concern Books, 2005.

Mitchell, I. What consumers really want and is the high street still relevant? Presentation at *ASBCI Conference, Fashion Impossible? What Consumers Really Want—On Trend, On Time, In Budget...Guilt Free!* Leeds, U.K., October 1, 2013.

Mitchell, J. The challenges of supplying the British high street. Paper presented at *Making it in the UK—Ready or Not? The Reality of Manufacturing Fashion in the UK, ASBCI Conference*, Hinckley, U.K., February 25, 2014.

Montagna, G., Carvalho, H., Catarino, A., and da Silva, F.M. A user centered design methodology for functional and smart garments. *CIAUD—Faculdade de Arquitetura, Universidade Técnica de Lisboa*, Lisboa, Portugal, October 2011. http://handle.net/1822/14464.

Mowbray, J. Inside track on outdoor retail. *Eco Textile News*, Issue No: 56 August/September 2013.

Nieper, C. British design and British made—The magic ingredients. Paper presented at *Making it in the UK—Ready or Not? The Reality of Manufacturing Fashion in the UK, ASBCI Conference*, Hinckley, U.K., February 25, 2014.

Ravasio, P. Business case grows for circular economy. *Eco Textile News*, Issue No: 58 December 2013/January 2014, p. 24.

Rhodes, C. Closed loop or wear nothing. Paper presented at *National Retail Consumer Conference*, Stratford-Upon-Avon, U.K., February 24–26, 2012. http://www.uk.coop/nrcc/closed-loop-or-wear-nothing-slides, accessed February 28, 2013.

Ridgeway, R. Circular economy. *Eco Textile News*, Issue No: 57 October/November 2013, p. 23.

Santen, M. and Kallee, U. Chemistry for any weather: Greenpeace tests outdoor clothes for perfluorinated toxins. 2012. Available via DIALOG. https://Greenpeace.org/romania/Global/romania/detox/Chemistry%20for%20any%20weather.pdf, accessed May 6, 2014.

Shankleman, J. UK throwing out £25m a year by sending clothes to landfill. *Business Green*, September 13, 2012. Available via DIALOG. http://www.businessgreen.com/bg/news/2204863/uk-throwing-out-gbp25m-a-year-by-sending-clothes-to-landfill, accessed February 28, 2014.

Sumner, M. Plan A—Consumer engagement; consumer perception. Presentation at *'Fashion Impossible? What Consumers Really Want—On Trend, On Time, In Budget...Guilt Free!' ASBCI Industry Conference*, Leeds, U.K., October 1, 2013.

Taylor, C. and Goodwin, D. The changing face of sourcing. Joint presentation at *Making it in the UK—Ready or Not? The Reality of Manufacturing Fashion in the UK, ASBCI Conference*, Hinckley, U.K., February 25, 2014.

Walker, S. Sustainable by Design: Explorations in theory and practice, Earthscan, UK and USA, 2006.

Willis, R. Clothes: A manifesto. *Intelligent Life, The Economist*, March/April 2013, pp. 48–54. Available via DIALOG. http://moreintelligentlife.com/content/lifestyle/rebecca-willis/clothes-a-manifesto?page=full, accessed February 28, 2014.

Wolfe, D.B. and Snyder, R.E. *Ageless Marketing*. Chicago, IL: Dearborn, 2003.

Wolff, M. Stop designing for only the young and able-bodied. *Guardian Sustainable Business*, February 13, 2014. http://www.theguardian.com/sustainable-business/michael-wolff-inclusive-design-seniors-disabled, last accessed September 6, 2014.

Wusaty, M. *Sustainability: A Dirty Word for a Clean Cause*. Brand Commercial, May 1, 2013. Available via DIALOG. http://brandandcommercial.com/articles/show/brand-strategy/109/sustainability-a-dirty-word-for-a-clean-cause, accessed February 28, 2014.

FURTHER INFORMATION

A New Dawn: Rebuilding UK Textile Manufacturing', *Conference*, London's Clothworkers Hall, London, U.K., November 2, 2012. http://www.clothworkers.co.uk/getattachment/Textiles/Textiles-Conference/A-New-Dawn-FINAL_LR.pdf.aspx, accessed February 28, 2014.

Best Ager, guide through the gallery: ISPO. *ISPO Magazine*, January 27–30, 2008. Available via DIALOG. https://www.ispo.com/sports/en/All-Sports/ISPO-NEWS/ISPO-MAGAZINE, accessed May 27, 2013.

Best Ager: ISPO, Munich, 2010. http://media.nmm.de/47/ispo10_bestager_presentation_eng_24372647.pdf, accessed January 10, 2015.

Brady, B. Boomers Are 'The Most Valuable Generation' For Marketers, Nielsen Report Finds, Posted: August 17, 2012, 12:37. pmhttp://www.huffingtonpost.com/2012/08/17/marketing-to-boomers-most-valuable-generation_n_1791361.html. accessed January 10, 2015.

Cline, E.L. (2013) Overdressed: The Shockingly High Cost of Cheap Fashion. Available via DIALOG. http://www.bbc.co.uk/northernireland/forteachers/linen/ethical/the_cost_of_cheap_fashion.shtml.

Closed-Loop Recycling System. Available via DIALOG. http://www.teijin.com/solutions/ecocircle/, accessed February 28, 2014.

Closed-Loop Recycling System: Eco Circle, 03, Solutions: Reaching beyond the known. Source: Fiber products 3R related research report, The Organization for Small & Medium Enterprises and Regional Innovation, Japan. http://www.teijin.com/solutions/ecocircle/. accessed January 10, 2015.

Common Threads Partnership. Available via DIALOG. http://www.patagonia.com/us/common-threads/, accessed February 28, 2014.

Design for Ageing Well. Improving the quality of life for the ageing population using a technology enabled garment system. Key Findings (2013). http://www.newdynamics.group.shef.ac.uk/assets/files/NDA%20Findings_20(1).pdf, accessed May 27, 2013.

Design for Ageing Well, Findings, 2013. http://www.newdynamics.group.shef.ac.uk/assets/files/NDA%20Findings_20(1).pdf

Fashion Impossible? What Consumers Really Want—On Trend, On Time, In Budget...Guilt Free! ASBCI *Conference*, Leeds, U.K. http://www.asbci.co.uk/news/archive-news/-fashion-impossible-press-review, October 1, 2013.

Has anyone tried the Warehouse Try Me On Thing?—Mumsnet: http://www.mumsnet.com/Talk/style_and_beauty/a1742349-Has-anyonetried-the-Warehouse-Try-Me-On-thing, (accessed January 10, 2015.)

Higg Index 2. Available via DIALOG. http://www.apparelcoalition.org/higgindex/, accessed February 28, 2014. http://www.ellenmacarthurfoundation.org/circular-economy/circular-economy, accessed February 28, 2014.

http://neweconomymanchester.com/stories/1933-the_alliance_project, accessed February 28, 2014.

Michael Wolff, designer and creative advisor—EU.com. http://www.wolff.eu.com (accessed January 10, 2015.)

Retail. *Retail Week*, 2014. http://www.kurtsalmon.com/UK/about-news-item/Retail-2014-identifies-the-key-challenges-for-UK-CEOs?id=423&language=en-uk#.UxG_wf3TRFI.

Sizemic, Fashion Technology, Size and Fit Solutions. http://www.sizemic.eu/. (accessed January 10, 2015.)

Sizemic Press release: Human-solutions and Sizemic—An exclusive collaboration. http://www.human-solutions.com/group/upload/Pressemitteilungen/Englisch/13_13_HS_Sizemic__eng.pdf. (accessed January 14, 2015.)

Sustainable Apparel Coalition, Higg Index FAQs, General Higg Index Questions: http://www.apparelcoalition.org/index-faqs/general-higg-index-questions/when-will-there-be-a-consumer-facing-score-or-label-based-on.html. (accessed January 17, 2015.)

Sustainable clothing commitment growth. *Eco Textile News*, Issue No: 58, December 2013/January 2014, p. 9.

Sustainable clothing roadmap: Progress report, 2011. https://www.gov.uk/government/uploads/system/uploads/attachment_data/file/69299/pb13461-clothing-actionplan-110518.pdf.

The shape of things to come—New thinking and innovations in clothing size and fit. *ASBCI Conference*, Huntingdon, U.K., April 10, 2013. http://www.asbci.co.uk/news/archive-news/asbci-sizing–fit-conference-huntingdon-april-2013.

Towards the Circular Economy. Available via DIALOG. http://www.ellenmacarthurfoundation.org/circular-economy/circular-economy/towards-the-circular-economy, accessed February 28, 2014.

World Health Organisation. Definition of an older or elderly person. http://www.who.int/healthinfo/survey/ageingdefnolder/en/. (accessed January 10, 2015.)

World Health Organisation. What is "active ageing"? http://www.who.int/ageing/active_ageing/en/ (accessed January 10, 2015.)

WRAP Textiles. http://www.wrap.org.uk/category/materials-and-products/textiles. (accessed January 10, 2015.)

12 Consumer Behavior in the Fashion Field

Kirsi Niinimäki

CONTENTS

12.1	Introduction	271
12.2	Background	271
12.3	Meaning of Consumption	272
12.4	Environmental Impact of Fashion Consumption	273
12.5	Current Practices in Fashion Consumption	274
12.6	Textile and Clothing Waste	276
12.7	Clothing Disposal	277
12.8	Consumer Values and Consumption	278
12.9	Increasing Environmental Information	279
12.10	Transforming Consumption	281
12.11	Whom to Blame?	281
12.12	Redirective Fashion for Sustainable Consumption	282
12.13	Fulfilling Consumers' Changing Needs	284
12.14	For Slow or Fast Consumption in the Future?	284
12.15	Conclusions	285
References		285

12.1 INTRODUCTION

This chapter presents perspectives on current consumer behavior in the fashion field. It describes and shows in figures the unsustainable practices of overproducing and consuming fashion and their consequences and impacts on the environment. The chapter describes the meaning of fashion consumption to Western consumers. Further, it explains how fast design and manufacturing practices lead to unsustainable consumption patterns: impulse shopping, overconsumption, very short use time of garments, and consumers' constant need for change and novelty. Overconsumption and constant growth in fashion consumption lead to an increase in textile waste in all Western countries. Further, the chapter discusses who to blame for the current unbalanced situation, whether green taxation should be developed or more information provided to consumers to influence their behavior, and what would be the best practices for transforming the current situation. Finally, the text shows how it is possible to change the situation and transform consumption toward sustainable practices through design strategies and rethinking value creation.

12.2 BACKGROUND

In recent years, the prices of fashion and garments have dropped, volumes have grown, and cycles of new trends and fashion have speeded up. Consumers in the Western world purchase clothing to fulfill not only their functional needs but also other emotional, cultural, and social needs.

Often, consumers use their clothes for only a very short time and replace them with new ones long before they are worn out.

The textile and clothing industry has a large environmental impact. For example, 25% of all chemicals produced worldwide are used in this industry. Additionally, this industry produces much waste and significant emissions to air, groundwater, and soil (Chen and Burns 2006). Health, safety, and ethical issues are also problematic, as cost efficiency has led this industry to move production to lower-cost countries, where workers toil for long hours in often poor conditions that are not well monitored and controlled (Niinimäki 2013). The current industrial system and consumption are thus not in a sustainable balance, and competition is heating up as companies aim only for easy profits and low costs.

Most garments are currently produced in lower-cost Asian countries, and there is no real option for a consumer to choose more sustainable clothes (Joergens 2006). Markkula (2011) points out those current practices in fashion manufacturing and marketing limit the actions available to fashion consumers to act as ecological citizens. She also argues that fashion marketing pushes consumers to engage in a certain kind of identity building through adopting fast-changing visual appearances and brand symbols.

Simultaneously, an increasing number of Western consumers are worried about the environmental load and ethical issues of low-cost, mass-manufactured garments, and they want to at least be given as much environmental information as possible to guide their choices. Ethical interest is also rising among consumers globally, for example, in 2004 in the United States, 80% of consumers considered themselves to be environmentalists and felt that environmental issues are important in clothing (Solomon and Rabolt 2004). However, consumers' interest in environmental issues has not affected their purchasing behavior, as sustainable fashion has remained a niche market and total clothing consumption is increasing all the time.

Mass manufacturing of clothing in low-cost Asian countries has led to a situation where cheap product prices encourage consumers to make impulse purchases and engage in unsustainable consumption behavior: overconsumption, very short use time of products, and premature disposal of products. Accordingly, the environmental impact of this industry is increasing constantly. Textile and clothing consumption is estimated to comprise about 5% of the environmental impact and carbon emissions of households (Niinimäki 2011). Even if that figure is relatively low, textile and clothing consumption is still on the rise, and the more recent shortening of the life spans of fast fashion items in particular increases the environmental burden of the industry: all those resources are wasted if the garments are worn for only a very short time or even not at all. Furthermore, textile waste poses growing problems in all Western countries, and additionally, the chemical burden of textile manufacturing is a huge problem in many Asian countries.

12.3 MEANING OF CONSUMPTION

Emotions lie at the center of human existence, and they influence most of our behavior, motivations, and thought processes (Desmet 2009, 379). Emotions also play a strong role in consumption. Richins (2009) argues that garments belong to the category of self-expressive products. Consumption-related emotions are important for a consumer of these products, at least immediately after the purchase event or even just before the purchase. According to Richins (2009), the purchase situation becomes a strong positive experience for a consumer, but this experience is very short-lived, disappearing soon after the purchase event is over. Richins (2009) further explains how the study of consumption-related emotions might be important from an environmental point of view if aiming for sustained positive emotions between a person and a product or when seeking solutions to replace consumption with other positive emotional states in order to extend the use time of garments and to decrease materialistic consumption.

Consumption has many meanings for the consumer. It can involve experience, integration, or classification, or it can mean play (Solomon et al. 2002, 15–16). Consumption as an experience colors the purchase situation with emotions; consumption is an emotional goal as such. When a consumer uses products in order to integrate their symbolic meanings with his or her own identity, consumption can be understood as integration. Classification refers to how consumers communicate symbolic meanings and status to others through objects. The element of play is important in contemporary consumer culture, and it can be seen in appearance and fashion, for instance. In consumption, the interplay between the consumer's wants, needs, values, attitudes, and experiences is emotionally meaningful for a Western consumer.

Consumer products that relate to our self-construction and identity are constantly evaluated on both aesthetic and social grounds. Products' symbolic meanings are connected to psychological satisfaction through an emotional response, for example, clothing and fashion enable consumers to gain social acceptance, affiliation with particular groups, and emotional experiences of beauty. When the product no longer offers a positive emotional response, for example, it falls out of fashion; the consumer feels a sense of psychological obsolescence and can easily replace the product with a new one, even if the original garment is still functional. Through the new purchase event, the consumer can again experience excitement, joy, and pleasure, at least for a moment (Niinimäki 2011).

12.4 ENVIRONMENTAL IMPACT OF FASHION CONSUMPTION

The textile and fashion industry is one of the biggest and oldest industrial sectors in the world. It uses more water in its processes than any other industry except agriculture and releases huge amounts of toxic chemicals into the environment (Fletcher 2008). The total volume of textile and fashion production at the global level is estimated to be more than 30 million tons annually, and therefore, the environmental impacts of this industry are remarkable and moreover increasing (Chen and Burns 2006).

It is estimated that clothing items, work wear, and household textiles have a large relative environmental impact during use, compared to their impact during production or disposal (Fletcher 2008). Mainly, the intensive laundering frequency causes this evaluation. Depending on the material and its need to be washed frequently, the impact of consumer care can be as high as 75%–80% of the total environmental impact of a, for example, cotton shirt (Lewis and Gertsakis 2001). Yet the rapid growth in textile and clothing consumption and disposal and simultaneously the growth in textile manufacturing are changing this evaluation.

On the consumption side, rapid growth is evident. For example, in the United Kingdom, the consumption of fashion has increased by 37% between 2001 and 2005 (the amount of clothes purchased per capita). In Sweden, the amount of clothes and home textiles released on the market rose by 40% between the years 2000 and 2009 (Carlsson et al. 2011 in Tojo et al. 2012). In Finland, the total volume of clothes sold in 2010 was 70,212 tons (Tojo et al. 2012), which, for a population of 5,426,674, equals an average of 13 kg of textiles per person. This Western overconsumption causes environmental and social problems on the other side of the globe, where the fibers are cultivated and fabrics are woven, dyed and finished with chemicals, and sewn into garments in often poor working conditions.

In Asian countries, the environmental laws and regulations are not so tight than in EU countries. Even if the regulations exist, they are not tightly followed. Therefore, the ever growing textile consumption in the Western world links to the growing manufacturing volumes in Asian countries, and this overgrown industrial system contributes to large environmental impact both in Asian countries and in Western countries. In Asian countries, the impact is on environmental pollution and causing people's health problems and in Western countries in increasing waste problem. Islam et al. (2011) argue that in Bangladesh, the most important industrial sector is the textile industry, and it accounts about 77% of the country's foreign exchange earnings. On the other hand, textile industry uses large quantities of freshwater and produces a large amount of highly toxic and polluted wastewater that goes into sewing system without any treatment (Islam et al. 2011). This causes a health risk to local people and further destroys the environment. While textile manufacturing has moved to lower-cost countries while seeking for the cheap end price of the product, so has also the most of the environmental impacts of textile and clothing manufacturing while the waste streams are growing in the West.

Through consuming more and more cheap textiles mass-produced in Asian countries, Western consumers affect the environment on the other side of the world. Lately, the ethical production of textiles has been a hot topic in public discussion. Western consumers have become informed about unfair production systems through the mass media. However, thus far, the environmental impacts that Western consumer habits have on other countries have been largely ignored.

There are several options for informing the end user about the global impact of textile production, that is, carbon footprint and water footprint. A carbon footprint is the estimated impact of a person's activities on the environment in quantitative terms: it includes all greenhouse gases. It is estimated that 5% of a typical person's total carbon footprint in the developed world is accounted for by clothing (Niinimäki 2011). The carbon footprint includes the primary footprint of direct CO_2 emissions (also including domestic energy consumption and transportation) and the secondary footprint (the whole life cycle of products we use). It is also possible to calculate the carbon footprint of individual products and place the calculation in context, thereby enabling the consumer to compare different products (Weidema et al. 2008). In cheap, mass-produced clothes, cotton cultivation and long logistic routes result in a very large footprint.

A water footprint is an indicator of both the direct and indirect water use of a consumer or a producer. Many countries, especially in the Western world, have externalized their water footprint by importing water-intensive products from elsewhere. Global international trade implies international flows of virtual water. For example, the water footprint of one cotton shirt is 2700 L (Chapagain et al. 2006; Water Footprint).

The carbon footprint and water footprint are also linked directly to the total consumption and welfare of nations. These are good indicators for evaluating consumption levels and the environmental load of consumption in total. Even though these figures are rather new, there has been public debate on topics such as carbon footprint–labeled clothes, for example, in the United Kingdom (Fletcher 2008; Niinimäki 2011).

There exist two trends on how to guide consumers toward more sustainable practices: one follows the path of voluntary change supported by labeling and increasing environmental information on products, while the other relies on the responsibility of authorities through better legislation, taking into account the environmental costs of the product and adding green taxation. It is quite obvious that current fashion consumption practices cause huge environmental problems, which are further discussed in the following texts.

12.5 CURRENT PRACTICES IN FASHION CONSUMPTION

Clothing prices are currently far cheaper relative to household incomes than a few decades ago. Textile and clothing prices have fallen, and currently the consumer possesses more and more cheap garments and low-quality textiles. According to Jackson and Shaw (2009, 146), in the 1950s, in the United Kingdom, 30% of a household's income went to clothing purchases; currently, that figure is 12%, with a higher amount of consumed items. Clothing prices have fallen, and currently the consumer possesses more and more cheap garments and low-quality textiles that are used for an extremely short time.

Examples of clothing and fashion consumption rates and increases in consumption are presented next. In Denmark, the domestic use of textiles and clothing amounted to about 89,000 tons in 2010 or 16 kg/capita. Danish consumers spending on textiles and clothing increased by close to 62% between 2003 and 2008. In Finland, textile imports amounted to 76,500 tons and domestic production to almost 2,400 tons in 2010. Accordingly, nearly all garments on the Finnish market (about 95%) are imported. The textile inflow per person was about 13 kg a year, and textile and clothing consumption increased by 35% between 2001 and 2006 (Tojo et al. 2012).

Textile and clothing consumption in Britain rose over 30% in the period 1995–2005 (Defra 2008). Fast fashion, marked by its low quality and very short product life span, is increasing its share of the market. In Britain, it is estimated that fast fashion currently accounts for 20% of the market, having doubled its growth between the years 1999 and 2006 (Defra 2008). This growth trend in fashion can also be seen in other countries.

Fast fashion with short life spans worries consumers. In a Finnish study, 78% of women and 68% of men were worried about the short life span of garments. When asked to estimate the life span of their oldest and most used garments, 74% of men and 66% of women answered over 5 years (see Table 12.1) (Niinimäki 2011).

When consumers estimate the shortest time they have used some of their garments, 10% of men and 22% of women answered less than 1 month (see Table 12.2). Moreover, the consumers' interest in environmental issues affects the life span of garments: 84% of those consumers with a high interest in environmental issues reported using garments for more than 5 years, while only 14% reported using some garment less than 1 month. Of the respondents who were less interested in the environment, 59% reported using some garment for over 5 years, and 22% said that they have used some garment for less than 1 month. In summary, the respondents' interest in the environment and their values are connected with the use time of their garments (Niinimäki 2011).

TABLE 12.1
Respondents' Estimation of the Longest Use Time of Their Garments

	Men (%)	Women (%)
7 months to 1 year	1.7	1.4
1–2 years	1.7	0
2–3 years	6.9	5.6
3–4 years	8.6	15.4
4–5 years	6.9	11.9
Over 5 years	74.1	65.7

Source: Niinimäki, K., From disposable to sustainable, The complex interplay between design and consumption of textiles and clothing, Doctoral dissertation, Aalto University, Helsinki, Finland, available at https://aaltodoc.aalto.fi/handle/123456789/13770.

TABLE 12.2
Shortest Use Time of Garments

	Men (%)	Women (%)
Less than 1 month	10.3	21.8
1–2 months	12.1	9.2
3–6 months	25.9	30.3

Source: Niinimäki, K., From disposable to sustainable, The complex interplay between design and consumption of textiles and clothing, Doctoral dissertation, Aalto University, Helsinki, Finland, available at https://aaltodoc.aalto.fi/handle/123456789/13770.

When consumers are shopping for garments, durability and long-term use are not important selection criteria: 33% of women and 22% of men have considered the long use time of the garment in their purchase choices. In addition, the consumers' value base can be seen in these responses; 43% of respondents with high environmental interest and 21% of respondents with low environmental interest have considered the long use time of the garment when making purchases (Niinimäki 2011).

Impulse shopping is one common practice in the field of fashion; 24% of all consumers felt that their latest garment purchase was based on impulse. Yet the value base of consumers also has an effect on this; 19% of consumers with a higher interest in environmental issues said that their latest garment purchase had been an impulse buy, compared to 31% of consumers with a lower environmental interest. Some consumers commented that they had bought a cheap, low-quality garment on impulse and had not even expected it to last for a long time. Consumers do not give as much thought to their purchases of cheap garments (Niinimäki 2011).

Very often, impulse shopping resulted in the wrong purchase decision. Consumers noticed that the garment did not fit in a satisfying way, the color was strange or wrong, the material felt uncomfortable in use, or the garment did not match their existing wardrobe. These kinds of garments purchased in error might end up not being used at all. Accordingly, cheap prices tempt consumers to make impulse purchases, which often result in dissatisfaction and quick garment disposal (Niinimäki 2011).

12.6 TEXTILE AND CLOTHING WASTE

The amount of textile waste is growing in all Western countries, which has a direct correlation with the increase in fashion consumption and the shorter use time of garments. Most consumers donate some of their old clothes to charity and do not consider this as disposal. However, we know that more and more of those donated garments end up in landfill due to their low quality. Charity shops receive more donations than they can actually use. Therefore, while retail markets are full of cheap garments, the secondhand markets for reused clothes are also saturated. In Finland, the secondhand charity shop U-landshjälp från Folk till Folk i Finland rf received over 8 million kilograms of old clothing in 2010 (Finland has a population of little over 5 million), an increase of 3.5% over the previous year (Keski-Uusimaa January 28, 2011; UFF 2011). The largest collection rate relative to the population was in the wealthy municipality of Kauniainen (UFF 2011). Mounting disposal costs pose a big problem for the charity organization, as does the cost of the extra work involved in sorting out the low-quality garments from the donations.

Donated garments end up in charity shops. Some of these garments are resold to consumers and some are transported to destinations such as Africa. In Africa, this invasion of Western garments has largely destroyed local textile and garment production. In this way, Western consumers have passed on their consumption problems to the developing world and simultaneously helped to destroy the cultural value of local production and textile traditions (Niinimäki 2013).

Estimates of the amounts of textile and clothing waste vary depending on calculation method. In Finland, it has been about 17 kg/person and in Sweden, around 24 kg/person (Tojo et al. 2012). Of course, this correlates directly to consumption levels in general. The American consumer disposes of 31 kg of textiles and clothing each year, most of which end up in landfill (about 85%), and it is estimated that in the United Kingdom, over 900,000 million items of clothing are thrown away each year (Brown 2010 in Tojo et al. 2012). Finnish consumers discard nearly 17 kg of textiles and clothing/person a year, 75% of which end up in landfill (Moilala 2013).

Fashion markets are oversaturated, and because of the extremely effective mass manufacturing system, the world is full of not only new fashion items and fashion shops but also unsold clothing. Discount sales in fashion shops seem to be a permanent phenomenon. In addition, not all new garments even enter the market; some go to landfill directly from the factory due to quality imperfections. Furthermore, some garments are never sold to consumers from the shop because there are too many offerings on the market. In fact, market oversaturation means that it is no longer possible to sell all produced garments to consumers. No one actually seems to know how much preconsumer waste and unsold garments end up in landfills. It is estimated that these unsold garments may account for as much as 5%–10% of total fashion production. Fashion companies try to maintain their brands' reputations in this risky business and would rather destroy the unsold garments than discount them and release them onto the market in overly large quantities.

12.7 CLOTHING DISPOSAL

Female consumers are generally more involved in fashion products (O'Cass 2004), and there exists a positive relationship between being female, fashion interest, and frequent clothing consumption and disposal (Lang et al. 2014), and therefore, it is worthwhile to study female consumers' reasons for disposing of clothing in more detail. A study of product–service systems (PSS) in fashion in Finland (Armstrong et al. 2014) included a questionnaire that investigated clothing disposal by inviting fashion-oriented females to fill out an online survey. Seventy-seven respondents between the ages of 24 and 67 filled out the questionnaire.

When the respondents were asked to estimate the amount of money they used on clothing per month, 41 answered less than EUR 100, 26 between EUR 100 and 200, and 10 between EUR 200 and 500. When asked about the frequency of clothing disposal, 3 reported that they did so monthly, 38 a couple of times a year, 20 once a year, and 16 respondents more seldom.

When the respondents were asked about the practical reasons for clothing disposal, the most important reason was that the clothing was damaged or had quality problems (see Table 12.3).

Furthermore, some of the survey respondents pointed out that they possessed so many garments that they had storage problems. Quotes from the questionnaire are presented in the following text:

> A real space problem. I have too many similar items of clothing, so I can't really keep track of what I have in the closet anymore. I should clear it out.
> No space in the closet anymore.
> Too little space to store all the clothing items I own.

When asked about other reasons for disposing of garments, the respondents stated that the most important reasons were "changes in own style" and "making space for new items" (see Table 12.4). Other studies point out the same. "We own more and more garments and use them less frequently." In the United Kingdom, a study showed that almost half of people's clothes have sat in a closet without being used during the last 12-month period. It is estimated that this means 2.4 billion items in the United Kingdom alone. Most of these unused clothes are owned by young consumers aged 25–34 (Belz and Peattie 2011, 125). Another study from the Netherlands showed that the average piece of clothing is owned for 3 years and 5 months. During that time, it is worn for only 44 days and laundered after being used for 2.4–3.1 days (Uitdenbogerd 1998, cited by Fletcher 2008).

TABLE 12.3
Reasons for Clothing Disposal

Reason for Clothing Disposal	Respondents
Damaged/quality problems	36
The lack of use	11
Wearing out	11
Uncomfortable in use	7
Needs too much maintenance	7

TABLE 12.4
Reasons for Clothing Disposal

Reasons for Disposing of the Garment	Most Important Reason 1	2	3	4	5	6	Less Important Reason 7	Mean
Changes in own style	18	15	11	5	8	9	2	3.07
Making space for new garments	8	18	16	11	14	3	0	3.2
Bored	6	15	12	21	8	5	4	3.58
Changes in fashion trends	2	8	5	10	14	16	10	4.75
Giving garment to charity	26	9	6	10	4	10	8	5.22
Season has changed	3	0	8	8	13	14	18	5.22
No emotional attachment to the garment anymore	4	4	7	3	7	14	31	5.44

Consumers dispose of their products not only because they want to make space for new purchases but also because new trends and fashions make products fall out of style, and additionally, consumers need to experience changes in their emotional levels. Consumers are actively seeking novelty and at the same time evaluating their appearance and the product world in a social context. The increase in waste streams can thus be understood as representing failed person–product relationships in the context of sustainable development (Chapman 2005, 20).

Other earlier studies have shown that low quality is one important reason for disposing of garments. In a study by Niinimäki (2014), the main determinants of clothing dissatisfaction were low durability and especially the poor quality of garments. Consumers pointed out that some garments might be used only until the first wash, after which the garment has lost its fit, size, or color or the material simply looks old after laundering. The first laundering seems to be the critical stage at which consumers experience the quality of the product. Accordingly, the experienced quality in use is one of the most important determinants of a garment's life span. A number of examples provided in Niinimäki's study (2014) show that sometimes contemporary garments do not even last to their first wash; some garments came apart even before that time. Garments' low durability and especially weak maintenance quality seem to be key determinants of consumer dissatisfaction, and they lead to product disposal. Also in this study, respondents had much to say about low quality, such as "the garments from fast fashion chains do not last even through the first laundering stage." Accordingly, it is of paramount importance to improve the quality and durability of garments to enable consumers to extend the life span of garments and make greater use of them.

12.8 CONSUMER VALUES AND CONSUMPTION

Most consumers in the Western world are well aware of the environmental impact of industrial production, but not of the impacts of current fashion consumption. Regarding sustainable development, consumer choices and the importance of environmental aspects in consumers' everyday purchasing decisions should be the center of attention (Jalas 2004). Yet a consumer's choices are somewhat irrational and not always well connected to his or her values. A consumer fulfills deep inner motivations and unconscious needs by consuming. Consumption has two kinds of functions when answering a person's needs, targets, and values: the consumer can try to achieve individual or collective benefits by consuming. Ethical products may be an answer to either individual motives or collective benefits for an ethical consumer. Individual benefits involve issues such as price, quality, saving of time, and purchase convenience (Moisander 1991).

Clavin and Lewis (2005) argue that a consumer who takes ethical issues into consideration behaves according to his or her ethical values, and he or she realizes these values in consumption behavior even if the behavior does not reflect well on him or her. This kind of consumer has committed himself or herself to a social value base. The consumer's ethical awareness is high, and he or she knows which enterprises function ethically. Ethical decision-making relates to the consumer's social orientation, ideals, and ideology. According to Freestone and McGoldrick (2007), social motivators represent a stronger lever for ethical behavior than personal ones. Ethical consumption can create an individual, symbolic feeling of advantage that is linked to a certain lifestyle or expression of personal identity and other social values (Moisander 1991).

Interest in environmental issues and ethical values are strong drivers for purchasing eco-clothes, eco-materials, recycled clothing, and ethically made garments. Among *ethical hard-liners*, a strong personal ideology is a prioritized value in clothing purchase decisions, even more important than one's own identity or aesthetic values. However, the *ethical hard-liners* represent only a niche market (about 9% of consumers). For all consumers, quality and aesthetics are highly important when purchasing clothes and fashion, even eco-clothes, and it is said that environmental aspects can only add value to an otherwise functional and aesthetically attractive product (Niinimäki 2010).

An ethical value base and an interest in environmental issues make consumers select garments more carefully, invest in good quality (Niinimäki 2009), use garments longer, and engage in less impulse shopping behavior (Niinimäki 2014). The most effective means of improving the current unsustainable situation might be to educate consumers to develop their own value base, as a strong environmental and ethical value base seems to be the best way of transforming consumer behavior toward sustainability.

12.9 INCREASING ENVIRONMENTAL INFORMATION

Consumers say that it is difficult to find information about products' environmental impacts and manufacturing ethicality (e.g., Niinimäki 2009; Oksanen 2002). Finding ethical information from the consumer point of view is problematic. Because consumers do not have access to this information, they still select products on the basis of price, appearance, design, convenience, ergonomics, and functionality (Worcester and Dawkins 2005). Younger consumers and the group of higher-educated consumers are also suspicious of the information companies give about their ethicality (Oksanen 2002).

In Niinimäki's study (2009), the respondents evaluated the best sources of environmental and ethical information in textiles and clothing. The most reliable were standardized environmental labels (Table 12.5). The second best source comprised information from the authorities,

TABLE 12.5
Best Information Sources

Information Source	Respondents (%) Totally Agreed
Standardized eco-labels	41.4
Consumer offices	22.5
Civic organization	19.7
Public authorities	19.3
Manufacturers	10.7
Media	5.8
Retailers	3.7
Importers	2.5

civic organizations, and the consumer authority. The respondents rated the reliability of the mass media as an information source only at the same level as companies, producers, importers, and retailers.

It confuses consumers that a company might have one ethical line and, at the same time, produce unethically. Accordingly, as producers offer contradictory information, it can be said that they do not help consumers to make ethical purchase decisions. Consumers cannot easily find ethical or environmental information on different products. At the same time, the vast amount of different kinds of information on textile products makes it very difficult to make informed comparisons. The public has recently shifted its focus toward the ethical production of clothes, and this information has become more common in the textile area.

Consumers have a strong need to acquire environmental and ethical information on the fashion field. Table 12.6 shows consumers' wishes to find environmental information in the future (Niinimäki 2009). They strongly supported all the existing labels: the Nordic swan and European environmental labels, as well as the fair trade logo and ethical production labels. This reveals that producers simply must provide more ethical and environmental information to consumers. At the same time, researchers have to develop further environmental labeling so that it is easy for the consumer to evaluate products on the basis of sustainability.

Even though consumers are worried about the environmental impacts of fashion production, these worries do not easily lead to action: actualization of ethical interest in the form of ethical purchasing decisions is complex, particularly so if consumers should decrease their total fashion consumption. The moral norm-activation theory of altruism by Schwartz (1973) defines a precondition in activating personal norms into action. In Schwartz's theory, the assumption is that environmental quality is a collective good, which activates consumers to act in a proenvironmental way. By acquiring information about the life cycles of different products and their environmental impact, consumers will wisely select products with a lower environmental load. This approach to sustainability also emphasizes the consumer's responsibility and hypothesizes that the number of ethical products available on the market will increase simply by offering more information to consumers. Hitherto, this hypothesis has not come true in practice (Niinimäki 2010). However, Western consumers are

TABLE 12.6
Information on Garments

Information on the Garments	Respondents (%) Totally Agreed
Production country	82
Information on ethical production	66
Fair trade label	63.5
Nordic environmental label (*the swan*)	57.6
Information on the environmental impact of production	57.5
Information on how to dispose of the garment	47.3
European environmental label (*the flower*)	46.9
Carbon footprint	41
LCA (production + use + disposal)	40.7
Water footprint	36.9
ISO14000 standards for minimizing environmental impacts of the company	35.5
Environmental impact of the use phase (e.g., laundering, energy use, and water use)	31.3

most eager to find more background information about fashion and have the means to do their own research on the background values of a product.

Consumers themselves point out that the best way to affect fashion consumption is to provide more environmental and ethical information (41.6%). That said, green tax increases and other control measures implemented by the public authorities also received strong support (37.4% agreed) in an earlier study (Niinimäki 2009). In conclusion, consumers still want to have the power and possibility to weigh their own individual purchase decisions, but at the same time, they wish to have as much help as possible from producers, legislation, and authorities.

12.10 TRANSFORMING CONSUMPTION

One of the most important factors in the environmental impact of the use phase is the garment's lifetime. Currently, garments are far cheaper relative to household incomes than a few decades ago. As textile and clothing prices have fallen, consumers possess greater amounts of cheap impulse-bought garments and low-quality textiles. These kinds of garments are often used for a very short time and are easy to discard. Extending the life span of garments is one of the most critical issues for sustainable development.

Contrary to the current economical and industrial system, product durability and long-term use are prerequisites for sustainable consumption (Cooper 2005). To slow down consumption, it is important to invest in good quality and durability as well as in lasting style and materials that retain their aesthetic appeal over time. In this regard, services that aim to extend the use time of garments offer value in the sustainable development context. One challenge in the current system is how to design products added with services that encourage consumers to adopt more environmentally responsible behavior, extending the use time of garments.

When focusing on sustainable consumption, the following issues should be considered:

- Purchasing fewer garments
- Investing in meaningful garments (promoting emotional bonding)
- Investing in durable garments, more classical style and high quality
- Investing in eco-materials and eco-labels
- Extending garments' owning time and using them more frequently
- Washing less, letting garments rest, and airing between uses
- Maintaining and repairing garments
- Using services to intensify use and to extend the use time (e.g., repair, upgrading) (Niinimäki 2013)

We should critically examine our own consumption habits and our relationships to the material world. We should build deeper relationships with the items we own and take good care of them. We should love our garments, keep them longer, maintain them, upgrade them, and simply cherish and love them. As one of the respondents commented in the 2009 survey (Niinimäki 2009),

> In our consumption behavior, we should go back to the time when we bought little, but the things we bought were expensive and good. Now, cheap products stand in the way of realizing this ideal.

12.11 WHOM TO BLAME?

Consumers do not necessarily see a connection between their own consumption patterns and environmental impacts and ethical problems in the textile and clothing business. Consumers think that the brand and the company behind it should take responsibility for environmental and ethical issues in the textile and clothing business: consumers do not differentiate between all the suppliers and subcontractors in the manufacturing chain, from the fiber producers to the manufacturers of the final product (Nordic Fashion Association 2009).

In the 1990s, the public authorities believed that increasing information about sustainable products and product life cycles would encourage consumers to act wisely and choose products with a lower environmental load. Providing more information would thus enable decreasing the environmental load of consumption and industrial production. A key responsibility was thereby shifted onto the shoulders of the individual consumer. It was also presupposed that this would serve to increase the number of ethical products on the market. A majority of consumers feel that their values are based on ethicality, but the authorities have given too much power and, at the same time, too much responsibility to individual consumers.

Many consumers feel that they now bear an overly heavy responsibility and that they are not to blame for the current unsustainability of the fashion field. Consumers even wish that products would be automatically optimized to minimize their environmental impact so that no extra environmental labeling would be needed (Niinimäki 2009).

The following is a quote from Niinimäki's 2009 study:

> Producers have to carry the responsibility. In shops there should only be ecological and ethical clothes and other products. It is incomprehensible that now the responsibility has been pushed onto consumers and while maximizing profits we have ended up in a situation where consumers have to separately demand ethicalness and "ecologicalness." Enterprises should somehow be forced to follow ethical and ecological principles with the help of legal institutions and laws, and these have to be strict enough.

When discussing the current situation with manufacturers, they complain that consumers are only seeking for low prices and trendy looks, and sustainability and quality issues are not at the core of consumer choices. This might be true. Although price is currently not as obviously connected to the quality of a product as it was earlier, consumers still expect higher prices to mean higher quality. Producers play with cost and quality in order to arrive at the desirable end price for the product, one hopefully cheaper than competing companies' prices, and this has created a situation where quality in the fashion field can only be average or lower (Jackson and Shaw 2009). Jackson and Shaw (2009, 125) claim that consumers are no longer seeking the best quality; they are quite happy with lower prices if the product performs at the minimum level and manufacturing standard. However, the authors (Jackson and Shaw 2009, 125) also point out that consumers forget about the price of the product shortly after purchase, while they remember its quality for a long time. This value resulting from the experienced quality of the product is reflected as trust in the producer's brand name and leads to repeat purchases of the brands that have given the consumer a quality experience.

Consumption habits and manufacturing practices are tightly linked to each other, and we have to change the whole system to create a new and more sustainable balance in this industry. As previously discussed, the value aspect of product satisfaction shows that there are nonmonetary values that are important for the consumer that could change the current business thinking in fashion. If businesses were to focus more on this value creation, they would have the opportunity to create new and more sustainable business models and thinking inside this industry.

12.12 REDIRECTIVE FASHION FOR SUSTAINABLE CONSUMPTION

How should we create and design fashion for sustainable consumption? How can we create farsighted and future-oriented sustainable design that can change consumers' behavior patterns toward more sustainable ones? In this way, design for sustainability can be a redirective practice that aims for sustainable consumption.

According to Fry, "every design decision is future decisive" (2009, 211). Every designed object transforms us: our knowledge, habits, and practices as well as our emotions and well-being.

Understood in this way, sustainable design could be or should be a redirective practice (Fry 2009, 53), which aims for sustainable practices in consumption.

A deeper understanding of people, consumption, and products and their complex meanings and interplay enables a better comprehension of how consumption today may affect tomorrow (Hodges et al. 2007, 342). The environmental and ethical issues in clothing manufacturing are complex and traditionally have been seen through a very limited lens by primarily focusing on eco-materials or ethical manufacturing principles. A more holistic approach and additional knowledge of consumers' relationships with products in the context of sustainable development are needed. Tukker et al. (2006) proposed that new knowledge about consumers' activities and choices and their connection to environmental issues is needed. They highlighted that this consumer-centered knowledge should be used in design to foster more sustainable tendencies in consumption (Tukker et al. 2006).

Products configure consumer needs and use patterns; hence, design can be said to be *practice oriented*, creating certain everyday practices and consumption behaviors (Shove et al. 2007, 134–136). Current design and manufacturing systems stimulate only consumerism and disposable products (Walker 2007, 5). In this way, designers and manufacturers are also responsible for the unsustainable consumption behavior of consumers. Therefore, designers have to evaluate how each design decision affects a consumer's consumption patterns. Understood in this way, sustainable design can be a redirective practice that aims for sustainable practices in consumption (Fry 2009, 53).

Consumption and consumerism lie at the core of Western societies. Consumption is an important function in people's everyday life. Current consumption patterns are strongly connected to current industrial manufacturing systems, economic systems, and the underlying economic values supporting this unsustainable system. Consumption and purchasing situations often involve a strong emotional experience for consumers. Therefore, we should also create systems that offer other kinds of emotional experiences and satisfaction than those provided by buying new stuff. For example, such a system could be introduced through a new kind of radical, holistic, and system thinking that includes PSS, for instance (Niinimäki 2011).

It is extremely difficult to sustain the value of a product in a temporary context. Walker and Chaplin (1997) identify the following values in artifacts: monetary or exchange value, use value (practical functions), artistic value (aesthetic quality), and personal or sentimental value (emotional). In some cases, the aesthetic value may last if the product is not too fashionable and it ages in an aesthetically pleasing way. Furthermore, personal memories can increase the emotional value of the product.

How then can a designer create a product or person–product relationship that has sustainable value? Positive emotions toward a product offer possibilities to elicit commitment to and bonding with this product, so that the consumer will cherish and take good care of it. This is a valuable aspect of sustainable development and provides the possibility to extend the product's life span.

The challenge in extending product lifetimes lies in achieving enduring satisfaction with the product also at the emotional level. Emotionally durable design aims for deeper understanding of consumer experiences as well as for satisfying use experiences. It is important to offer good product performance in those attributes that are important to the consumer and enable the long-term use of product. Deep emotional satisfaction is one way to postpone product disposal and thus extend the product's life span.

Consumer-based quality attributes, which are experienced in use situations, are most significant when defining the satisfaction attributes in long-term use. But merely increasing the intrinsic quality of the product is not enough. Designers have to fulfill consumers' emotional needs also. At this level, it is most important to identify the emotional and symbolic meanings of products. The key to facilitating deeper product relationships thus lies in a better understanding of the consumers, and an empathic approach is essential for success.

Products' value should be defined in the use context, not only at the point of purchase (Park and Tahara 2008). Products' prices should be connected to their quality and durability. Furthermore, the product value should be connected more deeply to consumer satisfaction and sustainability aspects, for example, to the experienced quality and long life span of the product. An earlier study has shown (Niinimäki 2011) that most of the consumers (83%) were ready to pay more for higher quality and a longer product lifetime in garments if it would be possible to estimate these aspects at the time of purchase. This consumer value base is an opportunity to create durable design.

12.13 FULFILLING CONSUMERS' CHANGING NEEDS

One of the biggest challenges in fashion is the consumers' constant need for newness and fast satisfaction of desires. Clothing choices are strongly linked to our identity building, and our needs with regard to our appearance are changing all the time. As consumers' aesthetic preferences change over time, the question is how to avoid the psychological obsolescence of garments. And how then can we offer change to consumers in a more sustainable way? At this level, it is not enough to merely provide better product quality and durability. Better product satisfaction can be achieved by meeting a consumer's individual needs, for example, through a unique design approach or *made-to-measure* strategies. Furthermore, designing, upgrading, modifying, and lending services can offer the needed emotional change experience for the consumer to postpone garment disposal.

Designing and constructing PSS create possibilities to fulfill consumers' different and changing needs in a less materialistic and more sustainable way. PSS even open up opportunities to dematerialize consumption by shifting the focus to satisfying consumer needs and wishes instead of manufacturing new products. For instance, the consumer can experience emotional satisfaction and change by borrowing the product instead of owning the product himself or herself. High-quality products can be targeted for intensive utilization, for example, renting and leasing. Recently, new kinds of clothing clubs have emerged where consumers can borrow a certain number of garments each month and change these garments next month. In this approach, the focus is not on owning the product, but rather on the function or the experience the product can offer to the consumer.

Products meant for intensive and long-term use (e.g., in renting services) must be made of high-quality and durable materials. Furthermore, they are investments for the renting company, and this approach decreases the total environmental impact of manufacturing and consumption while decreasing the material throughput in the system.

12.14 FOR SLOW OR FAST CONSUMPTION IN THE FUTURE?

When thinking about how to develop the current existing systems of design–manufacturing–consuming fashion, we can basically adopt two approaches. While it is true that nowhere near all consumers are willing to reduce their consumption and thereby create their identity without external symbols, perhaps, we even have to approach consumers with different strategies. Through material and design innovations and production processes, it is possible to produce clothes with different quality and life cycles and target these at different consumer groups according to their values and consumption habits. Perhaps there should be fast fashion and slow fashion production systems and different taxation and labels for these.

Slow fashion would be designed according to an ethical consumer's values. These high-quality, durable clothes would be made from sustainable materials. The production lines would comply with ethical principles and perhaps even employ local production. The style would be more classical and longer lasting in terms of design, colors, and print. The material choices would be optimized so that the clothes need very little maintenance, especially washing and ironing, and the materials and clothes could also be reused and even recycled into new textile material. Producers could also offer new service concepts, such as repairing, recycling, swapping, renting, and leasing clothes.

Fast fashion would be directed toward the younger generation, and it would be based upon their need to have new experiences and build their identity with a trendy look and fashionable items. This might mean new sustainable clothing materials that are optimized for the real lifetime of the product. Perhaps they are intensively used for only 6–12 months and then recycled into new materials. They are mainly made from recycled rather than virgin materials. A good recycling system could be introduced for these clothes, or they could be biodegradable. Perhaps they would not need to be washed at all during their short lifetime; airing would suffice.

12.15 CONCLUSIONS

This chapter presented perspectives on current consumer behavior in the fashion field and how it is tied not only to the current design and manufacturing system but also strongly to the unsustainable economical and marketing systems of fashion.

In the future, we all have to satisfy our needs in a longer-lasting manner instead of by consuming more and more products. This represents a huge step and involves changes not only in our behavior but also in industrial design and manufacturing practices. Balancing prices between sustainable and unsustainable products using green taxation might help consumers to behave more rationally. As one of the respondents commented, "cheap clothes encourage consumers to make less rational choices"; in other words, consumers do not invest in more expensive, higher-quality and more durable clothing and do not use garments longer. At the present, fashion marketing focuses strongly on consumers' unsustainable desires, rapid changes in fashion, and garments' short use times.

Furthermore, providing more environmental and ethical information to consumers would help them to compare products against their own value base and might help them to make wiser purchase decisions. That said, relying only on more information is a difficult task and might not help consumers to act more responsibly or, for example, consume less. Consumers still want to have the power and possibility to weigh their own individual consumption decisions, but at the same time, they wish to have help from producers, legislation, and authorities.

It is of paramount importance to rethink the industrial system in a totally radical way and create new designs and manufacturing–marketing strategies that aim for sustainable consumption practices. In this transformation process, some help from the public authorities is needed in the form of legislation or taxation. Furthermore, new research information from the consumer side helps manufacturers and marketers in this change process by pointing out consumers' values, expectations, and wishes. This research might even provide examples of how to fulfill consumers' desires in a more sustainable way. Therefore, new research information might open up new opportunities to do fashion business differently, in a more sustainable and redirective way. It is a challenging task to change the current industrial system, but by focusing on new value creation and consumers' deeper needs, we can create fashion that satisfies consumers longer—clothing that we can love, become attached to, and maintain in good condition.

REFERENCES

Armstrong, C., Niinimäki, K., Kujala, S., Karell, E., Lang, C. 2015. Sustainable product-service systems for clothing: Exploring consumer perceptions of consumption alternatives in Finland. *Journal of Cleaner Production* (in press).

Belz, F.M., Peattie, K. 2011. *Sustainability Marketing: A Global Perspective*. John Wiley & Sons, London, U.K.

Chapagain, A.K., Hoekstra, A.Y., Savenije, H.H.G., Gautam, R. 2006. The water footprint of cotton consumption: An assessment of the impact of worldwide consumption of cotton products on the water resources in the cotton producing countries. *Ecological Economics* 60(1):186–203.

Chapman, J. 2005. *Emotionally Durable Design: Objects, Experiences, Empathy*. Earthscan, London, U.K.

Chen, H., Burns, L.D. 2006. Environmental analysis of textile products. *Clothing and Textile Research Journal* 24(3):248–261.

Clavin, B., Lewis, A. 2005. Focus groups on consumers' ethical beliefs. In: Harrison R. et al. (eds.) *The Ethical Consumer*. Sage Publications, London, U.K., pp. 173–188.

Cooper, T. 2005. Slower consumption: Reflections on products' life spans and the 'throwaway society'. *Journal of Industrial Ecology* 9(1–2):51–67.

Defra. 2008. Sustainable clothing action plan. http://www.defra.gov.uk/environment/business/products/roadmaps/clothing/documents/clothing-action-plan-feb10.pdf (Accessed October 1, 2011).

Desmet, P.M.A. 2009. Product emotion. In: Schifferstein H., Hekkert P. (eds.), *Product Experience*, 2nd edn. Elsevier, San Diego, CA, pp. 379–397.

Fletcher, K. 2008. *Sustainable Fashion and Textiles*. Earthscan, London, U.K.

Freestone, O., McGoldrick, P. 2007. Motivation of the ethical consumer. *Journal of Business Ethics* 79:445–467.

Fry, T. 2009. *Design Futuring: Sustainability, Ethics and New Practice*. Berg, Oxford, U.K.

Hodges, N.N., DeLong, M., Hegland, J., Thompson, M., Williams, G. 2007. Constructing knowledge for the future: Exploring alternative models of inquiry from a philosophical perspective. *Clothing and Textiles Research Journal* 25(4):323–348.

Islam, K., Mahmud, K., Farui, O., Billah, M.S. 2011. Textile dying industries in Bangladesh for sustainable development. *International Journal of Environmental Science and Development* 2(6):428–436.

Jackson, T., Shaw, D. 2009. *Mastering Fashion Marketing*. Palgrave Macmillan, New York.

Jalas, M. 2004. Kuluttajat ympäristöjohtamisen kohteina ja osapuolina. [Consumers as subjects and participants of environmental management]. In: Heiskanen E. (ed.), *Ympäristö ja liiketoiminta. Arkiset käytännöt ja kriittiset kysymykset (Environment and Business—Everyday Practices and Critical Questions*, in Finnish]. Gaudeamus, Yliopistokustannus Oy, Tampere, Finland, pp. 211–226.

Joergens, C. 2006. Ethical fashion: Myth or future trend? *Journal of Fashion Marketing and Management* 10(3):360–371.

Keski-Uusimaa January 28, 2011. UFF kierrätti miljoonia kiloja vaatteita (UFF recycled millions of kilos of clothing).

Lang, C., Armstrong, C.M., Brannon, L.A. 2013. Drivers of clothing disposal in the United States: An exploration of the role of personal attributes and behaviors in frequent disposal. *International Journal of Consumer Studies* 37(6):704–714.

Lewis, H., Gertsakis, J. 2001. *Design + Environment: A Global Guide to Designing Greener Goods*. Greenleaf, Sheffield, U.K.

Markkula, A. 2011. Consumers as ecological citizens in clothing markets. Doctoral dissertation. Aalto University, Helsinki, Finland.

Moilala, S. 2013. *Tappajafarkut ja muita vastuuttomia vaatteita* (Killer jeans and other irresponsible garments). Into, Helsinki, Finland.

Moisander, J. 1991. Sosiaaliset arvot luomutuotteiden kulutuksessa (Social values in organic food consumption) Helsinki Business School, Consumption and Environment Project. Helsinki Business School publication D-144, Helsinki, Finland.

Niinimäki, K. 2009. Consumer values and eco-fashion in the future. In: Koskela M, Vinnari M (eds.), *Future of the Consumer Society. Proceedings of the Conference "Future of the Consumer Society"*, May 28–29, 2009, Tampere, Finland. Finland Futures Research Centre, Turku School of Economics, Turku, Finland, pp. 125–134. Available at http://www.tse.fi/FI/yksikot/erillislaitokset/tutu/Documents/publications/eBook_2009-7.pdf

Niinimäki, K. 2010. Eco-clothing, consumer identity and ideology. *Journal of Sustainable Development* 18(3):150–162.

Niinimäki, K. 2011. From disposable to sustainable. The complex interplay between design and consumption of textiles and clothing. Doctoral dissertation. Aalto University, Helsinki, Finland. Available at https://aaltodoc.aalto.fi/handle/123456789/13770 (Accessed January 1, 2014.)

Niinimäki, K. 2013. *Sustainable Fashion: New Approaches*. Aalto Arts Books, Helsinki, Finland. Available at https://aaltodoc.aalto.fi/handle/123456789/13769 (Accessed May 30, 2014.)

Niinimäki, K. 2014. Sustainable consumer satisfaction in the context of clothing. Vezzoli, C., Kohtala, C., Srinivasan, A., Xin, L., Fusakul, M., Sateesh, D. and Diehl, J.C. (eds.). In: *Product-Service System Design for Sustainability*. Greenleaf, Sheffield, U.K., pp. 218–237.

Nordic Fashion Association. 2009. This is NICE: NICE 10-year plan of action. December 2009. Available at http://nicefashion.org/files/10_Year_Plan.pdf (Accessed January 1, 2010).

O'Cass, A. 2004. Fashion clothing consumption: Antecedent and consequences of fashion clothing involvement. *European Journal of Marketing* 38(7):869–882.

Oksanen, R. 2002. Suomalaisten kuluttajien suhtautuminen eettiseen kaupankäyntiin ja reilunkaupan tuotteisiin (Finnish consumers' attitudes towards ethical markets and fair trade products). Jyväskylän yliopisto, Jyväskylä. Available at http://www.reilukauppa.fi/cms/img/text/283/Reetta_Oksanen,_pro_gradu_2002.pdf (Accessed November 1, 2008).
Park, P., Tahara, K. 2008. Quantifying producer and consumer-based eco-efficiencies for the identification of key ecodesign issues. *Journal of Cleaner Production* 16:95–104.
Richins, M. 2009. Consumption emotions. In: Schifferstein H, Hekkert, P (eds.), *Product Experience,* 2nd edn. Elsevier, San Diego, CA, pp. 399–422.
Solomon, M., Bamossy, G., Askegaard, S. 2002. *Consumer Behaviour. A European Perspective.* 2nd Ed. Pearson Education, Edinburgh, UK.
Solomon, M., Rabolt, N. 2004. *Consumer Behavior in Fashion*. Prentice Hall, Upper Saddle River, NJ.
Schwartz, S. 1973. Normative explanations of helping behaviour: A critique, proposal, and empirical test. *Journal of Experimental Social Psychology* 9(4):349–364.
Shove, E., Watson, M., Hand, M., Ingram, J., 2007. *The Design of Everyday Life*. Berg, Oxford, UK.
Tojo, N., Kogg, B., Kiørboe, N., Kjær, B., Aalto, K. 2012. *Prevention of Textile Waste: Material Flows of Textile in Three Nordic Countries and Suggestions on Policy Instruments*. Nordic Council of Ministers, Copenhagen, Denmark.
Tukker, A., Cohen, M., de Zoysa et al. 2006. The Oslo declaration on sustainable consumption. *Journal of Industrial Ecology* 10(1–2):9–14.
UFF, 1/2011 Uutiskirje (*UFF Newsletter*).
Walker, J.A., Chaplin, S. 1997. *Visual Culture: An Introduction*. Manchester University Press, Manchester, U.K.
Walker, S., 2007. *Sustainable by Design. Exploration in Theory and Practice* Earthscan, London, UK.
Water Footprint. Water footprint network. 2008. Available at http://www.waterfootprint.org/?page=files/home (Accessed April 25, 2009).
Weidema, B.P., Thrane, M., Christensen, P. et al. 2008. Carbon footprint. A catalyst for life-cycle assessment? *Journal of Industrial Ecology* 12(1):3–6.
Worcester, R., Dawkins, J. 2005. Surveying ethical and environmental attitudes. In: Harrison R et al. (eds.), *Ethical Consumer*. Sage Publications, Trowbridge, U.K., pp. 189–203.

13 Textile Sustainability
Major Frameworks and Strategic Solutions

Arun Pal Aneja and Rudrajeet Pal

CONTENTS

- 13.1 Introduction 289
- 13.2 Planetary Pulse 291
  - 13.2.1 The Great Unraveling 292
  - 13.2.2 The Great Warming 292
  - 13.2.3 The Great Inequities 293
- 13.3 Strategic Vectors of Sustainability 295
- 13.4 Major Sustainability Frameworks 296
  - 13.4.1 Ecological Footprint 297
  - 13.4.2 Natural Step 297
  - 13.4.3 Natural Capitalism 298
  - 13.4.4 Industrial Ecology 298
  - 13.4.5 Cradle-to-Cradle 299
  - 13.4.6 Biomimetics 299
  - 13.4.7 Zero Emissions Research and Initiatives 299
  - 13.4.8 Planetary Boundaries 300
- 13.5 Role of Textile in Sustainability 300
  - 13.5.1 Control of Growth of Greenhouse Gases 300
  - 13.5.2 Safeguard Water Quality 302
  - 13.5.3 Efficient Energy Utilization 302
  - 13.5.4 Biodiversity Protection and Desertification Prevention 303
  - 13.5.5 Waste Product Management 304
- 13.6 Moving into the Future: A Summary 304
- References 305

13.1 INTRODUCTION

Humanity is passing through tumultuous times particularly as it relates to our relationship with the entity we hold so dear—*Mother Earth*. Earth is unable to provide resources at the rate we are consuming them. Indeed, what we are doing now cannot go on. We live today by borrowing from the future. The living planet report makes clear that more than 30 years ago, the world moved beyond the point it could support its population sustainably. Since then, the population has continued to grow at an *explosive rate* (Figure 13.1), and each person consumes more. Textile consumption is about 10 kg fiber per year per person (Ashby 2012). Global resource depletion scales with the population and with per capita consumption. So today, humanity burns through the resources of 1.5 Earths in a year; thus, it takes the planet a year and a half to regenerate what we use annually.

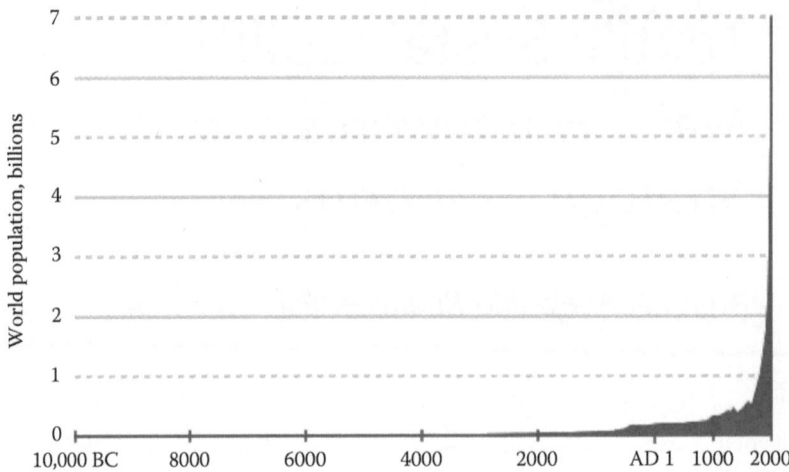

FIGURE 13.1 Graph of the global human population from 10000 BC to AD 2000. The graph shows the extremely rapid growth in the world population that has taken place since the eighteenth century. (From U.S. Census Bureau, Suitland, MD.)

At the current rate of growth, this will rise to 2 Earths by 2030 and 2.5 Earths by 2050—a clear impracticality (Global Footprint Network 2014). Viewed from the perspective of time scale, the planet is 4.5 billion years old (Bide 2012). The very concept of sustainability is about 50 years old or a blip in our collective thinking. Stated differently, we are destroying in an *instant* what has taken millennia to evolve. Future generations will ask, and justifiably so, "How did we let it happen?"

To combat the situation, sustainability has become the battle cry. It is an increasingly popular catchphrase and a movement that completes the spectrum of emotions from euphoria to depression. We know the narrative: ecosystems in decline, species loss, water scarcity, climate chaos, spread of diseases coupled with the resultant social and economic turmoil. However, it simultaneously provides the warm feeling that comes with being green, earth friendly, natural, and ethical. Several authors have tried to define sustainability, the most popular of which was coined in 1987 by the *United Nations' World Commission on Environment and Development—the Brundtland Commission*. In their definition, sustainability means "meeting the needs of the present without compromising the ability of future generations to meet their own needs" (Brundtland Report 1987). A more inclusive definition is provided by the American Institute of Architects. They envisioned sustainability as "the enduring prosperity of all living things."

So what is the role of textile and apparel industries in this debate? The textile and apparel industries operate on a global basis and are considered as one of the major product groups in world trade of manufactured goods as suggested by the World Trade Organization.

The future of textiles and apparel, much like other industries, is facing problems of limits to natural resources, global warming, sustainability issues, and social and political upheaval. The textile and apparel industries are considered among the most environmentally damaging industries being one of the major sources of greenhouse gas (GHG) emission (Eryuruk 2012). Furthermore, the industry is also one of the biggest consumers of water and energy. Additionally, it has been growing rapidly on a global scale, contributing to the environmental issues and wastes associated with it. Fast changing trends have led to premature product replacement. With a focus on growth and profits, it is now time to change that model and include global citizenship competencies to counter the problems around the world.

To become sustainable, the industry needs a dramatic change at all levels starting with design, production, marketing, sales, and promotion (Pal 2014). With growing awareness, sustainability has gained traction in the textile and apparel industries since 1994 and today is becoming even more important. There are several sustainability issues—along the three pillars of the *triple bottom line* approach—that specifically relate to the components of the supply chain for meeting the challenges of fashion logistics, overproduction, unsustainable consumption behavior, and social irresponsibility, hence cannot be ignored (Pal 2014).

In this context, the work in this chapter makes an assessment of a sustainable future for textile and apparel industries based on economic, social, and environmental dimensions along the major emergent patterns highlighted in eight critical sustainability frameworks (viz., ecological footprint, natural step, natural capitalism, industrial ecology, cradle-to-cradle, biomimetic, Zero Emissions Research and Initiatives [ZERI], and planetary boundaries). The chapter, at first, narrates the *big picture* that highlights the global-scale changes in the earth system due to dramatic increase in human activity forming the basis of our planetary pulse. It also highlights the contributions of the textile and apparel sectors in this *big picture*. The subsequent section briefly describes the goals and focus of the three strategic vectors (viz., economic, environmental, and social) of the triple bottom line approach. The next section summarizes the eight critical sustainability frameworks and identifies their strategic foci along the three pillars of the triple bottom line approach. Finally, the chapter concludes with a detailed discussion to motivate and achieve a fundamental mind shift in the textile and apparel industries by identifying various components of nonsustainability. The deeper insights and collective change suggested will not provide solutions to ensure success but rather provide insights and holistic and integrated systems perspective to answer the query in our head—"What must we do?"—to give rise to this major transformation. The narrative that currently frames this industry, and perhaps others, is summed up in a few simple words—"Why is enough for mankind never enough?"

This chapter positions its contributions as a *viewpoint* article. It proposes strategic directions and solutions that if adopted can manifest sustainability in the textile and apparel arena. A fundamental mind shift in these industries by identifying various components of nonsustainability is also suggested. Such deeper insights and collective changes will not only provide solutions to ensure success but will also provide a holistic and integrated systems perspective to give rise to this major transformation.

A major gap in the literature related to sustainability frameworks is its industry specificity—in terms of impacts, strategies, and solutions. In fact, all the frameworks provide the *big picture* in terms of the intrinsic components of nonsustainability; however, at the strategic level, a *one-size-fits-all* approach across industries does not work. As the contributions to nonsustainability are different across industries, the remedies proposed should also be different. Few works (e.g., Allwood et al. 2006; Eryuruk 2012) highlight textile industry–specific scenarios and solutions. However, they do not discuss the strategies and solutions in relation to the major sustainability frameworks. In this regard, the present work provides a holistic picture of the role and adoption of the major sustainability frameworks in the textile and apparel arena. Information gathered and recorded is an outcome of authors' long and enriched association with the textile industry, business, and education at various positions (aggregated over 40 years). The sustainability frameworks adopted in the work are deduced through extensive study of the publications of the originators of each of these.

13.2 PLANETARY PULSE

The growth-oriented society we live in is now out of control and destroying the bases of life itself. The globalized political economy, driven by its need to accelerate growth and measuring success by its rate of growth, is what systems theorists call a *runaway* and chaotic system. It is, in effect, a suicide economy. We're all part of it. What are the intrinsic components of this nonsustainability?

Quite clearly, the major causes of any imbalance in the earth's resilience are either due to (1) decrease in natural capital (discussed in Section 13.2.1), (2) increase in harmful substances in the ecosystem (discussed in Section 13.2.2), and (3) imbalance in resources (discussed in Section 13.2.3).

13.2.1 THE GREAT UNRAVELING

Human history marches on and continues to unfold its 0.15-million-year story of dazzling creativity. The genius of mankind has overcome numerous apparently insurmountable problems through this journey to meet its needs. The agricultural revolution, which took centuries, accomplished the basic task of providing food for an ever increasing number of inhabitants. The Industrial Revolution provided higher standards of living to a large group of people. Now, we confront another quandary; we are destroying the planet that provided us all of what we have today. Macy and Johnstone (2012) call this to be "The Great Unraveling" marked by economic decline, resource depletion, climate change, social division, war, and mass extinctions. Rockström and Klum (2012) term this as the *quadruple squeeze* on humanity's ability to secure long-term sustainable development on planet Earth, particularly the third squeeze resulting in a rapidly eroding resilience of the earth, where we have so far undermined 60% of the key ecosystem services in support of human well-being. This is reflected in

- Ecosystem decline
- Habitat loss
- Species extinction (Figure 13.2)
- Human body burden and chemical stressors of various types and quantity

In the textile and apparel sectors, pesticides and fertilizers are rampantly used in cotton farming and for dyeing of yarns and fabrics. Apart from that, the industry dominates the issue of climate change due to its huge requirement for burning fossil fuels for energy generation purposes. Water consumption due to extensive cotton cultivation can also lead to major environmental degradation and habitat loss (as was observed in the Aral Sea region). Moreover, the growing volume of wastes with the advent of *fast fashion* results in around 30 kg of textiles and apparel per capita to landfill each year (Allwood et al. 2006). In every way, this leads to various forms of ecosystem decline.

13.2.2 THE GREAT WARMING

The atmosphere has a natural supply of *GHGs*. It captures heat and keeps the surface of the earth warm enough for us to live on (Fagan 2008). Absence of greenhouse effect would lead to an uninhabitable planet.

Before the Industrial Revolution, the amount of carbon dioxide (CO_2) and other GHGs released into the atmosphere was in a rough balance with what was absorbed in natural sinks. During the

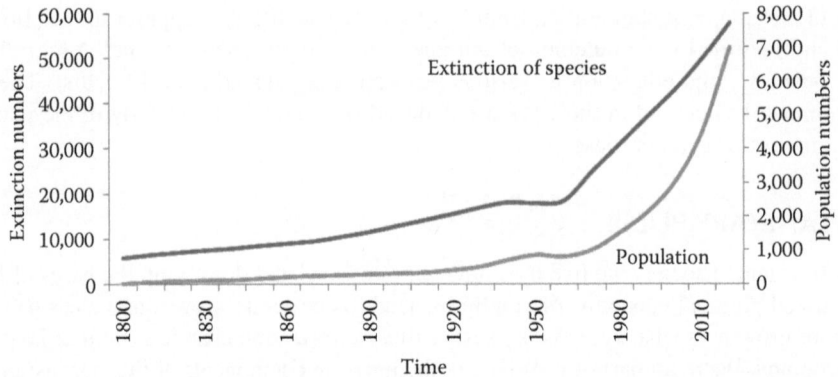

FIGURE 13.2 Price for human consumption. (From Biologicaldiversity.org, http://www.biologicaldiversity.org/programs/population_and_sustainability/extinction_and_population_graph.html, August 05, 2014.)

Textile Sustainability

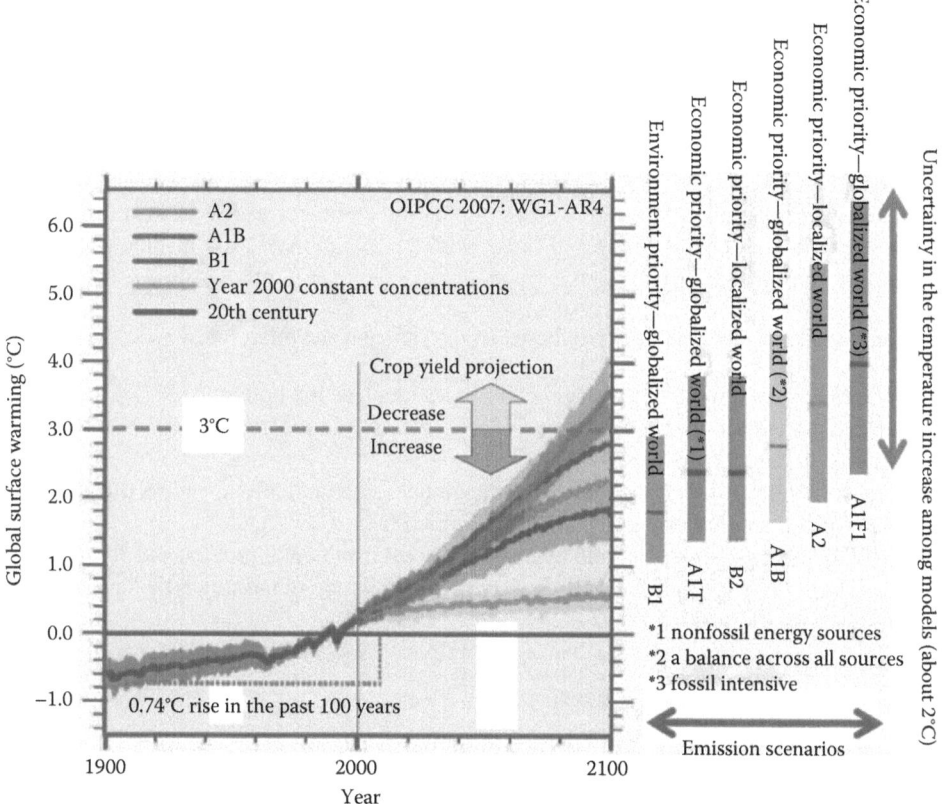

FIGURE 13.3 Average surface temperature. (From IPCC Fourth Assessment Report 2007, Summary for policy makers, https://www.ipcc.ch/pdf/assessment-report/ar4/wg1/ar4-wg1-spm.pdf, August 05, 2014.)

industrial age, large quantities of GHGs started to be emitted due to burning of fossil fuels. This has resulted in an increase in the GHG concentration (CO_2 eq) to 400 parts per million (ppm), with the possibilities of it going beyond the next threshold of 450 mark by 2060, crossing the Stern and Lomborg target (550 ppm) by 2090, and approaching 600 ppm by 2100 (Wheeler 2009). This precondition is risky as it may give rise to 2°C in average global temperature. The result is that the globe has heated up by about 1°F over the past century—and it has heated up more intensely over the past two decades (Figure 13.3). This is necessary to avoid melting of the big ice sheets, which can even raise the sea levels by 15 m (Mitrovica et al. 2009; Hansen et al. 2011).

This great warming is more than just about the heat. It is about the disruption of the global heat engine. More importantly, it is about how the planet responds with resultant economic dislocations creating climate refugees and impacting security of member nations. In this *big picture*, the textile industry is the fifth largest contributor of GHG emission accounting for around 4% of the secondary carbon footprint of an individual in the developed world (Dev 2009).

13.2.3 The Great Inequities

The current North–South disparity in economic development is extreme, while the gap associated with access to resources and information is narrowing. International equity—defined as the ratio of average income in non-OECD to OECD nations—was 0.13 in 2005 (Meadows et al. 1972). The World Institute for Development Economics at United Nations conducted a study that reports that the richest 1% of adults owned 40% of global assets in 2000 and richest 10% of adults accounted for

85% of global wealth. The bottom 50% had a meager 1% of global wealth. The primary measure of income inequality, the Gini coefficient, consists of two groups of countries:

- The first group with 13% of the world's population receives 45% of the world's purchasing power parity (PPP) income. This group includes the United States, Japan, Germany, the United Kingdom, France, Australia, and Canada and comprises 500 million people with an annual income level over 11,500 PPP dollars.
- The second group has 42% of the world's population and receives 9% of the PPP income. This group includes rural China and India, with a population of 2100 million people and income of about 1000 PPP dollars.

Here is another way to view the disparity between the rich and the poor:

- The richest 1% of people in the world receives as much as the bottom 57%.
- The three richest people possess more financial assets than the poorest 10% of the world's population combined.
- Of the world population, 6% own 52% of the global assets and 50% own less than 1%. The whole global assets volume is about 125 trillion USD.
- The 20/80 dilemma, characterized by bulk of the environmental problems so far, is due to the rich minority—20% of the world population—while the remaining 80% have the right to development (Figure 13.4).

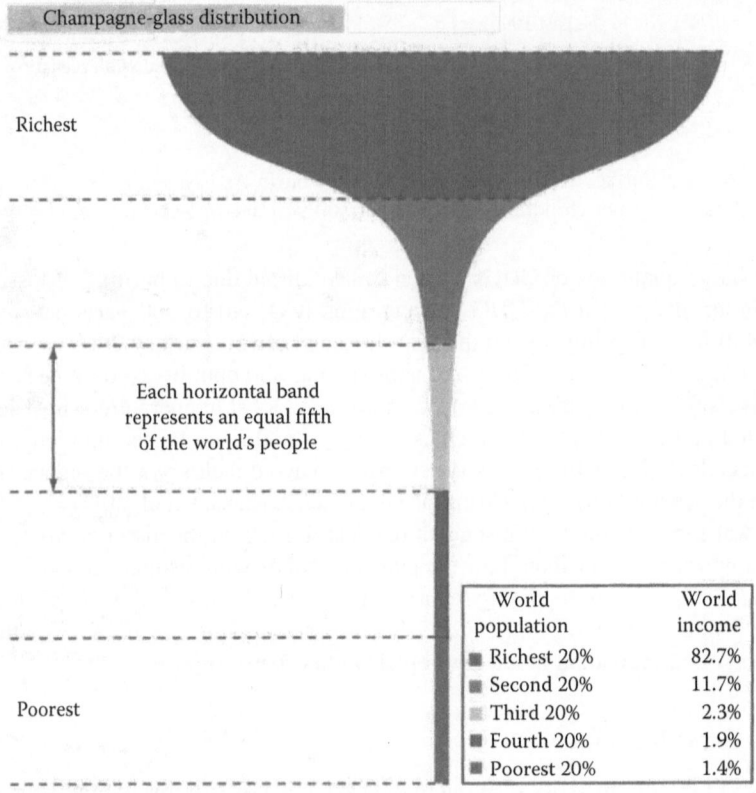

FIGURE 13.4 Global economic disparity. (Adapted from Wade, R.H., The rising inequality of world income distribution, *Finance and Development, International Monetary Fund*, 38(4), 2001.)

With the progressive polarization of the textile and apparel sectors between western countries (focused on innovation and brand retailing) and developing countries (focused on low-cost manufacturing), global economic disparity is increasingly catapulted by the sector as well. For instance, hourly wage in major producing countries in Asia ranges between 0.23 and 0.86 USD/h as compared to 10.03 in Germany and 11.16 in United States (Allwood et al. 2006). This income inequality is a source of social unsustainability cumulatively leading to several factors like nation's degree of income inequality and its rates of homicide, imprisonment, infant mortality, teenage births, and obesity (Meyer and Kirby 2014).

13.3 STRATEGIC VECTORS OF SUSTAINABILITY

Traditionally, we have emphasized economic development above all else in our pursuit to achieve prosperity. All indicators now suggest this to be shortsighted for long-term welfare of humanity—either we pay now or we pay later. As an example, too often in the past, costs of dealing with wastes, effluents, and emissions are not accounted for as costs of products, but instead pushed off onto the environment for people living near the exposure sites to pay later (much later). These costs include increased health risks and polluted air, water, and land. We only consider the lowest cost per unit produced instead of cost per unit consumed, which initiates more waste in the name of profits. One basic analytical approach is to view this process as an interaction among three systems: the environmental (and other resource) system, the economic system, and the social system forming the three strategic vectors of sustainability (Barbier 1987).

Hence, *real* improvement (economic development) cannot occur in the world unless the strategies that are being formulated and implemented are environmentally sustainable over the long term. In addition, they must be consistent with social values and institutions and encourage *grassroots* participation in the development process. Thus, "there will be no sustained development or meaningful growth without a clear commitment at the same time to preserve the environment and promote the rational use of resources." Similarly, to be socially and culturally sustainable, "development must be gauged by the values [which] a society itself or some member thereof, deems to be requisite for its health and welfare." Thus, we must not see environmental, social, and economic challenges as separate and competing but rather as an interconnected whole.

"Environmental sustainability is the process of making sure current processes of interaction with the environment are pursued with the idea of keeping the environment as pristine as naturally possible based on ideal-seeking behavior" (Barbier 1987). Thus, environmental sustainability demands that society designs activities to meet human needs while indefinitely preserving the life support systems of the planet. Hence, mankind must utilize resources (water, materials, energy, etc.) at a rate at which they can be replenished naturally. An unsustainable situation occurs when natural capital is used faster than it is replenished. The long-term results of environmental degradation may lead to catastrophe with possible extinction of human race. This is linked to the concept of carrying capacity, which is the maximum number of individuals of a given species that an area's resources can sustain indefinitely without significantly depleting or degrading those resources.

Social sustainability includes factors of human well-being. These consist of parameters of social justice such as human rights, labor rights, and corporate governance. In addition, social sustainability must provide future generations the same access to resources as the current generation. Thus, sustainable human development is the idea put forward by the Brundtland Commission that the development that promotes the capabilities of present people must not compromise the development of future generations. In the context of human development, environment and natural resources should constitute a strategy of better standards of living much like wealth in which the capability for an increase in social expenditure leads to the well-being of all.

Thus, each system has its own unique set of human-ascribed goals:

Environmental system goals:

- Genetic diversity
- Resilience
- Biological productivity

Economic system goals:

- Satisfying basic needs (reducing poverty)
- Enhancing equity
- Increasing useful goods and services

Social system goals:

- Equity and maturity
- Cultural diversity
- Institutional sustainability
- Social justice and quality of life
- Participation

The general objective of sustainable economic development, then, is to maximize the goals across all these systems through an adaptive process of trade-offs. In contrast, the current conventional consensus on economic development attempts to maximize only economic system goals, and Marxist economics maximize only environmental and social goals.

13.4 MAJOR SUSTAINABILITY FRAMEWORKS

No one is suggesting that for human well-being we stop all human activity. It does not mean no extracting of materials from the earth, or nonproduction of man-made substances or plundering of the natural landscape, but, rather, a need to turn back the current adverse trends of progressively greater environmental havoc by mankind. Our current industrial and economic systems have developed such that practices that have damaged the environment will continue to grow and have a greater impact over time such that it may reach an autononsustaining point, that is, a point where the conditions are nonsustainable. As a consequence of this behavior, we are at the cusp of human development where truly there are "not enough fish in the sea." This growth in the use of resources cannot be a sustainable strategy for a limited spaceship called Mother Earth. Along this context, Rockström and Klum (2012) highlight the risk of progressing beyond the tipping point, which will result in losing the earth's resilience, thus pushing the critical system into an undesired state.

These new pressures—predominantly acceleration of human demands and aspirations from a finite planet and acceleration of negative global environmental trends—have resulted in rejuvenating our mind-set toward a new paradigm that can unleash sustainable solutions and novel ways of implementing a new pathway for sustainable development (Rockström and Klum 2012). There are many thoughtful minds that have postulated a variety of viable alternatives to our current activities resulting in several frameworks as discussed briefly in this chapter. However, apart from the eight discussed in this chapter, there are few other frameworks (like the sustainability framework articulated by the International Finance Corporation and WaterAid sustainability framework) to address the sustainability issue. However, the justification to our selection of the eight is nonquantitative, aimed at providing an understanding to the readers that many frameworks are possible or available but we elicit some of the more common and perhaps widely used ones. Further, those discussed here were originated by noncommercial institutions mainly in universities and research institutes, unlike the rest being documented mostly by for-profit organizations.

13.4.1 Ecological Footprint

Conceived in 1990 by Mathis Wackernagel and William Rees at the University of British Columbia, the ecological footprint is widely used by scientists, businesses, governments, agencies, individuals, and institutions to monitor ecological resource use and advance sustainable development (Wackernagel and Rees 1996).

The ecological footprint has developed into one of the world's major measure of the stress imposed by mankind on nature. This system documents the demand side (footprint imposed by a population comprising an individual, city, business, nation, or all of humanity), that is, how much land and water human population uses including accommodation space, creating required infrastructure (roads, buildings, manufacturing, etc.) and a strategy to absorb the waste (solid, liquid, and gaseous carbon dioxide, methane, etc.). These calculations account for each year's prevailing technology, as productivity and technological efficiency change from year to year. The supply side documents how much biocapacity is available to provide all the services needed by mankind to accommodate the aforementioned demand. The system compares both the supply and demand to evaluate the stress on the plans imposed by human activity. This will help us manage our finite ecological assets judiciously and truly live within our means. This will lead to better management of our ecological assets wisely and take personal and cumulative commitment to provide a system where mankind lives within the natural limits of the bounty provided by the earth.

13.4.2 Natural Step

The Natural Step is a nonprofit organization. It was founded in 1989 by a Swedish cancer research physician Karl-Henrik Robert. It is currently active on a global scale (Robèrt 2002). The Natural Step approach to sustainability begins with four simple scientific principles:

"1. Matter and energy cannot be destroyed (the First Law of Thermodynamics).
2. Matter and energy tend to spread spontaneously (entropy, the Second Law of Thermodynamics).
3. Biological and economic value (quality) of matter is based on in concentration (availability) and structure.
4. Net increases in material quality on the earth are generated almost entirely by the sun-driven process of photosynthesis."

The Natural Step vision of sustainability was further defined to what are referred to as the *four system conditions*. These encompass

1. The systematic increase of concentrations of substances extracted from the earth's crust (e.g., heavy metals and fossil fuels)
2. The systematic increase of concentrations of substances produced by society (e.g., plastics, dioxins, PCBs, and DDT)
3. The systematic physical degradation of nature and natural processes (e.g., overharvesting forests, destroying habitat, and overfishing)
4. The conditions that systematically undermine people's capacity to meet their basic human needs (e.g., unsafe working conditions and not enough pay to live on)

13.4.3 Natural Capitalism

Natural capital consists of the earth's natural resources and ecological systems that provide society and all living things with all the goods and services vital for living (Hawken et al. 2000). Our current business practices abuse and waste the offerings of nature consisting of energy, materials, water, and soil. The next industrial revolution like the previous one will be a response to the developing scarcity and will create upheaval coupled with opportunities for new business model development called natural capitalism. *Natural capitalism* involves four interlinked business practice changes. These are as follows:

- *Dramatically increase productivity of natural resources.* Through use of emerging technologies in production, manufacturing, and product design, companies are developing ways to make natural resources—energy, minerals, water, and forests—stretch many times over than where we are currently. The savings achieved in investment, operational costs, and efficiency gains will lead to implementation of the other three principles.
- *Conversion to biologically inspired production models and materials.* Natural capitalism envisions a state where no waste is created. These closed-loop production systems are modeled on nature's strategies where the output of a process is either returned intelligently to the ecosystem as a nutrient or becomes an input for another manufacturing process. Industrial processes that mimic the benign chemistry of nature reduce dependence on nonrenewable inputs. This enables effective, efficient, and technologically elegant production systems.
- *Shift to a service-and-flow business model.* The traditional business model of manufacturing is governed by sale of goods. In the *service-and-flow* model, value is provided as a continuous flow of services. This is consistent with the profit potential of providers and customers by rewarding each one for resource productivity.
- *Focus and reinvestment in natural capital.* Consistent with the age-old adage that you need money to make money, any entity that consumes its capital is bankrupting future profit potential. With increasing needs of mankind, cost of deleterious impact on existing ecosystems, and consumer awareness of the environment, businesses are becoming more sensitive to restoring, sustaining, and expanding natural capitalism. Interestingly, all of these lead to new business opportunities and job creation.

This emerging industrial revolution is providing an opportunity to farsighted companies to have a higher profit potential and a competitive advantage by using these four principles. Management and employees have a *feel good* factor resulting in job satisfaction and future prosperity. These future state solutions are interconnected and as companies shut down unproductive assets and downsize, they tend to end up with higher payroll, which is productive, efficient, innovative, and more tuned to future growth.

13.4.4 Industrial Ecology

Industrial ecology is the systematic study of energy and material balance in a flow system through industrial processes. Hence, it transforms processes from linear systems consisting of an open loop. This involves movement of resources and capital eventually producing waste. However, we must move to a closed-loop system utilizing waste as input for new processes. The industrial economy of the world can evolve as a network of processes that utilize raw material from the earth and convert them into goods for global commercial transaction. Industrial ecology measures the flow of materials and provides assessment of their impact on the environment, the remaining extent of natural resources available, and issues related to disposal of waste. Industrial ecology is an emerging field comprising multiple disciplines consisting of biology, engineering, chemistry, economics, sociology, and toxicology (Frosch and Gallopoulos 1989). It addresses challenges associated with sustainability using a complex system involving integrated natural and human scientific disciplines to aid our thinking based on natural systems as a guidepost.

13.4.5 CRADLE-TO-CRADLE

The cradle-to-cradle concept uses *ecoeffective* strategies to generate products contributing to an overall sustainable prosperity. The concept was developed in 1995 by William McDonough and Michael Braungart to enhance *next industrial revolution* through intelligent design practices (McDonough and Braungart 2002). The concept is grounded in science with peer-reviewed process to evaluate and optimize materials used in products and manufacturing systems. The objective is the maximization of worker health, safety, efficiency, effectiveness, and reutilization of waste over multiple life cycles.

In applying this concept, materials used are evaluated for their impact on human health and environmental relevance under four categories (green, yellow, orange, or red). These are the following:

- Green: little or no risk. This chemical is acceptable for use in the desired application.
- Yellow: low to moderate risk. This chemical is acceptable for use in the desired application until a green alternative is found.
- Orange. There is no indication that this is a high-risk chemical for the desired application, but a complete assessment is not possible due to lack of information.
- Red: high risk. *Red* chemicals (also sometimes referred to as *X-list* chemicals) should be phased out as soon as possible. *Red* chemicals include all known or suspected carcinogens, endocrine disruptors, mutagens, reproductive toxins, and teratogens. In addition, chemicals that do not meet other human health or environmental relevance criteria are *red* chemicals.

After evaluation, chemicals/materials in the green subgroup are selected to replace existing high-risk substances. The certification provided (silver, gold, or platinum brands) suggests environmentally sound product and processes being utilized with socially responsible strategies.

13.4.6 BIOMIMETICS

Nature is always changing while local points of nonsustainability occur (Vincent et al. 2006). The system overall is generally stable and self-governing. So the idea of sustainability is the first and biggest lesson we get from nature, and it refers to sustainable systems. As the Kenyan proverb goes, "We do not inherit the land from our parents, we borrow it from our children." Hence, biomimetics forces us to look for answers to various problems related to our quest for new products, processes, or technology. Such complex human and engineering problems can be related to self-healing abilities, environmental exposure tolerance and resistance, hydrophobicity, self-assembly, and harnessing solar energy.

13.4.7 ZERO EMISSIONS RESEARCH AND INITIATIVES

The United Nations Development Program and Gunter Pauli created the ZERI Foundation in 1996. Its objective is to transform designs into revolutionary concepts of sustainability (Pauli 2002). ZERI brings together a global network of creative people who attempt to solve the principal challenges facing the world. The solutions obtained are derived from scientific principles and information available from open-source systems. The aim is to find sustainable solutions that can be used locally by communities and corporations.

For ZERI, it is important to work with many challenges at the same time in order to facilitate the synergy of multiple solutions and different organizational approaches. The sustainable solutions are based on what is locally available, building on local culture and expertise but also taking the inspiration of the ecosystem with what is available.

> Business must first and foremost respect the license under which it is granted a right to operate: respond to the needs of the client. Government is not exempt from this golden rule: respond to the needs of the tax payer and the needs in the community. Unfortunately, the current notions of leadership are based on power and control. Going beyond ZERI means focusing on creating a future generation and working with young minds which believe in 'impossible dreams'. Throughout the stimulation of creativity and innovation, ZERI believes in affirming the creative potential of each individual and their unique contribution towards the development of themselves and their communities.
>
> **ZERI (2013).**

ZERI utilizes scientific principles in creating fun educational materials using fairy tales to explore science through the workings of nature. This creates a new class of scientists and entrepreneurs who understand the strong connection between humanity and nature and of the importance of preserving this balance.

13.4.8 PLANETARY BOUNDARIES

The planetary boundaries framework is the central concept that sets safe limits to the critical processes or elements of the earth system that determine the stability of the Holocene (such as climatic system, how the ocean works, and chemical cycles)—aimed at lowering the environmental risks and avoiding tipping points at the regional and global levels (Rockström and Klum 2012). In 2009, a group of the earth system and environmental scientists led by Johan Rockström from the Stockholm Resilience Centre and Will Steffen from the Australian National University proposed the framework of *planetary boundaries* designed to define a "safe operating space for humanity" as a precondition for sustainable development. The nine planetary boundaries were identified for climate change, ozone depletion, land use change, freshwater use, biodiversity loss, ocean acidification, nitrogen and phosphorus discharges to the biosphere and oceans, aerosol loading, and chemical pollution (Rockström and Klum 2012).

The framework asserts that once human activity has passed certain thresholds or tipping points, defined as *planetary boundaries*, there is a risk of irreversible and abrupt environmental changes.

Summarizing the contributions of the eight sustainability frameworks along the key strategic vectors of environmental, economic, and social goals, Table 13.1 highlights the similar characteristics or arrangements to understand the scope and inspiration required for a fundamental mind shift in the textile and apparel industries, as discussed in detail in the following section.

13.5 ROLE OF TEXTILE IN SUSTAINABILITY

Textile as a sector of economic activity is not unique in its contribution to nonsustainability. Along with the rest of the world, it too is rapidly heading toward a precipice: an environmental, social, and political disaster that we can all see unfolding before our eyes. What will happen when we run out of resources? By the middle of the century, we will have 10 billion people competing for finite resources from clean water to energy to food. So what are the textile-specific issues and some logical solutions we should consider? Let us consider a limited number of components to understand the patterns of nonsustainability in textiles (Edwards 2010).

13.5.1 CONTROL OF GROWTH OF GREENHOUSE GASES

GHGs are predominantly oxides of carbon, nitrogen, CFCs, and methane. Their contributions are approximately 55%, 6%, 15%, and 24%, respectively. The major cause is our current technology of textile chemistry and long-distance transportation of textile-related goods.

TABLE 13.1
Eight Sustainability Frameworks along the Three Key Strategic Vectors

Major Sustainability Frameworks	Focus of Strategic Vectors			Fundamental Assumptions along the Strategic Vectors
	Social	Environmental	Economic	
Ecological footprint	✗	✓	✗	*Environmental*: Consumption of energy, biomass, building material, water, and other resources are converted into a normalized measure of biologically productive area necessary to supply the resources a human population consumes and to assimilate associated waste.
Natural step	✓	✓	✗	*Social*: Based upon the principle of "society moving toward more of well-being, happiness, and meeting human needs." *Environmental*: Based on the three principles of avoiding digging/making/destroying stuff from the earth at a rate faster than that of replenishment.
Natural capitalism	✓	✓	~	*Social*: Ecological economics to create sustainable prosperity "improving human well-being and social equity." *Environmental*: Redressing global inequities of material well-being, slowing down resource depletion, lessening pollution. *Economic*: Lower costs for business and society.
Industrial ecology	✗	✓	✓	*Environmental*: Concentrate on aspects, like dematerialization, decarbonization, environmental policy, technology, and life cycle planning. *Economic*: Concentrate on aspects, like ecodesign, product stewardship, ecoefficiency.
Cradle-to-cradle	~	✓	~	*Social*: Health of the overall system (human being, plants, and animals) is better maintained through removal of toxic nutrients. *Environmental*: A biomimetic approach to circulate nutrients in healthy and safe metabolisms. *Economic*: Lowers the financial cost of systems.
Biomimetics	✗	✓	✗	*Environmental*: Solve complex engineering problems by rethinking the approach to material development and reducing ecological degradation.
ZERI	~	✓	~	*Social*: ZERI provide entrepreneurs with new businesses (through improved material productivity, multidisciplinary research, etc.) generating jobs and building sustainable livelihoods. *Environmental*: Targeting zero emissions. *Economic*: ZERI builds on zero defects, zero inventory, and zero complaints.
Planetary boundaries	✗	✓	✗	*Environmental*: Maintaining the threshold of nine environmental earth system processes.

✗—not focused; ✓—focused; ~—partially focused.

The carbon footprint of the textiles is estimated based on the *embodied energy* in the fabric, comprising all of the energy used at each step of the manufacturing process from fiber to finished goods (Athalye 2012). In case of synthetics, the fibers are made from fossil fuels, where very high amount of energy is consumed in extracting the oil from the ground as well as in the production of the polymers. Further, the raw material used to produce a shirt—oil—is drilled in the Middle East, produced in China, and consumed in the United States. The location of the raw material, availability of cheap labor forces, and the consumer groups determine the development of these global supply chain patterns. A great amount of transport is needed to connect these different parts of the global network structure and make them work effectively. The textile industry, according to the U.S. Energy Information Administration, is the fifth largest contributor to CO_2 emissions (IEA 2007).

Much has been written about the adverse impact of these gases on the environment. The average surface temperature of the earth is likely to increase by $1.1°C–6.4°C$ by the end of this century relative to 1980–1990. The best estimate is $1.8°C–4.0°C$ based on the emissions of gases caused by our pursuits of growth (Chapman and Davis 2007). Even these seemingly low increases in global temperature will have a catastrophic impact on the world. Most coastal areas of the world will see increased flooding through rise of sea levels. In addition, there will be a sharp drop on all continents in agriculture and grain production.

What should be done? We must tirelessly work toward reducing CO_2 emissions both in terms of reducing consumption and conducting research toward sequestration of these gases. Sequestration of carbon can be carried out through various established processes, namely, biological (e.g., peat production, wetland restoration, and agriculture), chemical (e.g., scrubbing), and physical (e.g., biomass related) in nature. We must also move toward reducing the transportation distances between production, transformation, and consumption locations through various supply chain strategies and new business formats, for example, nearshoring. Spurred by these factors, the textile industry and players across the value chain have adopted various strategies for reducing the carbon footprint, for example, switching to more energy-efficient processes (fabrics made with Sorona—a renewable polymer-based fiber provides 30% CO_2 reduction, while the manufacturing process reduces GHG emissions by 63%) and developing innovative products with reduced carbon footprints (eco-friendly solutions by BASF) (Dev 2009).

13.5.2 SAFEGUARD WATER QUALITY

Clean water is becoming an increasingly precious commodity for the burgeoning global population. The breakdown of water consumption between agriculture, industry, and domestic household is 70%, 20%, and 10%, respectively. In addition, daily personal consumption is 700 L in North America, 200 L in Europe, and only 20 L in Africa. Furthermore, 1.5 billion people around the world do not have access to drinking water (Rost et al. 2008). To make matters worse, the annual depletion of underground drinking water in all corners of the world is increasing rapidly. This scarcity of water is the most important sustainability issue facing the textile industry because of the dependence on water to process, dye, and finish fabrics (Thiry 2011).

The industry must reduce its dependence on water, particularly by lessening the current water-based dyeing, replacing the inefficient and polluting processing technologies, and setting minimum standards for wastewater treatment. It is also about accounting and for the real price of water that has not been *accounted for*. Finally, with improvements in desalination technologies, it is possible we may reach unlimited supply of drinking water.

13.5.3 EFFICIENT ENERGY UTILIZATION

Energy availability is critical for industrial activity. However, oil-/coal-based sources of energy are not infinite, as fossil fuels are an unsustainable source. Price volatility, threat of global energy crisis, and the high cost of fuels have resulted in serious efforts to conserve energy maximum extent.

The textile industry has a dismal record with low energy-efficient utilization efficiency and is one of the major energy-consuming industries. About 34% of energy is consumed in spinning, 23% in weaving, 38% in chemical processing, and another 5% for miscellaneous purposes (Grady et al. 1989). Power dominates consumption pattern in spinning/weaving, while thermal energy is a major component for chemical processing (Izhar and Fahimuddin 2013).

The rule of thumb in textile industry is that it takes one truck of fuel to make one truck of fabric. It is known that thermal energy in textile mills is largely consumed in two operations, in heating water and drying water. Fuel consumption in textile mills is almost directly proportional to the amount of water consumed. This water/energy nexus is driven mostly by water utilization by the industry. Hence, if consumption of water can be reduced, it will also save energy. Conservation of energy can be affected through process and machine modification, proper chemical recipes, replacing water as a primary medium for dyeing, and new technologies. Some of these novel concepts of energy conservation include

- Supercritical dyeing technique
- Ultrasonic assisted wet processing
- Foam technology
- Global agreements on reduction of energy consumption

13.5.4 BIODIVERSITY PROTECTION AND DESERTIFICATION PREVENTION

Biodiversity is the degree of variation of life forms within a given species, ecosystem, or planet. Rapid environmental changes typically cause mass extinctions. One estimate is that <1%–3% of the species that have existed on the earth are extinct (Saier 2006). Since life began on Earth, five major mass extinctions and several minor events have led to large and sudden drops in biodiversity. Since humans have emerged, there has been ongoing reduction of biodiversity and genetic diversity. The extinction of species has accelerated in the past 40 years. Scientists estimate that 60,000 species have gone extinct in the past two centuries (Hannah 2012). This has resulted because of a combination of both consumption-driven growth and climate change, which is exacerbating the decline even faster. Furthermore, extinction of species could impact the food chain with adverse unknown consequences.

A companion issue is desertification, which is predominantly driven by intensive cotton farming. Desertification is a type of land degradation in which a relatively dry land region becomes increasingly arid, typically losing its bodies of water as well as vegetation and wildlife. It is caused by a variety of factors, such as climate change and other human activities. Desertification is a significant global ecological and environmental problem.

Cotton farming uses large quantities of pesticides, and scientists justifiably ask the question "Is cotton a real natural fiber?" The impact of cotton farming in the Aral Sea region in Uzbekistan has had profound impact on both desertification and loss of biodiversity. Aral Sea was a freshwater body with vibrance and diversity of species. It has seen a steady decline in biodiversity and a loss of water since the introduction of cotton farming, which has only accelerated. Now, only a small fraction of the freshwater remains. The same fate is predicted for Lake Chad and the basins of Rio Grande and Colorado River.

There are no simple answers to this very complex issue and there are many competing interest groups. However, we must find a solution to generate a fundamental mind shift to avoid catastrophe. Some strategies that have shown promise are

- Efficient cotton farming
- Development of new cotton variety
- Development of nonintrusive farming methods
- Farming and substitution by other natural fibers
- Development of new methods of transformation

13.5.5 WASTE PRODUCT MANAGEMENT

The increased awareness of consumers has contributed to an increased demand for products that has a minimal impact on the well-being of the world. There has been an increased demand for *green textiles* despite the higher cost to consumers. The use of recycled rather than virgin materials leads to conservation of natural resources and concomitant decrease in energy use. Furthermore, there is a reduction in carbon dioxide and other emission products coupled with less waste to landfill. Some of these recycled products are more expensive, partly because the practice of recycling has not reached a high level in the United States, further contributing to limited supply. As an example, only 25% of plastic bottles in the United States are recycled. The good news is that the percentages are on the rise every year. These recycled products are made with cleaner and more efficient processes. Hence, the overall higher cost is balanced by the advantage gained by living in a cleaner and healthier world with abundant natural (sustainable) resources (Welle 2011).

Depending on quality, condition, and fashion accuracy, there are four basic paths for used textiles:

- Used again, formally or informally
- Recycled, into new textile or other products
- Used for energy, incineration with energy recovery
- Landfilled, waste dumps

Landfill (or other means of disposal) is the least preferable option, and waste prevention is the most preferable option. Reuse of textiles is at the highest and second highest level in the hierarchy. The other waste hierarchy steps are self-explanatory. For a more sustainable waste product management strategy, we must develop improved reuse and recycling technologies. Some of the materials used in textiles today (e.g., cotton) are not sustainable even with a rather high grade of reuse and recycling and new materials therefore need to be developed. To reduce the environmental impact from textiles, the design of clothes need not only be focused on fashion but also on the life cycle of the apparel item. A short-lived apparel item (due to fashion, inherent properties, or others) needs to be designed with recycling in mind, while a long-lived apparel item should be designed to last long, perhaps with some parts interchangeable to enable easy repair and to make it suitable for the secondhand market.

13.6 MOVING INTO THE FUTURE: A SUMMARY

Sustainability is not about reaching a destination but rather a direction. There is no end point that, when reached, we can declare victory. We can always make our planet, through responsible care, a better and a safer place for all life forms. This can only be achieved through a collective mind-set where we see environmental, social, and economic challenges not as separate and competing but rather as an interconnected whole. Instead of keeping score as a gross cash flow, we must think of the whole system balance sheet. We must not consider profit making as an economy outside of nature but one where all is integrated with nature. The industrial processes must transform from an open-loop strategy where the resources and capital investment move through the system to produce waste to a closed-loop system where the waste can be converted in new processes fueled by evolving alternate energy technologies. This thinking by very nature is rooted not in our current mentality of short-term thinking but by looking long term for the developments that do not compromise the ability of future generation to meet their own needs.

So what answer is there to the question raised earlier, "Why is enough for mankind never enough?"

> There is enough in this world for every man's needs, but there is not enough in this world for every man's greed.
>
> **Mahatma Gandhi**

REFERENCES

Allwood, J., Laursen, S.E., De Rodriguez, C.M., and Bocken, N., 2006. *Well Dressed? The Present and Future Sustainability of Clothing and Textiles in the United Kingdom*. Cambridge, U.K.: University of Cambridge.

Ashby, M., 2012. *Materials and the Environment: Eco-Informed Material Choice*, 2nd edn. Oxford, U.K.: Butterworth-Heinemann.

Athalye, A., 2012. Carbon footprint in textile processing. *Indian Textile Journal*, 122(11), 20, http://www.indiantextilejournal.com/articles/FAdetails.asp?id=4652. (Accessed January 10, 2014.)

Barbier, E., 1987. The concept of sustainable economic development. *Environmental Conservation*, 14(2), 101–110.

Bide, M., 2012. Sustainability: A big picture. *American Association of Textile Chemists and Colorists International Conference*, p. 125, Charlotte, NC.

Biologicaldiversity, 2014, http://www.biologicaldiversity.org/programs/population_and_sustainability/extinction_and_population_graph.html, (Accessed August 5, 2014.)

Brundtland Report, 1987. *World Commission on Environment and Development: Our Common Future*. Oxford, U.K.: Oxford University Press.

Chapman, D.S. and Davis, M.G., 2007. Global warming—More than hot air. *Journal of Land, Resources, & Environmental Law*, 27(1), 59–77.

Dev, V., 2009. Carbon footprint of textiles, http://www.domain-b.com/environment/20090403_carbon_footprint.html, March 2014. (Accessed March 20, 2014.)

Edwards, A., 2010. *Thriving Beyond Sustainability: Pathways to a Resilient Society*. Gabriola Island, British Columbia, Canada: New Society Publishers.

Eryuruk, S.H., 2012. Greening of the textile and clothing industry. *Fibres and Textiles in Eastern Europe*, 20(6A/95), 22–27.

Fagan, B., 2008. *The Great Warming: Climate Change and the Rise and Fall of Civilizations*. New York: Bloomsbury Press.

Frosch, R.A. and Gallopoulos, N.E., 1989. Strategies for manufacturing. *Scientific American*, 261(3), 144–152.

Global Footprint Network, World Footprint: Do we fit on the planet? 2014. http://www.footprintnetwork.org/en/index.php/GFN/page/world_footprint/, February. (Accessed February 10, 2014.)

Grady, P.L., Mock, G.N., Pai, G.A., and Throneburg, K.W., 1989. A general purpose textile plant energy consumption model. *Textile Research Journal*, 59(3), 177–182.

Hannah, L.J., 2012. *Saving a Million Species: Extinction Risk from Climate Change*. Washington, DC: Island Press.

Hansen, J., Sato, M., Kharecha, P., and Von Schuckmann, K., 2011. Earth's energy imbalance and implications. *Atmospheric Chemistry and Physics*, 11, 13421–13449.

Hawken, P., Lovins, A.B., and Lovins, L.H., 2000. *Natural Capitalism: The Next Industrial Revolution*. New York: Little, Brown & Company.

IEA, 2007. Tracking industrial energy efficiency and CO_2 emissions. International Energy Agency, pp. 1–321. http://www.iea.org/publications/freepublications/publication/tracking_emissions.pdf, (Accessed January 10, 2014.)

IPCC Fourth Assessment Report, 2007. Summary for policy makers, https://www.ipcc.ch/pdf/assessment-report/ar4/wg1/ar4-wg1-spm.pdf, August 05, 2014.

Izhar, Q. and Fahimuddin, S., 2013. Analysis of global energy consumption patterns. *Journal of Environmental Research and Development*, 7(4A), 1761–1773.

Macy, J. and Johnstone, C., 2012. *Active Hope: How to Face the Mess We're in without Going Crazy*. Novato, CA: New World Library.

Mcdonough, W. and Braungart, M., 2002. *Cradle to Cradle: Remaking the Way We Make Things*. New York: North Point Press.

Meadows, D.H., Meadows, D.L., Randers, J., and Behrens III, W.W., 1972. *The Limits to Growth*. New York: Universe Books.

Meyer, C. and Kirby, J., 2014. Income inequality is a sustainability issue. *Harvard Business Review*, http://blogs.hbr.org/2014/01/income-inequality-is-a-sustainability-issue-2/, (Accessed August 16, 2014.)

Mitrovica, J., Gomez, N., and Clark, P., 2009. The sea-level fingerprint of West Antarctic collapse. *Science*, 323(5915), 753.

Pal, R., 2014. Sustainable business development through designing approaches for fashion value chains. *In* Muthu, S.S. (ed.), *Roadmap to Sustainable Textiles and Clothing*. Singapore: Springer Science+Business Media. pp. 227–261.

Pauli, G., 2002. *The Blue Economy*. Taos, NM: Paradigm Publications.
Robèrt, K.-H., 2002. *The Natural Step Story: Seeding a Quiet Revolution*. Gabriola Island, British Columbia, Canada: New Society Publishers.
Rockström, J. and Klum, M., 2012. *The Human Quest: Prospering within Planetary Boundaries*. Stockholm, Sweden: Langenskiöld.
Rost, S., Gerten, D., Bondeau, A., Luncht, W., Rohwer, J., and Schaphoff, S., 2008. Agricultural green and blue water consumption and its influence on the global water system. *Water Resources Research*, 44(W09405), 1–17.
Saier, M., 2006. Species extinction. *Environmentalist*, 26(2), 135–147.
Thiry, M., 2011. Staying alive: Making textiles sustainable. *AATCC Review*, Nov/Dec, 26–32.
Vincent, J., Bogatyreva, O., Bogatyrev, N., Bowyer, A., and Pahl, A.-K., 2006. Biomimetics: Its practice and theory. *Journal of the Royal Society Interface*, 3(9), 471–482.
Wackernagel, M. and Rees, W., 1996. *Our Ecological Footprint: Reducing Human Impact on the Earth*. Gabriola Island, British Columbia, Canada: New Society Publishers.
Wade, R.H., 2001. The rising inequality of world income distribution. *Finance and Development, International Monetary Fund*, 38(4).
Welle, F., 2011. Twenty years of PET bottle to bottle recycling—An overview. *Resources, Conservation and Recycling*, 55(11), 865–875.
Wheeler, D., 2009. Greenhouse emissions and climate change: Implications for developing countries and public policy. *Commission on Growth and Development*, 1, 247–283.
ZERI, 2013. What is ZERI?, http://zeri.org/ZERI/About_ZERI.html, August 2014. (Accessed August 16, 2014.)

Section IV

Assessment of Sustainable Apparel Production

14 Eco-Parameters and Testing of Sustainable Textiles and Apparels

Shanthi Radhakrishnan

CONTENTS

14.1 Introduction 310
    14.1.1 Industrial Processes in the Textile Industry 310
    14.1.2 Pollution Output in the Textile Industry 311
        14.1.2.1 Wastewater 311
        14.1.2.2 Air Emissions 313
        14.1.2.3 Other Wastes 314
14.2 Eco-Testing of Textiles 315
    14.2.1 Hazardous Substances and Mixtures 315
    14.2.2 Restricted Substances List 316
    14.2.3 RSL Testing 316
14.3 Pollution Prevention Opportunities 318
    14.3.1 Pollution Prevention Act 1990 319
    14.3.2 Methods for Pollution Prevention 320
        14.3.2.1 Quality Control of Raw Materials 320
        14.3.2.2 Chemical Substitution 320
        14.3.2.3 Process Modification 320
        14.3.2.4 Equipment Modification 321
        14.3.2.5 Good Operating Practices 321
14.4 Impact of Banned Substances on Textiles and Apparels 321
    14.4.1 Azo Dyes/Arylamines 321
    14.4.2 Formaldehyde 323
    14.4.3 Chlorinated Phenols: Pentachlorophenol, Trichlorophenol, and 2,3,5,6-Tetrachlorophenols 324
    14.4.4 Phthalates 324
    14.4.5 Heavy Metals 325
    14.4.6 Organotin Compounds Tributyltin, Dibutyltin, and Monobutyltin 325
    14.4.7 Alkylphenols, Alkylphenol Ethoxylates, Dimethyl Fumarate, Chlorinated Organic Carriers, Flame Retardants, Solvents (VOC), Perfluorooctane Sulfonates, Pesticides, and pH Value 325
14.5 Environmental Laws: Major Statutes for the Textile Industry 327
    14.5.1 Environmental Laws 327
References 329

14.1 INTRODUCTION

The textile and clothing industry forms an essential part of manufacturing, employment, and trade in developing countries with opportunities for export orientation and industrialization. This industry is a suitable workable tool for growth and economic development in many poor and developing countries to move them to a higher economic category of middle income countries like Vietnam and Mauritius. This industry has contributed to capital expenditure, exports, employment, and fiber consumption. Capital expansion, modernization, and automation have increased productivity while international trade and exports showed substantial improvement (Gereffi 2002).

In Mauritius, the textile and clothing industry generates around 9% of the GDP and is estimated to provide indirect employment for around 250,000 people and direct employment for around 78,000 people (ILO 2005). The Cambodian garment industry has been the key source of export growth and formal employment contributing to about 10%–12% to the country's GDP. Garments contribute to around 80% of Cambodia's exports and employ 65% of the manufacturing workforce in Cambodia. Around two-thirds of Cambodia's exports are for the U.S. market, while the remaining cater to the European Union (EU). In 2002, the ASEAN countries and the Asian industrialized countries were the main investors in Cambodia and have a strong foothold on the FDI flows and stock in the country (Keane and Velde 2008).

14.1.1 Industrial Processes in the Textile Industry

A general outline of the industrial processes in the textile industry is essential for understanding the processes involved and their impact on the environment that leads to terms like pollutant outputs, pollution prevention opportunities, and federal regulations. Basically, there are four main processes

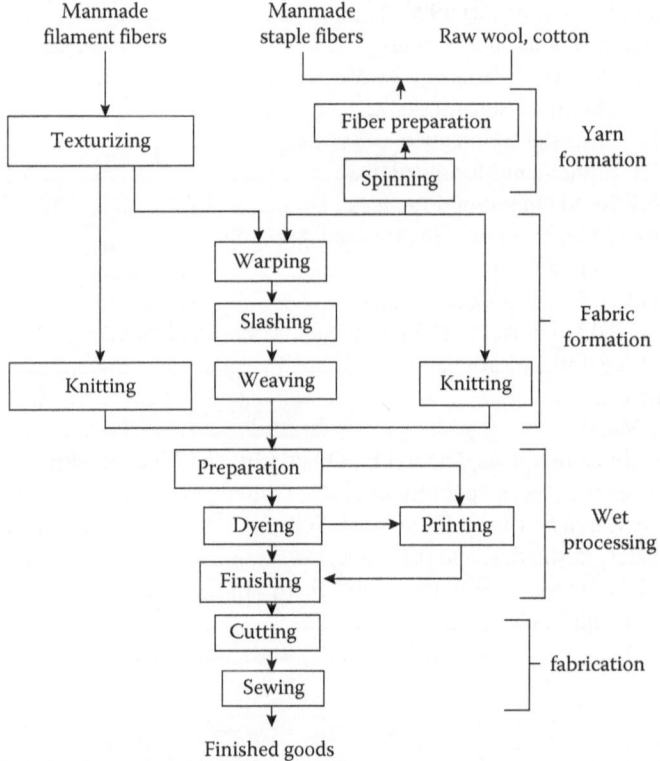

FIGURE 14.1 Typical textile processing flowchart.

in the textile industry that include yarn formation (spinning), fabric formation (weaving, knitting, or nonwovens), wet processing (dyeing, printing, and finishing), and fabrication (apparel or finished goods). Figure 14.1 represents the typical textile processing flowchart.

Textile manufacturers receive, prepare, and convert fibers into yarn, thread, or webbing. The yarn is used for the production of fabric by weaving or knitting. These fabrics are dyed and finished at different stages of production (Ghosh and Gangopadhyay 2000). At each stage of preparation and conversion, there are various types of waste generated that has to be handled with care to prevent environmental contamination.

14.1.2 Pollution Output in the Textile Industry

The textile industry has a high growth rate and production leading to a better economy and industrialization. The production processes in textile production has a high impact on environment in terms of wastewater, air emissions, and residual waste. The summary of the potential releases emitted during textile manufacture are given in Table 14.1.

14.1.2.1 Wastewater

The textile production is water based consuming high amounts of processed water and producing large quantities of polluted discharge. When the discharge is left untreated, it can cause extensive damage to the environment. The chemicals and treatment processes used for production play an important role in the content of the resulting wastewater. The types of wastewater include cleaning water, process water, noncontact cooling water, and storm water. The wet processing operations result in large volumes of effluent with alkali wastewater from preparation and wastewater from dyeing and finishing, which contain large amounts of salts, acid, or alkali. Another important feature of textile wastewater is the complex mixture of chemicals present that depends on the wide variety of process steps involved in the treatment of raw material (EPA 1997).

Textile dyeing effluents are extremely toxic in nature and contain a high concentration of various salts, heavy metals, and unused dyes. These contaminants are toxic to all living organisms. The hematological parameters, namely, hemoglobin, packed cell volume (PCV), RBCs, WBCs, and mean corpuscular hemoglobin concentration (MCHC), in freshwater fish exposed to various concentrations of untreated dyeing effluents were analyzed and found to vary from the results of the control fish to highlight the negative impact of textile dyeing effluents (Amte and Mhaskar 2013).

Pollutants in textile wastewater may include changes in pH of water, total dissolved solids, total suspended solids, chemical oxygen demand, biodegradable organic matter, total organic compounds, and heavy metals. The effects of these parameters have a derogatory outcome on fauna and flora of the environment. Table 14.2 shows the environmental impact of pH on fish and aquatic life.

The ideal pH of 6.0–9.0 is best suited for fish and aquatic life. Most wastewaters have a high pH value of 11–12 when the textile process is carried out in alkaline pH. Industrial wastewater may contain iron, lead, chromium, ammonium, mercury, and other elements. When the pH falls, the water becomes acidic in nature and increases the solubility of many substances and their absorption, for example, 4.8 mg/L of iron is not toxic in nature in pH 4.8 but 0.9 mg/L of iron at pH 5.5 can cause fish to die (RAMP 2000).

Similarly, the temperature of water has an important impact on aquatic life. Heat loadings that affect water temperature of streams include direct sun, industrial wastewater from cooling machinery, or high temperature industrial process, for example, bleaching and runoff water from sun-heated areas where there is no surrounding vegetation to lower water temperature. Water quality standards for temperature are seasonal, and temperature beyond 72°F–77°F could kill fish in ponds. Higher temperatures also create less dissolved oxygen. In water bodies, the dissolved oxygen is rich in the surface water due to photosynthesis of the large quantities of algae present. As the depth of

TABLE 14.1
Summary of Potential Releases Emitted during Textile Manufacture

Process	Air Emissions	Wastewater	Residual Wastes
Fiber preparation	Little or no air emissions generated	Little or no wastewater generated	Fiber waste, packaging waste, and hard waste
Yarn spinning	Little or no air emissions	Little or no wastewater generated	Packaging waste, sized yarn, fiber waste, cleaning and processing waste
Slashing and sizing	VOCs	BOD, chemical oxygen demand (COD), metals, cleaning waste and size	Fiber lint, yarn waste, packaging waste, unused starch-based sizes
Weaving	Little or no air emissions generated	Little or no wastewater generated	Packaging waste, yarn and fabric scraps, off-spec fabric, used oil
Knitting	Little or no air emissions generated	Little or no wastewater generated	Packaging waste, yarn and fabric scraps, off-spec fabric
Tufting	Little or no air emissions generated	Little or no wastewater generated	Packaging waste, yarn and fabric scraps, off-spec fabrics
Desizing	VOCs from glycol ethers	BOD from water-soluble sizes, synthetic sizes, lubricants, biocides, and antistatic compounds	Packaging waste; fiber lint; yarn waste; cleaning materials like wipes, rags and filters; cleaning and maintenance wastes containing solvents
Scouring	VOCs from glycol ethers and scouring solvents	Disinfectants and insecticide residues, NaOH, detergents, fats, oils, pectin, wax, knitting lubricants, spin finishes, and spent solvents	Little or no residue waste generated
Bleaching	Little or no air emissions generated	Hydrogen peroxide, sodium silicate or organic stabilizer, high pH	Little or no residual waste generated
Singeing	Small amount of exhaust gases from burners	Little or no wastewater generated	Little or no residual waste generated
Mercerizing	Little or no air emissions generated	High pH, NaOH	Little or no residual waste generated
Heat setting	Volatilization of spin finish agents applied during synthetic fiber manufacture	Little or no wastewater generated	Little or no residual waste generated
Dyeing	VOCs	Metals, salt, surfactants, toxics, organic processing assistants, cationic materials, color, BOD, COD, sulfide, acidity, alkalinity, spent solvents	Little or no residual waste generated
Printing	Solvents, acetic acid from drying and curing oven emissions, combustion gases, particulate matter	Suspended solvents, urea, solvents, color, metals, heat, BOD, foam	Little or no residual waste generated
Finishing	VOCs, contaminants in purchased chemicals, formaldehyde vapors, combustion gases, particulate matters	BOD, COD, toxins, suspended solvents, and spent solvents	Fabric scraps and trimmings, packaging wastes
Product fabrication	Little or no air emissions	Little or no wastewater generated	Fabric scraps

Source: American Textiles Manufacturers Institute (ATMI), Industrial processes comments on draft of this document, 1997, http://nepis.epa.gov/Exe/ZyPDF.cgi/50000HE9.PDF?Dockey=50000HE9.PDF.

TABLE 14.2
Environmental Impact of pH on Fish and Aquatic Life

Limiting pH Values		
Minimum	Maximum	Effects
3.8	10	Fish eggs hatch, but deformed young produced.
4	10.1	Limits for most resistant fish types.
4.3	—	Carp died in 5 days.
5.4	11.4	Fish avoided waters beyond these limits.
5.0	9.0	Tolerable range for most fish.
6.0	7.2	Optimum range for fish eggs.
7.5	8.4	Best range for growth of algae.

Source: Kentucky River Basin Assessment Report, Water quality parameters, August 2000, http://www.uky.edu/WaterResources/Watershed/KRB_AR/krww_parameters.htm.

the water body increases, the oxygen is low due to lack of sunlight and little or no photosynthesis. When dissolved oxygen move to below 2.0 mg/L or increase to 110%, large volumes of fish die, and aquatic invertebrates are also affected (RAMP 2000).

The total organic content (TOC) is another factor that affects the dissolved oxygen content of water bodies. The larger the TOC, the higher the growth of microorganisms that deplete the oxygen level, and when the carbon-containing compounds in the wastewater increases, toxicity results. When the total suspended solids increase, the viscosity of the water increases leading to clogging of fish gills and settling into the bottom of the water body decreasing the dispersion of oxygen and nutrients in the deeper layers (WRM 2000). Mineral contaminants like aluminum, barium, cadmium, chromium, lead, manganese, and zinc can be present in very small amount, but when in excess, it has a negative effect on aquatic life.

14.1.2.2 Air Emissions

The textile industry emits a wide variety of air emissions that make treatment and prevention very complex. Nitrogen and sulfur oxides are generated from boilers, resin finishing, and drying operations of different processes. Hydrocarbons are emitted from drying ovens and from mineral oils that are emitted, while the materials are dried and cured at a temperature of 200°C. It has been estimated that majority of resin finishing plants emit 1 ton/year of formaldehyde from storage tanks and fabric off-gassing. Fabric off-gassing is the evaporation of chemicals from the fabric at normal atmospheric pressure (ATMI 1997). Carriers used in disperse dyeing lead to volatilization of aqueous chemical emulsions during heat setting, drying, and curing; process chemicals like methyl naphthalene or chlorotoluene may exhaust into the fibers and are emitted as volatile organic compounds (VOCs) during the drying process (EPA 1996). Some of the common air emissions are given in Figure 14.2.

There are two categories in air emissions—fugitive emissions and point source emissions. Emissions that are due to the volatile nature of substances and emanate from roof vents, louvres, open door, and equipment leaks and from valves and flanges are called fugitive emissions. They may be as dust from piles of stock and volatilization of vapor from vats, open vessels, or spills or due to material handling. Point source emissions are those that are emitted through a vent or stack usually through a single point into the atmosphere. Air emissions are emissions due to combustion of natural gas, LPG, or oil that are very relevant to the textile industry that uses boilers and other heating equipment for steam generation and drying (NPI 1999).

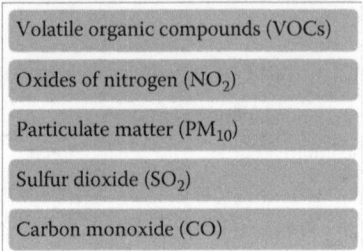

FIGURE 14.2 Common air emissions from textile and clothing manufacturing processes.

14.1.2.3 Other Wastes

Some of the wastes other than wastewater and air emissions are fiber wastes that may be soft fiber wastes and hard fiber wastes like yarn spinning wastes, beaming wastes, offcuts, packaging, spools, and creels. The major wastes generated by the textile and clothing sector have been classified by the CONSLEG system of the Official Publications of European Communities and are assigned a six-digit code that comprises two digits for the chapter, two for the subchapter, and two for the waste type. Table 14.3 shows the classification of the waste produced by the textile industry.

The textile and clothing industry has been continuously contributing to the pollution of the environment by means of wastewater, air emissions, and other wastes that are the results of mass production and lack of knowledge about the impact of the processes used for textile and apparel manufacture. Eco-friendly approaches and sustainable production methods coupled with strict

TABLE 14.3
Classification of Wastes According to the European Waste Codes (EWC)

Sl. No.	Classification	Wastes	EWCs
1.	Nondangerous packaging wastes	Paper packing wastes	15 01 01
		Plastic packing wastes	15 01 02
		Wooden packing wastes	15 01 03
		Metallic packing wastes	15 01 04
2.	Dangerous packaging wastes	Paper packing wastes	15 01 10
		Plastic packing wastes	15 01 10
		Metallic packing wastes	15 01 11
3.	Nondangerous packaging wastes	Textile waste	04 02 22
		Sludge from wastewater treatment	04 02 20
		Printing toners	08 03 18
4.	Dangerous wastes	Grease- and oil-impregnated rags	15 02 02
		Used oils	13 02
		Accidental leaks over sepiolite substratum	15 02 02
		Contaminated textile wastes with chemicals	15 02 02
		Solvent wastes	04 02 14
		Waste electronic and electric equipment with metals	16 02 13
		Batteries	16 06 01–03
		Chemical wastes, dyes, and print pastes	04 02 16

Source: European Commission, 2007. Textile waste minimization. http://ec.europa.eu/environment/life/project/Projects/index.cfm?fuseaction=home.showFile&rep=file&fil=LIFE05_ENV_E_000285_LAYMAN.pdf, Accessed August 29, 2014.

quality control and competition have brought about awareness among consumers, manufacturers, intermediate contractors, and retailers. The German ban on the use of nickel components in consumer articles followed by restrictions on pentachlorophenol (PCP), azo dyes, and carcinogens has highlighted the need for proper selection of raw materials, process methods, end-product testing, and certification. Eco-testing has aided retailers and manufacturers to prove the eco-friendliness of the product and to build the company's brand image in the consumer market.

14.2 ECO-TESTING OF TEXTILES

Today's designers and manufacturers have the responsibility to use sustainable materials and processes for the development of the eco-fashion industry. The impact of the choice of raw materials and processes involved to convert these raw materials to finished products is of primary concern and should be well understood for the reduction of waste, judicious utilization of natural resources, and safety of our planet. Carcinogenic pesticides used in agriculture have created a wide range of health problems to people using the end products, and many have been diagnosed with multiple chemical sensitivities. Consumers today are health conscious and prefer organic clothing to reduce the ailments due to chemical content of the end product (Sharma 2013). Growing ethical and environmental awareness calls for environmentally sustainable practices, fair trade and labor, contribution to community development, openness to public accountability, and building an ethical brand image among the consumers. To substantiate the environmental marketing claims on sustainability and eco-friendliness, eco-testing and certification have to be performed to prove to the consumer that all the salient features of the product advertised as *ethical* are true and genuine.

Some of the environmental concerns that have significant impact are energy use, resource depletion, greenhouse gas emissions, nutrient releases (leading to eutrophication), eco-toxicity from washing and dyeing of textiles, toxicity from fertilizer, pesticide and herbicide use, and emissions from fertilizer generation and irrigation systems related to production of fiber crops (RFS 2013). The concern for environment has brought about strict environmental legislations and clothing manufacturers, and retailers look for the complete absence of the restricted chemicals and substances in the end product. This has resulted in a demand for ecological testing for eco-parameters by various legislative and certification bodies at each stage of manufacturing. The testing of restricted substances in textiles is a highly specialized area and requires a specific procedure depending on the type of case administered.

14.2.1 Hazardous Substances and Mixtures

Products with EU Ecolabel would require that the product or any component should not contain substances that

- Are restricted or authorized in Article 57 of EC Regulation No. 1907/2006 and of the Council of December 2006 concerning the registration, evaluation, authorization, and restriction of chemicals (REACH)
- Appear on European Chemical Agency (ECHA) Candidate List as substances of very high concern (SVHCs) described in Article 59(1) of EC Regulation No. 1907/2006
- Are classified as carcinogenic, mutagenic, and reprotoxic (CMR) and toxic and hazardous to the environment as per EC Regulation No. 1272/2008 or Directive 67/548/EC that are identified in the form of hazard statements

It has been proposed that derogation shall not be given for substances that meet the criteria of Article 57 of EC Regulation No. 1907/2006 and Article 59(1) of the regulation and that are present in mixtures in an article or as part of a complex article in concentrations higher than 0.1% w/w (European Commission 2013a).

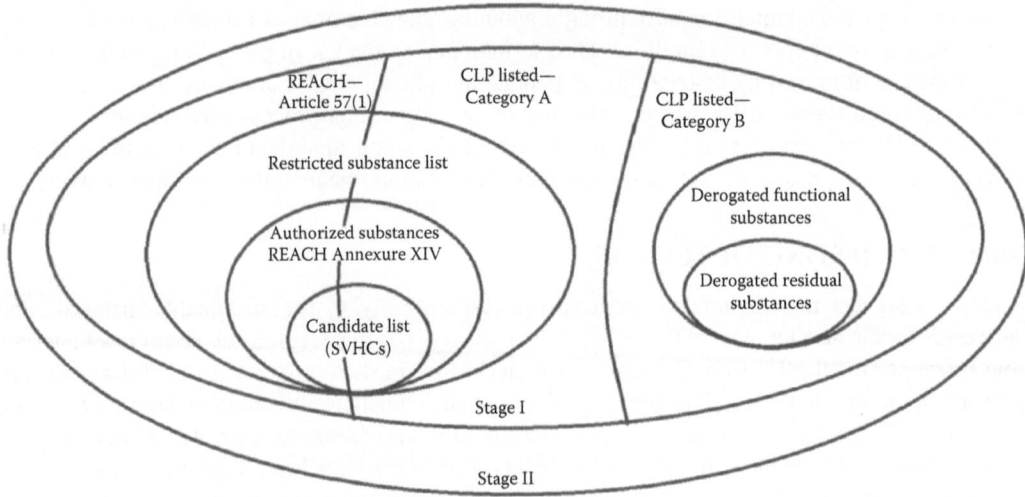

Stage I—Substances restricted or derogated
Stage II—All dyeing, printing and finishing substances

FIGURE 14.3 Schematic of substances restricted or derogated. (From Dodd, N., Revision of the EU ecolabel for textile products—Summary of proposed hazardous substances criteria, October 17, 2012, http://susproc.jrc.ec.europa.eu/textiles/docs/Ecolabel%20textiles_Hazardous%20substances%20summary.pdf.)

Hazardous substances are classified as CLP (classification, labeling, and packaging of substances and mixtures) Category A hazards (completely restricted) and CLP Category B hazards (those that could be used if they are derogated). Substances that are listed under hazardous substances may be used in dyeing, printing, and finishing processes. The derogated substances may be functional or residual. If these substances exist in the final product, they must be substituted in ecolabeled products (European Commission 2013b). Moreover, the concentration limits for each hazardous substance in the end product are stipulated, and if they are present above these levels, they would come under the category of restricted substances. Figure 14.3 represents the schematic of substances that are restricted or derogated.

14.2.2 Restricted Substances List

The EU Ecolabel for textile products has listed the textile restricted substance and prescribes that each restriction will be applied to specific groups according to the range of the restriction. An Oeko-Tex 100 certification would help to portray that the product has the requirements with reference to the restricted substances. Sometimes, other tests may also be undertaken apart from the Oeko-Tex 100 standard tests. Table 14.4 shows the restricted substance list (RSL) as per the EU Ecolabel and Oeko-Tex 100 along with the restricted substance group, scope of restriction, the limit values, and guidelines to applicants.

14.2.3 RSL Testing

The RSL testing is very complex with its roots in analytical chemistry as the substances tested need to be isolated and the precision with which the testing is done will be at the milligram/microgram/pictogram levels. Restricted substance testing requires special instruments like gas chromatography with mass spectrophotometer (GC–MS) or with electron capture detector (GC–ECD), high-performance liquid chromatography with diode array detector (HPLC/DAD) or with mass spectral detector (HPLC/MSD), atomic absorption spectrophotometer with graphite tube (AAS/graphite), inductively coupled plasma with mass spectrophotometer (ICP–MS), UV–visible spectrophotometer,

TABLE 14.4
Restricted Substances List with Limit Values and Guidelines

Substance Group	Scope of Restriction	Limit Values	Instructions to Applicants
SVHCs that are entered into the ECHA Candidate List	DMAC is subject to a higher limit value.	0.1% w/w in fiber prior to wet processing	If raw fiber testing meets the Oeko-Tex 100 limit value, final product testing not required.
	No specific requirement to consult the candidate list.	0.1% w/w for derogation	To follow RSL verification guidelines.
Biocides	Strictest EU Ecolabel requirement applies to Product Class I PCBs.	0.05% w/w sum total	The sum total value to be used where verification is required; additional testing shall be done for PCBs.
Auxiliaries and surfactants	Oeko-Tex 100 tests accepted for APs and APEOs on condition that the individual concentration of the APEO is provided.	100 ppm for each APEO	Individual APEO concentration should meet the 100 ppm limit value.
Dyes and carriers	Oeko-Tex 100 tests accepted for azo, CMR, sensitizing dyes, chlorinated benzenes, and toluenes (in relation to carriers).	NA	Separate EU Ecolabel verification required for nonuse of mordant dyes and metal complex dyes (as appropriate).
Heavy metals	Oeko-Tex 100 tests accepted for • Antimony, arsenic, cadmium, copper, lead, mercury • Class II–IV for chromium with Class I when metal complex dyes are used • Class I for cobalt with Class II–IV when metal complex dyes are used • Class I–IV for nickel with stricter Class I requirement when metal complex dyes are used	Limit values as applicable	Separate EU Ecolabel verification necessary for • *Chromium*: Class I (no metal complex dyes) • *Nickel*: Class I (metal complex dyes used) • *Cobalt*: Class II–IV (no metal complex dyes)
Printing	Oeko-Tex 100 tests accepted for print paste VOCs and plastisol printing based on: • Sum total of organic volatiles • Sum total of restricted phthalates	0.1% 0.5 mg/m^3	
Easy care	Oeko-Tex 100 tests accepted for formaldehyde.	Limit values as applicable	
Antifelting and shrink resistance	Wool OX levels not tested.	OX test methods	Verification requirement needs specific test method.
Water and oil repellent treatments	Oeko-Tex 100 tests accepted for PFOS; stricter limit values for PFOA.	PFOA 0.05 mg/kg	Separate EU Ecolabel verification for PFCA substance group according to stricter limit values.
Flame retardants	Oeko-Tex 100 tests accepted for flame retardants.		
Coatings, laminates, and membranes	Oeko-Tex 100 tests accepted for phthalates.	PFOA 0.05 mg/kg	Separate EU Ecolabel verification for PFCA substance group according to stricter limit values.

(Continued)

TABLE 14.4 *(Continued)*
Restricted Substances List with Limit Values and Guidelines

Substance Group	Scope of Restriction	Limit Values	Instructions to Applicants
Accessories—buttons, rivets, and zips	Oeko-Tex 100 tests accepted for phthalates in plastic accessories.	Nickel 0.5 µg/cm^2/week Chrome 60 mg/kg	Separate EU Ecolabel verification for metal accessories with nickel and/or chromium.

Source: European Commission, EU Ecolabel and Green Public Procurement—Stakeholders, Annexes 3,4,5 restricted substances list, Oeko-Tex 100 read across and risk matrix, February 2013a, http://susproc.jrc.ec.europa.eu/textiles/docs/130206%20Ecolabel%20textiles_Restricted%20Substance%20List_Draft%20v2.pdf.

and adsorbable organic halogen (AOX) analyzer (Jain and Easton 2010). Restricted substance testing consists of the following steps:

- Sampling: The sample should be representative of the test specimen, and the method of obtaining the sample should be determined.
- Various processes like extraction of digestion, concentration, cleanup, and derivatization if required may be adopted.
- Determination of restricted substances may be performed by chromatography for organic substances, atomic absorption or emission for metals, or spectrophotometry for both metals and certain organic substances.
- Evaluation of the obtained data. The experiments conducted must adhere to the standard procedures and reliable for repeatability and accuracy.

The general requirements for certification state that the final products and the production recipes used in the manufacture of the final product should not contain hazardous substances that are listed in the RSL at or above the specified concentration limits. The RSL should be communicated to the suppliers and agents who undertake spinning, weaving, printing, and finishing to the end product as verification and testing for RSL will be done for each production stage as well as for the final product. The applicant shall provide concrete evidence as documents from raw material and chemical suppliers and subcontractors with declarations and test results of samples of final product analysis. Laboratory analysis will be carried out at random for specific product lines according to the listed test methods. On application from the client, the testing will be done every year and the results informed to the competent body. Failure of test result will result in retesting for a specific product line. If the second test fails, the license shall be suspended for the specific product line. An evaluation report showing the cause for previous failure followed by the achievement of a compliant test result is mandatory for the reinstatement of the license (European Commission 2013b).

With the increase in green legislations worldwide and the rise of safety requirements in consumer products, eco-testing of textiles and products has become an essential part of product manufacture. Special attention should be paid for the selection of dyes and chemical auxiliaries to keep products free from hazardous substances as listed in the legislations and standards. Knowledge of evolving chemical restrictions, technical competence, and commitment to the understanding of requirements form the basis of essential eco-assurance and eco-certification.

14.3 POLLUTION PREVENTION OPPORTUNITIES

The textile industry releases or transfers toxic chemicals that are usually accompaniments of manufacturing or processing. Releases may be in the form of air emissions, water discharges, releases to land as landfills or as treated application farming, and underground injection that is a contained

TABLE 14.5
PBT Chemical Compound Categories

Sl. No.	Category Name	TRI Category	Reporting Threshold (lb or g)
1.	Dioxin or dioxin-like compounds	N150	0.1 g
2.	Lead compounds	N420	100 lb
3.	Mercury compounds	N458	10 lb
4.	Polycyclic aromatic compounds (PACs)	N590	100 lb

Source: EPA, Persistent Bioaccumulative Toxic (PBT) chemicals covered by the TRI program, March 16, 2014g, http://www2.epa.gov/toxics-release-inventory-tri-program/persistent-bioaccumulative-toxic-pbt-chemicals-covered-tri.

release of fluids into Class I or Class V wells. Transfers of toxic chemicals are to relocate or hand over the wastes to a facility that is separate from the manufacturing or processing industry. The toxic chemical waste may be transferred to publicly owned treatment works (POTWs); transferred to recycling units for recovery or regeneration; transferred to industrial furnaces for energy recovery or incineration; transferred for treatment like neutralization, biological destruction, or physical separation; and transferred to disposal as landfills or underground injection.

In order to provide background information on the pollutant releases of the textile industry, the best source is the Toxic Release Inventory (TRI). The TRI program conducts data quality checks and provides analytical support for efforts of enforcement taken up by the Environmental Protection Agency's (EPA) Office of Enforcement and Compliance Assurance (OECA). Reporting the TRI is a must if the industry is classified as a part of a specific industry sector (manufacturing, mining, power generation, etc.), employs 10 or more full-time employees, and manufactures or processes more than 25,000 lb of a TRI-listed chemical or uses more than 10,000 lb of the TRI-listed chemical in a given year (EPA 2014b). It is mandatory that the industry must submit an extended TRI Form R for each TRI-listed chemical; it manufactures or processes or uses in quantities above the limits prescribed by the EPA. A shorter version of the TRI form may be submitted by industries that does not use or produce a persistent bioaccumulative toxic (PBT) chemical; the chemical has not been manufactured, processed, or used more than 1,000,000 lb; and the total annual waste management (which include releases of the chemical during disposal, recycling, energy recovery, and treatment) of the chemical does not exceed 500 lb (EPA 2014b).

PBT chemicals are toxic in nature and also remain in the environment for long periods of time as they are not readily destroyed; they tend to build up or accumulate in the body tissue and lead to several health issues and hazards. The final list of 16 PBT chemicals along with their thresholds (limit values) and four PBT chemical compound categories are provided under Emergency Planning and Community Right-to-know Act (EPCRA) Section 313 (EPA 2014d,g). The four PBT chemical compound categories are given in Table 14.5.

14.3.1 Pollution Prevention Act 1990

The Pollution Prevention Act of 1990 established the national policy that pollution should be prevented or reduced at source where feasible. Pollution prevention avoids the need for investments in waste management or cleanup activities and will result in economic benefits. The current pollution prevention mandate in federal statutes is to eliminate pollutants in all media like water, air, and land. The Executive Order 1315, Federal Leadership in Environmental, Energy, and Economic

Performance states that the federal agencies shall increase energy efficiency; reduce greenhouse gas emissions from direct and indirect activity; conserve and protect water resources through efficiency, reuse, and storm water management; and eliminate waste, recycle, and prevent pollution. The order also promotes markets for sustainable technologies and environment-friendly materials, products, and services and also encourages federal employees in the achievement of these goals (U.S. Environmental Protection Agency 2012).

14.3.2 Methods for Pollution Prevention

14.3.2.1 Quality Control of Raw Materials

The quality of raw materials is bound to have an impact on the quality of the final product. In 1982, Cotton Inc. introduced the Engineered Fiber Selection Cotton Fiber Management System by which cotton mills were able to lower cost and improve product quality. Eco-Tex certificates were issued to textile manufacturers throughout the world to identify raw material that was less polluting, thereby reducing the use of chemicals, for example, organic cotton. The best method is to work with environmentally responsible purchasing policies and work with suppliers to obtain less polluting raw materials (Clapp et al. 2001).

It is advisable to perform tests on raw material shortly after receipt to determine the environmental effects, proper handling, and emergency procedures for chemicals. Promoting purchase of raw materials in returnable containers will help in transferring chemical wastes to production facility where off-site cleaning can be done.

14.3.2.2 Chemical Substitution

Substitution of harmful substances is an option to reduce the environmental impact of a process. It is the replacement or reduction of hazardous substances in products and processes by less hazardous substances. Easy-care agent applied to denim is formaldehyde based containing 0.1%–1% formaldehyde on weight basis. The risk was threat to workforce during the cutting process. The organization substituted the easy-care agent with a formaldehyde-free product that was three times expensive than the formaldehyde-containing easy-care agent. The results showed that there were no health hazards, and the productivity increased to compensate the price increase (Demirer 2008).

Waste can be reduced by replacing chemical treatments with other treatments. Implementation of cleaner production processes and pollution prevention measures can yield both economic and environmental benefits. A common wastewater treatment consists of screening, equalization, and settling to remove suspended solids. This can be followed by oxidation ponds or an aerobic process, instead of chemicals, which can remove up to 95% BOD (World Bank Group 1998).

14.3.2.3 Process Modification

Continuous preparation processes give many opportunities for wastewater reuse since the waste stream is continuous and has the same characteristics. Similarly, the exhausted dyebath can be reconstituted for the next dyebath. The spent dyebath is pumped into the dyeing machine, and the characteristics of the bath are analyzed. Most auxiliary chemicals will not be removed from the dyebath, and the makeup quantity of the dyebath can be estimated or determined analytically. Computer programs have been developed to analyze the composition of the spent bath, and then it is reconstituted with the required chemicals.

Exhausted reactive dyebath was ozonated, and the color was removed from the effluent. This effluent was reused for two successive dyebaths. Color removal of the effluent was achieved in 5 min contact time for yellow and blue shades at an ozone consumption of 37.5 and 36 g/L, respectively (Sundrarajan et al. 2007). Optimization of process conditions and combining of processes can also reduce pollution.

14.3.2.4 Equipment Modification

Modification and replacement of equipment can prevent pollution. Automated dosing systems in dyeing machines, continuous washing of dyeing machines, and use of low liquor ratio dyeing or processing machines can help to prevent pollution.

14.3.2.5 Good Operating Practices

Good operating practices can improve efficiency and maintain low operating costs by undertaking pollution prevention practices and managerial processes. Optimize cleaning and housekeeping practices and adopt training programs for workers.

Prevention is better than cure and this slogan is very ideal for the textile industry. Pollution prevention techniques not only improve efficiency but also increase profits by providing savings in the relevant field. Moreover, there is safety to the environment and humans. The amount spent for removal of hazardous chemicals can be spent for other beneficial causes and help the environment and industry. Reduction of waste at source followed by recycling and recovery of materials would bring about savings as well as reduce pollutant releases.

14.4 IMPACT OF BANNED SUBSTANCES ON TEXTILES AND APPARELS

The textile and clothing industry produces end products that pass through a long chain of processes like raw material production, spinning, weaving, processing, and end-product fabrication. All these essential processes of the industry have to become eco-friendly if the products are to be certified as eco-textiles or green products. The aim of green production is to adopt pollution control or prevention measures to manufacture products that do not cause any harm to human beings and environment. Apart from these principles, they target on economies in natural resource consumption by innovative technologies and reduction of waste and pollutants. Major textile and apparel buyers around the world are conscious of this awareness and are catering their products to meet the eco-requirements.

Textile products are evaluated in terms of comfort especially when they come in contact with the human body. The thermophysiological and sensorial aspects unite to create a subjective awareness of comfort (Barker 2008). In all cases, the textile or apparel products come in contact with the human skin and become an integral part of the wearer. This may not be so in technical textiles, but this concept suits all other categories of textile and apparel products. There are four textile ecology classes, namely, production ecology, human ecology, performance ecology, and disposal ecology. Human ecology studies the impact of textiles and the chemical content on the health and well-being of humans (Oeko-Tex 100). The more closely the product is in contact with the skin, the stricter the human ecology standards are to be maintained. To maintain ecology standards, eco-testing is required to cover legally banned and controlled substances, chemicals that harm human health, parameters for health protection, and substances targeted by the zero discharge of hazardous chemicals (Oeko-Tex 100a).

The four classes of products are Class I, textile items for babies and toddlers up to 3 years (clothing, toys, bed linen, and terry cloth items); Class II, textiles used close to the skin (underwear, bed linen, and T-shirts); Class III, textiles used away from the skin (jackets and coats); and Class IV, furnishing materials (curtains, draperies, upholstery) (Oeko-Tex 100b). To obtain Oeko-Tex Standard 100 Certification, all components of the textile products that include raw material as fabric, sewing threads, linings, prints, and nontextile accessories such as buttons, zips, and rivets should comply to the required eco-testing parameters. A few test parameters and their limits for garments and accessories are given in Table 14.6.

14.4.1 Azo Dyes/Arylamines

Azo dyes are used as both dyes and pigments on textile and leather products, on cotton, silk, viscose, synthetic fibers, and blue colorants used in wool and nylon furnishings and as alternatives

TABLE 14.6
Eco-Test Parameters and Limit Values in Garments and Accessories

Test Parameters	Product Class			
	I	II	III	IV
pH value	4–7.5	4–7.5	4–9	4–9
Formaldehyde (ppm)	20	75	300	300
Extractable heavy metals				
Lead (ppm)	0.1	0.1	0.1	0.1
Nickel	—	—	—	—
Cadmium (ppm)	0.02	0.01	0.02	0.02
Pesticides (ppm)	0.05	1.0	1.0	1.0
Chlorinated phenols (ppm)	0.05	0.05	0.05	0.05
Color fastness (staining)				
Acid/alkaline perspiration	3–4; 4	3–4; 4	3–4; 4	3–4; 4
OTCs				
TBT, DBT (ppm)	0.05; 1.0	1.0; NA	1.0; NA	1.0; NA

Source: Textile School, Testing in eco textiles, 2014, http://www.textileschool.com/articles/610/testing-in-eco-textiles.

ppm, parts per million; NA, not available.

for tinting denim. In the case of leather, they are found in shoes, watch straps, belts, and leather clothing and furniture.

Arylamines are organic compounds and functional groups that contain amines with an aromatic substituent. A simple structure of arylamine is given in Figure 14.4. There are 24 arylamines that are banned, and their source of origin may be azo dyes, PU materials, and other compounds. On decomposition, the azo dyes form the banned aromatic amines (Garment Tech 2013). The reductive cleavage of the azo dyestuff is shown in Figure 14.5.

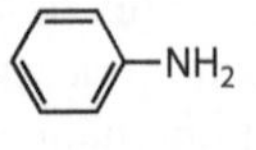

FIGURE 14.4 Chemical structure of arylamine. (From Garment Tech, Why need to be sure you foods, garment and textile fibre, apparel, toys, leather, paper and plastic product are azo dyes free, August 24, 2013, http://garmentstech.com/why-need-to-be-sure-your-gament-textile-fibre-apparel-toys-leather-paper-and-plastic-product-are-azo-dyes-free/.)

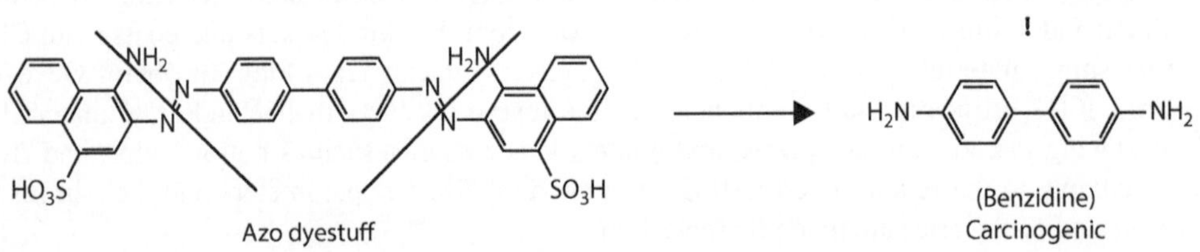

FIGURE 14.5 The reductive cleavage of azo dyestuff. (From Garment Tech, Why need to be sure you foods, garment and textile fibre, apparel, toys, leather, paper and plastic product are azo dyes free, August 24, 2013, http://garmentstech.com/why-need-to-be-sure-your-gament-textile-fibre-apparel-toys-leather-paper-and-plastic-product-are-azo-dyes-free/.)

The azo bond is the most accountable portion of the azo dye. This bond readily undergoes cleavage by biological degradation (biodegradation), enzymatic breakdown (metabolic cleavage) or abiotic degradation, and thermal or photochemical degradation (Øllgaard et al. 1998, Shu et al. 1995, Weber and Adams 1995).

When these compounds enter the human body either unintentionally or by design, the intestinal microbial azoreductase and the liver enzymes metabolize the azo dye compounds to aromatic amines. This has been confirmed by the presence of anaerobic bacteria isolated from cecal or fecal contents of experimental animals and humans showing evidence of the azo linkages to produce aromatic amines. In this study, 26 azo dyes have been included for the study of the reductive cleavage of the said dyes (Chung et al. 1992). The main routes of human exposure to azo dyes are oral ingestion, referring to the suckling nature of babies and young children and dermal absorption, when consumers wear azo-dyed textile products and inhalation, where workers handle newly dyed products at production centers. The report on the risk of cancer by the use of azo dyes on textile and leather goods, issued by the Laboratory of the Government Chemist (LGC), states that all the 22 amines classified as Category I and II carcinogens by EU (14) and by the MAK Commission (8) should be minimized to a negligible level or completely avoided in use as it causes the risk of cancer in consumers and among workers in the environment; further, when the exposure to these amines can be termed as *low*, the associated risks of the occurrence of cancer is very high to give cause for concern (CMPSGI 2012).

Two study reports by the Ecological and Toxicological Association of Dyes and Organic Pigments Manufacturers (ETAD) indicate that the leaching of the azo dyes from the garments to the simulated body fluids is possible, for example, wool and nylon socks leaching into alkaline perspiration simulant; estimations are also given that the total exposure of an adult over a lifetime use of an azo-dyed garment covering the whole body is calculated as 723 µg/kg (dermal exposure) while that of a child sucking a piece of azo-dyed garment (oral exposure) is 13 µg/kg (European Commission [EC] 1999). These are considered undesirable and have given the following recommendations:

- Restriction to the lowest possible limits or complete elimination of azo dyes that have the potential to give rise to aromatic amines that have been classified as Category I and II carcinogens
- Validated analytical methodology and evaluation of the socioeconomic impacts to be developed before implementation of the restrictions
- Data for assessment of mutagenic potential of the amines as well as for other dyes of interest to be formulated

14.4.2 Formaldehyde

Wrinkle-resistant finishes are based on formaldehyde and used commonly for cotton, viscose, linen, and its blends with synthetic fibers. In April 1992, Germany passed a legislation restricting the use of formaldehyde. The legislation states that textile products that are intended for use close to the skin, having a mass content of 0.15% or more free formaldehyde, must be labeled as "contains Formaldehyde – Recommended to wash garment before first use to avoid skin irritations" (Crem 2012).

The Australian National Industrial Chemicals Notification and Assessment Scheme has stated the health problems related to formaldehyde. Formaldehyde (CH_2O) is a VOC and is a colorless gas with a pungent smell. The vapors of formaldehyde are flammable, explosive, and toxic. Inhalation exposure to formaldehyde gas (vapor) as aerosol or mist may cause sensory irritation to carcinogenicity, and dermal exposure to formaldehyde solutions may result in skin sensitization. It can provoke the respiratory system causing lung edema and asthma. Formaldehyde has been classified as a hazardous substance under the Hazardous Substances and New Organisms Act that has hazardous properties that affect skin and eye irritation and carcinogenicity. The acceptable internationally recognized testing method is EN ISO 14184-1:1998 "Textiles, Determination of formaldehyde, Part 1: Free and hydrolysed formaldehyde (water extraction method)." It has been estimated that 28 million tons of formaldehyde solution was consumed across the world in 2006; 1 million European workers are exposed to some degree of formaldehyde, and 1.2%–2.3% of U.S. eczema sufferers have dermatitis caused by textile formaldehyde resin (Australian Competition and Consumer Commission [ACCC]).

Formaldehyde is considered as a textile allergen and attributed to formaldehyde-releasing perma press finishes. The eruptions in the skin is termed as clothing dermatitis and occurs in places where the garment fits snugly; allergy to finishes on bed linens or upholstery may worsen the dermatitis on the back, posterior legs, and other locations that come in contact with the linen or upholstery. Warmth and moisture from perspiration and wearing heavier clothing in winter can lead to textile acute clothing dermatitis (Fowler 2002).

The government product safety policy statement has set the acceptable limits of formaldehyde in clothing. The acceptable limits for infant and baby clothing and for people with sensitive skin are not more than 30 ppm (30 mg/kg); product Class II where a large portion of the textile comes in direct contact with the skin is not >100 ppm (100 mg/kg); product Class III for clothes that are not in direct contact with the skin is not >300 ppm (300 mg/kg) (Ministry of Business, Innovation and Employment).

14.4.3 CHLORINATED PHENOLS: PENTACHLOROPHENOL, TRICHLOROPHENOL, AND 2,3,5,6-TETRACHLOROPHENOLS

These are substances used to prevent fungal growth by bacteria in textiles and leather products. They are highly toxic and hazardous that REACH have limited use of the substance to 1000 ppm that is also specified as 0.1% limit in the European market (Commission Regulation EC 552/2009). Products containing more than 5 ppm PCP are prohibited in Germany. Short-term exposure to PCP causes harm to the liver, kidneys, lungs, gastrointestinal tract, and immune and nervous system and may lead to carcinogenic, renal, and neurological effects. These chemicals are highly stable and not decomposable, thereby causing harm to humans if they accumulate in the environment. The EPA has classified PCP in group B2—probable human carcinogen (Professional Testing and Consulting 2010). The presence of chlorinated phenols is analyzed through the GC–MS determination from the extracted solution. The sample size is 20×20 cm, and the turnaround time is 4 days (The Birmingham Assay Office 2014).

14.4.4 PHTHALATES

These are highly stable substances, low in cost, and ideal for large-scale production that are used as a softening agent of PVC for obtaining flexible PVC in floor coverings, clothing, tapes, polyester buttons, plastisol prints, and polyurethane coatings. About 85% of all flexible PVCs contain phthalates. There are six phthalate softening agents, DEHB, DBP, BBP, DINP, DIDP, and DNOP, which are strictly prohibited in the EU to a 0.1% limit banned under the REACH directive. Young children below 3 years of age are highly vulnerable to the phthalates released under stimulated mouthing conditions (Centexbel 2009). It has been reported that the class of phthalates called *ortho*-phthalates disrupt the endocrine system of laboratory test animals; hence, the second class terephthalates is more commonly used in the United States. The Consumer Product Safety Improvement Act of 2008 (CPSIA) has banned the *ortho*-phthalates, while the terephthalates are restricted to the use of 0.1% in products. The CPSIA aims at children and infant clothing especially those printed with plastisol inks that contain phthalate plasticizers. Phthalate plasticizers are not chemically bound to the PVC and may leach out. Young children and infants who chew the toy or clothing item may absorb some of the chemicals into their system and may be the reasons for many health problems (Branigan 2009). Determination of the presence of phthalates in the textile product is by GC–MS of an extracted solution of a sample 5×5 cm with turnaround time of 4 days.

14.4.5 Heavy Metals

Heavy metals are metallic elements that have relatively high density and are toxic in low concentrations. They are considered carcinogenic and found in dyes as metal complex dyes and in finishes (Rybicki et al. 2004, Tonetti and Innocenti 2009). Chromium-based dye CI Mordant Black 11 is highly toxic and used commonly for dyeing black color. Though metals and metal complex dyes are hazardous in nature, they are not completely prohibited since certain colors like turquoise blue, green, violet, and blue shades cannot be achieved without these dye components. Many national and European standards do not prohibit these dyes, but many voluntary organizations check these parameters for their ecolabeling and certification. Ecolabel and Oeko-Tex Standard 100 establish limits on permitted levels of pesticides, heavy metals, and toxic substances in both the raw material and products.

Toxic effects of heavy metals on humans are well known. They cause negative effects on metabolism, damage to organs, heart disease, disorder to nervous system, and allergies. Accumulation of heavy metals in body tissues and the binding of these metals to enzymes could disrupt the correct functioning of the cells with tumors and mutation development (Apostoli 2002, Muller et al. 2000). Determination of heavy metal content is essential for the safety of workers to exposure during production and for consumers. Several analytical methods like anodic stripping voltammetry, spectrophotometry, atomic absorption spectrophotometry, and x-ray fluorescence spectrophotometry may be used for determination of heavy metals. It has been reported that ICP–MS is a good instrument for measuring ultratrace metals from many samples in a few minutes (Pranaityte 2008).

14.4.6 Organotin Compounds Tributyltin, Dibutyltin, and Monobutyltin

Organotins are a family of substances in which organic (carbon based) groups are bonded to one or more of the available four sites on a tin atom. Generally, the mono-, di-, and trisubstituted organotins are used with the free sites occupied by the other chemical groups. These substances are toxic in nature and cause endocrine disruptive effects on aquatic organisms, can suppress the immune system (immunotoxicity) in humans, and also cause toxicity in reproduction. Biochemical processes like blood-forming mechanisms, disruption of the enzyme system, and damage to the liver and kidney have been identified (Intertek 2009). The toxic effects of organotin compounds, (OTCs) tributyltin (TBT), dibutyltin (DBT), and monobutyltin (MBT) in low concentrations (0.1–1 μM) evaluated in vitro in a neuroblastoma human cell line, were cell death, apoptosis, and DNA fragmentation (Ferreira et al. 2013).

These compounds, DBT, dimethyltin (DMT), and dioctyltin (DOT), are used extensively in the textile industry as heat stabilizers in the processing of PVC; as catalysts to speed up chemical reactions like polymerization of polyurethane, polyester, and silicones, for esterification and transesterification reactions in the production of polyesters; and as biocides where TBT is used for socks, shoes, and sportswear for its antimicrobial functions (HKTDC Research 2009). These compounds are used in a wide range of textile products like sanitary towels, nappies and diapers, carpets, tents, sportswear, socks, and underwear. The EU restrictions on the use of organotins was 0.1% by weight of tin from July 1, 2010, for trisubstituted organostannic compounds, January 1, 2012, for DOT and DBT compounds (Intertek 2009).

14.4.7 Alkylphenols, Alkylphenol Ethoxylates, Dimethyl Fumarate, Chlorinated Organic Carriers, Flame Retardants, Solvents (VOC), Perfluorooctane Sulfonates, Pesticides, and pH Value

The use of alkylphenols (AP) like nonylphenol and octylphenol, alkylphenol ethoxylates (APEOs), nonylphenol ethoxylates, and octylphenol ethoxylates, found in detergents, cleaning agents, or chemicals used for textile and leather production, has been restricted to not more than 0.1% in formulations (Eurofins 2012). These compounds often end up in wastewater and are highly persistent and nondegradable leading to problems to water bodies and environment.

Dimethyl fumarate (DMFU) is a biocide found in many consumer products. It is a skin sensitizer, capable of causing an allergic response affecting the skin and also the respiratory system. The restricted limit is 0.1 mg/kg above, which is banned. DMFU detection limit in articles has found to be in a range of 0.1 mg/kg to several thousand of mg/kg. Skin dermatitis and lesions have been caused by higher ranges of DMFU, namely, 0.47 mg/kg in chair and 3–95 mg/kg in shoes (ECHA 2011). These substances are used by spraying, or they are left in sachets along with the products.

Chlorinated organic carriers (COCs), like chlorobenzenes and chlorotoluenes, can accelerate the expansion of the fabric structure and facilitate the infiltration of the dyes into textiles. Excessive COCs can migrate from textiles to human bodies to cause problems to the central nervous system and skin allergies and are teratogenic and carcinogenic to humans. The restricted limit of use in textiles is 1.0 mg/kg as they are toxic, persistent, and possess bioaccumulative properties (Zhang et al. 2011).

Flame retardants used in textiles are very many. Hexabromocyclododecane (HBCD or HBCDD) used in upholstered furniture, automobile interior textiles, car cushions, and packaging materials causes extensive harm to the environment and living organisms. In October 2008, the ECHA has included HBCD in the substances of very high concern (SVHC) list within the REACH Framework (ECHA 2014). The identification of HBCD is usually difficult and is highly dangerous leading to a worldwide ban. They are fat soluble and tend to accumulate in the food chain. They decompose very slowly and can be easily transported to long distances. HBCDs have been detected in the Arctic. HBCDs are made up of stereoisomers that are identical in appearance but differ in their toxicity and environmental behavior. It has been reported that while flameproof polyesters are being processed, HBCDs can transform to unknown stereoisomers with potentially toxic derivatives. Representatives from 160 countries have passed a resolution at the UN Conference on Chemicals in Geneva that flame-retardant HBCD may not be produced or used for plastics, textiles, electronics, and interiors of buildings as it is included in the persistent organic pollutants (POPs) (Empa 2013).

Perfluorooctane sulfonate (PFOS) and its salts are fluorinated organic compounds that is a persistent bioaccumulative global contaminant that has been used for soil release and stain resistance and surfactant applications in various textiles, upholstery, carpeting, and speciality papers. Occurrence of PFOS in the tissues of humans and wildlife has been reported in Japan. All the fish samples taken from Tokyo Bay showed the presence of PFOS in the fish blood and liver, and the bioconcentration factors were ranging from 274 to 41,600. Concentrations of PFOS in the blood of Japanese human volunteers ranged from 2.4 to 14 ng/mL (Taniyasu et al. 2003). This global contaminant is a key ingredient for the fabric protector Scotchgard and numerous stain repellents. Some of the problems related to PFOS detection are chronic kidney problems and endocrine abnormalities. The restricted limit is 1 µg/m^2 as per the Restricted Substance List 2012.

The EPA defines a VOC as any compound of carbon that reacts with sunlight to create smog. However, there are many VOCs that have negligible photochemical activity. Under the European Law, a VOC is based on evaporation into the atmosphere and has been categorized into five levels or bands depending on the content present as minimal, low, medium, high, and very high. A Newsweek report states that the children who took refuge in the government trailer parks in Baton Rouge were found to be very sick and highly anemic. More than 42% have respiratory infections and high-risk health problems, while more than 50% had mental health problems (Ecotextiles 2010a,b). These problems were due to the presence of VOCs and other contaminants in the trailer parks.

The human skin is usually slightly acidic that inhibits the development of many diseases. Fabrics with pH as neutral (pH 7) or slightly acidic (below 7) are skin friendly, while extreme pH values may cause allergic reactions and damage the skin. Pesticides are used in cotton cultivation to ward off insects and as moth protection during storage. Herbicides are used as weed eradicators and as defoliant chemicals. They may be absorbed in the fibers and can remain in the finished products, though wet processing removes most of them. These substances are considered very toxic as they are easily passed through the skin (Intertek 2012).

The importance of eco-testing is enhanced by analyzing the impact of various chemicals and processes that are involved in the manufacture of product. The impact of using the product, the effect of the content of the product, and the impact of the chemicals involved in manufacturing the product play a vital role on the health and safety of all the living beings in the environment. Many new products enter the market, but the effect of all the contents and processes are brought to light after a certain period of time when manufacturers realize the toxic nature and its effect. This leads to certain restrictions and limitations that are very important in safeguarding the environment and resources from pollution. Thus, eco-testing plays a very essential role in all certifications and labeling process before they are termed as eco-*friendly or green* products.

14.5 ENVIRONMENTAL LAWS: MAJOR STATUTES FOR THE TEXTILE INDUSTRY

In a move to control pollution and protect plant and animal diversity, preservation, restoration, and improvement of the natural environment is important. Environmentalism is an attempt to provide stability between humans and the natural systems on which they rely, to accord a degree of sustainability. Many factors like overpopulation, hydrology, intensive farming and land use, and industrialization have led to severe degradation of environmental health, climate change, war over use of resources, resource depletion, and toxic pollution and waste. This has encouraged the need for conservation of ecosystems, natural resources, and habitat, which has culminated into environmental laws where the polluter pays for environmental crimes to uphold environmental justice.

A collective term that includes international treaties, statutes and regulations, and common laws in order to regulate and reduce the impact of human activity on environment is *environmental law*. It has two major segments, namely, pollution control and remediation and resource conservation and management. Environmental regulations that started to protect the regional and state problems regarding public health soon developed and spread wide to become international laws and multinational agreements (Cheever 2013).

The global nature of environmental challenges has resulted in the emergence of global regulatory regimes that help to provide appropriate global responses and solutions, for example, the United Nations Framework Convention on Climate Change (UNFCC) and the Convention on the Protection of the Ozone Layer (Vienna Convention). Most of the environmental laws are multilateral and are related to biodiversity, atmosphere, land, chemicals, and hazardous wastes and oceans, seas, and water. Multilateral environmental agreements (MEAs) are legal instruments that play an important role in international environmental protection. The role of conference of the parties (COPs) is crucial in shaping laws by interaction with the member states for guidance and consistency and in the implementation of the MEAs to meet the environmental challenges. Amendments are made, and they are adopted by two-thirds majority followed by a ratification process to come into force (Camenzuli).

14.5.1 ENVIRONMENTAL LAWS

- *Resource Conservation and Recovery Act*: It authorizes EPA to control hazardous waste from cradle to grave and sets a framework for the management of nonhazardous solid waste with a mission to protect human health and environment. The compliance and enforcement initiatives include a strategy for addressing environmental challenges and a compliance monitoring program including inspections (EPA 2014c).

- *The Comprehensive Environmental Response, Compensation, and Liability Act*: It authorizes EPA to respond to releases of hazardous substances that would endanger public health, welfare, or the environment. It also enables EPA to force parties responsible for environmental contamination to clean it up and reimburse the remediation costs incurred by EPA (EPA 2012).
- *The EPCRA*: It provides an infrastructure at the state and local levels to plan for chemical emergencies. Industries that store, use, or release chemicals require various reporting requirements that are made publicly available so that interested parties may be informed about dangerous chemicals in their community. This act includes emergency planning and release notification, hazardous chemical inventory reporting, and TRI (EPA 2013b).
- *Clean Water Act (CWA)*: It establishes the structure for regulating discharges of pollutants into the waters of the country and regulating quality standards for surface waters. According to the CWA, it is unlawful to discharge any pollutant from a point source into navigable waters unless permission is obtained (EPA 2014e, NRDC).
- *Toxic Substances Control Act*: It provides EPA the authority to require reporting, record maintenance and testing requirements, and restrictions relating to chemical substances. This act addressed the production, importation, use, and disposal of hazardous chemicals (EPA 2014f).
- *Clean Air Act*: It calls for states and EPA to solve multiple air pollution problems through programs based on the latest information in science and technology. It also has information on the role of science and technology; the role of state, local, tribal, and federal governments; public participation, flexibility and accountability, and clean air and economy (Bearden et al. 2013, EPA 2013a, 2014a).

The textile industry is a major water user, and the CWA is the most important regulation. The effluent guidelines for the point source are listed, as well as the effluent limitation, new source performance standards, and pretreatment standards. Under the Clean Air Act, the national uniform emissions for stationary industrial sources are given. Guidelines are given for reducing the emissions in a cost-effective manner. The Resource Conservation and Recovery Act states that all waste generated from textile organizations must classify solid waste as hazardous or not and the characteristics of the hazardous waste in terms of toxicity, ignitability, corrosivity, or reactivity. Some textile facilities may store hazardous waste in their facility, and when the waste is stored for 90 days, the unit becomes a storage facility under the RCRA and requires a permit in addition to contingency and emergency plans, record keeping and reporting, and use and management. Printing, dying, and fabric coating needs guidelines and is under development. The compliance and enforcement data include facility indexing system (FIS) and integrated data for enforcement analysis (IDEA). The enforcement of the regulations is maintained by inspections and reporting.

Today, ecology is an important agenda, and the environment is turning to green processes for safety and protection. Statutory laws and regulations have been established, and the textile industries and establishments need to conduct their business keeping in mind all the regulatory conditions to manufacture eco-friendly products. Eco-testing plays an important part in declaring and substantiating the authenticity of the characteristics of the product and processes to make environmental green claims. Many hazardous substances have been red listed and restricted to protect all living beings and to ensure the safety of the environment. The impact of these substances on environment and humans has been characterized to make people aware of their hazardous nature and need to take care when using them in the production processes. The need of the hour is to ensure that the environment is kept safe for future generations, and the textile industry must move toward a path of eco-friendliness and sustainability.

REFERENCES

American Textiles Manufacturers Institute (ATMI). 1997. Industrial processes comments on draft of this document. http://nepis.epa.gov/Exe/ZyPDF.cgi/50000HE9.PDF?Dockey=50000HE9.PDF.

Amte G.K. and Mhaskar T.V. 2013. Impact of textile-dyeing industry effluent on some haematological parameters of freshwater fish *Oreochromis mossambicus*. *Nat. Environ. Pollut. Technol.* 12(1), 93–98. http://isindexing.com/isi/papers/1406611680.pdf.

Apostoli P. 2002. Elements in environmental and occupational medicine. *J. Chromatogr. B* 778(1–2), 63–97. http://www.sciencedirect.com/science/article/pii/S037843470100442X.

Australian Competition and Consumer Commission. 2014. Formaldehyde in consumer products. https://www.productsafety.gov.au/content/index.phtml/itemId/973697.

Barker R.L. 2008. Multilevel approach to evaluating the comfort of functional clothing. *J. Fiber Bioeng. Inform.* 1(3), 173–176. doi:10.3993/jfbi2200801, www.jfbi.org/journal/PaperDownload.aspx?PaperID=232.

Bearden D.M., Copeland C., Luther L., McCarthy J.E., Tiemann M., Esworthy R., and Yen J.H. December 20, 2013. Environmental laws: Summaries of major statutes administered by the environmental protection agency. http://fas.org/sgp/crs/misc/RL30798.pdf.

Branigan Ed. October 1, 2009. Phthalates and why its impacting textile screen printing. http://internationalcoatings.wordpress.com/2009/10/01/phthalates-and-why-its-impacting-textile-screen-printing/.

Camenzuli L.K. The development of international environmental law at the Multilateral Environmental Agreements' Conference of the Parties and its validity. http://cmsdata.iucn.org/downloads/cel10_camenzuli.pdf.

Centexbel. 2009. Development of test methods to determine phthalates in textiles. http://www.centexbel.be/files/project-file/PN_phthalates_0.pdf.

Cheever F. November 9, 2013. Environmental law. http://www.britannica.com/EBchecked/topic/765435/environmental-law.

Chung K.T., Stevens S.E., and Cerniglia C.E. 1992. The reduction of azo dyes by the intestinal microflora. *Crit. Rev. Microbiol.* 18(3), 175–190. doi:10.3109/10408419209114557, http://informahealthcare.com/doi/abs/10.3109/10408419209114557.

Clapp T.G., Godfrey B.A., Greeson D., Johnson R.H., Rich C., and Seastrunk C. October 2001. Weaving a quality industry: Quality initiatives reshape the textile industry. http://www.qualitydigest.com/oct01/html/textile.html.

CMPSGI. July 2012. Aromatic azo- and benzidine-based substances. http://www.ec.gc.ca/ese-ees/9E759C59-55E4-45F6-893A-F819EA9CB053/Azo_Technical%20Background_EN.pdf.

Crem B.V. July 2012. German legislation: Formaldehyde in textiles (additional requirements). http://www.cbi.eu/system/files/marketintel/2012_Germany_legislation_Formaldehyde_in_textiles.pdf.

Demirer G.N. September 23–25, 2008. A chemical substitution study for a wet processing textile mill in Turkey. https://www.fona.de/pdf/forum/2008/beitrag/iii.s2_demirer_goeksel_01_presentation_forum_2008.pdf.

Dodd N. October 17, 2012. Revision of the EU ecolabel for textile products—Summary of proposed hazardous substances criteria. http://susproc.jrc.ec.europa.eu/textiles/docs/Ecolabel%20textiles_Hazardous%20substances%20summary.pdf.

ECHA. June 16, 2014. Candidate list of substances of very high concern for authorization. http://echa.europa.eu/web/guest/candidate-list-table.

Ecotextiles. October 27, 2010a. Lead and fabrics. https://oecotextiles.wordpress.com/tag/heavy-metals/.

Ecotextiles. March 17, 2010b. Volatile organic compounds. http://oecotextiles.wordpress.com/2010/03/17/volatile-organic-compounds-vocs/.

Empa. August 26, 2013. Worldwide ban on flame retardant. http://www.sciencedaily.com/releases/2013/08/130826105752.htm?utm_source=rss&utm_medium=rss&utm_campaign=worldwide-ban-on-flame-retardant-science-daily-press-release.

EPA. 1996. The US Environmental Protection Agency, EPA profile of the textile industry, September 1997. http://nepis.epa.gov/Exe/ZyPDF.cgi/50000HE9.PDF?Dockey=50000HE9.PDF.

EPA. September 1997. Profile of the textile industry. http://www.epa.gov/compliance/resources/publications/assistance/sectors/notebooks/textilsn.pdf.

EPA. June 27, 2012. Summary of the Resource Conservation and Recovery Act. http://www.epa.gov/agriculture/lcla.html.

EPA. August 15, 2013a. Air Pollution and Clean Air Act. http://www.epa.gov/air/caa/.
EPA. July 24, 2013b. The Emergency Planning and Community Right-to-know Act (EPCRA). http://www.epa.gov/superfund/contacts/infocenter/epcra.htm.
EPA. July 8, 2014a. Summary of the Clean Air Act. http://www2.epa.gov/laws-regulations/summary-clean-air-act.
EPA. April 22, 2014b. Toxic Release Inventory (TRI) program-basics of TRI reporting. http://www2.epa.gov/toxics-release-inventory-tri-program/basics-tri-reporting.
EPA. July 10, 2014c. Summary of the Resource Conservation and Recovery Act. http://www2.epa.gov/laws-regulations/summary-resource-conservation-and-recovery-act.
EPA. 2014d. PBT final rule summary. http://www2.epa.gov/sites/production/files/documents/pbtrulesum.pdf.
EPA. July 22, 2014e. Summary of the Clean Water Act. http://www2.epa.gov/laws-regulations/summary-clean-water-act.
EPA. July 10, 2014f. Summary of the Toxic Substances Control Act. http://www2.epa.gov/laws-regulations/summary-toxic-substances-control-act.
EPA. March 16, 2014g. Persistent Bioaccumulative Toxic (PBT) chemicals covered by the TRI program. http://www2.epa.gov/toxics-release-inventory-tri-program/persistent-bioaccumulative-toxic-pbt-chemicals-covered-tri.
Eurofins. 2012. Alkylphenol ethoxylates (APEO) in textiles. http://www.eurofins.com/media/17648/apeo%20in%20textiles%20-%20en.pdf.
European Chemical Agency (ECHA). March 8, 2011. Opinion on an Annex XV dossier proposing restrictions on dimethylfumarate (DMFu). http://echa.europa.eu/documents/10162/13641/dmfu_rac_restriction_opinion_20110308_en.pdf.
European Commission. January 18, 1999. Opinion on risk of cancer caused by textiles and leather goods coloured with azo-dyes. http://ec.europa.eu/health/scientific_committees/environmental_risks/opinions/sctee/sct_out27_en.htm.
European Commission, 2007. Textile waste minimization. http://ec.europa.eu/environment/life/project/Projects/index.cfm?fuseaction=home.showFile&rep=file&fil=LIFE05_ENV_E_000285_LAYMAN.pdf, Accessed August 29, 2014.
European Commission. February 2013a. EU Ecolabel and Green Public Procurement—Stakeholders. Annexes 3,4,5 restricted substances list, Oeko-Tex 100 read across and risk matrix. http://susproc.jrc.ec.europa.eu/textiles/docs/130206%20Ecolabel%20textiles_Restricted%20Substance%20List_Draft%20v2.pdf.
European Commission. May 2013b. Establishing the ecological criteria for award of the EU ecolabel for textile products. http://susproc.jrc.ec.europa.eu/textiles/docs/130522%20Textiles_EUEB%20Criteria%20document%20v2.pdf.
Ferreira M., Blanco L., Garrido A., Vietes J.M., and Cabado A.G. May 1, 2013. In vitro approaches to evaluate toxicity induced by organotin compounds of tributyltin (TBT), dibutyltin (DBT), and monobutyltin (MBT) in neuroblastoma cells. *J. Agric. Food Chem.* 61(17), 4195–4203. doi:10.1021/jf3050186, http://www.ncbi.nlm.nih.gov/pubmed/23534342.
Fowler J.F. 2002. Formaldehyde as a textile allergen. In: Elsner P., Hatch K.L., and Alberti W.W. (eds.), *Textiles andtheSkin*,Vol.31.Karger,NewYork,pp.156–165.http://books.google.co.in/books?id=QNabhdQIyzcC&pg=PA164&lpg=PA164&dq=%E2%80%9CFormaldehyde-related+textile+allergy:+an+update&source=bl&ots=wW8y9Iet2N&sig=gyiCShCaAFRQnbpX809PbrsBMBs&hl=en&sa=X&ei=NqEHVO_yNs-6ggSf44HgAQ&ved=0CDkQ6AEwBA#v=onepage&q=%E2%80%9CFormaldehyde-related%20textile%20allergy%3A%20an%20update&f=false.
Garment Tech. August 24, 2013. Why need to be sure you foods, garment and textile fibre, apparel, toys, leather, paper and plastic product are azo dyes free. http://garmentstech.com/why-need-to-be-sure-your-gament-textile-fibre-apparel-toys-leather-paper-and-plastic-product-are-azo-dyes-free/.
Gereffi G. April 4, 2002. Responding to globalization: Societies, groups, and individuals. http://www.colorado.edu/IBS/PEC/gadconf/papers/gereffi.pdf.
Ghosh P. and Gangopadhyay R. 2000. Photofunctionalization of cellulose and lignocellulose fibres using photoactive organic acids. *Eur. Polym. J.* 3, 625–634. http://www.sciencedirect.com/science/article/pii/S0014305799000932.
Hongkong Trade Development Council (HKTDC) Research. August 6, 2009. Organotin compounds in textile articles. http://product-industries-research.hktdc.com/business-news/article/Garments-Textiles/Organotin-Compounds-in-Textile-Articles/psls/en/1/1X000000/1X06LD5S.htm.

ILO. 2005. Better factories Cambodia: Facts and figures about Cambodia's garment industry. http://www.betterfactories.org/content/documents/Facts%20and%20Figures.pdf.

Intertek. June 2009. EU—Organotin restrictions extended to consumer goods. http://www.intertek.com/uploadedFiles/Intertek/Divisions/Consumer_Goods/Media/PDFs/Sparkles/2009/sparkle454.pdf.

Intertek. 2012. Textile testing. http://www.intertek.com/textiles/testing/apparel/.

Jain C. and Easton J. December 2010. Sustainability in textile processing. http://www.dystar.com/my_uploads/file/sustainability_textile_processing.pdf, 34—hazardous substances.

Keane J. and te Velde D.W. May 7, 2008. The role of textile and clothing industries in growth and development strategies. http://www.odi.org/sites/odi.org.uk/files/odi-assets/publications-opinion-files/3361.pdf.

Kentucky River Basin Assessment Report. August 2000. Water quality parameters. http://www.uky.edu/WaterResources/Watershed/KRB_AR/krww_parameters.htm.

Ministry of Business, Innovation and Employment. 2014. https://www.consumeraffairs.govt.nz/pdf-library/publications/Acceptable-Limits-of-Formaldehyde-in-Clothing-and-Other-Textiles.pdf.

Muller J., Sigel R.K.O., and Lippert B. 2000. Heavy metal mutagenicity: Insights from bioinorganic model chemistry. *J. Inorg. Biochem.* 79, 261–265. http://www.sciencedirect.com/science/article/pii/S0162013499001798.

National Pollution Inventory (NPI). July 1999. Emission estimation technique manual for textile and clothing industry. http://www2.unitar.org/cwm/publications/cbl/prtr/pdf/cat5/Australia_ftextile.pdf.

NRDC. 2014. Environmental laws and treaties. http://www.nrdc.org/reference/laws.asp.

Oeko-Tex 100. 2014. https://www.oeko-tex.com/en/manufacturers/textile_ecology/textile_ecology_start.html.

Oeko-Tex 100a. http://www.made-by.org/consultancy/standards/oeko-tex-100/.

Oeko-Tex 100b. https://www.oekotex.com/en/manufacturers/concept/oeko_tex_standard_100/oeko_tex_standard_100.xhtml.

Øllgaard H., Frost L., Galster J., and Hansen O.C. 1998. *Survey of Azo-Colorants in Denmark: Consumption, Use, Health and Environmental Aspects*. Ministry of Environment and Energy, Danish Environmental Protection Agency, Copenhagen, Denmark. www2.mst.dk/udgiv/publications/1999/87-7909-548-8/pdf/87-7909-546-1.pdf.

Pranaityte B., Padarauskas A., and Naujalis E. 2008. Determination of metals in textiles by ICP-MS following extraction with synthetic gastric juice. *Chemija* 19(3–4), 43–47. http://www.lmaleidykla.lt/publ/0235-7216/2008/3-4/43-47.pdf.

Professional Testing and Consulting. 2010. Pentachlorophenol (PCP). http://www.professional-laboratory.com/product_44.html.

River Assessment Monitoring Project (RAMP). 2000. Background information for watershed watch parameters. http://water.nr.state.ky.us/ww/ramp/.

Retail Forum for Sustainability (RFS). August 2013. Sustainability of textiles. http://ec.europa.eu/environment/industry/retail/pdf/issue_paper_textiles.pdf.

Rybicki E., Swiech T., Lesniewska E., Albinska J., Szynkowska M.I., Parjczak T., and Sypniewski S. 2004. Changes in hazardous substances in cotton after mechanical and chemical treatments of textiles. *Fibres Text. East. Eur.* 12(2), 67–73. http://www.fibtex.lodz.pl/46_18_67.pdf.

Sharma S. September 2013. Eco textile processing and its role in sustainable development. http://www.indiantextilejournal.com/articles/FAdetails.asp?id=5518.

Shu H.-Y. and Huang C.-R. 1995. Degradation of commercial azo dyes in water using ozonation and UV enhanced ozonation process. *Chemosphere* 31(8), 3813–3825. http://www.sciencedirect.com/science/article/pii/0045653595002557.

Sundrarajan M., Vishnu G., and Joseph K. 2007. Ozonation of light shaded exhausted reactive dye bath for reuse. http://www.sciencedirect.com/science/article/pii/S0143720806002269.

Taniyasu S., Kannan K., Horii Y., Hanari N., and Yamashita N. 2003. A survey of perfluorooctane sulphonate and related perfluorinated organic compounds in water, fish, birds and humans from Japan. *Environ. Sci. Technol.* 37(12), 2634–2639. http://www.chem.toronto.edu/coursenotes/CHM410/notes/Lecture%206%20extraction%20paper%203.pdf.

Textile School. 2014. Testing in eco textiles. http://www.textileschool.com/articles/610/testing-in-eco-textiles.

The Birmingham Assay Office. September 5, 2014. Textile testing. http://www.thelaboratory.co.uk/textiles_testing.html.

Tonetti C. and Innocenti R. 2009. Determination of heavy metals in textile materials by atomic absorption spectrometry: Verification of the test method. *AUTEX Res. J.* 9(2), 66–70. http://www.autexrj.com/cms/zalaczone_pliki/0301.pdf.

U.S. Environmental Protection Agency. February 16, 2012. Pollution Prevention (P2)—Laws and policy. http://www.epa.gov/oppt/p2home/pubs/laws.htm.

Water Resource Monitoring Site (WRM). 2000. Background information for watershed watch parameters. http://www.uky.edu/WaterResources/Watershed/KRB_AR/krww_parameters.htm.

Weber E.J. and Adams R.L. 1995. Chemicals and sediment-mediated reduction of the azo dye disperse blue 79. *Environ. Sci. Technol.* 29(5), 1163–1170. http://pubs.acs.org/doi/abs/10.1021/es00005a005.

World Bank Group. July 1998. Textiles pollution prevention and abatement handbook. http://www.ifc.org/wps/wcm/connect/55bd0c0048855a26853cd76a6515bb18/textile_PPAH.pdf?MOD=AJPERES&CACHEID=55bd0c0048855a26853cd76a6515bb18.

Zhang W.-Y., Sun Y., Wang C.-M., and Wu C.-Y. 2011. Solid-phase microfibers based on polyethylene glycol modified single walled carbon nanotubes for the determination of chlorinated organic carriers in textiles. *Anal. Bioanal. Chem.* 401, 1685–1693. doi:10.1007/s00216-011-5211-0, http://link.springer.com/article/10.1007/s00216-011-5211-0#page-1.

15 Functional Aspects, Ecotesting, and Environmental Impact of Natural Dyes

L. Ammayappan and Seiko Jose

CONTENTS

- 15.1 Introduction ... 334
- 15.2 Colors ... 334
  - 15.2.1 Dye ... 334
    - 15.2.1.1 Chromophore and Auxochrome ... 335
  - 15.2.2 Natural Dye ... 335
    - 15.2.2.1 Factors Responsible for Affinity of Natural Dyes ... 335
  - 15.2.3 Dyeing Mechanism ... 336
    - 15.2.3.1 Color of Natural Dye–Mordant Complex ... 337
    - 15.2.3.2 Chemical Bonding between Mordant, Fiber, and Natural Dye ... 337
- 15.3 Functional Properties of Natural Dyes ... 338
  - 15.3.1 Antimicrobial Property ... 338
    - 15.3.1.1 Antibacterial Property ... 338
    - 15.3.1.2 Antifungal Property ... 339
  - 15.3.2 Deodorizing Property ... 340
  - 15.3.3 UV Protection Property ... 341
  - 15.3.4 Antioxidant Property ... 342
- 15.4 Environmental Impacts of Natural Dyes ... 342
- 15.5 Ecotesting ... 343
  - 15.5.1 Ecolabeling ... 343
    - 15.5.1.1 Oeko-Tex ... 344
    - 15.5.1.2 Blue Sign ... 344
    - 15.5.1.3 International Association of Natural Textile Industry ... 344
    - 15.5.1.4 KRAV ... 344
    - 15.5.1.5 JOCA ... 344
    - 15.5.1.6 GOTS ... 344
    - 15.5.1.7 Ecolabel ... 344
- 15.6 Sustainable Advantages of Natural Dyes ... 344
  - 15.6.1 Sources ... 345
  - 15.6.2 Extraction Procedure ... 345
  - 15.6.3 Color ... 345
  - 15.6.4 Functional Property ... 345
  - 15.6.5 Eco-Friendliness ... 345
- 15.7 Future Perspectives ... 346
- 15.8 Conclusive Remarks ... 347
- References ... 347

15.1 INTRODUCTION

Humans are always fascinated with color. Throughout civilization, they invented hand spinning, knitting, handloom weaving, coloration, and finally fashion clothing with unique or various colors. Aristocrats felt that apparel with distinct color would symbolize their self-esteem, culture, identity, and social group as well as wealth. Designing of apparel in olden days was mainly dependent on the design, color, and end use. To impart different colors, craft dyers identified various sources for coloration of textiles from nature and termed the coloring matter as natural dye. Dyers extracted the natural dyes from locally available sources like vegetables, barks, roots, flower, dried insects, fruits, and even some colored soils. They classified each natural dye depending on its source, extraction procedure, and depth and tint of the color. For the development of fashion apparels for upper class and ruling people, traders imported natural dyes that had superior color value from other countries. So along with other agricultural commodities, business people traded natural dyes all over the world, which also enhanced the economy of countries (Ammayappan 2009).

During the 1920s, synthetic dyestuff and pigments came into the market in a big way to meet the requirement of dyeing and the apparel industry. Due to the exploitation of synthetic dyes, the natural dye market was invaded slowly. Most of the dyeing industries consumed synthetic dyes from the 1960s to 1990s, and as a result, gradually, natural dyes had lost its passion in the fashion market. Later on, it was found that azo-based (–N=N–) synthetic dyes had a carcinogenic effect on humans (Medvedev et al. 1988; Nadiger 2001; Umbuzeiro et al. 2005; Mathur and Bhatnagar 2007). Accordingly, in September 1994, Germany banned the use of azo dyes for textile dyeing, and the European Union (EU) imposed the ban on synthetic dyes responsible for carcinogenicity. Subsequently, the awareness on carcinogenic components of synthetic dyes present in the apparel was spreading well to many developing countries. The EU gradually established standards, certifying agencies, and labeling practices in order to prevent the application of banned dyestuffs and chemicals on apparel. Each certifying agency has its own protocols and standards to identify and evaluate the banned compounds. Buyers preferred to have any exportable apparel with eco-label i.e. it must be free from banned chemicals and dyes. Due to global warming, different associations around the world stressed that all textile industries should follow eco-friendly processing concept. Such awareness enabled industries to follow eco-friendly processing and to augment the reintroduction of natural dyes, since natural dyes are biodegradable, compatible with the environment, and nonallergenic to the skin (Bechtold et al. 2003; Ali and El-Mohamedy 2011).

This chapter deals with natural dyes, their dyeing theory, and functional aspects, the importance of ecotesting, and environmental impacts of natural dye processing. Due to the predicament on natural dyes for their less reproducibility and low productivity, their sustainability implication is also exploited.

15.2 COLORS

Nature is enriched with flora and fauna, so people can perceive color easily. Since color is associated with human emotions, qualities, seasons, festivals, passion, etc., it played a major role in the civilization of ancient people. Studies revealed that color can influence the behavior and mood of people since it induces a pleasant frame of mind. Generally, an apparel can be designed based on occasions, wealth of a wearer, and cost, that is, black dress for award, white for pure, and purple for luxury. Also, people believe that color is a powerful tool to attract a crowd's attention, to communicate any information politely, and to express working style.

15.2.1 Dye

An organic compound used to impart color on textile material is called a dye. Generally, dye has an affinity toward the functional groups of textile fiber, so it is firmly entrapped inside the fiber polymer through chemical bonding. The essential properties of the dye are intense color, water solubility, substantiveness toward fiber, thermal stability, and good fastness properties.

15.2.1.1 Chromophore and Auxochrome

Dye is composed of chromophores and auxochromes. According to Witt's theory, chromophore is a chemical group responsible for the color of the dye, which absorbs the visible region (400–700 nm) of the light and reflects its complimentary color, for example, azo group (–N=N) and phenoxy groups (Ph=C=O). Auxochrome is a functional group present in the dye molecule and induces the final color by delocalizing the free electrons present in the chromophore, for example, amino group (–NH_2) and hydroxyl group (–OH). Dyes may have either one or more than one chromophores and auxochromes (Trotman 1984).

15.2.2 NATURAL DYE

Ancient humans used fur, leather, and grass to protect themselves from changes in climatic conditions. Throughout civilization, gradually, they invented hand spinning and handloom weaving for making cloth in order to have protection against changing environmental conditions. They imparted different colors on cloth in order to develop designs and differentiate garments from one another. Literature information indicated that indigo-dyed linen cloth was used to wrap Egyptian mummies. Coloring with natural dyes was initially practiced during the Bronze Age in Europe. Primitive dyeing techniques were practiced by sticking plant fabric or rubbing crushed pigments on cloth. Among the natural dyes, tyrian purple– and crimson kermes–dyed apparels were considered as highly prized and luxurious items in the ancient and medieval ages. Blue dye derived from *Indigofera tinctoria* had been known in India for about 4000 years (Cardon 2007; Ammayappan 2009).

Artisans used natural dyes for a long time, even before its scientific basis has been known. Prior to the sixteenth century, natural dyeing was a closely guarded technology. For example, the production of Turkey red by the application of the madder root on alum-mordanted cotton was a secret. The dyeing technology had been unexposed for more than 250 years to other parts of the world. However to exploit this technology, many European industrial spies lost their lives. In countries like India, this traditional knowledge was intimately connected with culture, art, religious beliefs, and medicinal value. However, such knowledge rapidly disappeared with the encroachment of western lifestyles into the remotest parts of the globe and the loss of the last generation of *living treasures* among traditional dyers. The global demand for natural dyes in coloration of textiles is approximately 0.1 million tonnes, which is equivalent to only 1% of the world's synthetic dye consumption (Sivakumar et al. 2011; Anon 2013).

15.2.2.1 Factors Responsible for Affinity of Natural Dyes

Dyeing theory is initially formulated for synthetic dyes, since their structure, molecular weight, and characteristics are well known (Bancroft 1914; Jones 1967). Factors responsible for the exhaustion of synthetic dyes also resemble with natural dyes. Dye molecules are initially adsorbed on the surface of the fiber and then diffused inside the fiber matrix and fixed with functional groups of fiber polymer by means of chemical bonding. The affinity of natural dye on textile fiber depends on various factors like accessibility of fiber, preparatory process, chemical constitution of natural dye, mordanting, dyeing temperature, and pH (Ammayappan 2009; Ammayappan and Chakraborty 2013).

15.2.2.1.1 Accessibility of Fiber

Natural fibers have hydrophilic property with a sufficient amount of accessible regions (amorphous/disoriented region) and functional groups like hydroxyl, carboxyl, and amino groups. These factors are mainly responsible for the affinity toward natural dyes. Synthetic fibers do not have sufficient accessible regions and they are less hydrophilic in nature; natural dyes do not possess affinity toward them (Trotman 1984).

15.2.2.1.2 Preparatory Process

A natural dye needs an impurity-free and hydrophilic textile fiber for uniform fixation. Conventional preparatory processes like desizing, scouring, and bleaching are required to improve the exhaustion of natural dyes by removing the natural as well as added impurities present in the fiber.

15.2.2.1.3 Chemical Constitution of Natural Dyes

Each natural dye is different from one another by its chemical constitution. Natural dye with larger chromophore and more number of auxochrome groups can give better color yield with good fastness properties.

15.2.2.1.4 Mordanting

Mordanting can be done before dyeing (premordanting), along with dyeing (metamordanting), or after dyeing (postmordanting). A single natural dye can produce different colors with the same mordant at different mordanting methods. Generally, mordanting for longer duration at moderate temperature can give satisfactory fastness properties.

15.2.2.1.5 Dyeing Temperature

Heat of dyeing is the difference between the energy of bond formation and bond breaking during dyeing. Generally, the heat of dyeing of protein fiber with natural dyes is exothermic; that is, when there is an increase in dyeing temperature above 70°C, the dyeing affinity toward fiber will be decreased and so there will be reduction in chemical bonding between the dye and fiber at >80°C. Due to exothermic reaction, dyeing of wool fiber with ionic natural dyes is preferred at moderate temperature (70°C–80°C) in longer duration (2–6 h) (Temani et al. 2011). At high temperature, there is a possibility of uneven dyeing due to rapid migration of natural dye. Some natural dyes require high temperature to form a strong chemical bonding with functional groups of cellulosic fibres.

15.2.2.1.6 pH

The pH of dyeing bath can influence the dyeing rate of natural dye, since most of the natural dyes are ionic in nature. Similarly, pH can also induce the ionic nature of the natural fiber, that is, wool fiber will be amphoteric at pH 7, while in acidic pH, it forms additional cationic groups. The exhaustion of natural dye on wool fiber is higher at acidic conditions than neutral to alkaline pH. At pH > 4, natural dye can be easily chelated with metal ion, and so more insoluble complex would be formed in the dye bath and that will ultimately reduce the affinity of natural dye on fiber (Amarsinghee and Williams 2007). Due to the presence of delocalized electrons and aromatic moiety, some of the natural dyes are very sensitive to pH and form different colors in different mediums (Cardon 2007).

15.2.3 Dyeing Mechanism

Generally, dyeing of fibers with natural dyes is undertaken in the following steps: migration, adsorption, diffusion, and fixation. After dissolution of natural dye, it migrates from the aqueous bath to the fiber surface due to the concentration gradient, and the rate of migration depends on the concentration of natural dye, pH of dye bath, dyeing temperature, additives, mechanical agitation, and hydrophilicity of fibers. After migration, natural dye molecules adsorb on the fiber surface through physical adsorption and then gradually accumulate. Due to the concentration gradient between the fiber surface and fiber matrix, dye molecules diffuse inside the fiber, which is the rate-determining step. At high temperatures, natural dye molecules reach the activation energy required to react with the functional groups of fiber; as a result, chemical bonding will occur. Due to strong bonding between natural dye and fiber, it is fixed well inside the fiber. Limited research works have been done on the dyeing behavior of natural dye.

Dyeing of natural dye extracted from *Laccifer lacca* with cotton fibers followed the pseudo second order kinetic model. The dyeing rate of constant of *Laccifer lacca* inferred that the adsorption of this natural dye is faster in silk than cotton (Chairat et al. 2005). The activation energies of lac dye required to diffuse in cotton, nylon, silk, and wool fibers were 42.4, 45.31, 87.85, and 107.15 kJ/mol, respectively, which indicated that lac dye followed the chemisorption property, that is, they can form a chemical bond between dye and fiber (Chairat et al. 2008; Wei et al. 2013). However, the activation energy of natural dye can be reduced by ultrasonication, which will enhance the diffusion process by cavitational effect. It is observed that the reaction between natural dye molecules and aluminum-mordanted jute fibers is endothermic, so diffusion and fixation of dye molecules are increased with increase in dyeing temperature. After saturation of the functional groups present in the jute fiber with natural dye, further dye absorption is reduced (Samantha et al. 2012).

15.2.3.1 Color of Natural Dye–Mordant Complex

Elements from the d-block of the periodic table like Cr, Fe, Co, Ni, Cu, and Zn have d-orbits; when ligands, that is, coordination-forming groups –OH, –COOH, –NH$_2$, –CONH, –CN, and –Cl, are approaching the d-block ions, an octahedral field will be generated. Due to the octahedral ligand field, the d-orbits of metal ion may split into different energy levels, and a coordinate bond may be formed between ligand(s) and the d-orbits of the metal ion. The energy difference between the two energy sets is responsible for the blue-green absorption of metal complex. Each dye–metal coordination complex differs in shape, composition, geometry, and isomerism, and these factors decide the final color of natural dye on apparel (Trotman 1984; Ammayappan et al. 2014).

15.2.3.2 Chemical Bonding between Mordant, Fiber, and Natural Dye

Most of the natural dyes are not substantive toward natural fiber. To improve the fixation of natural dyes on textile fibers, traditionally, mordants like metallic salts, tannins, and biopolymers are used. Mordant forms a chemical bond with fiber and natural dye in different ways, and the extent of bonding depends on the functional and ionic groups present in the fiber and natural dye (Septhym et al. 2007).

Alum is used as metallic mordant. Aluminum ion is a trivalent cation and it initially binds with either the negative charge present on the surface of fibers or the natural dye anion (e.g., laccaic acid from lac dye). The functional groups present in the natural dye molecule (–OH, >C=O) and aluminum ion form the coordinated complex inside the fiber (Chairat et al. 2005). It is attributed that hydrogen atom of the phenolic hydroxyl group of the dye molecule (Ph–OH) is substituted by metal cation (Ph–OM) through coordinate bonding (Borges et al. 2012). The fiber–mordant–dye coordination complex can give different colors (Samantha and Agarwal 2009).

Tannins are water-soluble phenolic compounds with higher molecular weights and they can be used as natural mordants. They form the following chemical bonds with the fiber other than the coordinate bond (Ammayappan and Moses 2007):

- Inter- and intrahydrogen bonds between the hydroxyl/phenolic groups of mordant and amino/amido/hydroxyl/carboxyl groups of fiber.
- Ionic bond between charged anionic groups of mordant and cationic groups of fiber.

Nonpolar hydrocarbon regions of the protein fibers are responsible for the entrapment of the nonpolar natural dye inside the fiber matrix (Das et al. 2008). Mordanting method also plays a major role in the chemical bonding between fiber/dye molecule and mordant. Generally, premordanting gives better color yield than other mordanting methods.

Most of the natural dyes are ionic in nature and they are easily exhausted by protein fibers like wool and silk, while cellulosic fibers show moderate substantiveness. To improve the affinity of ionic natural dyes toward cellulosic fibers, anionic reactive groups like dichloro-s-triazinyl groups can be introduced. The reactive functional group is covalently bonded to the cellulosic fibers and gives anionic sites. The additional anionic groups can interact with cationic-based natural dye like berberine and improve its exhaustion as well as fixation on the cellulosic fibers. Similarly, cationic biopolymer like chitosan can also be applied on cellulosic fibers to improve the fixation of anionic dyes (Kim and Son 2005).

15.3 FUNCTIONAL PROPERTIES OF NATURAL DYES

Finishing is a textile process, used to enhance comfort, aesthetics, durability, and functionality of any textile product. Consumers prefer apparel which must be comfortable, good for traveling, and versatile with casual look. Finishing process also diversifies the use of textile product and satisfies the consumer demand as per end use. Functional finishing plays an important role for the development of apparels for specific end uses (Ammayappan et al. 2013). To impart functionality on apparels, specialty functional polymers can be applied after dyeing. However, some natural dyes have the potential to impart antimicrobial, UV protection, and deodorizing finishing in addition to color on apparels. The functional property of natural dye mainly depends on its chemical components (Feng et al. 2007). Some natural dyes have antimicrobial, deodorizing, UV protection, and antioxidant properties.

15.3.1 ANTIMICROBIAL PROPERTY

Natural fibers are susceptible to the attack of microorganisms, since they provide suitable conditions such as warmth, moisture, and nutrients for the growth of microorganisms (Purwar and Joshi 2004). Apparels used for underwear, sportswear, and medical textiles can be easily affected by microorganisms and will produce objectionable odor and cause skin infection as well as strength deterioration of apparels. Antimicrobial agents are used to either inhibit the growth or kill microorganisms by damaging their cell wall (Madigan and Martinko 2005). Antimicrobial agents such as nanosilver, quaternary ammonium salts, polyhexamethylene biguanide, triclosan, regenerable *N*-halamine compounds, and peroxyacids are used in medical textiles (Borkow and Gabbay 2008; Gao and Cranston 2008).

In olden days, dyers found that some natural dyes inhibited the growth of microorganisms, and so they used these dyes to impart antimicrobial finishing on fabrics. The antimicrobial activity of natural dyes mainly depends on the presence of functional groups like polyphenols (simple phenols, phenolic acids, quinines, flavonoids, flavones, flavonols, tannins, and coumarins), terpenoids, essential oils, alkaloids, lectins, polypeptides, and polyacetylenes (Cowan 1999; Gupta et al. 2004; Sarkar and Dhandapani 2009). Cationic groups like protonated amino groups present in a biopolymer can also inhibit the growth of gram-negative bacteria, while anionic groups like phenoxyl, carboxyl, and hydroxyl groups inhibit the growth of gram-positive bacteria. However, due to limited affinity toward the fiber, generally, natural dyes offer moderate antimicrobial property with respect to washing.

15.3.1.1 Antibacterial Property

Research works were carried on the antibacterial properties of natural dyes on textiles. The natural dye extracted from *Arnebia nobilis* was used to inhibit the growth of both *Staphylococcus aureus* and *Escherichia coli* on nylon, polyester, silk, wool, cotton, and acrylic fabrics. Due to the presence of alkannin β,β-dimethylacrylate, dyed textiles retained antibacterial activity up to 25 home launderings; however, it was reduced after exposure to light for 2 h (Aorara et al. 2012). In addition to antibacterial activity, the root bark of *Arnebia nobilis* showed antithrombotic activity, which are applicable for medicinal purposes (Khatoon et al. 1994). Quinone groups present in the natural dye can inhibit the growth of bacteria on fibers due to their oxidative stress and alkylation on bacteria. Natural dyes rich in naphthoquinones such as *Lawson* from henna, *juglone* from walnut, and *lapachol* from alkanet were reported to have antibacterial properties (Schuerch and Wehrli 1978; Wagner et al. 1989; Machado et al. 2003).

Eugenol is the major component present in the essential oil extracted from *Eugenia caryophyllus*, which inhibits the growth of some pathogenic bacteria (Agar et al. 1999). Curcumin is the dried

Functional Aspects, Ecotesting, and Environmental Impact of Natural Dyes

rhizome of the plant *Curcuma domestica* and chemically called as 1–7-*bis*(4-hydroxy-3-methoxyphenyl)-1,6-heptadiene-3,5-dione. The presence of methoxyl and phenoxyl groups of alkaloid, turmerol, and valeric acid components can inhibit the growth of *E. coli*, *Bacillus subtilis*, and *S. aureus* (Chandrana et al. 2005; Ammayappan and Jeyakodi Moses 2009; Joshi et al. 2009). Wool fabrics dyed with natural dye extracted from curcumin showed 45% and 30% inhibition rates against gram-positive and gram-negative bacteria, respectively, even after 30 cycles of home laundering (Han and Yang 2005).

Henna (*Lawsonia inermis*) contains hennotannic acid, which is chemically written as 2-hydroxy-1,4-naphthoquinone. It imparts orange hue to protein fibers with moderate fastness properties (Gulranjani et al. 1992). It is observed that chitosan biopolymer pretreated and henna-dyed woolen fabric showed 100% reduction in the growth of both *S. aureus* and *E. coli*, while only henna dyed woolen fabric reduced 40% of the growth of bacteria (Giridev et al. 2009). The methanolic extract of *Artocarpus heterophyllus* exhibits a broad spectrum of antibacterial activity on cotton fabric, while it is not active against fungi (Khan et al. 2003).

The bark of *Acacia catechu* contains catechin, phlobatannin, and quercitrin. These components are responsible for the inhibition of the growth of four strains of bacteria (Das et al. 2011). Neem extract was used along with polymeric resin glyoxal by pad–dry–cure method to impart anticrease as well as antibacterial finishing on polyester/cotton blended fabric. Results showed that antibacterial activity was higher against *B. subtilis* as compared to *Proteus vulgaris*. The inhibition of the growth of gram-positive bacteria by neem extract–applied fabric was more than 90% as compared to the untreated sample (Joshi et al. 2007). Woolen fabrics are dyed with extracts of green tea, madder, turmeric, and saffron petals in the presence of different metallic mordants. All of the naturally dyed woolen fabrics resisted the growth of *S. aureus*, *E. coli*, and *Pseudomonas aeruginosa*, out of which aluminum sulfate–mordanted woolen fabric has shown an antibacterial activity up to five cycles of washing and 300 min of light exposure (Ghaheh et al. 2014).

Some natural dyes like walnut extract can also be used to impart antibacterial finishing on synthetic fibers like polyamide fiber (Mirjalili and Karimi 2013). Nylon fabric was plasma-treated and then mordanted with copper sulfate and dyed with *Berberis vulgaris* extract. Berberis-dyed nylon fabric showed acceptable antibacterial activity against both gram-negative and gram-positive bacteria (Kim and Son 2005; Haji et al. 2013). Viscose fabric was dyed with extracts of walnut, turmeric, and henna under different pH with alum and potassium dichromate as mordants. Fabric dyed with turmeric under alkaline condition showed better reduction in bacterial growth than fabric dyed under acidic condition (Mirjalili and Abbasipour 2013).

15.3.1.2 Antifungal Property

Most of the fungi are beneficial for human life and are generally involved in the biodegradation of natural products. But some species of fungi can cause infections in the human body through skin, lung, and nasal passages (Espinel 1996). Antifungal agents such as azole, polyene, allylamine, and thiocarbamate are used to inhibit the synthesis of ergosterol, which is the predominant component of the fungal cell membrane (Parks and Casey 1996; Ghannoum and Rice 1999).

Natural dyes having naphthoquinone moiety such as Lawson from henna, juglone from walnut, and lapachol from alkanet inhibited the growth of bacteria as well as fungi (Harborne and Williams 2000; Singh et al. 2005). In inhibiting fungi growth, the minimum inhibitory concentration (MIC) of natural dyes ranges from 5 to 40 μg. It is reported that natural dyes like turmeric, terminalli, guava, and henna have an antifungal activity against *Cladosporium*, *Emericella*, *Candida*, and *Rhizopus* at a dosage level of 50 μL (Mariselvam et al. 2012).

Other than textile applications, some of the natural dyes are used as fungistatic and fungicidal agents. Red sandalwood contains components like santalin and deoxysantalin, and it is effective against certain classes of fungi. Marigold has a high content of flavonoids, and they are effective against 23 clinical fungi strains. Red cedar roots are rich in flavonols, procyanidins, quercetin, kaempferol, catechins, and thujaplicin; and the latter is responsible for the antifungal activity.

Most of the natural dyes did not have good antifungal activity when applied alone. It indicates that the antifungal activity of curcumin was sustained up to 25 washing cycles, when applied along with chitosan and curcumin by exhaustion method on peroxide-treated cotton and formic acid–treated wool/rabbit hair fibers (Ammayappan and Jeyakodi Moses 2009). However, all natural dyes will not give the required MIC on the fiber to inhibit the fungal growth; generally, they did not give good antifungal activity.

15.3.2 Deodorizing Property

The eccrine and apocrine glands secrete sweat from the skin to maintain the body temperature. Sweat contains low-molecular-weight fatty acids, aldehydes, ketones, sulfur compounds, and high levels of protein that can be easily broken by bacteria. The proteinaceous compounds of sweat are primarily responsible for body odor since they are degraded into odor-creating gases like ammonia by bacteria due to high humidity and warm temperature. Corynebacteria, streptococci, and propionibacteria are the main bacteria that cause malodor (Ohnuki 2006; Kanlayavattanakul and Lourith 2011). It was found that some natural dyes applied with mordants are able to absorb ammonia gas. The amount of absorption of ammonia gas by naturally dyed fabric is expressed as the deodorizing performance in comparison with the blank state.

The Japanese Society of Fibers has developed a method to determine the deodorizing performance of textiles by using ammonia gas detection tube. In this method, the amount of ammonia in the detection tubes with and without dyed textile materials, over a specified period of time is measured at 25°C (JED301 2001). The deodorizing property is evaluated using the following formula:

$$\text{Deodorization performance (\%)} = \frac{(C_b - C_s) \times 100}{C_b}$$

where
C_b is the gas concentration (ppm) of the test tube without dyed material (blank state)
C_s is the gas concentration of the test tube with dyed material (Huang and Yushan 2004)

The deodorizing properties of cotton, silk, and wool fabrics dyed with *Amur cork tree*, *Dryopteris crassirhizoma*, *Chrysanthemum boreale*, and *Artemisia* extracts were studied, and it was found that the deodorizing performance of these natural dyes on natural fiber ranged from 34% to 99% and increased in the order of cotton < silk < wool. Among them, the deodorizing performance of woolen fabrics dyed with the aforementioned natural dyes is higher (98%–99%) than cotton and silk fabrics (Lee 2007; Lee et al. 2010).

The deodorizing performance of these dyes is in the following order: *Gardenia* < *Cassia tora* < *Coffee sludge* < *Pomegranate*, in which the maximum deodorizing performance is observed in pomegranate dyed fabric (99%). The results showed that the deodorizing performance of the dyes is in the order gardenia < *Cassia tora* < coffee sludge < pomegranate, indicating that the maximum deodorizing performance is observed in pomegranate (99%). Ellagic acid present in pomegranate extract can easily react with ammonia by neutralization reaction, so it promotes an excellent deodorizing property. The deodorizing performance of woolen fabrics dyed with natural dyes extracted from sappan wood, black tea, peony, and clove is in the order of *peony* < *black tea* < *sappan wood* < *clove*. Clove gave a 98%–99% deodorizing performance due to the presence of eugenol, which is well known for its antiseptic property (Hwang et al. 2007, 2008).

It is inferred that metallic mordants also enhanced the deodorizing performance. Metal cations present in the fiber form the coordinate bond with unshared electron pairs of ammonia and absorbed additional odor molecules (Washino 1993; Lee et al. 1998). Similarly, the functional groups present in the natural fiber also improved the deodorizing performance. Wool fiber has more acidic functional groups than cotton and silk fiber, which can easily neutralize ammonia-based foul odors through the neutralization reaction; because of this, the deodorizing performance of wool fiber is higher than that of silk and cotton fibers (Hwang et al. 2008). Other than the properties of natural dye and fiber, some other factors are also greatly contributing to the deodorizing performance, which have to be studied in detail.

15.3.3 UV Protection Property

Sunlight has ultraviolet rays (UVRs) with a wavelength in between visible light and x-rays, that is, ranging from 100 to 400 nm (Gil and Kim 2000; Kullavanijaya et al. 2005). UVRs consist of the UV-A (320–400 nm), UV-B (290–320 nm), and UV-C (200–290 nm) regions. The UV-A region causes little visible reaction on the skin but can decrease the immunological response of skin cells, the UV-B region mainly causes skin cancers, and the UV-C region is totally absorbed by the atmosphere and does not reach the earth (Kaplan et al. 1988). Suitable clothing, hats and sunscreens, is required to protect the body against UVRs (Capjack et al. 1994; Hoffmann et al. 2001).

Ultraviolet protection factor (UPF) is the scientific term used to indicate the amount of ultraviolet (UV) protection provided by the apparel (Burgess 2009). The UPF of a textile material is the mean percentage ratio between transmission in the UV-A region (320–400 nm) and transmission in the UV-B region (280–320 nm). It is estimated using a UV transmission analyzer according to the standard AS/NZ 4399:1996 or AATCC 183-2004 method by the following equation (Gies et al. 2000):

$$\text{UV Protection Factor (UPF)} = \frac{\sum_{290}^{400} E_\lambda S_\lambda \Delta_\lambda}{\sum_{290}^{400} E_\lambda S_\lambda T_\lambda \Delta_\lambda}$$

where
 E_λ is the erythemal spectral effectiveness (unitless)
 S_λ is the solar UVR spectral irradiance (W/m²/nm)
 T_λ is the spectral transmittance of the fabric
 Δ_λ is the bandwidth (mm)
 λ is the wavelength (nm)

Apparels having UPF values of 15–24, 25–40, and >40 are already considered to have good, very good, and excellent UV protection property respectively; hence, apparels are not assigned with a UPF value >50. The UV protection property of apparels depends on the nature of the fiber, fabric construction, finishing processes, chemical constituents of dye, and concentration of dye and its UV-absorbing property (Grifoni et al. 2009). Compact weave and hosiery apparels can protect the body from UVRs better than plain weave apparel. Dyed fabrics can protect against UVRs better than undyed fabric, and its protection is directly proportional to the concentration of dye (Reinert et al. 1997; Driscoll 2000). Tannin-based-mordant-treated fabric has shown good UV protection, while after dyeing with natural dye, UV protection level is enhanced (Grifoni et al. 2009).

Cotton, hemp, and flax fibers have antiallergenic properties, and so they are mainly used in making summer clothes. Lignocellulosic fibers show good UV protection due to the presence of lignin, waxes, and pectin that act as excellent UV absorbents (Gambichler et al. 2001). Flavonoids and phenolic compounds present in the natural dye extract play a major role in the reduction of free radical species formed by UV radiation. It is also observed that metallic mordanting also significantly enhanced the UV protection properties of dyed apparels (Feng et al. 2007).

Cotton fabrics dyed with cochineal and indigo natural dyes gave 25–40 UPF, while madder-dyed fabric gave 15–24 UPF, which is mainly due to chemical component of natural dye (Sarkar 2004).

Chitosan biopolymer was used as a natural mordant for dyeing cotton fabrics with green tea. It is found that chitosan-mordanted + green tea–dyed cotton textiles showed higher UPF than unmordanted green tea–dyed cotton (Kim 2006). The UPF of silk fabrics dyed with eucalyptus leaf extract is less than 15 and is increased with increasing dye concentration. However, after mordanting with $CuSO_4$, silk fabrics attained very good UV protection, and $FeSO_4$ mordanting also gave an excellent UV protection (Mongkholrattanasit et al. 2011).

The UV protection properties of nylon fabric dyed with madder and safflower dyes using alum, zinc sulfate, and tannic acid mordants were studied. It is inferred that premordanting followed by natural dyeing has shown improvement in the UV protection against the harmful UV-B radiation due to the good light absorption characteristics of the madder/safflower dye–mordant complex (Ibrahim et al. 2013). Jute fabrics have shown good UV protection property, after mordanting with biomordant and/or metallic mordants followed by dyeing with natural dyes extracted from manjistha roots (*Rubia cordifolia*), annatto seeds (*Bixa orellana*), ratanjot barks (*Onosma echioides*), and babool barks (*Acacia arabica*). The UPF is in the following order: babool > annatto > manjistha > ratanjot; the variation in UPF is due to the chemical constitution of natural dyes (Chattopadhyay et al. 2013).

15.3.4 ANTIOXIDANT PROPERTY

Some natural dyes have carotenoids, flavonoids, phenols, and indigoids, and these groups posses an antioxidant property, that is, ability to quench the singlet oxygen formed in food, so they are also used in foods, cosmetics, etc. (Ramamoorthy et al. 2011). The literature information on the antioxidant properties of apparel dyed with natural dye is scanty.

15.4 ENVIRONMENTAL IMPACTS OF NATURAL DYES

Natural dyes are fascinatedly termed *green chemicals*, and their application on apparels has a high demand due to its biodegradability and unique colors. However, during extraction, dyeing, and aftertreatment of natural dyes, textiles face the following important problems, which pretense an impact on the environment:

- Pesticides are generally applied on plants during cultivation in order to protect them against pests. Natural dyes extracted from these plants may have traces of pesticides, which may cause problem during ecotesting.
- All natural dyes cannot be extracted by conventional aqueous extraction method; some natural dyes need either acidic or alkaline medium or solvent for efficient extraction. For example, extraction of natural dye from madder root needs sulfuric acid, which can also increase the pollution load indirectly.
- Some natural dyes need chemicals other than mordants for their application, which may cause effluent problem, like indigo dye needing sodium hydroxide and sodium hydrosulfite for dyeing cotton apparels.
- To match the commercial 1% shade developed by the synthetic dye, 10%–25% natural dye is required, which might increase the pollution load in effluent treatment plant (Chen and Burns 2006).
- Chan et al. (2002) inferred that the COD/BOD ratio of effluent from natural dyes is approximately 2. They implied that this effluent is highly biodegradable and treatable by effluent treatment plant. However, they indicated that the amount of copper, iron, and chromium present in the dye effluent exceeded the acceptable limit, which required more attention since copper, iron, and chromium metals were banned.
- Dyers in New Mexico are extracting natural dyes from certified organic plants grown in farms in the United States or Canada and used the extracts for coloration of woolen

Functional Aspects, Ecotesting, and Environmental Impact of Natural Dyes

materials in primary colors. They have a standard market and cultivate specific natural dyes yielding in a wide area, while in other parts of the world like India, cultivation area is still a constraint.

- The natural dyeing process consumes more energy and time than conventional dyeing process, since mordanting and dyeing need a longer duration for good fixation of the natural dye–mordant complex.
- Generally, natural dyeing units are run by small-scale industries, and so it is not economically feasible for them to set up an effluent treatment plant, since it will increase the cost of production. They are disposing the dye effluent to nearby canals or rivers or any other natural water bodies, which will further contaminate water sources.
- Some natural dyes need high amounts of metallic mordants like Cu, Sn, and Fe for their fixation, and eco-friendliness of the process is really questionable.

Nevertheless, naturally dyed apparels are gaining popularity day by day, due to their fanciness. It is high time for government authorities as well as scientific communities to join hands for collective actions to solve the issues.

15.5 ECOTESTING

Before the 1990s, there was no restriction on synthetic dye application. Later on, it was found that people working in the synthetic dye manufacturing industries were affected by bladder cancer. Scientists investigated that dyes based on benzidine, auramine, 2-naphthylamine, and some of the dyeing auxiliaries were carcinogenic to the human body. They inferred that such carcinogenic dyes contain electrophilic groups like R_2N^+ or R_3C^+ and would attract the nucleophilic site of DNA to form a covalent bond, thus making the structural changes in the DNA of cells. The structural changes in DNA will cause skin-related diseases like cancer. To prevent the application of such synthetic dyes on the apparel, in 1996, the EU banned a series of carcinogenic synthetic dyes. Similarly, certification standards such as the Global Organic Textile Standard (GOTS) and Oeko-Tex were established to verify the presence of carcinogenic dyes in any textile products. These agencies will evaluate the amount of banned dyes/chemicals present in the apparel as per their standards. After testing, they will provide suitable certificate, if the apparel does not contain any carcinogenic dyes and such certified apparels are eligible for exporting to the EU. For example, GOTS restricted dyes particularly azo and disperse dyes that can release >30 mg/kg amine from an apparel (Nadiger 2001). In 2008, the International Wool Textile Organization introduced the *eco-wool* label for woolen products, which satisfies the requirements of the EU ecolabel, and it should be free from banned chemicals/dyes at all stages of production and processing (Anon 2014a).

15.5.1 ECOLABELING

Ecolabeling is introduced in the late 1990s to confirm the absence of carcinogenic and toxic dyes/chemicals in any textile products. Oeko-Tex is an ecolabeling agent. The main objective of the ecolabeling is to provide the information about the textile value chain and assure the safety of apparel. Now for customers' point of view, ecolabeling is more important than price, functionality, or brand of the apparel.

Globally, there are two types of ecolabeling system used for textiles. The first type is privately owned certification like Oeko-Tex 100; this mainly concentrate on the analysis of residual harmful chemicals present in the final product, that is, in garments. The second type is association based like the EU ecolabel; evaluation starts from fiber harvesting to labeling of garments in order to inform the sustainability and environmental impact of the final product, process, and raw material (Anon 2014b). Simply, they brief the life history of a textile product. Some of the important certification agencies are given as follows.

15.5.1.1 Oeko-Tex

Oeko-Tex, which is issued by the *International Association for Research and Testing in the Field of Textile Ecology*, was introduced by the Hohenstein Institute and the Institute for Ecology, Technology and Innovation, Vienna, in 1992. Products that carry the Oeko-Tex Standard 100 label are free from more than 100 banned substances. Oeko-Tex is testing the textile material in four different categories based on intensive skin contact of a product (Anon 2014c).

15.5.1.2 Blue Sign

This Switzerland-based standard organization defines their standard as follows: "It certifies products that have been rigorously tested against harmful effects on humans and the environment and for efficient consumption of relevant resources" (Anon 2014d).

15.5.1.3 International Association of Natural Textile Industry

This Germany-based association makes awareness on eco-friendly textile products among the consumers and the retailers. The International Association of Natural Textile Industry (IVN) issues the quality standard, that is, *Naturtextil IVN certified BEST*, which will inspect the entire textile chain in terms of ecology and social accountability, so it is considered as one of the strictest ecolabel in the EU (Anon 2014e).

15.5.1.4 KRAV

This Sweden-based organization develops organic standards, inspects those standards, and promotes the KRAV label. The standards are adapted from the International Federation of Organic Agriculture Movements (Anon 2014f).

15.5.1.5 JOCA

The Japan Organic Cotton Association (JOCA) was established in 2000. It certifies and promotes organic cotton products as well as garments in Japan.

15.5.1.6 GOTS

The GOTS is recognized worldwide as the leading textile processing standard for organic fibers governed by an international working group comprised of the OTA (United States), iVN (Germany), Soil Association (United Kingdom), and JOCA (Japan). The awareness on GOTS started in the Intercot Conference held at Düsseldorf, Germany, on 2002 and established in 2006 with an objective to stipulate the basic requirements throughout the supply chain in the manufacturing of a textile product using organically produced raw materials. Recently, GOTS approved the *Rubia Pigmenta Naturalia*–dyed fabrics based on long-term research. Saco River Dyehouse is using mainly natural dyes, and it is the only GOTS-certified dyeing factory in Biddeford, Maine, United States (Anon 2014g).

15.5.1.7 Ecolabel

Ecolabel can be given to the apparel products that satisfied the criteria of less environmental impact. For certification, the testing starts from raw material to its disposal. It is given to the products that satisfied the criteria of less environmental impact, and the testing starts from the raw material to its disposal (Anon 2014b).

15.6 SUSTAINABLE ADVANTAGES OF NATURAL DYES

Each technology has its own merits and demerits. The sustainability of the technology is mainly dependent on its social, economic, and environmental responsibility. A sustainable technology should be cost-effective, meet customer demands without deteriorating their basic requirements, have less impact on the environment, and be easily adaptable (Anon 2014h). During 1900s, the consumption of synthetic dyes has increased rapidly in lieu of natural dyes due to its wide range of colors, ease of application, and cost-effectiveness; however, natural dyes are sustained due to its eco-friendliness and inherent functional properties. After reporting the carcinogenicity of azo-based synthetic dyes during the 1990s and compulsion of ecotoxicity tests for apparels, now, dyers are showing more interest on natural dyes (Hancock and Boxworth 1997; Thiry 2011).

Literature information inferred that nearly two million tonnes of apparel are imported to the United Kingdom, mainly from India and China. Among them, 50% are thrown away every year and they end up in landfills. For sustainable apparels, branded companies are now adopting the holistic green approach for sourcing in all possible ways. They prefer that apparels should be made from either organic cotton or lyocell, motivate the cotton farmers to reduce the application of pesticide, and suggest eco-friendly dyes/processing methods for garment manufacturing. Some companies prefer to harvest genetically modified cotton, which has an artificially built-in blue color. Such production of GM cotton may increase the cost of production and mainly face the adaptability issues among farmers (Anon 2014i,j). Being an alternative to synthetic dyes, these situations led to revival of the application of natural dyes on apparels. Natural dyes have sustainable advantages like gorgeous colors, eco-friendliness, functional properties, agricultural residual mass utilization, and ease of application.

15.6.1 Sources

The sources of natural dyes are either from primary agriculture products or from by-products/waste products of agriculture biomass like bark from eucalyptus/timber industry and leaves from teak (Bechtold et al. 2007). Such agriculture waste/by-products may provide additional income for the farmers as well as lead to the effective utilization of agriculture waste. However, research interventions are required to optimize the supply chain system and standardization of extraction variables (Bechtold et al. 2006).

15.6.2 Extraction Procedure

Natural dyes can be extracted by simple aqueous extraction method, which can be easily adapted by small-scale industries. Recently, ultrasonicator-based extraction process is used to extract more dye from the source than conventional methods, which can reduce the cost of dyeing and effluent load (Bechtold et al. 2008).

15.6.3 Color

Generally, natural dyes have more than one auxochrome, and they are sensitive to changes in pH and metallic mordants. This property can be used to develop a wide range of elegant colors from a single natural dye with different mordants (polygenetic behavior). In addition to colors, eco-friendly metallic mordant also imparts good light and washing fastness to naturally dyed apparels (Hancock and Boxworth 1997).

15.6.4 Functional Property

Natural dyes are also preferred for food and cosmetic coloration, other than coloration of textiles, due to their functional properties like antibacterial and UV protection properties. Such functional properties of natural dyes can diversify their application such as in protective apparels.

15.6.5 Eco-Friendliness

The source of natural dyes is mainly from agriculture products. The application of natural dye in the presence of biomordant is considered as an eco-friendly process, and the respective effluent will be easily biodegraded. To reduce the effluent load on the environment, natural dyes are still considered as an alternative to synthetic dyes. Natural dyes have good chelating property with metallic mordants, so natural dye waste can also be used as heavy metal adsorbents in chromium-contaminated areas (Shahid et al. 2013).

However, to sustain the natural dye application on apparels in the future, it is necessary to study the life cycle assessment of the whole process starting from harvesting, land use, extraction process, dyeing, and end use with proper ecolabeling and disposal (degradability).

15.7 FUTURE PERSPECTIVES

Shoppers are generally showing a great deal of interest in fancy textile products. They usually update the current trends in the apparel and fashion market regularly due to swift communication between people from one corner to another in the world. Recently, there is more awareness on ecolabeling of apparels; consumers prefer to wear *ecolabel*-tagged apparels like *Oeko-Tex 100*, which is given by certifying agencies. Since synthetic dyes have their own demerits, it is believed by the consumers that apparels dyed with natural dyes are environment friendly. So now, small- and medium-scale industries are regaining interest in the use of natural dyes for apparels. Globally, farmers are still cultivating natural dye–yielding plants and herbs traditionally for their income. In the United States, Native American and Hispanic weavers are cultivating regularly Navajo tea (*Thelesperma gracile*), big sage (*Artemisia tridentata*), golden rabbitbrush (*Chrysothamnus nauseosus*), snakeweed (*Gutierrezia sarothrae*), and mullein (*Verbascum thapsus*) for dyeing cotton- and wool-based textile products. In India, the woolen carpet industry of north India, small-scale silk industries in Karnataka and West Bengal, and cotton industries in Rajasthan and Andhra Pradesh are using vegetable dyes like indigo (*I. tinctoria*) and Kum (*Strobilanthes flaccidifolius*) (Gokhale et al. 2004; Siva 2007). Natural dyes like persimmon fruits are used traditionally to dye cotton garments at Jeju Island in Korea, and such garments have been preferred by farmers and fishermen, because they show improvement in air permeability, water repellency, and stiffness in touch (Park et al. 2005).

Scientific communities did commendable research and development works on natural dyes for their exploitation, starting from its extraction, characterization, application using advanced technology, and evaluation of their functional properties; however, majority of the research outputs has still not reached to the commercial level. The future of natural dyes for the application of apparel products can be flourished only by the interventions of the scientific bodies and government agencies to fill the gap between lab and field in the following ways:

1. Utilization of natural dyes is limited due to the nonfrequent supply of natural dyes in the market. Suitable natural dye–yielding plants can be identified and cultivated for the regular supply to the cottage industries.
2. Color yield of natural dyes from its sources is generally low and it ranges from 0.5% to 20% by the conventional extraction method. To enhance the extraction of natural dyes, ultrasonicator-/supercritical solvent–based extraction technology could be disseminated Sivakumar et al. (2011).
3. In place of metallic mordants like $Cu^{2+}/Cr^{2+}/Sn^{2+}$, biomordants like tannic acid, citric acid, tartaric acid, and chitosan can be used either alone or in combination with eco-friendly metallic mordants.
4. Knowledge of color theory and color matching can help to produce variety of shades by using combination of natural dyes. For example, by mixing turmeric extract with a little amount of madder, the color hue will shift from yellowish to reddish. Similarly, *Alkanna tinctoria* with turmeric can change the hue toward green (Seiko 2006).
5. Some of the natural dyeing units run by small-scale industries did not have any awareness on ecolabeling. Governments should take necessary steps to make a national policy for compulsion of ecolabeling.

15.8 CONCLUSIVE REMARKS

Natural dyes have some distinct advantages over synthetic dyes; however, the main constraint about its adaptability by the industry for commercialization is due to less availability, nonreproducibility, and moderate fastness properties. To sustain the natural dyes in the apparel industry, particularly for small-scale and cottage industries, it is the right time to identify suitable natural dyes that can give high color yield, very good fastness properties, and matching ecostandards. Governments should take necessary actions and policies to exploit the adaptability of natural dyes, which can make a huge impact on the economy of the farmers as well as its stakeholders. Being a widespread communicative world, each government should initiate awareness programs on eco-friendly textile processing, natural dyes, and ecolabeling of apparels in public, which can surely sustain the quality of naturally dyed apparels.

REFERENCES

Agar, E., S. Şaşmaz, and A. Agar. 1999. Synthesis and properties of phthalocyanines containing eugenol (4-allyl-2-methoxyphenol). *Turk. J. Chem.* 23: 131–137.

Ali, N.F. and R.S.R. El-Mohamedy. 2011. Eco-friendly and protective natural dye from red prickly pear (*Opuntia lasiacantha Pfeiffer*) plant. *J. Saudi Chem. Soc.* 15: 257–261.

Amarasinghe, B.M.W.P.K. and R.A. Williams. 2007. Tea waste as a low cost adsorbent for the removal of Cu and Pb from wastewater. *Chem. Eng. J.* 132: 299–309.

Ammayappan, L. 2009. Application of natural colours on woollen materials, in: S.A. Karim, D.B. Shakyawar, and A. Joshi (eds.) *Wool Technology: Innovations in Wool Production and Technologies for Value Addition*, 1st edn. Udaipur, India: Agrotech Publishing Academy, pp. 336–348.

Ammayappan, L. and S. Chakraborty. 2013. Natural colourants and their applications: An overview. *Man-Made Text. India* 56: 427–430.

Ammayappan, L. and J. Jeyakodi Moses. 2007. An overview on application of tanning in textile processing. *Man-Made Text. India* 50: 293–297.

Ammayappan, L. and J. Jeyakodi Moses. 2009. Study of antimicrobial activity of aloe vera, chitosan, and curcumin on cotton, wool and rabbit hair. *Fibres Polym.* 10(2): 161–166.

Ammayappan, L., L.K. Nayak, D.P. Ray, S. Das, and A.K. Roy. 2013. Functional finishing of jute textiles—An overview in India. *J. Nat. Fibres* 10: 390–413.

Ammayappan, L., D.B. Shakyawar, C. Lal, M. Sharma, and S.A. Wani. 2014. Extraction of natural colourants from agricultural residues and their application on woolen fabric: Part 1. *Man-Made Text. India* 57: 182–185.

Anon. 2013. http://dsir.csir.res.in/webdsir/#files/reports/reports.html. (accessed October 20, 2010.)
Anon. 2014a. https://oecotextiles.wordpress.com/category/fibres/wool/. (accessed January 2, 2014.)
Anon. 2014b. http://ec.europa.eu/environment/ecolabel/. (accessed January 2, 2014.)
Anon. 2014c. https://www.oeko-tex.com/en/manufacturers/concept/oeko_tex_standard_100/oeko_tex_standard_100.xhtml. (accessed January 2, 2014.)
Anon. 2014d. www.naturtextil.com/profile/quality-seals/best.html. (accessed December 20, 2013.)
Anon. 2014e. http://oecotextiles.wordpress.com/tag/ivn/. (accessed January 2, 2014.)
Anon. 2014f. www.krav.se/about-krav. (accessed January 2, 2014.)
Anon. 2014g. http://naturallysaferproducts.com/tag/global-organic-textile-standard/#sthash.yXDRHcM9.dpuf. (accessed January 2, 2014.)
Anon. 2014h. http://www.sda-uk.org/issues/issues.htm. (accessed January 2, 2014.)
Anon. 2014i. http://www.organicconsumers.org/clothes/color090804.cfm. (accessed January 2, 2014.)
Anon. 2014j. http://www.theecologist.org/green_green_living/clothing/881447/greening_the_high_street_marketing_trick_or_real_deal.html. (accessed January 2, 2014.)

Arora, A., D. Gupta, D. Rastogi, and M.L. Gulrajani. 2012. Antimicrobial activity of naphthoquinones extracted from *Arnebia nobilis*. *J. Nat. Products* 5: 168–178.

Bancroft, W.D. 1914. The theory of dyeing. *J. Phys. Chem.* 18: 1–25.

Bechtold, T., A. Mahmud-Ali, E. Ganglberger, and S. Geissler. 2008. Efficient processing of raw material defines the ecological position of natural dyes in textile production. *Int. J. Environ. Waste Manage.* 2: 215–232.

Bechtold, T., A. Mahmud-Ali, and R.A.M. Mussak. 2007. Reuse of ash-tree (*Fraxinus excelsior* L.) bark as natural dyes for textile dyeing: Process conditions and process stability. *Color. Technol.* 123: 271–279.

Bechtold, T., R. Mussak, A. Mahmud-Ali, E. Ganglberger, and S. Geissler. 2006. Extraction of natural dyes for textile dyeing from coloured plant wastes released from the food and beverage industry. *J. Sci. Food Agric.* 86: 233–242.

Bechtold, T., A. Turcanu, E. Ganglberger, and S. Geissler. 2003. Natural dyes in modern textile dye house—How to combine experiences of two centuries to meet demands of the future? *J. Cleaner Prod.* 11: 499–509.

Borges, M.E., R.L. Tejera, L. Díaz, P. Esparzab, and E. Ibáñezc. 2012. Natural dyes extraction from cochineal (*Dactylopius coccus*). New extraction methods. *Food Chem.* 132: 1855–1860.

Borkow, G. and J. Gabbay. 2008. Biocidal textiles can help fight nosocomial infections. *Med. Hypotheses* 70: 990–994.

Burgess, C.M. 2009. Chapter 70: Cosmetic products, in: A.P. Kelly and S.C. Taylor (eds.) *Dermatology for Skin of Color*. New York: McGraw-Hill Medical.

Capjack, L., N. Kerr, S. Davis, R. Fedosejevs, K.L. Hatch, and N.L. Markee. 1994. Protection of humans from ultraviolet radiation through the use of textiles: A review. *Fam. Consum. Sci. Res. J.* 23: 198–218.

Cardon, D. (ed.). 2007. *Natural Dyes: Sources, Tradition, Technology and Science.* London, U.K.: Archetype Publications Ltd.

Chairat, M., S. Rattanaphani, J.B. Bremner, and V. Rattanaphani. 2005. An adsorption and kinetic study of lac dyeing on silk. *Dyes Pigm.* 64: 231–241.

Chairat, M., S. Rattanaphani, J.B. Bremner, and V. Rattanaphani. 2008. Adsorption kinetic study of lac dyeing on cotton. *Dyes Pigm.* 76: 435–439.

Chan, P.M., C.W.M. Yuen, and K.W. Yeung. 2002. The effect of natural dye effluent on the environment. *Res. J. Text. Apparel.* 6: 57–62.

Chandrana, H., S. Baluja, and S.V. Chanda. 2005. Comparison of antibacterial activities of selected species of Zingiberaceae family and some synthetic compounds. *Turk. J. Biol.* 29: 83–97.

Chattopadhyay, S.N., N.C. Pan, A.K. Roy, S. Saxena, and A. Khan. 2013. Development of natural dyed jute fabric with improved colour yield and UV protection characteristics. *J. Text. Inst.* 104: 808–818.

Chen, H.L. and L.D. Burns. 2006. Environmental analysis of textile products. *Cloth. Text. Res. J.* 24: 248–261.

Cowan, M.M. 1999. Plant products as antimicrobial agents. *Clin. Microbiol. Rev.* 12: 564–582.

Das, D., S.R. Maulik, and S.C. Bhattacharya. 2008. Dyeing of wool and silk with *Rheum emodi*. *Indian J. Fibre Text. Res.* 33: 163–170.

Das, P.K., A.K. Mondal, and Mondal, S.P. 2011. Antibacterial activity of some selected dye yielding plants in eastern India. *Afr. J. Plant Sci.* 5: 510–520.

Driscoll, C. 2000. Clothing protection factor. *Health Prot. Ag. Radiol. Bull.* 222: 1–4.

Espinel, I.A. 1996. History of medical mycology in the United States. *Clin. Microbiol. Rev.* 9: 235–272.

Feng, X.X., L.L. Zhang, J.Y. Chen, and J.C. Zhang. 2007. New insights into solar UV-protective properties of natural dye. *J. Cleaner Prod.* 15: 366–372.

Gambichler, T., S. Rotterdam, P. Altmeyer, and K. Hoffmann. 2001. Protection against ultraviolet radiation by commercial summer clothing: Need for standardized testing and labeling. *BMC Dermatol.* 1: 6.

Gao, Y. and R. Cranston. 2008. Recent advances in antimicrobial treatments of textiles. *Text. Res. J.* 78: 60–72.

Ghaheh, F.S., S.M. Mortazavi, F. Alihosseini, A. Fassihi, A.S. Nateri, and D. Abedi. 2014. Assessment of antibacterial activity of wool fabrics dyed with natural dyes. *J. Cleaner Prod.* 72: 139–145.

Ghannoum, M.A. and L.B. Rice. 1999. Antifungal agents: Mode of action, mechanisms of resistance, and correlation of these mechanisms with bacterial resistance. *Clin. Microbiol. Rev.* 12: 501–517.

Gies, P.H., C.R. Roy, and G. Holmes. 2000. Ultraviolet radiation protection by clothing: Comparison of in vivo and in vitro measurements. *Radiat. Prot. Dosim.* 91: 247–250.

Gil, E.M. and T.H. Kim. 2000. UV-induced immune suppression and sunscreen. *Photodermatol. Photoimmunol. Photomed.* 16: 101–110.

Giridev, V.R., J. Venugopal, S. Sudha, G. Deepika, and S. Ramakrishna. 2009. Dyeing and antimicrobial characteristics of chitosan treated wool fabrics with henna dye. *Carbohyd. Polym.* 75: 646–650.

Gokhale, S.B., A.U. Tatya, S.R. Bakliwal, and R.A. Fursule. 2004. Natural dye yielding plants in India. *Nat. Prod. Radiance* 3: 228–234.

Grifoni, D., L. Bacci, G. Zipoli, G. Carreras, S. Baronti, and F. Sabatini. 2009. Laboratory and outdoor assessment of UV protection offered by flax and hemp fabrics dyed with natural dyes. *Photochem. Photobiol.* 85: 313–320.

Gulranjani, M.L., D. Gupta, A. Varsha, and J. Manoj. 1992. Some studies on natural yellow dyes: Part III quinones: Henna, Dolu. *Indian Text. J.* 102: 77–83.

Gupta, D., S.K. Khare, and A. Laha. 2004. Antimicrobial properties of natural dyes against gram-negative bacteria. *Color. Technol.* 120: 167–171.

Haji, A., A.M. Shoushtari, and M. Mirafsharb. 2013. Natural dyeing and antibacterial activity of atmospheric-plasma-treated nylon 6 fabric. *Color. Technol.* 130: 37–42.

Han, S. and Y. Yang. 2005. Antimicrobial activity of wool fabric treated with curcumin. *Dyes Pigm.* 64: 157–161.

Hancock M. (ed.). 1997. *Potential for Colourants from Plant Sources in England and Wales.* Boxworth, Cambridge, U.K.: Arable Crops Horticulture Division, ADAS, vol. 2, pp. 12–19.

Harborne, J.B. and W.A. Williams. 2000. Advances in flavonoid research since 1992. *Phytochemistry* 55: 481–504.

Hoffmann, K., J. Laperre, A. Avermaete, P. Altmeyer, and T. Gambichler. 2001. Defined UV protection by apparel textiles. *Arch. Dermatol.* 137: 1089–1094.

Huang, H. and A.N. Yushan. 2004. Development of deodorizing finish. *Dyeing Finish* 11: 39–41.

Hwang, E.K., Y.H. Lee, and H.D. Kim. 2007. Dyeing and deodorizing properties of cotton, silk, and wool fabrics dyed with various natural colorants. *J. Korean Soc. Dyers Finish* 19: 12–20.

Hwang, E.K., Y.H. Lee, and H.D. Kim. 2008. Dyeing, fastness, and deodorizing properties of cotton, silk, and wool fabrics dyed with gardenia, coffee sludge. *Cassiatora L*, and pomegranate extracts. *Fibres Polym.* 9: 334–340.

Ibrahim, N.A., W.M. El-Zairy, M.R. El-Zairy, and H.A. Ghazal. 2013. Enhancing the UV-protection and antibacterial properties of polyamide-6 fabric by natural dyeing. *Text. Light. Indus. Sci. Tech.* 2: 36–41.

JED301. 2001. *Deodorizing Performance of Textiles Certification Standards.* Osaka, Japan: Japan Textile Evaluation Technology Council.

Jones, F. 1967. The theory of dyeing. *Rev. Prog. Color. Relat. Top.* 1: 15–22.

Jose, S. 2006. Eco-friendly dyeing of silk using combination of natural dyes. MSc dissertation, Alagappa University, Karaikudi, India.

Joshi, M., S. Wazed Ali, R. Purwar, and S. Rajendran. 2009. Eco-friendly antimicrobial finishing of textiles using bioactive agents based on natural products. *Indian J. Fibre Text. Res.* 34: 295–304.

Joshi, M., S. Wazed Ali, and S. Rajendran. 2007. Antibacterial finishing of polyester/cotton blend fabrics using neem (*Azadirachta indica*): A natural bioactive agent. *J. Appl. Polym. Sci.* 106: 793–800.

Kanlayavattanakul, M. and N. Lourith. 2011. Body malodours and their topical treatment agents. *Int. J. Cosmet. Sci.* 33: 298–311.

Kaplan, D.L., S.J. Moloney, and S.R. Pinnel. 1988. A new stabilized ascorbic acid solution: Percutaneous absorption and effect on relative collagen synthesis. *J. Cutaneous Aging Cosmet. Dermatol.* 1: 115–121.

Khan, M.R., A.D. Omoloso, and M. Kihara. 2003. Antibacterial activity of *Artocarpus heterophyllus*. *Fitoterapia* 74: 501–505.

Khatoon, S., S. Mehrotra, and V.K. Bajpai. 1994. Ultramorphology of some boraginaceous taxa used as Ratanjot. *Feddes Repert.* 105: 61–71.

Kim, S. 2006. Dyeing characteristic and UV protection property of green tea dyed cotton fabrics. Focusing on the effect of chitosan mordanting condition. *Fibres Polym.* 7: 255–261.

Kim, T.K. and Y.A. Son. 2005. Effect of reactive anionic agent on dyeing of cellulosic fibres with a berberine colorant. Part 2: Anionic agent treatment and antimicrobial activity of berberine dyeing. *Dyes Pigm.* 64: 85–89.

Kullavanijaya, P., W. Henry, and H.W. Lim. 2005. Photo protection. *J. Am. Acad. Dermatol.* 52: 959–961.

Lee, H.S., J.H. Chang, I.H. Kim, and S.W. Nam. 1998. Dyeing of cotton with clove extract. *J. Korean Soc. Dyers Finish* 10: 29–35.

Lee, Y.H. 2007. Dyeing, fastness, and deodorizing properties of cotton, silk, and wool fabrics dyed with coffee sludge (*Coffea Arabica* L.) extract. *J. Appl. Polym. Sci.* 103: 251–257.

Lee, Y.H., E.K. Hwang, Y.J. Jung, S.K. Do, and H.D. Kim. 2010. Dyeing and deodorizing properties of cotton, silk, wool fabrics dyed with *Amur Cork tree, Dryopteris crassirhizoma, Chrysanthemum boreale, Artemisia* extracts. *J. Appl. Polym. Sci.* 115: 2246–2253.

Machado, T.B., A.V. Pinto, M. Pinto, I.C. Leal, M.G. Silva, A.C. Amaral, R.M. Kuster, and K.R. Netto-dos Santos. 2003. In vitro activity of Brazilian medicinal plants, naturally occurring naphthoquinones and their analogues, against methicillin resistant *Staphylococcus aureus*. *Int. J. Antimicrob. Agents* 21: 279–284.

Madigan, M.T. and J.M. Martinko (eds.). 2005. *Brock Biology of Microorganisms*, 11th edn. Lebanon, PA: Pearson Prentice Hall, Inc.

Mariselvam, R., A.J.A.R. Singh, and K. Kalirajan. 2012. Antifungal activity of different natural dyes against traditional products affected fungal pathogens. *Asian-Pac. J. Trop. Biomed.* 2: S1461–S1465.

Mathur, N. and P. Bhatnagar. 2007. Mutagenicity assessment of textile dyes from Sangner (Rajasthan). *J. Environ. Biol.* 28: 123–126.

Medvedev, Z.A., H.M. Crowne, and M.N. Medvedeva. 1988. Age related variations of hepato carcinogenic effect of azo dye (3′-MDAB) as linked to the level of hepatocyte polyploidization. *Mech. Ageing Dev.* 46: 159–174.

Mirjalili, M. and M. Abbasipour. 2013. Comparison between antibacterial activity of some natural dyes and silver nanoparticles. *J. Nanostruct. Chem.* 3: 37–39.

Mirjalili, M. and L. Karimi. 2013. Antibacterial dyeing of polyamide using turmeric as a natural dye. *Autex Res. J.* 13: 51–56.

Mongkholrattanasit, R., J. Krystufek, J. Wiener, and M. Vikováa. 2011. UV protection property of silk fabric dyed with eucalyptus leaf extract. *J. Text. Inst.* 102: 272–279.

Nadiger, G.S. 2001. Azo ban, eco-norms and testing. *Indian J. Fibre Text. Res.* 26: 55–60.

Ohnuki, T. 2006. Recent study of body odor generating mechanisms and development of antiperspirants. *Frag. J.* 34: 15–23.

Park, J.H., C.H. Kim, H.D. Suh, D.S. Kim, and K.S. Hwang. 2005. Cultivar and maturity effect on dyeing properties of persimmon fruit juice. *Acta Hort.* 685: 143–148.

Parks, L.W. and W.M. Casey. 1996. Fungal sterols, in: R. Prasad and M. Ghannoum (eds.) *Lipids of Pathogenic Fungi*. Boca Raton, FL: CRC Press, pp. 63–82.

Purwar, R. and M. Joshi. 2004. Recent developments in antimicrobial finishing of textiles—A review. *AATCC Rev.* 4: 22–25.

Ramamoorthy, S., M.G. Palackan, L. Maimoon, T. Geetha, D. Bhakta, P. Balamurugan, and S. Rajanarayanan. 2011. Evaluation of antibacterial, antifungal, and antioxidant properties of some food dyes. *Food Sci. Biotechnol.* 20: 7–13.

Reinert, G., F. Fuso, R. Hilfiker, and E. Schmidt. 1997. UV-protecting properties of textile fabrics and their improvement. *Text. Chem. Color.* 29: 36–43.

Samanta, A.K. and P. Agarwal. 2009. Application of natural dyes on textiles. *Indian J. Fibre Text. Res.* 34: 384–399.

Samanta, A.K., A. Konar, and S. Datta. 2012. Dyeing of jute fabric with tesu extract: Part II—Thermodynamic parameters and kinetics of dyeing. *Indian J. Fibre Text. Res.* 37: 172–177.

Sarkar, A.K. 2004. An evaluation of UV protection imparted by cotton fabric dyed with natural colorants. *BMC Dermatol.* 4: 1–8.

Sarkar, A.K. and R. Dhandapani. 2009. Study of natural colorants as antibacterial agents on natural fibres. *J. Nat. Fibres* 6: 46–55.

Schuerch, A.R. and W. Wehrli. 1978. Beta-Lapachone, an inhibitor of oncornavirus reverse transcriptase and eukaryotic DNA polymerase-alpha. Inhibitory effect, thiol dependence and specificity. *Eur. J. Biochem.* 84: 197–205.

Septhum, C., V. Rattanaphani, and S. Rattanaphani. 2007. UV-Vis spectroscopic study of natural dyes with alum as a mordant. *Suranaree J. Sci. Tech.* 14: 91–97.

Shahid, M., Islam, S., and F. Mohammad. 2013. Recent advancements in natural dye applications: A review. *J. Clean. Prod.* 53: 310–331.

Singh, R., A. Jain, S. Panwar, D. Gupta, and S.K. Khare. 2005. Antimicrobial activity of some natural dyes. *Dyes Pigm.* 66: 99–102.

Siva, R. 2007. Status of natural dyes and dye-yielding plants in India. *Curr. Sci.* 92: 916–925.

Sivakumar, V., J. Vijaeeswarri, and J.L. Anna. 2011. Effective natural dye extraction from different plant materials using ultrasound. *Indian Crops Prod.* 33: 116–122.

Temani, P., D.B. Shakyawar, L. Ammayappan, V. Goyal, and S.A. Wani. 2011. Standardization of dyeing condition of cochineal extract on Pashmina yarn. *J. Text. Assoc.* 72: 90–92.

Thiry, M.C. 2011. Staying alive: Making textiles sustainable. *AATCC Rev.* 11: 26–32.

Trotman, E. (ed.). 1984. *Dyeing and Chemical Technology of Textile Fibres*, 6th edn. London, U.K.: Wiley, p. 387.

Umbuzeiro, G.A., H.S. Freeman, S.H. Warren, D.P. Oliveira, Y. Terao, T. Watanabe, and D.L. Claxton. 2005. The contribution of azo dyes to the mutagenic activity of Cristais River. *Chemosphere* 60: 55–64.

Wagner, H., B. Kreher, H. Lotter, M.O. Hamburger, and G.A. Cordell. 1989. Structure determination of new isomeric naphthol (2,3,-b) furan- 4, 9-diones from Tabebuia avellanedae by the selective INEPT technique. *Helv. Chim. Acta* 72: 67–659.

Washino, Y. (ed.). 1993. Functional fibres: Trends, in: *Technology and Product Development in Japan*. Tokyo, Japan: Toray Research Center, Inc. (Asahi Kosoku Insatsu K.K.), pp. 216–245.

Wei, B., Q.Y. Chen, G. Chen, R.C. Tang, and J. Zhang. 2013. Adsorption properties of lac dyes on wool, silk, and nylon. *J. Chem.* 2013: Article ID 546839.

16 Test Methods Related to Characteristics, Performance, and Ecological and Safety Parameters of Textiles

Luis Almeida

CONTENTS

16.1 Introduction: Standardization	352
16.1.1 International Standardization Organization	352
16.1.2 European Committee for Standardization	353
16.1.3 Other Standardization Bodies	354
16.2 Quality Control Standards for Textiles	354
16.2.1 Fiber Tests	354
16.2.2 Yarn Tests	355
16.2.3 Fabric Tests: Basic Characteristics	356
16.2.4 Tests for Mechanical Resistance of Fabrics	358
16.2.5 Properties Related to Fabric Structure and Finishing	361
16.2.6 Dimensional Changes of Fabrics	364
16.2.7 Color Fastness	365
16.2.7.1 Part A: General Principles	365
16.2.7.2 Part B: Color Fastness to Light and Weathering	366
16.2.7.3 Part C: Color Fastness to Washing and Laundering	366
16.2.7.4 Part D: Color Fastness to Dry-Cleaning	366
16.2.7.5 Part E: Color Fastness to Aqueous Agents	366
16.2.7.6 Part F: Standard Adjacent Fabrics	367
16.2.7.7 Part G: Color Fastness to Atmospheric Contaminants	367
16.2.7.8 Part J: Measurement of Color and Color Differences	367
16.2.7.9 Part N: Color Fastness to Bleaching Agents	367
16.2.7.10 Part P: Color Fastness to Heat Treatments	367
16.2.7.11 Part S: Color Fastness to Vulcanization	367
16.2.7.12 Part X: Miscellaneous Tests	367
16.2.7.13 Part Z: Colorant Characteristics	368
16.2.8 Care Labeling	368
16.3 Sustainability Parameters Related to Product Ecology	368
16.3.1 pH of Aqueous Extract	369
16.3.2 Free Formaldehyde	369
16.3.3 Metals	369
16.3.4 Dyes and Pigments	370
16.3.5 Phthalates	370

	16.3.6	Flame Retardants	370
	16.3.7	Alkylphenolethoxylates	371
	16.3.8	Other Harmful Substances	371
16.4	Safety of Clothing		372
16.5	Conclusion		373
References			373

16.1 INTRODUCTION: STANDARDIZATION

Standardization is essential to maximize compatibility, interoperability, safety, repeatability, or quality.

Standardization started mainly in the beginning of the nineteenth century, with the need of the practical interchangeability of screws and later with other machine parts (Ping, 2011).

Standards are essential for society. There are different types of standards. These can include standards for

- Terms and definitions
- Specifications
- Management systems
- Test methods

Standards can be internal of an organization, but for trade, it is important that they are accepted for a large community. There are standards used in a private or local community or in a certain region. Standardization has had many developments at national levels, under the supervision of national standardization bodies. But the need for the interchange of products and services at global level has led, especially since the second half of the twentieth century, to a strong development of international standards.

16.1.1 INTERNATIONAL STANDARDIZATION ORGANIZATION

International standardization started in 1926 with the former ISA—the International Federation of the National Standardizing Associations. This federation stopped its activities in 1942. After the Second World War, ISA was approached by the recently formed United Nations Standards Coordinating Committee. The history of International Standardization Organization (ISO) itself started in London, where in 1946 delegates from 25 countries met to discuss the future of international standardization. In 1947, ISO officially comes into existence with 67 technical committees. The first offices were located in a private house in Geneva, Switzerland, in 1949. Since then, ISO has developed gradually, joining today more than 160 countries all over the world. The new offices in Geneva employ almost 150 people directly (compared with 5 in the early 1950s).

The first ISO standard has been published in 1951: ISO 1—Geometrical Product Specifications—standard reference temperature for geometrical product specification.

The standard number 2 of ISO in fact is related to textiles, ISO 2 (last version from 1973, confirmed in 2011), and concerns the designation of the direction of twist in yarns and related products.

The development of the standards is based on technical committees. One of the technical committees, created since the beginning of ISO, is ISO TC38 "Textiles." The first subcommittee of this technical committee is ISO/TC 38/SC 1 Tests for colored textiles and colorants, which met regularly since 1951 and has published already 111 standards, under the ISO 105 series (textiles—tests for color fastness). Further reference of this standard will be presented later in this chapter.

The most well-known and popular of ISO standards are of the 9000 family (quality management), published since 1987. There are also many other management systems standards (environmental management occupational health and safety, food safety, risk, energy, innovation, information security, etc.). Social responsibility is also especially relevant to the textile and garment industry and is a requirement for many big retailer chains and some ecolabels (see standard ISO 26000:2010 "Guidance on social responsibility").

Today, there are more than 19,000 international standards published. Out of these, about 370 standards are related to ISO TC 38 (textiles), most of them concerning textile test methods. There are also other textile-related standards developed by other ISO technical committees. This include, for instance, ISO/TC 45 "Rubber and rubber products (involving coated fabrics)," ISO/TC 94 "Personal safety—Protective clothing and equipment (involving protective clothing)," and ISO/TC 133 "Clothing sizing systems—Size designation, size measurement methods and digital fittings."

ISO standard documents include also technical specifications (ISO/TS), technical reports (ISO/TR), and international workshop agreements (IWA).

ISO standards are normally published in two languages: English and French. Some of them are also published in Russian language.

ISO works in close cooperation with International Electrotechnical Commission (IEC), an organization created in 1906 that prepares and publishes international standards for all electrical, electronic, and related technologies. There are several relevant common ISO/IEC standards.

16.1.2 EUROPEAN COMMITTEE FOR STANDARDIZATION

The European Committee for Standardization (CEN) has been created in 1961, as a nonprofit organization whose mission is to foster the European economy in global trading, the welfare of European citizens, and the environment by providing an efficient infrastructure to interested parties for the development, maintenance, and distribution of coherent sets of standards and specifications.

The European standardization became more important after the Maastricht Treaty (or treaty of the European Union [EU]), signed in 1992, which includes the free circulation of goods and services. Common standards are essential for the single market, so, once approved, all the European (EN) standards must be adopted by all the CEN members.

CEN is officially recognized by the EU and by the European Free Trade Association (EFTA) as being responsible for developing and defining voluntary standards at European level. CEN country members include all the 28 EU members (Austria, Belgium, Bulgaria, Croatia, Cyprus, Czech Republic, Denmark, Estonia, Finland, France, Germany, Greece, Hungary, Ireland, Italy, Latvia, Lithuania, Luxembourg, Malta, the Netherlands, Poland, Portugal, Romania, Slovakia, Slovenia, Spain, Sweden, and the United Kingdom) as well as the 3 countries that are members of EFTA (Iceland, Norway, and Switzerland) and 2 EU candidate members (former Yugoslav Republic of Macedonia and Turkey). There are also several observer members.

European standardization plays an important role in the development and consolidation of the European Single Market. The fact that each EN standard is recognized across the whole of Europe, and automatically becomes the national standard in 33 European countries, makes it much easier for businesses to sell their goods or services to customers throughout the European Single Market.

In the case of textiles, the most relevant technical committee is CEN/TC248 (Textiles and Textile Products), with already almost 400 standards published. There are still several other TCs also dealing with textile-related standards. This is the case of CEN/TC162, "Protective clothing including hand and arm protection and lifejackets," with almost 200 standards published, relevant for the CE marking of protective clothing.

CEN and ISO have signed in 1991 the so-called Vienna Agreement, to avoid duplication of (potentially conflicting) standards between CEN and ISO. Especially since 2000, CEN has adopted a number of ISO standards that replaced the corresponding CEN standards. In the case of textiles, most of the CEN/TC248 published standards are in fact ISO/EN standards.

All EN standards are published in the three official languages of CEN: English, French, and German. National standardization bodies also publish often versions in the corresponding languages. These versions are translations and have the same statute as the official versions.

16.1.3 Other Standardization Bodies

Apart from ISO, EN, and national standards, there are also some organizations that have relevant standards for the textile trade. These include the following:

American Society for Testing Materials (ASTM). ASTM publishes a comprehensive set of textile standards that provide the specifications and test methods for the physical, mechanical, and chemical properties of fibers, yarns, fabrics, and garments, used worldwide.

The American Association of Textile Chemists and Colorists (AATCC) is also worldwide known for developing and publishing test methods for textiles, especially related to dyeing and finishing. Many of them have served as a basis to the present ISO textile test methods. AATCC publishes annually the AATCC Technical Manual, which includes all the updated test methods and is a useful tool for all textile testing laboratories.

The International Wool Textile Organization (IWTO) publishes specifications that include all test methods developed within the committees of IWTO for the measurement of wool fiber, yarn, and fabric properties. Some of these methods served as a basis and are equivalent to ISO standards.

16.2 QUALITY CONTROL STANDARDS FOR TEXTILES

In order to have reproducible and intercomparable results, it is important that all the stakeholders in the textile-clothing chain use the same test methods. Most of the standards developed within ISO/TC38 are related to textile testing, and many of them have been adopted by CEN as EN ISO standards, according to the Vienna Agreement.

Test methods include fiber, yarn, fabric (woven, knitted, nonwoven), and garment testing.

In terms of fabrics, specific tests for testing coated fabrics and for nonwovens are not mentioned here.

This section includes a brief overview of the most relevant standards and tests.

16.2.1 Fiber Tests

In terms of textile fibers, the two following standards specify the names of all the major textile fibers that are used worldwide, both natural and man-made:

ISO 6938:2012—Textiles, Natural fibers, Generic names and definitions
ISO 2076:2013—Textiles, Man-made fibers, Generic names

Within the EU, in 2011, a specific regulation has been published related to the textile fiber names and related labeling and marking of the fiber composition of textile products (Regulation EU No. 1007/2011 of the European Parliament and the Council of September 27, 2011). There are similar regulations in other regions or countries.

In terms of fiber composition of textile products, the following standards are relevant:

ISO/TR 11827:2012—Textiles, Composition testing, Identification of fibers. This technical report describes procedures for the identification of natural and man-made fibers and may be used, when necessary, to coordinate with methods for the quantitative analysis of fiber blends.

ISO 5089:1977—Textiles, Preparation of laboratory test samples and test specimens for chemical testing. This standard specifies methods of obtaining laboratory test samples of textile materials from laboratory bulk samples taken from a bulk source and gives general directions for the preparation of test specimens of convenient size for chemical tests and, namely, for quantitative chemical analysis of fiber blends.

ISO 1833 series—Textiles, Quantitative chemical analysis. This is a series of more than 20 standards that specify test methods for the analysis of fiber blends. Although most of the standards refer to chemical analysis, manual separation of the fibers should be used whenever it is feasible.

Most of the methods of the ISO 1833 series are also included in the Regulation EU No. 1007/2011 mentioned earlier.

There are also many standards related to test methods for the quality of textile fibers: fiber length, fiber diameter, fineness, moisture regain, mechanical properties, chemical characteristics (namely, for wool fibers), impurities, etc. These test methods are important for fiber producers, traders, and yarn manufacturers but not so relevant for the garment manufacturing industry.

16.2.2 Yarn Tests

There are several standards related to yarns.

The standard ISO 2, already mentioned, refers to the twist of yarns:

ISO 2:1973—Textiles, Designation of the direction of twist in yarns and related products
This standard, reconfirmed in 2011, is applicable to yarn intermediates such as slivers, slubbings, or rovings; to single yarns, plied yarns, and cabled yarns; and to threads, twine, cordage, and rope. The direction of the twist is indicated by the capital letters S and Z.

ISO 1139:1973—Textiles, Designation of yarns
This standard, confirmed in 2012, describes two methods of indicating the composition of yarns, whether single, folded, cabled, or multiple wound. The notation comprises linear density indicated in the Tex System and number of filaments in filament yarns of all kinds except special kinds of yarns, for example, fancy yarns, textured yarns, bulked yarns, or yarns produced by wrapping a textile or nontextile material round a core.

In terms of yarn numbering system, ISO has published the following standards:

ISO 1144:1973—Textiles, Universal system for designating linear density (Tex System)
This standard gives the principles and recommended units of the Tex System for the expression of linear density and includes conversion tables for calculation the Tex values of numbers or counts in other systems together with a statement of the procedure for the implementation of the Tex System in trade and industry.

In order to facilitate the changeover by industry and commerce from traditional yarn numbering systems to the Tex System, the following standard provides a range of recommended rounded linear densities in the Tex System to replace the yarn numbers in the six main traditional numbering systems that are still used by the industry:

ISO 2947:1973—Textiles, Integrated conversion table for replacing traditional yarn numbers by rounded values in the Tex System

The following two ISO standards concern textured filament yarns:

ISO 8160:1987—Textiles, Textured filament yarns, Vocabulary
ISO 10132:1993—Textiles, Textured filament yarn, Definitions

There are also several standards that relate to test methods to evaluate the quality of yarns. The most important are

> ISO 2060:1994—Textiles, Yarn from packages, Determination of linear density (mass per unit length) by the skein method
>> This standard, confirmed in 2013, specifies a method for the determination of the linear density of all types of yarn in package form. It includes seven optional procedures based on different methods of conditioning and preparation. This method is applicable to single yarns, folded yarns, and cabled yarns, but it is not applicable to highly stretchable yarns or to yarns having a linear density greater than 2000 tex.
>
> ISO 2061:2010—Textiles, Determination of twist in yarns, Direct counting method
>> This standard specifies a method for the determination of the direction of twist in yarns; the amount of twist, in terms of turns per unit length; and the change in length on untwisting, by the direct counting method. It is applicable to single yarns (spun and filament), plied yarns, and cabled yarns. Another method applicable to single yarns is
>
> ISO 17202:2002—Textiles, Determination of twist in single spun yarns, Untwist/retwist method

The following standard specifies methods for the determination of the breaking force and elongation at break of textile yarns taken from packages:

> ISO 2062:2009—Textiles, Yarns from packages, Determination of single-end breaking force and elongation at break using constant rate of extension (CRE) tester

Four methods are given:

A. Manual—Specimens are taken directly from conditioned packages.
B. Automatic—Specimens are taken directly from conditioned packages.
C. Manual—Relaxed test skeins are tested after conditioning (this method should be used in cases of dispute regarding elongation at break of the yarn).
D. Manual—Specimens are tested after wetting.

An important quality parameter of a yarn is its evenness. The following standard describes the most widely used method, developed by Zellweger Uster, using capacitance measuring equipment, for determining the unevenness of linear density along the length of textile strands:

> ISO 16549:2004—Textiles, Unevenness of textile strands, Capacitance method

The method is applicable not only to yarns (spun and continuous filament) but also to tops, slivers, and rovings, made from either natural or man-made fibers.

In the case of yarns for knitting or for sewing threads, an important parameter is the coefficient of yarn friction to metal: if the yarn friction is high, there is an increase of temperature of the needles, leading to its damage or break. For determination of this parameter, this is an ASTM standard:

> ASTM D3108—Standard Test Method for Coefficient of Friction, Yarn to Solid Material

16.2.3 Fabric Tests: Basic Characteristics

ISO has published, under the ISO 7211 series, several standards related to the construction of woven fabrics. All the six standards have been published in 1994 and confirmed in 2010.

ISO 7211-1:1984—Textiles, Woven fabrics, Construction, Methods of analysis, Part 1: Methods for the presentation of a weave diagram and plans for drafting, denting and lifting

According to this standard, the weave repeat shown on design paper is adopted as the means for showing the weave of fabric, and provision is made in the repeat for showing the disposition of different yarns in relation to the weave when there is more than one yarn in the warp or weft.

ISO 7211-2:1984—Textiles, Woven fabrics, Construction, Methods of analysis, Part 2: Determination of number of threads per unit length

This standard includes three methods of determining the number of threads per centimeter. The principles are as follows:

Method A: A section of fabric of dimension specified is dissected and the number of threads counted. The threads that are to be counted are preferably short, 1 or 2 cm being suitable.

Method B: The number of threads visible within the aperture of a defined counting glass is determined.

Method C: The number of threads per centimeter of the fabric is determined with the aid of a traversing thread counter.

There is a EN standard (EN 1049-2:1993) with slight differences (it does not include method B of ISO 7211-2:1984).

ISO 7211-3:1984—Textiles, Woven fabrics, Construction, Methods of analysis, Part 3: Determination of crimp of yarn in fabric

Threads are removed from a strip of fabric of known length, straightened by a tension that is varied according to the nature and linear density of the yarn, and measured in the straightened state. The difference between the straightened length of the thread and the distance between the ends while in fabric is expressed as a percentage of the latter.

ISO 7211-4:1984—Textiles, Woven fabrics, Construction, Methods of analysis, Part 4: Determination of twist in yarn removed from fabric

A length of yarn is removed from the fabric and, while under tension, is secured in two clamps that are at a known distance apart. One of these clamps is then rotated until all twists are removed from the length of yarn. This method is only applicable to yarns spun on conventional systems and not applicable to open-end spun or interlaced yarns, for example.

ISO 7211-5:1984—Textiles, Woven fabrics, Construction, Methods of analysis, Part 5: Determination of linear density of yarn removed from fabric

Threads are removed from rectangular strips of fabric, the straightened length of a portion of them is determined, and their mass is determined. Linear density is calculated from the mass and the sum of the straightened length.

In the case of knitted fabrics, there is the following EN standard:

EN 14970:2006—Textiles, Knitted fabrics, Determination of stitch length and yarn linear density in weft knitted fabrics

The measurements should be applied to yarns from each knitting machine feeder and/or different patterning courses. The results can be used for the analysis of fabric fault, for example, barré analysis. The method can also be applied to warp knitted fabrics if it is possible to de-knit the fabric.

ISO 7211-6:1984—Textiles, Woven fabrics, Construction, Methods of analysis, Part 6: Determination of the mass of warp and weft per unit area of fabric

This standard establishes two methods:
Method A: The outline of the fabric specimen to be dissected is marked in the form of a square or rectangle, and the nonfibrous matter is removed while the marked area still forms part of a larger sample.
Method B: A specimen of known area is dissected, and the nonfibrous matter is removed from the warp and weft threads.

An important parameter for fabrics is the mass per unit area. For the determination of this property, these are the following standard methods:

ISO 3801:1977—Textiles, Woven fabrics, Determination of mass per unit length and mass per unit area
EN 12127:1997—Textiles, Fabrics, Determination of mass per unit area using small samples

This last standard includes the most common method used in the industry: cutting of circular samples of 100 cm^2 and determination of the weight.

For the determination of thickness of normal fabrics, this is the following standard:

ISO 5084:1996—Textiles, Determination of thickness of textiles and textile products

This standard describes a test method for the determination of the thickness of textiles and textile products under specified pressure. It is not applicable to textile floor coverings (see ISO 1765), nonwovens (see ISO 9073-2), geotextiles (see ISO 9863, parts 1 and 2), and coated fabrics (see ISO 2286-3).

In woven fabrics, the weft yarns are, in principle, perpendicular to the warp yarns. The following standard allows determining the skew (related to the obliquity of the weft) and the bow (related to the concavity/convexity of the weft):

ISO 13015:2013—Woven fabrics, Distortion, Determination of skew and bow

Specifically for knitted fabrics, ISO 23606:2009 (confirmed in 2014) specifies various systems of symbolic notation and pattern design for knitted fabrics.

ISO 8499:2003 describes defects that commonly appear during the inspection of knitted fabrics. This standard makes special emphasis on the defects that, if appearing in a prominent position in an article made from the fabric, would readily be seen and rejected by a prospective purchaser.

16.2.4 Tests for Mechanical Resistance of Fabrics

ISO 13934-1:2013—Textiles, Tensile properties of fabrics, Part 1: Determination of maximum force and elongation at maximum force using the strip method
This method specifies the determination of the maximum force and elongation at maximum force of test specimens in equilibrium with the standard atmosphere for testing and of test specimens in the wet state. The method is mainly applicable to woven textile fabrics, including fabrics that exhibit stretch characteristics imparted by the presence of an elastomeric fiber, mechanical, or chemical treatment.
ISO 13934-2:2014—Textiles, Tensile properties of fabrics, Part 2: Determination of maximum force using the grab method
This method is similar to ISO 13934-1:2013, but it is not necessary to fray the samples. It is used mainly for internal control of the companies.

In terms of elasticity of fabrics, there are three EN standards of the series:
EN 14704: Determination of the elasticity of fabrics:

Part 1 (2005): Strip tests
Part 2 (2007): Multiaxial tests
Part 3 (2006): Narrow fabrics

In terms of tear resistance of woven fabrics, ISO has published four standards, with the reference ISO 13937-x:2000 "Textiles—Tear properties of fabrics" (all the standards reconfirmed in 2011):

Part 1: Determination of tear force using ballistic pendulum method (Elmendorf)
Part 2: Determination of tear force of trouser-shaped test specimens (single-tear method)
Part 3: Determination of tear force of wing-shaped test specimens (single-tear method)
Part 4: Determination of tear force of tongue-shaped test specimens (double-tear test)

The first standard (Elmendorf apparatus) is the most commonly used. The method describes the measurement of the tear force required to propagate a single-rip tear of defined length from a cut in a fabric when a sudden force is applied.

The aforementioned methods are not suitable for evaluating the mechanical resistance of knitted fabrics. For these fabrics, bursting strength is normally used. ISO provides two test methods, with the reference ISO 13938-x:1999 "Textiles—Bursting properties of fabrics" (both standards reconfirmed in 2010):

Part 1: Hydraulic method for determination of bursting strength and bursting distension
Part 2: Pneumatic method for determination of bursting strength and bursting distension

Both methods are applicable to all kinds of fabrics. From the available data, there appears to be no significant difference in the bursting strength results achieved using hydraulic or pneumatic burst testers, for pressures up to 800 kPa. This pressure range covers the majority of performance levels expected of general apparel. For specialty textiles requiring high bursting pressures, the hydraulic apparatus is more suitable.

For abrasion resistance, there are several test methods with different severity.

The Martindale abrasion tester is the more frequent test for fabrics for clothing and is based on friction against a standard wool fabric. ISO provides four standards, with the reference ISO 12947-x:1998 "Textiles—Determination of the abrasion resistance of fabrics by the Martindale method" (all the standards reconfirmed in 2010):

Part 1: Martindale abrasion testing apparatus
Part 2: Determination of specimen breakdown
Part 3: Determination of mass loss
Part 4: Assessment of appearance change

There are many other tests for abrasion resistance, for instance:

AATCC 93 (Accelerotor method): The fabric specimen is driven by an impeller (rotor) along a zigzag course in a generally circular orbit within a cylindrical chamber, so that it repeatedly impinges on the walls and abradant liner of the chamber while at the same time being continually subjected to extremely rapid, high velocity impacts.
ASTM D 3885 (Stoll flexometer): The fabric is submitted to a simultaneous flexion and abrasion test.

For coated fabrics, the following ISO standard can be used:

ISO 32100:2010—Rubber- or plastics-coated fabrics, Physical and mechanical tests, Determination of flex resistance by the flexometer method.

By the effect of a slight abrasion, pilling can occur.

Pills are formed when fibers on a fabric surface *tease out* and become entangled during wear. Such surface deterioration is generally undesirable, but the degree of consumer tolerance for a given level of pilling will depend on the garment type and fabric end use.

Generally, the level of pilling that develops is determined by the rates of the following parallel processes:

a. Fiber entanglement leading to pill formation
b. Development of more surface fiber
c. Fiber and pill wear off

The rates of these processes depend on the fiber, yarn, and fabric properties. Examples of extreme situations are found in fabrics containing strong fibers versus fabric containing weak fibers. A consequence of the strong fiber is a rate of pill formation that exceeds the rate of wear off. This results in an increase of pilling with an increase of wear. With a weak fiber, the rate of pill formation competes with the rate of wear off. This would result in a fluctuation of pilling with an increase of wear. There are other constructions in which the surface fiber wear off occurs before pill formation. Each of these examples demonstrates the complexity of evaluating the surface change on different types of fabric.

For the evaluation of fuzzing and pilling propensity of fabrics, ISO has three test methods, of the series ISO 12945 "Textiles—Determination of fabric propensity to surface fuzzing and to pilling":

Part 1 (2000): Pilling box method
Part 2 (2000): Modified Martindale method
Part 3 (2014): Random tumble pilling method

There are several test methods to evaluate the behavior of fabrics under the effect of a seam.

Seam slippage is the movement, in a woven fabric, of weft yarns over warp yarns (or warp yarns over weft yarns) as a result of a pulling action. Seam slippage is a fabric property and should not be confused with seam strength.

The sewn fabrics may be obtained from previously sewn articles or may be prepared from fabric samples, as agreed by the parties interested in the results.

To evaluate seam slippage, there are three standards of the series ISO 13936 "Textiles—Determination of the slippage resistance of yarns at a seam in woven fabrics":

Part 1 (2004): Fixed seam opening method
Part 2 (2005): Fixed load method
Part 3 (2005): Needle clamp method

The seam tensile properties can also be determined at break. There are in this case two standard methods, of the series ISO 13935:2014 "Textiles—Seam tensile properties of fabrics and made-up textile articles":

Part 1: Determination of maximum force to seam rupture using the strip method
Part 2: Determination of maximum force to seam rupture using the grab method

In terms of mechanical stresses, all the tests mentioned previously are destructive. Professor Sueo Kawabata has been a pioneer in developing low mechanical stress tests, namely, related to the handle of fabrics (see, for instance, Kawabata et al., 1982). These tests have not been adopted by ISO standards.

16.2.5 Properties Related to Fabric Structure and Finishing

The *thermal insulation* provided by woven and knitted fabrics and nonwoven fiber aggregates is of considerable practical significance in connection with the use of these materials as clothing (especially cold-weather clothing) and other textiles. The following standard applies:

ISO 5085-1:1989—Textiles, Determination of thermal resistance, Part 1: Low thermal resistance
Part 2, concerning high thermal resistance materials, has been withdrawn in 2005.

Air permeability of fabrics is related to the velocity of an air flow passing perpendicularly through a test specimen under specified conditions of test area, pressure drop, and time. Air permeability is an important parameter not only for technical textiles such us air filters or airbags but also for comfort. A relevant test method is included in the following standard:

ISO 9237:1995—Textiles, Determination of the permeability of fabrics to air

Water permeability of fabrics is essential for dyeing/printing/chemical finishing processes and for many end uses. On the other hand, in many cases, it is required to have water repellent properties. In order to evaluate water repellency, ISO proposes several test methods:

ISO 811:1981—Textile fabrics, Determination of resistance to water penetration, Hydrostatic pressure test
ISO 4920:2012—Textile fabrics, Determination of resistance to surface wetting (spray test)
ISO 9865:1991—Textiles, Determination of water repellency of fabrics by the Bundesmann rain-shower test
ISO 18695:2007—Textiles, Determination of resistance to water penetration, Impact penetration test
ISO 18696:2006—Textiles, Determination of resistance to water absorption, Tumble-jar.
ISO 22958:2005—Textiles, Water resistance, Rain tests: exposure to a horizontal water spray

There are also several test methods for evaluating water penetration in nonwovens (series ISO 9073).

Water vapor permeability is an important property related to the physiological properties of the textiles. It is also a requirement for protective clothing. The most important standard method, which also measures thermal resistance, is

ISO 11092:1993—Textiles, Physiological effects, Measurement of thermal and water-vapor resistance under steady-state conditions (sweating guarded-hotplate test)

ISO also provides a simpler test for quality control:

ISO 15496:2004—Textiles, Measurement of water vapor permeability of textiles for the purpose of quality control

There are other test methods for measuring thermal resistance and water vapor permeability, not included in ISO standards (see, for instance, Hes and Loghin, 2009).

Oil repellency is a parameter that allows to evaluate the stain resistance of textiles.

ISO 14419:2010—Textiles, Oil repellency, Hydrocarbon resistance test

This method is applicable to the evaluation of a substrate's resistance to absorption of a selected series of liquid hydrocarbons of different surface tensions, providing a guide to oil stain resistance. It can provide a rough index of oil stain resistance. Results are given in a scale of 1–8. ISO 14419:2010 can also be utilized in determining if washing and/or dry cleaning treatments have any adverse effect on the oil repellency characteristics of a substrate.

In terms of *aqueous liquid repellency*, the following test, similar to ISO 14419, can be used:

ISO 23232:2009—Textiles, Aqueous liquid repellency, Water/alcohol solution resistance test

Crease recovery is an important parameter for textiles, in terms of evaluating the crease propensity and easy care properties. There are several ISO standard test methods in this area:

- ISO 2313:1972—Textiles, Determination of the recovery from creasing of a horizontally folded specimen of fabric by measuring the angle of recovery. This test method is very time demanding and is normally used in research works.
- ISO 9867:2009—Textiles, Evaluation of the wrinkle recovery of fabrics, Appearance method. The appearance of textile fabrics after induced wrinkling is evaluated by comparison with plastic replicas (scale 1–5).

The following standards are related to the evaluation of the appearance of fabrics after cleansing treatments:

- ISO 7768:2009—Textiles, Test method for assessing the smoothness appearance of fabrics after cleansing.
- ISO 7769:2009—Textiles, Test method for assessing the appearance of creases in fabrics after cleansing.
- ISO 7770:2009—Textiles, Test method for assessing the smoothness appearance of seams in fabrics after cleansing.
- ISO 15487:2009—Textiles, Method for assessing appearance of apparel and other textile end products after domestic washing and drying. This standard is a combination of the three methods 7768, 7769, and 7770. Each property is evaluated in a scale of 1–5. Smooth aspect and creases are evaluated by comparison with plastic replicas, while for seams, there are standard photographs. Evaluation should be made under specified viewing conditions.

There is a wide variety of test methods to evaluate the *burning behavior* or *flame-retardant* properties of textiles. In terms of ISO methods, the following can be mentioned:

- ISO 6940:2004—Textile fabrics, Burning behavior, Determination of ease of ignition of vertically oriented specimens
- ISO 6941:2003—Textile fabrics, Burning behavior, Measurement of flame spread properties of vertically oriented specimens
- ISO 10047:1993—Textiles, Determination of surface burning time of fabrics

Specifically for bedding items, ISO provides two test methods:

- ISO 12952:2010—Textiles, Assessment of the ignitability of bedding items:
   Part 1: Ignition source—smouldering cigarette
   Part 2: Ignition source—match-flame equivalent

CEN has also published several standards related to the burning behavior of curtains and drapes and technical textiles.

The flammability of textile materials can be affected by a combination of different aspects that occur during washing: shrinkage of the material, causing an increase in mass per unit area; abrasion of the material, causing a decrease in mass per unit area; removal of finishes; chemical modification of the fiber or finish; deposition of hard water salts; and application of softeners in the rinse cycle. ISO provides the two following laundering procedures to be performed prior to flammability testing:

- ISO 12138:1996—Textiles, Domestic laundering procedures for textile fabrics prior to flammability testing
- ISO 10528:1995—Textiles, Commercial laundering procedure for textile fabrics prior to flammability testing

In terms of *effects related to microorganisms*, there are two types of aspects to be considered: tests concerning the degradation of the fibers and tests concerning the antibacterial activity of textiles.

Cellulose is sensible to the action of microorganisms, when stored in wet. The following standard is applicable in this case:

- ISO 11721—Textiles, Determination of resistance of cellulose-containing textiles to microorganisms, Soil burial test. This standard has two parts:
  Part 1 (2001): Assessment of rot-retardant finishing
  Part 2 (2003): Identification of long-term resistance of a rot-retardant finish

The following standard in two parts is related to the evaluation of antifungal activity of textiles:

- ISO 13629—Textiles, Determination of antifungal activity of textile products:
  Part 1 (2012): Luminescence method
  Part 2 (2014): Plate count method

ISO presents two standards for the determination of antibacterial activity of textiles:

- ISO 20645:2004—Textile fabrics, Determination of antibacterial activity, Agar diffusion plate test

This method is applicable to testing hygienic finishes of hydrophilic, air-permeable materials or antibacterial products incorporated in the fiber.

- ISO 20743:2013—Textiles, Determination of antibacterial activity of textile products

This standard specifies quantitative test methods to determine the antibacterial activity of all antibacterial textile products including nonwovens. It is applicable to all textile products, including cloth, wadding, thread, and material for clothing, bedclothes, home furnishings, and miscellaneous goods, regardless of the type of antibacterial agent used or the method of application. Based on the intended application and on the environment in which the textile product is to be used and also on the surface properties of the textile properties, the user can select the most suitable of the following three inoculation methods on determination of antibacterial activity:

1. Absorption method (an evaluation method in which the test bacterial suspension is inoculated directly onto specimens)
2. Transfer method (an evaluation method in which test bacteria are placed on an agar plate and transferred onto specimens)
3. Printing method (an evaluation method in which test bacteria are placed on a filter and printed onto specimens)

In terms of *UV protection* of textiles, there is at present no specific ISO standard. The most commonly used standard methods are the following:

> AS/NZS 4399:1996—Sun protective clothing, Evaluation and classification (Australian/New Zealand standard).
> AATCC Test Method 183-2010—Transmittance or blocking of erythemally weighted ultraviolet radiation through fabrics.
> EN 13758-1:2001—Textiles, Solar UV protective properties, Part 1: Method of test for apparel. Part 2 (2003) is devoted to classification and marking.

16.2.6 Dimensional Changes of Fabrics

Dimensional stability is essential for fabrics for apparel. There are several ISO standards to cover this topic:

> ISO 3759:2011—Textiles, Preparation, marking and measuring of fabric specimens and garments in tests for determination of dimensional change

This standard specifies a method for the preparation, marking, and measuring of textile fabrics, garments, and fabric assemblies for use in tests for assessing dimensional change after a specified treatment, for example, washing, dry cleaning, soaking in water, and steaming.

In terms of *washing*, the following standards can be considered:

> ISO 5077:2007—Textiles, Determination of dimensional change in washing and drying.
> ISO 6330:2012—Textiles, Domestic washing and drying procedures for textile testing.
> ISO 15797:2002—Textiles, Industrial washing and finishing procedures for testing of workwear.
> ISO 23231:2008—Textiles, Determination of dimensional change of fabrics, Accelerated machine method. This standard provides a quick test, useful for internal control of the companies.

Laundering can also induce *spirality*, especially in knitted fabrics. For evaluating this phenomenon, ISO provides three standards of the series.

> ISO 16322:2005—Textiles, Determination of spirality after laundering:
> Part 1: Percentage of wale spirality change in knitted garments
> Part 2: Woven and knitted fabrics
> Part 3: Woven and knitted garments

The following method is applicable to fabrics which, in use, are subjected to *cold water* without agitation:

> ISO 7771:1985—Textiles, Determination of dimensional changes of fabrics induced by cold-water immersion

In terms of *dry-cleaning* and *wet-cleaning*, ISO provides four standards of the series:

> ISO 3175—Textiles, Professional care, drycleaning and wetcleaning of fabrics and garments:
> Part 1 (2010): Assessment of performance after cleaning and finishing. This part includes a complete assessment of the behavior of fabrics and garments, not only dimensional changes.
> Part 2 (2010): Procedure for testing performance when cleaning and finishing using tetrachloroethene.

Part 3 (2003): Procedure for testing performance when cleaning and finishing using hydrocarbon solvents.

Part 4 (2003): Procedure for testing performance when cleaning and finishing using simulated wetcleaning.

To evaluate the dimensional changes due to the effect of *steam* (important, for instance, in garment manufacturing), there is the following test method:

ISO 3005:1978—Textiles, Determination of dimensional change of fabrics induced by free-steam

16.2.7 COLOR FASTNESS

The need for standards for testing color fastness dates already from the nineteenth century. German dyestuff producers created in 1911 Deutsche Echtheitskommission (DEK, in English *German Color Fastness Committee*), which led to the development of standardized and simple tests, performed in small samples, essential for the suppliers, for the dyehouses, and for the retailers to obtain reproducible and intercomparable results. DEK created the five-grade gray scales, still used today all over the world, for evaluating the results of color fastness tests, a good example in terms of international standardization (Schiller, 2011). DEK has promoted the creation of the European Colorfastness Establishment (ECE, in German *Europäisch-Continentalen Echtheitskommission*), joining the main European manufacturers of dyestuffs.

Within ISO, the work related to color fastness tests has been developed by subcommittee ISO/TC38/SC1, since 1951, in collaboration with ECE. The first standard ISO 105/R has been published, and since then, all the ISO color fastness tests have the number ISO 105 (Valldeperas, 2011).

In the United States, AATCC has also developed in parallel many test methods for color fastness. Although there are some differences with ISO standards, there is a tendency to make AATCC and ISO standards more similar.

Standards of the series 105 are classified in the following manner:

105-Y-xx

where
Y is a letter from A to Z
xx is a digit starting from 1 by chronological order of the development of the standards in each group

There are at present more than 100 standards included in the series ISO 105. Only the most relevant for the garment industry will be mentioned here.

16.2.7.1 Part A: General Principles

The standards of this section describe the general principles and testing conditions, the gray scales for change in color and staining, and the instrumental assessment of color fastness:

Part A01 (2010): General principles of testing
Part A02 (1993): Grey scale for assessing change in color
Part A03 (1993): Grey scale for assessing staining
Part A04 (1989): Method for the instrumental assessment of the degree of staining of adjacent fabrics
Part A05 (1996): Instrumental assessment of change in color for determination of grey scale rating

16.2.7.2 Part B: Color Fastness to Light and Weathering

Color fastness to light and weathering is evaluated by comparing the performance with blue wool standards, in a scale of 1–8 (or L2–L9).

Part B01 (1994): Color fastness to light, Daylight.
Part B02 (2013): Color fastness to artificial light, Xenon arc fading. This standard is the most commonly used, as xenon arc allows a reproducible and much quicker fading than daylight.
Part B03 (1994): Color fastness to weathering, Outdoor exposure.
Part B04 (1994): Color fastness to artificial weathering, Xenon arc fading lamp test.
Part B05 (1993): Detection and assessment of photochromism.
Part B06 (1998): Color fastness and ageing to artificial light at high temperatures, Xenon arc fading lamp test.
Part B07 (2009): Color fastness to light of textiles wetted with artificial perspiration.
Part B08 (1995): Quality control of blue wool reference materials 1 to 7.

16.2.7.3 Part C: Color Fastness to Washing and Laundering

The most traditional color fastness tests are the test to washing of the series 105 C01 to C05. Since 2006, these five tests have merged in only one:

Part C10 (2006): Color fastness to washing with soap or soap and soda

Today, the following are most important tests:

Part C06 (2010): Color fastness to domestic and commercial laundering. This standard provides a series of alternatives, at different temperatures, with the possible addition of a bleaching agent and different mechanical actions.
Part C08 (2010): Color fastness to domestic and commercial laundering using a nonphosphate reference detergent incorporating a low-temperature bleach activator. This standard provides tests with the addition of tetraacetylethylenediamine (TAED), common in most washing detergents used nowadays.
Part C09 (2001): Color fastness to domestic and commercial laundering, Oxidative bleach response using a nonphosphate reference detergent incorporating a low temperature bleach activator. This test corresponds in fact to test the behavior of dyeing to nonchlorine bleaching processes.
Part C12 (2004): Color fastness to industrial laundering.

16.2.7.4 Part D: Color Fastness to Dry-Cleaning

Part D01 (2010): Color fastness to drycleaning using perchloroethylene solvent.
Part D02 (1993): Color fastness to rubbing, Organic solvents. This test uses the same equipment—crockmeter—as test X12 (dry and wet rubbing).

16.2.7.5 Part E: Color Fastness to Aqueous Agents

Part E01 (2013): Color fastness to water
Part E02 (2013): Color fastness to sea water
Part E03 (2010): Color fastness to chlorinated water (swimming-pool water)
Part E04 (2013): Color fastness to perspiration

Tests E01, E02, and E04 use the same test equipment—perspirometer—and are performed for 4 h at 37°C. Test E04 prescribes two artificial sweat solutions: acid and alkaline. E03 includes the immersion of the test specimens in solutions with three different active chorine concentrations, at 20°C.

Part E07 (2010): Color fastness to spotting, Water (action of a drop of water)

Textile Test Methods

16.2.7.6 Part F: Standard Adjacent Fabrics

Part F specifies the adjacent fabrics that are necessary for all the tests that require the evaluation of staining. F01–F09 concern monofiber adjacent fabrics, and F10 concerns multifiber adjacent fabrics, which allow to evaluate simultaneously the staining in six different fibers.

16.2.7.7 Part G: Color Fastness to Atmospheric Contaminants

This section is devoted to the exposure of several atmospheric contaminants (G01, nitrogen oxides; G02, burnt-gas fumes; G03, ozone in the atmosphere; G04, oxides of nitrogen in the atmosphere at high humidities).

16.2.7.8 Part J: Measurement of Color and Color Differences

The four standards included in Part J are not really color fastness tests but methods to quantify color measurement, including relative whiteness.

16.2.7.9 Part N: Color Fastness to Bleaching Agents

The most important standard for garments is as follows:

> Part N01 (1993): Color fastness to bleaching, Hypochlorite. This standard evaluates the action of bleaching baths containing sodium or lithium hypochlorite in concentrations normally used in commercial bleaching, including home treatments. It is applicable mainly to natural and regenerated cellulose materials.

The four other parts concern bleaching in textile processes using peroxide (N02), sodium chlorite (N03 and N04), and sulfur dioxide (N05).

16.2.7.10 Part P: Color Fastness to Heat Treatments

This part contains two standards, allowing to evaluate color fastness to dry heat, excluding pressing (P01) and steam pleating (P02).

16.2.7.11 Part S: Color Fastness to Vulcanization

This part contains three standards that concern the effect of three different processes of vulcanization on color. These tests are rarely used.

16.2.7.12 Part X: Miscellaneous Tests

This part includes different test methods. Among these, the most relevant for garments are

> Part X11 (1994): Color fastness to hot pressing

This method concerns hot-pressing (namely, ironing) when the textile is dry, when it is wet, and when it is damp. The standard considers three different temperatures: 110°C, 150°C, and 200°C.

> Part X12 (2001): Color fastness to rubbing

This standard uses a crockmeter and describes two tests to evaluate staining, one with a dry rubbing cloth and one with a wet rubbing cloth.

> Part X18 (2007): Assessment of the potential to phenolic yellowing of materials

Phenolic yellowing of textile materials is caused by the action of oxides of nitrogen and phenolic compounds during storage, namely in plastic bags, which lead to the production of a yellow color.

16.2.7.13 Part Z: Colorant Characteristics

Most of the standards including in this section are useful mainly for dyestuff manufacturers.

16.2.8 CARE LABELING

The variety of fibers, materials, and finishes used in the production of textile articles, together with the development of cleansing and care procedures, makes it difficult and often impossible to decide on the appropriate cleansing and care treatment for each article simply by inspecting the textile article. To help those who have to make such a decision (principally, the consumer but also launderers and dry cleaners), care labeling of garments is very important. The following standard presents graphic symbols that are widely used:

> ISO 3758 (2012)—Textiles, Care labeling code using symbols
> This standard has been developed based on the GINETEX (International Association for Textile Care Labeling) system, based on five graphic symbols for washing, bleaching, drying, ironing, and professional textile care. ISO 3758 also presents specific requirements for Japan and the United States. This standard includes reference to test methods for the correct selection of care symbols.

For industrial laundering, the following standard should be used:

> ISO 30023:2010—Textiles, Qualification symbols for labeling workwear to be industrially laundered
> This standard establishes a system of graphical symbols, intended for use in the marking of workwear articles and protective clothing providing information on the suitability for professional industrial laundering, including washing, bleaching, tunnel finishing, and tumble drying after washing.

16.3 SUSTAINABILITY PARAMETERS RELATED TO PRODUCT ECOLOGY

In order to answer to the needs of the consumers for eco-friendly and more sustainable products, several ecolabeling systems have been developed in the last two decades (Almeida, 2014).

It is worth mentioning that within CEN/TC 248, the following document is being developed as a technical report:

prCEN/TR 16741—Textiles and textile products, Guidance on health and environmental issues related to chemical content of textile products intended for clothing, interior textiles, and upholstery

In Europe, several of the substances mentioned as follows are restricted, according to the REACH Regulation (Regulation No 1907/2006 of the European parliament and of the Council of December 18, 2006) and namely its annex XVII (which includes all substances included in previous legislation). Annex XVII is regularly updated, taking into account the inclusion of substances of very high concern (SVHC) in the candidate list. Several amendments have already been published in 2014. Most of the ecolabels take into account the restriction of these substances (Almeida, 2014).

In the following sections, several test methods to evaluate the compliance of the textiles to these criteria are presented. Tests that concern production processes and the corresponding emissions are not included here, as they are not made in the textile itself. ISO or EN test methods are mentioned as far as possible. For many parameters, there are still no international tests. For instance, for testing according to OEKO-TEX 100, OEKO-TEX Association uses specific test methods that are not publically disclosed. In the case of the *clear to wear* label, Inditex group uses tests developed by the University of Santiago de Compostela, reserved to laboratories accredited by Inditex's Corporate Social Responsibility Department.

16.3.1 pH of Aqueous Extract

The pH is an important parameter, especially for textiles that are in skin contact. For these textiles, it is normally recommended that pH is between 4.0 and 7.5. ISO specifies the following test method:

ISO 3071:2005—Textiles, Determination of pH of aqueous extract. The determination is based on the extraction and measurement in a KCl solution.

16.3.2 Free Formaldehyde

ISO prescribes two methods for the evaluation of free and releasable formaldehyde in textiles:

ISO 14184-1:2011—Textiles, Determination of formaldehyde, Part 1: Free and hydrolyzed formaldehyde (water extraction method)

ISO 14184-2:2011—Textiles, Determination of formaldehyde, Part 2: Released formaldehyde (vapor absorption)

The first standard is based on the so-called Japanese method and is used to simulate the effect of skin contact. It is the test method referred in most of the ecolabels. A quantity below the detection limit, which is 16 ppm (or mg/kg), is a demand for several labels, namely, for baby articles.

The second standard is more adequate for occupational exposure to formaldehyde, namely, in a garment manufacturing company.

16.3.3 Metals

Metals and particularly several heavy metals are restricted by legislation and by most ecolabels. There are at present four relevant DIN standards devoted to methods for the determination of metals, of the series DIN 54233 "Testing of textiles—Determination of metals":

Part 1 (2010): Determination of metals using microwave digestion
Part 2 (2014): Determination of extractable metals with hydrochloric acid
Part 3 (2010): Determination of extractable metals acid synthetic sweat solution
Part 4 (2014): Determination of extractable metals with synthetic saliva solution

CEN/TC248 is developing the following two standards to determine metals in textiles, based on DIN standards:

prEN 16711-1 Determination of metals using microwave digestion
prEN 16711-2 Determination of metals extracted by acidic artificial perspiration solution

Nickel can be present especially in metal accessories. The following EN standard specifies a method for accelerated wear and corrosion, to be used prior to the detection of nickel release from coated items that come into direct and prolonged contact with the skin:

EN 12472:2005—Method for the simulation of wear and corrosion for the detection of nickel release from coated items

Nickel release can be detected by the following EN standard:

EN 1811:2011—Reference test method for release of nickel from all post assemblies which are inserted into pierced parts of the human body and articles intended to come into direct and prolonged contact with the skin

In the case of leather, ISO has the following standard for determining chromium(VI) in solutions leached from leather under defined conditions:

ISO 17075:2007—Leather – Chemical tests, Determination of chromium(VI) content.

16.3.4 DYES AND PIGMENTS

In terms of dyestuffs, there are several test methods related to the extraction and identification, especially devoted to trace dyestuffs that are potentially harmful for the human health.

The following are the standards of the series ISO 16373 "Dyestuffs":

- Part 1 (Still under development): General principles of testing colored textiles for dyestuff identification.
- Part 2 (2014): General method for the determination of extractable dyestuffs including allergenic and carcinogenic dyestuffs (method using pyridine–water). This standard lists the allergenic and carcinogenic dyestuffs that can be analyzed using this method.
- Part 3 (2014): Method for determination of certain carcinogenic dyestuffs (method using triethylamine/methanol). This standard specifies a method for the detection and quantitative determination of the presence of carcinogenic dyestuffs in dyed, printed, or coated textile products by chromatographic analysis of their extracts.

Certain azo colorants can cleave into aromatic amines that are restricted by most ecolabels or forbidden by legislation, namely, in the EU. The following standards, recently published, are relevant to detect the potential release of arylamines:

ISO 24362 (2014)—Textiles, Methods for determination of certain aromatic amines derived from azo colorants:

- Part 1: Detection of the use of certain azo colorants accessible with and without extracting the fibers
- Part 3: Detection of the use of certain azo colorants, which may release 4-aminoazobenzene

These standards correspond to the EN standards EN 14362, Parts 1 and 3 (2012), presently under revision. Note that Part 2 does not exist, as it has been merged with Part 1.

16.3.5 PHTHALATES

Phthalates, used as plasticizers, are an issue for textile manufacturers and retailers due to their use within motifs, coated fabrics, plastisol prints, buttons, etc. Phthalates present a health risk, especially for babies and children. The presence of several phthalates is therefore forbidden.

ISO 14389:2014—Textiles, Determination of the content of phthalates

This standard specifies a method of determining phthalates in textiles with gas chromatography–mass spectrometry (GC-MS) with mass selective detector. It is applicable to textile products where there is a risk of the presence of some phthalates.

16.3.6 FLAME RETARDANTS

Several chemicals used as flame retardants can be harmful for human health. The following test methods are at present under development:

ISO 17881 series—Textiles, Determination of certain flame retardants:

Part 1 Brominated flame retardants
Part 2 Phosphorus flame retardants
Part 3 Short-chain chlorinated paraffins (SCCP)

16.3.7 Alkylphenolethoxylates

Nonylphenol ethoxylates belong to the nonionic surfactant category and are of particular concern. The biodegradation of nonylphenol ethoxylate releases the branched nonylphenol, which is difficult to biodegrade. Nonylphenol is a hormonal acting substance that is toxic for waterborne organisms and many other organisms. For this reason, the release of nonylphenol ethoxylate into the environment should be avoided.

The following test method, based on test by HPLC-MS, is at present under development within ISO:

ISO/DIS 18254—Textiles, Method for the detection and determination of alkylphenolethoxylates (APEO)

This method detects extractable alkylphenolethoxylates (nonylphenol ethoxylates and octylphenolethoxylates) in textile products, based on liquid chromatography with mass spectrometry system, after extraction with methanol.

16.3.8 Other Harmful Substances

There are many other harmful substances that are limited by legislation or ecolabels. In most of the cases, the detection is based on an extraction, followed by an appropriate chromatographic test: liquid chromatography or gas chromatography. Mass spectrometric detection or electron capture detection methods can be used.

The following compounds are normally controlled:

- Pesticides
- Chlorinated phenols (namely, pentachlorophenol [PCP] and its salts and esters, as well as tri- and tetrachlorophenols)

In the case of leather, the following is the test method for determining the content of PCP, its salts and esters:
ISO 17070:2006—Leather, Chemical tests, Determination of PCP content:

- Organostannic (or organotin) compounds
  Note: for footwear, the following is the test method:
  ISO/TS 16179:2012—Footwear, Critical substances potentially present in footwear and footwear components, Determination of organotin compounds in footwear materials:
- Organochlorinated compounds (used as carriers in polyester dyeing)

The following is the DIN method:
DIN 54232:2010—Textiles, Determination of the content of compounds on the basis of chlorobenzene and chlorotoluene:

- Dimethyl fumarate
- Perfluorinated compounds (especially perfluorooctane sulfonate [PFOS] and perfluorooctanoic acid [PFOA])
- Polycyclic aromatic hydrocarbons (PAH)

ISO provides the following test method for the quantitative determination of 16 PAH in soil:

ISO 18287:2006—Soil quality, Determination of PAH, Gas chromatographic method with mass spectrometric detection (GC-MS):

- Engineered nanomaterials

There is an increased concern related to the presence of nanoparticles in textile materials, as long-term effects on human health are not known. At present, finishing with nanoparticles is already quite frequent, for instance, in terms of UV protection, antibacterial finishing, and antisoiling/self-cleaning of surfaces. For instance, GOTS organic labeling simply forbids the use of nanoparticles (Almeida, 2014). The development of test methods to evaluate the transfer of nanoparticles from textiles to the skin, for instance, by the effect of sweat, is foreseen within CEN. There are already several research works published in this area (see, for instance, von Götz et al., 2013).

16.4 SAFETY OF CLOTHING

The previous section covered chemical aspects related to the presence of harmful substances in textiles.

In terms of flame retardancy, there are no general international or European regulations for garments. Nevertheless, there are different national regulations that must be taken into account by producers and importers. For instance, in the United Kingdom, the nightwear safety regulations impose severe restrictions for children's nightwear.

In terms of mechanical aspects of garments, special care must be taken in the design and manufacture of children's clothing. The following EN Standard is very relevant:

EN 14682 (2014)—Safety of children's clothing, Cords and drawstrings on children's clothing, Specifications

This standard covers all children's clothing including disguise costumes and skiwear intended to be worn by children up to 14 years of age. Specifications are included for a wide range of items including drawstrings, functional cords, decorative cords, loops, zip pullers, adjusting tabs, shoulder straps, halter necks, tied belts and sashes, and belt loops. It also includes guidance on and should be used to inform the process of risk assessment of children's clothing.

EN 14682 is a harmonized EN standard and is related to the General Product Safety Directive (Directive 2001/95/EC). It is used by the authorities to check the security of children's clothing. It must be noticed that a significant number of garments, especially import from outside Europe, not complying with this standard, are detected every week and are reported by RAPEX (rapid alert system for nonfood dangerous products).

This standard has been recently revised and includes now an informative Annex to clarify several doubts that appeared in the market.

In terms of safety of children's clothing, the following technical report has recently been published:

CEN/TR 16792 (2014) Safety of children's clothing, Recommendations for the design and manufacture of children's clothing—Mechanical safety. This Technical Report is applicable to clothing including bonnets, hats, gloves, scarves, socks and other clothing accessories intended for all children up to 14 years of age.

There are other two documents in preparation:

Safety of children's clothing, Security of attachment of buttons, Specifications and test methods

Safety of children's clothing, Security of attachment of metal press fasteners, rivets, eyelets, and similarly applied components.

In terms of risks in the sleeping environment for children, the following standards are being developed within CEN, following a mandate from the European Commission:

prEN 16779-1—Textile child care articles, Safety requirements and test methods for children's cot duvets—Part 1: Duvet (excluding duvet covers)
prEN 16780—Textile child care articles, Safety requirements and test methods for children's cot bumpers
prEN 16781—Textile child care articles, Safety requirements and test methods for children's sleep bags

16.5 CONCLUSION

The textile and clothing are probably the most globalized industries at world level. In order to evaluate the performance and safety of apparel, it is important to use as far as possible internationally recognized test methods. The ISO has published since the 1950s many standards, especially in terms of test methods. Most of them have been adopted as EN standards, according to the Vienna Agreement.

Recently, a lot of effort is being put in the development of standard test methods related to the safety of textile products, with special emphasis on the detection and quantification of harmful substances. These aspects are very important for sustainable apparel production.

In this chapter, a survey of the present version of the standards, especially ISO standard test methods, has been presented. It is important to note that standards are in constant evolution. ISO standards are submitted to a review every 5 years; often, they are confirmed but in some cases, they are updated or even withdrawn. There are also a number of new standards in development related to textile testing and linked, namely, to the functionality of textiles and to ecological and safety parameters. It is therefore essential for all the stakeholders to follow regularly the evolution of the standards, using the more updated versions.

REFERENCES

Almeida, L. (2014). Ecolabels and organic certification for textile products. In: *Roadmap to Sustainable Textiles and Clothing: Regulatory Aspects and Sustainability Standards of Textiles and Clothing Supply Chain*. Textile Science and Clothing Technology Series, S.S. Muthu (ed.). Springer, Singapore, Singapore, ISBN: 978-981-287-163-3.
Hes, L., Loghin, C. (2009). Heat, moisture and air transfer properties of selected woven fabrics in wet state. *Journal of Fiber Bioengineering and Informatics*. 2(3): 141–149.
Kawabata, S., Postel, R., Niwa, M. (1982). *Proceedings of the Japan–Australia Joint Symposium on Objective Specification of Fabric Quality, Mechanical Properties and Performance*. Textile Machinery Society of Japan, Kyoto, Japan.
Ping, W. (2011). A brief history of standards and standardization organizations: A Chinese perspective. East-west Center Working Papers, Economic Series, No.117. East-West Center (EWC), Honolulu, HI.
Schiller, W. (2011). 100 Jahre Deutsche Echtheitskommission DEK—100 Jahre Farbechtheitsprüfung. In: *Proceedings of the Festive Symposium on Occasion of the 100th Anniversary of Deutsche Echtheitskommission DEK*. Erding, Germany, October 10–12, 2011.
Valldeperas, J. (2011). History of 50 years of ISO meetings—Evolution of ISO 105. In: *Proceedings of the Festive Symposium on Occasion of the 100th Anniversary of Deutsche Echtheitskommission DEK*, Erding, Germany, October 10–12, 2011.
von Götz, N., Lorenz, C., Windler, L., Nowack, B., Heuberger, M.P., Hungerbuehler, K. (2013). Migration of Ag- and TiO_2-(nano)particles from textiles into artificial sweat under physical stress: Experiments and exposure modeling. *Environmental Science and Technology*. 47(17): 9979–9987.

17 Environmental Communication and Green Claims of Textile Products

Shanthi Radhakrishnan

CONTENTS

17.1 Introduction to Environmental Communication 376
    17.1.1 Importance of Environmental Communication 376
    17.1.2 Core Essentials of Environmental Communication 376
    17.1.3 Environmental Communication in Textile and Clothing Industry 377
17.2 Green Claims of Textile Products 378
    17.2.1 Meaning and Importance 378
    17.2.2 Green Guides 379
    17.2.3 Product and System Certifications 380
17.3 Eco-Promising and Sustainability Communications 381
    17.3.1 Background 381
    17.3.2 Current Trends in Eco-Promising 382
    17.3.3 Salient Features of Eco-Promising 382
    17.3.4 Guidelines for Crafting Eco-Promises 383
17.4 Environmental Claims for Marketing 384
    17.4.1 Standards for Environmental Labeling 385
    17.4.2 Standard for Environmental Claims in Marketing 386
17.5 Certification as Environmental Communication Tools for Green Claims 387
    17.5.1 Product Certifications 388
        17.5.1.1 EU Ecolabel 388
        17.5.1.2 Global Organic Textile Standard 389
        17.5.1.3 OEKO-TEX Standard 100 390
        17.5.1.4 Textile Exchange Labeling System 391
    17.5.2 System Certifications 392
        17.5.2.1 OEKO-TEX Standard 1000 392
        17.5.2.2 ISO 14001 Environmental Management Systems 393
        17.5.2.3 EMAS 393
        17.5.2.4 ISO 9001 Quality Management System 394
        17.5.2.5 SA8000 Standard 394
17.6 Future Directions 395
References 396

17.1 INTRODUCTION TO ENVIRONMENTAL COMMUNICATION

All living beings communicate in some form or another. Information is exchanged between individuals or groups through certain means, for example, movements, sound, reactions, and transformations. Usually, communication creates or shares information that conveys some meaning and helps to link people and places. It becomes a key function of management without which no system can perform, improve, or attain any height. According to the business dictionary, communication is a two-way process of reaching mutual understanding in which participants not only exchange information, news, or ideas but also create and share meaning (Business Dictionary 2014). In order to perpetuate living, the simplest to the most complex organisms communicate with their environment and other beings. This is termed as a critical function because it is very essential for the survival of living systems. The study and practice of how individuals, institutions, societies, and cultures receive, understand, and use message about environment and the human interaction with the environment is known as environmental communication. From the perspective of practice, this concept has been defined as the application of communication approaches, principles, and strategies to environmental management and protection (Flor 2004).

17.1.1 IMPORTANCE OF ENVIRONMENTAL COMMUNICATION

Research scholars are anxious about the ways in which people communicate with the natural world as it has far-reaching effects especially during human-caused environmental problems. There are certain assumptions around which environmental communication systems evolve. The means of powerful communication affect our perceptions of the living world and these perceptions tend to shape how we interact with nature. Hence, environmental communication not only reflects but also constructs, produces, and naturalizes human relations with environment. It has also been supported that human representations and communications of nature are affected by social, economic, and political contexts and interests, which lend their way to direct humans to see nature in a particular context. It has been stated that the term *environment* reflects anthropocentric or human-centered cultural views of and relations with the living Earth. It also highlights the exploitive and destructive actions that continuously and materially shape the biosphere. Environmental communication scholars critique and raise awareness about the enormous damage to environment by the materialistic actions of the human race (Stephen and Foss 2009).

The importance of environmental communication is best described in the case of the Danish textile industry during the period between 1991 and 1997, when high environmental awareness created a massive change in the perception of a group of frontrunners, leading to a collective learning process by which good housekeeping and substitution moved to environment management systems, product-oriented LCA, and ecolabeling schemes. The industry and local environmental authorities developed a high degree of consensus on environmental perception, leading to the adoption of clean technologies in the area of textile finishing. This wave of change encouraged the Danish Ministry of Environment to take up new environmental problems and initiatives from research results, public debates, and business technology options at very short notice and has resulted in finding environmental protection measures from which the industry can benefit. Thus, ecological modernization and institutional transformation was strongly influenced by the change in environmental perceptions within the Danish textile industry. Hence, it can be understood that environmental communication, which is a tool for perception change, is very important in building a rapport between the human and nature to develop eco-friendly process systems for the textile industry (Hansen et al. 2001).

17.1.2 CORE ESSENTIALS OF ENVIRONMENTAL COMMUNICATION

The core messages of environmental communication are four in number. They are *everything is connected to everything else*, so harming one part of the ecosphere will bring consequences to the other parts; *everything must go somewhere* deals with waste management, implying that when waste is discarded without proper treatment its impact will affect all living organisms; *nature knows best* means that nature has its own ways of maintaining equilibrium and no technology can completely solve the problems of society; *there is no such thing as free lunch* and exploiting natural resources comes with a price.

Apart from the aforementioned core messages, there are six essential elements that should be understood by those involved in environmental communication. A knowledge of Barry Commoner's ecological laws is of utmost importance; sensitivity to cultural dimension is necessary as nature and culture coevolve and cultural dimension may be used as a tool for environmental management and protection; the ability to network can help in advocacy, which is necessary for environmental communication; efficiency in using media for setting environmental agenda; appreciation and practice of environmental ethics will help people to follow the environmental ideals; conflict resolution, mediation, and arbitration is very essential in environmental communication as innumerable conflicts arise when any projects related to environment is launched. Lack of communication between the project proponents and the conflicting entities like the local community is the main reason for the conflict and this can be sorted out through arbitration and mediation (Flor 2004).

17.1.3 ENVIRONMENTAL COMMUNICATION IN TEXTILE AND CLOTHING INDUSTRY

The global consumers today are conscious and demand strongly that the manufacturers respect ethical conduct, people, and natural environment. Many buyers have started including ethical garment lines to their fashion collections and fashion shows are based on themes like *environment friendly* and *sustainability*. Social and environmental reports and certifications serve as tools for communication between the consumers, the stakeholders, and management and also give them the access to use labels to give information on features like raw material composition, the country of origin, and the effects of process parameters on environment. The rules for eco-certification have been created by a group of international organizations like European Community, World Trade Organization (WTO), United Nations Environmental Program (UNEP), and the International Organization for Standardization (ISO) (Koszewska 2011).

Environmental labeling of textiles is a means of communication to inform the consumer about the impacts of the product on the environment. Several labels are available, which may be based on the life cycle assessment of the product or on issues like quality or recyclability. The aim of these labels is to increase the sales of products that are environment friendly and to decrease the emissions from the product chain. A communication of the important environmental effects of a product is known as environmental product declaration (EPD) and the technical parameters used for evaluating these effects are being developed by ISO under the ISO Type III environmental declaration development work. The technical requirement for an environmental label is to determine the possible harmful effects of the finished product across the entire supply chain. The OEKO-TEX standardization process and their database on numerous chemicals used for the manufacturing of textiles will be of great help to formulate these technical parameters for environmental labeling. The market for textile products is global and hence a system that can be applied on a global scale in a universal manner would be useful. It has been reported that in 2001, the COST Action 628 *Life cycle assessment of textile products, eco-efficiency, and best available technology (BAT) of textile processing* was launched with the involvement of 16 European Union (EU) research institutes and universities in textiles and EURATEX as action members. A universal clear and explicit mode of communication regarding the environmental impacts of products will inform the consumers about the product and help in the sales of products with a clean background. To ensure a uniformity to ecolabeling, the ISO has published a series of standards (ISO 14020, ISO 14021, ISO 14024, ISO 14025) (Kalliala 2003).

Certain mechanisms are available for transmitting information about chemicals in textile products. *CiP information exchange* is explained as the passing on of information about some or all of the chemicals in a product through some stage of the product life cycle. *Negative content information* is the passing on of information about the chemicals that are not in the product. The exchange of information can be of four types, namely, the information on chemicals within the production chain, from producer to consumer, from producer to end of life of product, and from external stakeholders to consumers and general public. The restricted substances list is used as a communication tool by which the company expresses to the consumer that certain chemicals that harm humans and environment are not used in the production process and assure that they will not be present in their products. Periodic product testing is a means of assuring that the product does not contain the chemicals in the Restricted Substances List.

Another system of environmental communication is the EPD, which communicates the environmental performance of a product or system. The framework for developing the EPD is found in ISO 14025:2006 and the procedure builds upon other ISO standards. Information about the environmental impacts associated with a product or service can be as raw material procurement; material content and chemical composition; energy consumption and efficiency; air, soil, and water emissions; and waste generation. Since there are no standard levels in determining environmental performance, the ISO stresses the need that the EPD is built on well-organized quantitative data, which can be used for business-to-business (B2B) communication or business-to-consumer (B2C) communication (UNEP 2011, UNEP/DTIE 2011).

Environmental communication has gained immense importance as it serves and supplies relevant information to consumers thereby encouraging them to make the right choice of products with great concern for the environment. Developments in this field are aimed as standardizing the environmental information presented to consumers to prevent misleading conceptions and to meet the global demands. Efforts are directed towards full disclosure of material composition and transparency in the communication to make it visible to the consumer that no harmful substances exist in their products. This serves to build confidence in the consumer to choose the products with ample information about the extent of friendliness towards the environment, people, and planet.

When transparency and material composition are informed to the customer, inclusion of banned substances to the extent possible in terms of limits of use can be identified easily. The consumer is bound to know that such substances have not been used in the product and this concept will build confidence in him or her to purchase the product or article. Second, ecolabeling is given to products only after there is assurance from the manufacturer that the product has been manufactured with the least impact on the environment and certified by third-party organizations. Ecolabels tend to declare that the raw materials chosen for the product, the process optimization, and the systems developed for manufacturing the product are least harmful to the environment. If the manufacturer is able to show some information on the impact of the product and maintenance when in use and disposal when compared to the other products in the market in terms of numerical values, then this information will help the consumer to reach out for the most environment-friendly product and gain satisfaction that he or she has contributed to the safety of the environment. When such declarations are made, it not only helps the customer in choosing the correct products but it also helps in brand building and service to society. Environmental claims with data on the impact of the product would serve to provide the essential information to the consumer to enable choice of product and brand preference.

17.2 GREEN CLAIMS OF TEXTILE PRODUCTS

17.2.1 MEANING AND IMPORTANCE

The well-being of planet Earth and human health are the primary concerns in today's world. The consumer of this era has immense regard for the raw materials and processes involved in manufacturing a product and they also favor products that do not have any detrimental effects on the environment. The terms like cradle to gate, cradle to grave, and cradle to cradle serve as banners for the sale of products and sustainability is the word that has caught the attention of one and all. The manufacturers today are taking steps to reduce the environmental impacts of their products due to the expectations of the consumer, the responsibility inculcated by the environmental communication, and due to the realization that these techniques are economical. Products enter the market with different connotations like green, eco-friendly, sustainable, renewable, recyclable, organic, and a host of related terms implying that they are environment friendly.

Green claims suggest that the product or process is beneficial or has minimal impact on environment with regard to resource, energy, raw material usage, greenhouse gas, toxic emissions, and waste generation. Green claims found all around are so varied and many that it does not provide a clear picture of the true environmental impact of the product. The claim "100 percent organic cotton garment" is a general phrase that has no certifiable meaning but just conveys the idea of environmental responsibility or can relate to any one aspect of the product. Such claims if deceptive and misleading will make consumers doubtful and result in mistrust of all valid claims made for other products. To address this problem of green marketing issues, the Federal Trade Commission (FTC) in 1992 developed the *Guides for the Use of Environmental Marketing Claims* also known as green guides to provide a baseline for voluntary agreement with the FTC regulations with regard to environmental marketing and advertising practices (Rodie 2008).

17.2.2 GREEN GUIDES

The core principle of the FTC green guides is to substantiate or qualify all green marketing claims, whether specified or implied, in such a way that it can be understood by the consumer. Examples of general and specific claims are given in the guide to highlight the degree of accuracy and clarity. Another point to note is that the green claim varies from product to product and requires careful coining of the claim otherwise it could be misleading. To quote an example, the National Organic Program states that if a textile material is made up of two materials, say organic cotton and spandex/nylon, both the components with the percentage content need to be mentioned in the label and not only the organic cotton content.

Green claims must be very specific when a blend of two materials is used, organic cotton and recycled polyester. Here, the raw materials are sustainable individually but when they are blended, they are nonrecyclable. Similarly, the green claims on the raw material bamboo are controversial. Bamboo is a renewable fast-growing resource that requires little water and pesticides for its cultivation and is considered sustainable. The mechanically extracted bamboo from the stalk of the plant is antimicrobial and has moisture transport performance as apparel manufacturer's claim but these claims do not suit regenerated solvent-spun fiber that may be either rayon or lyocell. The lyocell fiber is made using recoverable solvent in a sustainable closed loop process. With an increase in green marketing claims, especially in the field of textiles and building, the FTC is planning to make further revisions in the green guides (Green Guides, Miles 2011).

Transparency in product information is the key to earn customer loyalty. When a customer learns about the product components, how they have been manufactured, the persons who made it and their work environment, and the life after death of the product, a sense of bonding and loyalty remains etched in his mind, which will help him or her to choose the products with correct certification and satisfaction.

Third-party certifications and seals are used by many marketers to substantiate their green marketing claims. The FTC has devoted an entire section to third-party certification and seals in the revised issue. Third-party certifications help the consumers to differentiate the sustainability aspect of the product in comparison with the other competitive products available in the market. These certifications and seals should have a concrete scientific meaning, and the certified products should be supported by competent, reliable scientific evidence. If these seals are not legitimate and are in violation of Section 5 of the FTC, it is very important that the advertising and marketing green claims are substantiated by environmental benefits or attributes, whether the claim is general or specific in nature (Sullivan and Worcester 2011).

17.2.3 PRODUCT AND SYSTEM CERTIFICATIONS

Certifications are credibility boosters for products or systems and are useful to increase its marketability as it creates reliability in the minds of the consumer. Certifications are useful in the achievement of goals rather than a tool to ward off competition. It is a sound check to assure that the product or system meets the required standards or specifications for which it is intended. Certification helps in attesting and validating the authenticity or truth of something or someone. In the case of green claims, two main certifications are important, namely, product certification and system certification. Product certification endorses that a certain product meets the qualification criteria mentioned in contracts or regulations as it has passed performance and quality assurance tests. System certification validates the process or method that helps to build and maintain green claims based on certain evaluation criteria and recommendations.

Certification of the sustainability aspects of products are by means of ecolabels in the field of textiles. The most popular labels are OEKO-TEX Standard 100, Global Organic Textile Standard (GOTS), and community ecolabel mark:

- *OEKO-TEX Standard 100*: About 7000 companies around the world and innumerous products use this standard that certifies the compatibility of textiles with the intended use and its impact on human health. It certifies that the product is free of harmful substances based on certain performance and eco-tests.
- GOTS certifies that the textile products are made from organically grown natural fibers. This certification starts analyzing the growing environmental, social, and health problems associated with the production and use of traditional process for growing cotton.
- *Community ecolabel mark*: Guarantees eco-friendly characteristics throughout the life cycle of the product and this are proved through some performance requirements.

Environment management systems are a group of processes and practices that enable the organization to reduce its environmental impact and increase efficiency. National or international acclaimed environment management systems that accredits by an independent third party are sought after by industries as they are more structured environmental impact mechanisms. Though a number of ecological system certifications are available, two most important environmental management systems the ISO 14001 standard and the Eco-Management and Audit Scheme (EMAS) regulation are discussed in the following:

- *ISO 14001 Standard*: It is most widely used and well known as the international standard 14001. It provides a framework that helps an organization to develop and implement an environmental policy that addresses the environmental impacts and satisfies legal requirements. The highlights of this system are widespread adoption and international recognition that leads to business opportunities; it has a simple implementation profile with low human resource requirements leading to cost-effectiveness; since it is an independent accreditation process, its credibility has been vastly improved and it provides assurances to organizations and interested parties. However, a few points prove to be disadvantageous like the organization has no requirement to assess its environmental impact as it is based on the fact that once the EMS is established in the organization, it leads to continuous improvement with regard to environmental impacts. Another point to note is that the performance information is not available for publicity.
- *EMAS*: The next standard designed to improve an organization's environmental performance is the EU EMAS. It recognizes and rewards organizations that perform beyond the minimum legal compliance and show continuous improvement in environmental performance. Though similar to ISO 14001, it has a plan–do–check–review cycle and the focus is different. This system requires organizations to publish their environmental performance

statement annually. The accuracy and reliability of the report is independently verified by a third-party environmental assessor. Since the process is strict and costly, many small organizations find it difficult to register under this system and the reporting system and strict compliance requirements also prevent organizations to enroll into the system as minor breaches of standards may exist. The absence of worldwide recognition and its focus in the EU and European Economic Area is a limitation (TexEASTile Sustainable Innovation for Textile in South East Europe).

Both the aforementioned organizations serve public or private business enterprises and are responsible for the assessment of environmental performance. These organizations are not specific to the textile sector nor are they a product mark to proclaim that the organizations are environment friendly but formulate and ensure that the certificate holder complies with mandatory environmental requirements and sets goals for continuous improvement. Another latest development that requires mention are the lines of recycled or recyclable clothing that have self-declared environmental claims but no third-party verification. However, the organization should procure evidence to substantiate its green claims with regard to recycled or recyclable clothing. The consumer should be aware of these points before he or she purchases a product with any ecological label.

17.3 ECO-PROMISING AND SUSTAINABILITY COMMUNICATIONS

The world we live in today is highly competitive and has innumerable products and services for the customer's choice. The consumer can be tempted to pick a product produced by a specific manufacturer or industry by communicating the special features of the product and also by highlighting in what way the product is better than its peers in the market. The method of advertising has turned to informing the consumer about the environmental impacts of the product, and bold commitments are being issued by high-profile companies to showcase their efforts towards the safety of the planet. The increased drive to communicate to the consumers involve claims of improved environmental credentials for product lines with regard to product raw material scouring, clean technologies in manufacturing, and lesser use of harmful chemicals. Sincere efforts are directed by manufacturers towards reducing the environmental footprints of products with emphasis on energy efficiency during use coupled with a take-back policy at the end of life of the product to enable recycling or any other eco-friendly approach.

Apart from these efforts, technology tools are available to track supply chains and enhancing methods by which environmental information is shared by several business enterprises. Consumers inform their needs by using online resources and demand for more detailed product information. However, new ecolabels are multiplying in the market and they vary in the methodology and the type of assurance they offer. New approaches have been developed to further the scope of environmental communications like cross-industry voluntary standards (specifications for product greenhouse gas emissions), industry-specific tools (environment tools for processing industries), and company-specific methodologies (Timberland's Green Index) (CU 2014).

17.3.1 Background

Eco-promising started with Blue Angel certification, developed by Germany in 1977 with the innovation of combining life cycle analysis, to estimate the environmental performance of the product, with third-party assurance. By 1986–1990, national level ecolabeling schemes like AENOR in Spain, Norme Francaise Environnement in France, Eco Mark in Japan, and Green Seal in the United States were launched. Nongovernmental ecolabel schemes came into existence such as OEKO-TEX with focus on textiles and Blue Flag beach certification for tourism. By the turn of this century, there was a wide range of goods and a range of labels that encompassed many meanings on certification. Companies now use ecolabels as a component of broad eco-promising approaches.

Walmart is the largest buyer of organic cotton, and it has been advertised that organic sourcing has reduced the quantity of chemical fertilizers and pesticides by 22,000–27,000 kg (the International Trade Centre).

The term eco-promising is the practice of making environmental claims and communication for products and services. They are assurances made by the manufacturers and retailers that their products maintain certain environmental claims and this is substantiated by third-party certification and labels. Opportunities arise with eco-promising, to increase ethical purchasing and improve brand differentiation but if these environmental claims are misleading the risk of negative consumer attitude and reporting cannot be avoided. Eco-promises can cover a number of activities and may take up many forms. They may occur as advertisements on websites or on the products sold as pictures, text, symbols, and labels. Whatever may be the mode of operation and information, the aim of eco-promising is to attract customers and seal the purchase.

17.3.2 CURRENT TRENDS IN ECO-PROMISING

The latest trend in eco-promising is to develop in-house methodologies and messaging to the people. Nike's Considered Index is a sustainable product design tool used to evaluate the environmental footprint of the product (Green Chemistry and Commerce Council). Timberland has developed the Green Index that evaluates the greenhouse gas emissions, recycled content, and chemical use. So retailers are extending their special stamp in the market for recognition of consumers through their eco-promising values. Many manufacturers and retailers are stretching beyond the label information's and connecting the consumer to more information through special links and electronic tags. This can include the carbon credentials, the list of contractors, and suppliers or environmental credentials.

The focus now is on customer use of a product. Apart from providing information about the raw materials used in the product, the sustainable production process incorporated, a detailed analysis is given with regard to the impact of the raw material use and the end of life disposal. The use of implied attributes is highlighted by the retailers and manufacturers by suggestive language and images that imprints the attributes in the minds of the people in an indirect way. Many companies cannot stay idle about their environmental promises as third-party ratings are being done for all the companies in the sector and rankings are given to inform the consumers about the status of different companies and institutions with regard to the environmental impacts and promises. The Carbon Monitoring for Action database provides the carbon emission data of about 50,000 global power plants to make the consumers aware of the status and ratings (Ummel 2012).

17.3.3 SALIENT FEATURES OF ECO-PROMISING

The motivation behind eco-promising is very many. The life cycle analyses reveal useful information on waste in production and distribution, supply chain risks, and procurement opportunities thereby improving risk and operational management. Second, these promises improves the reputation of the company's image among the general public like the campaign of Mark and Spencer, *Look Behind the Label*, produced a huge impact on its brand image than any other activities carried out in the past (Butler 2007). Eco-promising has increased sales percentage and the markets for environmentally based products are expanding rapidly. Environmental messaging on clothing has offered a competitive edge and price premium. The growing demand for eco-friendly products has put organizations into the forefront when they strive to move ahead of the potential regulations and take up these activities voluntarily rather than an enforced compliance. It also tends to shape the regulatory framework and puts business in a much stronger negotiating position with a wide range of stakeholders.

Eco-promising can also cause detrimental effects when they are overdone. Many consumers are not able to differentiate between fair trade, ethical, and organic products. Most of the logos and claims are vague and meaningless, lacking in transparency and lacking in standards and verification. Consumers look for vital information on attributes of the product, and the additional information on the environmental impacts of the products make them feel that there is too much of information that it may be neglected causing a term called *green overload*.

Green claims will usually call for more cost on the product and the willingness to pay extra on the part of the customer is to be evaluated. The *value-action* gap demonstrates the stated values of the consumer and his buying behavior for environmental cost additions to the product. Further, too much of environmental information and claims will cause a lack of understanding of concepts among customers. Hence, producers and retailers must be able to subtly convey the information through a range of complex issues.

Effectiveness to pass on information through the supply chain calls for special attention to significant features that are to be highlighted as more systems and data can add to the costs in the supply chain. Constraints in certification include environmental labels that cause distraction from the brand and placement of increased limitation on innovative product development. However, consumers have more trust and regard for third-party labels and certification rather than in-house messages. Charges on *green washing* whether true or false can cause severe damage to the credentials of the company as it will be taken up by the media, investors, and other stakeholders. This situation is nonreversible, and many consumers will lose faith on the green claims projected by different companies (Schuchard et al. 2008).

17.3.4 Guidelines for Crafting Eco-Promises

Creating eco-promises is an art and has to be captivating to catch the attention of the customers. Managers in consumer-based companies are responsible to communicate the environmental credentials of products honestly and convincingly and must understand the risks and challenges associated with the issue of environmental claims. Certain guidelines to craft effective eco-promises are discussed in the following:

- *Knowledge of the biggest impacts of products*: Life cycle analysis gives information about the environmental impacts of the manufacture and use of products and forms the base for shaping reliable communication about environmental credentials. This information opens up new challenges and also helps in providing data for further improvements in environmental performance. The data available can be compared to the national or international targets for environmental outcomes and identify potentially high-impact products and services to formulate effective green claims.
- *Transparency in industrial practices and impacts*: Customers and stakeholders want companies to express transparency in the environmental and societal impacts of the products manufactured and used as it can improve relationships with peer groups and can deliver a competitive edge for securing brand loyalty. The Green Index is a communiqué to the consumer, which is displayed as a simple box sticker showing overall product rating as a number followed by more detailed performance measures. After this, background information is provided in small print followed by a web address for further information.
- *Substantiate green claims with independent verification*: Third-party verification can bring legitimacy to eco-promises. The verification may range from qualitative assurance in general terms to detailed verification of the different stages of the life cycle of the product. One should analyze the level of investigation in terms of the type of product and its impact, the market it caters, and the brand loyalty and integrity that exists.
- *Encourage consumers to act*: Consumers in different parts of the world are concerned about environmental issues, but they need encouragement and constant communication to enable this concern into changes in purchasing behavior. This can be enabled by using category management techniques and buying criteria to rule out unsustainable products, by

the use of promotions and reward schemes, by supplying information beyond the labels at important customer contact points like point of sale, magazines, websites, roadshows, etc., and by relating these issues to current trends and political movements.
- *Avoid making empty claims*: Empty promises or highlighting only one advantage while all the rest lead to negative impacts is a high risk factor. Without proper information and useful defense, the questions that arise from the consumer cannot be addressed and the brand image will be affected. Claims that give personal benefit like nontoxic cleaning agents will first catch the attention of the consumer.
- *Formulate different communications for different target markets*: Based on the target markets and customers, different methods of expression are necessary to initiate action. Strong eco-promises can be used for markets that have importance for environmental issues and vice versa. Different groups are motivated in different ways and once success is achieved this methodology can be used as an overall strategy if required.
- *Technology as a tool for communication*: Latest advances in technology can be used to spread the message. RFID tags can provide ample information about the product and can also help in sharing this information to different customers via information databases. Similarly, satellite imaging, chemical markers, and interactive labels can provide effective environmental communication to consumers.
- *Industry participation in rule making*: Currently, many new environmental standards are coming into the environment impact measurement tools and it is a must for every industry to study them and select the one that are most suitable in highlighting the products integrity and building the brand image of the company (Consumers International and Accountability 2007).

These guidelines serve to help organizations to carefully craft the eco-promises to suit the need of the product and enable the consumers to get attracted to the products based on the environmental green claims and eco-promises.

17.4 ENVIRONMENTAL CLAIMS FOR MARKETING

Information about the environmental characteristics of products is being conveyed to the consumer in a variety of ways like ecolabels, claims, declarations, and other modes. For many years, environmental labels and information schemes (ELIS) have been used with the aim of providing assurance on products with regard to best environmental characteristics. These certifications have multiplied manyfold varying in scope, size, nature, and effectiveness. This increase and diversity of ELIS has created confusion in the minds of the consumer and has also led to reporting of exaggerated environmental performance of products called *green washing*. In order to differentiate their products from their competitors, many business organizations report their environmental footprints to suppliers and investors, and ELIS forms a link in environmental communication between B2B, B2C, business to government (B2G), and government to consumer (G2C). The means of communication may be as environmental labels, reports, or declarations, and the claim content may include information about natural resources, energy, sources of pollution, bio diversity, climate, waste, and other issues (EPOC 2013).

Companies use environmental and ethical claims as marketing tool and may contain generally accepted positive statements or properties. When such marketing techniques are used by a trader, the issue must be assessed by the Marketing Practices Act to protect consumers against misleading information and green washing, to promote real and fair competition among traders, and to make it clear that traders may use lawful ethical marketing statements for marketing. While marketing to consumers, the environmental and ethical claims must be in accordance with the requirements stipulated in the Marketing Practices Act or if these claims do not come under the purview of the

Act, then special legislation that contain special circumstances should be additionally observed, for example, special rules apply for chemical substances. Violation of regulations is liable to litigation and severe punishment.

The principal rule for liability for marketing rests on the trader who may be a manufacturer or retailer or both. The trader must ensure that the product complies with the marketing claims as any negligence may lead to criminal liability. If a third-party is involved, for example, an advertising agency may assist the trader to draft the message or provide advisory service, then both may incur joint liability. In some cases, the editors of the media are also liable for the content of the advertisements and may incur liability for involvement as per the Criminal Code. In general, while using environmental and ethical claims for marketing, the following requirements are to be followed:

- All claims, including environmental and ethical claims, must be correct, accurate, relevant, balanced, clearly worded, and easy to understand to avoid misleading consumers.
- The wording, layout, color choice, images, sound and symbols, etc., should all convey a true, balanced, and loyal overall impression that the product or service possesses environmental and ethical benefits.
- All environmental and ethical marketing claims must be substantiated wherever necessary with relevant documentation and this documentation must be available before the claims are used for marketing purpose for the first time.
- The access to such documentation must be accessible to the recommended supervisory authority as the public is entitled to relevant explanation and on the method of documentation (DCO 2011).

17.4.1 Standards for Environmental Labeling

A voluntary environmental performance certification, practiced around the world, is ecolabeling. These are claims developed by manufacturers and service providers that are based on life cycle considerations, awarded by impartial third-party organization for products and services that meet transparent environmental leadership criteria. The standard classification is listed in the following:

- Type I (ISO 14024:1999) is an international standard that awards a license that authorize the use of environmental labels for products and services that shall meet a set of predetermined environmental requirements to make them environmentally preferable within a particular product category. Type I environmental labeling program are voluntary, multiple criteria based, and can be operated by public agencies in a regional, national, or international level (ISO 14024:1999).
- Type II (ISO 14021:1999 en) is an international standard that describes the general as well as the specific evaluation and verification methodology for self-declared environmental claims without independent third-party certification. The evaluation methodology should be clear, transparent, scientifically sound, and documented to give assurance of the validity of the claims (ISO 14021:1999).
- Type III (ISO 14025:2006 en) are environmental declarations that give quantified information on the life cycle of a product to make comparisons between products and services that fulfil the same functions. The criteria may be provided by one or more organizations and based on independent LCA data and inventory analysis in accordance with the ISO 14040. The organization making the environmental declaration should ensure that the data is independently verified by internal or external agencies; hence, the ISO uses the term *third-party verification* instead of *certification*. The common terms used for Type III environmental declarations are ecoleaf, eco-profile, environmental profile, and EPD (ISO 14025:2006).

Although the aforementioned standards are important and has wide reportage, this standard does not represent the full diversity of environmental labeling and information schemes. Various consumer products are certified by third-party audit without life cycle base and such schemes do not come under these standards. ISO 14000 has covered such labels but the efforts are not yet completed.

17.4.2 Standard for Environmental Claims in Marketing

The FTC has developed guides for environmental claims to help marketers avoid making environmental claims that are unfair or deceptive under Section 5 of the FTC Act. If the FTC proves that the claim is deceptive, then the organization or individual is liable to criminal offence. The general principles are as follows:

- *Qualification and disclosures*: Simple plain language in large type would make qualifications and disclosures to be understandable and clear. Inconsistent statements and distracting elements that would undermine or contradict the disclosure should be avoided.
- *Distinction between benefits of product, package, or service*: The marketing claim should specify whether the claim is pertaining to product, package, or service or all inclusive. If a claim is made in general and it may not pertain to a part/component in the product, package, or service, the claim becomes deceptive.
- *Overstatement of environmental attribute*: Marketers should not state or imply environmental benefits if they are negligible.
- *Comparative claims*: When making comparative claims between products, there should be enough substantiation for the claim.
- *General environmental benefit claims*: Unqualified general environmental benefit claims are difficult to interpret and hence it is better for a marketer to make claims highlighting the specific benefits with substantiation.
- *Carbon offsets*: Claims that indicate reduction in carbon emissions should clearly mention if the reduction will occur in the next 2 years or longer.
- *Certifications and seals of approvals*: A marketer can use a name, logo, or seal of approval only when a third-party certification agency has assessed the product or service and authenticated that it has met the criteria for endorsements provided in the FTC endorsement guides, which include definitions, general considerations, expert endorsements, endorsements by organization, and disclosure of material connections. Certification and seals cannot be used as an assurance for general environmental benefits; when environmental seals or certificates are obtained on scientific basis, then the claim should clearly mention the specific environmental benefits to carry the correct message to the consumer (FTC-a).
- *Other claims*: This includes claims like compostable claims, degradable claims, free-of claims, nontoxic claims, ozone safe and ozone-friendly claims, recyclable claims, recycled content claims, refillable claims, renewable energy claims, renewable materials claims, and source reduction claims. A product should have all items to break down into or otherwise become part of usable compost in a safe and timely manner in an appropriate composting facility to use the term *compostable claims*. The term *degradable claims* can be used only when the marketer has substantial evidence that the product or package will completely break down and return to natural elements within a reasonably short period of time after customary disposal. In case it is not so, then the marketer should clearly mention the rate and extent of degradability and it is stipulated that the product should degrade completely within 1 year and not land up into the solid waste stream. Terms like degradable, biodegradable, oxo-degradable, oxo-biodegradable, or photodegradable cannot be used without scientific evidence.

Messages like product, package, or service is *free of* certain harmful substances only if the presence of the substance is within the norms specified as trace contaminant or background level, the substance if present should not cause harm to the consumers using it, and the substance should not be intentionally added to the product, for example, claims like chlorine-free bleaching in shirts. A *nontoxic claim* can be used only if the product, package, or service has scientific evidence that the substance is nontoxic to both humans and environment. Direct or indirect implications that a product, package, or service is ozone safe or friendly to the ozone layer is considered deceptive if it contains any ozone-depleting substances mentioned in the Class I and II chemicals in Title VI of the Clean Air Act of 1990.

A product, package, or service is considered recyclable only if it can be collected, separated, and recovered from the waste stream through an established recycling program for reuse in manufacturing or assembling another item. It is considered deceptive or misleading if *recyclable claims* are attached to products that do not have the required specifications. *Recycled content claims* are mentioned when products are made from recycled material that is recovered from the waste stream during manufacturing (preconsumer) or after use (postconsumer). The percentage content of the recycled material and substantiation for the recycled material should be given. Direct or indirect implications that a product or package has *refillable claims* is considered deceptive if the marketer does not make provision for refilling or the life of the product does not last for the required refills stated in the claims (FTC-b).

Renewable energy claims cannot be issued if fossil fuels or electricity derived from fossil fuel is used to manufacture any part of the product or package or is used to power any part of the service undertaken. Renewable energy certificates authenticating the use of renewable energy for the product or service have to be given to use such claims. Renewable materials claims can be used only when the product, package, or service is entirely made with renewable materials. This claim must be validated by identification of the renewable material, the percentage of renewable material in the product, and reasons for considering it as renewable material. Claims that directly or indirectly imply that a product or package has reduced in terms of weight, volume, toxicity, or waste generation are termed as *source reduction claims* and may be beneficial to the environment in a specific way. The extent of source reduction and the basis of comparison that leads to reduction at source are of utmost importance (FTC-c).

Environmental claims have become an important feature in marketing of products to the conscious consumers who have a major role in making the right decisions to purchase eco-friendly products thereby contributing to the safety aspect of humans and nature. Claims for marketing with regard to environment have to be carefully framed so that they are not misleading or deceptive and the product should meet the promises and assurances declared by the marketer or manufacturer. Deviations from the claims and lack of substantiation will lead to criminal offence, and there are many standards and rules that have been adhered to while making environmental claims. One notable recent development in advertising is the proliferation of green claims, and it is appropriate that existing claims are reviewed and interpreted in the right manner.

17.5 CERTIFICATION AS ENVIRONMENTAL COMMUNICATION TOOLS FOR GREEN CLAIMS

The environmental performance of products is of primary importance in the textile and apparel industry. Communication of environmental benefits is by environmental labeling and use of claims that serve as advertisements to promote the purchase of eco-friendly products by the consumer. There are a wide range of laws, labels, and logos that serve as key drivers to describe environmental claims to consumers, which is termed as *green marketing*. These environmental claims allow consumers to differentiate between products and also encourage purchasing decisions in relation to environment. This in turn will be a key driver for industries to invest and introduce sustainable environmental practices. The value of the environmental claims lies on the guarantee that the information provided is valid, credible, and easily distinguishable by the consumers. Standards and certifications play an important role in providing guidance to ensure that authenticated claims are issued in green advertising in the industry. Standards for environmental claims help the consumers, industry, and advertisers by providing a common ground coupled with reliability and uniformity in terms, regulations, and applications. When environmental claims form the background for industrial products and processes, new techniques, situations, and systems will evolve, and continuous improvement and update of these standards will help to maintain the level and scope of environmental benefits (Canadian Standards Association 2008).

A written assurance in the form of a certificate that the product, system, or service in question meets certain specific requirements is termed as certification. Certification is an important tool to add credibility as it demonstrates the product or service meets the requirement of the consumer. In some cases, certification is legal or part of the requirements of the contract. Certification may be given in two major forms as product certifications and system certification. Product certification in terms of environmental standards is a process of endorsing that the product has performed well in the performance and quality assurance tests and also meets the criteria specified in contracts/regulations or specifications (ISO). System certifications are structured environmental impact assessment mechanisms that help to evaluate the process and systems existing in the industry to meet the requirements and norms specified by the certifying body. Generally, an environment management system is adopted in the organization to reduce its environmental impacts and improve efficiency. Some of the most important product certifications and system certifications are discussed here (TexEASTile Sustainable Innovation for Textile in South East Europe).

17.5.1 PRODUCT CERTIFICATIONS

17.5.1.1 EU Ecolabel

The EU Ecolabel launched in 1992 is a commitment to environment sustainability. It has grown immensely, and by the end of 2011, more than 1,300 licenses have been awarded and more than 17,000 products use these labels. The license gives the organization the right to use the Ecolabel flower logo for specified product groups (Eco Label 2014a). The Ecolabel is a voluntary scheme that when used on any product guarantees limited use of substances harmful to environment and health, reduced air and water pollution, resistance of textiles to shrinkage during washing and drying, and color fastness to perspiration, washing, rubbing, and exposure to light. Further, this label gives assurance for nonuse of inorganic fibers like mineral fibers, glass fibers, metal fibers, carbon fibers, lead-based pigments, plasticizers and solvents, zinc and copper, azo dyes, or any dyes classified as carcinogenic, mutagenic, and toxic for reproduction. It also guarantees that the residues of harmful pesticides are negligible (Eco Label 2014b).

The focus of the label is to analyze the product life cycle stages to find out the stage that gives the highest environmental impact. All efforts are maintained to reduce the environmental impacts. Each product group criteria elaborates the compliance declarations, testing, and verification required. The main criteria for textiles are textile fibers criteria, processes and chemicals criteria, and fitness for use criteria. Care is taken to see that the product is of good quality with high performance using the fitness for use criteria. The product-specific criteria for the development of the EU Ecolabel are developed and revised by a group of experts and stakeholders in a transparent manner taking the life cycle of the product into account—from extraction of raw materials for production, packaging, transport, consumption, and the recycling bin. The current criteria are valid up to July 2013 after which the Joint Research Centre of the EU is working on the revision and development of essential criteria for different products. This label has been awarded to many products across Europe and has caught the attention of consumers who look out for the Ecolabel logo making green choices easy among the competitive market (Eco Label 2014c).

From the business point of view, the Ecolabel is advantageous by promoting sales right across the EU as this label signifies eco-friendliness, which is the main motto for product selection in the minds of the consumers. It serves as a marketing tool and provides a competitive edge for green procurement. The application is a simple online process and special discounts are given for SMEs, microenterprises, and applicants from developing nations. This label is not a certification given by the industry but signifies rigorous certification procedures and compliance checks by independent qualified scientists and hence it builds trust in the minds of the consumers to make green selection easy. The Ecolabel portrays eco-reputation and high order in corporate social responsibility and increases sales, which helps business organizations. They serve as communication tools as Ecolabeled products are promoted at international fairs and newsletters of EU Ecolabel where list of awarded companies are published and distributed, Ecolabeled products are listed online in the Green Store, and meeting guides and brochures are available in the Ecolabel website (Eco Label 2014d).

The quality assurance system is ensured through certain requirements to achieve certification, which include chain of custody, site visits, and other metrics and audit and surveillance requirements and specifications for duration of certification, time to achieve certification, reviewing, and retesting. In order to obtain certification for one or more products, the said products are to be selected and a dossier is built for each product. The dossier should contain information about the selected product (raw materials, life cycle stages from concept to supply and use) along with technical information like the general description of the company submitting the application, description of the manufacturing process, the list of chemicals used, list of suppliers, results of the testing of product carried out by ISO 17025–certified professional bodies, and additional supporting information if any. All information about the product and organization must be submitted to a competent body who will evaluate the ecological and performance criteria and make a visit to the manufacturing site.

The contract holders must be ready with a journal of information regarding test results and relevant documentation for the inspection by the competent body and testing of products. The members of the national competent bodies are responsible for the drafting of Ecolabel criteria, assessing applications, awarding the label and provision of technical support, and consulting. On compliance with the specified standards, an official certificate is sent after which the flower Ecolabel logo can be used. The validity of the certification is for 3 years until which the organization has to follow and maintain relevant criteria, test products periodically and keep records for annual inspection, inform the certification body on any changes in supplier list or manufacturing process, and pay relevant fee to the national certification body (Eco Label 2014d,e).

17.5.1.2 Global Organic Textile Standard

GOTS is a worldwide leading textile standard for organic fibers, which is independently certified for the entire textile supply chain with checks on ecological and social issues. The GOTS Organic Textile Standard International Working Group is composed of four reputed member organizations, OTA (United States), IVN (Germany), Soil Association (United Kingdom), and JOCA (Japan), which contribute to GOTS along with international stakeholder organizations and experts in organic farming and environment and socially responsible textile processing. The vision of the organization is to work for organic textiles to become part and parcel of the day-to-day life of the people and environment. The mission is to develop, implement, verify, and promote the GOTS. The standard specifies that organically produced raw materials are to be used in textile and apparel manufacturing and stipulates requirements for ecology and labor conditions throughout the supply chain. It also states that organic production is based on a system of farming that replenishes soil fertility without using toxic pesticides and fertilizers. Organic production that requires raw material from animals also requires adequate animal husbandry and does not include genetic modification (GOTS 2013).

The latest revised Version 4.0 of this standard was published in March 2014. This standard aims at outlining the worldwide requirements for certification of organic status of textiles and providing assurance to the consumer in the form of labels that authenticate its friendliness to environment. An organization that participates in the GOTS certification scheme should work in compliance with the criteria specified by the standard. The criteria mentioned in the standard are criteria for fiber production and criteria for processing and manufacturing (environmental criteria, technical quality and human toxicity criteria, minimum social criteria).

The key criteria for fiber production include organic certification of fibers based on international and national standards, for example, IFOAM Family of Standards, EEC 834/2007, USDA NOP; certification of fibers from conversion period is the standard permits such certification and the minimum percentage content for Label I is 95% certified organic fibers and for Label II 70% certified organic fibers.

The environmental criteria include differentiating the organic fiber products from conventional ones and prohibition of chlorine bleaching and use of materials like synthetic sizing agents; heavy-metal-containing oils; carcinogenic azo dyes; printing auxiliaries like aromatic solvents; phthalates and PVC; accessories made of PVC, nickel, and chrome; PVC packaging material; and critical inputs like toxic heavy metals and functional nanoparticles, genetically modified organisms, and their enzyme production. Chemical inputs must meet the basic requirements for toxicity and biodegradability, environmental policy to be formulated with goals and procedures for minimizing waste and discharges, record keeping of the use of chemicals, energy, water consumption, waste water treatment, and sludge disposal. Recycled paper certified as per FSC and PEFC could be used for packing material.

The technical quality and human toxicity criteria include product as per specifications with respect to testing like color fastness test and shrinkage values. Residue testing forms important criteria for raw materials, intermediaries, finished products, and accessories. The minimum social criteria are based on International Labour Organization (ILO) terms and requirements. Social requirements include specifications for freely chosen employment without forced labor, freedom for association and right to collective bargaining, safe and hygienic working environment, prohibition of child labor, living wages, working hours specifications, hours of rest and work, indiscrimination in work environment, humane treatment and dignity of labor, and regularized employment.

The standard provides two label grades known as label grade I and label grade II. Label grade I is termed as *organic*, and according to this label, the product should contain ≥95% certified organic fibers and ≤5% natural or synthetic fibers. Label grade II is termed as *made with X% organic*, and this label stipulates that the product should contain ≥70% certified organic fibers and ≤30% nonorganic fibers. Of the 30% nonorganic fibers, a maximum of 10% can be allocated to synthetic fiber component, as long as the raw materials used are not from certified organic origin, sustainable forestry program, or recycled. Blending of conventional fibers with organic fibers of the same type in the same product and the use of conventional cotton, angora, and virgin polyester in the remaining balance of the fiber composition are not permitted. If the raw fibers used are not certified organic fibers but are *organic in conversion*, then the label grade will read *organic in conversion* and made with X% *organic-in-conversion materials*.

The revision process includes the active participation of the international stakeholder organizations in the process of review and revision of GOTS. The review process is continuous and the revisions are anticipated once in 3 years. The GOTS revision process also includes the GOTS-approved certification bodies, which has revisions made through the *Certifier's Council* (GOTS 2014).

17.5.1.3 OEKO-TEX Standard 100

OEKO-TEX Standard 100 is a testing and certification system for textile raw materials, intermediate, and end products at all stages of production. It is based on the four sectors of textile ecology, which includes production ecology (Sustainable Textile Production [STeP] by OEKO-TEX), which examines the impact of processes in production on people and environment; human ecology (OEKO-TEX Standard 100) studies the impact of textiles and chemical ingredients on health and well-being of humans; performance ecology deals with the impact of the use of textile products on environment; and disposal ecology deals with problems related to disposal, reuse, recycling, and removal of textile products (OEKO-TEX-a).

Since 1992, the product label "Confidence in Textiles—Tested for harmful substances according to OEKO-TEX Standard 100" is a global certified system for all types of textiles that provides guarantee to retailers and customers that the labeled products are free of harmful substances. The label "OEKO-TEX Standard 100plus" represents the highest level of the system and can be applied to products that have been tested for harmful substances as per the requirements of the OEKO-TEX Standard 100 as well as produced under sustainable production conditions (OEKO-TEX-b,c).

OEKO-TEX testing focuses on the actual use of textile and stipulates that the closeness of the product to the human skin will determine the extent to which the human ecological regulations are to be met. According to this standard, there are four product classes (which includes textile items for babies and toddlers up to 3 years), namely, product class II (textiles close to human skin), product class III (textiles away from human skin), and product class IV (furnishing materials). The testing of harmful substances includes illegal substances, legally regulated substances, and harmful, yet legally not regulated, chemicals and parameters for health care. Items eligible for certification are gray or unfinished yarns, fabrics and knits, textile and non-textile accessories, ready-to-use furnished products, and all types of ready-made articles (OEKO-TEX-d).

While undertaking this certification, the benefits attained are to be analyzed. It helps in building a transparent supplier relationship and facilitates the flow of information regarding harmful substances. This certification has become mandatory for many organizations and has become an integral part of the organization's image. The costs of certification are distributed throughout the companies along the textile supply chain. Tests are usually done on the materials that are added at each individual stage of production. When certified source material is used, duplication of testing is not carried out.

It is mandatory that the organization should comply to certain prerequisites for certification. OEKO-TEX Standard 100 certification will be provided only if all the constituents of the product meet the required criteria without omission. Certification is facilitated on receiving a written application from the business organization sent to any one of the authorized test institutes or global certified representative offices of the OEKO-TEX Standard. To guarantee consistency in the testing results, the submitted sample materials are sent for testing to the member institutes in Japan or Europe. The prerequisite to enable certification by the testing institute or certification center is a declaration on the part of the manufacturer stating that at all times the quality of the products manufactured and sold shall correspond to the tested textile samples throughout the license period of 12 months. The next prerequisite for product certification is the conduct of company audits to examine the quality assurance and production processes to create the most promising certification conditions and to conform to sound human ecology product quality for the stipulated certification period. The certification includes regular controls of the product as an important part of the system, which is carried out in the market by the OEKO-TEX institutes to ensure that the required criteria are maintained on a regular basis. When the products have passed all these stages, the certification is given and the label can be used as labels, hang tags, and prints on products. Every year, control tests are carried out, which amount to 20% of all the certificates issued, and any deviations found in market testing will lead to the withdrawal of certificates if the manufacturer is unable to clarify the results (OEKO-TEX-e).

Consumers are increasingly sensitive and enquire about the safety of the product in use in terms of the product components to health or test results displaying safety to humans. The OEKO-TEX label is a source of value addition, extensive safety, transparency, and reliability and is independent and verifiable, up to date, and distributed across all types of products including technical textiles. The OEKO-TEX label becomes an important tool for closing a sale after price discussions and conveys a feeling of memorability and good feeling among consumers.

17.5.1.4 Textile Exchange Labeling System

Textile Exchange is a global nonprofit organization that promotes organic cotton growth in conjunction with farmers, manufacturers, brands, and retailers, from $240 million in 2001 to $6.8 billion in

2011 (Textile Exchange-a). This organization has its headquarters in United States and is committed to the responsible expansion of textile sustainability across the global textile value chain. The vision of the organization is to develop a global textile industry that enhances lives and protects and restores the environment. Textile Exchange focuses on acceleration of sustainable practices in the textile value chain, minimizing harmful impacts and maximizing positive effects (Textile Exchange-b). Textile Exchange standards have been developed through multistakeholder approach to safeguard the product claims through verification by an independent third party. The Organic Content Standard (OCS) checks the accurate amount of organic content in the final product by using two tools: Content Claim Standard (CCS) and CCS implementation. It does not address issues like use of chemicals and social and environmental impacts of production. The Recycled Claim Standard (RCS) identifies the recycled raw materials through the supply chain of the product. This standard was developed by the Materials Traceability Working Group of the OIA's Sustainability Working Group. The Responsible Down Standard (RDS) ensures that the down used in products comes from ethically reared geese. This standard was developed by the joint effort of North Face, California, United States; Control Union, Argentina; and Textile Exchange, EU, and is backed by the Textile Exchange CCS. The CCS is the base for all the standards by Textile Exchange and provides a tool to verify the content of specific input material and to check the supply chain of the organization under study by independent third-party verification. The Global Recycled Standard Version 3 provides brands with a tool for precise correct labeling, for encouraging the use of reclaimed materials, for establishing transparency in the supply chain, and to offer valid information to consumers (Textile Exchange-c). The OE 100 Standard (OE 100) 2009, Version 1.3, is a standard that tracks and documents the purchase, handling, and use of 100% certified organic cotton or organic-in-conversion cotton fiber used in yarns, fabrics, and finished goods. The OE Blended Standard, Version 1.2, provides standards for fiber-only claims in situations for products containing organic cotton that has been processed to third-party standards, for products containing organic cotton that has not been processed to third-party standards, and for products containing a percentage blend of organic cotton. The OE Blended Standard tracks and documents the purchase, handling, and use of certified organically farmed cotton fiber used in blended yarns, fabrics, and finished goods. This standard is applicable for all goods containing a minimum of 5% organic cotton content (Textile Exchange 2008, 2009).

The core to successful growth of the sustainable textile industry is integrity as perceived by Textile Exchange, and this aspect is promoted through tool development, education, information sharing, and collaboration with key stakeholders. Promotional activities are undertaken at conferences, trade shows, and individual companies for spreading the information about integrity, standards, and certification methods adopted by Textile Exchange.

17.5.2 System Certifications

17.5.2.1 OEKO-TEX Standard 1000

OEKO-TEX Standard 1000 is a certification system that includes testing and auditing of all the operations along the textile chain with the underlying concept of environment friendliness. This standard stipulates that an organization will be qualified for certification by OEKO-TEX Standard 1000, only when the organization has met environment-friendly and socially acceptable production processes at all the different stages of the manufacturing and provides evidence that at least 30% of the total production is already certified according to OEKO-TEX Standard 100. An independent certified auditor from OEKO-TEX International–Association for the Assessment of Environmentally Friendly Textiles will inspect the company and issue a certificate that is valid for 3 years and can be renewed on request. The central idea of the standard is the redefinition of targets and measures and development of criteria and limit values to enable continuing optimization of environmental performance and provision for benchmarking.

Some of the advantages of this standard are cost savings by efficient production processes, protection of resources, waste minimization, recycling of raw materials and auxiliaries used, improved employee identification by increased responsibility for employees, occupational safety and health management, strong market position due to improved image and credibility, certification by independent bodies or agencies, requirements and criteria suitable for the textile and apparel industry, significant contribution to environmental protection and additional marketing for own products.

The other environmental management systems like ISO 14001 or the EMAS system along with the in-house quality assurance system or ISO 9001 are taken into account for the certification. The management systems provide the necessary structure and data that form the basis for the starting point for an OEKO-TEX certification. The quality management process may analyze the innovative technical solutions, ecological improvements, savings in consumption, fair business practices, and employee satisfaction, which will form the foundation of future sustainable goals (OEKO-TEX Association).

17.5.2.2 ISO 14001 Environmental Management Systems

This standard aims to execute an effective management system that will remain commercially successful without compromising environmental responsibilities. It provides a framework for corporate responsibility as well as legal or regulatory requirements with focus on reducing environmental impact leading to sustainable success. The developed systems are suitable for all types and sizes of organizations and help participants to reduce waste and energy use leading to better environmental management, reduce costs and improve efficiency, meet legal obligations and win stakeholder and consumer trust, and face the changing business environment confidently. Its widespread adoption and international recognitions leads to global business opportunities and enhanced national and regional prospects; simple implementation profile, low human resource requirement, makes it a cost-effective solution; this standard has an independent accreditation process that showcases good credibility that reassures organizations and business enterprises. However, some points in this system require improvement. This standard does not call for continuous assessment of actual environmental performance to maintain accreditation, though a guideline ISO 14031 exists. The assumption that effective environmental management system will automatically lead to continuous improvement in environmental performance is highly criticized. In this system, there is no compulsion to make environmental performance information publicly available, and there is no requirement to show how the compliance has been achieved although the organization is required to publish its environment policy (BSI 2014).

17.5.2.3 EMAS

The EU EMAS is a management instrument for any organization that is eager to improve its environmental performance. This standard has been developed by the European Commission for companies in order to evaluate, report, and improve environmental performance. This premium environmental management tool leads to enhanced performance, credibility, and transparency and is suitable to organizations that are willing to improve their environmental and financial performance and to communicate their environmental performance to the stakeholders and society at large. The third revision of the EMAS regulation has brought about improved schemes applicability and credibility to strengthen its visibility and outreach.

An organization must adopt an environmental policy to comply with the relevant environmental legislations in order to achieve improvements in environmental performance. Second, an environmental review must be conducted whereby all environmental aspects of the activities, products and services, assessment methods, legal and regulatory framework, and the existing environmental management practices and procedures are to be taken into consideration. Based on the results of the review, an environmental management system must be established in line with the organizations environmental policy as specified by the management of the institution. The management system is required to establish responsibilities, objectives, operational procedures, training needs, monitoring and communication systems, and resources. This will be followed by an environmental audit that assesses the conformance of the system with the ideals of the management of the organization as well as the relevant environmental regulations. The next step is to provide a statement of the environmental performance of the organization against the objectives and the steps to be taken in future to continuously improve the environmental performance of the said organization. Finally, the environmental review, the EMS, the audit procedure, and the environmental statement must be approved by an accredited EMAS verifier and the validated records are to be sent to the EMAS competent body for registration. All the information must be made public before an organization can use the EMAS logo (EMAS 2014).

17.5.2.4 ISO 9001 Quality Management System

The quality management system in any organization is based on requirement for the products and services offered by the organization. Its design and implementation is influenced by certain factors like the changes and risks associated with the environment, the needs and objectives, the products and processes involved, and the size and structure of the organization. The international standard can be used by both internal and external parties and certification bodies to assess the organization's ability to meet customer, statutory, and regulatory requirements with regard to the requirements of the organization. The ISO 9000 and the ISO 9004 have been used for the development of this international standard.

While developing and improving the effectiveness of the quality management system, the process approach has been adopted to fulfil customer requirements and to promote customer satisfaction. The first step in process approach is to determine and manage linked activities. A process is a set of activities where resources are used and managed to enable the conversion of inputs to outputs. Usually, the output of one process may be the input of another process. The term *process approach* is the identification and interaction of a set of processes and its application and management in the said organization to produce the desired outcome. This system helps to gain ongoing process control over the individual processes as well as control over their combination and interaction. When this approach is used in the quality management system, requirements can be understood and met, processes are to be considered in terms of added value, results are to be obtained on process performance, and effectiveness and continuous improvement of processes can be achieved based on objective measurement.

While evaluating the processes in an organization, the objectives and the processes required to produce results as per consumer needs and management policies are to be established. The processes are to be implemented and the processes are monitored and results measured in terms of output against policies, objectives, and requirements for the product are to be reported. This should be followed by appropriate action to achieve continuous process performance (ISO 2008).

17.5.2.5 SA8000 Standard

SA8000 Standard is the first global auditable social certification standard for decent working environment across all the industrial sectors. It is based on the UN Declaration of Human Rights and conventions of the ILO, UN, and National Law to create a common platform to measure social performance. Organizations that require certification of SA8000 should have policies and procedures that safeguard the basic human rights of workers, and this standard specified structures and procedures to ensure compliance that is continuously reviewed. By January 2016, SA8000 will be completely transitioned to SA14000. The key elements of SA14000 are

- *Child labor*: Use and support of child labor is prohibited. Remediation and support provided to children at work to attend school. Employment of young workers conditional.
- *Forced and compulsory labor*: No use and support of forced and compulsory labor inclusive of prison labor; no financial labor deposits for work; no enforcement of personnel to work by withholding salary, benefits, property, and documents; freedom to leave premises after work and to terminate their employment; and no support to human trafficking.

- *Occupational health and safety (OHS)*: Provision of safe and healthy workplace to meet the basic needs of the workers, safety measures as personal protection equipment, and medical attention during emergencies; establishment of H&S Committee with management representatives and workers; and appointment of senior manager to ensure OSH and orientation of the features of OSH to all the personnel in the industry.
- *Freedom of association and right to collective bargaining*: The workers should have the freedom to join trade unions or to create one through which bargaining for necessary rights can be carried out in a collective manner.
- *Discrimination*: No discrimination in areas of recruitment, remuneration, access to training, promotion, termination, and retirement, based on race, origin, caste, birth, religion, disability, gender, sexual orientation political opinion, and age.
- *Disciplinary practices*: Zero tolerance to corporal punishment, mental and physical abuse, and harsh or inhumane treatment; all personnel to be treated with dignity and respect.
- *Working hours*: Compliance with laws and industry standards, normal work week, overtime shall not exceed 48 h, 1 day off after every 6 working days, voluntary overtime not more than 12 h per week.
- *Remuneration*: Right to living wage, all workers to be paid at least minimum wage to meet basic needs and provide discretionary income; prohibited use of labor on long-term contracts.
- *Management systems*: Integration of standards to maintain certification followed by merging them into management systems and practices (SAI 2014).

Social Accountability International offers global training and capacity building programs like the SA8000 Auditor Training Social Fingerprint, implementation programs for UN Guiding Principles for Business and Human Rights, custom training and professional development and technical assistance, corporate programs, and guidance and standard interpretation (SAI 2012).

17.6 FUTURE DIRECTIONS

The most notable developments in advertising are the increase of *green claims* and the emerging interest in the concepts of environmental sustainability. Sustainability comprises aspects like economic and social values, action by public and private institutions, and environmental aspects, but in the market place, the focus is on environment. Many cases of environmental communication of green claims are usually exaggerated and misled the consumers but many organizations have set up standards for preventing *green washing*.

There are many challenges facing manufacturers, retailers, and consumers to improve the current situation in environmental communication and green claims. They include the improvement of the social and working conditions of the workers, offering organic textiles at affordable prices to consumers, changing the attitude of consumers to judiciously purchase durable clothing that lasts longer, and providing relevant information about the environmental footprint of products to consumers to make wise decisions. The role of retailers include the promotion of environment-friendly textiles, demand environment and social accountability from producers, communicate and highlight to consumers the value of sustainability and environment-friendly behavior, encourage recycling by promotion of locally produced clothes banks, use of eco-friendly textiles for staff uniforms in retailing, and include sustainability in staff training programs. Manufacturers can use best practices in technological innovations and contribute to the reduction of environmental impacts, encourage recycling, remanufacturing, and fashion upgrades, and encourage research in developing production processes and innovative materials with sustainability as the motto (Retail Forum for Sustainability 2013).

On the whole, changes can occur in the direction of sustainability among manufacturers, retailers, policy makers, and other organizations only when the awareness towards such issues are kindled by planned and well-directed environmental communications that reach the end users to shop wisely. These communications become well balanced and far reaching when products display the *green claims* to catch the attention of consumers in the competitive market. Truthful and substantiated claims build the confidence of the consumer who will move towards products that are sustainable in all aspects and will be willing to pay the price for showing their contribution and willingness to safeguard the environment for the forthcoming generations.

REFERENCES

BSI. 2014. ISO 14001 environmental management. http://www.bsigroup.com/en-GB/iso-14001-environmental-management/.

Business Dictionary. 2014. Communication. http://www.businessdictionary.com/definition/communication.html#ixzz38YiCiMY6.

Butler S. 2007. How the M&S boss turned green and decided that his Bentley just had to go. *Times* [Online, U.K.]. http://business.timesonline.co.uk.

Canadian Standards Association. June 2008. Environmental claims: A guide for industry and advertisers. http://www.competitionbureau.gc.ca/eic/site/cb-bc.nsf/eng/02701.html.

Consumers International and AccountAbility. 2007. Press Release: US/UK consumers want tougher government action on products that cause climate change. Consumers International and AccountAbility. www.consumersinternational.org.

CU. 2014. Eco labels—Here you will find out what the labels on your favorite products mean. http://www.greenerchoices.org/eco-labels.

Danish Consumer Ombudsman (DCO). January 2011. Guidance from the Consumer Ombudsman on the use of environmental and ethical claims, etc., in marketing. http://www.consumerombudsman.dk/~/media/Consumerombudsman/dco/Guidelines/The%20Use%20of%20Environmental%20and%20Ethical%20Claims%20etc%20in%20Marketing%20%20Guidance%20from%20the%20Consumer%20Ombudsman.pdf.

Eco Label. July 31, 2014a. Facts and figures. http://ec.europa.eu/environment/ecolabel/facts-and-figures.html.

Eco Label. July 31, 2014b. More about the EU ecolabel. http://ec.europa.eu/environment/ecolabel/the-ecolabel-scheme.html.

Eco Label. July 31, 2014c. EU ecolabel for consumers. http://ec.europa.eu/environment/ecolabel/eu-ecolabel-for-consumers.html.

Eco Label. July 31, 2014d. EU ecolabel for businesses. http://ec.europa.eu/environment/ecolabel/eu-ecolabel-for-businesses.html.

Eco Label. July 31, 2014e. EU ecolabel—Information and contacts. http://ec.europa.eu/environment/ecolabel/information-and-contacts.html.

EMAS. August 7, 2014. About EMAS. http://ec.europa.eu/environment/emas/about/index_en.htm.

Environment Policy Committee (EPOC). 2013. A characterization of environmental labelling and information schemes. http://www.oecd.org/officialdocuments/publicdisplaydocumentpdf/?cote=ENV/EPOC/WPIEEP(2013)2/FINAL.

Federal Trade Commission-a. 2014. Part 260—Guides for the use of environmental marketing claims. http://www.ftc.gov/sites/default/files/attachments/press-releases/ftc-issues-revised-green-guides/greenguides.pdf.

Federal Trade Commission-b. Commercial practices. http://www.ecfr.gov/cgi-bin/text-idx?tpl=/ecfrbrowse/Title16/16cfr260_main_02.tpl.

Federal Trade Commission-c. Guides and trade practice rules. http://www.ecfr.gov/cgibin/text-idx?tpl=/ecfrbrowse/Title16/16cfr260_main_02.tpl.

Flor A.G. 2004. Environmental communication: Principles, approaches and strategies of communication. https://www.academia.edu/181519/Environmental_Communication.

GOTS. October 10, 2013. Global organic textile standard—Ecology and social responsibility: About us. http://www.global-standard.org/about-us.html.

GOTS. March 20, 2014. Global organic textile standard—Ecology and social responsibility: General description. http://www.global-standard.org/the-standard/general-description.html.

Green Chemistry and Commerce Council. Considered chemistry at Nike: Creating safer products through the evaluation and restriction of hazardous chemicals—Case study for the Green Chemistry and Commerce Council (GC3). http://www.greenchemistryandcommerce.org/downloads/Nike_final.pdf.

Green Guides. Part 260—Guides for the use of environmental marketing claims. http://www.ftc.gov/sites/default/files/attachments/press-releases/ftc-issues-revised-green-guides/greenguides.pdf.

Hansen O.E., Holm J., and Sondergard B. 2001. Ecological modernization and institutional transformation in the Danish textile industry. *2001 Ninth International Conference of Greening of Industry Network*, Bangkok, Thailand, January 25. www.researchgate.net/...and...textile.../72e7e521b23786afc6.pdf, http://ec.europa.eu/environment/industry/retail/pdf/issue_paper_textiles.pdf, http://www.unep.org/chemicalsandwaste/Portals/9/CiP/CiPWorkshop2011/CiP%20textile%20case%20study%20report_21Feb2011.pdf, https://www.oeko-tex.com/en/retailers/basics_1/basics.html.

ISO. Certification. 2014. http://www.iso.org/iso/home/standards/certification.htm.

ISO. 2008. ISO 9001—Quality management systems—Requirements. https://www.iso.org/obp/ui/#iso:std:iso:9001:ed-4:v1:en.

ISO 14025:2006. Environmental labels and declarations—Type III environmental declarations—Principles and procedures. https://www.iso.org/obp/ui/#iso:std:iso:14025:ed-1:v1:en.

ISO 14024:1999. Environmental labels and declarations—Type I environmental labelling—Principles and procedures. https://www.iso.org/obp/ui/#iso:std:iso:14024:ed-1:v1:en.

ISO 14021:1999. Environmental labels and declarations—Self-declared environmental claims (Type II environmental labelling). https://www.iso.org/obp/ui/#!iso:std:23146:en.

Kalliala E.N. 2003. Environmental indicators of textile products for ISO (Type III) environmental product declaration. *AUTEX Research Journal* 3(4):206–211. http://www.autexrj.org/No4-2003/0056.pdf.

Koszewska M. 2011. Social and eco-labelling of textile and clothing goods as means of communication and product differentiation. *Fibres and Textiles in Eastern Europe* 19(4): 20–26. www.fibtex.lodz.pl/article536.html.

Miles M. May 20, 2011. Labeling of textiles that contain organic ingredients. http://www.ams.usda.gov/AMSv1.0/getfile?dDocName=STELPRDC5090967.

OEKO-TEX-a. 2014. Textile ecology. https://www.oeko-tex.com/en/manufacturers/textile_ecology/textile_ecology_start.html.

OEKO-TEX-b. OEKO-TEX Standard 100—What is that?

OEKO-TEX-c. Concept. https://www.oeko-tex.com/en/retailers/basics_1/oets_concept_1/oets_concept.html.

OEKO-TEX-d. Concept—OEKO-TEX Standard 100. https://www.oekotex.com/en/manufacturers/concept/oeko_tex_standard_100/oeko_tex_standard_100.xhtml.

OEKO-TEX-e. Certification—Prerequisites. https://www.oeko-tex.com/en/manufacturers/certification/prerequisites/prerequisites.html.

OEKO-TEX Association. Concept of OEKO-TEX Standard 1000. https://www.oekotex.com/en/manufacturers/concept/oeko_tex_standard_1000/concept_oets_1000/concept_oets1000.html.

Retail Forum for Sustainability. August 2013. Sustainability of textiles.

Rodie J.B. November–December 2008. Going Green: Beyond marketing hype textile world. http://www.textileworld.com/Issues/2008/November-December/Features/Going_Green-Beyond_Marketing_Hype.

SAI. 2012. About SAI. http://www.sa-intl.org/index.cfm?fuseaction=Page.ViewPage&pageId=1365.

SAI. 2014. SA8000 Standard. http://www.sa-intl.org/index.cfm?fuseaction=Page.ViewPage&pageID=937.

Schuchard R., Berry T., Skinner C., Stewart E., and Uren S. April 2008. Eco-promising: Communicating the environmental credentials of your products and services. http://www.bsr.org/reports/BSR_Eco-Promising_April_2008.pdf.

Stephen W.L. and Foss K.A. 2009. *Encyclopedia of Communication Theory*. SAGE Publications, Inc., Thousand Oaks, CA. DOI:10.4135/9781412959384, https://theieca.org/sites/default/files/Milstein_Enviro_Com_Theories.pdf.

Sullivan & Worcester. 2011. Federal trade commission and consumers seek to curb potentially misleading green marketing claims. http://www.sandw.com/assets/htmldocuments/CLIENT%20ADV.%20%20FTC%20Commission%20and%20Consumers%20Seek%20B1494837.PDF.

TexEASTile Sustainable Innovation for Textile in South East Europe. 2014. Green tools handbook—A guide on main product and system certifications focused on ecological improvement of textile products. http://www.uctm.edu/uctm-uploads/labs/GTOOLS_Handbook_FINAL.pdf.

Textile Exchange. 2008. Organic exchange blended standard. http://old.icea.info/Portals/0/OE%20Blended%20Standard%20Aug%202008_Revisioni-Approvate.pdf.

Textile Exchange. 2009. OE 100 Standard. http://textileexchange.org/sites/default/files/te_pdfs/integrity/OE100%20Standard_v1.3.pdf.

Textile Exchange-a. History. http://textileexchange.org/about-us/history.

Textile Exchange-b. About us. http://textileexchange.org/about-us.

Textile Exchange-c. Standards—TE Standards. http://textileexchange.org/standards.

The International Trade Centre. 2014. The organic cotton market. http://www.intracen.org/The-organic-cotton-market/.

Ummel K. October 15 2012. CARMA notes: Future data. http://carma.org/.

UNEP/DTIE. February 2011. A synthesis of findings under the UNEP/IOMC project on information on chemicals in products. http://www.unep.org/chemicalsandwaste/Portals/9/CiP/CiP%20Project%20synthesis%20report_Final.pdf.

UNEP. January 2011. The chemicals in products project: Case study of the textiles sector.

18 Supplier Assessment in Global Apparel Supply Chains

Shams Rahman and Aswini Yadlapalli

CONTENTS

18.1 Introduction 399
18.2 Apparel Industry in Developing Nations: Overview 400
    18.2.1 Global Sourcing in Apparel Industry 400
    18.2.2 Apparel Supply Chain... 401
    18.2.3 Apparel Industry: Unethical Practices.. 402
18.3 Supplier Assessment 403
18.4 Conceptual Framework for Supplier Assessment.. 404
    18.4.1 Supplier Selection 404
    18.4.2 Supplier Selection Criteria Used in the Apparel Industry 405
        18.4.2.1 Economic Criteria.. 405
        18.4.2.2 Environmental Criteria 406
        18.4.2.3 Social Criteria.. 406
        18.4.2.4 Methods for Decision Making in Supplier Selection........................ 408
    18.4.3 Supplier Development... 408
        18.4.3.1 Supplier Evaluation 409
        18.4.3.2 Collaboration... 410
    18.4.4 Continuous Improvement.. 410
18.5 Case Study 411
18.6 Conclusion 413
References.. 414

18.1 INTRODUCTION

In spite of their commitment for corporate social responsibility (CSR), many apparel manufacturers and retailers in the past have abused human rights by operating sweatshops in developing nations. The first incident that attracted significant media attention is the working conditions of a factory producing Kathie Lee Gifford's line of clothing for WalMart in 1996 (Park-Poaps and Rees 2010). Since then, there has been a steady increase in tragic incidents in this industry arising out of work rights abuse and human exploitation to safety that resulted in huge death tolls (Perry and Towers 2013). Some of the more recent incidents include fires in Ali Garment and Tazreen Fashion factory in September and November 2012, respectively. In these two devastating fire incidents, more than 400 workers were killed and over 500 workers suffered injuries. However, the incident that shocked the apparel supply chains was the collapse of Rana Plaza in May 2013, which killed more than 1130 workers and injured more than 2500 (Lund-Thomsen and Lindgreen 2013). These incidents have raised concerns regarding social and environmental aspects of the global apparel supply chain among stakeholders such as customers, competitors, regulators, industry peers, and media (Park-Poaps and Rees 2010, Gallear et al. 2012). The usual problem that the Western garment retailers encounter is with the visibility of the complex network of global supply chains (Park et al. 2010).

This implies that the retailers are prone to greater risks associated with the transfer of social responsibility (SR) beyond their boundary of ownership and control (Gimenez and Tachizawa 2012). Since "a company is no more sustainable than its supply chain" (Krause et al. 2009, p. 18), addressing sustainability aspects in supply chains is a key to successful apparel retailers.

Globalization and outsourcing from low-cost producing nations have significantly impacted the dynamics of business practices and forced the senior executive of the big retailers to examine their relationships with suppliers (Lamming et al. 1996). The literature indicates that the supplier relationship management is critical for supply chain performance. It is estimated that manufacturers spend more than 60% of their total sales on purchasing items such as raw materials, parts, and components and 70% of the product cost consists of purchases of goods and services (Liao and Kao 2011). This indicates the importance of supply management and procurement process in the global apparel supply chain. Therefore, proper assessment of suppliers from the low-cost producing nations plays a crucial role in the procurement process. Literature suggests that the two key elements of a procurement process are supplier selection and supplier development (Kannan and Tan 2002). Supplier assessment is used for supplier selection and segmentation of suppliers to enhance the supplier development process (Watts and Hahn 1993, Tan and Wisner 2003, Park et al. 2010). It focuses on evaluating suppliers' products and processes based on certain specific criteria. Traditionally, costs, quality, service, and performance are used for supplier assessment. In the recent times, the stakeholder pressure forced many retailers to include social and environmental aspects in the supplier assessment process. The objective of this chapter is to provide an overview of the apparel industry in developing nations, develop and discuss a conceptual framework for supplier assessment, and apply this framework to a real-world case study.

The remainder of the chapter is organized as follows. Section 18.2 provides an outlook on apparel supply chains and the role of developing nations in apparel manufacturing. Section 18.3 introduces various aspects of supplier assessment processes and methods. An integrative conceptual framework for supplier assessment and development is suggested in Section 18.4. The proposed framework is used to illustrate an Australian case study in Section 18.5. The chapter ends with a brief conclusion in Section 18.6.

18.2 APPAREL INDUSTRY IN DEVELOPING NATIONS: OVERVIEW

Since the 1960s, Western retailers have been using offshore garment manufacturing facilities in developing nations to capitalize on the low-cost manufacturing (Singleton 1997). Increased pressure on retailers to reduce price further led to greater use of offshore manufacturing facilities. Under this circumstance, the apparel industry is not only competing against the price but also the quality and speed (Bruce and Daly 2011). Recent disasters in this industry shifted retailers toward ethical sourcing. Until 2005, the textile and garment manufacturing industry used to be governed by quota restrictions under multifiber arrangement (MFA), which led to the internationalization and use of more developing nations in garment manufacturing and supply. Even after the abolition of the MFA, the industry experienced a steady increase in the export value (World Bank 2012).

18.2.1 Global Sourcing in Apparel Industry

According to World Trade Organization (WTO), clothing and textile exports registered a significant growth since 2000 and reached US $706 billion in 2011. In 2011 alone, the world clothing export increased by 17% and the top 10 exporting countries experienced a growth over 13% (WTO 2012). Figure 18.1 indicates the trade shifts among the world's leading exporters between 2000 and 2011. It appears that the share of world exports from United States and Mexico decreased considerably from over 4% to 1%. On the other hand, Asian countries such as Bangladesh, Vietnam, and Cambodia emerged as new sourcing destinations with dramatic increase in the export share. India and Turkey have also experienced a moderate growth in export, increasing from 3.0% to 3.5% and 3.3% to 3.4%,

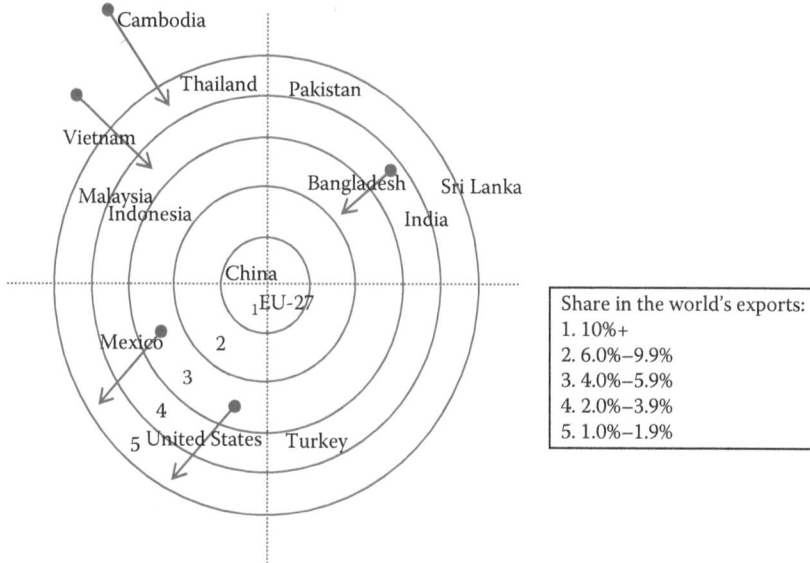

FIGURE 18.1 Shifts in the market share of leading clothing exporters, 2000–2011. (From World Trade organisation [WTO], Available at: http://www.wto.org/english /res_e/statis_e/its2012_e/its12_highlights2_e.pdf.)

respectively. However, there is a modest drop in export from Indonesia, Thailand, and Pakistan. Overall, the export share of the Asian region has increased during the period between 2000 and 2011. China remained as the leading exporter of cloths with over 10% of the global export (WTO 2012). Low capital requirements and emphasis on labor-intensive manufacturing of apparel industry acted as a gateway for countries toward export-oriented industrialization (Gereffi and Frederick 2010).

18.2.2 APPAREL SUPPLY CHAIN

Apparel supply chains are globally dispersed with products designed in one country and manufactured in another, whereas raw materials are sourced from a different country. The two most value-adding processes in apparel supply chains are marketing and branding. The responsibilities for such activities are vested with big retailers (Gereffi and Frederick 2010). So large retailers, marketers, or branded manufacturers orchestrate the apparel supply chain, whose reputation and competitiveness are based on the performance and actions of the whole apparel supply chain.

The apparel supply chain is long and complex and involves various parties. Figure 18.2 illustrates a higher-level version of an apparel supply chain. Each block in the supply chain may represent several members. For example, a retailer represents multiple channels of retail outlets such as department stores, specialty stores, discounted stores, and factory outlet or e-commerce websites where the product is sold to final consumers. Similarly, an export network may include brand-named apparel companies, overseas buying houses, and trading companies. Apparel manufacturers can be classified into domestic and overseas based on their geographic locations and assembler, original equipment manufacturing (OEM), and original brand manufacturing (OBM) based on their production systems.

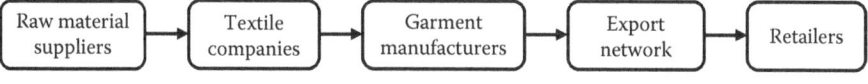

FIGURE 18.2 A generic global apparel supply chain.

Apparel industry is characterized by three types of production models. An assembly system is a type of model in which sewing plants are provided with imported inputs for assembly of the final garment, whereas an OEM is a production model in which OEM manufactures products that are ordered and purchased by another company and retailed under that purchasing company's brand name. Finally, OBM is the most sophisticated form of production model where manufacturers possess capability to design and develop new products and sell these products under their own brand name (Gereffi and Memedovic 2003). The shift in the production model from OBM to OEM and assembly has led to increase in complexity of the global sourcing networks with lack of visibility especially in the upper echelon of the supplier tiers.

Gereffi and Memedovic (2003) identified two types of economic supply chain networks, namely, producer-driven and buyer-driven. Producer-driven supply chains are technology-intensive industry with huge capital investments where manufacturers play a vital role in the network. On the other hand, buyer-driven supply chains are dominated by branded retailers or marketers. Apparel supply chains are predominantly buyer-driven with power asymmetries between the manufacturers and global retailers. In most cases, the lead retail firm outsources manufacturing system to global network of suppliers. Intense competition among apparel manufacturers for foreign investments and contracts leaves them with little leverage to negotiate. Thus, large retailers, marketers, or branded manufacturers setting up decentralized production networks in developing nations with no manufacturing facilities of their own characterizes a buyer-driven chain. In buyer-driven supply chains, buyers act as intermediaries between retailers and manufacturers. Due to the transition of apparel industry toward buyer-driven chains and involvement of the increased number of parties, retailers can no longer control manufacturers. This scenario tends to create an opportunistic behavior among the members of supply chains that leads to unacceptable practices along the apparel supply chains.

18.2.3 APPAREL INDUSTRY: UNETHICAL PRACTICES

The apparel industry is known for its behavior of shopping around the globe for cheap labor and unethical practices. The industry is accused of not paying living wages to workers; using underage labor; discriminating against gender, religion, and social class; abusing human rights; preventing employees to join the unions; and failing to provide minimum labor standards that led to some of the worst industrial disasters (Awaysheh and Klassen 2010). Table 18.1 shows a list of some of the incidents in apparel manufacturing and the corresponding death toll signifying the magnitude of these disasters. These incidents questioned the fire safety and structural integrity of the apparel factories. Lack of emergency fire evacuation plans, blocked fire exists and stair cases, and locked doors and windows are some of the unacceptable working conditions.

TABLE 18.1
Major Incidents in Apparel Manufacturing

Company, Country	Year	Number of Deaths	Cause
Triangle shirt waist factory, United States	1911	100	Fire
Ali Garment factory, Pakistan	2012	289	Fire
Tazreen Fashions, Bangladesh	2012	112	Fire
Rana Plaza, Bangladesh	2013	1132	Building collapsed

Characteristics that favor unfair employment practices within the apparel industry are (1) labor-intensive production with limited automation, (2) competitive pressure to lower production costs, and (3) transparency issues in supply chains having several subcontractors (Park-Poaps and Rees 2010). Perry and Towers (2013) suggested that high product variety, high volatility, low predictability, seasonality, and intense competition are the other factors that greatly influence this industry. For instance, retail buying practices such as shorter lead times, variability in order quantities, and pressure to lower prices force garment workers to work overtime without proper remuneration (Ruwanpura and Wrigley 2011).

18.3 SUPPLIER ASSESSMENT

Although the studies on criterion-based supplier selection go back in the 1960s (Dickson 1966), research on supplier assessment is relatively new. Ellram (1995) highlighted the supplier assessment task that includes supplier selection, establishment of relationship, and evaluation of relationships as the critical mechanism for the procurement process. Generally, supplier selection and development (supplier collaboration and evaluation) practices are associated with each other and must be examined together (e.g., see Kwong et al. 2002, Park et al. 2010, Gualandris et al. 2014). Often monitoring and evaluation are used as alternatives or synonyms to assessment (Ittner et al. 1999, Vachon and Klassen 2006, Park et al. 2010). Assessment refers to "the activities of collecting and processing data and information about different entities in the supply chain" (Gualandris et al. 2014, p. 2). On the other hand, monitoring is defined as "activities using markets or arms-length transactions conducted by the buying organization in order to evaluate and control its suppliers" (Vachon and Klassen 2006, p. 798), whereas "supplier evaluation essentially involves the buying company measuring the performance of suppliers" (Johnsen et al. 2008, p. 275). In the literature, too many terminologies and systems are used for supplier assessment that often create confusion for researchers (Lamming et al. 1996). In this chapter, we use the definition of supplier assessment given by Lamming et al. (1996, p. 174), which is defined as a "formal process that combines the vendor selection stage (to aid decisions on which supplier best suits the customer's requirements) with a vendor performance measurement stage (to monitor the performance of a supplier and compare it with customer's expected level of performance)."

Supplier assessment is a form of rating of supplier value based on a number of selected capabilities. It can be formal or informal. A formal assessment provides both subjective and objective ratings, whereas an informal assessment leans more toward subjective ratings. A variety of scorecards are available for organizations to assess suppliers. A scorecard system is a formal tool for evaluating suppliers to build long-term relationships. Generally, scorecards are based on factors and criteria, weighting, rating scale, and ease of use and effectiveness:

- *Factors and criteria*: Criteria are used to evaluate the supplier based upon buyers' requirements.
- *Weighting*: Industry determines the weights given to each factor. The weighting scales include percentage per criterion and numerical values.
- *Rating scales*: The suppliers are rated against the criteria using the scales. Scales can be either objective or subjective and the methodology being adopted determines the rating scale.
- *Ease of use and effectiveness*: The instructions on how to use a scorecard for supplier evaluation and the process of gathering information needed to evaluate the supplier using the scorecard determine the ease of use and effectiveness.

Supplier certification is another form of supplier assessment. A detailed questionnaire or audits of supplier operations are some of the common tools used for supplier assessment (Park et al. 2010). Through extensive literature review, Kwong et al. (2002) identified various assessment practices, whereas Briggs (1994) demonstrated a supplier assessment model using a questionnaire designed incorporating components such as quality, culture, commercial issues, material logistics, health and safety, and design engineering. Full business assessment, quality system survey, and quarterly assessment are some of the supplier rating schemes representing human judgment. A common form of supplier classification is a performance-based measure where suppliers are assessed based on outcomes and their process capability (Johnsen et al. 2008). Although many methods of supplier assessment are available, generally they are adopted depending upon the local and global business environment (Kwong et al. 2002).

The results of supplier assessment can be used for supplier selection and to support capacity development of suppliers (Park et al. 2010). From the buyer's perspective, the possible benefits of supplier assessment include improved overall quality, better all-round service, improved delivery performance, improved relationships, and reduced costs. From the supplier's perspective, improved relationship is the main benefit of supplier assessment (Lamming et al. 1996).

18.4 CONCEPTUAL FRAMEWORK FOR SUPPLIER ASSESSMENT

Performance or process-based assessments alone are no longer relevant in relationship-focused paradigm of supply management (Johnsen et al. 2008). An alternative performance measurement of supplier relationship has become prominent. Lamming et al.'s (1996) relationship assessment program model (RAP model) is considered as one of the earliest model on vendor relationship assessment. Since then, research has continuously adopted supplier selection, supplier development, and supplier performance evaluation as a mechanism for supplier relationship management (see, e.g., Kannan and Tan 2002). More recently, Park et al. (2010) developed an integrated framework for supplier assessment and development. Gualandris et al. (2014) advanced supplier assessment model by integrating verification element to validate the information emerging from assessment.

Briggs (1994) suggested that the supplier assessment process can be comprehensive when all aspects of supplier's business are incorporated. Based on the research of Vachon and Klassen (2006) and Park et al. (2010), a comprehensive framework of supplier assessment is suggested that includes supplier selection and supplier development as the elements of supplier assessment. Figure 18.3 shows the structure of the supplier assessment model and the elements of the model are discussed in the following subsections.

18.4.1 SUPPLIER SELECTION

Supplier selection is a major function of acquiring required materials, services, and equipment in a business (Weber et al. 1991). This element of supplier assessment is considered as the first stage of establishing relationship among supply chain partners. A right selection mechanism ensures that supplier capabilities will address buying firms' current and future challenges (Paulraj 2011). It is also critical for the enhancement of performance and gaining and maintaining competitiveness

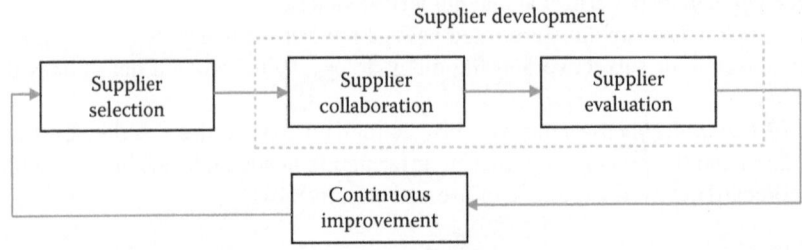

FIGURE 18.3 Supplier assessment framework. (Adapted from Park, J. et al., *Indust. Manage. Data Syst.*, 110(4), 495, 2010; Vachon, S. and Klassen, R.D., *Int. J. Operat. Product. Manage.*, 26(7), 795, 2006.)

TABLE 18.2
Studies Using Different Criteria for Supplier Selection

Criteria	Source
Environmental	Zhu and Sarkis (2007), Liao and Kao (2011), Paulraj (2011), Gimenez and Sierra (2013), Kannan and Tan (2002), Handfield et al. (2002), Bai and Sarkis (2010), Hsu et al. (2013), Shaw et al. (2012), Baskaran et al. (2012), Kannan et al. (2014), Govindan et al. (2013a), Xu et al. (2013), Kumar et al. (2014)
Social	Bai and Sarkis (2010), Paulraj (2011), Baskaran, Nachiappan and Rahman (2012), Kannan and Tan (2002, 2003), Govindan et al. (2013a), Xu et al. (2013), Kumar et al. (2014)
Economic	Chan et al. (2008), Dickson (1966), Liu and Hai (2005), Paulraj (2011), Carr et al. (2008), Kannan and Tan (2002, 2003), Weber et al. (1991), Kahraman et al. (2003), Liu et al. (2000), Sarkis and Talluri (2002), Kannan et al. (2014), Govindan et al. (2013a)

(Ittner et al. 1999, Krause et al. 2000, Park et al. 2010). Hence, supplier selection element, once an operational level function, has become an element of strategic issue in the context of supply management (Weber et al. 1991).

Supplier selection requires the assessment of suppliers based on specific selection criteria (Govindan et al. 2013a). Traditionally, it focused on suppliers that are favorable in terms of price and then shifted to quality of products and services (Kwong et al. 2002). In other words, the selection process of suppliers needs a multicriteria decision approach (Kwong et al. 2002). According to Kannan and Tan (2002) and Park et al. (2010), the selection process of supplier requires not only information on criteria but also on the process of implementing the decision under consideration.

Lamming et al. (1996) emphasized that the future assessment should incorporate a wide range of criteria into the existing ones. It is being suggested that the inclusion of social criteria in supplier selection process will enhance the sustainable performance of supply chains (Gallear et al. 2012). In recent studies, the use of green and sustainability–based selection criteria has become quite common (Handfield et al. 2002, Humphreys et al. 2006, Lu et al. 2007, Tsai and Hung 2009, Kannan et al. 2013, 2014). However, studies addressing social criteria in supplier selection are still scarce (Dai and Blackhurst 2012). A study by Baskaran et al. (2011) highlighted that the selection of suppliers based on social criteria will manifest buying firms' commitment toward social issues. A list of studies using different criteria for supplier selection is shown in Table 18.2.

18.4.2 Supplier Selection Criteria Used in the Apparel Industry

While selecting a global apparel supplier, the two most commonly used criteria are cost of product and quality and delivery time. Reliability and flexibility regarding product design and quantity are the two specific selection criteria considered in apparel supply chains. With greater emphasis given to SR in the recent past, social and environmental criteria have gained more prominence within the selection criteria (Ahsan and Azeem 2010). Table 18.2 provides a summary of studies using different supplier selection criteria in general. In the following subsection, we discuss studies that used social, environmental, and economic criteria in the supplier selection process within the context of the apparel industry.

18.4.2.1 Economic Criteria

Dickson's (1966) study is one of the earliest studies on criterion-based supplier selection. This study identified 23 different criteria of supplier selection primarily related to economic aspects. Earlier studies that have empirically examined selection criteria (e.g., Sibley 1978, Billesbach et al. 1991,

TABLE 18.3
Studies Using Economic Criteria in Apparel Supply Chains

Criteria	Subcriteria	References
Quality	• Quality assurance • Process capabilities • Reject rate • Warranties and claim policies	Chan and Chan (2010), Koprulu and Albayrakoglu (2007), Thaver and Wilcock (2006), Åkesson et al. (2007)
Price	• Purchasing price • Price performance value • Transportation costs	Chan and Chan (2010), Koprulu and Albayrakoglu (2007), Thaver and Wilcock (2006), Åkesson et al. (2007)
Delivery performance	• Assurance of supply • Flexibility responsiveness • Reliability • Service quality	Chan and Chan (2010), Koprulu and Albayrakoglu (2007), Thaver and Wilcock (2006), Åkesson et al. (2007)

Walton et al. 1998, Simpson et al. 2002, Lehmann and O'Shaughnessy 1982) suggest that not only costs but also quality, delivery, and service are important selection metrics (Kannan and Tan 2002). Quality, delivery, performance, and service are listed as the most important traditional selection criteria (Govindan et al. 2013a). Liao and Kao (2011) conducted an extensive literature review on selection criteria and grouped costs, quality, and delivery performance under economic criteria. Due to the emergence of fast fashion, criteria related to flexibility, reliability, and order quantities gained importance in the apparel supply chain. Table 18.3 provides the list of economic criteria in context to apparel supply chains.

18.4.2.2 Environmental Criteria

Since early 1990, there has been an increase in the use of green practices in production, which also has given importance to green selection (Liu et al. 2000). Research also started focusing on environmental criteria for supplier selection (Kannan et al. 2014). An environmental criterion varies from quantitative articulated in monetary terms to qualitative ones focusing on company's image (Humphreys et al. 2003). Noci (1997) developed an assessment model for supplier selection based on environmental criteria and identified green competencies, environmental efficiency, supplier's green image, and net life cycle cost as the criteria. Handfield et al. (2002) listed criteria for environmentally conscious purchasing and broadly grouped into product attributes, waste management, labeling/certification, packaging/reverse logistics, compliance to government regulations, and environmental programs at supplier facilities. In literature, most of these criteria are narrowed down under broad categories to facilitate the ranking process. Compared to other industries, literature focusing on supplier selection in the apparel industry based on the green criteria is limited and focused on a limited number of criteria. The criteria that are specific to apparel manufacturers are listed in Table 18.4.

18.4.2.3 Social Criteria

Several recent disasters in the garment industry are related to human rights violation, long working hours, child labor, discrimination, and lack of health and safety standards (OXFAM 2013). In the context of developing nations, it is very common to see organization practices such as bribery, excessive gift giving, and unethical marketing to gain advantage over competitors (Baskaran et al. 2011). Despite the long history of SR, application of social concepts has only emerged in recent years (Ciliberti et al. 2011). Carter and Jennings (2002) were one of the first to define purchasing SR as an umbrella term with diversity, environment, human rights, philanthropy/community, and safety dimensions. Since then, these dimensions are used for socially responsible supplier selection.

TABLE 18.4
Studies Using Environmental Criteria in Apparel Supply Chains

Criteria	Subcriteria	References
Pollution emissions	• Air emissions • Waste water • Solid wastes • Costs of waste and pollution treatment • Use of harmful materials	Baskaran et al. (2012), Cervellon and Wernerfelt (2012), Lo et al. (2012)
Environmental management systems	• Environmental policies and planning • ISO 14001 certification • Management commitment • Reverse logistics	Baskaran et al. (2012), Lo et al. (2012)
Resource consumption	• Energy consumption • Material consumption • Recycle material consumption	Baskaran et al. (2012), Cervellon and Wernerfelt (2012)

TABLE 18.5
Social Selection Criteria Used in Apparel Supply Chains

Criteria	Subcriteria	References
Human rights	• Rights of stakeholders • Interests and rights of employee	Baskaran et al. (2012), Bremer and Udovich (2001), Cooke and He (2010)
Child labor	• Identification procedure • Verification of employment	Baskaran et al. (2012), Bremer and Udovich (2001)
Discrimination	• Policies against discrimination • Procedures to prevent discrimination	Baskaran et al. (2012), Bremer and Udovich (2001), Cooke and He (2010)
Health and safety standards	• Training to handle equipment and hazardous substances • Restriction on working hours • Health checkups • Safety procedures (fire exits, fire extinguisher, emergency evacuation plans)	Baskaran et al. (2012), Bremer and Udovich (2001)
Unfair competition	• Bribery • Advertisement • Ethical violations • Nondisclosure of privacy information	Baskaran et al. (2012)

Govindan et al. (2013b) classified social measures for supplier selection into internal measures with employment practices and health and safety; and external measures with local communities influence and contractual stakeholders. Due to the prevalence of unethical workplace practices in the garment industry, supplier selection based on social criteria is a contested topic. Table 18.5 provides a summary of the list of social criteria used for apparel supplier selection.

Literature suggested that in addition to economic, environmental, and social criteria, several other factors such as purchase conditions, purchase type (e.g., routine and nonroutine orders), and location of purchasing should also be considered while selecting suppliers (Kannan and Tan 2003).

TABLE 18.6
Methods for Supplier Selection

Decision-Making Tools	Source
Analytic hierarchy process (AHP)	Sarkis and Talluri (2002), Chan et al. (2010), Handfield et al. (2002), Deng et al. (2014), Noci (1997), Lee et al. (2009), Chan and Chan (2010), Xu et al. (2013)
Analytic network process (ANP)	Zhu et al. (2010), Sarkis (1998)
Data envelopment analysis (DEA)	Kuo et al. (2010)
Importance–performance matrix (IPM)	Khan and Rahman (2014)
Case-based reasoning (CBR)	Humphreys et al. (2003)
Interpretative structural modeling (ISM)	Kannan and Haq (2007), Kannan et al. (2008), Kumar et al. (2014)
Fuzzy set theory	Chan et al. (2008), Lee et al. (2009), Govindan et al. (2013a)
Rough set theory	Bai and Sarkis (2010)

18.4.2.4 Methods for Decision Making in Supplier Selection

The extensive nature and complexity involved in the selection process makes it heavily dependent on multicriteria decision models. Numerous multicriteria decision-making tools have been developed by adopting methods from a wide range of mathematical practices and approaches (Wu et al. 2010). Recently, the supplier selection process has integrated the social and environmental dimensions with more and more intangible factors that posed challenges for assessment. To overcome the shortcomings of intangible criteria, researchers have utilized fuzzy logic (Kwong et al. 2002). Given the complexity of the real-world decision process, researchers have started integrating more than one approach to achieve a more realistic result (Kannan et al. 2013).

Linear weighting model is the most simplistic approach used for supplier selection, whereas analytic hierarchy model is considered the most advanced scoring model (Khan and Rahman 2014). Through the review of supplier selection literature, Ho et al. (2010) identified data envelopment analysis (DEA) as the most popular approach followed by the analytic hierarchy process (AHP), case-based reasoning (CBR), and analytic network process (ANP). Kannan et al. (2013) reviewed green supplier selection literature and identified AHP as the most popular method followed by ANP and DEA. Mathematical programming techniques such as linear programming, integer programming, nonlinear programming, goal programming, and multiobjective programming are also widely used in supplier selection models (Ho et al. 2010). Recently, Khan and Rahman (2014) used the importance–performance matrix for supplier selection in foreign-aid funded procurement. Table 18.6 summarizes the various methodologies adopted in supplier selection literature.

18.4.3 SUPPLIER DEVELOPMENT

The term supplier development was first introduced by Leenders in 1966 to address the efforts of manufacturers to improve supplier capabilities (Lu et al. 2012). Supplier development is defined as "any effort of buying firm with its suppliers to increase the performance and/or capabilities of the supplier and meet the buying firm's short-term and/or long-term supply needs" (Krause and Ellram 1997, p. 21). Organizations consider the supplier development as a mechanism to assist suppliers in developing their capabilities. A recent survey suggested that the top priority of the chief procurement officers is the supplier development (Checketts and Bartolini 2006). Several studies revealed that it has also gained attention from researchers as an effective mechanism for supplier relationship management (Sucky and Durst 2013).

A variety of supplier development mechanisms are available to improve the performance and capabilities of suppliers (Krause et al. 2000). Supplier development–related activities vary widely and include evaluation/assessment, feedback of evaluation, education/training, direct capital investments in the supplier, supplier dependency, and buyer dependency (Krause and Ellram 1997, Humphreys et al. 2004, Modi and Mabert 2007, Carr et al. 2008, Lu et al. 2012). Wagner and Boutellier (2003) identified three aspects that determine supplier development activities, namely, nature of the supplier (new versus existing), motivation of the customer (reactive versus proactive approaches), and role of the customer (indirect versus direct supplier development). Although supplier assessment schemes are generally applied to the entire supply base (Johnsen et al. 2008), nevertheless, it is not feasible to apply supplier development programs to the entire supply base. So a buyer needs to choose suppliers strategically to build capabilities of those suppliers that generate the highest return. Segmentation of the supply base provides the buying firm an opportunity to create value by allocating its limited human, technical, and financial resources to selective suppliers (Day et al. 2010).

Literature suggested various purchasing portfolio models to differentiate different types of suppliers (Johnsen et al. 2008). Generally, evaluation of supplier is reliant on some type of matrix-based design and hence, portfolio modeling or portfolio matrix is considered as an important assessment approach (Day et al. 2010). Kraljic (1983) is one of the first researchers who proposed purchasing portfolio matrix based on the risks and profit impact. Olsen and Ellram (1997) proposed a relationship attractiveness matrix with strength of relationship on one axis and relative supplier attractiveness on the other. Segmentation of suppliers for relationship management is often based upon the assessment criteria (Day et al. 2010). Park et al. (2010) segmented suppliers into the following four groups for development programs:

1. *Prime group*—For constructing long-term relationships; very high incentives are provided.
2. *Collaboration group*—Reinforces the collaboration between two parties and provides mutual benefits.
3. *Maintenance group*—Maintains the existing state of relationship and pursues a mutual benefit.
4. *Improvement group*—Focuses on inspecting and improving supplier activities.

What constitutes the supplier development activity differs from author to author. For example, Sucky and Durst (2013) classified activities into direct and indirect based on human and financial resources. Other studies categorized supplier development activities into reactive and proactive based on the strategic orientation of firms. Several authors have used supplier evaluation and collaboration classifications (Klassen and Vachon 2003, Lee and Klassen 2008, Large and Gimenez 2011, Gimenez and Tachizawa 2012). Notwithstanding these differences in classification, evaluation and collaboration are considered as crucial mechanisms for relationship management (Park et al. 2010). In this chapter, we apply the same approach and consider supplier evaluation and supplier collaboration as mechanisms of supplier development. Table 18.7 provides a summary of the literature referring to these mechanisms.

18.4.3.1 Supplier Evaluation

Evaluation is considered as the first step to identify actions required for improvement and collaboration is the follow-up step to make the necessary improvement (Gimenez and Tachizawa 2012). Supplier evaluation communicates customer expectations of social and environmental practices (Vachon and Klassen 2006). It also helps to identify the suppliers development needs (Gimenez and Sierra 2013) and reinforces the competitive pressure by comparing with the competitors performance (Krause et al. 2000). Supplier evaluation through formal processes, management systems, and feedback about results are the activities of the supplier assessment (Krause 1997). Feedback provides

TABLE 18.7
Studies Using Evaluation and Collaboration as Criteria for Supplier Development

Criteria	Source
Evaluation	Cao and Zhang (2011), Gimenez et al. (2012), Gimenez and Sierra (2013), Gimenez and Tachizawa (2012), Gualandris et al. (2014), Humphreys et al. (2004), Krause (1997), Krause et al. (2000), Klassen and Vachon (2003), Large and Gimenez (2011), Lu et al. (2012), Vachon (2007), Vachon and Klassen (2006, 2008)
Collaboration	Cao and Zhang (2011), Gimenez et al. (2012), Gimenez and Sierra (2013), Gimenez and Tachizawa (2013), Krause (1997), Klassen and Vachon (2003), Large and Gimenez (2011), Lu et al. (2012), Vachon (2007), Vachon and Klassen (2006, 2008)

the sense of guidance to the supplier firms on the direction of improvement (Krause et al. 2000). Information from supplier evaluation can be used for supplier segmentation and supply base reduction (Krause and Ellram 1997, Park et al. 2010).

18.4.3.2 Collaboration

In the context of supply chains, collaboration is regarded as a dominant mechanism to implement SR (Gimenez and Tachizawa 2012). According to Florida (1996), collaboration in SR is considered as the direct involvement of an organization with its suppliers and customers to plan jointly for SR management. It provides a platform for suppliers and buyers to learn from each other (Bjorkuld 2009). Collaborative activities including joint planning and knowledge sharing regarding products and processes, site visits, training and/or education, and technical assistance are suggested in the literature (Klassen and Vachon 2003). Collaboration with suppliers can be established either in product development stage or production stage. In line with previous studies, our study considers site visits, joint planning, technical assistance, and training/education as factors of collaboration.

18.4.4 CONTINUOUS IMPROVEMENT

Continuous improvement is a strategic activity that helps a company to maintain competitive position by continually achieving key performance indicators (KPIs). It is a key concept for many schools of management philosophy such as lean, quality management, and business process reengineering (Bond 1999). Since stakeholders' demand changes over time, the assessment of supplier requirements needs to be adjusted and updated. Until late 1980, supplier assessment was primarily based on cost and performance. Later the requirements for green and social aspects of business became critical. Therefore, a continuous update of criteria for supplier selection is crucial. Deming's plan–do–check–act (PDCA) virtuous cycle of improvement is a starting point for any improvement process. The PDCA cycle may be implemented as follows:

- *Plan*—To understand the current assessment criteria and systems and develop plans for improvement.
- *Do*—Introduce the improvements in the criteria or methodology on a trial basis.
- *Check*—Examine the changes and validate the results against the anticipations.
- *Action*—Standardize the changes to the assessment model on a permanent basis.

In the next section, we apply the proposed conceptual framework for supplier assessment using an Australian case study. The purpose is to analyze and assess the importance of criteria for supplier assessment using the AHP approach.

18.5 CASE STUDY

The apparel industry in Australia is dominated by the continued growing market penetration from imports. The advantages of low-cost manufacturing along with the scarcity of the skilled workforce promoted the influx of clothing imports in Australia. According to WTO statistics, the clothing imports to Australia increased from $711 million to $5839 million (an increase of over eight times) between 1990 and 2011 (WTO 2013) (see Figure 18.4). China is the leading apparel exporter to Australia with over 70% of the import share (DFAT 2004). During the past few years, Bangladesh emerged as a new clothing sourcing destination for Australia. Figure 18.5 illustrates that the clothing export from Bangladesh to Australia increased by almost tenfold between 2008–2009 and 2012–2013 (BGMEA 2013). The prevalence of issues related to work rights, child labor, and health and safety at manufacturers' facilities in developing nations raised concerns among Australian retailers. The challenge for many Australian organizations, including case company, is to assess the supplier's behavior based on social and environmental aspects.

To keep the identity of the case firm anonymous, it has been identified as "Smart Retailing Group." It is a leading Australian garment retail chain with a portfolio of 10 brands and 1030 stores operating in 14 countries. The company started in 1991 as a family-run business and now it has emerged as a global company with over 17,000 employees across all brands. The firm's business is based on a powerful idea of "delivering the right product at the right price and ensures that customers receive what they want at a great value." The Smart Retailing Group is known for its fast fashion for men, women, teenagers, and children. It focuses on factors such as fast, flexibility, and reliability so that they can be a dominant player in any chosen market. With the intense competition and rivalry among the global retailers, the Group is aiming for the title of world's fastest growing

FIGURE 18.4 Apparel imports in Australia. (From World Trade organisation [WTO], Available at: http://www.wto.org/english/res_e/statis_e/its2013_e/its13_merch_ trade_product_e.pdf.)

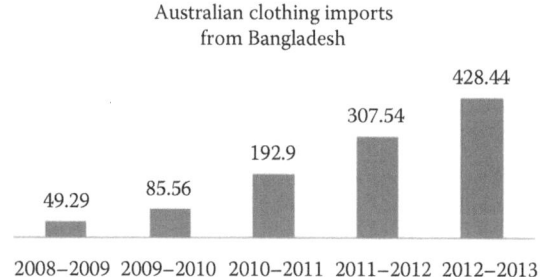

FIGURE 18.5 Bangladesh apparel exports to Australia. (From Bangladesh Garment Manufacturers Export Association [BGMEA], Available at: http://www.bgmea.com.bd/home/pages/AboutGarmentsIndustry#.UkofhzAybaM, accessed on August 10, 2013.)

retailer with a strategy of further expanding its operations into another half a dozen countries by 2014.

As of 2013, the Group outsourced most of its production to 125 factories in Asia. In the financial year 2012–2013, the Group sourced 11% of its global total purchase from Bangladesh. The Group's trust in suppliers is considered as the backbone of their success. Suppliers are considered as an extension of their workforce and the Group strives for the development of their suppliers. In order to facilitate better understanding between supply chain partners, the Group invests in long-term relationships. The Group's strategy of sourcing 65% of the products from the top 20 suppliers exhibits their commitment toward a long-term relationship.

In the wake of the recent Rana Plaza incident, the Smart Retailing Group is one of the first few Australian companies that signed the Bangladesh fire and safety accord. It has defined their role as a good corporate citizen by developing an ethical framework that provides guidelines to assess suppliers and their relationships. Their policy of zero tolerance to bribery, corruption, and forced and child labor forms the basis of assessment guidelines. The Group engages both internal and external auditors to assess the suppliers and strives for ways to continually improve the safety of the supplier's manufacturing sites. Their policy of supplier selection and development along with continuous improvement reflects the assessment framework proposed in the study.

The Group's practice of assessing an apparel manufacturer from developing nations is considered in the study. The criteria used for supplier selection can be broadly classified into economic, environmental, and social. Criteria such as supplier evaluation and supplier collaboration are considered as the development mechanism of the supplier. Figure 18.6 illustrates the hierarchical

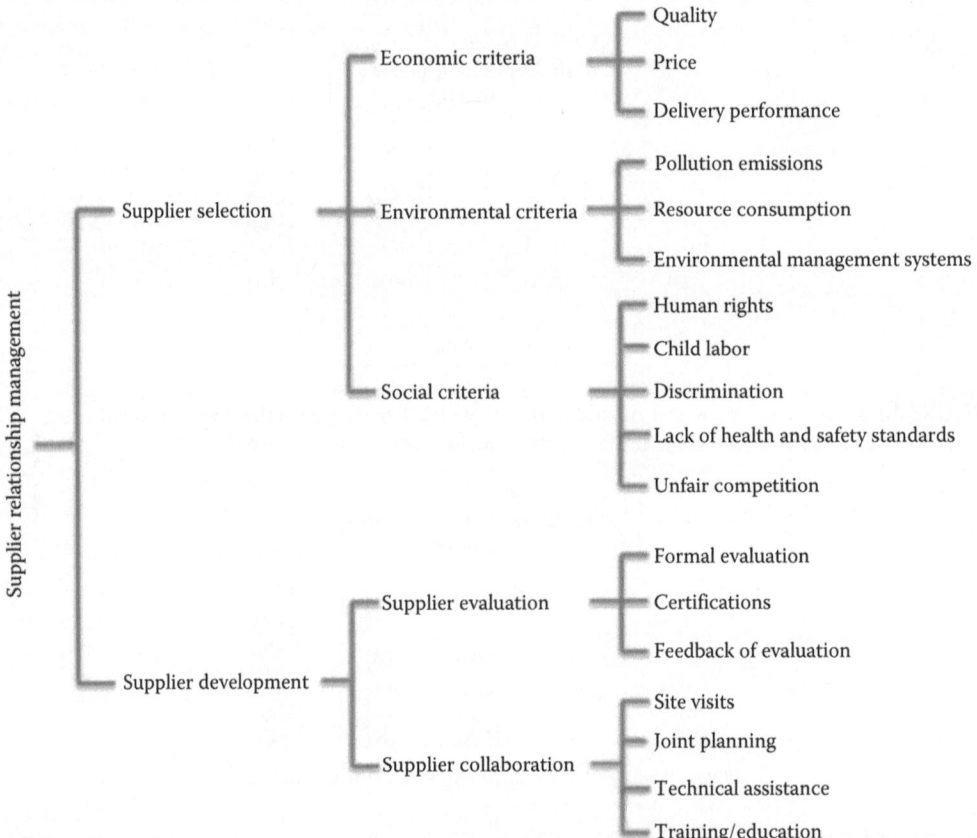

FIGURE 18.6 Hierarchical structure of supplier relationship management.

TABLE 18.8
Supplier Relationship Mechanisms and Corresponding Rankings

Mechanism	Relative Weight	Criteria	Relative Weight	Factor	Relative Weight	Ranking
Supplier selection	0.750	Economic	0.094	Quality	0.094	3
				Price	0.627	1
				Delivery performance	0.280	2
		Environmental	0.280	Pollution emissions	0.268	2
				Resource consumption	0.117	3
				Environmental management systems	0.614	1
		Social	0.627	Human rights abuse	0.112	3
				Child labor	0.549	1
				Discrimination	0.064	4
				Lack of health and safety	0.229	2
				Unfair competition	0.046	5
Supplier development	0.250	Supplier evaluation	0.800	Formal evaluation	0.818	1
				Certifications	0.091	2
				Feedback of evaluation	0.091	2
		Supplier collaboration	0.200	Site visits	0.216	2
				Joint planning	0.079	4
				Technical assistance	0.116	3
				Training/education	0.590	1

representation of the criteria used. In this study, supplier selection and supplier development together are considered as the supplier assessment. Table 18.8 provides the priority weights given to each factor with respect to the corresponding higher-level construct. Factors such as price, environment management system, and child labor are given the highest priority in the economic, environmental, and social criteria, respectively. Within the supplier development mechanism, highest priorities are given to aspects such as formal evaluation and training/education. From the analysis, it is also clear that the Group gives higher priority to supplier selection compared to supplier development. Its policy against child labor and lack of health and safety standards receives higher priority while making supplier assessment (see Figure 18.7). Smart Retailing Group's procedure of formally evaluating manufacturers is also emphasized. Overall, while assessing the suppliers the top five factors considered by the Group are child labor, environment management systems, formal evaluation, health and safety, and human right abuse.

18.6 CONCLUSION

Globalization and off-shore sourcing have significantly impacted the dynamics of business practices in the apparel industry and forced managers to reinvestigate their relationships with suppliers. The chapter provides an overview of the apparel industry in developing nations in terms of business and competitiveness. Based on an extensive literature review, this chapter develops an integrative conceptual framework for supplier assessment that consists of 2 elements, 5 criteria, and 17 factors. The proposed framework was applied in a leading Australian garment retail chain with a portfolio of 10 brands and 1030 stores operating in 14 countries. The elements, criteria, and factors of supplier assessment were subjected to analysis using the AHP. The results indicate that the Smart Retailing Group places higher priority on the supplier selection element of supplier assessment. Among the selection criteria, highest priority is given to the social aspects, and among

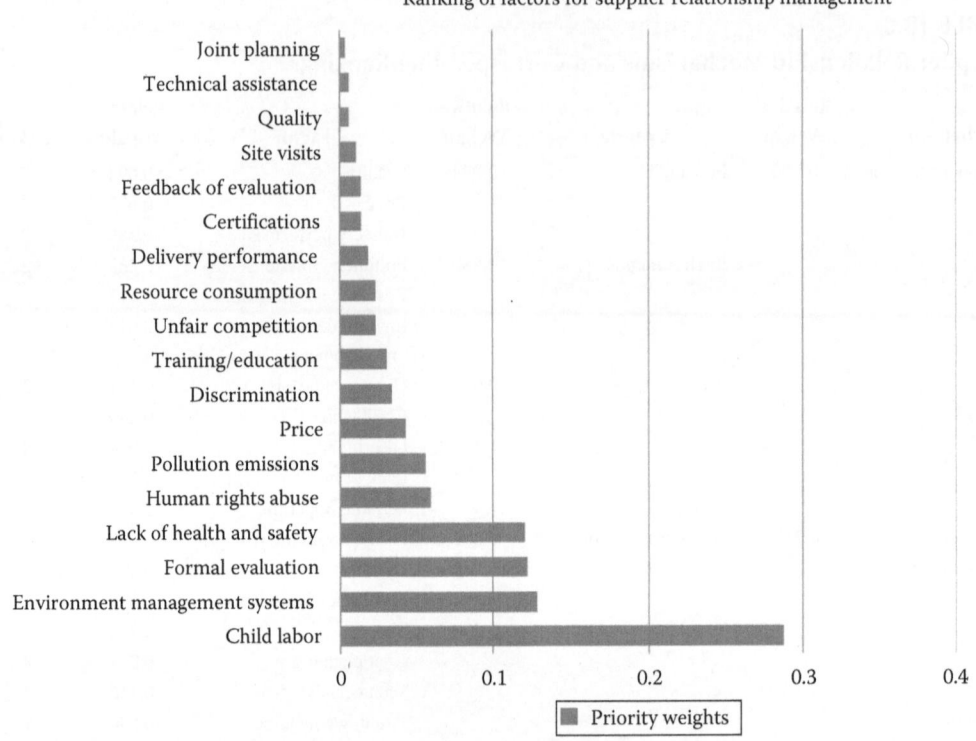

FIGURE 18.7 Overall priority ranks for supplier relationship mechanisms.

the criteria of supplier development, higher priority is given to supplier evaluation compared to supplier collaboration. Overall, top five factors considered by the Group while assessing suppliers are child labor, environment management systems, formal evaluation, health and safety, and human right abuse.

REFERENCES

Ahsan, K. and A. Azeem. Insights of apparel supply chain operations: A case study. *International Journal of Integrated Supply Management* 5(4) (2010): 322–343.

Åkesson, J., P. Jonsson, and R. Edanius-Hällås. An assessment of sourcing strategies in the Swedish apparel industry. *International Journal of Physical Distribution & Logistics Management* 37(9) (2007): 740–762.

Awaysheh, A. and R. Klassen. The impact of supply chain structure on the use of supplier socially responsible practices. *International Journal of Operations & Production Management* 30(12) (2010): 1246–1268.

Bai, C. and J. Sarkis. Green supplier development: Analytical evaluation using rough set theory. *Journal of Cleaner Production* 18(12) (2010): 1200–1210.

Bangladesh Garment Manufacturers Export Association (BGMEA). Available at: http://www.bgmea.com.bd/home/pages/AboutGarmentsIndustry#.UkofhzAybaM, accessed on August 10, 2013, 2013.

Baskaran, V., S. Nachiappan, and S. Rahman. Supplier assessment based on corporate social responsibility criteria in Indian automotive and textile industry sectors. *International Journal of Sustainable Engineering* 4(4) (2011): 359–369.

Baskaran, V., S. Nachiappan, and S. Rahman. Indian textile suppliers' sustainability evaluation using the grey approach. *International Journal of Production Economics* 135(2) (2012): 647–658.

Billesbach, T.J., A. Harrison, and S. Croom-Morgan. Supplier performance measures and practices in JIT companies in the US and UK. *International Journal of Purchasing and Materials Management*, 21(4) (1991): 24–28.

Björklund, M. Benchmarking tool for improved corporate social responsibility in purchasing. *Benchmarking: An International Journal* 17(3) (2010): 340–362.

Bond, T.C. The role of performance measurement in continuous improvement. *International Journal of Operations & Production Management* 19(12) (1999): 1318–1334.

Bremer, J. and J. Udovich. Alternative approaches to supply chain compliance monitoring. *Journal of Fashion Marketing and Management* 5(4) (2001): 333–352.

Briggs, P. Vendor assessment for partners in supply. *European Journal of Purchasing & Supply Management* 1(1) (1994): 49–59.

Bruce, M. and L. Daly. Adding value: Challenges for UK apparel supply chain management—A review. *Production Planning & Control* 22(3) (2011): 210–220.

Cao, M. and Q. Zhang. Supply chain collaboration: impact on collaborative advantage and firm performance. *Journal of Operations Management* 29(3) (2011): 163–180.

Carr, A.S., H. Kaynak, J.L. Hartley, and A. Ross. Supplier dependence: Impact on supplier's participation and performance. *International Journal of Operations & Production Management* 28(9) (2008): 899–916.

Carter, C.R. and M.M. Jennings. Logistics social responsibility: An integrative framework. *Journal of Business Logistics* 23(1) (2002): 145–180.

Cervellon, M.-C. and A.-S. Wernerfelt. Knowledge sharing among green fashion communities online: Lessons for the sustainable supply chain. *Journal of Fashion Marketing and Management* 16(2) (2012): 176–192.

Chan, F.T.S., N. Kumar, M.K. Tiwari, H.C.W. Lau, and K.L. Choy. Global supplier selection: A fuzzy-AHP approach. *International Journal of Production Research* 46(14) (2008): 3825–3857.

Chan, F.T.S. and H.K. Chan. An AHP model for selection of suppliers in the fast changing fashion market. *The International Journal of Advanced Manufacturing Technology* 51(9–12) (2010): 1195–1207.

Checketts, V. and A. Bartolini. *The CPO's Strategic Agenda: Managing People, Managing Spend.* Aberdeen Group, Boston, MA, 2006.

Ciliberti, F., J. De Haan, G. De Groot, and P. Pontrandolfo. CSR codes and the principal-agent problem in supply chains: Four case studies. *Journal of Cleaner Production* 19(8) (2011): 885–894.

Cooke, F.L. and Q. He. Corporate social responsibility and HRM in China: A study of textile and apparel enterprises. *Asia Pacific Business Review* 16(3) (2010): 355–376.

Dai, J. and J. Blackhurst. A four-phase AHP–QFD approach for supplier assessment: a sustainability perspective. *International Journal of Production Research* 50(19) (2012): 5474–5490.

Day, M., G.M. Magnan, and M.M. Moeller. Evaluating the bases of supplier segmentation: A review and taxonomy. *Industrial Marketing Management* 39(4) (2010): 625–639.

Deng, X., Y. Hu, Y. Deng, and S. Mahadevan. Supplier selection using AHP methodology extended by D numbers. *Expert Systems with Applications* 41(1) (2014): 156–167.

Department of Foreign Affairs and Trade (DFAT). Australia–China free trade agreement feasibility study. Available at: https://www.dfat.gov.au/fta/acfta/submissions/cfta_submission_4ma18.pdf, accessed on August 2014, 2004.

Dickson, G.W. An analysis of vendor selection systems and decisions. *Journal of Purchasing* 2(1) (1966): 5–17.

Ellram, L.M. A managerial guideline for the development and implementation of purchasing partnerships. *International Journal of Purchasing and Materials Management* 31(1) (1995): 9–16.

Florida, R. Lean and Green: The Move to Environmentally Conscious Manufacturing. *California Management Review* 39(1) (1996): 80–105.

Gallear, D., A. Ghobadian, and W. Chen. Corporate responsibility, supply chain partnership and performance: An empirical examination. *International Journal of Production Economics* 140(1) (2012): 83–91.

Gereffi, G. and S. Frederick. The global apparel value chain, trade and the crisis: Challenges and opportunities for developing countries. World Bank Policy Research Working Paper Series, 2010.

Gereffi, G. and O. Memedovic. *The Global Apparel Value Chain: What Prospects for Upgrading by Developing Countries.* United Nations Industrial Development Organization, Vienna, Austria, 2003.

Gimenez, C. and E.M. Tachizawa. Extending sustainability to suppliers: A systematic literature review. *Supply Chain Management: An International Journal* 17(5) (2012): 531–543.

Gimenez, C. and V. Sierra. Sustainable supply chains: Governance mechanisms to greening suppliers. *Journal of business ethics* 116(1) (2013): 189–203.

Govindan, K., R. Khodaverdi, and A. Jafarian. A fuzzy multi criteria approach for measuring sustainability performance of a supplier based on triple bottom line approach. *Journal of Cleaner Production* 47 (2013a): 345–354.

Govindan, K., S. Rajendran, J. Sarkis, and P. Murugesan. Multi criteria decision making approaches for green supplier evaluation and selection: A literature review. *Journal of Cleaner Production* (2013b): 1–18.

Gualandris, J., R. Klassen, S. Vachon, and M. Kalchschmidt. Building credibility into sustainable supply chains: towards a conceptual framework. In Proceedings of 20th European Operations Management Association (EurOMA) Conference, UCD and TCD Schools of Business, Dublin. (2013): 293–303.

Handfield, R., S.V. Walton, R. Sroufe, and S.A. Melnyk. Applying environmental criteria to supplier assessment: A study in the application of the analytical hierarchy process. *European Journal of Operational Research* 141(1) (2002): 70–87.

Ho, W., X. Xu, and P.K. Dey. Multi-criteria decision making approaches for supplier evaluation and selection: A literature review. *European Journal of Operational Research* 202(1) (2010): 16–24.

Hsu, C.-W., T.-C. Kuo, S.-H. Chen, and A.H. Hu. Using DEMATEL to develop a carbon management model of supplier selection in green supply chain management. *Journal of Cleaner Production* 56 (2013): 164–172.

Humphreys, P., A. McCloskey, R. McIvor, L. Maguire, and C. Glackin. Employing dynamic fuzzy membership functions to assess environmental performance in the supplier selection process. *International Journal of Production Research* 44(12) (2006): 2379–2419.

Humphreys, P.K., W.L. Li, and L.Y. Chan. The impact of supplier development on buyer–supplier performance. *Omega* 32(2) (2004): 131–143.

Humphreys, P.K., Y.K. Wong, and F.T.S. Chan. Integrating environmental criteria into the supplier selection process. *Journal of Materials Processing Technology* 138(1) (2003): 349–356.

Ittner, Christopher D., David F. Larcker, V. Nagar, and Madhav V. Rajan. Supplier selection, monitoring practices, and firm performance. *Journal of Accounting and Public Policy* 18(3) (1999): 253–281.

Johnsen, T.E., R.E. Johnsen, and R.C. Lamming. Supply relationship evaluation: The relationship assessment process (RAP) and beyond. *European Management Journal* 26(4) (2008): 274–287.

Kahraman, C., U. Cebeci, and Z. Ulukan. Multi-criteria supplier selection using fuzzy AHP. *Logistics Information Management* 16(6) (2003): 382–394.

Kannan, D., K. Govindan, and S. Rajendran. Fuzzy Axiomatic Design approach based green supplier selection: A case study from Singapore. *Journal of Cleaner Production* (2014).

Kannan, G. and A. Noorul Haq. Analysis of interactions of criteria and sub-criteria for the selection of supplier in the built-in-order supply chain environment. *International Journal of Production Research* 45(17) (2007): 3831–3852.

Kannan, G., A. Noorul Haq, P. Sasikumar and S. Arunachalam. Analysis and selection of green suppliers using interpretative structural modelling and analytic hierarchy process. *International Journal of Management and Decision Making* 9(2) (2008): 163–182.

Kannan, V.R. and K.C. Tan. Supplier selection and assessment: Their impact on business performance. *Journal of Supply Chain Management* 38(3) (2002): 11–21.

Kannan, V.R. and K.C. Tan. Attitudes of US and European managers to supplier selection and assessment and implications for business performance. *Benchmarking: An International Journal* 10(5) (2003): 472–489.

Kannan, D., R. Khodaverdi, L. Olfat, A. Jafarian, and A. Diabat. Integrated fuzzy multi criteria decision making method and multi-objective programming approach for supplier selection and order allocation in a green supply chain. *Journal of Cleaner Production* 47 (2013): 355–367.

Khan, S. and S. Rahman. An importance-performance analysis for supplier assessment in foreign-aid funded procurement. *Benchmarking: An International Journal* 21(1) (2014): 3–28.

Klassen, R.D. and S. Vachon. Collaboration and evaluation in the supply chain: The impact on plant-level environmental investment. *Production and Operations Management* 12(3) (2003): 336–352.

Koprulu, A. and M.M. Albayrakoglu. Supply Chain Management in the Textile Industry: A Supplier Selection Model with the Analytical Hierarchy Process. ISAHP, Vina Del Mar, Chile, 2007.

Kraljic, P. Purchasing must become supply management. *Harvard Business Review* 61(5) (1983): 109–117.

Krause, D.R. Supplier development: Current practices and outcomes. *International Journal of Purchasing and Materials Management* 33(1) (1997): 12–19.

Krause, D.R. and L.M. Ellram. Critical elements of supplier development: The buying-firm perspective. *European Journal of Purchasing & Supply Management* 3(1) (1997): 21–31.

Krause, D.R., T.V. Scannell, and R.J. Calantone. A structural analysis of the effectiveness of buying firms' strategies to improve supplier performance. *Decision Sciences* 31(1) (2000): 33–55.

Krause, D.R., S. Vachon, and R.D. Klassen. Special topic forum on sustainable supply chain management: Introduction and reflections on the role of purchasing management. *Journal of Supply Chain Management* 45(4) (2009): 18–25.

Kumar, D. Thresh, M. Palaniappan, D. Kannan, and K. Madan Shankar. Analyzing the CSR issues behind the supplier selection process using ISM approach. *Resources, Conservation and Recycling* 92 (2014): 268–278.

Kuo, R.J., Y.C. Wang, and F.C. Tien. Integration of artificial neural network and MADA methods for green supplier selection. *Journal of Cleaner Production* 18(12) (2010): 1161–1170.

Kwong, C.K., W.H. Ip, and J.W.K. Chan. Combining scoring method and fuzzy expert systems approach to supplier assessment: A case study. *Integrated Manufacturing Systems* 13(7) (2002): 512–519.

Lamming, R.C., P.D. Cousins, and D.M. Notman. Beyond vendor assessment: Relationship assessment programmes. *European Journal of Purchasing & Supply Management* 2(4) (1996): 173–181.

Large, R.O. and C.G. Thomsen. Drivers of green supply management performance: Evidence from Germany. *Journal of Purchasing and Supply Management* 17(3) (2011): 176–184.

Lee, Amy HI., H-Y. Kang, C.F Hsu, and H.C. Hung. A green supplier selection model for high-tech industry. *Expert systems with applications* 36(4) (2009): 7917–7927.

Lee, Su-Yol, and Robert D. Klassen. Drivers and Enablers That Foster Environmental Management Capabilities in Small-and Medium-Sized Suppliers in Supply Chains. *Production and Operations Management* 17(6) (2008): 573–586.

Lehmann, D.R. and J. O'Shaughnessy. Decision criteria used in buying different categories of products. *Journal of Purchasing and Materials Management* 18(1) (1982): 9–14.

Liao, C.-N. and H.-P. Kao. An integrated fuzzy TOPSIS and MCGP approach to supplier selection in supply chain management. *Expert Systems with Applications* 38(9) (2011): 10803–10811.

Liu, J., F-Y. Ding, and V. Lall. Using data envelopment analysis to compare suppliers for supplier selection and performance improvement. *Supply Chain Management: An International Journal* 5(3) (2000): 143–150.

Liu, Fuh-Hwa F. and H.L. Hai. The voting analytic hierarchy process method for selecting supplier. *International Journal of Production Economics* 97(3) (2005): 308–317.

Lo, C.K.Y., A.C.L. Yeung, and T.C.E. Cheng. The impact of environmental management systems on financial performance in fashion and textiles industries. *International Journal of Production Economics* 135(2) (2012): 561–567.

Lund-Thomsen, P. and A. Lindgreen. Corporate social responsibility in global value chains: Where are we now and where are we going? *Journal of Business Ethics* (2013): 1–12.

Lu, Louis YY., C.H. Wu, and T-C. Kuo. Environmental principles applicable to green supplier evaluation by using multi-objective decision analysis. *International Journal of Production Research* 45(18-19) (2007): 4317–4331.

Lu, R.X.A., P.K.C. Lee, and T.C.E. Cheng. Socially responsible supplier development: Construct development and measurement validation. *International Journal of Production Economics* 140(1) (2012): 160–167.

Modi, S.B. and V.A. Mabert. Supplier development: Improving supplier performance through knowledge transfer. *Journal of Operations Management* 25(1) (2007): 42–64.

Noci, G. Designing 'green' vendor rating systems for the assessment of a supplier's environmental performance. *European Journal of Purchasing & Supply Management* 3(2) (1997): 103–114.

Olsen, R.F. and Lisa M. Ellram. A portfolio approach to supplier relationships. *Industrial marketing management* 26(2) (1997): 101–113.

OXFAM. Available at https://www.oxfam.org.au/my/act/bangladesh-fire-and-safety-accord/, accessed on August 6, 2014, 2013.

Park, J., K. Shin, T.-W. Chang, and J. Park. An integrative framework for supplier relationship management. *Industrial Management & Data Systems* 110(4) (2010): 495–515.

Park-Poaps, H. and K. Rees. Stakeholder forces of socially responsible supply chain management orientation. *Journal of Business Ethics* 92(2) (2010): 305–322.

Paulraj, A. Understanding the relationships between internal resources and capabilities, sustainable supply chain management and organizational sustainability. *Journal of Supply Chain Management* 47(1) (2011): 19–37.

Perry, P. and N. Towers. Conceptual framework development: CSR implementation in fashion supply chains. *International Journal of Physical Distribution & Logistics Management* 43(5/6) (2013): 478–501.

Ruwanpura, K.N. and N. Wrigley. The costs of compliance? Views of Sri Lankan apparel manufacturers in times of global economic crisis. *Journal of Economic Geography* 11(6) (2011): 1031–1049.

Sarkis, J. and S. Talluri. A model for strategic supplier selection. *Journal of supply chain management* 38(4) (2002): 18–28.

Sarkis, J. Evaluating environmentally conscious business practices. *European journal of operational research* 107(1) (1998): 159–174.

Shaw, K., S. Ravi, S.Y. Surendra, and L.S. Thakur. Supplier selection using fuzzy AHP and fuzzy multi-objective linear programming for developing low carbon supply chain. *Expert Systems with Applications* 39(9) (2012): 8182–8192.

Sibley S.D. How interfacing departments rate vendors. *Journal of Purchasing and Materials Management* (1978) Summer.

Simpson, Penny M., Judy A. Siguaw, and Susan C. White. Measuring the performance of suppliers: An analysis of evaluation processes. *Journal of Supply Chain Management* 38(4) (2002): 29–41.

Singleton, J. *Singleton the World Textile Industry*. Routledge, London, U.K.,1997.

Staritz, C. *Making the Cut? Low Income Countries and the Global Clothing Value Chain in a Post-Quota and Post-Crisis World*. The World Bank, Washington, DC, 2012.

Sucky, E. and S.M. Durst. Supplier development: Current status of empirical research. *International Journal of Procurement Management* 6(1) (2013): 92–127.

Tan, K.C. and J.D. Wisner. A study of operations management constructs and their relationships. *International Journal of Operations & Production Management* 23(11) (2003): 1300–1325.

Thaver, I. and A. Wilcock. Identification of overseas vendor selection criteria used by Canadian apparel buyers: Is ISO 9000 relevant? *Journal of Fashion Marketing and Management* 10(1) (2006): 56–70.

Tsai, W-H, and Shih-Jieh Hung. A fuzzy goal programming approach for green supply chain optimisation under activity-based costing and performance evaluation with a value-chain structure. *International Journal of Production Research* 47(18) (2009): 4991–5017.

Vachon, S. and R.D. Klassen. Extending green practices across the supply chain: The impact of upstream and downstream integration. *International Journal of Operations & Production Management* 26(7) (2006): 795–821.

Vachon, S. and Robert D. Klassen. Environmental management and manufacturing performance: The role of collaboration in the supply chain. *International journal of production economics* 111(2) (2008): 299–315.

Vachon, S. Green supply chain practices and the selection of environmental technologies. *International Journal of Production Research* 45(18–19) (2007): 4357–4379.

Wagner, S.M. and R. Boutellier. Capabilities for managing a portfolio of supplier relationships. *Business Horizons* 45(6) (2002): 79–88.

Walton, Steve V., Robert B. Handfield, and Steven A. Melnyk. The green supply chain: Integrating suppliers into environmental management processes. *International Journal of Purchasing and Materials Management* 34(1) (1998): 2–11.

Watts, C.A. and C.K. Hahn. Supplier development programs: An empirical analysis. *International Journal of Purchasing and Materials Management* 29(1) (1993): 10–17.

Weber, C.A., John R. Current, and W.C. Benton. Vendor selection criteria and methods. *European journal of operational research* 50(1) (1991): 2–18.

World Trade Organization (WTO). Available at: http://www.wto.org/english/res_e/statis_e/its2012_e/its12_highlights2_e.pdf, accessed on August 6, 2013, 2012.

WTO (World Trade Organization). 2013. Available at: HYPERLINK "http://www.wto.org/english/res_e/statis_e/its2013_e/its13_merch_trade_product_e.pdf" http://www.wto.org/english/res_e/statis_e/its2013_e/its13_merch_trade_product_e.pdf, (accessed on January 12, 2015.)

Wu, C. and D. Barnes. A literature review of decision-making models and approaches for partner selection in agile supply chains. *Journal of Purchasing and Supply Management* 17(4) (2011): 256–274.

Xu, L., D. Thresh Kumar, K. Madan Shankar, D. Kannan, and G. Chen. Analyzing criteria and sub-criteria for the corporate social responsibility-based supplier selection process using AHP. *The International Journal of Advanced Manufacturing Technology* 68(1–4) (2013): 907–916.

Zhu, Q. and J. Sarkis. The moderating effects of institutional pressures on emergent green supply chain practices and performance. *International Journal of Production Research* 45(18–19) (2007): 4333–4355.

Zhu, Q., Y. Dou, and J. Sarkis. A portfolio-based analysis for green supplier management using the analytical network process. *Supply Chain Management: An International Journal* 15(4) (2010): 306–319.

19 Sustainable Measures Taken by Industry Affiliates, Nonprofit Organizations, and Governmental and Educational Institutions

Thilak Vadicherla and D. Saravanan

CONTENTS

19.1	Introduction	420
	19.1.1 Impact Measures	420
	19.1.2 Sustainable (Eco-) Standards	420
19.2	Sustainable Measures: Role of Industry Affiliates	421
	19.2.1 American Apparel and Footwear Association	421
	19.2.2 Control Union	422
	19.2.3 Cotton Incorporated	422
	19.2.4 European Outdoor Group	422
	19.2.5 Flo-Cert	423
	19.2.6 International Wool Textile Organization	423
	19.2.7 Oeko-Tex	423
	19.2.8 Bluesign® Technologies	424
	19.2.9 Cradle to Cradle Products Innovation Institute	424
	19.2.10 GreenEarth® Cleaning	425
	19.2.11 Zero Discharge of Hazardous Chemicals	425
19.3	Sustainability Measures: Role of NPO and Governmental and Educational Institutions	427
	19.3.1 Aid by Trade Foundation	427
	19.3.2 Better Cotton Initiative	428
	19.3.3 Fairtrade Foundation	428
	19.3.4 Made-By®	429
	19.3.5 Textile Exchange	430
	19.3.6 U.S. Environmental Protection Agency	430
	19.3.7 U.K. Department for Environment, Food and Rural Affairs	432
	19.3.8 National Resource Defense Council	432
	19.3.9 Sustainable Apparel Coalition	433
	19.3.9.1 Higg Index	433
19.4	Summary	434
References		436

19.1 INTRODUCTION

Sustainability is not a problem to be solved; it is a future to be created.

—**Peter Senger, Leadership and Sustainability at MIT**

Everyone accepts that sustainable practices have to be created, but the onus of creating those sustainable practices remains largely unaddressed. Should it be associated with the manufacturers or brands or retailers or consumer or the people on the planet? A holistic view envisages that not only the manufacturers or brands or retailers have a role to play in sustainability but also a greater and vital role lies in the hands of industry affiliates, nonprofit organizations (NPOs), and governmental and educational institutions to take the lead and make the change possible. However, in reality, everyone needs to contribute to the sustainable practices. Starting from the farmers, cultivating the fibers or manufacturers who manufacture the fibers to consumers who play a vital role in providing the raw materials for recycling should be involved actively in the sustainability cycle. It is interesting to note that many industry affiliates, nongovernmental organization (NGO) or NPO, and governmental institutions have taken the lead in ensuring the sustainability in the textile value chain through policies, regulatory issues, strategic processes, audits, avoiding certain chemicals in the production, benchmarking the products and processes, adopting the best practices, ranking systems, and certification processes. Though many of these agencies started the operation in a small province, their efforts are visible across the globe, today. The role of all these agencies will be best understood with the understanding of *impact measures and sustainable (eco-) standards* and the kind of support these organizations offer to the public.

19.1.1 IMPACT MEASURES

Impact measure is the term associated with the materials/resource that indicates the amount of impact the material/process/resource has on the environment. Impact measures for each and every material or resources have to be available across a variety of platforms. Unless and until these data are available, it is difficult for the designer/manufacture what certain material/process/resource to be used. All the stakeholders involved with the product are expected to have the awareness on the basic terminology available with regard to sustainability. The terminology includes carbon footprint, carbon neutral, carbon credit, water footprint, and life-cycle analysis.

A carbon footprint [1] is *the total sets of greenhouse gas emissions caused by an organization, event, product, or person*. It is a measure of the total amount of carbon dioxide (CO_2) and methane (CH_4) emissions of a defined population, system, or activity. Carbon footprint is calculated as carbon dioxide equivalent (CO_2e). Carbon neutral [2] may be defined as the balancing of measured amount of carbon dioxide released with an equivalent amount sequestered or offset, or buying enough carbon credits to make up the difference. Carbon credit [3] indicates a tradable certificate that permits the right to emit one tonne of carbon dioxide or the mass of another greenhouse gas with a carbon dioxide equivalent (CO_2e) equivalent to one tonne of carbon dioxide. Water footprint [4] can be defined as the total volume of freshwater used to produce the goods/services consumed by the individual/community/business. Utilization of water is measured in water volume consumed (evaporated) and/or polluted per unit of time. Life-cycle assessment (LCA) is a technique to assess environmental impacts associated with all the stages of a product's life from cradle to grave (i.e., from raw material extraction through materials processing, manufacture, distribution, use, repair and maintenance, and disposal or recycling) [5].

19.1.2 SUSTAINABLE (ECO-) STANDARDS

Private organizations, NGOs, NPOs, educational institutions, and governmental institutions and industries are promoting sustainable/eco-design standards and the awareness on these standards among the stakeholders involved with the products/services. Certain mandatory compliances imposed by various countries include EU Flower, Green Mark (China, Taiwan Province), Eco Mark (Japan, India, Africa), Ecoseal (the Netherlands), Ekolabel (Indonesia), Thai Green Label (Thailand), Korean Eco-label (Korea), Austrian Ecolabel (Austria), Good *Environmental Choice Australia* (Australia/New Zealand), Environmentally Friendly Label (Croatia), and Milieukeur (the Dutch environmental quality label); compliance with third-party proprietary standards include Oeko-Tex® and AKN; and compliance with proprietary company standards include Coop Naturaline, Hessnatur, Green Cotton, Steilmann, and Ottoversand. Environmental management systems (EMS) followed by apparel manufacturing companies include international systems like ISO 14000, European Management and Audit Scheme (EMAS), REACH, Oeko-Tex (Hohenstein Institute, Germany), and certain noncertified systems [6]. Social accountability that supports sustainability issues includes International Labor Organization (ILO), Social Accountability (SA) 8000, Worldwide Responsible Apparel Production (WRAP), Fairtrade International (FLO), and the campaigns Clean Clothes Campaign, Let's Stitch Together, and Better Cotton Initiative (BCI).

A variety of roles like compliance monitoring, trademark and labeling, benchmarks and standards certification, assessment tools and services, certification process, responsible production method, surveys, workshops, best practices, LCA models and farm research, input stream management, quality categories and awards, license, environmental performance, and chemical safety are being handled by the industry affiliates, NPOs, and governmental and educational institutions.

19.2 SUSTAINABLE MEASURES: ROLE OF INDUSTRY AFFILIATES

19.2.1 AMERICAN APPAREL AND FOOTWEAR ASSOCIATION

The American Apparel and Footwear Association (AAFA) that represents more than 1000 world-renowned brands has the trusted public policy and reflects the voice of the apparel and footwear industry and its management and contribute to more than USD 350 billion annually in U.S. retail sales. AAFA delivers a unified voice on key legislative and regulatory issues related to apparel and footwear industry including sustainable issues [7].

AAFA advocates and offers a wide variety of tools, services, and resources in bringing the awareness and implementing the sustainability aspects related to the apparel and footwear industry. Following are the some of the tools suggested for the implementation, while considering the sustainability issues [8]:

- Business intelligence
- Publications and statistics
- Trade agreement tools and resources
- Corporate social responsibility
- Partnership with industries [9]
- Country profiles
- Restricted substances lists and their updates
- Reasonable testing guidance
- Voluntary Product Environmental Profile (VPEP)—a web-based collaboration solution designed to exchange important environmental and chemical management information through a voluntary profile along the supply chain
- State-specific resources
- Drawstrings
- Labeling
- U.S. labeling resources
- Radio-frequency identification
- AAFA-endorsed colleges and universities

19.2.2 CONTROL UNION

Control Union offers independent inspection services, certification, and collateral services on a global level and offers two certification programs, namely, Global Organic Textile Standard (GOTS) and Organic Exchange (OE) certification, for sustainable textile production activities [10]. It also offers chemical input assessment, a conformity assessment procedure to the sustainable textile programs meant for the specific dyes and processing aids used in the production of textile and textile products.

Sustainable fiber program (SFP) of Control Union aims to safeguard the livelihood and well-being of the people by reducing poverty and, at the same time, minimizing the impacts on the environment in various activities of fashion and textile industries. SFP also offers innovative and economically viable opportunities for plant growers, employees, consumers, policymakers, and others involved in the value chain. Procedures of sustainable fiber certification ensure sustainability in terms of environmental protection, social equity, and economic viability and cover all the natural fibers. The standard focuses on the adoption of the best practices in agriculture and animal farming, lists of approved chemicals with the recommended dosages, training and emphasis on sustainable productivity from fiber to retail preferably using traceability software such as *Trace N Trade*, and laboratory analyses.

The standard addresses the social issues like the welfare of the farming community and fair pricing system; environmental concerns including greenhouse gases, conservation of the water and energy for a safe and hygienic, better-working environment, and profit in a fair and transparent manner are also considered.

19.2.3 COTTON INCORPORATED

Cotton Incorporated was established in the year 1970 through the Cotton Research and Promotion Act of 1966 in the United States [11], with an aim to increase the presence, demand for, and profitability of cotton textiles through research and promotion. It ensures the assessment of every bale of cotton produced in the United States, together with USDA.

Cotton LEADS™ is a program committed to responsible cotton production, consistent with the best sustainability practices and traceability in the supply chain. It has attracted many manufacturers, brands, and retailers in Australia and the United States to implement a reliable supply chain solutions and confidence that the raw material, cotton, is responsibly produced and identified. This program shares five core principles that include commitment, recognition, understanding, belief, and confidence for the common benefit of manufacturers, brands, and retailers.

19.2.4 EUROPEAN OUTDOOR GROUP

The usage of textiles in outdoor applications has undergone a phenomenal growth, and the outdoor activities have become one of the major contributors in the sporting goods business. The European Outdoor Group (EOG) undertakes numerous efforts to benefit the European industry through market surveys, workshops and forums, trade shows, cooperation with trade associations and decision makers, and promotion of best practices to popularize sustainable measures [12].

The Sustainability Working Group, an initiative of EOG, started in the year 2007, promotes and adopts the best standards of sustainability and incorporates comprehensive life-cycle perspectives with the key focus areas of

- Improving the ecological/environmental impact
- Updating the relevant environmental, health, and safety legislations related to products and services
- Maintaining the dialogues among the relevant stakeholders

- Providing the facts-based communication of sustainability credentials, certification, and labeling systems
- Addressing common issues on sustainability topics for the benefit of EU outdoor industry

The *State of Trade report*, prepared by the EOG, delivers facts about the European outdoor industry, based on wholesale figures for outdoor products, size, and scope of the markets in terms of season, country, and categories [13]. The main product categories highlighted in the report include apparels, footwear, backpacks, sleeping bags, tents, and climbing and outdoor accessories, further divided into 48 subcategories, in an online platform.

19.2.5 FLO-CERT

Flo-Cert is an approved social certification body for Fairtrade® [14] that offers solutions to supply chains challenges related to social norms, environmental demands, finance, or other risk-related issues. Flo-cert offers a wide variety of services that include (1) climate change and environmental services, novel business solutions that help the customers to measure and communicate the environmental performance of the organization or products in terms of carbon footprint (the concept of *insetting*, offsetting within the company's own supply chain, provides a measure to create a sustainable supply chain); (2) video collation, prepared by *Flo-Cert*, of the experiences of fair trade auditors and producers in the area of climate change; (3) supply chain services; and (4) fair trade services promoted by *Flo-Cert* to help implement the values like fairness and good social conditions for workers and the environment in day-to-day operations in the form of certification, training, and gap analysis.

19.2.6 INTERNATIONAL WOOL TEXTILE ORGANIZATION

Wool organizations from across the world have united to research the environmental credentials of this natural fiber, wool. The International Wool Textile Organization (IWTO) is an international and independent organization, which focuses on world wool trade regulations [19]. The IWTO has partnered with *The Woolmark* Company to produce a better-informed assessment of the environmental attributes of wool researching the world's sustainable premium fiber. Nine current LCAs of wool through various stages of its production and use have been assessed for their values and critical recommendations made in relation to the next steps toward creating a categorical study of wool's environmental credentials. The recommendations include (1) consolidating existing data and address data gaps, (2) developing globally applicable guidelines for conducting wool LCAs, (3) developing a communication strategy for the wool industry, and (4) engaging key stakeholders.

IWTO is working through a process to encourage collaboration between wool LCA researchers across the world to ensure the quick aggregation of data, allocation methods resolved, data gaps filled, and new comprehensive wool LCA data for carpet and apparel published [15]. The Woolmark Company is currently funding on-farm research in which more on-farm data are collected including data on allocation methods and greenhouse gas emissions. Further research has been initiated in the areas of the consumer phase, disposal, and recycling. Partnerships and working groups are currently being established with the relevant stakeholders within the field of apparel and carpets.

19.2.7 OEKO-TEX

Sustainable Textile Production (*STeP*) is the new certification system introduced by the Oeko-Tex for brands, retailers, and manufacturers operating in the textile chain to communicate their achievements regarding sustainable production to the public and others involved in the supply chain. The main objective of STeP is the permanent implementation of environmentally friendly production processes, optimum health and safety, and socially acceptable working conditions. Certification system encompasses production facilities of all processing stages from fiber manufacturing, spinning, weaving, and knitting to finishing facilities and manufacturers of ready-made textile items. The STeP standard provides the benchmarks and allows certified companies to continuously improve their environmental protection, social responsibility, and efficiency [16].

Oeko-Tex has also instituted the Sustainability Award under five different categories, environmental management, quality management, social responsibility, safety management, and product innovation to acknowledge the Oeko-Tex-certified companies to distinguish the efforts and special commitment to sustainability.

19.2.8 BLUESIGN® TECHNOLOGIES

Bluesign Technologies AG, one of the promoters of the sustainable textile industry, was founded in Switzerland in the year 2000 with a view of uniting the textile supply chain, eliminating restricted substances, and ensuring the responsible use of resources and safety for people and the environment. bluesign Technologies aims to link suppliers, textile manufacturers, and brands together to foster a profitable textile value chain.

Sustainability with bluesign advocates the elimination of the harmful substances from the beginning of the process through the *input stream management*, which addresses the environmental issues [17]. The input stream management suggests the preaudited components and the processes in the production that guarantee the use of sustainable ingredients in a clean process that ensures the safely manufactured products, irrespective of the number of process steps or manufacturers involved in the process. The input stream management with an appropriate network of bluesign system partners provides an unquestionable chemical expertise and implements the existing restrictions and bans.

The concept proposed by the bluesign revolves around five principles, namely, resource productivity, consumer safety, water emission, air emission, and occupational health and safety. The bluesign also proposes an evaluation and rating system for chemical components based on risk assessments and complex eco-toxicological information according to the principles of sustainability, known as homologation (Table 19.1). The criteria of the rating systems are based on the *best available technology* (BAT), and aim of the system partner criteria is the continuous development and improvement with a focus on ecological efficiency, uncompromising functionality, quality, or design. The consumer safety limits are specified in the bluesign system substances list (BSSL). Textile waste processing units are expected to adopt verifiable procedures; processes for such labeling standard and bluesign labels are issued to socially conscious versions of jackets, pants, shirts, sweaters, and accessories like hats and gloves [17].

19.2.9 CRADLE TO CRADLE PRODUCTS INNOVATION INSTITUTE

Cradle to Cradle Products Innovation Institute, headquartered in San Francisco, is an NPO that administers the publicly available Cradle to Cradle Certified™ Product Standard that aims to promote the commitment to sustainability in the manufacturing processes [18,19].

TABLE 19.1
Component Categories Suggested by bluesign

Category	Significance
Blue	Components meet all of the bluesign criteria and requirements.
Gray	Components shall only be used under certain appropriate conditions.
Black	Components do not meet the bluesign criteria.

Source: Adapted from http://www.bluesign.com/.

FIGURE 19.1 Cradle to Cradle certified product scorecard. (From http://www.c2ccertified.org/.)

The standard rates different products across five quality categories and recognizes the achievements and commitments to continuous improvement, with five different awards/levels (Figure 19.1), namely, basic, bronze, silver, gold, and platinum, based on the material health, material reutilization, renewable energy and carbon management, water stewardship, social fairness, and a final cumulative rating.

The institute administers the Cradle to Cradle Certified Product Program, which trains and accredits the assessors for the certification process. The institute has also developed an open, public database of *preferred* alternative chemicals, materials, and processes to ensure the sustainable processes and products.

19.2.10 GreenEarth® Cleaning

GreenEarth Cleaning, Kansas, was founded in the year 1999 and is the world's largest brand of environmental dry cleaning. *GreenEarth* has become synonymous with the environmentally safe dry-cleaning process that replaces petroleum-based solvents with liquid silicone, a gentle solution made from one of the Earth's safest and most abundant natural resources, silica or sand [20]. GreenEarth Cleaning focuses on issuing licenses to qualified dry cleaners and partnering with a number of leading operators, including Procter & Gamble and General Electric.

With chemical inertness and low surface tension, the liquid silicone gently penetrates the fabrics and removes the dirt, as efficient as that of peroxides or water. As it cleans the fabrics without being abrasive or aggressive, leaching dyes, or damaging trims, it is a safe and effective green cleaner for even the most delicate garments and accessories like beads, delicate lace, silk, and cashmere without affecting the dimensional stability of the materials.

19.2.11 Zero Discharge of Hazardous Chemicals

Zero Discharge of Hazardous Chemicals (ZDHC) group was founded in the year 2011 by major apparel and footwear brands and retailers with a shared commitment to help lead the industry

TABLE 19.2
Sustainability Measures of Industry Affiliates

No.	Industry Affiliate	Logo	Support Provided	Methods/ Process	Reference
1.	AAFA		Legislative and regulatory issues	Assessment tools and services	[8]
2.	Control Union		Independent assessment	Certification process	[10]
3.	Cotton Incorporated		Promotion of cotton fiber usage	Responsible production method	[11]
4.	EOG		Practices for outdoor products	Surveys, workshops, best practices	[12]
5.	Flo-Cert		Fair trade	Assessment and certification	[14]
6.	IWTO		Popularizing environmental credentials of wool fiber	LCA models and farm research	[15]
7.	Oeko-Tex		Processing audit	Certification and awards	[16]
8.	bluesign		Preaudited products and processes	Input stream management and BS certification	[17]
9.	Cradle to Cradle Products Innovation Institute		Training and assessment	Quality categories and awards	[18]
10.	GreenEarth Cleaning		Sustainability in dry cleaning	GEC license	[20]
11.	ZDHC		Elimination of hazardous chemicals	Environmental performance and chemical safety	[21]

toward zero discharge of hazardous chemicals by the year 2020. adidas Group, C&A, Esprit, G-Star Raw, H&M, Inditex, Jack Wolfskin, Levi Strauss and Co., Li Ning, M&S, New Balance Athletic Shoe, Inc., Nike, Inc., and PUMA SE are the current signatory members of this ZDHC group. The Joint Roadmap, a ZDHC program, is a highly ambitious plan that aims to set a new standard of environmental performance for the global apparel and footwear industry [21]. This group has published so far two joint roadmaps, version one in November 2011 and version two in June 2013. This group works on the key principle of *transparency*, through which the group of brands report regularly and publicly on the progress against the published Joint Roadmap timeline (quarterly in 2012, annually from 2013 to 2020). The mission of the ZDHC program is to transform the global apparel and footwear industry by improving environmental performance and chemical safety, thereby delivering a safer and cleaner environment as we work toward zero discharge of hazardous chemicals in the life cycle of all products by 2020.

By implementing tasks in seven workstreams defined in the roadmap, specifically chemical hazard assessment, prioritization and action: training; right to know; assessment and auditing; management systems approach, structure, and documentation; stakeholder partnering; and chemicals management best practices pilot, the ZDHC group will develop and promote industry best practices to deliver a safer and cleaner environment. The workstreams assist in measuring short- and long-term goals in addition to interim 2015 goals.

Table 19.2 gives a summary of the sustainability measures taken by the various industry affiliates, as discussed earlier, for the benefit of the readers.

19.3 SUSTAINABILITY MEASURES: ROLE OF NPO AND GOVERNMENTAL AND EDUCATIONAL INSTITUTIONS

19.3.1 AID BY TRADE FOUNDATION

The philosophy of the *Aid by Trade* Foundation is *social business* and offering help by means of commercial activities with the win–win situation for clothing suppliers and cotton farmers. The intriguing aspect of this initiative is that the African farmers are the partners of this foundation and the foundation works with *Cotton Made in Africa (CmiA) Initiative*, a PPP model, for the improvement of conditions of life of the African cotton farmers. The Aid by Trade Foundation defines sustainability indicators: 3P, *profit* (*economic element*), income and assets of the farmers; *people* (*social element*), percentage of primary school-age children in school; and *planet* (*ecological element*), water use and soil fertility [22].

CmiA Initiative has a considerably smaller ecological footprint as conventionally produced cotton, that is, more than 70% less greenhouse gas emissions per kilogram of cotton. By building an alliance of international retail companies that uses sustainably produced cotton, members of the demand alliance pay license fee to the foundation. The license earnings are reinvested in the project regions like agricultural training courses for farmers and efficient and environmental-friendly methods of cultivation with the *CmiA* verification criteria. Even though cotton made in Africa is not organic cotton, cultivation is ensured through modern, efficient, growing methods, reduced pesticide use, rain-fed cultivation, and crop rotation. Compliance is regularly monitored by means of verification by independent organizations. *CmiA Initiative* exclusion criteria and other sustainable indicators are elaborated:

- *CmiA*'s exclusion criteria include slavery, human trafficking, child labor, and hazardous pesticides.
- Sustainability indicators and their compliances are assessed by a traffic light system, with the ratings *red*, *yellow*, and *green*, whereby green stands for sustainable management.
- Every 2 years, cotton companies and smallholder farmers producing cotton made in Africa are checked by independent *verification* companies.

The social, ecological, and economic values generated by *CmiA* is determined in *impact monitoring*, the long-term impact on the lives of the participating farmers' families and their local conditions. Both qualitative and quantitative methods are used to determine changes in indicators that include social aspects (percentage of children that attend school or complete at least primary school), ecological aspects (checking the quality of the farming soils), economic aspects (development of incomes of farmers' families), and food security (sufficient supply of food in terms of quality, quantity, and nutritional variety).

19.3.2 BETTER COTTON INITIATIVE

Cotton is one of the world's most widely used important natural bras from the renewable natural resource. In the year 2005, a group of organizations made an attempt to establish the ways to safeguard the future of cotton and evolved a popular slogan, *There has to be a better way*, which turned out to be *Better Cotton*. BCI stewards the global standards for *Better Cotton* and addresses everyone involved in the cotton's complex supply chain, from the farmers to the retailers so that *Better Cotton* becomes a sustainable mainstream commodity [23].

A holistic approach to sustainable cotton production that covers all three pillars of sustainability, environmental, social, and economic, is suggested through Better Cotton Standard System (BCSS). BCI systems involve six major principles and are mentioned in Table 19.3.

19.3.3 FAIRTRADE FOUNDATION

FLO sets and maintains the fair trade standards for both the producers and traders [24]. FLO is owned jointly by 21 national labeling initiatives covering 22 countries and producer networks. These networks represent the certified farmers, employees, and organizations across Asia, Africa, Latin America, and the Caribbean with the standards and supports to meet the requirements; a separate certification organization, *Flo-Cert*, regularly inspects and certifies producers against these standards. While the exporters in the producer country are certified by Flo-Cert, the importers are certified either by Flo-Cert or by local labeling initiatives. In fair trade cotton, the person at the very bottom of the textile supply chain who has grown the cotton has been paid a fair price for his or her crop. It is not the finished fabric or item of clothing but the cotton itself is Fairtrade certified.

The *FAIRTRADE Mark* products include both nonfood (cotton, cut flowers, ornamental plants, and sports balls) and food products (bananas, cocoa, coffee, cotton, dried fruit, fresh fruit and fresh vegetables, honey, juices, nuts/oil seeds and purees, quinoa, soybeans and pulses, rice, herbs and spices, sugar, tea, and wine). FLO has developed distinct standards for two major groups that relate to their different ownership structures and other characteristics like (1) small

TABLE 19.3
Principles for Sustainability Achievements

BCI Principle	Description
Production principles and criteria	Providing a global definition of better cotton through sustainable and ethical farming practices
Capacity building	Supporting and training farmers in growing BC with experienced partners
Assurance program	Regular farm assessment and measurement through various indicators
Chain of custody	Incorporates demand and supply in BC supply chain
Claims framework	Propagating through data, information and facts
Results and impact	Monitoring and evaluation mechanism to ensure the intended impacts

farmers' organizations and (2) commercial farms and other companies that permanently employ hired labor.

The generic standards apply to all producers within the FLO geographical scope regardless of the products or specific requirements for individual products covered in the product-specific Fairtrade standards. Product-specific standards include additional social, economic, and environmental criteria, related to a specific product, which must also be met over time. Trade standards and Fairtrade Foundation's standards are a set of criteria that also cover the terms of trade that traders who buy Fairtrade-certified products must comply with. They include technical requirements relating to quality, shipment conditions, terms of payment, and other commercial provisions. Salient features of the Fairtrade standards include

- Paying a price to producers that at least covers the costs of sustainable production (Fairtrade minimum price)
- Paying a premium that producers can invest in the developments (Fairtrade premium)
- Partially paying in advance, if required
- Signing contracts that allow for long-term planning and sustainable production practices
- Traceability and record keeping systems
- Subcontractors' compliance with the relevant requirements

Traders are audited against these standards to ensure the use of Fairtrade products from Fairtrade-certified producer groups and manage the systems in their factories and warehouses that enable the accuracy in the traceability.

19.3.4 Made-By®

Made-By, started in the year 2004, is a European NPO with the mission to improve the environmental and social conditions in the fashion industry with the sustainability practices. *Made-By* is known for its tools and benchmarks that include *Made-By* benchmark for social standards, *Made-By* environmental benchmark for fibers, and *Made-By* social policy for developed countries [25]. *Made-By* partner brands commit to the social and environmental progress, published annually in the form of transparent, *Made-By* scorecards, based on the benchmarks that rank and classify fibers and social compliance initiatives in terms of sustainability, making the efforts visible to the public. It also emphasizes sustainable and ethical production methods used in the supply chains. The scorecards also illustrate the percentage of the supplier's factories that are socially certified, in addition to the quantum of organic cotton and environment-friendly materials used to produce their collections.

The Benchmark for Social Standards is based on the six internationally recognized codes for monitoring and certifying improvements to employment conditions (for employees at garment manufacturing units), which include business social compliance initiative (BSCI), ethical trading initiative (ETI), fair labor association (FLA), fair wear foundation (FWF), SA8000, and WRAP. These are based on the standards of ILO conventions in relation to child labor, salary, working hours, union membership, and safety in the workplace. A special ranking is given to suppliers located in countries, which *Made-By* has deemed, as *low risk*. These countries include EU countries (excluding Romania and Bulgaria), Japan, Canada, Australia, and New Zealand. In essence, social policy distinguishes between certain developed countries, where workers' rights are adequately protected, and the rest of the world, ranked from Class A to Class C.

Environmental Benchmark for Fibers supports the brands in limiting the impact of the materials on the environment and advises on sustainable alternatives based on emissions and their use of land space, pesticides, chemicals, water, and energy. *Made-By* has ranked 28 fibers into five classifications based on six parameters with different weightages, like greenhouse gas emissions (20% weightage), level of human toxicity (20% weightage), level of eco-toxicity involved in the fiber production (20% weightage), the total energy (13.33% weightage), water input needed (13.33% weightage), and the land use required in the fiber production (13.33% weightage), due to the difficulty in measuring these impacts directly. Based on these parameters, the fibers are classified into five classifications: Class A to Class E or an additional category, *unclassified* (fibers not listed). Recently, in the year 2013, the benchmark has been updated to include three fibers, flax, bamboo lyocell, and spandex.

Made-By has developed a list of countries that can be classified as *low risk*, where production factories in these countries do not require social audits to be carried out based on the Human Development Index (HDI) of the United Nations Development Program (UNDP), members of the Organisation for Economic Co-operation and Development (OECD), members of the ILO, and country reports of the U.S. Department of State.

19.3.5 TEXTILE EXCHANGE

Textile Exchange, formerly known as Organic Exchange, is an NPO incorporated in the year 2002 with a vision of *responsible expansion of textile sustainability* across the global textile value chain. Though the Textile Exchange is headquartered in the United States, its contractors located in other countries advocate product and industry integrity, help to bring positive innovations, improve organic farmers' visibility and access to stable markets, and develop business capabilities through education and create partnerships that accelerate sustainable practices across the global textile industry through partnership programs. Textile exchange offers a wide variety of standards [26] as shown in Table 19.4.

19.3.6 U.S. ENVIRONMENTAL PROTECTION AGENCY

Environmental Protection Agency (EPA) of the United States was established as early as 1970 in response to the elevated concern about environmental pollutions and to consolidate, in one agency, a variety of federal research, monitoring, standard setting, and enforcement activities to ensure environmental protection [27]. Ever since its inception, U.S. EPA has been working toward ensuring a cleaner, healthier environment for the people. The Earth Day celebrated by the EPA in the United States in the year 1970, subsequently, has become an international event from the year 1990 onward, and more than 192 countries demonstrate support for environmental protection, every year. Notwithstanding the pressures, many communities have started celebrating *Earth Week*, an entire week of activities focused on environmental issues. EPA's themes for the better future include

- Making a visible difference in communities across the country
- Addressing climate change and improving air quality
- Taking action on toxics and chemical safety
- Protecting water
- Launching tribal and local partnerships
- Embracing EPA as a high-performing organization
- Working toward a sustainable future, incentive-based efforts to complement the base of solid regulations and reviewing of new and key existing regulations to examine sustainable enhancements, which are important actions

U.S. EPA has an action plan accomplishing the mission, based on the following strategies:

- Develop and enforce the regulations.
- Give grants to state environmental programs, nonprofits, educational institutions, community cleanups, and others.
- Study environmental issues.

TABLE 19.4
Textile Exchange: Standards, Scopes, and Certification

TE Standard	Scopes	Certification
Content Claim Standard (CCS)	Provides a tool to verify the content of specific input materials through third-party checks. It is the foundation for other standards.	
Organic Content Standard (OCS)	Third-party verification to verify a final product to assess the use of organically grown materials.	
Recycled Claim Standard (RCS)	A chain of custody standards to track recycled raw materials through the supply chain.	
Responsible Down Standard (RDS)	Allows ensuring down the products that come from ethical treatment.	
Global Recycle Standard (GRS)	Intended for making and/or selling products with recycled content, applies to the full supply chain.	
OE Standard	OE 100 and OE Blended track the purchase, handling, use of 100% certified organic cotton, and goods that contain a minimum of 5% organic cotton as blends with any fiber (replaced by OCS).	

- Sponsor partnerships in working with business, NPOs, and state and local governments in the areas like conserving water and energy, minimizing greenhouse gases, reusing solid waste, reviewing pesticide risks, and actively sharing information among the public and partners.
- Educate the people about the environment.

ENERGY STAR, a voluntary program of EPA, helps businesses and individuals save money and protect climate through superior energy efficiency [28]. Energy use in homes, buildings, and industry account for two-thirds of greenhouse gas emissions in the United States, and *ENERGY STAR* has been an instrument in reducing the energy use in order to realize significant reduction of greenhouse gas emission.

19.3.7 U.K. DEPARTMENT FOR ENVIRONMENT, FOOD AND RURAL AFFAIRS

Legislative controls on pollutant emissions from the industry in the United Kingdom originated as early as in the mid-nineteenth century. The Environmental Protection Act 1990 was the immediate action that consolidated controls on emissions to air from smaller installations and introduced integrated pollution control for the larger installations with greater pollution potential [29]. U.K. Department for Environment, Food and Rural Affairs (U.K. DEFRA) is responsible for the policy and regulations on environmental, food, and rural issues whose priorities are to develop the rural economy and suggest the improvements in environment. U.K. DEFRA works directly in England, has the administrative activities in Wales, Scotland, and Northern Ireland, and negotiates powers in the EU and internationally. U.K. DEFRA is responsible for policy and regulations on

- Natural environment, biodiversity, plants, and animals
- Sustainable development and the green economy
- Food, farming, and fisheries
- Animal health and welfare
- Environmental protection and pollution control
- Rural communities and issues

The current framework manifests the concept of BAT for dealing with potential pollutions. The *Responsible Care* initiatives of the European Chemical Industry strongly encourage the development of the BAT and the attempts to bring awareness among all the stakeholders.

The term *best available technology* provides the guidelines for sustainable manufacturing in terms of technology, processes, and chemistry as well as the recommendations. Compared to traditional approaches and processes, the *BAT* significantly helps to save resources manyfold. Low-temperature treatments, reduced water consumption, enhanced dye exhaustion and fixation, efficient washing systems, and easy care and maintenance of fabrics and garments with better fastness properties are the major focus of the *BAT*.

Besides carbon footprint and water footprint, machinery footprint is also becoming fast-accepted phenomenon, which emphasizes on versatility of the machines to process with wide range of process parameters like temperature, speed, and multipass facilities. Intelligent design of the machines allows the processing of both woven and knitted fabrics without compromising the quality of the finish imparted. BAT is not independent of product quality and must therefore be judiciously adapted in each case.

19.3.8 NATIONAL RESOURCE DEFENSE COUNCIL

National Resource Defense Council (NRDC) is the U.S.-based environmental action group that includes online activists, lawyers, scientists, and other related professionals [30] with the following priorities:

- Curbing global warming
- Creating the clean energy future
- Reviving the world's oceans
- Defending endangered wildlife and wild places
- Protecting people health by pollution prevention
- Ensuring safe and sufficient water
- Fostering sustainable communities

19.3.9 SUSTAINABLE APPAREL COALITION

The Sustainable Apparel Coalition (SAC) was founded by a group of sustainability leaders from global apparel and footwear industries with an understanding that the industry's current social and environmental challenges are both a business imperative and an opportunity [31]. The SAC seeks to identify common metrics and approaches to reduce the social and environmental impacts of apparel and footwear products.

The SAC attempts to lead the industry toward a shared vision of sustainability, formulated upon a common approach for measuring and evaluating apparel and footwear product sustainability performance that in turn leads to the priorities for action and opportunities and finally to the technological innovation. The major focus of the SAC concentrates on the Higg Index that measures the environmental impact of apparel and footwear products.

19.3.9.1 Higg Index

The Higg Index 1.0 is an indicator-based tool for footwear and apparel that enables the companies to evaluate material types, products, facilities, and processes based on a range of environmental and product design choices, validated through rigorous pilot testing. The Index gauges the environmental sustainability performance and the improvements from a set of practice-based, qualitative questions, based on the Outdoor Industry Association's Eco Index and Nike Environmental Apparel Design Tool.

In the Higg Index 1.0, materials are partially scored (50%) using the Materials Sustainability Index (MSI) that helps product teams to select materials and engage the suppliers having lower environmental impacts (high MSI scores).

The MSI is a cradle-to-gate index informed by LCA-derived inventory data that span from the origin of raw materials to a finished textile or component parts ready to be shipped, to engage designers and the global supply chain of apparel and footwear products in environmental sustainability. The MSI influences the design and make phases of the product life cycle and does not include consumer use or end-of-life or reuse phase.

The MSI balances scoring based on the combination of base material score and qualitative questions regarding material environmental attributes and supplier practices. This leads to a robust scoring framework and comprehensive materials assessments with evenly weighting four environmental impact areas across the MSI scoring framework, chemistry, energy and greenhouse gas intensity, water and land use intensity, and physical waste. The base material score examines publicly disclosed LCA studies, industry reports, and supplier information to assess the impact areas and calculate the score for every raw material, using a representative supply chain.

Higg Index 2.0 is an indicator-based assessment tool for apparel and footwear products, based largely on the Eco Index, Nike Environmental Apparel Design Tool, Global Social Compliance Program (GSCP) reference tools, Social/Labor Best Practice Tools, and intensive pilot studies. The index ascertains the environmental sustainability performance and drive behavior for improvements. The Higg Index 2.0 is a standardized tool to measure and evaluate environmental performance of apparel products across the supply chain at the brand, product, and facility levels. Higgs Index 2.0 includes three individual tools, namely, facility, brand, and product tools, as detailed in the Table 19.5.

The Higg Index 2.0 is based on life-cycle thinking and spans the apparel life cycle (materials, manufacturing, packaging, transportation, use, and end of life), but retail activities are not included. Table 19.6 gives the summary and comparison between the Higg Index 1.0 and Higg Index 2.0.

Table 19.7 summarizes the roles of various NPOs and governmental institutions in implementing the sustainability measures in manufacturing of textile materials and the methods of implementing such measures.

TABLE 19.5
Tools and Modules of Higg Index Assessment

Higg Tools	Modules	Assessment
Facility tools	Facility environmental module and facility social/labor module (Beta)	Environmental performance of material, packaging, manufacturing facilities, and their social and labor performance.
Brand tools	Brand environmental apparel and footwear modules, brand social/labor apparel/footwear module (Beta)	Product-specific environmental practices at brand levels. Beta assesses practices in relation to social and labor.
Product tools	Rapid design module (RDM, Beta)	Guidance to designers to select sustainable product design and MSI data explorer.

TABLE 19.6
Comparison of Higg Index Parameters

Parameter	Higg Index 1.0	Higg Index 2.0
User interface	Excel	Web tools and excel
Assessment	Qualitative indicators	Both qualitative and quantitative
Sustainability topics—environment	✓	✓
Sustainability topics—social/labor	×	✓
Product categories—apparel	✓	✓
Product categories—footwear	×	✓
Value chain area—Higg index tools	Brand module (environment)	Brand module (environment, social/labor)
	Facility module (environment)	Facility module (environment, social/labor)
	Product module	RDM—Beta
Validation	None	Validation protocol for environmental facility module
Materials assessment	44 materials	46 materials with updated version
Chemistry	Basic indicator questions, MSI	Chemical management module for full assessment and social/labor—chemical hazards

✓, available and ×, not available.

19.4 SUMMARY

Changing global situations, demanding pollution-free living conditions, climate change, and global warming are expected to alter the way in which the industries are operating including textile and fashion, in the near future. Sustainable future is in the minds of all the stakeholders involved in the well-being of the planet. Responding to the situation, many organizations voluntarily come forward to initiate the required measures to ensure the sustainability across the length and breadth of the manufacturing processes including cultivation of fibrous materials. Though sops are available in many occasions, the ways and means to available such things are not propagated among the manufacturers, traders, and consumers. This issue has been addressed by many industry affiliates, NGO, NPO, governmental agencies, and educational institutions with different levels of emphasis. Everyone involved in the textile value chain is motivated to adhere to sustainable practices by a variety of assessment tools, labeling and certification of processes and products, readily available best practices, awards, and benchmarks and improve the existing processes to address the three vital issues of sustainability, environment, economy, and the society. Technology will continue to transform the lives of the people

TABLE 19.7
Sustainability Measures by NPO and Other Institutions

No.	NPO/Govt. Institutions	Logo	Principle	Methods/Modes	References
1.	Aid by Trade Foundation/ *CmiA* Initiative		PPP Model initiatives	Compliance monitoring	[22]
2.	BCI		Safeguard future of cotton	Six principles of BCI	[23]
3.	Fairtrade Foundation		Standards for producers and traders	Trademark and labeling	[24]
4.	Made-By		Improving environmental/social conditions	Benchmarks and standards	[25]
5.	Textile Exchange		Responsible expansion of sustainability	Certification	[26]
6.	U.S. EPA		Environmental protection	ENERGY STAR certification	[27,28]
7.	U.K. DEFRA		Emission control	BAT	[29]
8.	NRDC		Environmental protection	Addressing priority areas	[30]
9.	SAC		Reduction of social and environmental impact	Quantitative metrics	[31]

and businesses and create new opportunities; the attitude of the people toward resource shortages, climate change, and sustainability is also expected to change in the coming days.

REFERENCES

1. http://timeforchange.org/what-is-a-carbon-footprint-definition. (accessed on April 23, 2014.)
2. http://en.wikipedia.org/wiki/Carbon_neutrality. (accessed on April 23, 2014.)
3. www.sxcs.edu.in/model-united.../Study%20Guide%20-%20UNCCC.pdf. (accessed on April 23, 2014.)
4. www.waterfootprint.org/. (accessed on April 23, 2014.)
5. http://unterm.un.org/dgaacs/unterm.nsf/8fa942046ff7601c85256983007ca4d8/331914bc525f87fd85256a01000064cc?OpenDocument. (accessed on April 5, 2014.)
6. M.I. Tobler-Rohr, *Handbook of Sustainable Textile Production*. Woodhead Publishing Limited, Cambridge, U.K., 2011.
7. https://www.wewear.org/. (accessed on April 5, 2014.)
8. www.vpepxchange.com. (accessed on April 13, 2014.)
9. www.amefird.com/wp-content/.../2013–2014-Sustainability-Report.pdf. (accessed on April 11, 2014.)
10. http://www.controlunion.com/en. (accessed on April 11, 2014.)
11. http://www.cottoninc.com/. (accessed on April 11, 2014.)
12. http://www.europeanoutdoorfacts.com. (accessed on April 11, 2014.)
13. http://www.fairtradeafrica.net/uncategorized/measuring-producers%E2%80%99-carbon-footprint-flo-cert-offers-a-new-service/. (accessed on April 11, 2014.)
14. http://www.flo-cert.net/flo-cert/. (accessed on April 16, 2014.)
15. http://www.iwto.org/. (accessed on April 16, 2014.)
16. https://www.oeko-tex.com/. (accessed on April 16, 2014.)
17. http://www.bluesign.com/. (accessed on April 16, 2014.)
18. http://www.c2ccertified.org/. (accessed on April 16, 2014.)
19. http://www.treehugger.com/corporate-responsibility/cradle-to-cradle-announces-green-products-innovation-institute.html. (accessed on April 16, 2014.)
20. http://www.greenearthcleaning.com/. (accessed on April 16, 2014.)
21. http://www.roadmaptozero.com/. (accessed on April 16, 2014.)
22. www.cotton-made-in-africa.com/. (accessed on April 16, 2014.)
23. http://bettercotton.org/. (accessed on April 16, 2014.)
24. http://www.fairtrade.org.uk. (accessed on April 16, 2014.)
25. http://www.made-by.org/. (accessed on April 16, 2014.)
26. http://textileexchange.org. (accessed on April 16, 2014.)
27. http://www.epa.gov/. (accessed on April 22, 2014.)
28. http://www.energystar.gov. (accessed on April 23, 2014.)
29. http://www.defra.gov.uk/. (accessed on April 22, 2014.)
30. http://www.nrdc.org/. (accessed on April 23, 2014.)
31. http://www.apparelcoalition.org/. (accessed on April 16, 2014.)

Section V

Sustainability and Fashion

20 Exploring a Framework for Fashion Design for Sustainability

Alison Gwilt

CONTENTS

20.1 Introduction and Context	439
20.2 Bringing Sustainability into the Fashion Design Studio	441
20.2.1 Tools and Checklists	441
20.2.2 Design Scenario	442
20.3 Sustainable Design Strategies for the Fashion Design Process	443
20.3.1 Principles of Sustainable Design Strategies	443
20.4 Introducing the Fashion Design for Sustainability Model	445
20.4.1 Background: Developing the FDS Model	445
20.4.2 Preparations for Use	445
20.4.3 Applying the FDS Model	447
20.5 Challenges of Integrating Sustainable Strategies in the Fashion Design Process	449
20.6 Conclusion	451
References	451

20.1 INTRODUCTION AND CONTEXT

The fashion designer discussed within the context of this chapter is defined as the person charged with leading the development of the new seasonal collection in a micro-, small-, or medium-sized business. In this context, the designer is typically considered to be a significant and central member of a team and is chiefly responsible for guiding the collection from early design ideas to the production of the sample range (Renfrew and Renfrew 2009). At the center of the creative process, the fashion designer is expected to direct and work with a number of specialist people and teams, with the aim of producing a collection on time, within budget and to market expectations. However, it is from this position of influence that the fashion designer can steer and guide the team in making considered decisions, which can positively affect the way that garments are designed and produced.

Understanding the role of the designer and the context within which the designer operates is fundamental in developing a framework to support during design inception. The global production of fashion involves numerous companies, manufacturers, and retailers who develop products for specific sectors and market levels. Although the sectors of the fashion industry are broad (ranging from sportswear to lingerie), a company will typically operate within one sector with corresponding womenswear, menswear, or childrenswear products (Stecker 1996). However, while the fashion industry is separated into levels and sectors, in general terms the system of design and production utilized is universal across its different levels and sectors. Well documented in design texts, this system broadly includes a range of stages and activities, which Sinha suggests can be classified into five distinct phases that comprise the research and analysis phase, the synthesis phase, the selection phase, the manufacturing phase, and the distribution phase (Sinha 2002, p. 7). Typically the fashion designer is most engaged in two of these phases, the research and analysis phase and the synthesis phase; however, the activities undertaken by the designer in these two phases may differ between companies depending largely on the scale of the business. The research and analysis phase usually involves conducting trends, marketing and design research, and designing and editing the collection. Whereas the synthesis phase involves the designer engaging with or overseeing the pattern making, toiling, and production of the sample range. Within the industry, fashion design educational institutes and fashion design texts, these activities are commonly grouped together and referred to as the fashion design process (Sorger and Udale 2006; Renfrew and Renfrew 2009). With some variation, it is reasonable to suggest that this description of activities accurately describes the current fashion design process for the majority of micro and SME business.

Across all sectors of the fashion industry, from lingerie to high-street brands to luxury labels, there is a need to develop sustainable products and services, which consider economic, environmental, social, and ethical issues throughout the product life cycle (Allwood et al. 2006; Black 2008; Fletcher 2008; Hethorn and Ulasewicz 2008; Gwilt and Rissanen 2011). In contemporary society, the obsession for consuming fashion goods has seen an enormous growth in the "fast fashion" market, a sector that is typically responsible for low-quality, trend-driven products that utilize "just in time" technology to develop and ship garments to the retail store in the quickest possible time. However, while across the industry prices for fashion goods are expected to continue to drop as competition increases, and new production technologies force a reduction in labor needs, there is mounting pressure from the consumer, alongside increasing legislation and international ethical campaigns, that is driving the demand for positive change (Allwood et al. 2006). Disappointingly, while those external to the fashion industry look for change, the industry itself appears to be responding slowly, or in some quarters not at all.

For some fashion designers, the principles of sustainability can be perceived as an obstruction to good design: a set of restrictions that can incapacitate the process of design and innovation. Rather than see sustainability as an opportunity to engage in new creative design practices, a designer with this negative perspective regards sustainability as an optional method of practice that can inhibit design choices. In part, this view may arise from a distant, yet still dominant, belief that it is only through careful material selection that a designer can make a difference. This view contributes further to the common misconception that sustainability limits creative opportunities, which has created disengagement in some areas of the fashion industry. Fashion designers who are unfamiliar of working within a sustainable framework typically share negative comments about sustainability. This point was highlighted in a 2008 report from the Centre for Sustainable Fashion; one London-based designer claimed that while the company recognized their responsibility to the environment, the designer had "...no idea of how to go about doing anything or if we can afford to spend the time and resources on this subject" (Centre for Sustainable Fashion 2008, p. 22). However, it is reasonable to suggest that fashion designers are the critical link in the chain in the development of new products and they have the ability and the opportunity to improve the environmental, ethical, and social performance of fashion garments. To produce sustainable fashion garments means changing the way that fashion designers do things, particularly since it is widely acknowledged that almost "...60–80% of the (life cycle) environmental impacts from products are determined at the design stage" (UNEP 2002, p. 11). It follows that in order to stimulate design students and designers in industry to reflect upon and improve their approach to the fashion design process, it is necessary to challenge the current model of fashion designing.

This chapter argues that the designer can make a significant contribution in developing garments that can lessen the impact on the environment, while balancing economic, social, and ethical concerns. Existing studies have revealed that within the fashion design process there is often no consideration for the life cycle of a garment, garment end-of-life strategies, or any sense of responsibility for the textile waste generated through pattern making, manufacture, or use (Black et al. 2009;

Gwilt and Rissanen 2011). The chapter explores the notion of a design practice that integrates sustainable design strategies across the life cycle of garments as a means of developing improved products/services, and introduces the fashion design for sustainability (FDS) model as a framework that aims to support the designer to make considered choices.

20.2 BRINGING SUSTAINABILITY INTO THE FASHION DESIGN STUDIO

While this chapter posits a new holistic framework for fashion design practice, there are a number of existing studies that have promoted the importance of sustainability within the fashion industry. In recent years, researchers (Black 2008, 2012; Fletcher 2008; Hethorn and Ulasewicz 2008; Gwilt and Rissanen 2011; Fletcher and Grose 2012; Gwilt 2014) have critically investigated the possibilities for sustainable design within the fashion and textiles disciplines. Although investigations often highlight positive case studies of best practice drawn from the contemporary fashion industry, it is often argued that the fashion industry needs to investigate/adopt a wider range of strategies in order to encourage the use of multiple approaches to sustainable fashion. Since Black (2008) identified a collection of sustainable strategies that demonstrated alternate methods of practice are possible, it appears that across the industry designers may be unaware that such a wide variety of design-led approaches to sustainable fashion exist. At the same time for those designers that are aware, there can be confusion or misunderstanding as to how to integrate one or more strategies within their fashion design practice. This section begins with a review of a variety of qualitative and quantitative methods and approaches that the fashion designer can engage with to improve the environmental/ethical performance of a fashion product.

20.2.1 TOOLS AND CHECKLISTS

In exploring the topic of fashion and sustainability, inevitably most research tends to draw on existing models that are used within the product design discipline (Black 2008; Fletcher 2008). Although different types of tools, models, and software programs exist to assess the environmental impact of a product, the life-cycle assessment (LCA) is an analytical model typically used to provide a quantitative measure of environmental performance. Beginning with a set brief of defined parameters that are concerned with the environmental impacts of a specific product, an LCA helps determine the scope of negative impacts associated with the life cycle of the product. The assessment usually concentrates on gathering and evaluating data concerned with the inputs, outputs, and potential environmental impacts of a product throughout its life cycle. The results are then quantified into a single figure or an index, where the environmental issues are weighted according to their performance. The results then reveal areas for improvement and at the same time can be used to compare the performance of one product with another. Although it is a useful exercise, it can be a complex process that often involves extensive data collection, which can be costly if specialists are employed to undertake the study. For the fashion industry, where a wide range of products are created each season, the process of developing a fashion garment, from design to product, may occur in a matter of weeks within the mass-manufacturing sector. In this situation, although an LCA may be a suitable option for a generic garment, for example, a blouse, a pair of jeans, or a tailored skirt (Allwood et al. 2006), conducting an LCA for each individual fashion garment in a collection may be considered difficult and time consuming. While an LCA report is likely to be comprehensive, a blouse that is assessed once may be radically redesigned for the following season, incurring new and different design features that might require changed amounts of materials; be constructed from different fabric; involve different production steps and surface treatments; and so on. This redesign then might occur three or four times a year, which in essence would require further LCA studies.

Abbreviated/simplified versions of the analytical framework often form the basis of computerized tools such as the Higg Index, EcoMetrics™ calculator, Nike Considered Design Index, and others, which have been developed to help the design and production team to reduce the negative environmental impacts associated with the production and use of a garment. These tools usually assist the designer during the production phase but while useful for the quick responses that are required in large organizations, the use of these impact calculators can lead a "novice" or unsuspecting designer to consider sustainability as an (complicated) add-on, rather than as an integral part of the fashion design process (Black et al. 2009). This could also be said of reference guides and checklists, which can be nondiscipline specific or may be biased toward one set of strategies and solutions. It is also apparent that sustainability texts and guides can date extremely quickly. For the sustainably minded designer, this is obviously a concern if he/she is to remain engaged in the debate since continuous research and the updating of personal knowledge is required to capitalize on new information as it appears. More broadly, a solution-focused approach does not challenge or encourage designers to seek alternative approaches to the creation of garments during design inception, concept, or research stages. While tools and checklists can be helpful in certain circumstances, they can at the same time be a distraction from the fundamental problem, which is that the current fashion design paradigm does not encourage a philosophical approach to the integration of sustainability in the fashion design process.

20.2.2 DESIGN SCENARIO

Rather than becoming reliant on solution-focused texts, guides, and tools, an alternative approach is to focus on revisiting the way that the designer engages in the design process. This moves the notion of sustainable design from a "results"-based paradigm toward a philosophical viewpoint, which suggests that a rethinking of the framework under which designing takes place needs to occur. This meta-level, conceptual approach can be witnessed in the nondiscipline-specific design guide produced by Edwin Datschefski (2001). Datschefski's model was created to aid designers in the development of products that were "…100 per cent cyclic, solar and safe" (2001, p. 154). The framework is based on the employment of five basic principles in the design and manufacture process: cyclic, solar, efficient, safe, and social. Developed so that a designer could quickly assess a product for its environmental and social performance, the redesign of a product using this system would address the five principles to help effectively improve a product's "sustainable" performance. While acknowledging the framework is a guide (requiring some data collection), and not a replacement for a thorough LCA, Datschefski's model has one crucial advantage over other examples: it is a simple visual tool that designers can refer to and apply with some ease. Although it was not designed to replace other more complex assessment tools, as a directional guide it offers a methodology that can be applied to a broad range of design disciplines.

Alongside models for designing that are dependent upon LCAs, data collection, or texts and guides, the notion of the "design scenario" has been advocated as a means to engage with the theories and principles of sustainable design within design practice (Bras-Klapwijk and Knot 2001; Manzini and Jégou 2003; Fry 2009). The design scenario allows the designer to explore a specific current issue or problem, or propose a future prototype, by developing a response to a defined narrative or hypothesis. According to Thorpe, this provides the designer with the opportunity to think beyond the present situation and "…avoid the trap of short termism" (2007, p. 166). For example, within fashion design this extended "design thinking" (moving beyond the focus of designing a single item or look) could allow the designer to develop a response to a narrative that may, for example, be based on a wearer in a particular context or environment. However, these extended design scenarios should still have a clear motivation or goal and a defined method of implementation. Once defined, these scenarios can be used by a single designer or by a small group, which might provide a range of different outcomes from the same set of criteria. Manzini and Jégou (2000) advocate a design orientating scenario (DOS) that is centered on the changes in a product system rather than on policy or political intervention. However, some critics see design scenarios as forms of "idealized fictions" (Fry 2009, p. 155). While the DOS model can be an enabler to rethink design practice, it may not be considered a guide that easily steers the fashion designer toward an alternate model of fashion practice, or be a tool to reference daily in an industry context. The DOS relies on developing a response to a situation, rather than changing the fashion design process itself. Fry argues instead for a scenario of design, which aims to explore how design practice itself can change rather than focusing on the "...desired destination..." (Fry 2009, p. 154). Nevertheless, the use of a scenario in fashion design can provide the designer with the means to reflect on the current conventional model of fashion designing, and in this context can be useful in allowing the designer to see new ways of engaging in sustainable design strategies.

20.3 SUSTAINABLE DESIGN STRATEGIES FOR THE FASHION DESIGN PROCESS

An alternative approach for the fashion designer would be to work within a framework that promotes an engagement with sustainable strategies for fashion design for sustainability. Rather than relying on responding to detailed LCA reports, there is an opportunity to encourage behavioral change that would enable the fashion designer to approach design practice from a new perspective. Through a new framework for design that supports the integration of sustainable strategies within the design process, the fashion designer may be able to identify negative impacts and employ appropriate strategies, according to a proposed garment life cycle. However, it is imperative that the designer is informed of the possible sustainable strategies that are available for fashion design. This section introduces the function of sustainable design strategies and their application within fashion design practice.

20.3.1 PRINCIPLES OF SUSTAINABLE DESIGN STRATEGIES

A sustainable design strategy is considered a framed approach that is typically employed by a designer to minimize or avoid specific environmental and/or social impacts associated with the life cycle of a product (Gwilt 2014). Research relating to the use of sustainable strategies within design practice initially appeared within the context of the Industrial design discipline, and in recent years, a number of the strategies have been explored and adopted within fashion design practice. Although sustainable design strategies have appeared on the landscape across the design disciplines—from product design to architecture, textiles to fashion—the terminology used to describe the different strategies can vary. This complex terrain of differing terms and language has perhaps created a barrier to engagement for fashion designers, and a process of education, and possibly consistency across the disciplines, is required to communicate common aims and objectives. At the same time, new approaches continue to emerge as manufacturing technologies, social attitudes and practices, and economic factors change, which increases the ongoing collection of strategies.

There are numerous strategies available for use in fashion design practice, and through the research activities conducted in academic institutes and in the industry new strategies are frequently being identified and trialed. Within the fashion industry there are brands and labels that have been keen to adopt approaches from *slow fashion*, where the focus is placed on producing garments that are durable and/or designed for an extended life cycle, to *upcycling* that places an emphasis on adding value to discarded or damaged garments or materials so they may become wearable again. However, the sustainable design strategies that are applicable to the design and production of a garment generally intend to meet one or more of the following principles (Crul and Diehl 2006):

- Minimize the consumption of resources: aim to reduce material and energy/water consumption across the life cycle of a fashion garment
- Choose low-impact processes and resources: aim to reduce the use of pollutant/harmful materials, production and distribution processes; use energy/water efficient processes and materials

- Improve production techniques: aim to reduce production steps and waste; adopt lower and cleaner energy and material use; work with ethical/fairtrade manufacturers
- Improve distribution systems: aim to use less/cleaner packaging; use efficient transport and logistics; use local suppliers
- Reduce the impacts created during use: aim to use less/cleaner energy and water during laundering and care
- Improve product lifetime: aim to develop durable garments; improve repair and maintenance opportunities
- Improve the use of end-of-life systems: aim to employ reuse, remanufacture or recycling strategies; use closed loop systems

Each principle is typically addressed through the use of an ever-growing set of individual sustainable design strategies, which promote particular goals and intentions. Many of the strategies focus on making environmental improvements, which is an outcome of the extensive work that has been conducted within the area of eco or green design. However, design for sustainability is also concerned with social and ethical issues in balance with economic needs, and strategies that aim to make improvements in these areas must be considered to be equally as important. Dusch et al. (2010) surveyed a selection of research studies that explored sustainable design strategies applied in disciplines such as architecture and product design. In particular, the authors focused on seven different studies that emerged between 1993 and 2009. Within these studies, Dusch et al. (2010) found that the scope of approaches being proposed within a single study ranged from between 47 sustainable design strategies (Keoleian and Menerey 1993 cited Dusch et al. 2010, p. 3) to the 210 strategies that were advocated by Vezzoli and Manzini (2009 cited Dusch et al. 2010, p. 3). While this clearly demonstrates how research in the field of sustainable design has expanded through the last decade, it also reveals the complexity facing the designer in industry. As the scale and diversity of information has risen, fashion designers have become overwhelmed with the deluge of discussions, which has in turn fuelled confusion and/or resistance among the practicing fashion design community.

The adoption of single or multiple strategies by fashion labels and brands within the industry has been patchy; however, it is argued that if the link between sustainable strategies and life cycle thinking was established in a framework to support engagement, then participation may increase. This approach has to some extent been explored within the product design discipline where comprehensive, and sometimes complex, models have emerged. For example, the design for sustainability (D4S) framework extensively explores improvements in efficiency, product quality, market opportunities, and environmental performance through assessment/redesign approaches that draw on select strategies relevant to different phases within a life cycle of a product (Crul and Diehl 2006). The rationale for applying this type of approach in fashion design lies in the opportunities that are provided by addressing specific impacts at particular points in the life cycle of a product. A sustainable design strategy can be applied at one or extend across several phases in the life cycle of a garment, but the integration is at its most successful when the life cycle of a garment is fully understood. For example, improvements can be made in one specific phase through a strategy such as *design for disassembly*, which concentrates on creating a garment that can easily be disassembled at the end of its useful life in order to reuse or recycle the material components. Alternatively, if the fashion designer wishes to minimize the waste of resources (such as materials and/or water or energy), this can be approached during a number of phases including at the outset of design, during manufacturing processes and at the disposal phase, creating benefits within many phases in the life cycle of a garment. By aligning sustainable design strategies with life cycle thinking, a clear framework emerges that builds upon and extends the familiar supply chain model that is understood by the fashion designer (Gwilt 2011). Moreover, if the new framework is applied at the early, formative stages of the design brief the designer is able to see which strategies and where they can be integrated within the manufacturing phases to reduce environmental and ethical impacts across the entire life cycle of a garment.

20.4 INTRODUCING THE FASHION DESIGN FOR SUSTAINABILITY MODEL

20.4.1 Background: Developing the FDS Model

One way forward is to develop a taxonomy of sustainable strategies that is set within a framework, which can be used and referenced in the fashion studio environment. However, in analyzing the points given earlier, there are a number of emergent issues that need to be considered in the development of an alternate model of fashion design practice, including the need to generate a model that is instantly recognizable to the fashion designer.

The FDS model has been developed over a 5-year period and research conducted through different modes of enquiry including contextual research; empiric interviews and examination of garments; and practice-led enquiry. In particular, through a series of practice-led components, it has been possible to examine how the integration of sustainable design strategies within the creative process might unfold, from the perspective of the fashion designer. The FDS model has two main purposes: first, to encourage the fashion designer to see the design and production phases in the context of life cycle thinking; second, the model attempts to reveal how and where relevant sustainable strategies can be integrated within fashion design practice. The process of development began with the mapping of the activities within the supply chain, as discussed in Sinha's model of the design and production process, and then extended in the context of the life cycle of a garment (Figure 20.1). Additionally, the phases of design and production have been divided further so that the fashion designer can recognize the conventional activities, as established by Stecker (1996), Jenkyn-Jones (2002), Jackson and Shaw (2006), Sorger and Udale (2006), and Renfrew and Renfrew (2009). With a basic framework for the FDS model in place, the structure has been overlaid with appropriate sustainable strategies that are now aligned with the relevant life cycle phases of a garment. Critically, the fashion designer is placed at the heart of the FDS model, and the sustainable design strategies that have been suggested are those that are relevant and accessible for the designer to employ during design inception. However, the FDS model is considered as a flexible framework that can be adapted to meet the needs of designers located in different sectors of the fashion industry.*

Although the fashion design and production process in the FDS model is presented from a life cycle perspective, it has been important to identify the types of sustainable design strategies that may be appropriate for fashion clothing. The FDS model currently highlights a select number of design strategies that are available from a range of strategies applied across the design disciplines. The selected strategies detailed in the FDS model are not considered a definitive list, rather they aim to suggest approaches that have been or can currently be applied within contemporary fashion design practice. It is, however, the alignment between the selected strategies and the appropriate life cycle phase that provides a foundation from which the designer can begin to explore and activate in the design studio. By referring to models such as the D4S framework and existing research studies, it has been possible to select and suggest design strategies that have been highlighted by Fuad-Luke (2002), Crul and Diehl (2006), Black (2008, 2012), Fletcher (2008), and Gwilt (2014) and others, which support the particular needs of the fashion designer.

20.4.2 Preparations for Use

In the studio, as with other models, guides, and checklists, the first step to engaging with the FDS model is to conduct an assessment exercise. The fashion designer needs to establish the inputs and outputs of design and production decisions, along with those connected to use and disposal, at the design inception phase of a collection or garment. Although an LCA can be useful, it is suggested that conducting a basic assessment exercise will help the fashion designer to understand,

---

* See Gwilt (2014) for an expanded entry level introduction to the FDS model.

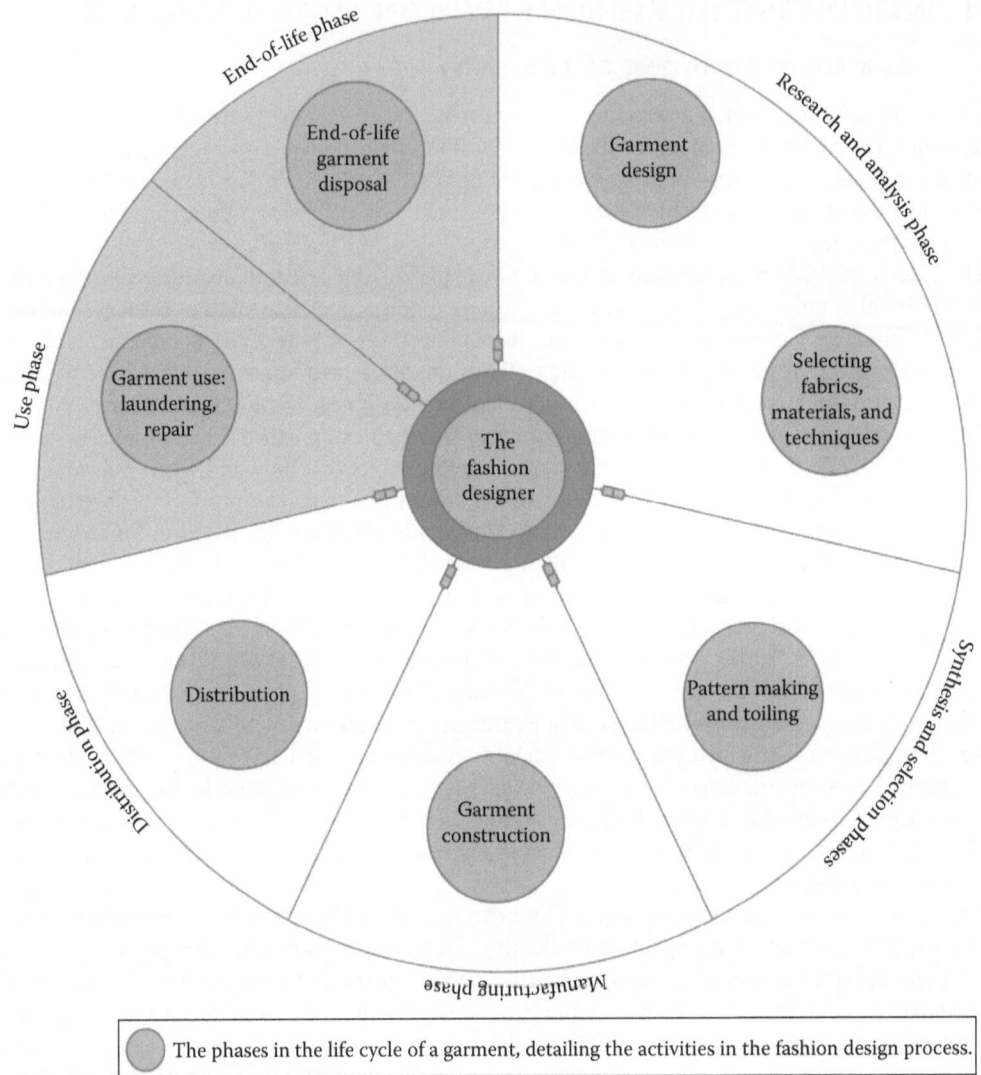

FIGURE 20.1 The FDS model template: Sinha's model (2002) has been extended to reflect the fashion design and production process when linked to life cycle thinking.

philosophically and practically, where the negative/positive impacts exist. For this, the designer needs to begin by mapping the impacts associated with a specific life cycle phase, and a mind-mapping exercise will help the designer think through the relevant inputs and outputs involved. This type of exercise can assist in identifying issues that exist at different levels. For example, the assessment may reveal impacts that occur in response to decisions made at an operational level, in the development of the collection, and in the creation of a specific or complex garment. The assessment tool itself can take many forms, including impact matrices, checklists, eco-wheels, and inventory/ analysis software; however, a basic assessment tool can be used to map and measure the negative and positive impacts using a scale system between 1 and 10, or a scale ranging from excellent to bad, which is then translated, for example, into a visual spider or wheel diagram. While the assessment will highlight problem areas, significantly it provides a clear picture of the current situation, and it is an exercise that should be revisited frequently so that the current situation is always understood.

This is just as important when determining the strategies to be employed to improve problem areas, when it becomes important to see whether making improvements actually creates other impacts/problems within the life cycle. Overall conducting an assessment, whether executed simply and often, or in greater detail, provides the groundwork for improvement, after which the designer can refer to the FDS model and select and integrate appropriate strategies that aim to minimize or avoid the identified impacts.

20.4.3 Applying the FDS Model

To apply the FDS model, the designer needs to see how sustainable strategies can link to the activities of the fashion design and the production process (when seen from a life cycle perspective). This can be best understood when using parallel lines of thought (Lawson 2006), which enables the designer to work out different aspects of the design at the same time. Using the concept of parallel lines of thought within the fashion design and production process, the designer is able to design, plan, and create a new garment design while integrating sustainable strategies. Through these conversations the fashion designer can begin to identify and respond to areas within the current paradigm that require positive intervention. The FDS model has been visually developed to encourage the designer's parallel lines of thought and to create an open-ended means of engagement with sustainable design strategies during the design process (Figure 20.2). As discussed earlier, in applying the FDS model it is important to reflect on the impact of engaging change, and as sustainable strategies are identified for use (to tackle specific problem areas) the assessment exercise should be revisited to see if/where new impacts may arise as a consequence change. Figure 20.3 reflects an appraisal of a small designer/owner fashion label, which following an assessment exercise, identified that the garment use phase was an area for improvement. The assessment exercise helped the designer to establish a design brief that was focused on integrating the most relevant sustainable design strategies suggested in the FDS model.

Following the designer's decision to engage with the strategy "Design for low-impact care" that promotes a water/energy-efficient approach to clothing care as a means of reducing some of the impacts created during laundering, the diagram in Figure 20.3 shows where effects may be felt as a consequence of change. The diagram clearly demonstrates that integrating one or more strategies is likely to create a secondary effect at other points within the garment's life cycle and reiterates the point that revisiting the assessment exercise is a crucial step in creating positive change. In this example, the designer's goal is to optimize efficiency during the garment's use phase—for example, reducing the energy used during the laundering process, while conserving water and reducing harmful pollutants. In the diagram, it is apparent that one effect of implementing change is reflected in the choice of fabric to be used in the garment's design since the specific properties of a fabric require different laundering methods, with some more inefficient than others. Moreover, the designer also has to consider fabric choice in relation to the range of different activities in the laundering process—pretreatment, washing, tumble drying, ironing, dry cleaning—that each create different types/levels of impacts. At the same time, the designer can look at improving the way the garment is designed to enhance efficiency. For example, through "Design for modularity" garments can be developed so that the wearer may easily remove heavily soiled areas (e.g., cuffs, collars) for washing, without the need to disrupt the rest of the system. In order to develop modular garments, the designer will have to consider impacts that may be created during the pattern making and toiling phase, which will incur increased time/thought/testing to meet the technical needs of modular design concepts, and their ability to function in the specified fabric. Moreover, a key contributor to improvements in the use phase relies on the participation of the wearer, and so communicating care information to engage the wearer is critical—if the wearer does not adhere to the principles deployed in the garment, then the designer's intentions will be ineffective. The points highlighted in this example demonstrate

Sustainable strategies

1
- Design for empathy
- Design for well-being
- Design for closed-loop systems

2
- Design for low-impact materials and processes
- Design for monomaterials

3
- Design for zero waste
- Design for efficient use of materials and resources

4
- Design for durability
- Design for ethical and fair trade production

5
- Design for need
- Design to minimize transportation
- Design to reduce/reuse packaging
- Design to engage with local communities

6
- Design for multifunction
- Design for modularity
- Design for low-impact care
- Design for customization
- Design for repair
- Design for product/service systems

7
- Design for reuse
- Design for remanufacturing
- Design for recycling/upcycling
- Design for disassembly

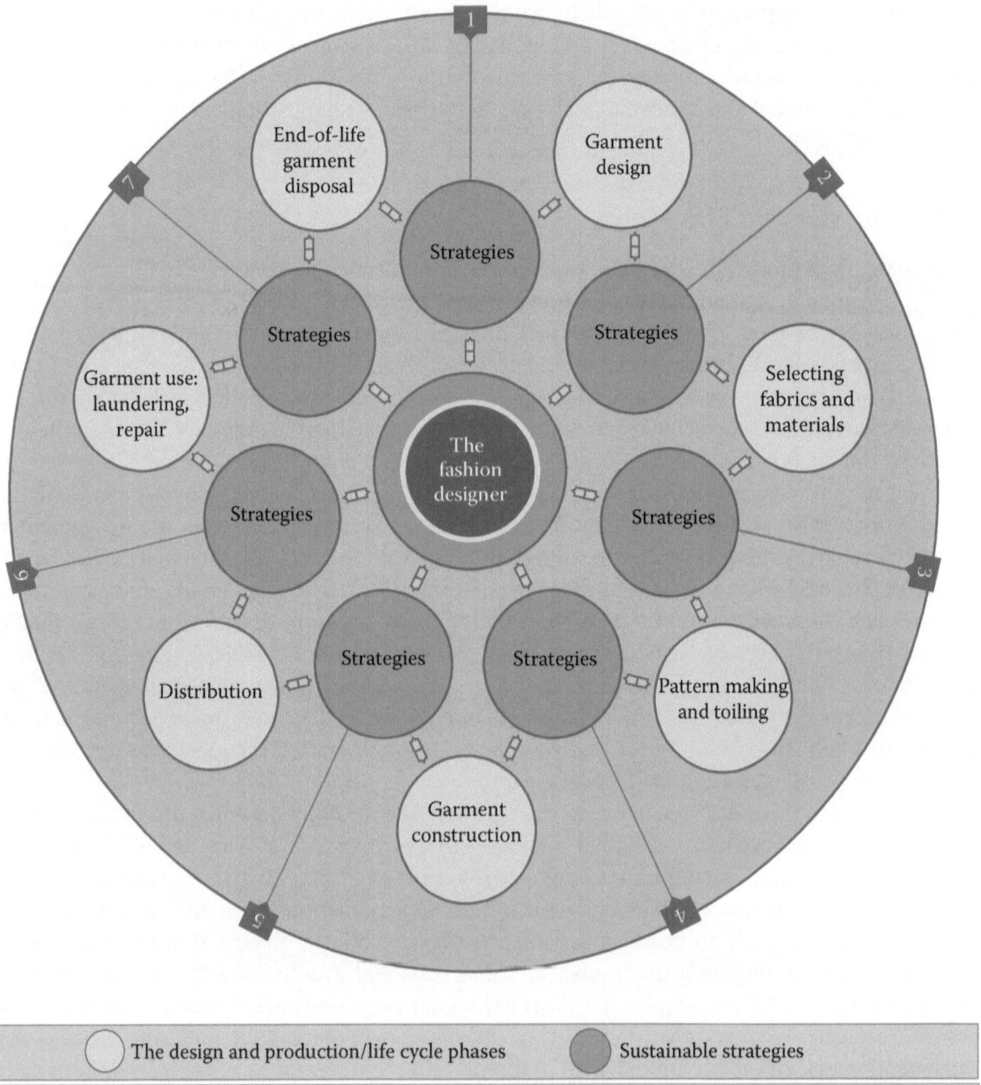

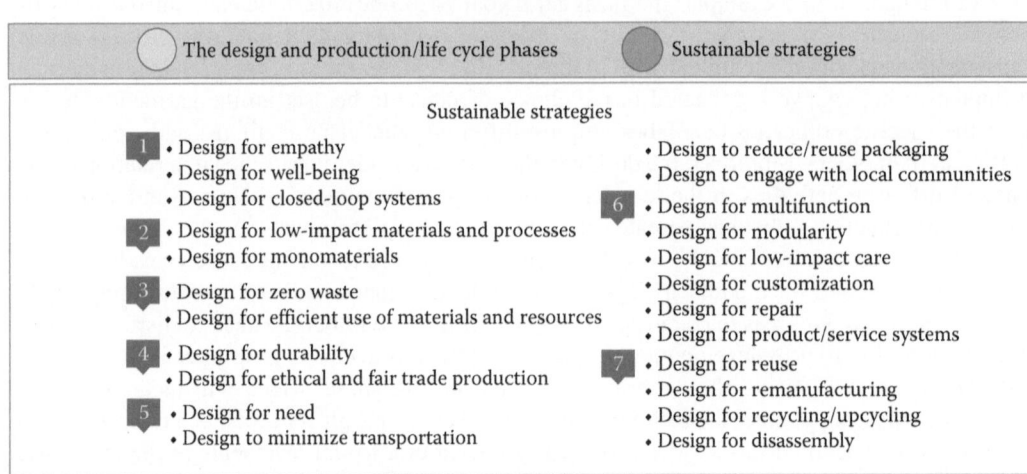

FIGURE 20.2 The FDS model presents the fashion design and production process from a life cycle perspective and aligns key phases with sustainable design strategies.

Exploring a Framework for Fashion Design for Sustainability

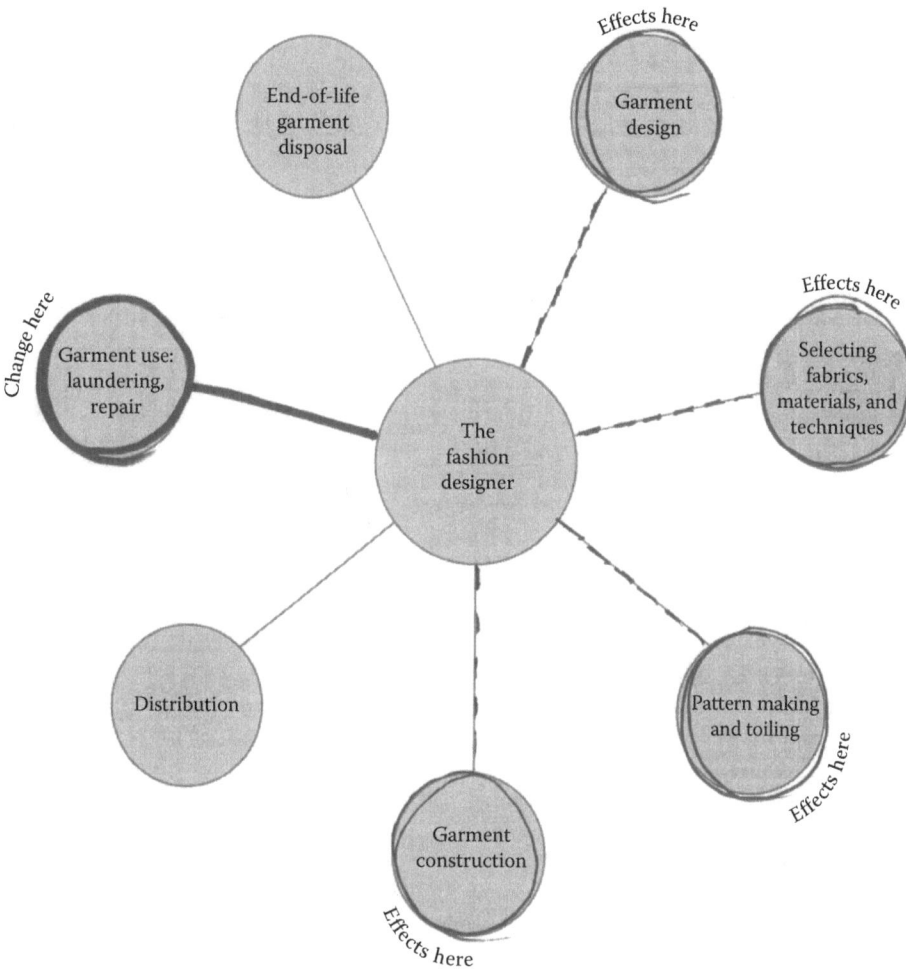

FIGURE 20.3 Diagram reflecting the designer's goal to make improvements to garments during the use phase. The diagram reveals other phases in the life cycle that can be affected by the improvements.

the need to reflect and revisit the assessment exercise in order to trace any potentially unwanted/negative consequences that may be detrimental to the best intentions of the designer.

20.5 CHALLENGES OF INTEGRATING SUSTAINABLE STRATEGIES IN THE FASHION DESIGN PROCESS

While engaging with the FDS model can help lead the designer to develop improved products and services, it does highlight the complexities involved with influencing change across the various activities within the design and production process and beyond into the garment use phase. The cooperation of other people including, for example, the sourcing/buying team; the pattern maker; and the wearer is vital if designers want to create real change. At the same time, it is important to recognize how a designer, when placed in a central role, can influence the use of sustainable design strategies across the design and production process that can create positive change across the entire life cycle of a garment. By initially engaging with one or two strategies, the designer can maximize positive impacts in phases that are identified as areas in need of improvement. But with confidence, the designer can then begin to expand upon the range of strategies to include and can further minimize negative impacts that appear within other life cycle phases.

Despite the potential for the FDS model to increase the fashion designer's engagement with sustainable design strategies, there may be, as with other tools and guides, some issues that impede a wider implementation. First, it is acknowledged that there is a need for continuing research to broaden and expand the range of strategies suitable for fashion design practice. For this to occur, it is important to appreciate that currently the use of different terms and descriptions applied to design strategies has created confusion and resistance among the practicing fashion design community. It is then essential for the fashion design community to establish a common set of terms that is relevant and accessible to fashion design practitioners.

Second, although it is apparent that there are problems associated with fashion clothing, the real issue lies with the model of design practice being used to create garments and the cultural desire for these products. The designer in most small- and medium-sized companies is responsible for developing a garment that is driven primarily by aesthetic and economic factors. Furthermore, the conventional model of fashion design practice applied within many SMEs and larger manufacturers does not engender a relationship between the producer and consumer; the relationship typically ends when the garment reaches the shop floor (Black et al. 2009). This is in contrast to the methodology applied in the made-to-measure sectors of tailoring and haute couture where the relationship with the wearer is fundamental to the model of production (Gwilt 2011). Although the FDS model provides a framework for the fashion producer, it is evident that both the fashion designer and the wearer need to engage with sustainable fashion in order for sustainable garments to "work." This is highlighted in the example associated with Figure 20.3, where the designer has elected to integrate a water/energy-efficient clothing care strategy in the design and production of new garments. In developing a garment for low-impact care, the fashion designer may need the wearer to change ritualistic laundering practices to render the objective successful. But while the issue of responsibility may be problematic, considering a holistic approach to the design and production of garments provides a wide range of positive opportunities across the whole garment life cycle.

Considering the use and disposal phases, as integral to the design and production process at the outset of design will enable designers to explore a wider range of strategies that reduce/avoid impacts across the whole garment life cycle. For example, designers can begin to develop garments specifically for end-of-life disassembly and recycling as a method of material reuse. Moreover, as producers begin to employ strategies such as product services systems (PSS) with the aim of providing products with services/or services for products as a method of reducing material consumption, then the wearer has to revise and rethink attitudes toward product ownership. By extending the fashion and production process to include the use and disposal phases, the designer can begin to see the relationship between the designed garment and its journey, which goes beyond the retail store. Life cycle thinking allows the designer to appreciate that the impacts of fashion clothing are felt in phases that are normally considered outside of their jurisdiction. Furthermore, with such an understanding the designer can then reflect and act upon their own design practice and begin to find ways to improve the activities within the company and the sustainable performance of its products.

This leads toward a third issue that within the typical fashion magazine the reader may see one or two pages of eco/ethical fashion garments, but the majority of fashion spreads in the publication are dedicated to conventional types of fashion garments—that is, fashion garments that do not claim to or appear to address any environmental, social, or ethical issues. This apparent lack of press coverage devoted to sustainable fashion garments ensures that neither the fashion designer nor the wearer is being exposed to or educated about how fashion garments can be created and used in accordance with sustainable criteria. Importantly then, while sustainable fashion design practices in the industry may improve and indeed need to continue to improve, the industry and design press also need to find ways to encourage the sector as a whole, and the public's expectations of the sector, to engage in change.

20.6 CONCLUSION

The chapter has argued that a new model of fashion design practice can exist if the process of fashion design and production is considered from a life cycle perspective. Furthermore, once a life cycle approach to fashion design is accepted, then sustainable strategies can be integrated within the design methodology. Many advocates endorse a wide variety of approaches and strategies for use in design practice (Manzini and Jégou 2003; Chapman and Gant 2007; Black 2008; Fletcher 2008; Fuad-Luke 2009; Gwilt and Rissanen 2011). For these advocates, sustainable strategies can, and should, improve the designer's ability to innovate in design practice. Moreover, if fashion designers perceive sustainable design strategies in terms of the opportunities for innovation they might offer, rather than as a barrier to design practice, then a wider engagement with the principles of sustainability might be commonplace. While there are a wide variety of design strategies that can be applied specifically within fashion design (and in multiples), often strategies are considered in isolation, which gives an impression that only one phase or impact can be tackled at a time. This approach is often reiterated through the use of case studies of best practice that commonly fail to promote the use of multiple strategies during design and production, which can tackle a range of impacts across the garment life cycle.

In developing the FDS model, it was important to analyze, reflect upon, define, and arrange the activities and phases within the fashion design process. This points to a need to critique fashion design practice in order to prepare for a constantly changing industry. Fashion producers need to embrace change, and the FDS model intends to show the designer how positive contributions can be made in, across and beyond the design and production phases. While the FDS model is not considered a replacement model for LCA exercises, or other detailed data collective tools, its primary function is to encourage a change in attitude and behavior. The FDS model is not prescriptive rather it aims to motivate designers to take tentative steps toward change and to seek further support, advice, and guidance. In particular, the FDS model is useful at the inception stage of the fashion design process, when the designer is engaged in the research and analysis phase, identified by Sinha (2002). At this point in the design and production process, the design brief is established, which draws on market, trend, and resources research. The use of the FDS model at this stage would assist the designer to also include life cycle thinking within the design brief and conceptual development and design decision-making processes. Moreover, at this point the entire life cycle of the garment can be imagined allowing for the incorporation of positive interventions at all stages of the life cycle.

It can be argued that perceptually there is little need/desire for the fashion designer to take a life cycle approach to design practice, and this is the crux of the problem. Whether through lack of knowledge or willingness to change/challenge existing practices, or other economic/resource issues, without a consideration for the entire life cycle of the garment it becomes impossible to make a genuine sustainable transformation. The fashion designer needs to acknowledge that life cycle thinking requires an understanding of the impacts—positive and negative—across all aspects of the garment's life cycle. With a good understanding of sustainable design strategies it becomes possible, at the design inception phase, to select the most beneficial strategy for a particular point of intervention in the design and production process. Although a perfect, all encompassing model for creating sustainable fashion garments does not as yet exist, the objective should be to minimize/avoid negative impacts created by the industry and to capitalize on the positive impacts of good practice, and this has been a key motivation for the FDS model.

REFERENCES

Allwood, J.M., Laursen, S.E., Maldivo de Rodriguez, C., and Bocken, N.M.P. *Well Dressed? The Present and Future Sustainability of Clothing and Textiles in the United Kingdom.* Cambridge, U.K.: Institute for Manufacturing, University of Cambridge, 2006.

Black, S. *Eco-Chic: The Fashion Paradox*. London, U.K.: Black Dog Publishing, 2008.

Black, S. *The Sustainable Fashion Handbook*. London, U.K.: Thames and Hudson, 2012.

Black, S., Eckert, C., and Eskandarypur, F. The development and positioning of 'the considerate design tool' in the fashion and textiles sector. In: The Centre for Sustainable Design (ed.) *Sustainable Innovation 09: Towards a Low Carbon Innovation Revolution*. Farnham, U.K.: University of the Creative Arts, 2009, pp. 68–75.

Bras-Klapwijk, R.M. and Knot, J.M.C. Strategic environmental assessment for sustainable households in 2050: Illustrated for clothing. *The Journal of Sustainable Development*, 9(2) (2001): 109–118.

Centre for Sustainable Fashion. *Fashion and Sustainability: A Snapshot Analysis*. London, U.K.: London College of Fashion, 2008. Retrieved October 12, 2008, from: http://www.sustainable-fashion.com/resources/signposts/our-publications-2/.

Chapman, J. and Gant, N. *Designers, Visionaries and Other Stories*. London, U.K.: Earthscan, 2007.

Crul, M. and Diehl, J.C. Design for sustainability: A practical approach for developing economies. United Nations Environment Program/DELFT University of Technology, 2006. Retrieved December 10, 2009, from: www.d4s-de.org/manual/d4stotalmanual.pdf.

Datschefski, E. *The Total Beauty of Sustainable Products*. Crans-Pres-Celigny, Switzerland: Rotovision, 2001.

Dusch, B., Crilly, N., and Moultrie, J. Revisiting sustainable design strategies. In: The Centre for Sustainable Design (ed.) *Sustainable Innovation 2010: Creating Breakthroughs: Green Growth, Eco-Innovation, Entrepreneurship and Jobs*. Rotterdam, the Netherlands: University of the Creative Arts, November 8–9, 2010.

Fletcher, K. *Sustainable Fashion and Textiles: Design Journeys*. London, U.K.: Earthscan, 2008.

Fletcher, K. and Grose, L. *Fashion & Sustainability: Design for Change*. London, U.K.: Laurence King, 2012.

Fry, T. *Design Futuring: Sustainability, Ethics and New Practice*. Sydney, Australia: UNSW Press, 2009.

Fuad-Luke, A. *The Eco-Design Handbook*. London, U.K.: Thames and Hudson, 2002.

Fuad-Luke, A. *Design Activism: Beautiful Strangeness for a Sustainable World*. London, U.K.: Earthscan, 2009.

Gwilt, A. Producing sustainable fashion: The points for positive intervention by the fashion designer. In: Gwilt, A. and Rissanen, T. (eds.) *Shaping Sustainable Fashion*. London, U.K.: Earthscan, 2011.

Gwilt, A. *A Practical Guide to Sustainable Fashion*. London, U.K.: Fairchild Books, 2014.

Gwilt, A. and Rissanen, T. (eds.) *Shaping Sustainable Fashion*. London, U.K.: Earthscan, 2011.

Hethorn, J. and Ulasewicz, C. (eds.) *Sustainable Fashion: Why Now? A Conversation About Issues, Practices, and Possibilities*. New York: Fairchild Books, 2008.

Jackson, T. and Shaw, D. (eds.) *The Fashion Handbook*. Oxon, U.K.: Routledge, 2006.

Jenkyn-Jones, S. *Fashion Design*. London, U.K.: Laurence King Publishing, 2002.

Lawson, B. *How Designers Think: The Design Process Demystified*. Oxford, U.K.: Architectural Press, 2006.

Manzini, E. and Jégou, F. The construction of design orientating scenario, Final Report. Delft University of Technology, 2000. Retrieved November 12, 2011, from: www.solutioning-design.net/li/pdf/SOL_038.pdf.

Manzini, E. and Jégou, F. (eds.) *Sustainable Everyday: Scenarios of Urban Life*. Milan, Italy: Edizioni Ambiente, 2003.

Renfrew, E. and Renfrew, C. *Basics Fashion Design 04: Developing a Collection*. Lausanne, Switzerland: AVA publishing SA, 2009.

Sinha, P. Creativity in fashion. *Journal of Textile and Apparel, Technology and Management*, 2(4) (2002). Retrieved May 06, 2006, from: www.tx.ncsu.edu/jtatm/volume2issue4/.../sinha/sinha_full_25_02.pdf.

Sorger, R. and Udale, J. *The Fundamentals of Fashion Design*. Lausanne, Switzerland: AVA Publishing SA, 2006.

Stecker, P. *Fashion Design Manual*. Melbourne, Australia: Macmillan Education Australia, 1996.

Thorpe, A. *The Designer's Atlas of Sustainability*. Washington, DC: Island Press, 2007.

UNEP. *Sustainable Consumption and Cleaner Production: Global Status 2002*. Paris, France: UNEP, 2002.

21 Fashion Industry and New Approaches for Sustainability

*Kirsi Niinimäki, Esben Rahbek Gjerdrum Pedersen,
Kerli Kant Hvass, and Lisbeth Svengren-Holm*

CONTENTS

21.1 New Business Models for Sustainable Fashion 453
    21.1.1 What Is a Sustainability Business Model? 453
    21.1.2 Collaborative Consumption for Sustainable Fashion 456
    21.1.3 Postretail Business Models for Sustainability 457
    21.1.4 Supply Chain Solutions: New Partnerships for Sustainable Fashion 458
21.2 New Approaches for Sustainability in the Fashion Industry through PSS and Communication 459
    21.2.1 Innovation in the Fashion Business 459
    21.2.2 Fashion Service Concepts 460
    21.2.3 Service Concepts in Selected Swedish Fashion Design Companies 461
        21.2.3.1 Filippa K 461
        21.2.3.2 Uniforms for the Dedicated 462
        21.2.3.3 Nudie Jeans 463
        21.2.3.4 Boomerang 464
    21.2.4 Communication Strategies of Sustainability Policies and Actions 464
21.3 PSS and Consumer-Based Eco-Efficiency 466
    21.3.1 New Business Thinking for Consumer Satisfaction 466
    21.3.2 Consumer-Based Eco-Efficiency 466
    21.3.3 Product Satisfaction 467
    21.3.4 Product Service Systems 468
    21.3.5 PSS for Product Satisfaction through Perfect Fit 468
    21.3.6 PSS to Fulfill Fashion Needs for Change 469
    21.3.7 PSS to Fulfill Consumers' Cognitive Needs 470
    21.3.8 Shifting Business toward Service Thinking and Consumer-Based Eco-Efficiency? 470
21.4 Conclusions 471
References 471
Online References 474

21.1 NEW BUSINESS MODELS FOR SUSTAINABLE FASHION

21.1.1 What Is a Sustainability Business Model?

The term business model emerged during the dotcom-boom in the 1990s when new online businesses entered the markets with characteristics differing significantly from traditional bricks-and-mortar industries. Since then, scholars and practitioners alike have increasingly adopted the business model concept that has now become an integrated part of the mainstream business/management literature (Zott et al., 2010). However, despite its fast-growing popularity, the concept remains fluffy and ill-defined, which implies that it has been difficult to get the precise fix on the core characteristics of a business model (Morris et al., 2005). However, there seems to be consensus that value is crucial to the understanding of a business model. For instance, Osterwalder and Pigneur (2010) define business models as follows: "(…) the rationale of how an organization creates, delivers, and captures value" (Osterwalder and Pigneur, 2010, p. 14). In a similar vein, Chesbrough (2007, p. 12) argues the following:

> At its heart, a business model performs two important functions: value creation and value capture. First, it defines a series of activities, from procuring raw materials to satisfying the final consumer, which will yield a new product or service in such a way that there is net value created throughout the various activities. This is crucial, because if there is no net creation of value, the other companies involved in the set of activities won't participate. Second, a business model captures value from a portion of those activities for the firm developing and operating it. This is equally critical, for a company that cannot earn a profit from some portion of its activities cannot sustain those activities over time.

The mainstream business model literature is commercial in nature and gives little priority to ethics, responsibility, and sustainability. Recently, however, a number of scholars have tried to integrate social and environmental responsibility into the discussions of business model theory of practice (see, e.g., Lüdeke-Freund, 2009; Schaltegger et al., 2011; Stubbs and Cocklin, 2008). The new stream of literature aims at supplementing the existing economic and customer-centric focus with the inclusion of, for example, social and environmental costs/benefits (Osterwalder and Pigneur, 2010), stakeholders (Stubbs and Cocklin, 2008), nonmarket aspects (Lüdeke-Freund, 2009), and the ecosystem (Michelina and Fiorentina, 2012). Overall, what seem to tie this literature together is that business models must be viewed from a more holistic perspective that give priority to a broader set of stakeholders and more dimensions of performance (people, profit, planet).

How does a sustainability business model look like in the fashion industry? As a starting point, sustainability business models in the fashion industry is about looking for new ways of value creation, value delivery, and value capture that maximizes positive and minimizes negative societal impacts within the fashion life cycle. The definition is in line with DEFRA (2010, p. 5), which argues that sustainable clothing "(…) maximizes positive and minimizes negative environmental, social and economic impacts along its supply and value chain. Clothing that is sustainable does not adversely impact people or the planet in its production, manufacture, transport, retail or end of life management." Moreover, the definition bears similarities within existing attempts to conceptualize sustainability business models (see, e.g., Bocken et al., 2014, p. 44).

There is ample evidence of a need for new and more sustainable business models in the fashion industry. Massive amounts of pesticides and water are used in conventional cotton production, toxic chemicals from fashion manufacturing cause serious harm on the environment, and poor labor conditions persist throughout the fashion supply chain (Pedersen and Gwozdz, 2014). Moreover, the current fashion consumption is far from sustainable: an average consumer discards more than 20 kg of clothes and textiles every year (MISTRA, 2010). Moreover, approximately 30% of the clothes owned by consumers have not been worn for a year or more (WRAP, 2012). Last, current fashion consumption (washing, tumble drying) is a key contributor to the overall environmental footprint of fashion.

In response to the sustainability challenges in the fashion industry, a number of new initiatives have been introduced to lower the social and environmental impact of fashion manufacturing and consumption. For instance, at present there is experimentation with new fibers (paper, milk, seaweed, soy, etc.), which can lower the dependence on conventional cotton and polyester (Fletcher, 2008). Likewise, a wide range of new tools and technologies reduce their social and environmental impact in manufacturing processes, packaging, and transportation. For instance, Puma's Clever Little Bag attempts to minimize the use of packaging materials, and DyeCoo provides industrial

Fashion Industry and New Approaches for Sustainability

CO_2 dyeing equipment that eliminates water consumption. Last, a number of organizations have developed standardizations (e.g., GOTS) and systems to better assess the environmental impact of various garments (e.g., Higg index), which will hopefully promote the development of an institutional infrastructure for sustainability in the fashion industry (Pedersen and Andersen, 2014; Waddock, 2008).

Still, however, evidence indicates that only a minority of fashion companies are proactive when it comes to sustainability/corporate social responsibility (CSR). For instance, a study among 400 Nordic fashion companies concludes that most companies conform to social and environmental demands when they emerge (conformers), whereas fewer try to foresee future stakeholder requirements (anticipators) or move beyond expectations (definers) (Pedersen and Gwozdz, 2014, see Figure 21.1). Moreover, evidence from a survey among 492 Swedish fashion companies indicates that sustainability is rarely seen as a core success factor (Pedersen and Laursen, 2014).

In realization of the significant sustainability challenges in the fashion industry and the reactive approach of most fashion companies, it becomes very relevant to discuss new business models for sustainable fashion. In the following sections, we will therefore discuss three new fashion business models to address sustainability: collaborative consumption, postretail business models, and cross-sector partnerships in the supply chain. The choice of business models is highly selective and makes no claim of covering the entirety of new, innovative sustainability initiatives within the fashion industry (see, e.g., Pedersen and Andersen, 2014). Moreover, the topics addressed in this section

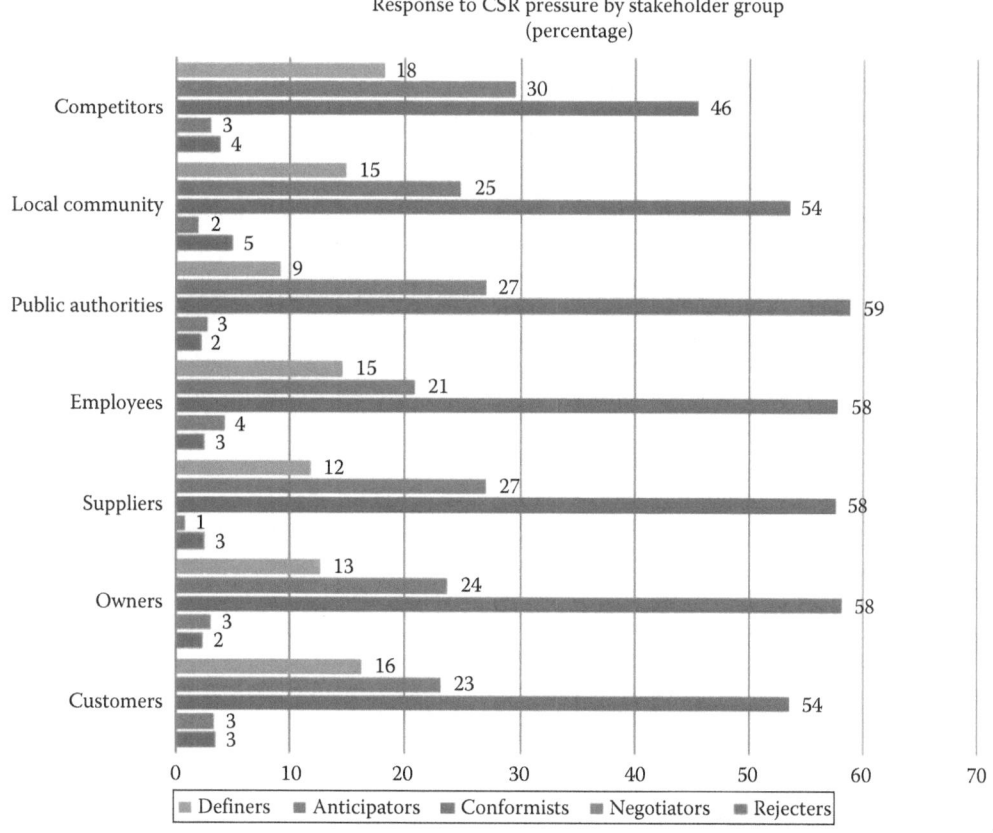

FIGURE 21.1 Response to CSR requirements from stakeholders, evidence from a study among Nordic fashion manufacturers and retailers (Denmark, Norway, Sweden, Finland, and Iceland). A more comprehensive analysis of the findings can be found in Pedersen and Gwozdz (2014).

deal with some parts of the fashion lifecycle even though transformative change toward sustainability necessitates a rethinking of the entire industry: from extraction of raw materials and all the way to consumption and subsequent recycling/upcycling/downcycling.

21.1.2 COLLABORATIVE CONSUMPTION FOR SUSTAINABLE FASHION

Collaborative consumption and related ideas like the sharing economy have come to the fore as a game changer that will radically transform current business structures and consumer behavior (Walsh, 2011; WEF, 2013). As noted by Rachel Botsman and Roo Rogers (2010, p. 30a): "Collaborative consumption is not a niche trend, and it's not a reactionary blip to the recession. It's a socioeconomic groundswell that will transform the way companies think about their value propositions—and the way people fulfill their needs." Ultimately, collaborative consumption is about new business models based on access, reuse and redistribution and represent and broader development from ownership and possession towards sharing, collaboration and participation, whether it concerns cars or clothes (Bardhi and Eckhardt, 2012). Some authors even talk about an emerging *sharing economy* and *access regime* (Bardhi and Eckhardt, 2012; STWR, 2012).

Collaborative consumption is often presented as a commercial opportunity as well as an environmental solution although the latter is often an indirect benefit from these business models (Botsman and Rogers, 2010b). More specifically, collaborative consumption is said to fight overconsumption by enabling a better utilization of resources. However, the actual environmental impacts from collaborative consumption are still debated as they depend e.g. on the extent to which sharing replaces or supplements ownership (Armstrong and Lang, 2013; Tukker, 2004). To give an example from the fashion industry, clothes sharing only reduces overconsumption if it replaces conventional, possession-based shopping. (EMF, 2013a).

New business models for collaborative consumption initiatives are also starting to take hold in the fashion industry. Traditionally, there has always been a significant secondhand market for fashion and other products. For instance, the UK spend on charity shops and secondhand outlets is approximately GBP 719 million (The Co-Operative, 2012). Moreover, traditional one-off rentals of formal wear have been a niche market in the fashion industry for centuries (WRAP, 2013). However, what we are seeing now is the emergence of new business models that experiment with access-based rather than ownership-based business models. For instance, small, subscription-based fashion libraries have emerged in a number of countries offering their members access to a shared wardrobe (Pedersen and Netter, 2013). Moreover, Mud Jeans has recently launched a leasing system that allows consumers to lease new jeans for a deposit fee and a monthly rate. Other examples include Le Tote, Girl Meets Dress, Thredup, Lütte Leihen, and Resecond (WRAP, 2013; Zelwak and Pedersen, 2014).

When assessing the sustainability potentials of collaborative fashion consumption, it is also relevant to look at some of the related industries. After all, a significant part of the environmental footprint from fashion is related to the consumption phase that makes it interesting to explore environmental-friendly alternatives to conventional washing and drying (e.g., Fletcher, 2008). For instance, already in the late 1990s, the washing machine provider Electrolux tested a pay-per-wash system in Gotland, Sweden, as an alternative to the ownership-based business models (Brass, 2012). Although the test can hardly be perceived as a success, a redesigned pay-per-wash system could nonetheless still be a relevant business model to explore in the future if it is designed to provide consumers with financial incentives to be sustainable (e.g., higher washing temperature—higher price).

Looking into the future, collaborative consumption business models are likely to become more common in the fashion industry. However, a lot of these initiatives are also likely to remain niche phenomena due to resource constraints, scalability problems, and limited consumer demand. Moreover, collaborative consumption initiatives often also face the problem that the products offered are designed for ownership rather than sharing. In the future, it will be relevant to explore new fashion designs for collaborative consumption that give more priority to longevity and flexibility.

21.1.3 POSTRETAIL BUSINESS MODELS FOR SUSTAINABILITY

In the context of increasing amounts of textile waste, the world's decreasing natural resources and the industry's growing need for raw materials it is increasingly apparent that fashion companies need to transition beyond their traditional business models and incorporate end-of-life (EoL) perspectives in their businesses. A recent UN Global Compact study among world business leaders regarding sustainability demonstrates that closed-loop business models, with their aim to decouple growth from resource use, environmental and social impact, are increasingly attractive to companies seeking disruptive innovation in the search for both positive sustainability impact and business value (Accenture and UN Global Compact, 2013).

Until recently, fashion companies mainly focused on creating and capturing value from the sale of new products, and once the garments were sold these products were not regarded as part of their business model. This situation is currently changing and integrating product reuse and EoL aspects in companies' business models while taking responsibility for the postretail phases of products is an emerging phenomenon in the fashion industry. In parallel with policy-driven extended producer responsibility (EPR) initiatives for textiles, pioneered by French self-financing entity, EcoTLC (Kelly, 2012), the last years have also brought several business model innovations to the market that focus on EoL management of garments and closed-loop supply chains (Wells and Seitz, 2005). Closed-loop business models focus on taking products back from customers for reuse, repair, recycling, and remanufacturing, and they link the traditional forward supply chain activities with reverse supply chain activities (Morana and Seuring, 2007).

In general, two broad strategies can be distinguished in how companies address product EoL issues through business model innovation, namely developing resell/reuse platforms for prolonging the life of garments and thereby capturing the resell value they offer and secondly, implement in-store product take-back schemes (Kant Hvass, 2014). Developing new resell/reuse channels is mainly chosen by premium and high-end fashion brands with higher-quality products as this strategy requires the highest quality possible to ensure that garments retain their value and can be rebought many times (Fletcher and Grose, 2012). For example, the Swedish store, Filippa K Second Hand, in collaboration with a local consignment store where preowned Filippa K clothes and collection samples are sold on consignment, or American Eileen Fisher with the Green Eileen secondhand store concept where preowned garments are obtained through donations for further resell. A similar concept is Boomerang Effect by the Swedish brand, Boomerang, where consumers can donate their used garments, which are then resold in Boomerang stores or upcycled into Boomerang Home Collection products.

In-store product take-back schemes or direct donations to charities are the other broad postretail strategies chosen by fashion companies. These are often large-scale, market-driven brands with complex distribution channels (Kant Hvass, forthcoming). Either the take-back initiatives are often organized by entering a donation partnership with a well-known charity organization to facilitate the collection of used clothes (e.g., Marks and Spencer's Schwopping initiative in collaboration with charity organization Oxfam), partnering with a third-party professional collector (e.g., H&M's Long Live Fashion Garment Collecting initiative in collaboration with global collector I:Co), or companies organize their own take-back scheme (e.g., Uniqlo's All-Product Recycling Initiative or Patagonia's Common Threads Initiative). The previously described broad strategies for business model innovations are not clear-cut and reuse, recycling, and reduce approaches are often intertwined and behind these initiatives are several varieties of implementation.

Innovating business models based on the circular economy (Ellen MacArthur Foundation, 2013), where companies are aiming to close the material loop of their products and recover resources of postconsumer textile waste, is also an emerging phenomenon. For example, companies such as Patagonia and Nike, who collect their postconsumer products, ship them back to a fiber mill for chemical recycling into new fibers, which are then manufactured into new products again to be sold to Nike and Patagonia customers (Ulasewicz and Baugh, 2013). PUMA provides another good example with their in-store take-back initiative and new InCycle closed-loop collection of biodegradable or recyclable shoes (Ecotextile News, 2013). However, closing the loop of textiles is still in its experimental phase and scaling up the practices requires overcoming many barriers. For example, the ability to guarantee sufficient volume of product returns, preferably of own brand products due to chemical concerns, is an important aspect (Guide and Van Wassenhove, 2009). For example, Morana and Seuring (2007) studied the recycling network, ECOLOG, a closed-loop supply chain for polyester apparel for corporate wear that operated in the 1990s. The case study showed that even though the closed-loop system was technically sound and economically viable if used products were obtained, customer return behavior was not appropriately considered and the initiative failed. Returning used clothes to the store is a new behavior expected of end consumers, and even though the return system may be convenient, incentivized, and well communicated, collecting necessary volumes for closing the loop may be challenging. In addition, technological innovation that improves the cost/performance relation associated with the remanufacturing of used fabrics is currently lacking (Ravasio, 2013/2014), which is an important factor to scale up the practices and develop markets for recycled fibers. Finally, for a well-functioning closed-loop fashion system, it is important to have transparent and open collaboration between designers, producers, consumers, and EoL value chain recyclers in order to produce products that are recyclable at the end of their lives, to collect them when they have lost their use value and remanufacture them into new products.

21.1.4 SUPPLY CHAIN SOLUTIONS: NEW PARTNERSHIPS FOR SUSTAINABLE FASHION

The recent factory collapse in Bangladesh, killing more than 1000 workers, represented the tragic climax of a decades-long criticism of the fashion supply chain (Pedersen and Andersen, 2014). Politicians, media, environmentalists, human rights groups, and researchers have repeatedly highlighted the massive social and environmental problems (e.g., pollution, sweatshop conditions, health and safety risks, child labor, underpayment), which have also been documented in a number of publications and reports (Pedersen and Gwozdz, 2014). According to Brito et al. (2008, p. 538): "Fashion SC (Supply chain, our amendment) is particularly sensitive to sustainability due to its inherent characteristics and some specific trends. The production process makes intense use of chemical products and natural resources (land and water), generating a high environmental impact. Furthermore, the search for lower cost production has led to a dramatic relocation of production sites towards the Far East (...)." However, despite the widespread recognition of the massive social and environmental problems in the fashion supply chain, the current responses of the industry can be expected to have only limited sustainability impacts (Deloitte, 2013).

The global, complex, and fragmented supply chain structure has made the fashion industry the epitome of unsustainable capitalism. However, there are signs that more actors are beginning to step up on the sustainability agenda when it comes to the supply chain. For instance, new materials and manufacturing principles are introduced to reduce the environmental footprint. An example is Kuyichi, a brand introduced by the NGO Solidaridad, which intents to create a more sustainable supply chain by producing garments of organic cotton, recycled plastic bottles, hemp, etc. Moreover, a number of new standards, tools, and systems for transparency and traceability have been developed (DEFRA, 2010; Pedersen and Andersen, 2014). For instance, GoodGuide ranks fashion companies based on their ability to address sustainability in the supply chain. Moreover, individual companies such as Patagonia, Icebreaker, and Rapanui have also introduced initiatives that indicate commitment to transparency in the supply chain (Deloitte, 2014). More transparency is in itself not a business model (except for those organizations providing transparency solutions), but transparency may be the first step to rethink the current layout of the fashion supply chain. Without transparency, it will be difficult to create awareness and change. An interesting example of an alternative supply chain is the *IOweYou (IOU)* Project, which has replaced the term supply chain with prosperity chain. The initiative tries to connect the actors in the chain by enabling the consumer to trace the fabric all the way back to the weaver.

Another interesting niche phenomenon is the cross-sector partnerships between fashion companies and local communities. In recent years, a number of fashion brands have made special collections by making use of the local handcraftsmanship (Pedersen and Andersen, 2014). For instance, the knitting company Gudrun & Gudrun has employed Faroese and Jordanian knitting artisans to create modern hand-knitted garments. Moreover, a number of fashion brands have partnered up with COOPA-ROCA, a nongovernmental organization (NGO) that works with local artisans in the Brazilian Favela, Rocinha. Last, the company Indigenous has developed a comprehensive network with more than 300 artisan cooperatives or knitting groups (ibid).

A related development is the so-called slow fashion movement that attempt to address some of the more fundamental root problems in the fashion supply chain. The current fast fashion business model, emphasizing speed and efficiency, is simply difficult to combine with the basic principles of sustainability that necessitates a search for alternative approaches to fashion manufacturing and consumption. Therefore, it is necessary to develop more durable, and less seasonal, fashion products, which can be repaired and recycled.* Part of the slow fashion movement is also looking at locally made garments; something that differ significantly from the present dominance of anonymous, global supply chains. A recent example is Manufacture NY, which intends to promote transparency and local production by uniting the whole supply chain under the same roof (Pedersen and Andersen, 2014).

21.2 NEW APPROACHES FOR SUSTAINABILITY IN THE FASHION INDUSTRY THROUGH PSS AND COMMUNICATION

21.2.1 Innovation in the Fashion Business

The fashion industry needs to reduce its negative impact on the environment and have better conditions for those working in the production. Briefly, we can say that in order to become more sustainable, changes both in production and consumption are needed. In some discussions, the volume of the production is referred to as one of the major problems and the logical solution is that it has to be reduced. With the increasing number of people that need to have clothes for pure functional reasons and the increasing demand of fashionable clothes by those who can afford it, it is questionable if it is possible to reduce the volumes as this will have an immediate effect on companies' revenues. It is generally argued that changes are needed on a systemic level—to make the whole fashion system more sustainable and achieve a cradle-to-cradle approach (Carbonaro and Votova, 2009; Fletcher, 2008; McDonough and Braungart, 2002). This will however not happen very quickly. Especially not in an industry that is based on a material dominated growth logic.

The fashion industry is based on the physical products with very little service as part of the revenues. Would it be possible to transfer the money consumers spend on physical products to immaterial services? Maybe, but fashion as a concept seems so bounded in its materiality that services in the context of fashion are rarely discussed—with some exceptions (Fletcher, 2008). Only recently some researchers have started to discuss whether there could be new business models based on services that consumers would be willing to pay for (Armstrong and Lang, 2013; Niinimäki, 2009). This would be a way to reduce the volumes of physical products, but still provide an income for the companies (Niinimäki and Hassi, 2011). Innovative sustainable ideas that can change the system are needed because innovations will attract followers, which will further move the industry toward sustainable solutions.

There are indeed many innovative ideas for new fashion concepts and business models that could lead to a more sustainable fashion industry, even if these solutions are just small steps on a thousand-mile march. Innovations are traditionally thought of as new technologies, design, and materials.

---

* http://en.wikipedia.org/wiki/Slow_Movement.

New fibers, especially cellulose-based ones, and digital printing that needs less water and chemicals, are examples of new technologies that contribute to more sustainable solutions (Loker, 2008).

There are however also new business concepts, which are actually not new but have been lost in the industrialization of fashion: for instance, tailor-made clothes (which is now related to slow fashion), redesign or upcycling of existing clothing, repair of and rent of clothing, and reuse, that is, channels offering secondhand or vintage clothing. These concepts are back, even if in new versions adapted to current technologies, design concepts, and trends. But these new concepts are still very marginal for the fashion industry, also for most of those companies that have launched them. If these concepts are going to take off, and grow they need to be of interest for the last part in the fashion system—the consumers—and the consumers first of all need to become aware of them. This also includes becoming more knowledgeable about why they should embrace them, and what benefits them as consumers. Therefore, it is also important how the companies that have developed these concepts communicate them.

Consumers' attitudes to sustainability usually rank sustainability quite low when buying clothes, and consumers do not want to pay a higher price just because the clothes are more sustainable or ethical (Svengren Holm and Holm, 2010). Consumers' view is that sustainability is the companies' responsibility. Fashion companies are indeed taking a greater responsibility to become more sustainable even if it is done under pressure, for instance, from the media but also from several political and nonpolitical organizations. One example is the Better Cotton Initiative that was the result of a roundtable meeting organized by the WWF with the goal to find more sustainable solutions for farmers. Many of the most well-known retailers and brands are members of this program, for example H&M, Gap, IKEA, and Marimekko, just to mention a few. Many of these companies are also members of another initiative, the Sustainable Apparel Coalition (SAC) with aim to create a tool for designers, the so-called Higg index tool. With the help of this tool, designers can learn about the environmental and social impact of, for instance, material choices and as a result take more sustainable decisions. The Higg-index is built on the Nike index for materials, and further developed to make it easy to use. Nike is also one of the members alongside more than 80 of the world's leading companies, for instance, Patagonia, H&M, Zara, Puma, Adidas, and Burberry. This could be one basis in the future to also inform and enable consumers to make better choices. But we are not there yet.

We need to influence consumers' behavior both when shopping and in the use phase, that is, washing the clothes. We need to reduce the production of new materials, of using the land for fiber production. Is it possible to find new, nonmaterial products that maintain the joy people seem to have in relation to clothes? Few, but still some, companies try to be proactive and develop service concepts that support sales and efficient use of existing materials. As these focus on supporting the fashion products, these concepts are referred to as product service systems (PSS) (Armstrong and Lance, 2013; Niinimäki and Hassi, 2011). Redesign is becoming more common (McDonough, 2013), also in fashion (Brown, 2013; Ericsson, 2014; von Busch, 2008), but seems not to have moved out of the workshop to the regular market and to the established brands. In following, we will describe and discuss some service concepts that Swedish fashion design companies have embraced in order to act more sustainably as part of their business model and how this is communicated to customers.

21.2.2 Fashion Service Concepts

Selling products in several cycles has normally been the role of nonprofit organizations such as the Red Cross or Oxfam. Reuse, or secondhand, sometimes phrased as vintage if we mean high-quality branded items, is quite a large and growing business (e.g., Allwood et al., 2006; Fletcher, 2008; Scott, 2008). This is confirmed by nonprofit organizations like the Red Cross and in Sweden *Myrorna* (the secondhand store of the Salvation Army). Approximately half of the turnover of these nonprofit organizations in 2011, 890 million SEK, was reported to be clothes (ASFB, 2013). The private secondhand trade on the online Swedish secondhand site *Blocket* increased by 23% in the third quarter of 2012 compared to the same period in 2011 (ASFB, 2013). Whereas secondhand is dominated by nonprofit organizations, there is an interest in secondhand/vintage by established designers and fashion design companies (e.g., Scott, 2008). The Swedish fashion brands Filippa K and fashion retailer Boomerang both have a secondhand selection in their stores.

Redesign has become a popular concept, for instance, for the nonprofit organizations that see redesign as a way to utilize all those garments and textiles that they receive. The nonprofit organization can use their whole network of voluntary work that is part of their network. Also, many workshop activities are organized where people gather to make new products from old ones in a very creative way. But very few commercial companies do it because it requires a lot of manual work and this is just too costly. Boomerang, which has made this an important part of its concept, is a rare example.

A third way to reduce production and consumption is to provide a repair service and thereby prolong their life (Fletcher and Grose, 2012). What used to be a necessity for most people was abandoned in the wave of mass production of clothes where the price fell to such a level that nobody considered it worthwhile to repair old clothes, at least not in the affluent Western parts of the world. Older people are however still more likely to mend clothes (Ekström et al., 2012). Exceptions are only those clothes that we truly like so much that we indeed invest time and money to mend them (Niinimäki, 2012). For most clothes, this is no longer considered worthwhile. As a consequence, there are also very few tailors left where you can hand in the clothes that need to be mended. Some laundry services run by people with tailoring skills have repairs and mending of clothes as an additional service. Nudie Jeans, a Swedish fashion company with a focus on denim products, offers this service for its customers and their concept is described next.

A fourth concept is to rent the clothes and in that way satisfy the consumers' need for variety without frequent buying of new clothes. Renting would decrease the material consumption and material flows through slowing down the system. To rent formal dresses is an old concept, but few fashion brands do it, most of all because there is an assumption that we want to own—and control—the clothes we wear. One company that has is testing this concept is Uniforms for the Dedicated, which will be described later. Please address the implications of these four concepts in detail with respect to fashion sector.

In the next section, some companies who have developed these kinds of services as part of their sustainability policy are presented and how they have communicated it on their website and in the store. This is followed by a discussion of different communication strategies for sustainability actions in relation to the brands.

21.2.3 Service Concepts in Selected Swedish Fashion Design Companies

21.2.3.1 Filippa K

Filippa K is a fashion company founded in 1993 and today one of the largest Swedish fashion design companies. Filippa K has a CSR policy and states the alignment between the policy and its general business idea. According to its history, Filippa K took some strategic decisions to act with the aim to become more sustainable in 2008 (Filippa K, Sustainability Report, 2012). This year the company became a member of the Fair Wear Foundation and opened a secondhand store for its own branded clothes.

21.2.3.1.1 Communication of the Secondhand Store

The Sustainability Report for 2012 reports many activities regarding how the negative impact on the environment is being reduced as well as the social conditions in production. However, the secondhand store is not mentioned on Filippa K's website in the list of stores, nor do the regular Filippa K stores in Stockholm have any information about it. When a shop assistant was asked, she knew of it but stated: "It is not part of us, it does not belong to Filippa K." There is also nothing on the company's website about the possibility to hand in used clothes.

One needs to read the report on the company's responsibilities to finally find information about the secondhand store as one of the company's many initiatives. If one is lucky enough to find this information or know there is a secondhand store, an Internet search reveals its website.* There could be several reasons for this opaqueness. It could be a different kind of store, not included in the regular store concept, and possibly Filippa K indeed does not own the store. The secondhand store emerged in collaboration with Judit's secondhand, a well-known secondhand store in Stockholm who also owns the Filippa K secondhand store. The store is located almost beside Judit's secondhand store in Södermalm, in Stockholm. The website for the Filippa K secondhand store is informative, but minimalistic, giving only the most basic needed information: a short description of the store, general information about what assortments to expect, what is asked for, opening hours, where to find, and how to contact the store (Filippa K, secondhand).

A visit to the store reveals that it is nothing like the regular Filippa K's store design; on the contrary, it has a typical secondhand store interior: old items as décor and rather small, but still well organized and pleasant. The clothes sold however are not only secondhand; the selection also includes new clothes from Filippa K, for instance, their sample collections from this year's launch. According to the shop assistant, demand is much greater than supply. On the website there is a tab "searched for" where it is stated that there is demand especially for larger sizes. Filippa K is well known for its slim design, and the shop assistant states that it is common that those who hand in their used Filippa K clothes have outgrown them; hence, the most common clothes in the store are in small sizes. This is of course also the case for the sample collections that were launched for the catwalk. Compared to the other activities that Filippa K conducts in design and production for sustainability purposes, the store has a marginal effect. But it has been mentioned as an example of an innovative concept for enabling more sustainable fashion consumption and is something visible for the consumers, those who find it and are interested.

21.2.3.2 Uniforms for the Dedicated

Behind Uniforms for the Dedicated is a group of creative people who shared interests in snowboarding, mountain climbing, surfing, and travelling. The business started with a T-shirt they designed to wear themselves in the first place—to communicate their identity as a social community—but they also sold the T-shirts in friends' stores. They started as a group of designers, artists, filmmakers, and musicians with a common interest in sustainability—and in working in the creative sector. They all felt dedicated to the idea. In 2007, they decided to try and make a business of what they were doing, not only with clothes, but also with designed furniture, music, and films. Uniforms for the Dedicated was born. Their success is men's clothes. After only a couple of years, the clothes are sold by more than 150 retailers worldwide, including online stores such as Asos and Zalando. The brand has received awards for its creativity. In 2008, Swedish Fashion Council awarded it Rookie of the Year (Swedish Fashion Council), and in 2009 it was awarded *Newcomer of the Year* by the magazine Café (Café Magazine).

In September 2012, Uniforms for the Dedicated opened their first store in Stockholm, in Södermalm. This stays in line with their philosophy of acting sustainably; the materials used for the store are materials they have upcycled themselves. The store is for their range of clothes but also pieces from their furniture collection (Uniforms for the Dedicated).

In the autumn of 2013, Uniforms for the Dedicated selected some parts of its collection for short-term lease in their store in Stockholm. The idea is that renting should provide an alternative to just buying—and thereby contribute to more considerate and careful shopping overall.

21.2.3.2.1 Communicating the Policy and the Concept

On the company's website the philosophy is described under different headlines, such as "Design Philosophy" and "Change-Earth," and the renting concept is labeled "The Collection Library" found

---

* http://www.filippaksecondhand.se.

Fashion Industry and New Approaches for Sustainability

on the bottom lines on the first web page. Here the interested person can read about the idea, why it exists, and the practical details of how to rent the clothes. But the most important communication tool is, according to one of the owners, Mike Lind, to use the store as the platform to show and talk about the concept. One reason for this is that is such a new concept for these kinds of brands that the consumers are not familiar with and needs to experience it and discuss it with the store personnel. So far the number of customers who have rented any apparel is not large, but according to the store manager, the people who enter the store are curious and people have started to talk about it *on the street* and to some extent in social media. It takes time for customers to get used to such a new concept in a regular fashion store. According to Mike Lind, it has been encouraging so the company will continue with the concept and expand it.

21.2.3.3 Nudie Jeans

Nudie is a Swedish jeans and casual wear label founded in 2001 by Maria Erixon Levin, a Swedish designer and entrepreneur, and her ex-husband Joakim Levin (Nudie Jeans). Besides a stated passion for denim and jeans, the owners show a strong commitment to sustainability from all perspectives. Sustainability is the core of its brand. The company has seen positive growth, and 10 years later it is a medium-sized company employing around 50 persons. The owner stresses the philosophy to grow organically without any external investors and therefore has chosen to rely on reinvesting the profits (SVD). According to the two owners, this is because they want to have total control themselves and no one else should tell them what to do.

The company is specialized in unwashed jeans and has received several awards for its ecological and ethical work. There is no doubt that Nudie Jeans is committed to act in a sustainable way. The price level of the products is a consequence of that good quality and ethical behavior throughout the supply chain.

21.2.3.3.1 Communication of the Sustainable Concept and the Repair Store

Nudie Jeans is recognized for its communication of what it does and is ranked as an "A" company on the "Rank a Brand" list, a website ranking the company from a sustainability perspective.* Nudie Jeans' philosophy is that jeans are something very personal, *not just a piece of clothing*—more like *a second skin* for the wearer according to the website. Nudie has thus described their jeans as the *naked truth about denim*. As the wearer is supposed to have an intimate relationship with and passion for his/her jeans, they are intended to be worn for a very long time. But this also means that the jeans become worn out and parts need to be repaired. As a consequence of this philosophy, Nudie Jeans therefore offers a repair service for free in several of its stores around the world. If the wearers do not have access to a Nudie Jeans store, a sewing kit especially for jeans repair can be ordered—also free of charge.

Nudie Jeans' store in Stockholm is located in Södermalm, an area well known as a hipster community where the jeans density is high. A visit to the store was made to see how the repair concept is realized. The store is rather spacious, and there are two sewing machines in the sales room. On the table with the sewing machines are some jeans obviously in the process of being repaired. On the wall behind the cashier hangs around 20 repaired jeans waiting to be picked up by their owners. It is clearly visible that the jeans have been mended although it is nicely done. Worn seams have been patched and holes have been mended with zigzag stitches—further contributing to the personalization of the jeans and the relationship with the wearer. On the shelf opposite, in 4 piles, at least

---

* http://www.rankabrand.org/index.php/ranker/Niels (accessed January 16, 2014). Rank a Brand is a website that, with the help of consumers and volunteers, assesses how companies communicate what they do to become more sustainable. All product categories are considered, also fashion companies. The argument for this ranking is the need for information, according to Niels Oskum, a Dutch engineer and CSR expert: "Most products in our shops are made in low wage countries. When buying stuff, brands are guiding signposts. I like to know what those brands mean in terms of carbon emissions, environment and labour conditions."

40 jeans lie waiting to be repaired. Between five and seven jeans are handed in each day for repair according to the shop assistant.

Nudie Jeans has also expanded its line of denim products with a new category: denim rags. The rags are made from worn-out Nudie Jeans denims that were beyond repair. The rags are referred to as slow craft as they are woven by hand on manual shuttle looms. Pictures on the Nudie Jeans website show how this is done to further communicate the value of these rags (Nudie Jeans).

21.2.3.4 Boomerang

Two entrepreneurs who wanted to create a Scandinavian brand for casual wear founded the Swedish fashion retailer Boomerang in 1976. The first store opened in Örebro and a couple years later, a store opened in Stockholm, which was quickly followed by Göteborg, Malmö, and Jönköping. From 1999, the stores also offered women's wear in the same casual style, followed by collections for children and teenagers in 2008. This year Boomerang also launched the concept *Boomerang effect* with an ambition to be a sustainable actor in the fashion sector. Quality, long-term use and sustainability are mentioned as ingredients of the brand when it is described on the company web page (Boomerang).

The concept *Boomerang effect* includes a return system, vintage, and redesign. The consumer can bring back the clothes that are no longer wanted to the store and receives a 10% discount on the next item bought. The idea behind the junior collection is claimed to be an example as it should be possible for the parents to bring back the outgrown children's clothes and to receive a discount when buying a new one. If the garment is OK, it is then marked as "Good Environmental Choice," a symbol from the Swedish organization Naturskyddsföreningen (Swedish Society for Nature Conservation) and as Boomerang Vintage and sold as *vintage* in some of the regular stores. Those products that do not qualify as vintage are cut and the material, together with other spill material in the production, are used to design interior products that are sold in the Boomerang Home section, a concept created in 2009.

21.2.3.4.1 Communication of the Boomerang Effect

The concept of Boomerang effect is found on the bottom of the first web page with the subheadline Boomerang Vintage, Boomerang Home, and Boomerang Effect Collection. The philosophy of the founders and their view upon quality and sustainability is told in the history of the brand.

Boomerang Vintage is sold in a number of their own regular stores, three in the Stockholm region, in Malmö, Linköping, and Karlstad. In the store, the vintage clothes hang in a special section clearly marked and easy to recognize when you walk around in the store.

21.2.4 COMMUNICATION STRATEGIES OF SUSTAINABILITY POLICIES AND ACTIONS

Historically, we used to repair clothes and shoes, but when prices went down and we could afford it, we bought new things. And the prices for clothes have continued to decrease. We do not know whether the generations that have become used to shopping and buying new and cheap fashion frequently will endorse repair, renting, or buying secondhand to a larger extent. But there is an awakening among people in general about environmental problems (Hethorn and Ulasewicz, 2008). The interest in these issues among students, for instance, at the Swedish School of Textiles, University of Borås, is much larger compared to only 3 or 4 years ago. This is evident also in their choice of subjects for theses. It might be that PSS (Niinimäki, 2012, 2014) and the RE-concepts (REpair, REuse, REturn, REnt) could be the basis for developing new services that people are also willing to pay for DG. The young generation is knowledgeable about climate problems, and the RE concepts could also support their desire for fashion on an individual level, that is, they can see more benefits than just the contribution to a sustainable fashion industry.

The fact that Filippa K engages in sustainability and opens a secondhand store is a positive activity, as it contributes to the reuse of existing clothes and gives consumers an opportunity to clean out their wardrobe before buying new clothes. A bulging closet is a problem for many fashion-interested people, but this is an activity on the margin of the company's business and not supported by its regular brand or business activities. Uniforms for the Dedicated is dedicated to creating a new lifestyle and embracing sustainability. The renting service is in line with this strategy. Nudie Jeans also has sustainability as a core part of its brand and has developed repair and recycle as a service contributing to a more sustainable industry. Nudie Jeans is however the only company that highlights this on the top of the first page of their website. It is easy to understand why the company receives an "A" ranking on the www.rankabrand.org website.

Frequent shoppers and those who are actively interested in sustainability and new lifestyles will learn about the services, but it requires this special interest to find information about them on the company websites. In the stores where decisions on purchase are made, most companies do not communicate sustainability, so we are expected to look at the company's websites to find information about what the company is doing—or be knowledgeable about the brand stories. To be fair, many companies now have training and education of their staff, including shop personnel, about sustainability. Nevertheless, with no special campaign and no immediate first-page information they will not capture the attention of those consumers who only have a tepid interest in sustainability and never ask questions.

There are different strategies for how companies communicate their sustainability to consumers. Some companies have sustainability as part of their operations, like Filippa K, but not as core of the brand, as, for instance, Nudie Jeans, Boomerang, and Uniforms for the Dedicated. Instead, companies like Filippa K seem to be very careful not to mix the core brand with anything other than what they stand to in relation to fashion. Logically, Filippa K is also not mixing its secondhand with its regular store, as, for instance, Boomerang does.

The Nudie Jeans website gives no doubt about the passion for jeans—*and* for sustainability. Actions for sustainability are communicated as a way to endorse the brand. Services like the repair of damaged jeans support the brand strategy. All stories of the brand of Boomerang are related to the Boomerang Effect concept. Compared to most fashion companies, the extensive information about Nudie Jeans' and Boomerang's policy and activities is however rather an exception.

We can therefore conclude that companies have chosen different strategies for communicating how they perceive and act upon sustainability. These strategies are either direct or indirect communication. The companies that communicate directly on the front page of the website also use their regular stores for the services, which further strengthen the sustainability aspects of the brand. Companies that only indirectly communicate the services and other action for becoming more sustainable have externalized the services in relation to the brand. Those companies who do have a sustainability policy but only tells about it in their policy document have indeed integrated the activities in the design and production of the clothes but do not communicate it to others than those who are really interested.

This overview of services and communications strategies does not reveal anything about consumers' response. There is also no evidence that companies who do not communicate are not actually doing anything. Some of these companies might have strict policies, CSR managers, and sustainability actions in many aspects. The question is why do they not communicate it on their websites or in the stores to their consumers? Another question that also still needs to be answered is what prevents direct communications that could clarify for consumers what really is going on in terms of sustainability in those companies who almost hide their services or sustainability actions on a lower hierarchy of their web page tabs? It seems that many companies do not trust the consumers to be interested in sustainability activities or sustainability dimensions of the clothes and the brand. If the system is going to change, also the consumers need to be included. Only when the consumers embrace the services these will take off and grow into new businesses.

21.3 PSS AND CONSUMER-BASED ECO-EFFICIENCY

21.3.1 New Business Thinking for Consumer Satisfaction

The economic and industrial systems of the fashion industry are currently based on fast-changing trends and the planned obsolescence of products. Thus, we need to find new radical ways to create a new sustainable win–win situation for both consumers and manufacturers (Burns, 2010). We need more knowledge about consumers and the consumption side to initiate a sustainable transformation process in the fashion industry and business (Throne-Holst et al., 2007). Without a deeper understanding of the consumption side, it is not possible to achieve a business transformation that would offer value for both consumers and producers—and even for the environment—by reducing the material throughput in the system.

Thus, the question to be addressed is how fashion offerings should be designed and manufactured to better suit consumer preferences in a more sustainable way. At present, the business models are linked to the volume of sales and production alone. Therefore, more sustainable consumption, that is, a decrease in purchasing, production, and sales, is seen merely as leading to reduced volumes and lower profitability in production, not as an opportunity for a new kind of green business (Allwood et al., 2008). A radical new mindset among designers, manufacturers, and consumers is needed in order to find more sustainable ways to fulfill consumer needs and to attain sustainable improvements in the relationship between production and consumption. The manufacturing side is not the only cause of the textile industry's environmental load. To a large extent, sustainable development hinges on change, not only in production systems but also in consumption patterns (Perrels, 2008).

The fashion industry has placed an emphasis on keeping the price of the final product low and increasing efficiency in production. Designers, manufacturers, and retailers have paid less attention to other dimensions of the offerings and therefore consumers' deep wishes and values are not at all in the core of the fashion industry. Fashion development does not listen to the views of consumers. Hence, the products are designed and produced to cater to regularly changing trends that enable quick profit, rather than radically rethinking design and manufacturing in order to base them on consumer needs and sustainability. Thus, the question to be addressed is how textile and clothing offerings should be designed and manufactured to better suit consumer preferences in a more sustainable way.

21.3.2 Consumer-Based Eco-Efficiency

The eco-efficiency principle means *doing more with less*. This has meant improving technologies and systems to be less resource-intense, which has yielded cost savings for industry, and hence it can be said to have involved a techno-economic approach that attempts to decrease the environmental impact of industrial manufacturing. Industry's interest in the eco-efficiency approach since the 1990s has not led to significant advances in sustainability, as this approach does not account for the dimension of human needs in the current unsustainable consumption patterns. Even while industry has moved toward more effective processes, the volumes of production and consumption have increased due to the cheap prices of commodities (Michaelis, 2001).

Park and Tahara (2008) argue that eco-efficiency can be used only as an evaluation tool for design alternatives, not to identify key sustainability problems. They propose that producer-based eco-efficiency has to be combined with *consumer-based* eco-efficiency in order to better identify key eco-design issues. This process involves not only analyzing the environmental aspects of a product, but also assessing product quality and consumer satisfaction. Consumer satisfaction is best addressed by offering good performance in those dimensions that are important to the consumer. Focusing on consumer satisfaction opens up possibilities to create a new kind of green business thinking in the area of fashion. Furthermore, consumer-based eco-efficiency can create value simultaneously for both the consumer and company.

21.3.3 Product Satisfaction

Burns (2010) proposes that a new sustainable balance could be driven by social and environmental goals such as reducing environmental impacts and waste, increasing consumer satisfaction and encouraging emerging technological evolution and positive social change. To date, consumer satisfaction has not been widely researched. Hence, knowledge on how to deliver sustainable satisfaction for consumers in different product groups is still largely lacking.

Product satisfaction in the fashion field is not well known or studied. On the other hand, earlier studies on product attachment have shown that positive emotions, memories, special meanings, and reflective levels operating between the product and user create deep attachment (e.g., Csikszentmihalyi and Rochberg-Halton 1981; Mugge et al., 2005; Niinimäki, 2010). The reflective level of product attachment comprises feelings, emotions, self-image, personal satisfaction, memories and cognition, and it is constructed during a longer period (Norman, 2005). Therefore, these emotional attachments are very individual and related mainly to old garments that may no longer even be in use (Niinimäki, 2010). We have less knowledge about the consumer–product relationship during ownership or use, even though this phase is critical for understanding product replacement (Mugge et al., 2010).

In contemporary society, consumers become attached to some objects while easily disposing of others. From a sustainable development viewpoint, it is important to lengthen the lifespan of the product.

Designers must strengthen consumers' attachment to products in order to lengthen the psychological lifespan of products (Mugge et al., 2005; van Hinte, 1997). Simultaneously, when consumers develop deep product attachment with some garments, they may still continue to purchase new ones. On the other hand, Belk (1991) argues that attachment is one way to foster caring toward the product and postpone its replacement. When we deeply value products and they are meaningful and precious to us, they deserve to be well taken care of (Walker, 2007). Deep attachment toward a product prevents the disposal of this object. Earlier studies have shown that consumers mend and take good care of cherished garments, which is one reason for the long life of meaningful clothing (Niinimäki and Koskinen, 2011).

The following attributes determine the longevity of and satisfaction with clothing: quality, functional and esthetic aspects, and the values of the product or company. Quality means durable materials, durability in use, durability in laundering and high manufacturing quality. Furthermore, functional aspects are important to the consumer: that is, easy maintenance, suitability in use, and satisfying use experience. Easy maintenance means that garments are suitable for water washing (no chemical laundering) and they do not need frequent washing, as some synthetic fibers do. Furthermore, esthetic attributes are most important in the context of garment satisfaction: that is, beauty, style, color, fit, and tactility (comfortable materials) (Niinimäki, 2014).

For those consumers who have a strong interest in ethical and environmental issues, the following values are important in clothing choices and also play a role in clothing satisfaction: local or ethical production, eco-materials, and long lifespan of garments. These environmental attributes are linked to a consumer's individual values and are defined to be credence quality attributes (ibid).

In the study of Niinimäki and Armstrong (2013), long-term use of garments was investigated among American consumers. The most important attributes that led consumers to keep a garment for a long time were memories, comfortable use experience, and good fit, in this order. Memories correlate to product attachment, but comfortable use and good fit are linked to the use situation and functional experiences.

Further, the same study points out that in long-term ownership of garments, there are four different stages of attachment: use enjoyment, liking and loving, reflective stage, and finally cemented memento. In the first stage, the most important aspect is active use (owned for 0–6 years). In the second stage (owned for 7–18 years), active use decreases. In the third stage (owned for 19–21 years), all use categories are as strong and stable (active, seldom, and in storage), but the memory aspect is also present and increasing. Finally in the memento stage (owned for over 22 years), the garment is used seldom or not at all; it has become a memento that is mainly kept in storage. Accordingly, the active use period and positive use experiences in the early period of ownership are most significant when aiming for product attachment and product satisfaction in the fashion field.

Designers cannot really offer the memory aspect to a consumer through a product, as memories are very personal and linked mainly to individual reasons and events in one's life. For product satisfaction, the active use period is most important, as in this early phase the pleasurable use experience is the most dominant attribute. Accordingly, comfortable use and good fit are the most important factors in providing product satisfaction to consumers.

21.3.4 Product Service Systems

In the context of sustainable development, two polar positions can be identified: from end-of-pipe strategies that address waste in a traditional production consumption system to those in which new qualitative circumstances and criteria for sustainable lifestyles are identified (Vezzoli and Manzini, 2008). This wide variety of sustainable approaches offers different design and manufacturing strategies and even different levels of dematerializing consumption, from very little to significant. When evaluating previous approaches and questioning business thinking in fashion, PSS offer possibilities to create value in a new way. There are three different ways to create a PSS: product-oriented, use-oriented, and results-oriented (Tukker, 2004). Product-oriented is the easiest to build by adding a service aspect to the actual product, and it is also the easiest for consumers to accept and even easiest to build from a business point of view. The goal when constructing this kind of system might be to add some new business possibilities to the traditional fashion system or to develop better consumer satisfaction or even to extend the use time of garments. Still, the product represents the main focus and biggest value offering.

The most radical approach is a result-oriented PSS, which concentrates on delivering satisfaction to consumers in a totally dematerialized way by, for example, offering some functional activity without a physical product. Here, the value offering is wholly provided in the service aspect. One example of a service without any product offering is stylist consulting, where a stylist comes to your home and goes through your closet to see what you should keep, what items still fit and how to create a new look with old garments.

Use-oriented PSS, in which the value is as much in the product as in the service, lies between these two approaches. For example, lending and renting are these kinds of services that offer temporal use and satisfaction to the consumer without the owning of the product. In such an offering, the meaning of the product is as important as that of the service aspect.

The interesting thing in PSS is that it offers an opportunity to disconnect the value aspect of the offering from the product and production and even materialistic consumption and therefore makes it possible to dematerialize consumption (Mont, 2002). PSS thinking offers possibilities to concentrate more on consumer satisfaction issues, desires, and wishes than traditional mass manufacturing. When developing PSS strategies, a company has to develop a longer and deeper contact with the consumers and end users to understand their needs. Further, new kinds of networks could be created around the service aspect.

21.3.5 PSS for Product Satisfaction through Perfect Fit

The garment's good fit is one of the most important attributes in the experienced quality of clothing, something that is tested and experienced during the use phase and that is hard to estimate when purchasing the garment. Further, good fit is the best way to offer product satisfaction to the consumer.

NOMO Jeans is a company that offers made-to-measure jeans. According to the company's research, the average consumer tries on approximately 12 pairs of jeans before finding suitable ones.

NOMO Jeans has identified their main customer groups and their main purchase drivers in the jeans sector as follows:

- Women aged 30–60, who are fit-seekers
- Men aged 30–40, who seek individuality
- Men aged 40–60, who want to invest in convenience

In NOMO Jeans' system, each customer's measurements are taken with a 3D body scanning machine. The jeans are then made to measure. The customer can select fabrics, details, and decorations from the offered selection and this process results in individualized jeans. This offers deep customer satisfaction through perfect fit and an individual look.

The company's basic idea is to produce high-quality products that are guaranteed to fit. Through this strategy the company is offering a satisfaction guarantee for the customer; the customer has 30 days to return the jeans to the company if he or she is not happy with them and especially with the fit. Once a customer's measurements have been taken, he or she can also order new jeans online, which is a convenience factor for the customer. The company aims to create a solid customer relationship based on product satisfaction (NOMO Jeans).

Another example of a business with PSS offering product satisfaction is Anna Ruohonen. Ruohonen's business is based on high-quality design, quality materials, and good fit. The products are made to order, which minimizes the amount of inventory on hand. The company creates its own collections, produces a couple of sample collections, and makes it possible for consumers to specify all orders and measurements on an individual basis. Garments can then be created based on each wearer's measurements, thus enabling him/her to experience greater satisfaction through perfect fit. This design and manufacturing strategy may also help producers avoid the problem of overproduction. Small enterprises could offer their collections in small shops carrying a sample collection and obtain orders directly from consumers, allowing them to avoid extra production.

21.3.6 PSS to Fulfill Fashion Needs for Change

The fashion system is based on fast-changing trends, constant change, and the unsustainable desires of consumers. Consumers seek novelty and change through fashion. Fashion has a strong connection to our emotional side, and we are looking for social acceptance and emotional stimulation through our appearance and clothing choices. Fashion and garments show and hide our identity, and garments are part of the social code and communication with others.

How then could a business offer change and the fulfillment of fashion needs in a more sustainable way? Recently, several companies have started creating redesign fashion from old textiles and clothing. One pioneer in the field of redesign is Globe Hope in Finland. This company was established in 2003 and focuses on using recycled textile material to create redesign and upcycled products that mainly appeal to young consumers. Globe Hope fulfills the emotional aspect of satisfaction through trendy fashion, but operates in a more environmentally responsible way than an ordinary fashion company. The company bases its offering on the statement that eco-fashion does not have to be ugly and boring, but can instead be fun and trendy. Redesign fashion is a better way to offer fashion change to consumers than using virgin materials in production, yet it does not require consumers to change their behavior and is therefore easy for them to accept.

A more radical way of offering change in fashion is fashion libraries. There are several such libraries in the Nordic countries. In Finland, the Fashion Library (Nopsa vaatelainaamo) lends out clothes, shoes, and apparel to its members. The collection consists of new Finnish design and vintage clothing. The fashion collections are from Finland's most inspirational young designers. The membership is valid for 6 months and is reasonably priced. Students get a discount. Members get to borrow products for 2 weeks at a time and can also purchase the pieces they have borrowed. The collection grows and changes each week, which keeps up the interest of members (http://www.nopsatravels.com/nopsa/klubi/).

Borrowing garments is a good way to consume fashion in a more ecological way. Consumers can decrease their impulse purchasing by trying out different styles without actually purchasing and owning the garments. Additionally, the use frequency of the Fashion Library garments is high, which increases the eco-efficiency of the service.

The Fashion Library is a radical example of how to dematerialize consumption. Even though these existing fashion libraries are probably yet not profitable, they do show that some consumers are ready to accept collaborative consumption and decrease their own fashion consumption. They are experiments in how to fulfill consumers' fashion needs in a sustainable, fun, and social way. A customer can easily try out different fashion styles and expensive brands without purchasing anything. This offers value to the customer by enabling her/him to save money.

21.3.7 PSS to Fulfill Consumers' Cognitive Needs

Consumers in the Western world are worried about environmental issues in fashion and have a high need for environmental and ethical information about this sector. A web portal focusing on green and ethical clothes can provide basic information and also critical commentary concerning the fashion sector. The Internet can thus help consumers to make better choices as well as fulfill their cognitive information needs. In the best cases, offering environmental information on different brands might even satisfy the customer's emotional and cognitive needs with no need to make purchases.

Green Garments (Vihreät vaatteet) is a web portal that provides objective information on ethical and sustainable fashion. This service is provided by Anniina Nurmi from Nurmi Design and is aimed at both professional designers and consumers. In addition to information, the page offers links to sustainable fashion brands' webpages and online shops, which makes it easy for consumers to reach these companies and their offerings. This is a convenience factor for consumers. Anniina Nurmi also designs and produces her own brand, Nurmi Design, which is based on sustainable and ethical principles. One of the main values behind the company is transparency in all aspects. The company discloses a great deal of information on all aspects of its design, manufacturing and materials, and in this way offers cognitive satisfaction to ethical consumers.

As consumers say that finding ethical and environmental information about garments and brands is difficult, services to offer this information would help make life easier for consumers. The Green Garments webpage and Nurmi Design satisfy the need for information and knowledge, which represents the intellectual aspect of ethical consumption. The Green Garments web portal also provides consumers with easy access to other green brands and in this way fulfills customers' fashion and aesthetic needs.

21.3.8 Shifting Business toward Service Thinking and Consumer-Based Eco-Efficiency?

Services that aim for deeper product satisfaction, dematerializing consumption or fulfilling fashion needs in a more sustainable way offer not only environmental benefits but also new business opportunities in fashion. Yet product service thinking is novel and needs a creative way of fundamentally rethinking the business and offerings and further focusing on consumer needs and wishes. Product service systems offer business possibilities both at the local (e.g., renting, upgrading, made-to-measure) and the global level (e.g., mass-manufacturing, online exchange stocks, and do-it-yourself concepts). Product service systems allow manufacturers to be in direct contact with the consumers and better address their wishes and develop a connection to the end user and even to collect information from consumers to create better-fitting services and products for better fulfilling consumers' needs and deep wishes.

In PSS thinking, companies have a possibility to construct their business offering according to consumers' needs. Therefore, they provide opportunities to offer consumer-based eco-efficiency and in so doing aim for more satisfied consumers, who are ready to decrease their own fashion consumption.

21.4 CONCLUSIONS

To transform the current fashion system toward sustainable development, radical thinking and other dimensions of the fashion offering have to be developed: for example, more radical business models, rethinking value creation, as well as concentrating on consumer wishes and values. Sustainable fashion business models must be viewed from more holistic perspective that prioritizes a broader set of stakeholders and more dimensions of performance (people, profit, planet).

This chapter introduced new sustainable approaches and initiatives in the global fashion industry. Additionally, it presented green business thinking for sustainable fashion through theory and real-life cases, which showed that we need brave designers and companies who question today's practices and examine how to do things differently in the fashion field. These forerunners can demand sustainable change from their suppliers, subcontractors, or manufacturers and through this pressure create change in the fashion industry. They can even educate and show consumers how to be a more ethical and responsible consumer and therefore these forerunners can show examples, how to do transformation in fashion.

The examples presented in this chapter highlighted some of the sustainability initiatives in the upstream and downstream supply chain that may eventually inspire the development of more sustainable business models in the fashion industry. Depending on perspective, there are reasons to be both optimistic and pessimistic about the future of sustainability in the fashion industry. On the one hand, we are seeing an increasing interest in various types of recycling, upcycling, and downcycling, which ultimately challenge the linear system view of today's fashion industry. Moreover, a mushrooming of standards, systems, and tools offer guidelines for businesses, promote transparency, and help institutionalizing sustainability within the fashion industry. On the other hand, we are still to see a major shift in dominant manufacturing and consumption patterns. Still, the fast fashion sector has grown dramatically, and although consumers overall express an interest in sustainability, these interests are rarely translated into everyday consumption behavior (Allwood et al., 2006; DEFRA, 2010). Therefore, it is too early to celebrate the success of sustainable fashion. Concerted action is required from businesses, policymakers, researchers, and consumers to address the fundamental sustainability challenges of the fashion industry. Moreover, these sustainability challenges call for new business models that provide environmentally sound and financially viable offerings to customers and consumers.

REFERENCES

Allwood, J., Ellebaek Laursen, S., Malvido de Rodriguez, C., Bocken, N. 2006. Well dressed? The present and future sustainability of clothing and textiles in the United Kingdom. University of Cambridge, Institute for Manufacturing, Cambridge, U.K.

Allwood, J. M., Laursen, S. E., Russell, S. N. et al. 2008. An approach to scenario analysis of the sustainability of an industrial sector applied to clothing and textiles in the UK. *Journal of Cleaner Production* 16:1234–1246.

Armstrong, C., Lang, C. 2013. Sustainable product-service systems: The new frontier in apparel retailing? *Research Journal of Textile and Apparel* 17(1):1–12.

ASFB. 2013. Modebranschen i siffror—statistik och snalys. (Fashion Industry in Figures—Statistic and Analysis). Report 13:03. Association of Swedish Fashion Brands and Fashion Incubator in Borås Volante, Stockholm, Sweden.

Bardhi, F., Eckhardt, G. W. 2012. Access-based consumption: The case of car sharing. *Journal of Consumer Research* 39(4):881–898.

Belk, R. W. 1991. The ineluctable mysteries of possessions. *Journal of Social Behaviour and Personality* 6(6):17–55.

Bocken, N. M. P., Short, S. W., Rana, P., Evans, S. 2014. A literature and practice review to develop sustainable business model archetypes. *Journal of Cleaner Production* 65(15):42–56.

Botsman, R. and Rogers, R. 2010a. Beyond Zipcar: Collaborative Consumption, Harvard Business Review, 88(10):30–30.

Botsman, R. and Rogers, R. 2010b. What's Mine is Yours, Collins, London.

Brass, C. 2012. What if the washing machine became a service and not a product. In Black S (ed.), *The Sustainable Fashion Handbook*. Thames and Hudson, London, U.K., p. 61.

Brito, M. P., Carbone, V., Blanquart, C. M. 2008. Towards a sustainable fashion retail supply chain in Europe: Organisation and performance. *International Journal of Production Economics* 114:534–553.

Brown, S. 2013. *Refashioned: Cutting-Edge Clothing from Upcycled Materials*. Laurence King Publishing, London, U.K.

Burns, B. 2010. Re-evaluating obsolescence and planning for it. In: Cooper, T. (ed.), *Longer Lasting Products: Alternatives to the Throwaway Society*. Gower Publishing, Farnham, U.K., pp. 39–60.

Carbonaro, S., Votava, C. 2009. The function of fashion? The design of new styles of thoughts. *Nordic Textile Journal* 30–45.

Chesbrough, H. 2007. Business model innovation: It's not just about technology anymore. *Strategy & Leadership* 35(6):12–17.

Csikszentmihalyi, M., Rochberg-Halton, E. 1981. *The Meaning of Things: Domestic Symbols and the Self*. Cambridge University Press, Cambridge, U.K.

DEFRA. 2010. Sustainable clothing action plan (Update February 2010). Department for Environment, Food and Rural Affairs, London, U.K.

Deloite. 2014. Modeanalysen. Deloitte.

Deloitte. 2013. Fashion sustainability 2013. Deloitte.

Ecotextile News. 2013. Puma to debut cradle to cradle collection. Available: http://www.ecotextile.com/2013021211911/fashion-retail-news/puma-to-debut-cradle-to-cradle-collection.html (January 15, 2013).

Ekström, K., Gustafsson, E., Hjelmgren, D., Salomonson, N. 2012. Mot en mer hållbar konsumtion. En studie om konsumenters anskaffning och avyttring av kläder (Towards a more sustainable consumption. A study of consumers' purchase and disposal of clothes). University of Borås, School of Business and IT, Borås, Sweden.

EMF. 2013. Towards the circular economy 1: Economic and business rationale for an accelerated transition. Ellen MacArthur Foundation, Cowes, U.K.

Ericsson, A. 2014. The life of a dress. Master Thesis. Swedish School of Textiles, University of Borås, Borås, Sweden (forthcoming).

Fletcher, K. 2008. *Sustainable Fashion & Textiles: Design Journeys*. Earthscan, London, U.K.

Fletcher, K., Grose, L. 2012. *Fashion and Sustainability: Design for Change*. Laurence King Publishing, London, U.K.

Guide, V. D. R., Van Wassenhove, L. N. 2009. OR FORUM—The evolution of closed-loop supply chain research. *Operations Research* 57(1):10–18.

Hethorn, J., Ulasewicz, C. 2008. Sustainable fashion: Why now? A conversation about issues, practices and possibilities. Fairchild Books, New York.

Kant Hvass, K. (forthcoming). Post-retail responsibility of garments—A fashion industry's perspective. *Journal of Fashion Marketing and Management*.

Kant Hvass, K. 2014. Post-retail responsibility of garments—A fashion industry perspective, *Journal of Fashion Marketing and Management*. 18(4):413–430.

Kelly, M. 2012. In the loop. *Ecotextile News*. June/July edition.

Loker, S. 2008. A technology-enabled sustainable fashion system: Fashion's future. In: Hethorn, J., Ulasewicz, C. (eds.), *Sustainable Fashion: Why Now? A Conversation about Issues, Practices, and Possibilities*. Fairchild, New York.

Lüdeke-Freund, F. 2009. Business model concepts in corporate sustainability contexts. Centre for Sustainability Management, Leuphana Universität Lüneburg, Lüneburg, Germany, http://pure.leuphana.de/ws/files/1174317/Luedeke_Freund_Business_Model_Concepts_in_Corp._Sust._Contexts.pdf. Accessed January 12, 2015.

McDonough, W., Braungart M. 2002. *Cradle to Cradle: Remaking the Way We Do Things*. North Point Press, New York, USA.

Michaelis, L. 2001. Sustainable consumption and production. In: Dodds, F., Middleton, T. (eds.), *Earth Summit 2002: A New Deal*. Earthscan, London, U.K., pp. 264–277.

Michelini, L., Fiorentino, D. 2012. New Business Models for Creating Shared Value. *Social Responsibility Journal* 8(4), 561–577.

MISTRA. 2010. Mistra future fashion research program: Background paper. The Foundation for Strategic Environmental Research (MISTRA), Stockholm, Sweden.

Mont, O. K. 2002. Clarifying the concept of product-service system. *Journal of Cleaner Production* 10(3):237–245.

Morana, R., Seuring, S. 2007. End-of-life returns of long-lived products from end customer—Insights from an ideally set up closed-loop supply chain. *International Journal of Production Research* 45(18–19):4423–4437.

Morris, M., Schindehute, M., Allen, J. 2005. The entrepreneur's business model: Toward a unified perspective. *Journal of Business Research* 58:726–735.

Mugge, R., Schifferstein, H., Schoormans, J. 2010. Product attachment and satisfaction: Understanding consumers' post-purchase behavior. *Journal of Consumer Marketing* 27(3):271–282.

Mugge, R., Schoormans, J., Schifferstein, H. 2005. Design strategies to postpone consumer's product replacement: The value of a strong person–product relationship. *The Design Journal* 8(2):38–48.

Niinimäki, K. 2009. Consumer Values and Eco-Fashion in the Future. In: Koskela, M., Vinnari, M. (eds.). *Future of the Consumer Society*. Proceedings of the Conference, 28–29 May, 2009, Tampere, Finland, pp.125–134. Finland Futures Research Centre, Turku School of Ecnomics. http://www.tse.fi/FI/yksikot/erillislaitokset/tutu/Documents/publications/eBook_2009-7.pdf, Accessed January 1, 2013.

Niinimäki, K. 2010. Forming sustainable attachment to clothes. *Seventh International Conference on D&E Conference in IIT*, October 4–7, 2010, Chicago, IL.

Niinimäki, K. 2012. Proactive fashion design for sustainable consumption. *Nordic Textile Journal* 1:60–69.

Niinimäki, K. 2014. Sustainable consumer satisfaction in the context of clothing. In: *Product-Service System Design for Sustainability*. Vezzoli, C., Kohtala, C., Srinivasan, A. (eds.) Greenleaf, Sheffield, U.K., pp. 218–237.

Niinimäki, K., Armstrong, C. 2013. From pleasure in use to preservation of meaningful memories: A closer look at the sustainability of clothing via longevity and attachment. *International Journal of Fashion Design, Technology and Education* 6(3):190–199. doi: 10.1080/17543266.2013.825737.

Niinimäki, K., Hassi, L. 2011. Emerging design strategies in sustainable production and consumption of textile and clothing. *Journal of Cleaner Production* 19:1876–1883.

Niinimäki, K., Koskinen, I. 2011. I love this dress, it makes me feel beautiful: Emotional knowledge in sustainable design. *Design Journal* 14(2):165–186.

Norman, D. 2005. *Emotional Design. Why We Love (or Hate) Everyday Things*. Basic Books, New York.

Osterwalder, A., Pigneur, Y. 2010. *Business Model Generation*. John Wiley & Sons, Inc., Hobroken, NJ.

Park, P., Tahara, K. 2008. Quantifying producer and consumer-based eco-efficiencies for the identification of key ecodesign issues. *Journal of Cleaner Production* 16:95–104.

Pedersen, E. R. G., Andersen, K. R. 2014. The SocioLog.dx experience: A global expert study on sustainable fashion, MISTRA Future Fashion. http://www.mistrafuturefashion.com/en/publications/Documents/CBS%202014-01-23%20Report%20Project%201.pdf. Accessed January 12, 2015.

Pedersen, E. R. G., Gwozdz, W. 2014. From resistance to opportunity-seeking: Strategic responses to institutional pressures for corporate social responsibility in the Nordic Fashion Industry. *Journal of Business Ethics* 119:245–264.

Pedersen, E. R. G., Laursen, L. M. 2014. Back to basics? Exploring the stakeholder orientation among managers in the Swedish Fashion Industry. Work in progress document.

Pedersen, E. R. G., Netter, S. 2013. Collaborative consumption: Business model opportunities and barriers for fashion libraries. CBS working paper: In peer review.

Perrels, A. 2008. Wavering between radical and realistic sustainable consumption policies: In search for the best feasible trajectories. *Journal of Cleaner Production* 16:1203–1217.

Perry, P., Towers, N. 2009. Determining the antecedents for a strategy of corporate social responsibility by small- and medium-sized enterprises in the UK fashion apparel industry. *Journal of Retailing and Consumer Services* 16(5):377–385.

Ravasio, P. 2013/2014. Business case grows for circular economy. *Ecotextile News*. December 2013/January 2014 edition.

Schaltegger, S., Lüdeke-Freund, F., Hansen, E. G. 2011. Business cases for sustainability and the role of business model innovation—Developing a conceptual framework. Centre for Sustainability Management, Leuphana Universität Lüneburg, Lüneburg, Germany.

Scott, W. 2008. Recycle and reuse as design potential. In: Hethorn, J., Ulasewicz, C. (eds.), *Sustainable Fashion: Why Now? A Conversation about Issues, Practices, and Possibilities*. Fairchild, New York.

Stubbs, W., Cocklin, C. 2008. Conceptualizing a "sustainability business model". *Organization & Environment* 21(2):103–127.

STWR. 2012. Financing the global sharing economy. Share the World's Resources (STWR), London, U.K.

Svengren Holm, L., Holm, O. 2010. Sustainable fashion and new business models. *Nordic Textile Journal* 1:30–39.

The Co-Operative, 2012. Ethical consumer markets report 2012. http://www.co-operative.coop/PageFiles/416561607/Ethical-Consumer-Markets-Report-2012.pdf. Accessed January 12th, 2015.

Throne-Holst, H., Sto, E., Strandbakken, P. 2007. The role of consumption and consumers in zero emission strategies. *Journal of Cleaner Production* 15:1328–1336.

Tukker, A. 2004. Eight types of product-service system: Eight ways to sustainability? Experiences from SusProNet. *Business Strategy and the Environment* 13(4):246–260.

Ulasewicz, C., Baugh, G. 2013. Creating new from that which is discarded. The collaborative San Francisco Tablecloth Repurposing Project. In: Gardetti, M. A., Torres, A. L. (eds.), *Sustainability in Fashion and Textiles: Values, Design, Production and Consumption*. Greenleaf, Sheffield, U.K., pp. 165–181.

Vezzoli, C., Manzini, E. 2008. *Design for Environmental Sustainability*. Springer, New York.

van Hinte, E. (ed.). 1997. *Eternally Yours: Visions on Product Endurance*. Rotterdam, the Netherlands.

von Busch, O. 2008. Fashion-able: Hacktivism and engaged fashion design. Art Monitor. A Publication Series from the Board for Artistic Research (NKU) of the Fine. Applied and Performing Arts University of Gothenburg, Gothenburg, Sweden.

Waddock, S. 2008. Building a new institutional infrastructure for corporate responsibility. *Academy of Management Perspectives* 22(3):87–108.

Walker, S. 2007. *Sustainable by Design: Exploration in Theory and Practice*. Earthscan, London, U.K.

Walsh, B. 2011. 10 Ideas that Will Change the World, http://content.time.com/time/specials/packages/article/0,28804,2059521_2059717_2059710,00.html. Accessed January 12, 2015.

WEF. 2013. Circular economy innovation & new business models initiative. World Economic Forum (WEF) Young Global Leaders Taskforce.

Wells, P., Seitz, M. 2005. Business models and closed-loop supply chains: A typology. *Supply Chain Management: An International Journal* 10(4):249–251.

WRAP. 2012. Valuing our clothes. Waste & Resources Action Programme (WRAP). http://www.wrap.org.uk/sites/files/wrap/VoC%20FINAL%20online%202012%2007%2011.pdf. Accessed January 30, 2012.

WRAP. 2013. Evaluating the financial viability and resource implications for new business models in the clothing sector. Waste & Resources Action Programme (WRAP). http://www.wrap.org.uk/sites/files/wrap/Clothing%20REBM%20Final%20Report%2005%2002%2013_0.pdf. Accessed January 30, 2013.

Zelwak, R., Pedersen, E. R. G. 2014. Collaborative consumption: New model for fashion Consumption, Unpublished report. To be published on MISTRA Future Fashion website Summer 2014.

Zott, C., Amit, R., Massa. L. 2010. The business model: Theoretical roots, recent developments, and future research. IESE Business School, University of Navarra, Barcelona, Spain.

ONLINE REFERENCES

Accenture and UN Global Compact. 2013. *The UN Global Compact-Accenture CEO Study on Sustainability—Architects of a Better World*. Accenture http://www.accenture.com/SiteCollectionDocuments/PDF/Accenture-UN-Global-Compact-Acn-CEO-Study-Sustainability-2013.PDF Accessed March 15, 2014.

Deloitte 2014: http://www2.deloitte.com/content/dam/Deloitte/dk/Documents/consumer-business/Modeanalyse-2014.pdf (accessed January 12, 2015).

Café Magazine, http://cafe.se/cafes-stora-modepris-2009/. Accessed January 31, 2014.

EU Waste Hierarchy, http://en.wikipedia.org/wiki/European_Waste_Hierarchy. Accessed January 25, 2014.

Fashion Library, Nopsa Vaatelainaamo, http://www.nopsatravels.com/nopsa/klubi/.

Filippa K, 2012, Sustainability report, http://www.filippa-k.com/wp-content/uploads/2008/06/Sustainability-report-2012.pdf. Accessed January 31, 2014.

Filippa K, Secondhand, http://www.filippaksecondhand.se. Accessed January 31, 2014.

Globe Hope, http://www.globehope.com/. Accessed January 1, 2013.

Green Garments, Vihreät vaatteet, http://www.vihreatvaatteet.com/.

NOMO Jeans, http://www.nomojeans.com/.

Nudie Jeans, http://www.nudiejeans.com. Accessed January 31, 2014.

Rank a Brand, http://rankabrand.org. Accessed March 19, 2014.

Sustainable Apparel Coalition, http://www.apparelcoalition.org. Accessed February 10, 2014.

SVD, http://www.svd.se/naringsliv/branscher/handel-och-tjanster/kreativt-kaos-pa-vag-mot-jeanstoppen_7406842.svd. Accessed February 10, 2014.

Swedish Fashion Council, http://www.stockholmfashionweek.com/rookiesdesigners.php. Accessed January 25, 2014.

Uniforms for the Dedicated, http://se.uniformsforthededicated.com/pages/stockholm-store. Accessed March 19, 2014.

WEF 2013: http://www.collaborativeconsumption.com/2013/06/13/report-world-economic-forum-sharing-economy-position-paper/ (accessed 12 January, 2015)

22 Eco-Design/Sustainable Design of Textile Products

Thilak Vadicherla and D. Saravanan

CONTENTS

22.1 Introduction 475
22.2 Sustainable Design.. 476
22.3 Fashion versus Sustainability 476
22.4 Sustainable Design Principles 477
    22.4.1 Sustainable Fibers.. 479
    22.4.2 Low-Impact Materials... 481
    22.4.3 Renewable Sources of Energy 482
    22.4.4 Energy Efficiency.. 482
    22.4.5 Water Efficiency.. 484
    22.4.6 Waste Management (Reduction of Material and Recycling)............................... 487
    22.4.7 Biomimicry.. 488
    22.4.8 Easily Biodegradable Products 489
    22.4.9 Pollution Prevention.. 490
    22.4.10 Design for Reuse and Recycling 491
    22.4.11 Service Substitution 492
    22.4.12 Community Couture 492
    22.4.13 Do-It-Yourself and Patchworks 494
    22.4.14 Fair Trade/Ethical Practices 494
    22.4.15 Near-Sourcing... 495
22.5 Best Available Techniques.. 496
22.6 Conclusion 497
References... 497

22.1 INTRODUCTION

Sustainability is on the lips of almost every fashion designers these days. Whenever a new design/product/service is launched, many people raise the questions, "Is this design/product/service sustainable?" "Is it eco-friendly?" "Are low impact materials, energy and water efficient processes used?" "What about the ease of recyclability?" etc. All these questions indicate the consumer's increased awareness and the concern over the sustainability practices. Textiles/apparels are no exception to these. In the context of surge in fast fashion and increased amounts of textiles/apparels, wastes have forced the designers and manufacturers focus their efforts toward sustainable design throughout the processes and the supply chain as the whole. The aim of this chapter is to provide the awareness on sustainable design principles through which sustainability can become a routine rather than a luxury.

Conventional 3R (reduce, recycle, reuse) or 4R (reduce, recycle, reuse, rebuy) concepts have clear focus toward the sustainability; however, newer principles have also been evolved over a period of time. Principles based on 10 point or 12 point or 14 point approaches have been postulated for introducing the sustainability into the fashion and fashion items. All these approaches converge toward advocating the use of less energy, fewer or limited use of resources, safeguarding the natural resources from depletion, not involving in polluting the environment and recommend the reuse or recycling at the end of life cycle of the products. Key priority areas and the roadmap for eco/sustainable design of textile products include (1) enhancing the performance and reducing the environmental impact of the products through sustainable design, selection of fibers and fabrics, maximizing the benefits of 3R/4R approaches and cleaning or maintenance of textiles and clothing; (2) analyzing the consumption trends and attitude of the consumers; (3) bringing the awareness through education and other social networks; (4) creating the market with niche products; and (5) the measures to improve the entire supply chain. Worldwide Responsible Accredited Production (WRAP) actively popularizes the sustainable clothing action plan (SCAP) to improve the sustainability of clothing throughout the life cycle through creating awareness among the industry, government, and corporate performance.

22.2 SUSTAINABLE DESIGN

Sustainable design is the philosophy of designing the physical objects, built environment, and services to comply with the principles of social, economic, and ecological sustainability [1]. The term sustainable design is also interchangeably used with eco-design, ecological design, environmental design, environmentally sensitive design, environmentally conscious design, environmentally sustainable design, environmentally responsible design, design with nature, green design, sustainable development and holistic resource management, and all these terms aim at sustainability, the capability of environmental, economical, and social systems over time, without any adverse effects.

Sustainable designs are often demonstrated by the less or prohibited use of nonrenewable resources, minimal impact on the environment and providing natural environment to the people to the maximum possible extent. In the fashion industry, creative designers and fashion illustrators play the most challenging role in terms of redefining the fashion in terms of upcycling and creating newer products. Very often, accessories, appliqués form the major focus to impart the eco-design inputs in the garments. In the recent years, manufacturers have started reusing and recycling the wastes into various raw materials and thus preventing them from going as mere landfills. However, the commitment to the green designs not only comes at a cost but also limits the esthetic options available to a manufacturer or designer.

22.3 FASHION VERSUS SUSTAINABILITY

Since the immemorial, the fashion industry is widely lauded for the creativity and culture that the fashionable products can propagate. Nevertheless, most of the fashion garments also generate textile wastes, water pollution, carbon load, and higher-energy use in the manufacturing process. It has been estimated that nearly 10 million tons of textile materials are discarded every year in Europe and the United States alone, originally produced using large amounts of resources. Many adverse effects are apparent while manufacturing or while using (life cycle) the different textile and fashion items. This necessitates the designers and manufacturers to look for selection of *eco-materials* and making *eco-decisions* to use or consume the textile and fashion materials sustainably and address the environmental issues [2]. Fundamental rethinking in design and the value creation offer opportunities to develop the sustainable options in production as well as in consumption, and through this knowledge it is possible to redesign the business also [3].

The moment when one thinks of fashion and sustainability, few questions that might arise in the minds could include, Is there any relation exist between fashion and sustainability? Are these two terms contradictory to each other? Is sustainability in fashion, a marketing gimmick and nothing else to offer? Fashion has been denoted as an expression that is widely accepted by a group of people over time and has been characterized by several market factors such as low predictability, high impulse purchase, shorter life cycle, and high volatility of the market demand [4]. Fashion industry's popularity and success had a focus on low-cost mass production of standardized styles till mid-1980s with the exception of haute couture [5]. From 1990s onward, fashion shows and catwalks have become the public phenomenon, where photographs of the fashion shows could be seen in the magazines and also on the worldwide web, leading to delineating the fashion process [6]. Information and trends are moving around the globe at tremendous speeds, resulting in consumers' ability to exercise more options and purchase more often than ever before [7]. Consumers are becoming more demanding and fashion savvy that force the fashion retailers to provide the right products at the right time in the market—in other words, provide quick (fast) fashion [8]. Fashion-conscious consumers are continuously exposed to exclusive designs and styles inspired from runways. Retail chains such as Zara, H&M, Mango, New Look, and Top Shop adopt new designs rapidly to attract consumers and introduce the interpretations of the runway designs to the stores in short periods [9].

The changing dynamics of the fashion industry, increase in the number of fashion seasons, and modified structural characteristics in the supply chain, all, have forced the retailers to desire low cost, modular and flexible designs, quality, and speedy delivery to the markets [10]. Buying practices of fast fashion, namely, a combination of global and local suppliers, a lean approach, trust and integration of key internal activities and processes to facilitate the speed of buying decisions that may be required [11].

Fashion is about being trendy, up-to-date, and the latest while sustainability is about long-lasting and durable, low impact, and eco-friendly. The simplest understanding of these two terms indicates a conflict between them as if these two are diametrically opposite and can never go hand in hand and the responsibility of correcting the misconception lies with all the stakeholders involved in fashion process, that is, designer, manufacturer, distributor, retailer, and customers.

22.4 SUSTAINABLE DESIGN PRINCIPLES

Even though the classical 3R approach, that is, reduce, reuse, and recycle, still holds good for sustainability, sustainable designs can be best understood using the principles mentioned later [12,13]. Most of the sustainable designs of textiles and apparels revolve around these principles depending upon the nature of the products and processes. Many of the principles are incorporated in a single process or product simultaneously rather than relying on a single aspect.

1. Sustainable fibers
2. Low-impact materials
3. Renewable sources of energy
4. Energy efficiency
5. Water efficiency
6. Waste management (reduction of material and recycling)
7. Biomimicry
8. Biodegradable products
9. Pollution prevention
10. Design to reuse and recycle
11. Service substitution
12. Community couture
13. DIY (do-it-yourself) and patchwork
14. Fair trade/ethical practices
15. Near-sourcing

A pictorial representation of these principles is shown in Figure 22.1.

Any process/product involves inputs and corresponding outputs. Inputs can be in the form of materials like fibers, fiber assemblies, dyes, chemicals and auxiliaries, water, energy, etc., and

478   Handbook of Sustainable Apparel Production

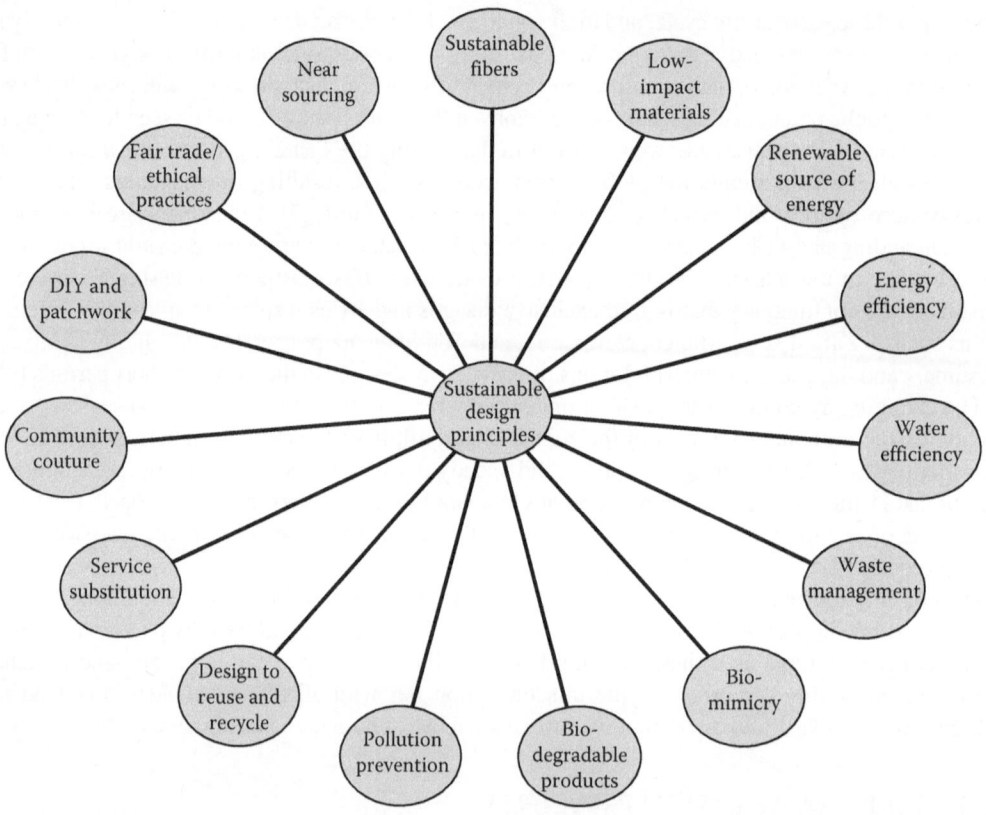

FIGURE 22.1 Sustainable design principles.

outputs can be product, emissions to air, water, land, etc. Sustainable design begins with the selection of sustainable fibers, using low-impact materials like low-impact dyes, chemicals, auxiliaries and processes, and water and energy efficiency during processing. Not only energy efficiency but also whether the source of energy is renewable or nonrenewable is also significant and renewable source of energy is considered in sustainable design. During product development, invariably waste happens in a process/product that gives the scope for waste management at various levels like reduction of materials, modification of process, and recycling. Pollution in the form of emission to air, water, land, etc., is another eco-factor that must be taken into account while designing the sustainable product. Whether a product can be reused as it is or how quickly and easily it could be recycled is also of great importance as extending the life of a product is one of the parameters in sustainable design. Biomimicking also helps in redesigning systems/products on biological lines that enable the constant reuse of materials in continuous closed cycles. Service substitution that promotes minimal resource use per unit of consumption is of pivotal importance that yields good results in sustainability. Community's continuous focus on self-sufficiency like community laundries, disposal of clothing for monetary benefits and local community providing energy and materials are few of the examples that fall under this category. Encouragement for the secondhand/*preloved* clothing; secondhand markets with tailors/stylists on hand, in the retail stores; and DIY and patchwork will improve the sustainability prospects in the long run. Sourcing of materials at the nearby place rather than at a longer distance is obviously useful in promoting sustainability. Degradability of the product, once it is used and disposed, also a vital concern and the products that can be decomposed quickly in the presence of air, water, and commonly present soil organisms are preferred. Sustainability becomes complete if and only, with the addition of ethical practices, as sustainability is a triple bottom line that includes *environment, economy, and society.* Many leading textile and garment manufacturers across the world employ these principles to improve their *image*, to meet the norms/standards enforced by the local government and the magnanimous aims to save the environment.

22.4.1 SUSTAINABLE FIBERS

A careful selection of raw materials is the first and essential step in achieving the sustainability. Selection of raw materials based on life-cycle assessment or life cycle thinking enables the designers to choose the raw materials with less impact on the environment as the whole. A wide range of natural fibers, regenerated fibers, and synthetic fibers offer a great challenge and options for the designers to select suitable fibers for the sustainable design. An insight into the different fibers and manufacturing practices is of huge benefit to the designers.

Natural fibers are those fibers that are produced from natural sources such as plants, animals, and geological processes. The unique advantages of these fibers are they are safe to use and are biodegradable over the time. They can be classified according to their origin as vegetable fibers (cotton, hemp, jute, flax, ramie, sisal, and bagasse), animal fibers (silkworm silk, spider silk, sea silk and hairs such as cashmere, wool, mohair, angora, and furs), and mineral fibers include asbestos and basalt. Notwithstanding the demands of the consumers, newer fibers are being explored that are available in specific regions like agave, banana, coir, and kenaf. Designers of various places make use of these fibers to reflect the sustainability and the values in the designs and products.

Approximately 25 million tons of cotton fibers are grown annually [71] and cotton is the largest natural fiber used across the world; however, from the life cycle perspective around 39% of water [72] and huge amounts of energy is spent on harvesting the cotton itself. This necessitates certain alternative fibers in making the products sustainable ones. Cotton-harvesting practices like integrated pest management (IPM), conservation tillage, and organic cultivation are better known sustainable measures [14]. The primary goal of the *Organic*, as per the Organic Trade Association (OTA), is to "optimize the health and productivity of interdependent communities of soil life, plants, animals and people." Even though organic agricultural practices do not ensure that products are completely free from polluting residues, specific methods can be used to minimize the pollution from the air, soil, and water. IPM, an initiative extensively practiced by farmers of Texas State, focuses on the long-term prevention or suppression of pest problems with minimum impact on the human health, the environment, and the nontarget organisms. IPM is best practiced by the application of less harmful pesticides, biological control, adaptation of agricultural practices to avoid pests, and changes of habitats for pests [15]. Conservation tillage focuses on leaving the most of the previous crop residue on the soil surface to provide mulch for the soil, increase water infiltration rates into the soil, and decrease wind and water erosion compared to conventional methods of seedbed preparation and thereby enhance the cultivation. The best ecological solution for natural fiber cultivation would obviously be a combination of *organic, conservation tillage* and reduced dependence on the synthetic support of rain-grown cotton crops [16].

Wool is regarded by many retailers and consumers as a premium fiber and accounts for <3% share of the total fiber market. Annually, greasy wool fiber is produced nearly 2.1 million tons [17], which gives less than 1 million tons of clean wool. Recent developments in processing of wool fabrics with chlorine and polymers have resulted in the machine washable materials. Water vapor–absorbing properties of merino wool fibers have led to the development of *sports wool*, which is a much sought-after material in the sportswear manufacturing [17].

Man-made-regenerated fibers are not directly available from natural sources but manufactured using regeneration process of the natural sources like cellulose (viscose, modal), protein (casein), and alginates. Synthetic fibers are derived from synthetic chemicals obtained from the petroleum refinery products and fibers like polyester, acrylic, polyvinyl, and polyolefin belong to this category. Other man-made fibers include metallic fibers, glass fiber, and carbon fiber. Table 22.1 summarizes various fibers available for different fashions along with their merits and demerits.

TABLE 22.1
Fibers for Fashion

Fiber	Merits	Demerits
Cotton	Biodegradable and renewable.	Water-intensive cultivation.
	Organic cotton reduces use of pesticides.	Use of more fertilizers and pesticides (conventional cotton).
	Enhances comfort.	
	Waste fibers used for regenerated fibers.	Ironing requirement.
Wool	Biodegradable and renewable.	Sheep breeding and processing of wool are water- and energy-intensive ones.
	Longer fiber length.	
	Flame resistant.	
	Retains warmthness.	
	Multiple clippings from the sheep.	
Silk	Biodegradable and renewable.	Silkworm breeding is water- and energy-intensive process.
	Mildew and mold resistant.	Delicate and expensive.
Flax/linen	Biodegradable and renewable.	Region-specific crops.
	High durability.	Limited availability.
	Blended with wool/silk.	Low spinnability.
Bamboo	Biodegradable and renewable.	Water- and fertilizer-intensive cultivation.
	Organic bamboo is also possible.	Fiber extraction involves environmentally harmful chemicals.
Viscose	Renewable cellulose raw material.	Fiber manufacturing is a water-, energy-, and chemical-intensive process.
	Low maintenance.	
	Wash at lower temperatures and air dry.	Use of highly corrosive and toxic chemicals.
	Biodegradable and renewable.	
Modal	Renewable cellulose raw material.	Fiber manufacturing is water intensive.
	Low maintenance.	
	Wash at lower temperatures and air dry.	
	Biodegradable and renewable	
Polyester	Longer life and durable.	Petroleum based and nonrenewable.
	Recyclable.	
	Low water requirement.	
	Low laundry requirements and no ironing required.	
Nylon	Tough and durable.	Petroleum based and nonrenewable.
	Can be recycled.	
Acrylic	Synthetic version of wool.	Petroleum based and nonrenewable.
	Lightweight, soft, and warm.	Requires dry cleaning or specific laundry processes.
	Resistant to moths, oils, and acidic chemicals.	
Polypropylene	Recyclable.	Petroleum based and nonrenewable.
	High performance to weight ratio.	

22.4.2 Low-Impact Materials

Materials selection for garments shall be based on the overall impact of those materials on the three sustainable parameters such as the environment, economy, and the society. Materials with nontoxic or less-toxic levels, which require little energy to process, and use of recycled materials, have less impact on the environment. Avoiding petroleum-based plastics and thinner or lighter materials have a huge impact on sustainability, while the uses of materials that have high impact on sustainability are to be prohibited.

More developments in the area of coloration, printing, and finishing are witnessed frequently where the usage of auxiliaries, dyes, pigments, and finishing chemicals play the major roles. Many dye-leveling agents are suggested in dyeing [18] for improving the diffusion of dyes into the fibers and thus reducing the impact on effluent generated in the process. Also, such agents warranty the reduced consumption of dyestuffs in the process. Besides, color deepening, water-based formulations are often advocated for reducing the use of dyes and auxiliaries, for example, Dark Knight 101 of Sarex Chemicals, which is known for enhancing the perception of dyed fabrics [19]. Seventeen percent of global cotton production is used in denim production [73] using synthetic indigo dyes. Most of the process houses rely on reuse of dye bath to put a check on the effluent generated in the process and also to optimize the consumption of expensive dyes. Indigo dyes used in denim dyeing process demand high quantities of hydrose for reducing that is also slowly replaced with low-impact substances like glucose to reduce the COD loads in the effluents [20]. Notwithstanding the pressures, many dyestuff manufacturers also make attempts to extract dyestuffs from various natural resources and offset the environmental impacts.

Many manufacturing processes are redesigned to incorporate the sustainability measures and considering the extent of the impact that can be reduced on the environment. AirDye, a waterless dyeing technology, developed by the Colorrep of USA, is based on the subliming properties of the dyestuffs that are normally exploited in the case of printing the synthetic fabrics, like transfer printing. Besides waterless dyeing process, the fabrics do not require post-was treatments after the dyeing process that also results in much less impact on the water loads [19].

Worldwide, there is a shift in the use of various high-impact ingredients used in the printing processes, especially in the case of the garments meant for kids. Heavy dependence on alkyl phenol ethoxylates, phthalates, PVC is slowly replaced with water-based inks. Water-based quick drying inks are strongly recommended in the ink-jet printing in place of high VOC solvents. Low-impact dendrimers and emulsifier-free silicone technology are advocated in the case of water repellent finishes of fabrics in the place of fluoropolymers [20].

Recently published list of the substances of very high concern (SVHC) by the European Chemical Agency provides an easy guide to identify the substances with high impact on the environment and to scout for the alternatives. Bluesign™ also has released the substances' list for the certified manufacturers and retailers to follow during the different processes. Some of the substances that can be avoided in the garment manufacturing are listed in the following.

- Allergenous dyestuffs and pigments
- Azo dyes and azo pigments that can be reduced to carcinogenic amines
- Carcinogenic auxiliaries, dyestuffs, and pigments
- Dyestuffs and pigments containing heavy metals like lead, cadmium, and chromium
- Dyestuffs and pigments with acute toxicity ($LD_{50} < 100$ mg/kg)
- Printing systems based on heavy benzene
- Printing with high VOC inks
- Dichromate as oxidizing agent to improve color fastness
- Chlorinated organic carriers (chlorobenzene, chlorotoluene, chlorophenol)
- FRs containing toxic metals, brominated flame retardants
- Chlorinated and fluoro chlorinated organic solvents in open systems

- Multiple halogenated hydrocarbons in biocides, solvents
- Less degradable tensides and complexion agents
- Cross-linking agents with a high content of free formaldehyde
- Wet adhesives
- Use of pesticides for storing and transportation

22.4.3 RENEWABLE SOURCES OF ENERGY

Fossil fuels use nonrenewable resources and cause toxic emissions into the air and have become a threat to the earth. Oil, gas, and coal, the most important nonrenewable resources, are used as the fuels and also in the production of synthetic fibers and auxiliary chemicals. The use of renewable sources of energy is less polluting compared to that of nonrenewable sources [74].

A threat to the nonrenewable resources is estimated in which most of the nonrenewable resources will be depleted in the next 50 years if the production continues at current rates and it will happen sooner in the next 25 years, if the production grows at current rates [75]. It means we are heading toward a grave and alarming scenario. To rescue ourselves and the earth, it is high time we start looking at renewable resources. Technologies based on the utilization of renewable sources for energy such as solar, wind, hydro, bioenergy, geothermal, and hydrogen are recommended to greater extent as the sources of energy.

22.4.4 ENERGY EFFICIENCY

Conversion of raw materials into a final product involves the utilization of energy at various stages of the manufacturing process. Processes, products, and services that require less energy often translate to the sustainability of the processes and products. It makes sense to optimize the energy consumption for the sake of economic viability also.

A wide variety of operations [21] in the cultivation of fiber crops are recommended like switching from conventional tillage to no-till practices result in fuel saving; nutrient management and crop rotations to reduce the use of synthetic fertilizers, lesser irrigation timing, implementation of integrated pest management strategies, precise applications of the resources, efficient weed control and lower herbicide usage, switching to highly energy-efficient equipment like irrigation pumps, introducing perennial crops and rotational grazing reduces energy. Conversion of plant residues to biofuels like ethanol or biodiesel, recycling of lubricants and other petroleum-based materials, utilizing alternative energy sources like anaerobic digestion, solar and wind that reduces carbon and greenhouse gas emissions.

Energy efficiency in staple fiber spinning can be achieved with the optimization of beating points (or cleaning efficiency) in the blowroom, proper design of chute feed, synchronous card cylinder speed and output, ideal total draft in ring frame, optimum humidification, proper maintenance and avoiding/minimizing the reprocessing of the intermediate products to avoid energy losses.

Wet processing of textile materials includes many high-energy-consuming operations and consumes approximately 80% of total energy requirement of all the operations [65,66], out of which, about 66% of the energy are consumed in heating and evaporation of water from the fibers. In spite of improved performances of many drying machines and new techniques adopted for drying the textile materials, the increase in energy cost makes every manufacturer to think in terms of conserving the energy required in these areas. Though intermediate drying is not carried out in many fabric preparation stages, the individual processes also consume more amount of energy since these processes are carried out, generally, at higher temperatures. Attempts have been made, in the past, to optimize the individual unit operations and to derive the advantages in the respective unit operations only [21–26]. Manufacturers of batch processing machines are cleverly helping the processors to reduce the energy consumption by reducing the material-to-liquor ratio required for various processes, which in turn requires less energy for heating the process liquor, less effluent treatment also.

Provision of smaller pad troughs in the padding mangles ensures minimal chemical and energy requirements for processing of the fabrics [27]. Savings in the energy may also be realized with frequency converters and variable speed drives to the motors, plugging the steam and air leaks in the process line offer significant reduction in the energy consumed in the process.

Various approaches to promote low-impact laundering, in terms of washing temperature, washing frequency, and size of load, have also been suggested in the literature [28]. Energy efficiency in garmenting can be achieved through maximizing the capacity utilization of manufacturing, high pattern or marker efficiency, light reflectors in shop floor, low-energy lights, auto-switch off facilities based on motion sensors, minimized internal transport of goods, and minimization of rework.

Retailing is one of the crucial areas that need to be considered for energy efficiency. Unfortunately, very few retailers employ technology experts with hands-on experience on various processes involved in the fabric and garment manufacturing. Nevertheless, retailers can insist the suppliers of the garments and fabrics to follow sustainable routes in their manufacturing. Major retailers, across the world, follow the codes of practice relating the ecological aspects by insisting not to use certain chemicals in their products. Besides, types and ways of energy used in the processes and effluent disposal methods followed in the manufacturing firms may be audited and certified by the retailers to ensure the sustainable and ethical work practices.

Many of the retailers take measures to follow the sustainable options in establishing the retail stores. One can take a cue from the Puma's Sustainable Store [29], Bangalore, India, started in the year 2013. The sustainability measures implemented in the store include construction of building with recycled steels from old DVD players, bicycles and tiffin boxes; porotherm blocks used in the shell of the building have been made up of silt; furniture and fixtures are made up of recycled wood, coated with the low VOC paints, need for artificial lighting is reduced with more than 90% of the structures with access to natural lighting in the interior space with reduced use of the artificial lighting; foam roof of the building for insulation, stack effect natural cooling, under floor air distribution systems combined with air passing through underground tunnel to cool the retail showroom without air-conditioning, 100% solar-powered energy and occupancy sensors are some of the sustainable measures that give the clue for others also. Puma also has set up a trend by launching the event in the carbon neutral way as the energy is generated by the pedal powers of customers entering the store.

Levi's life-cycle assessment of 501 jeans and Dockers, carried out in the year 2007, paved the sustainability roadmap for the coming years [30]. The surprising results indicate that nearly 50% of water is consumed in cotton harvesting, another 45% of water is used by the consumers during washing, and nearly 60% of the energy is used in making and taking care of a pair of jeans. International Association for Textile Care Labelling (GINETEX) recommends a *Clevercare*™ symbol (Figure 22.2) that advises the consumers to reduce the environmental impacts during the garment use and care by providing suitable recommendations [31].

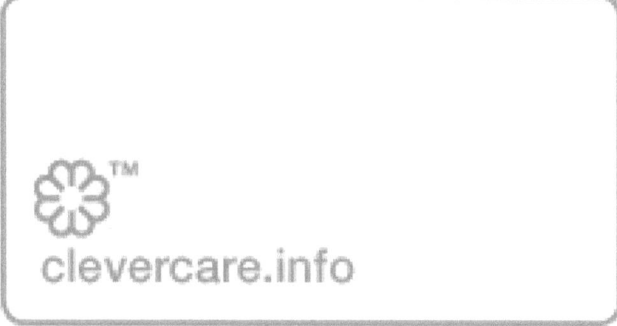

FIGURE 22.2 Clevercare symbol. (From http://www.ginetex.net/clevercare/symbols.)

Various maintenance practices recommend to reduce the energy consumption, and the environmental impacts in the *Clevercare*™ symbols for textiles/apparels care are listed in Table 22.2.

Even though the *Clevercare* symbol is not a technical symbol, it advises the consumers to consider a wide range of actions that include recommendation for a *clever care*, tips for wash and care systems. Besides, efficient washing is carried out by the use of efficient washing machines (star rating), detergent dosage recommendations according to the level of water hardness and degree of soiling, usage of prewash programs only when the articles are very dirty, usage of quick wash programs for the lightly soiled laundry, follow the care label instructions before using tumble dryer, and use of washing machine with warm and cold water in takes. Natural drying methods in the fresh air are encouraged; tumble dryer fitted with moisture sensor and proper filling reduces the environmental impacts. Simple and quick actions like removing garments from the washing machine or dryer as soon as it stops, use of the fabric conditioner to avoid creases and ironing at appropriate temperatures also reduces environmental impact.

22.4.5 Water Efficiency

Water, one of the five vital elements for the survival of species on the earth, needs to be utilized in an utmost conscious way. Not only the water preservation but also water consumption plays a major role in the sustainable process. It is estimated that demand for water, worldwide, would increase by 40% by 2030. Adding to the woe, many experts predict that wars will erupt in various parts of the world on water-rich countries to meet the demands. Careful planning to reduce the water consumption can be thought of from the fiber cultivation onward, which obviously might result in the substantial saving of water in the entire process chain. Technologies like row (furrow) irrigation, low-energy precision application (LEPA) or drip irrigation [17]. The water required in the textile and garment manufacturing needs to be of very high quality and is at least as clean as drinking water. Almost nearly 150–200 L of water is used for every kilogram of cotton or regenerated cellulosic fibers dyed and saving potential per kilogram of material processed has been estimated to be at least 25 L [33]. All the manufacturing processes, products, services, etc., must lay emphasis on the reduction of water usage or to find the alternatives to the usage of water [34].

In most of the countries, at least a part of the water used by the processing houses is expected to be recycled for reusing in the process. Very sadly, there is no quick fix solution to reduce the use of water, especially in wet processing of textile materials.

Water Framework Directive (WFD) of the European Union has been designed to drive sustainable water management in the years to come. Many researchers have advised combined preparatory process in fabric preparation instead of individual unit processing for desizing, scouring, and bleaching processes. Such combined preparatory processes reduce the amount of water required for intermediate washing and overall water requirements for the preparation [35–37].

In the case of wet processing, right first dyeing that has been advocated for more than three to four decades achieves considerable savings in the water used in the processing and following practices need to be considered for right first time dyeing: process control at every stages in preparatory processes, highly standardized products, shortest possible dyeing process, high fastness level, simple procedures, and easy to implement in lab and bulk practices [38,39].

In the normal wet processing methods, the wet pickup at the end of conventional padding process is, normally, about 60%–100%, depending on the type of the fiber employed in the fabric. Eco padding systems have been developed to reduce the pickup levels in the past [40]. Also, moisture measurement systems in the conventional application processes have been developed continuously for monitoring the pickup of dyes and chemicals in the fabric during the processing stage itself [41]. Low add-on techniques, though they offer the possibility of reducing the water and energy requirement, should not affect the distribution of chemical on the fabric [18]. The optimum level of liquor in the fabric is to be ensured to have uniform distribution of the chemicals, expressed in terms of critical add-on value. Foam application techniques can be recommended in the preparatory

TABLE 22.2
Clevercare Symbols for Textiles and Apparel Care

Symbol	Explanation	Remark
Washing process		
[95]	Boiling wash (normal process)	—
[60]	60°C colored wash (normal process)	—
[60 underlined]	60°C mild process	Drum should be not more than half full. Avoid spinning to minimize the risk of creasing.
[40]	40°C colored wash (normal process)	—
[40 underlined]	40°C mild process	Fine laundry, for example, in modal, viscose, synthetic fibers.
[40 double underlined]	40°C very mild process (wool wash cycle)	Machine-washable wool. The drum should be not more than a third full.
[30]	30°C colored wash (normal process)	—
[30 underlined]	30°C mild process	Drum should be not more than half full. Avoid spinning to minimize the risk of creasing.
[30 double underlined]	30°C very mild process (wool wash cycle)	Machine-washable wool. Drum should be not more than a third full.
[hand wash symbol]	Wash by hand	Washing by hand is recommended instead of hand wash programs.
[crossed out]	Do not wash	—
Bleaching		
△	Chlorine and oxygen bleach	—
△ with lines	Oxygen/nonchlorine bleach	—
△ crossed out	Do not bleach	—
Natural drying		
□ with vertical line	Line drying	—
□ with vertical line and diagonal	Line drying in the shade	—
□ with two vertical lines	Drip line drying	—
□ with two vertical lines and diagonal	Drip line drying in the shade	—
□ with horizontal line	Flat drying	—
□ with horizontal line and diagonal	Flat drying in the shade	—
□ with two horizontal lines	Drip flat drying	—
□ with two horizontal lines and diagonal	Drip flat drying in the shade	—

(Continued)

TABLE 22.2 (*Continued*)
Clevercare Symbols for Textiles and Apparel Care

Symbol	Explanation	Remark
Tumble drying		
	Normal drying process	Drying in the tumbler is under normal load and temperature (80°C) without limitations.
	Mild drying process	Select mild process with reduced action (temperature 60°C, duration of treatment).
	Do not tumble dry	
Ironing		
	Hot iron	Iron with maximal sole plate temperature at 200°C (cotton/linen setting).
	Iron at moderate temperature	Iron with maximal sole plate temperature at 150°C (wool/silk/polyester/viscose setting).
	Iron at low temperature	Iron with maximal sole plate temperature at 110°C (nylon, acetate setting).
	Do not iron	
Professional dry cleaning		
P	Professional dry cleaning in tetrachloroethene and/or hydrocarbons	Normal cleaning process with no restrictions.
P	Professional dry cleaning in tetrachloroethene and/or hydrocarbons	Mild cleaning process with stringent limitation of added humidity and/or mechanical action and/or temperature.
F	Professional dry cleaning in hydrocarbons (heavy benzene)	Normal cleaning process with no limitations.
F	Professional dry cleaning in hydrocarbons (heavy benzene)	Mild cleaning process with stringent limitation of added humidity and/or mechanical action and/or temperature.
	Do not dry clean	—
Professional wet cleaning		
W	Professional wet cleaning	Normal wet cleaning process with no restrictions.
W	Professional wet cleaning	Mild professional wet cleaning for sensitive textiles with reduced mechanical action.
W	Professional wet cleaning	Mild wet cleaning process for very sensitive textiles and apparels with much reduced mechanical action.
	Do not wet clean	—

processes [43,44], dyeing, durable press (DP) finish, softening, soil-release finish, water, oil repellant finish, flame retardant (FR) finish, and antistatic finish [45–47]. Low wet pickup techniques such as saturation–removal (vacuum extraction, porous bowl techniques, air-jet ejectors, transfer padding, i.e., squeeze–suck technique) and topical application (kiss-roll, loop transfer system, engraved roll, nip padding system, spray system, and foam application) need to be encouraged without the focus lost on the quality of the process. Tandem wet-on-wet foam application of both crease-resist and antistatic finishes is also explored [48].

Textile Systems BV, the Dutch company, has developed the DyeCoo process that eliminates the use of water in dyeing of textile materials, at lower costs compared to conventional methods of dyeing. DyeCoo does not necessitate the consumption of water, chemicals, and drying procedures that make the process energy attractive also. Dyeing process utilizes the carbon dioxide heated to above 31°C and a pressure of 74 bar, where it reaches the super critical fluid state, is an attractive state for dissolving hydrophobic dyes and thereafter used on hydrophobic fibers with relatively shorter dyeing time. At the end of the process, gaseous carbon dioxide is released in the process, which eliminates the need for drying the materials. With gas-like low viscosities, better diffusion of dyestuffs is expected, which in turn enhances the fastness properties also. Extraction of excess dyes and removal of spinning oils can also be carried out using the same equipment under different conditions [49].

Adidas DryDye technology [50] is a waterless polyester fabric dyeing process that requires 50% fewer chemicals and less energy than the traditional fabric dyeing processes. This process uses no water; instead dye is injected into the fabric using compressed carbon dioxide. Company boasts of greater dye retention than traditionally dyed fabrics, saves water, chemicals, and energy and thereby calling it as a truly eco-friendly process.

Levi's Waterless™ denims [50] aim at the reduction of water consumption during manufacturing, are produced using ceramic stones and rubber balls, and change the filtration systems in the washing machines. A distressed look of the denim is achieved with just 4 L of water, while the normal styles consume up to 45 L of water per pair of a garment.

Plasma processing method, another dry-treatment technology, has been postulated by many researchers in improving the desizing and scouring, dyeing and printing, and also in imparting various functional finishes to the textile materials [51–53].

On a wild thinking, it is also suggested to label the garments as *no wash*, *low wash*, surfaces with wipe-clean and enhanced ventilation in the garments depending upon their functional use and applications [28].

The proposed international standard of International Organisation for Standardisation (ISO), ISO 14046, aims to clarify the principles, requirements, and guidelines for assessing the water footprint of the products, processes, and organizations. It proposes to measure the water footprint through the four-stage approach that includes clear definition of the goal and scope of the study, inventory analysis (amount of water used, consumed, and polluted), assessment of the impacts associated with water, and interpretation of the results.

22.4.6 Waste Management (Reduction of Material and Recycling)

Concept of waste management includes the reduction of waste during the process or enhancing the utilization of resources and recycling the materials/resources used in the process. *Reduction of material consumption* during sampling and product development or prototype development, manufacturing, packing, and selling eliminates the pressure on the environmental impact, and extends the benefits of manufacturing more products with lesser environmental impacts. Adidas, the leading sportswear manufacturer, represents a suitable example for this case, which successfully reduced the amount of materials through process modification, color palette reduction, and virtual product sample catalogues.

Modification in the process like *For Motion Soles* [54], introduced by Adidas, uses nearly 50% less material in production than that required for a typical sports shoe soles. Adidas's *Element Soul*, an element introduced during Spring/Summer 2013 Collection, is truly a sustainable development, made up of the combination of recycled polyesters, soybean-based foams, a one-piece mid-sole and high pattern efficiency. The shoe has been designed using only essential items that resulted in a lighter shoe with about two-thirds the weight of a standard shoe. Adidas has also achieved 19% reduction in color palette used in the last 2 years and has set an ambitious target of reducing to the extent of 50% within the Adidas sports performance division in the coming years. Virtual technology can help in the reduction of the physical samples required to design and sell new products. Adidas has been able to produce 600,000 lesser samples in the years 2011 and 2012 compared to 2010 just because of the benefit of the virtual technology. Adidas's consolidation of various hand tags across the genders and all business units from the Spring/Summer 2012 Collection and the usage of single-wall transportation cartons that are thinners and use less paper than double-wall cartons have enabled a huge reduction of the material usage. Starting with the raw materials to the final products, many attempts have been taken by various manufacturers and agencies across the world.

Sustainability is a matter of how fibers and fashion materials are produced, it also includes how much waste is produced with pattern cutting of a garment. Zero-waste fashion design often advocates converting the "negative space of the pattern into a positive ones" and thereby reducing the wastes generated in the process. Most of the garment has around 15% of the fabric lost in the process. What would ordinarily be considered remnant fabric to be discarded by other designers becomes a luxurious embellishment that adds value to the garment. Zero-waste fashion design is about constructing garments without wasting fabric, usually making good use of a single piece of fabric [64]. Seamless garments knitted directly on a knitting machine also eliminate the cutting and sewing wastes and considered as a novel zero-waste design concepts [65].

Popular brands like Levi's, Nike, and Adidas are the leading companies in the *sustainable front*. Levi's WasteLess™ jeans consist of at least 29% postconsumer recycled plastic, which is made from 8 plastic bottles. Nike Flyknit is an innovative manufacturing process [55] that reduces waste in knitting together the upper of the shoe. Adidas Fluid Trainer, the most sustainable shoes ever as claimed by the company, is designed in such a way that it reduces waste generated in the upper design, with over 50% recycled content in the upper, and incorporates 20% and 10% regrind into the sock liner and soles.

22.4.7 BIOMIMICRY

The term biomimetics is derived from the Greek word bios, *life,* and mimetics, which means *having an aptitude for mimicry*, and thus translating to imitation of nature or copying or adaption or derivation from the biology/nature. Different terminologies used in imitating the biological systems and their principles are shown in Table 22.3. Biomimicry is the imitation of the products/models/systems/elements of nature for the purpose of solving complex human problems. Modern biomimetic exercises attempt to systematically develop an interdisciplinary approach for comprehending biological structures, processes, and functionality that provide the technical solutions to complex problems. Designing of airplanes based on flying kinetics of birds, design inputs for tsunami warning systems from sensory systems of dolphins, and adapting diving characteristics of kingfisher to introduce the quietest bullet trains are some of the classic examples of biomimicking. Biomimicking also helps in redesigning systems/products on biological lines, which enables the constant reuse of materials in continuous closed cycles.

Learning from nature inspires everyone and stimulates the endless creativities and those concepts could be used as the solutions to difficulties encountered in the process of fashion design [56]. Biomimetics or bionics or biomimicry is not a newer term for textiles and apparels that are as old as humankind and at every stage of civilization, people have been trying to imitate certain aspects of nature into their costumes and apparels. Applications of biomimicry observations on fashion designs enable the designers to learn from two different subjects, design and nature. All the features of the biological systems could also be mixed to create the new styles, fusion designs. The outline, color, material, and form of the fashion design could imitate all kinds of living beings from nature. From the time immemorial, camouflage fabrics and textiles have been used in the warfare in the form of concealments using flexible nets, covers, accessories, and disruptive print patterns in the uniforms. Besides, camouflage principle is applied to ultraviolet and IR regions also. Color-changing behavior of squids using muscular movements has also been delineated and incorporated in the textile materials [57]. Recombinant DNA methods [23] are successfully applied

TABLE 22.3
Biological Systems in Design and Development of Fashion

Terminology	Principle/Concept
Bionics	It is a science modeled or imitated on the coexistence systems whose characteristics have been presented by imitation. Bionic engineering is derived from a biological concept that has been promoted and assisted the human body organ or replaced body parts as a device.
Biomimetics	Bios means *unit of life* and mimesis refers to imitate, emulate, and mimicry. It is also considered as the synonym of bionics.
Biomimicry	Study of specific organisms whose behavior, shape, color, or behavioral simulation enhances favorable situation for human being.
Biodesign	A form or structural characteristics of biological shape.
Bio-inspired design	Designs inspired by the behavioral characteristics of natural species.
Bionical creativity engineering	It is the scientific basis for technology development, in particular nanoscience and technology. It is inspired by evolutionary nature and seeks solution to the problems to complex problems with similar concepts.
Biomorphology	Biomorphic curves are considered as the source of inspiration for creating designs and the appearance of the objects.

Source: Adapted from Chen, T.Y. and Peng, L.H., Nature-inspired fashion design through the theory of biomimicry, http://design-cu.jp/iasdr2013/papers/1512–1b.pdf.

to the production of protein-based fibers as biomaterials and fibers are spun from protein solutions. Biomimetic principles of spider silks are effectively used in the development of high-performance fibers. Development of self-cleaning textiles using the lotus effect is also another example of biomimicry in the introduction of high performance and functional textile materials. Biomimetic approach is currently analyzed for the production of sustainable structural composites using plant fibers and plant stems.

22.4.8 Easily Biodegradable Products

IUPAC defines biodegradation is the degradation caused by enzymatic process resulting from the action of cells, at least in part [59]. Sustainable or eco-design of the products or raw materials demand the design in such a way that they can be decomposed quickly in the presence of air, water, and commonly present soil organisms. Mostly, biodegradable substances are organic materials that also include textile and fashion accessories and other materials originating from the living organisms. Biodegradable substances break down into more than one set of chemicals, which are usually called primary and secondary degradation products. However, biodegradation does not warranty safer primary and secondary degradation products alone. Time period required for the biodegradation varies from 1 week to many years (Table 22.4) depending upon the structure and properties of the materials under degradation [60].

Biocoutre represents the types of fibers that represent biodegradable substances and since the biocoutre are produced without damaging the environment and sustainable solutions. Survival of microorganisms on the fibers and textile materials can be incorporated in the design, a fact that is known as *biomimicry*.

Natural fibers are naturally biodegradable and regenerated fibers like lyocell, poly(lactic acid), and poly(hydroxyalkanoate) are also biodegradable. More research works are being carried out in the development of biodegradable natural fiber green composites, biodegradable nonwovens natural geotextiles, etc. Use of unconventional fibers like Agave, Abaca, coir, and palm fibers is also explored in the regions where they are grown abundantly.

TABLE 22.4
Time Required for Biodegradation

Substance	Time Required for Biodegradation
Paper towel	2–4 weeks
Cardboard box	2 months
Cotton glove	1–5 months
Wool glove	1 year
Plywood	1–3 years
Plastic bag/bottles	100 years
Disposable diapers	50–100 years
Aluminum can	200 years
Monofilament fishing line	600 years

Source: Adapted from http://cmore.soest.hawaii.edu/cruises/super/biodegradation.htm.

YKK of Japan's *ReEarth®* zippers [61] are made up of corn and other plant materials, which has the unique advantage of having the ability to decompose in the appropriate composting environment. Adidas, *Green Silence Shoe*, boasts of soy-based inks, biodegradable midlayer soles, and recycled materials. Other biodegradable products that could find among the textile manufacturers include easily degradable sizing agents manufactured from biological polysaccharides, natural dyes, and finishing chemicals. Size recovery using ultrafiltration makes synthetic sizing equally attractive.

22.4.9 POLLUTION PREVENTION

Pollution prevention programs that focus on the reduction of water use and efficient use of process chemicals that have adverse effects on the environment are suggested. Manufacturers of fibers, dyes, and chemicals often provide modified process routes to prevent the pollution levels and *quality* use of the dyes and chemicals.

Huntsman has developed the *Gentle Power Bleach*, jointly with Genecor, based on peroxide-catalase enzyme system that converts the peroxide into the active species in a controlled manner at neutral pH (in contrast to the alkaline pH) and relatively low temperature, 65°C, compared to conventional peroxide bleach. Gentle Power Bleach preserves more strength and allows deeper shades, which in turn reduces the dye consumption also.

Many dyestuff manufacturers are voluntarily withdrawing their manufacturing facilities of after-chrome dyes and concentrate on metal-free dyes for wool fibers. However, such initiatives necessitate the process changes that might include the following:

- Avoid nondegradable or less degradable surfactants in washing and scouring.
- Substitute the substances and chemicals that deplete atmospheric ozone.
- Minimize the use of volatile organic solvents.
- Process variables need to be matched to type and nature of fabrics.
- Production batches to be managed for minimizing waste at the end of cycles.
- Encourage the use of transfer printing for synthetic fiber fabrics or water-based printing pastes, wherever feasible.
- Encourage jet dyeing machines with liquid-to-fabric ratio of 4:1 to 8:1 rather than machines that demand higher material-to-liquor ratio.
- Encourage peroxide-based bleaches instead of sulfur- and chlorine-based bleaches.
- Avoid chlorine-based bleaches, benzidine-based azo dyes, and dyes containing heavy metals.

Eco-Design/Sustainable Design of Textile Products

- Encourage the use of less toxic dye carriers and finishing agents. Avoid carriers containing chlorine, chlorinated aromatics.
- Encourage peroxide oxidation for vat and sulfur dyes instead of dichromate oxidation.
- Replace the nondegradable spin finish and size with degradable alternatives.
- Use biodegradable preservatives instead of polybrominated diphenylethers, Hg, As, or PCP in mothproofing, carpet backing, and other finishing processes.
- Adoption of low-heat and low-temperature processes.
- Employ countercurrent rinsing systems.
- Program to reuse the dye bath wherever possible.
- Reuse and recover the process chemicals such as caustic and size.
- Recovery of heat from wash water, boilers, and other heating systems.
- Improve cleaning and housekeeping measures.

A wide range of eco-friendly options are available treating the effluent including enzyme oxidization, enzyme hydrolysis, aerobic and anaerobic biological digestion methods, and bioscrubbing for the color removal are suggested as the safe methods of treating the effluents generated in the various textile process houses.

Many manufacturers are adopting the Bluesign standard, one of the voluntary standards for environmental sustainability, and Oeko-Tex Standard 1000 has recently been replaced by STeP, a testing auditing and certification system for eco-friendliness in various operations and processes. Developing countries are also attempting to bring awareness among the manufacturers to prevent the pollution in the manufacturing processes. Partnership-for-Cleaner Textiles (PaCT) of Bangladesh in support with the World Bank and leading brands like H&M, C&A, and G Star is a project initiated to develop and adopt the best practices to ensure the cleaner production.

22.4.10 DESIGN FOR REUSE AND RECYCLING

Sustainable designs should focus on products/processes/systems that are adoptable for reusing and recycling of various waste materials. Various designs may be divided into five different types, based on the increasing order of the recycling–friendliness that include single-material system, single-material composite systems, multimaterial composite system with detachable connections, multimaterial composite system with compatible materials, and multimaterial composite systems with permanent fixed connections. Products made up of only one material in a single system (single-material system) are the easiest option to reuse and recycle. Popular examples include the PET bottles, which are recycled in the manufacturing of polyester fibers. Most of the needle-punched nonwovens are single-polymer systems. Single-material composite system provides easy recycling options by means of agglomeration, for example, 100% PE-based structure into a granulate, unipolymer composite.

Reversible clothing that consists of wide variety of materials and detachable connections like Velcro/zippers/fasteners are good examples for multimaterial composite system with detachable connections. Upholstery and most of the furniture also come under this category. Separable interlocking composites use Velcro™ strip principle and Clamp-type assembling.

Automotive industry generally uses the conventional seat materials that are multimaterial composite systems that cannot be separated, which can be regarded as the *worst case* considering the design of easily recyclable products. In general, these composite systems consist of thermal lamination of a decorative material, PUR foam, and a thin textile bottom layer. Such composites need to be substituted by the easy-to-recycle structures. Warp-knitted nonwovens Malivlies, Kunit, and Multiknit are bonded by means of meshing and do not use threads and can be the appropriate substitute for seats that use thin layer of PUR foam. A warp-knitted spacer nonwoven Multiunit developed using Caliweb process is used instead of the layer of foam. A flat-bed laminating is used to carry out the composite formation between decorative material and the nonwoven with the use of thermo-sensitive gluing substances.

TABLE 22.5
Material Systems

S. No.	Material System	Examples
1	Single-material system	PET bottles
2	Single-material composite system	Unipolymer composites
3	Multimaterial composite system with detachable connection	Fashion accessories
4	Multimaterial composite system with compatible materials	Reversible clothing
5	Multimaterial composite systems with permanent fixed connection	Automobile seat materials

Multimaterial composite system with permanent fixed connections is the most difficult to recycle and from the reuse and recycle perspective these kinds of designs shall be given the last priority. Table 22.5 summarizes the different types of material systems with suitable examples.

Modular designs of garments with multimaterial composite systems with detachable parts facilitate the laundering processes easier and reduced load during washing and laundering process since heavily soiled parts can be separated and taken for laundering treatment [28,62].

Groz Beckert has introduced the concept of sustainable shoe [63], which attempts to replace the multimaterial composite system of making shoe uppers with a single-component in-form shoe production by suitably modifying the needles in the knitting machine. Conventionally, the sports shoes have been made using different materials and components combined together by thermal sealing or suitable adhesive systems. On the other hand, Groz Beckert's knitted single-component shoe upper provides a fashionable look with perfect fit. This also facilitates the reduction in the weight of the shoes and the waste materials generated in the process. This novel design is expected to reduce the waste generation to the extent of ~2000 tons, an estimate based on the global shoe production statistics. Shifting from multimaterial composite systems into the single-component/material systems also enhances the possibilities of recycling such materials after the use.

22.4.11 SERVICE SUBSTITUTION

Shifting the mode of consumption from the personal ownership of products to provision of the services that provide similar functions, for example, from a private automobile to a car sharing service and mass transport systems like bus, train, etc. Such a system promotes minimal resource use per unit of consumption (e.g., per trip driven). Community's continuous focus on self-sufficiency will encourage the sorts of clothing made at home or community-run recycling centers linked to local, hyper-efficient factories. Secondhand/*preloved* clothing; secondhand markets with tailors/stylists on hand, in the retail stores; community laundries; disposal of clothing for monetary beneficial; and local community providing energy and materials are needed to be encouraged. It is also a better idea to encourage DIY and patchwork for better sustainability.

22.4.12 COMMUNITY COUTURE

Communities, across the world, in the longer run may be able to adapt secondhand/*preloved/pre-used* clothing in the same manner as it is being used for automobiles. Secondhand markets with tailors and stylists on hand, in retail stores, may be a positive phenomenon. Clothes may be manufactured and laundered in community-run factories/organizations. Secondhand clothing is encouraged and donated to orphanages and old-age people.

The terms secondhand, pre-owned, preloved, vintage, reused, and recycled are used interchangeably. In fact, the industry for secondhand existed in the eighteenth century itself in the United Kingdom with considerable market potential. In those days, it appeared that used garments were sold after cleaning with the turpentine. During the World War II, the British government propagated the "Make Do and Mend" campaign, featuring Mrs Sew-and-Sew [57]. Used clothing unsuitable for sale in a premium market may still find a buyer or end-user in different market, and sometimes, secondhand clothing is sorted, restored, and redistributed to other nations in many occasions.

The word *used* clothing often gives the portraits of the dirty rags that are stained, damaged, or smelly to ever wear in public; contrary to these notions more pre-owned clothing are available in highly esthetic conditions with higher esteem values. Many online dealers and traders are mushrooming on a large scale to deal with the pre-used garments.

Unlike newly tailored garments, pre-owned clothing offers the classic and vintage styles that fit the personal taste and style of an individual regardless to whatever the current fashion trend is. Availability of the pre-owned clothing freely in the market also helps to avoid making any huge investment for stocking the wardrobe with a wide range of attires (Figure 22.3).

A quick guide for selecting the secondhand/*preloved*/*pre-used* clothing includes size of the garments; measurements of the clothing; color and cut of the items and closer-to-the reality photographs or pictures; condition of the garments including rips, tears, holes, and missing fasteners; and finally fabric content and care instructions [65]. In many occasion, brand names of the garments and accessories also play the major role in selecting the pre-used garments by many consumers. Obviously, purchasing the pre-used garments could facilitate the *dream come true* for many of the middle-class population, in particular possessing highly acclaimed brands like Levi's, Wallmart, Calvin Kline, etc. Many freight carriers offer flat price shipping to such clothing and accessories, which in turn reduces the transport costs also. Besides, many traders also offer attractive return policies, insurance, and credits to the purchaser coupled with warranty claims. One unwritten rule that needs to be practiced by the consumers while purchasing a secondhand garment is that one should never hesitate to ask a question about the missing information on the piece of garments or any accessories.

However, the used garments frequently suffer the *image* problems in spite of the widespread sustainability or reuse/rebuy concepts. Many buyers lodge the complaints about the smell of *old age*, mustiness or mothballs and unpleasant overtones, in the case of secondhand garments. Concerns related to hygiene and pathological proliferations need greater concern while preserving the used garments [66].

FIGURE 22.3 Local market for secondhand garments. (From http://www.bergfashionlibrary.com/static-files/Encyclopedia/Secondhand-Clothing-Global-Fashion.pdf.)

22.4.13 Do-It-Yourself and Patchworks

DIY aims at building, modifying, or repairing the products without the aid of experts or professionals, who otherwise would be required for such purposes. Practices like DIY and patchwork or appliqués can be encouraged. DIY is best practiced for its economic benefits, lack of product availability and product quality, urge for customization, craftsmanship, empowerment, community seeking, uniqueness, etc. DIY is very popular among the craft work, making and repairing the broken fasteners, and fashion fraternity. Patchwork may involve needlework for sewing the pieces of fabric together into a larger design or fabric. Larger designs include repeat patterns of various fabric shapes or colors and final geometric shape is achieved by piecing together. An Appliqué, usually one-piece article, is a surface ornamentation technique applied onto a different surface. Appliqués are extensively used in textiles and garments as the value addition measure. Of late, appliqués are considered as a remarkable technique for mending different garments also. Patchwork and appliqués together might contribute to DIY (Figure 22.4).

22.4.14 Fair Trade/Ethical Practices

Favorable growth conditions that prevailed in the Western economies, in the past, witnessed the shifting of textiles and apparels manufacturing facilities to the so-called developing countries. Unfortunately, the required levels of infrastructure, legal framework to protect the environment and employees are also not sufficient enough in many of the developing countries to ensure the sustainability measures. Always, there has been a need for the textile industry to take the responsibility and ensure the sustainability measures to reduce the adverse effects on global environmental issues like GHGs, global warming, etc. All these necessitated many of the international agencies to stipulate the standards and norms to ensure the sustainability in the manufacturing operations.

Sustainability, as many people understand, is not only about environment considerations but also includes the social aspects. Organizations are insisted to adhere to the norms prescribed by the renowned international organizations like ILO (International Labor Organization), WRAP (Worldwide Responsible Apparel Production), SA 8000 (Social Accountability), and other provincial and international norms prescribed for ethical and sustainable work cultures. In general, most of the norms include the 10 basic principles that include prohibition of child labor, prohibition of forced labor, prohibition of harassment or abuse, prohibition of discrimination, health,

FIGURE 22.4 Do-it-yourself appliques and patchwork with appliques. (From http://www.ggdesignsembroidery.com/and jenniferjangles.blogspot.com.)

safety and security, freedom of association and collective bargaining, hours of work, remuneration, compensation and benefits, disciplinary practices, and management systems.

Besides, many organizations are making innumerous efforts to make their industrial practices ethically acceptable with the inclusion of choosing the responsible partners and award more business opportunities, promoting diversity among employees, awareness programs on sustainability among employees into their standard code of conduct, supplier relationship programs, audits, ratings, analysis of supplier management systems for capacity building, training programs for the suppliers on sustainable practices, human rights policy based on UN guiding principles, grievance redressal system, improved fire safety, etc.

Adidas trains the cotton farmers through Better Cotton Fast Track Program (BCTFP), encouraging suppliers to become members of the Better Cotton Initiative (BCI), establishing traceability code for the sustainable materials used in the apparel products and offer different ratings for the leather products.

Levi's has taken a ground-breaking approach of combining sustainable design and environmental practices with the emphasis on "supporting the well-being of the apparel workers" who make the garments and produce Dockers® Wellthread khakis, jackets, and T-shirts. Forum for the Future, the sustainable development charity initiated by the Levi Strauss, works in tandem with the manufacturers and service providers and helps them to devise sustainable strategies and deliver new products and to enhance people's lives and are better for the environment [67].

22.4.15 Near-Sourcing

In spite of many odd situations that prevail in the businesses, the manufacturers are forced to react to the trivial issues of rising costs in the supply chain and employee's cost-to-company. This forced many manufacturers/countries to look for cheaper alternatives to achieve the cost benefits. In the later part of the year 2011, China alone contributed to nearly 45% of the total apparel market of the developed countries, regardless of its geographical location or disadvantages. Until recently, there has been an unstoppable shift of textile and clothing production facilities from developed countries to developing countries due to favorable situations that could exist in those countries. However, this flux has witnessed a slowdown, a trend that is attributable to the concept *near-sourcing*. The cost advantages offered by those countries are slowly vanishing due to less availability of workforce available for the skilled operations and uncertain political and economical situations encountered by those countries and inability of the manufacturers to follow the stringent sustainable measures imposed by the importers causing the major reasons for reversal of the outsourcing trend in turn favor near-sourcing [68].

Near-sourcing is a term used to describe a business strategically placing some of its operations closer to the location where the end products are sold and this, essentially, is in contrast to the activities like outsourcing the low-wage manufacturing activities [69]. One of the ways that many companies are trying to make their supply chain's leaner through *near-sourcing* concept. Near-sourcing helps to address the problems related to cultural barriers, proximity, time zone constraints, and skill surplus available in a specific region. Though near-sourcing does not mean bringing back the operations to the headquarters of the firm, it could possibly mean locating operations in the neighboring nations that offer some incentives for new businesses. Such moves are expected to improve the cost efficiency of the supply chain, delivery schedule, and transport costs.

A five-step process has been recommended to assess the viability of near-sourcing process that includes the following: (1) calculate the costs associated with switching (like all of the costs and benefits including fuel costs, quality, productivity, technical expertise, etc., to potentially offset higher material and/or labor costs), (2) analyze your delivery needs, (3) factor in the potential risks of using overseas vendors in particular to the intellectual property–related products and processes, (4) rate of change in the product mix, and (5) a smart study about the competitors. Exhaustive reports have also been published on different strategies and opportunities available on near-sourcing [70].

Other intangible benefits associated with the near-sourcing include quick response of the suppliers to the fast fashion and reduced *carbon footprint* since lower quantities of the garments would be shipped for longer distance exports. Manufacturers who are unable to implement near-sourcing have the option to think about partner-sourcing also.

22.5 BEST AVAILABLE TECHNIQUES

The term best available techniques (BATs) provides the guidelines for sustainable manufacturing in terms of technology, processes, and chemistry, as well as recommendations. Compared to traditional approaches and processes, the best available technology can significantly help to save resources in many folds. Low-temperature treatments, reduced water consumption, enhanced dye

TABLE 22.6
Best Available Techniques/Practices

Process	Best Available Technique/Process
Cotton cultivation and ginning	Water irrigation systems.
Steam production	Low flow temperature.
	Prime source to be natural gas for high heating value and low emissions.
	Avoidance of oil and diesel.
Wet processing	Exhaust technology for wet-in-wet processing.
	Accurate and precise dosage of ingredients.
	Closed systems in the application of finishes with emission like FR and antibacterial.
	Incineration of discharged air with VOC in coating and finishing.
	Good insulation of HT equipment (polyester).
Process efficiency	Processes with low temperatures, short reaction time, reduced volume of liquor ratio in batches.
	Reduction of water consumption.
Drying equipment	Good insulation and reduced surface.
	Automated shutdown.
	Heat recovery of exhausted air (saves up to 50%).
	Neutralization of effluents with CO_2 emissions.
	Fast motion of fabric with high temperatures for reduced heat loss.
Domestic laundry	Choose machinery with options for different laundry quantities.
	Sort laundry according to temperature, color, and delicacy.
	Run machines with a full load.
	Choose low temperatures (in accordance with hygiene requirements).
	Apply compact laundry formulas.
	Apply kit for different temperatures and fibers.
	Adapt dosage to hardness grade of water.
	Apply prewashing only for very dirty laundry.
	Apply pretreatment for specks on laundry.
	Do without softener where possible.
	Use a clothes line or a fan system.
	Reduce ironing.
Water management	Reduction of rinsing and drying processes (more wet-in-wet processes).
	Reuse of baths (countercurrent), heat exchange (temperatures 40°C–60°C).
	Heat exchange from effluents to fresh water (up to 80% feasible).
	Pumps with slower recirculation (20% slower, 50% less energy).

Source: http://eippcb.jrc.ec.europa.eu/reference/txt.html.

exhaustion and fixation, efficient washing systems, and easy care and maintenance of fabrics and garments with lasting colors (better fastness properties) are the major focus of the best available technology. The "Responsible Care" initiatives of the European Chemical Industry strongly encourage the development of the best available technology and attempt to bring the awareness among all the stakeholders [18].

Besides carbon footprint and water footprint, machinery footprint is also becoming a fast accepted phenomenon, which emphasizes on versatility of the machines to process with a wide range of process parameters like temperature, speed, and multipass facilities. Intelligent design of the machines allows the processing of both woven and knitted fabrics without compromising the quality of the finish imparted [18]. BAT is not independent of product quality and must therefore be judiciously adapted in each case. Some of the recommendations are shown in Table 22.6. Every convinced producer and user is ready to adopt the best available technology and pave the way for the sustainability practice.

22.6 CONCLUSION

Changing global situations like explosion of population growth, climate change, and global warming are expected to alter the production of the natural and man-made fibers and their proportions significantly in the near future. However, fashion goes beyond the simple clothing to express identity, create well-being, embrace creativity, and connect global communities. A fundamental change is needed with the strategic innovations to create a sustainable way of doing business in the form of the user experience and rethinking value creation. The apparel industry has caused significant environmental problems, including large quantities of harmful wastes that are generated at every stage of the apparel manufacturing process, especially during laundering. Sustainability should not merely be considered as the reduced volumes or usage, but as an opportunity to do a kind of green business.

However, with the kind information available at the click of the worldwide web, the design industry is well placed to create design solutions that are both creative and desirable, coupled with sustainability. This emphasizes the need to create awareness on sustainability issues among the consumers including cultivars and manufacturers of different fibers. Technology will continue to transform the lives and businesses, and create new opportunities, and the attitude of the people toward resource shortages, climate change, and sustainability is also expected to change in the coming days.

REFERENCES

1. McLennan J.F., *The Philosophy of Sustainable Design*, Ecotone Publishing, Kansas, MO, 2004.
2. Slater K., *Environmental Impact of Textiles*, Woodhead Publishing, Cambridge, U.K., pp. 17–18, 2003.
3. Niinimäki K., Hassi L., Emerging design strategies in sustainable production and consumption of textiles and clothing, *Journal of Cleaner Production*, 19, 1876–1883, 2011.
4. Fernie J., Azuma N., The changing nature of Japanese fashion: Can quick response improve supply chain efficiency?, *European Journal of Marketing*, 38 (7), 749–769, 2004.
5. Brooks J., A friendly product, *New Yorker*, November 12, 58–94, 1979.
6. http://www.sydneylovesfashion.com/2008/12/fast-fashion-is-trend.html. (accessed on February 22, 2014.)
7. Anon, The future of fast fashion: Inditex, *The Economist*, 375 (8431), 63, 1979.
8. Hoffman W., Logistics get trendy, *Traffic World*, 271 (5), 15, 2007.
9. Barnes L., Greenwood G., Fast fashioning the supply chain: Shaping the research agenda, *Journal of Fashion Marketing and Management*, 10 (3), 259–271, 2006.
10. Doyle S.A., Moore C.M., Morgan L., Supplier management in fast moving fashion retailing, *Journal of Fashion Marketing and Management*, 10 (3), 272–281, 2006.
11. Bruce M., Daly L., Buyer behaviour for fast fashion, *Journal of Fashion Marketing and Management*, 10 (3), 329–344, 2006.
12. http://archive.defra.gov.uk/environment/business/products/roadmaps/clothing/documents/clothing-map-intro.pdf. (accessed on February 22, 2014.)

13. http://www.wrap.org.uk/content/sustainable-clothing-action-plan-1. (accessed on February 18, 2014.)
14. Hauser L., Evaluation of cotton production systems in the Texas high plains upon their sustainability, ETH thesis, 2000.
15. Flint M.L., Daar S., Molinar R., Establishing integrated pest management policies and programs: A guide for public agencies, University of California IPM Publication, Davis, CA, Compendium of IPM Definitions (CID), 1991.
16. Matocha J.E., Keeling J.W., Conservation tillage systems and research in Texas, *Proceedings of the Beltwide Cotton Conference*, Texas, Vol. 1, pp. 665–666, 1998.
17. Duffield P., Wool—The naturally sustainable fibre, *The Colourist*, 1, 5–7, 2010.
18. Anon, *The Textile Dyer*, 2, 2–8, 2010.
19. Anon, *The Textile Dyer*, 2, 2–8, 2009.
20. Meichsner R., Schmick W., Water based ink for ink jet printing, US Patent Office, US 5183502A dated February 2, 1993.
21. http://cottontoday.cottoninc.com/natural-resources/energy. (accessed on February 14, 2014.)
22. Harrison P.W., Low-liquor dyeing and finishing, *Textile Progress*, 14 (2), 17–34, 1986.
23. Energy Conservation Center (ECC) of Japan, *Seminar on Energy Conservation in Textile Industry*, The Energy Conservation Center (ECC), Japan, pp. 1–56, 1992.
24. Dickinson K., Oxidative desizing, *Review Progress in Coloration*, 17, 1–6, 1987.
25. Trotman E.R., *Dyeing and Chemical Technology of Textile Fibres*, 6th edn., Charles Griffin and Co., London, U.K., pp. 187–218.
26. Girsoy N.C., Hall M.E., Optimisation of hydrogen peroxide bleaching, *International Textile Bulletin*, 5, 80–86, 2001.
27. Anon, Environmental matters and energy savings, *The Colourist*, 4, 6–7, 2009.
28. Hu Y., A study on the sustainable fashion design in the process of use, *International Journal of Arts and Commerce*, 1 (4), 54–59, 2012.
29. www.puma.com/. (accessed on January 28, 2014.)
30. http://www.levistrauss.com/sustainability. (accessed on February 25, 2014.)
31. www.ginetex.net. (accessed on February 27, 2014.)
32. Duffield P., Wool—The naturally sustainable fibre, *The Colourist*, 1, 5–7, 2010.
33. Welham A., Water—Our responsibility, *The Colourist*, 2, 5–6, 2010.
34. Welham A.C., The global crisis for drinking water or how dyers can be transformed from villains to heros, *The Colorists*, 2, 6–8, 2011.
35. Ammayappan L., Muthukrishnan G., Saravana Prabhakar C., A single stage preparatory process for woven cotton fabric and its optimisation, *Man Made Textiles in India*, 47 (1), 29–35, 2003.
36. Gulrajani M.L., Development, optimisation and soarisation of combined preparatory process, *Colourage*, 2, 19–28, 1989.
37. Gulrajani M.L., Sukumar N., Optimisation of a single-stage preparatory process for cotton using NaOCl, *Textile Research Journal*, 55 (10), 614–619, 1985.
38. Bille H.E., Correct pre-treatment—The first step to quality in modern textile processing, *Journal of Society of Dyers and Colourists*, 103, 427–434, 1987.
39. Parton K., Right-first-time dyeing–The dye manufacturer's role, *Journal of Society of Dyers and Colourists*, 1, 4–5, 1994.
40. Srivastava C.P., Eco-pad system for low add-on, *Man Made Textile in India*, 12, 513–516, 1993.
41. Pleva R., Dye liquor pick-up and application moisture, *International Textile Bulletin*, 2, 62–64, 2002.
42. Anon, *The Textile Dyer*, 2, 2–8, 2010.
43. Bandyopadhyay B.N., Pradhan S.S., Scouring and bleaching by foam application—A novel approach, In Gulrajani M.L. (ed.), *23rd Technological Conference Jointly Sponsored by SITRA, ATIRA, BTRA, and NITRA*, Coimbatore, February 12–13, 1982.
44. Kumar R., Sarma T.S., Srivastava H.C., Conservation of energy in textile wet processing—Use of Foams, In Gulrajani M.L. (ed.), *23rd Technological Conference Jointly Sponsored by SITRA, ATIRA, BTRA, and NITRA*, Coimbatore, February 12–13, 1982.
45. Thomas H.L., Casey B., Richard C., Robert W., Method of obtaining color effects on fabric or garments using foam carriers and cellulase enzymes, US Patent 5,435,809, July 25, 1995.
46. Lambert A.H., Harper R.J., Single side cross-linking via foam finishing to produce garment dyeable cotton fabrics, *Journal of Coated Fabrics*, 19 (1), 1990, 169–180.
47. Dawson T.L., Foam dyeing and printing of carpets, *Conference on Foam Processing for Dyeing and Finishing*, Shirley Institute, Manchester, U.K., No. 4, pp. 29–38, 1981.

48. Pearson J., Elbadawi A., Tandem wet-on-wet foam application of both crease-resist and antistatic finishes, In *Ecotextiles: The Way Forward for Sustainable Development in Textiles*, Woodhead Publication, Cambridge, U.K., 2007.
49. Anon, DyeCoo: Waterless dyeing, *The Colourist*, 3, 8–9, 2010.
50. http://www.adidas-group.com/en/sustainability/welcome.aspx. (accessed on February 22, 2014.)
51. Yip J., Chan K., Sin K.M., Lau K.S., Low temperature plasma treated nylon fabrics, *Journal of Materials Processing Technology*, 123, 5–12, 2002.
52. Canup L., McCord M., Hauser P., Qiu Y., Cuomo J., Hankins O., Bourham M.A., Modification of nylon fabrics with atmospheric pressure plasmas, www.ntcresearch.org/pdf-rpts/Bref0602/C99-NS09-02.pdf. (accessed on February 22, 2014.)
53. Chaivan P., Pasaja N., Boonyawan D., Suanpoot, Vilaithong T., Low temperature plasma treatment for hydrophobicity improvement of silk, *Surface and Coatings Technology*, 193 (1–3), 356–360, 2005.
54. http://www.adidas-group.com/en/sustainability/welcome.aspx. (accessed on February 22, 2014.)
55. http://www.nikeresponsibility.com/report/content/chapter/our-sustainability-strategy. (accessed on February 18, 2014.)
56. Chen T.Y., Peng L.H., Nature-inspired fashion design through the theory of biomimicry, http://design-cu.jp/iasdr2013/papers/1512–1b.pdf. (accessed on February 22, 2014.)
57. http://www.sffashtech.com/2012/06/05/from-biorythms-to-biomimicry-the-future-of-textiles-is-here/. (accessed on February 9, 2014.)
58. *Seminar on Energy Conservation in Textile Industry*, The Energy Conservation Center (ECC), Japan, pp. 1–56, 1992.
59. Hatada K., Fox R.B., Kahovec J., Marikhal E., Shibaev V., Definition of terms relating to degradation, aging, and related chemical transformations of polymers, *Pure and Analytical Chemistry*, 68 (12), 2313–2323, 1996.
60. http://cmore.soest.hawaii.edu/cruises/super/biodegradation.htm. (accessed on February 11, 2014.)
61. http://www.ykkfastening.com/products. (accessed on February 1, 2014.)
62. http://www.bsr.org/reports/BSR_Sustainable_Fashion_Design.pdf. (accessed on February 22, 2014.)
63. Anon, *Sustainable Shoes, Impact: Textiles and Environment*, 2, 37, 2014.
64. http://www.mysomapatterns.com/why-zero-waste-fashion.html. (accessed on February 22, 2014.)
65. http://myweb.polyu.edu.hk/~tcshyam/8ISS/8-Panel%20on%20Environment%20and%20sustainability.pdf. (accessed on February 22, 2014.)
66. http://moreintelligentlife.com/content/arts/rebecca-willis/second-hand-clothes. (accessed on February 22, 2014.)
67. http://www.forumforthefuture.org/project/fashion-futures-2025/overview. (accessed on February 22, 2014.)
68. http://jolien.keablogs.dk/tag/near-sourcing/. (accessed on February 22, 2014.)
69. http://www.wisegeek.com/what-is-near-sourcing.htm. (accessed on March 14, 2014.)
70. http://www.tradecard.com/a-guide-to-apparel-near-sourcing/. (accessed on March 14, 2014.)
71. http://www.naturalfibres2009.org/en/fibres/cotton.html. (accessed on March 14, 2014.)
72. www.waterfootprint.org/Reports/Report18.pdf. (accessed on March 14, 2014.)
73. http://www.theguardian.com/lifeandstyle/2008/sep/03/denim. (accessed on March 14, 2014.)
74. http://www.unep.org/climatechange/mitigation/RenewableEnergy/tabid/29346/Default.aspx. (accessed on March 14, 2014.)
75. http://www4.ncsu.edu/~kpadia/CS895/HW5/. (accessed on March 14, 2014.)

23 Fashion Industry and Sustainability

Shanthi Radhakrishnan

CONTENTS

23.1 Introduction to Sustainable Fashion 502
    23.1.1 Concept of Sustainability 502
    23.1.2 Sustainability in Fashion 503
    23.1.3 Salient Features of Sustainable Fashion 504
    23.1.4 Myths on Green Fashion 505
    23.1.5 Sustainable Elements in Apparel Manufacture 506
23.2 Key Issues in the Fashion Industry 507
    23.2.1 Scenario of the Fashion Industry 507
    23.2.2 Environmental Issues Related to Fashion Industry 508
    23.2.3 Social Issues Related to Fashion Industry 509
    23.2.4 Metrics on Impact of the Fashion Industry 510
23.3 Framework for Sustainability 511
    23.3.1 Principles Underlying Framework of Sustainability 511
    23.3.2 Framework for Sustainable Fashion Consumption 511
23.4 Sustainability in Design and Production 513
    23.4.1 Meaning and Importance 513
    23.4.2 Sustainable Design Principles 513
    23.4.3 Product Life Cycle 513
    23.4.4 Future Scenario for Sustainable Design and Production 514
23.5 Application of Sustainable Materials and Technologies 516
    23.5.1 Sustainable Fibers 516
    23.5.2 Environment-Friendly Materials for Fashion 518
    23.5.3 Sustainable Technologies 520
        23.5.3.1 Use of Enzymes in Textile Processing 520
        23.5.3.2 Natural Dyeing 521
        23.5.3.3 Clean Technologies 521
        23.5.3.4 Recycling and Upcycling 522
23.6 Future Trends 523
    23.6.1 Trends in Sustainable Sourcing 523
        23.6.1.1 Trends in Cotton and Natural Fibers 523
        23.6.1.2 Trends in Silk and Luxury Fabrics 523
        23.6.1.3 Trends in Innovative Fabrics, Dyeing, and Design 523
        23.6.1.4 Trends in Wool and Heavyweights 523
        23.6.1.5 Trends in Components and Accessories 524
    23.6.2 Trends in Sustainable Retailing 524
    23.6.3 Fashion Shows with a New Facelift 525
    23.6.4 Implementation of Sustainable Fashion 526
References 527

23.1 INTRODUCTION TO SUSTAINABLE FASHION

23.1.1 CONCEPT OF SUSTAINABILITY

Sustainability was the term first coined in the field of ecology where it was used to define how biological systems had the strength to withstand and remain varied and productive. To quote a few examples are forests, seas, and wetlands. However, it now covers four domains, namely, ecology, economics, politics, and culture, and sustainability is the long-lasting endurance of social welfare, durability, and flexible living. According to another school of thought, the three main pillars of sustainability are social equity, environmental resilience, and economics. Despite the various meanings, ecology remains to be the basic block of sustainability as safe ecosystems and environment are required for the well-being and existence of all living materials.

Some of the negative impacts of globalization and industrialization are deforestation, mass production, and pollution. Clearing of forests and wildlife to serve the expanding need of industries has resulted in lesser trees that are a source of maintaining the oxygen balance of the atmosphere. The unethical use of chemicals and the emissions of toxins into the environment have given rise to global warming and depletion of ozone layer. However, there are a number of ways to reduce these negative effects, namely, environmental protection, environmental resource management, and environment-friendly chemical engineering. All the three methods work on data from green chemistry, conservation biology, environmental science, cultural and political concerns, and management of resources and human consumption. Though many societies work for sustainability, this appears to be a dream in the current situation that seems to be ruled by environmental degradation, climate change, overconsumption, and overpowering nature of society to achieve unlimited economic growth (Admin in Environment 2012; Bakari 2014).

The fashion industry is the most vibrant industry that is susceptible to change. The term *change* is important as fashion-conscious customers aspire for something new, unique, and exciting that would complement their looks and would lend to the attention of their peers and society. To cater to this basic need, the fashion industry has been working traditionally with large-scale manufacture of clothing with long lead times of around 6 months for the next collection. However, this trend changed and the so-called fast fashion principle was adopted by which retail stores changed designs on show every few weeks. This change was implemented by speeding up production by the integration of textile finishing and dyeing processes with textile weaving factories along with integration of clothing manufacture and distribution networks. This was accompanied by the introduction of new industrial robotics that substituted expensive labor and helped in accelerating the production process. The fashion industry focused on the demand and supply but operated to the detriment of social and environmental aspects in the context of sustainability. Waste generation was the result of the production and use of fashion garments as fully functional garments were discarded just for change in fashion. The textile manufacturing chain from spinning to finishing uses fossil fuels, resources, and chemicals resulting in nonbiodegradable wastes and hazardous effects on the environment. The marketing and sales processes have carbon emissions and waste through packaging, distribution systems, and transport (Joy et al. 2012). Sustainability measures that can be followed in the fashion industry are given in Figure 23.1.

The process of transforming this industry into a more sustainable one requires long-term commitment and change in personal, social, and institutional levels. The traditional processes in the textile and fashion field have to be revamped right from agricultural practices, consumption patterns, and ecological concepts to international energy policies. On the whole, the sustainability of this sector necessitates the participation and involvement that extends beyond boundaries and calls for the establishment of links with other disciplines, industries, societies, and international groups. Sustainability principles have the potential to transform the fashion industry, and the emerging trends adopted by the leaders in this industry are moving ahead in this direction.

Fashion Industry and Sustainability

FIGURE 23.1 Fashion industry and sustainability.

23.1.2 Sustainability in Fashion

The Ethical Fashion Forum (EFF), noted for its sustainable efforts, has described sustainable fashion as a methodology or a system that is conscious of the designing, sourcing, and manufacturing of clothing such that it maximizes the benefits to the people and communities coupled with minimizing the effects on the environment. The concept of sustainability in fashion has grown in the last decade and is high in the last few years. The idea of sustainability should have a three-branch support: from the government through legislations; from the manufacturers and retailers who take up the initiatives in designing, raw material sourcing, and manufacture; and from the consumers who are the users of the products and who can state their choice by using the products.

The three main pillars of sustainability, social equity, environmental resilience, and economics are grouped together under the name triple bottom line (Figure 23.2). These attributes must be incorporated at the center of policy and business practices and should extend from the board room to the shop floor or factory. The social factor should enhance the well-being and ability of the workforce of any industry. Poverty and excessive utilization of the human workforce in the fashion industry affect sustainability. The environmental line calls for monitoring all business operations throughout the supply chain for reducing the environmental impact. The company should go beyond the immediate target of minimizing environmental issues in operations and get involved in creating awareness about sustainability among different classes of the fashion fraternity and support all initiatives to safeguard the environment. The economics of the fashion business should be strong to face the stiff competition and to change policies and accompanied production technologies. This can be achieved when the sustainable business model includes quality products and services to meet the market needs that form the basis of sustainability.

Many companies and industries have been moving toward ecofashion by streamlining their business approach and setting it for the future in the competitive market and also using this principle for building trust and relationship with consumers. Here are a few initiatives to show the movement

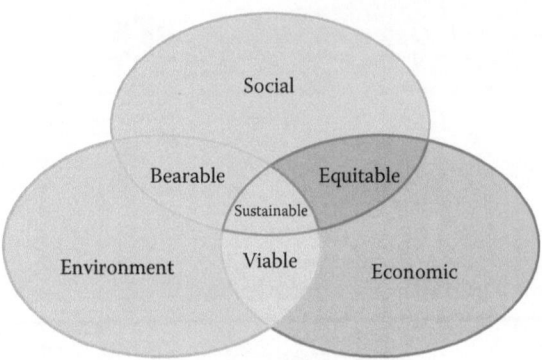

FIGURE 23.2 Pillars of sustainability.

toward sustainability in fashion. Hennes and Mauritz have used recycled polyesters, linens, and cottons to create a trendy summer wear "Conscious Collection," sustainable yet affordable for customers. Mark and Spencer encourage their customers to recycle their old clothing by their latest concept "Shwopping," where customers can hand over their old clothes of any brand to the shwop points across the United Kingdom. This was an extension of "plan a program" where people who donated their clothes as Oxfam could get a £5 voucher for their thought. This program was based on the policy of making the customer hand an item of clothing every time they buy one. Levis Strauss has joined the ethical group by introducing their "Waterless Jeans Collection," recuing the water footprint by 20%. The new collection is available in a variety of styles and comes with a label encouraging customers to wash their jeans infrequently (Smith 2012).

The EFF has launched the global platform for sustainable fashion called "Source" in 2011. The aim of this platform is to facilitate the fashion professionals and businesses work on the principles of sustainable fashion and also create an awareness among all down the fashion value chain. This online platform works in three major areas, namely, source intelligence, source database, and source network. Source intelligence is a global leader for sustainable fashion business intelligence from information about brand leaders to suppliers, market reports, and supply chain innovation to be used according to the needs of fashion specialists. Source database has created a sourcing directory that gives contact information of ethical suppliers and manufacturers and is also developing a database for the fashion sector. Source network unites thousands of individuals and businesses across the supply chain to connect and spread ethical practices and projects in fashion (EFF).

23.1.3 Salient Features of Sustainable Fashion

In order to understand the sustainability aspect of fashion, all the processes and systems involved in fashion should be under consideration. The role of the fashion designer is highly important and forms a link between humans and environment and is inclusive of all related components. Further, the fashion industry has a global impact and is complex and dynamic. Sustainability in fashion should be investigated in terms of its relationship with processes, life cycle of the product, the personnel of the system, the global impact, the importance of attitudes and results, and the value of wholesome solutions. Trend is a process of change and starts by-product development. This may result in new products or as a forecast of some change and adoption. Sustainability became a trend in the late 1980s after the environmental crisis that companies like Patagonia (1985) and Komodo (1988) promoted ethical fashion. This trend became popular and was predicted by Helen Job, editor of the trend forecasting service, World Global Sourcing Network (WGSN), United States, in 2007. Her forecast analysis was based on ecocities, consumer products, technology, and architecture. This trend has grown by leaps and bounds and works its ways into many issues like climate change, food, architecture, home products, and individuals who seek healthy lifestyles.

According to WGSN, it is very important to analyze the trends in sustainable consumption of fashion, indicating that the challenges that arise by social and environmental impacts of fashion consumption can be taken into account by sound design innovation; designers are going beyond the boundary of the product, and this attitude results in new processes, services, and systems; underlying concepts in fashion are integrated with the safety, longevity, and versatility of the product coupled with ethical production methods; the impact of the products at the end of the life phase is assessed and is significant in encouraging sustainable fashion consumption; the transparency in production methods and education have opened opportunities to create changes in the attitudes of the consumer. The essence of the analysis has been reported as design longevity, disassembly, ethical production for design, cradle to grave of products and services, transparency in production, and consumer education (Obregon 2012).

Sustainable design reduces the negative impacts on environment, health, and comfort of the consumer and aims at improving the performance of the products. The objectives of sustainable design are decreasing the consumption of nonrenewable resources, creating healthy and productive environments, and minimizing waste. While going for sustainable designs, principles like minimum energy consumption, use of natural materials, selection of eco-friendly production processes, conservation of water, and optimization of operational and maintenance procedures are essential. Taking up a sustainable design philosophy calls for decisions at each phase of the design process to reduce negative effects on environment and consumers without compromising in the important basics. It is an integrated approach that has an impact on all phases of the product life cycle that encompasses design, construction, and operation and postconsumption. When designers use the intelligence of natural systems, like the nutrient cycling or the energy of the sun, they can develop designs that will allow nature and trade to grow together (Horn and Davis 2014).

Fashion consumption is usually governed by the need to be distinctive and the need to conform. In order to appear distinctive, the individual uses fashion apparel showing his individuality and visual identity within the context of social relationships. In the second case, fashion is all about the acceptance of style by the consumers and hence youngsters would buy to show their conformity to the latest fashion (Pears 2006). To promote sustainable consumption, the culture of consumption must be analyzed in different prospects. These include consumption to care, consumption to utilization of services, and consumption to nonconsumption. Another important aspect that has been overlooked is the person and product relationship that was dominated by increased production and consumption. The role of the designer is to reach the consumer to create a sense of belonging with the garment to minimize consumption. There should also be a radical change in the minds of the designers and manufacturers to create sustainable fashion rather than regularly changing trends to increase profit. Clothing consumption has increased greatly over the past decade. The textiles going for incineration and landfills are vast, and the increased consumption and disposal have been the root cause of intensifying waste management systems. However, the current trend has highlighted the transformation of the industry and consumption patterns toward a more sustainable approach (Netter 2013).

23.1.4 MYTHS ON GREEN FASHION

Many professionals in the field of fashion have debated on the meaning of sustainable or green fashion. The creative director of Gucci, Frida Giannini, feels that sustainability deals with products of quality that will endure for a long period of use such that it can be used over and over again and can be passed on. Brand founder and designer Oscar de la Renta expressed that sustainable fashion is an attachment and commitment to tradition and culture; the sewing personnel must be skilled to make clothes that are extremely beautiful and professionally excellent. Designer Anya Hindmarch describes green fashion as locally found material that is nonpolluting from design to grave and requires minimum resources to obtain the finished product. Designer Dries van Noten refers to sustainable fashion as the analysis of methods followed for the production of raw materials used to develop the new product, for example, carbon footprint.

It would seem that sustainable fashion means different for different people though all work for the same industry. Labels and terms like sustainable, green, organic, eco, and ethical are increasingly used in the world of fashion. These terms may be looking attractive and catchy but the percentage of eco-friendliness in the product is very important. Terms like *recyclable* and *preorganic* mean that it can be disposed easily or it is made in a natural way with natural material. On investigation, the cotton that is claimed to be organic may not be organic in the farm; shoes may be termed *recyclable* by the manufacturer but the recycling authorities may not accept them for recycling. Efforts are being made to make production sustainable, but the extent of sustainability in the product is very crucial as it may mean something different in the minds of the consumers. The World Wide Fund for Nature published a report "Deeper Luxury," where grading was done for the 10 biggest luxury brands in 50 different eco and ethical categories, and it was found that none of the products got higher than a C+ grade. The world's premier jeweler and America's designer Tiffany was rated at D+ grade. In fashion, sustainability is not just a fashion trend that comes in and goes out with the move of seasons, but it requires a fundamental change in approach and attitudes, which needs to be explained and experimented (Friedman 2010).

23.1.5 Sustainable Elements in Apparel Manufacture

While manufacturing apparels, various elements that contribute to sustainability need to be analyzed in order to reduce problems. Fiber is the cradle stage of apparel production as it is the raw material for yarn and fabric. The use of eco-friendly fibers is recommended as they are easily recyclable. In case of blending of fibers for yarn development, it is advisable to choose eco-friendly organic fibers that are biodegradable. The next element is textile production. Use of efficient processes for pretreatment, dyeing, and finishing and in energy and water consumption can help in reducing the effluent load as well as the quantity of effluent produced. Eco-friendly alternatives and substitution of chemicals with enzymes can help in making conventional processes environment friendly.

The design for clothing should be focused on disassembly that facilitates the dismantling of clothing and reintroducing it in the garment life cycle resulting in zero waste as shown in Figure 23.3. The carbon footprint of the manufacturing of apparel, the distribution to the different retail centers, and the retailing business followed by the use phase of the consumer should be accounted and evaluated. If the product ends up in a landfill after use, it is considered to be not sustainable and

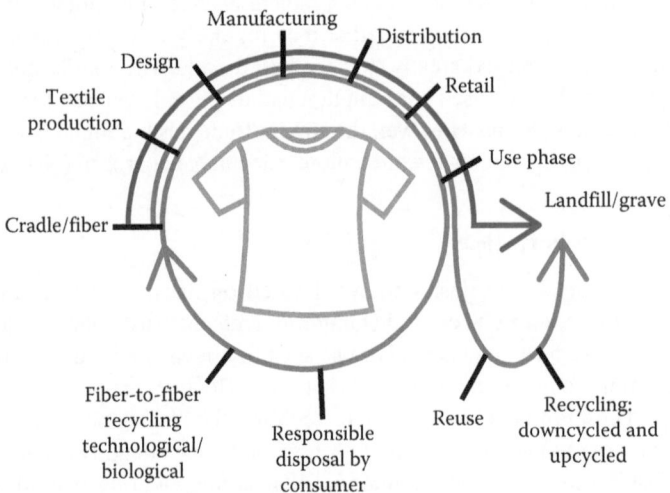

FIGURE 23.3 Sustainable elements in apparel manufacture.

Fashion Industry and Sustainability

requires modification in design and manufacturing. Partial sustainability is achieved if the product can undergo recycling and reuse. The green closed loop implies recycling and reuse of product followed by reuse as raw material in manufacturing and negligible or no wastage (Undress Runways Inc.). This is the aim and objective of sustainability, and many measures are undertaken by associations, government, and law to subject the products to assessment by the manufacturers to indicate the level of sustainability for certification to be offered in the market. The stamp of sustainability will ensure that the products have been manufactured in safe conditions by eco-friendly processes. The fashion and eco-conscious consumer will be ready to pay the price for the label and ensure his accountability for retaining sustainability by proper selection, use of product, and postuse phase. Eco-friendly production with minimal waste coupled with social accountability like safe working environment and fair wages is the most important part of production. Serving the domestic market and enabling production with local raw materials help to reduce the carbon footprint as traveling is avoided in distribution. Concepts like reuse, reassembling, downcycling and upcycling, vintage garments, and secondhand clothing will reduce the burden on retailing. Use clothes with easy maintenance and good care characteristics so that they will last longer. After use of garment, it should be recycled. The use of biodegradable products would minimize the load on landfills.

The different stages of garment manufacture have shown the areas in which sustainability can be enabled. Sustainability in apparel production is a complex issue involving many contributors, and it is not possible to create a 100% sustainable garment. An awareness of sustainability will send feelers to the consumers who will in turn demand from designers and manufacturers. Today's consumer is technologically equipped and is in a position to dictate the production centers and designers to fulfill his needs. The government regulations for production and export harness sustainable practices to develop new products and practices. Hence, sustainability enhances the ongoing capacity of living systems to maintain life in a healthy and safe way.

23.2 KEY ISSUES IN THE FASHION INDUSTRY

23.2.1 Scenario of the Fashion Industry

The textile and the clothing sectors represent around 7% of the global exports with about 2.6 million people who are involved in the production of apparel to suit the needs of the consumer. The textile industry undertakes the manufacture of fabric from fiber to finish, which makes the fabric ready for apparel conversion, while the clothing sector deals with the production of apparels for consumers. The fashion industry encompasses both the textile and clothing sectors, dealing with the materials of fashion (fibers, fabrics, and trimmings) at the primary level and the production of apparels and accessories in the secondary level. The developing countries account for three-fourths of the world's clothing exports, while the United States is a leader in cotton and Australia for wool. There have been rapid changes in international trade agreements and the removal of the quota system and tariffs have opened the market to efficient players (Allanwood et al. 2006). Despite these developments, the rules of trade remain complicated and continue to change rapidly. This sector is becoming highly integrated. Traditionally, fashion changed twice a year, while *fast fashion* has introduced patterns where stores change designs every 2 weeks. Textile dyeing and finishing industries are integrated with weaving, factories, clothing manufacture, and distribution networks. Industries are investing in increased capacity building with robotics substituting costly labor with innovative technologies. The development of business clusters is evident as they give rise to support and regional integration.

The shifting boundaries of the fashion industry call for great attention. In early days, the fashion industry was a section of the clothing industry and associated with haute couture or the exclusive design elements of fashion. Next came the luxury goods that were goods having a higher selling price when compared to the manufacturing cost ratio since the most important part was the design. This context was shifted toward utilitarian clothing where mass production and mass market is the concern. There is a unification of designs and manufacture to cater to the mass market. The expansion of the existing market and the growth in the developing world have opened up the world of fashion (Pratt et al. 2012).

The technological advances have brought about a revolutionary change in the fashion industry. This enables quick and efficient output with reduced man power. However, all processes need to be manned by the technical operators. Technological innovations have replaced manual operations and surveillance with microprocessor-controlled equipment and evaluation methodologies that have fastened the production processes leading to new design development, process optimization, and production technologies. The drive to undertake fast production and fast fashion is encouraged by the huge orders gained through access to global markets. Thus, the fashion industry has geared itself to face the challenges of high-speed production and changes in the fashion market.

23.2.2 ENVIRONMENTAL ISSUES RELATED TO FASHION INDUSTRY

However, there are major environmental issues associated with the fashion industry. Extensive use of energy is found in the production of primary materials, which include yarn development of natural fibers, production of man-made fibers, and laundry. Toxic chemicals are being used in the production metrics, for example, conventional cotton production, which are harmful to man and the environment. The wastewaters of pretreatment, dyeing, finishing, and laundry release chemicals into the environment that harm aquatic life. Solid waste arising from manufacturing, wastewater treatment, and disposal of nonbiodegradable products after use is alarming.

It has been reported that textiles and clothing made in Cambodia, India, and China are produced using dangerous chemicals that can cause birth defects and cancer. The pesticides used for cotton cultivation have been banned in Europe. The UN Division for Sustainable Development has reported that from the year 1997 to 2006, one farmer committed suicide every 32 min in India (Hough 2012).

Fashion industries use animal fur that involves inhumane treatment of animals, either trapping and skinning wild animals for the fur or raising the animals domestically for their fur. It has been estimated that more than 1 billion animals are slaughtered for hides and skins and 30 million animals are raised in cages for their fur. The killing of animals may be breaking of the animal's neck, gassing, harmful injections, and genital or anal electrocution (Allanwood et al. 2006). Fur is the covering of animals and has the tendency to rot. Manufacturers prefer to avoid decay of fur by using chemicals to prevent putrefaction. Formaldehyde and chromium are the main chemicals that are being used and can affect the workers in the processing plants and the people who will wear them. The World Bank has stated that the fur-dressing process is one of the world's worst industries for toxic metal pollution. When these chemicals leak into waterways, the effects can be extremely harmful to aquatic life and other living forms that are dependent on the water for day to day use (Hoskins 2013). Some high-end fashions use *blood diamonds* in their designs. These are diamonds mined in poor conditions by immoral people who use the funds from these diamonds for corrupt activities (Xaxx 2014).

The wastewater generated from textile processing can cause damage if it is discharged without proper treatment. The aquatic toxicity of the textile effluent will vary according to the production facilities. Materials like salt, surfactants, ionic metals, toxic organic chemicals, biocides, and toxic anions are sources of aquatic toxicity. Other pollutants are detergents, emulsifiers, and dispersants that contribute to BOD, foaming, and aquatic toxicity. Air emissions are abundant in textile operations like fabric preparation, dyeing, printing, resin finishing, and wastewater treatment plants. High-temperature drying or curing and drying ovens emit hydrocarbons. These processes can also release formaldehyde, acids, softeners, and other volatile compounds. Sulfur dioxide, emitted during textile processing and by transportation vehicles, when released into the atmosphere can cause acid rain, which has harmful effects on plants, animals, and infrastructure (Parvathi et al. 2009).

The Chinese textile industry creates about 3 billion tons of soot each year; in 2010, the textile industry in China ranked the third for wastewater discharge of 2.5 billion tons per year. The China Pollution Map Database in 2012 had 6000 records of textile industries that violated the environmental regulations causing excessive damage to the ecosystem. The issues were discharge of wastewater from hidden pipes, discharge of untreated pollutants, misuse of wastewater treatment facilities, discharging water with a pollution load that was greater than the standard for discharging water, and using the facilities that were shut down by the authorities for a variety of reasons (Breyer 2012).

23.2.3 Social Issues Related to Fashion Industry

The social implications of this industry are of great concern. The elimination of child labor is the law, but it remains a challenge as there are many intermediaries in the production chain like subcontractors, indirect workers, and homeworkers. The industry has a workforce of young unskilled women who are subjected to various forms of abuse and who cannot maintain their rights as workers. Further, there are various issues that affect the workers in the industries like exposure to hazardous chemicals, fiber dust, noise, and monotonous repetitive processes (Allwood et al. 2006).

The work practices in Asia and India are very alarming. In Cambodia, women workers are forced to work for 100 h a week with €0.28 as wages. China produces more than one-fourth of the world's textile and clothing supply and has the largest labor force earning very low wages (Hough 2012). The unethical working conditions were highlighted in Rana Plaza factory collapse report that said that 1129 Cambodian textile workers were killed, of which 33% were malnourished and medically underweight (Borromeo 2013).

The fashion industry is complex and works in union with other allied fields like manufacturing, raw material production, advertising, transportation, and retailing. Profits are huge and the thirst for money may lead to unethical practices. Conditions need to be handled better to save the industry from disreputable methodologies. To maintain the aesthetic expectations of the public, the models in the fashion houses are extremely thin in stature to match with the idealized image. This leads to eating disorders and poor body image with very low body mass index (Xaxx 2014). Fashion retailers have an excessive thirst for fresh new designs. They release up to 300 new styles per week and these styles are presented in a fashion show with lead times as short as 13 days. A section of the fashion personnel are involved in sourcing raw materials from farmers and cultivators for bulk production. The styles are mass produced in factories identified in the developing countries. This chain of design to manufacture is usually very tedious and faceless (Pratt et al. 2012).

The search in fashion for new and different looks calls for borrowing special elements from varied cultures. This leads to accusations of cultural appropriation by the members of the culture or clan. Further, the materials, production, and transportation have a negative impact on the environment. Natural materials used for food production or synthetic materials derived from petroleum products are generally used in the fashion industry (Xaxx 2014).

The role of advertising is of prime importance in the fashion industry. The role of advertisement is to lure consumers to buy as much as possible. The concept of change in the arena of fashion does not allow the consumer to use the apparel for a long time but make them ready for the next purchase, which may not be necessary. Some people look at this consumer behavior as harmless or a boon to economy, while others feel it instigates mindless consumerism. Exclusiveness of the product is the largest appeal in fashion clothing, which is accompanied with higher status and glamour. A majority of people may not be able to buy it, leading to fakes and imitations. Consumers are carried away by the brand names and this encourages forgers to create cheap knockoffs with labels of leading fashion houses. There is a well-known conviction that the fashion industry promotes the livelihood of the poor and downtrodden in developing countries. The unethical practices, poor environmental and social standards, and employment of child labor have given rise to the term *sweatshop labor*, which has been criticized by many (Xaxx 2014).

23.2.4 METRICS ON IMPACT OF THE FASHION INDUSTRY

The raw material used in the fashion industry is natural and synthetic fibers. Fossil fuels are being used for the production of cotton, wool, and polyester. About 60% of the crude oil consumed for the production of cotton fabric is in the pretreatment, dyeing, and finishing area, while 21% is used in the cultivation of cotton. The global warming potential is higher due to the methane emissions from the sheep until the manufacturing stage. The utilization of hard coal in the wool scouring and yarn development is about 46% and 43% of the total usage. It should be noted that polyester manufacturing consumes the most energy from fossil fuels coupled with greenhouse gas (GHG) emissions and resource depletion when compared with the natural fibers (Norwegian Fashion Institute).

Since manufacturing, production of the end product, and target market have been grouped together from different parts of the world, it becomes necessary to dispatch products over long distances by land, sea, or air. During the life-cycle assessment, the emissions produced by transportation will clearly indicate the environmental impact due to transportation. The carbon footprint will add the emission details to the country that benefits in the trade, for example, the exporter. Another area of concern is the consumer handling process where two-thirds of the energy consumption is due to laundering and one-third for drying. Here, fabrics from natural fibers use more energy than synthetic materials.

It has been roughly estimated that 90% of the waste produced is from consumer use and garment disposal as a majority of clothing is not reused or recycled but ends at landfills. Materials that are nonsynthetic in nature can release methane when deposited into the environment and form one of the reasons for climate change impact during the clothing life cycle. The development of raw materials and production are reasons for deforestation, soil erosion, and water contamination, which have an adverse effect on fauna and flora. The toxins of manufacturing processes are harmful for workers, consumers, and the environment. Millions of gallons of toxic dye effluent are discharged into lakes, rivers, and oceans.

The fashion industry wastes 70 million gal of water each year. About 700–2000 gal of water is required to produce cotton necessary to manufacture one T-shirt. About 3.5% of the total volume of water used for crop production worldwide, that is, 210 billion m^3 of water, is used for cotton production. This production cycle pollutes 50 billion m^3 of water every year (Dietz 2013).

Pesticides used in raw material growth and water toxicity affect biodiversity. Conventionally grown cotton uses a high quantity of insecticides when compared to the other crops. Cotton cultivation employs 25% of the world's insecticides and 10% of the world's pesticides. These pesticides are carried along with the runoff rainwater polluting lakes, rivers, and waterways. Cotton that is grown in the environment of pesticides, insecticides, herbicides, and genetically modified organisms is harmful to the living organisms and entire ecosystems (Dietz 2013).

It is common knowledge that GHG emissions are caused primarily by burning of fossil fuels leading to global warming. It has been estimated that 2 kg of carbon dioxide is released into the atmosphere for every kilogram of textile produced since fossil fuels are utilized for all stages of clothing manufacture. The main GHG emissions are carbon dioxide (CO_2) from usage of energy and methane (NH_4) and nitrous oxide (N_2O) due to cotton cultivation. This industry uses 0.6 kg of oil equivalent primary energy for every kilogram produced. In 2004, the energy use was 5.217 million tons of oil equivalents. It is also said that 1.5 billion gal of oil is required to manufacture 1 million ton of clothing (Dietz 2013). Statistics reveal that the textile and clothing industry has very high energy requirements that increase steadily year after year (Norwegian Fashion Institute).

While formulating policies for innovation, it is essential to understand the basics of clothing production in the fashion industry. Each stage of production has to be analyzed on its own as well as in the context of the other stages of the life cycle. In order to receive the bulk of the orders as against competitors, developing countries do not give the accurate information about the production procedures. Clothes meant for the short life cycles are considered the most harmful ones with regard to its impact on environment.

23.3 FRAMEWORK FOR SUSTAINABILITY

The World Commission for Environment and Development has rightly coined the concept of sustainable development that has become increasingly popular on a global scale. There is a definite relationship between the population and the carrying capacity of the environment. Since a number of measureable variables exist, it offers superior leverage to estimate the human activity on the planet. The activities of the population bear a demand on resources, and also by use of these resources, by-products are generated. These two factors have an impact on the carrying capacity, which is the ability to continue in a healthy manner. To put this concept in a nutshell, it is the organization of alignment between individuals, society, economy, and the regenerative capacity of the life-supporting ecosystems of the planet.

23.3.1 Principles Underlying Framework of Sustainability

There are five basic principles that govern the framework of sustainability, namely, the material domain, the economic domain, the domain of life, the social domain, and the spiritual domain. These are generalized principles that can be catered to specific sectors of economy, developmental issues, business strategies, and individual initiatives.

The material domain forms the basic regulation of the flow of materials and energy that forms the basis of the existence. Energy conservation is vital and there should be productive use of the resources. By use of energy, certain disorderliness exists, which will be made orderly in both individual and complex systems. The main objective will be highest resource productivity and recycling of nonregenerative resources for better stability and continued availability. The economic domain highlights the accounting framework that misrepresents values and does not take into account factors like depletion of resources and pollution into the cost components. Regulations and subsidies offered continue to foster the negative effects of the economy and stunt the scope of initiatives that aim at best practices. The only approach for sustainability with regard to economy is to take into account natural, human, social, manufactured, and financial elements as key forms of capital and to relate them to the regenerative capacity of nature in order to impose a measure of well-being and human development.

The domain of life deals with a very important aspect of maintaining all forms of life in the biosphere. A trend that has come into existence in recent times is the destruction of individual animals, species, habitats, and whole ecosystems due to the actions of human population. The quick propagation of human life in all parts of the planet shows their adaptive success at the expense of many other forms of life. The fundamental sacred forms of life are to be protected. Responsibility to conserve the existing life by sustainable living is a must and human encroachment on other forms of life should be avoided to save all forms of life. The social domain indicates tolerance for social interactions and development of social forms for realizing the need to live socially without adversely affecting others. Provision should be made for incorporation of universal rights and equitable access for nurturing resources, cooperation for managing global issues and planetary commonalities, and help in bringing sustainability into the legislation. The principle of spiritual domain is to understand the power of nature and wisdom of the outer reaches of the cosmos that links our solar system, planet, and its biosphere. The consciousness about the mystery of the underlying aspects of existence and the intricate ecology of the earth of which humans form an important part will promote the respect for nature and all living beings (Ben-Eli 2006).

23.3.2 Framework for Sustainable Fashion Consumption

In order to develop and put into practice design ideas and business practices and to manage the economic, social, and environmental factors, sustainable ideas are important. These are usually linked to the life cycle of apparel, footwear, accessories, and other fashion goods. Fashion consumption involves varied interactions between the consumption needs of the individuals and the productions output of the fashion system. The basic need for the use of clothing is utility. When people use clothes for reasons beyond utility that is fulfilled while keeping resources for the needs of the future generations, the process is called *sustainable fashion consumption*.

The visualization of any organization is that the consumers and society as a whole must be able to enjoy the new design ideas, products, and services of the fashion industry. This should have a two-way approach where contributions for sustainable development process should reach across the globe while reducing the social, environmental, and economic impacts. Thus, the qualities associated with the fashion industry should not be lost but should highlight the positive impacts.

There are four focus areas in the framework of sustainable fashion consumption, namely, awareness of sustainable fashion, plan and acquire, wear care and share, and end of life.

1. *Awareness of sustainable fashion*: Changes in attitude and behavior that should be linked with fashion consumption are of primary importance. The most significant is to spark the awareness of sustainable consumption and the economic, environmental, and social impacts of the consumption and production of fashion. Consumer studies and behavioral studies may be used to gauge the level of awareness among customers. Many such studies include Mistra Future Fashion Project in Sweden, Textile Waste as Resource in Norway, and Sustainable Clothing Action Plan in the United Kingdom. These studies will serve as a good foundation research for identifying the levels of awareness among customers and measures can be formulated for increasing the awareness and a change in behavior can be predicted and developed.
2. *Plan and acquire*: Consumers should start analyzing their wardrobe and purchase sustainable goods and services related to fashion. They must be produced in a sustainable environment and must have good quality to last long and certified for sustainable standards. Attitude and behavior related to planning, searching for, and acquiring fashion can be converted to numeric measures by following details like the number of garments an individual has in possession, online websites and searches for sustainable fashion items and the type of purchasing behavior in retail items, the sections of the retail shop that promote sustainable fashion, and secondhand shops.
3. *Wear care and share*: Sustainable thinking also includes individuals to maintain and keep garments for more seasons and occasions and go in for garments that require less care like low-temperature washing and line drying. Garments are to be mended and used for longer periods. This sustainable attitude or trend in consumers can be measured by modern tools like mobile technology and social media that work and give new opportunities to motivate and track customer behavior. The data to be collected include the average use of garments, the wash and drying methods, the extent to which clothes are repaired at home or in mending shops, and the practices followed for caring or maintenance and sharing of garments with others.
4. *End of life*: It is ideal for individuals to give garments for second use, recovery, reuse, and recycling that includes recovery or upcycling of fabrics and fibers. Information collected from municipal waste agencies will estimate the number and weight of garments that walk into landfills. Studies on the consumer behavior will help in tracking the fate of garments when not wanted, while the recycling industry can provide statistics on the weight, value, and fate of postconsumer materials.

On the part of the industry, more transparency regarding design choice and sustainability features of products and services like life-cycle assessments is useful to understand the negative impacts of products on the environment. Similarly, corporate governance and accountability are responsible for the current and future statures of the fashion industry (Nice Fashion 2012).

23.4 SUSTAINABILITY IN DESIGN AND PRODUCTION

23.4.1 Meaning and Importance

Sustainable design is also known as environmental design. The philosophy of designing objects or products, services and environment is based on the principles of social, economic, and ecological sustainability. Sustainable designs are skillfully and intensively planned so as to eliminate any negative environmental impacts. These designs do not require nonrenewable resources but help to integrate people with the environment. Sustainable design must create a balance between economy and society and create a long-term relationship between the user and the object and also have an awareness of environmental and social differences.

Sustainable design strategies for textile and fashion designers are necessary to help them to create designs with the sustainability issues in mind. As more and more discussions related to the unsustainable products and systems in the textile and fashion field are being held all over the world, the responsibilities of the creators of the design are greater and the need for frameworks is essential. Much of the research work projects like London Sustainable Innovation are doing work on the revival of traditional textile crafts, MISTRA Future Fashion aims at innovation and analysis of the life cycle of the product in the cradle-to-cradle environment, (D)urability Day has developed practical strategies to reuse garments by whitewash and overprinting, and Digital Craft works on digital textile design and print and their impact. Extensive work is being carried out to develop a methodology for application of practice-based sustainable design strategies by budding designers to create new designs that have reduced or no impact on the environment (Politowicz 2010).

23.4.2 Sustainable Design Principles

There are certain principles that govern sustainable design creation. The raw materials chosen for the design should be from renewable sources and nontoxic, sustainably produced, or recycled materials that need comparatively little energy to process. These materials are known as low-impact materials as they do not harm the environment. The energy requirement to create the new product and to produce merchandise should be low using lesser fossil fuels. The design should create an emotional bond between the product and the user, thereby increasing the durability of product. The product, processes, and systems should perform in commercial settings till the end of its life. It should be biodegradable and compatible for reuse and recycling. The impact of the design should be measurable for total carbon footprint and life-cycle assessment. These calculations should be available for every resource used in the design. Many of these measures are complex but they help in estimating the impact of resources on the environment.

Many professional design guides and design standards are available that make it easy and user friendly to calculate the impact of the design on the environment. This new technology is termed as *sustainability science* and is being promoted by a wide variety of educational and governmental institutions. For sustainable designs, the principle of *biomimicry*, which is based on designing industrial systems on biological lines, is very useful. This principle enables the constant reuse of materials in closed-loop cycles. The principle of pooling of work and substitution of personal ownership to sharing of services creates minimal utilization of resources. The raw materials should be locally sourced and must be compostable after use. Strong design principles should be utilized for any design that is going to address a pollution source (Hawkin et al. 2008).

23.4.3 Product Life Cycle

The product life cycle begins with the raw material and its origin. The energy needed for the manufacture of the product is an important factor and should be taken into consideration, as shown in Figure 23.4. The next step is to gather information about the production techniques, distribution,

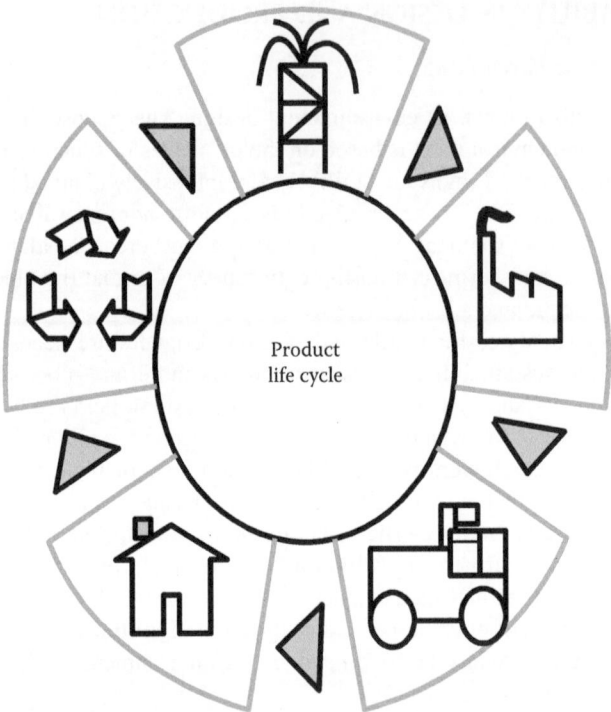

FIGURE 23.4 Life cycle of a product.

and use of the product that includes the utilization of the product, generation of products for reuse or recycling, and the ultimate disposal. The environmental effects at each stage of the product life cycle are accountable in a unified manner. The most important issues include consumption of the input materials at each stage of the life cycle of the product, which includes water, nonrenewable resources, and energy. The output materials that may be in the form of water, heat, emissions, and waste must be taken into account. This may also include noise, radiation, electromagnetic fields, and vibration.

The life cycle of a shirt is taken as an example for analysis. Shirts are made of blended material of natural and synthetic fibers. The production of natural fiber includes energy, fertilizers, water, and pesticides as in the case of cotton, while the synthetic fiber requires fossil fuels. The raw materials as fibers are combined to produce yarn and fabric. During this stage, water, energy, and chemicals are used to process and dye the fabric and to impart other characteristics. Shirts are produced from the developed fabric, packaged, and distributed to retail stores. This requires labor and energy. After the consumer has purchased the shirt, he or she will discard the packaging and will use the shirt. During the use phase, the shirt may be used about 100 times and washed, dried, and ironed. The environmental impacts include water, energy, and detergent. Finally, when the shirt is worn out, it will be disposed. It will not be compostable because of the synthetic parts and may not be recyclable since it is a blended fabric. During its manufacture, the shirt may have traveled many kilometers since the fabric production may have been in Asia, production in North Africa, and retail in Europe (Crul and Dieh 2005).

23.4.4 Future Scenario for Sustainable Design and Production

Knowledge of the future scenario is necessary to develop and create sustainable design and production methods. Innovative thinking coupled with an open mind to foresee things and explore the means to bring about sustainable successful designs and production systems is the need of the hour. These scenarios will help and guide the fashion and textile experts to develop a vision for innovation and organizational development. The future scenarios are *slow is beautiful, community couture, techno chic*, and *patchwork planet* (Levi Strauss and Company 2010a).

Slow is beautiful is the theme of this scenario. The attitude is risk averse, moralistic, low carbon, and highly regulated with sustainable lifestyles and mind-sets. Fashion is either sustainable or cool with consumers willing to pay high-quality sustainable clothing. The choices are durable fabrics from organic natural fibers or man-made fibers from renewable sources, handcrafted, vintage second hand clothes, or smart clothes that monitor health. Clothes are taken from small or virtual stores with efficient logistics. Clothes are manufactured in different parts of the world according to the manufacturing required, and the workers will have a decent wage and working environment. Clothes are washed without harmful chemicals and they last longer and are washed less frequently at low temperature. Clothes after use can be returned to the retailer and these will be remanufactured in other places. Sustainable labeling and digital tagging will ensure that the consumers know the origin of their clothing and the impact on the environment. The economy will be dull with poor labor standards as there are people who cannot get used to the new slower world. Sustainable fashion businesses are transparent and most sustainable for the best value.

In the community couture scenario, the strength is the strong bondage of the community. Here, the world is facing resource shortages and the impact of climate changes, but the community culture is strong and promotes self-sufficiency. Two types of fashion exist, namely, one that is expensive and new and the other that is cheap and secondhand. A dramatic fall is foreseen in the production and sale of new clothing due to high cost of raw materials and disrupted supply chains. Secondhand, preloved ethical clothing and community-grown locally available materials are used, and only the rich can afford certified new clothes made from virgin raw materials. Clothes will be available secondhand, in retail stores, and from clothing libraries. Clothes may be made domestically or in community-monitored recycling centers that are linked with the local factories. Care of clothes includes community laundries and low-volume-water washing machines. Used clothes are resold for an additional income. Nothing is disposed as it will be used for secondhand clothing. Successful business organizations will have a strong community bond and will provide energy supply, education, and subsidized rates or free food for employees.

Techno chic is a scenario that focuses on technology and its impact. The world is ultrahigh tech, healthy, and wealthy with an aim of lightweight living. Materialism is not in favor but people will believe in consumerism. Fashion will be fast paced, low in carbon footprint, and cheap, and the fabrics used will be manufactured from new high-tech low-impact biodegradable fibers to be converted to biodegradable, nanotechnology fabrics with nontoxic sprays on clothing. The manufacture will include high-definition programmable clothing. Machine-made clothes will cater to the consumers with the help of online stores. The care for clothing will be high tech with valet services, finishes will reduce the number of washes, and they will be recyclable. According to the design, clothes will be composted, disassembled, remanufactured, or reused. The industry will be financially viable due to low-carbon low-impact production and sustainable solutions but may not be able to keep up with the pace. Successful business organizations are those who find creative ways to maintain the customer loyalty and anticipate a correct demand to avoid waste.

Planet patchwork is based on the concept of world-level activities and occurrences. The world is divided into cultural blocks with unequal economic performance. It has been defined that Asia will be the economic and cultural powerhouse with conflict over the dwindling resources. The fabrics are sourced from local manufacturers and the clothes are made in regional factories to reach the consumers quickly. Fashion is strongly influenced by regional trends and celebrities and is highly personalized. The care for clothing has brought about the waterless washing machines or nanocoatings that require no cleaning. After use, the clothes are thrown away and there are many agencies that are involved in clearing the waste and have to face mounting tensions and environmental constraints. Successful business organizations are those that have a strong local heritage. This is the scenario for

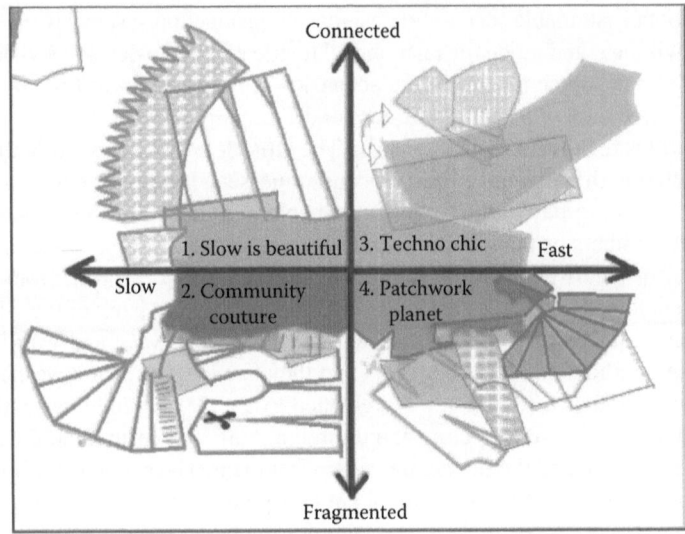

FIGURE 23.5 Scenarios for future fashion.

planet patchwork and the designers are to work with these ethics in mind for developing sustainable designs (Levi Strauss and Company 2010b).

These scenarios were based on two major issues that shape the fashion world as in Figure 23.5. The issues relate to the extent of connectivity or fragmentation that exists in the world and the swiftness with which society and fashion change. The issues are explained with the changes that occur in society and its effect on the attitude of the people. When the term is connected, globalization has moved further and expanded, trade barriers are lowered, and communications are united. This results in similar global cultures that keep the world united and similar. On the other hand, where globalization has reversed, long-distance trade is declining and local and regional identities are stronger, and a closed community results, which is termed as *fragmented*. The second issue is how fast the change occurs in society and fashion, which depends on communication and media, flow of financial capital, and the pace with which change occurs.

23.5 APPLICATION OF SUSTAINABLE MATERIALS AND TECHNOLOGIES

Sustainability of materials is an area that is widely debated while evaluating fashion products. Sustainability evaluation is done by tracing the sourcing and fabrication pathway of the product right from raw material, production techniques, product manufacture, and consumption of the product till the end of life of the product or reuse. The basic nature and properties of the raw material that comprise the product are of utmost importance. In addition to this, the techniques and technologies involved in the manufacture of these products will rate the quality of the product. While considering the sustainability of the material, many factors should be considered, namely, renewability and source of a fiber, the process involved in converting fiber to fabric, the environment and working conditions of the people involved in the production of the material, and the total carbon footprint of the material.

23.5.1 SUSTAINABLE FIBERS

Natural fibers may be divided into animal and plant fibers. Fibers may also include regenerated fibers that are obtained from natural sources and converted by chemical processing to produce continuous filaments that can be spun into yarns. Regenerated fiber from cellulose are rayon, which is produced using harsh chemicals and is not sustainable, while Tencel is considered to be an eco-friendly fiber, which is sourced from the forests certified by the Forest Stewardship Council and processed in a closed-loop system whereby very little effluent escapes into the environment. Regenerated fibers from protein origin are called *azlons* and are sourced from soy, corn, milk, and peanuts. Synthetic fibers are polymers made from petroleum-based chemicals and are not sustainable. New developments in biotechnology have seen the use of agricultural sources for the production of eco-friendly synthetic fibers like nylon, polyester, acrylic, olefin, and spandex.

Organic cotton is grown as per the certified organic standard without the use of chemicals, fertilizers, sewage sludge, irradiation, or genetic engineering and should be certified by an independent organization. The harmful effects on people and environment are considerably less when compared to conventional cotton. Recycled cotton may be from spinning waste or from scrap yarn, cut waste from garments, and scrap fabric that is processed and recycled into yarn. Organic certified linen and hemp are also available for yarn development. Organic livestock raised as per the certified organic standard will be able to produce organic wool, and Ahimsa silk produced from cocoons that allow the moth to finish its full cycle is considered as a sustainable fiber.

Recycled synthetic fibers may be from postindustrial or postconsumer waste. Mechanical recycling involves the melting of the waste and reextruding it into a yarn. There is a loss in strength and streakiness in dyeing due to impurities. This process can be done a few times before the molecular structure of the polyester breaks down and becomes unsuitable for textiles. Chemical recycling is based on the depolymerization of waste polyester and is equivalent to virgin polyester. The material can be recycled multiple times and is similar to the original fiber. Recycling is similar for nylon fiber and is from postindustrial waste. Several standards have been developed for recycled fibers and third-party certifications are also available. The life-cycle assessment of the fiber includes the energy, water, and land use, GHG emissions, toxicity to humans and ecosystems, usefulness of the product, and disposal techniques (Organic Exchange [OE]).

Bio-based man-made fibers are from renewable sources like plant sugars or oils. These polymers have desirable properties and are also environmentally safe. Ingeo™ polylactide, a biopolymer produced by NatureWorks, Minnetonka, uses lactic acid from corn sugars. The fiber and yarn producers of Ingeo include Fiber Innovation Technology Inc., Johnson City, TN; O'Mara Inc., Rutherford College; and Palmetto Synthetics LLC, Kingstree, South Carolina. DuPont™ Sorona polytrimethylene terephthalate is made up of 37% annually renewably sourced plant-based ingredients by weight and 28% bio-based carbon. Rilsan® PA11 is produced by Arkema, which is a France-based chemical company, and derived from the oil of castor beans. These bio-based fibers are very good in performance with regard to mechanical and comfort properties and dyeability and are suitable for high-performance technical textiles. The GHG emissions are comparatively reduced and they do not contain any heavy metals. They are biodegradable and recyclable as they serve to reduce the burden on the ecosystems (Rodie 2011).

Unconventional fibers are also used these days to produce eco-friendly fabrics. Hemp belongs to the family of cannabis and requires less water for cultivation when compared to other fibers. It also helps to rejuvenate the soil and no rest periods are required for the soil where hemp is grown. Jute is an eco-friendly fiber as one hectare of jute absorbs 15 tons of carbon dioxide and releases 11 tons of oxygen. The net carbon released is zero and it has a purifying effect on the atmosphere. A combination of jute and hemp is the hessian cloth used by the military in Germany and is suitable for the forest and tough terrain assignments. Ramie is a natural fiber eight times stronger than cotton and is commonly known as China grass. It can withstand hot temperatures and does not shrink and is ideal for technical textiles like fire hoses, canvas, and paper. It is usually blended with cotton or wool and is used for apparel. Nettle fiber is a weed and is a natural antimicrobial agent and moth repellant. It provides good insulation properties and is used for fine clothing and sail cloth. Bamboo fiber is used extensively because of its antimicrobial property and moisture absorbency. It releases 35% more oxygen into the atmosphere than the regular trees and helps to prevent soil erosion. Fibers are produced from bamboo pulp and are used for all types of apparel (Image Credit).

23.5.2 Environment-Friendly Materials for Fashion

Sustainability has called attention to alternatives for the existing materials as there has been extensive depletion of resources that have been constantly taken from the environment due to the demands of the population. Fish skin is a waste material in an industrial waste that is separated from the meat and discarded. It is a durable, warm, thin, and soft material when compared to leather, and the surface of fish skin can be cleaned manually without the use of chemicals. It ages well like leather and can be colored using any dye. A tannery in Iceland, Atlanta Leather, specializes in fish leather production from species of fish like salmon, perch, wolfish, and cod. Exclusive niche products like wolfish leather and washable salmon leather are available here. A few images of fish leather are shown in Figure 23.6 (Smithsonian Institution National Museum of National History 2013; Pinterest.) Abundant renewable resources like hot water from geothermal sources and hydroelectric power have given Atlanta Leather a sustainable advantage to be world leaders in fish leather

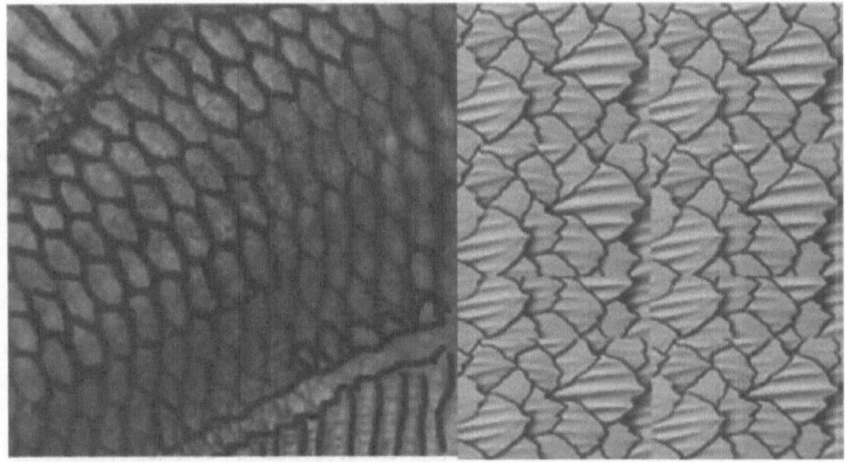

FIGURE 23.6 Fish leather.

FIGURE 23.7 Cocona jacket from coconut husk fibers.

Fashion Industry and Sustainability

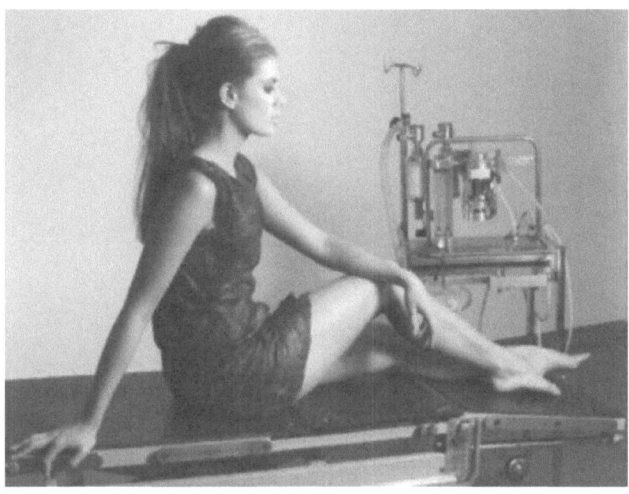

FIGURE 23.8 Garment from fermented wine.

FIGURE 23.9 Naoron bag, fabric from wood pulp.

production. This organization is a member of the BLC Leather Technology Centre Limited and is ECHA tested and REACH certified (The Sustainable Angle).

The tough fibrous coconut husks disposed by the food service industry have been converted to Cocona fabric, which is made into a lightweight, breathable, and durable insular jacket (Figure 23.7). A group of scientists in the University of Western Australia have used microbes *Acetobacter* to work on fermented wine to produce the fabric that has a definite odor (Figure 23.8). A water-resistant textile Naoron, derived from wood pulp and recycled polyester, is an alternative for conventional leather. This textile is animal friendly and also requires none of the chemicals used to create faux leather (Figure 23.9). A company called Qmilch has developed fabric from the proteins of soured secondary milk that is not suitable for human consumption. This fabric requires no harsh chemicals and requires less water than normal milk-based fabrics (Figure 23.10). Novel materials are being used to produce innovative fabrics for apparel development (Steph 2012).

FIGURE 23.10 Garment from spoilt milk fabric.

23.5.3 Sustainable Technologies

The increasing global consumption of major resources like water and energy, the rising costs, and the impact of processes on the world climate have brought about stringent regulations and limits and environmental obligations for water and air. This has made the textile industry take up measures to save heat energy and freshwater consumption. New breakthrough techniques in processing have been undertaken to bring about reduction in the use of chemicals, alternatives to harsh chemicals, use of novel environment-friendly gentle processes, and use of machinery to take up processing in different media other than water. This requires immense changes in the mind-set of the processors who have been using extensive quantity and quality of chemicals for decades. Apart from this the challenge will be to convert the current textile facilities into a new technology that uses lesser resources.

23.5.3.1 Use of Enzymes in Textile Processing

Cotton fabrics require pretreatment that involves singeing, desizing, scouring, bleaching, peroxide removal biopolishing, mercerizing, and optical brightening. The objectives of pretreatment are to make the fabric suitable for further processing by removing natural impurities and by making it hydrophilic. Enzymes have been used in ancient times in a small way, but with mass production and chemical processing, not much attention was paid for the incorporation of enzymes into processing. Amylase for biodesizing, pectinases for bioscouring, xylanase and glucose oxidases for bleaching, catalases for peroxide killing, and cellulases for biopolishing are some of the enzymatic processes used in the pretreatment area. In finishing, laccases are used for color removal in jeans production and cellulases are also used for stone-washed effects. There are many advantages while using enzymes, which include savings in time, energy, water, and weight and a smooth surface profile of the material. There is also a reduction in cost and use of lesser auxiliaries resulting in reduced effluent load and in turn reduction of effluent treatment costs (Auterinen 2006; Senthilkannan and Nithyanandhan 2008a,b).

The sticky insect secretions from the silk fibers are removed by enzymes by a process called *degumming with enzymes*. Protease enzymes are used for all protein-based fibers. Biostoning is used for stone-washed denim, while biopolishing is used to defibrillate cotton to give a smooth finish. Shrink proofing of wool is also effectively done with enzymes. Research has been carried out to combine different enzymatic processes in the same bath as enzymes are target specific and will get activated only when the right environment occurs. When two enzymes work under the same conditions, they can be combined together in the same bath in order to save more water, time, energy, and costs, for example, combined bleach cleanup and dyeing, combined bioscouring and biopolishing, and combined desizing and bioscouring. This requires a lot of research and careful handling of the different enzymes. Research in the field of biotechnology is giving rise to different varieties of enzymes that would help to streamline the processing methods of the textile industry (Blackburn 2009; Periyasamy 2010; Menezes and Choudhari 2011).

23.5.3.2 Natural Dyeing

Natural dyeing of different textiles and leathers has been used for specialty products in the decentralized sector owing to specific advantages and limitations. Due to increased environmental awareness, this art is gaining ground since it is nontoxic and eco-friendly and is safe to the environment. This method of dyeing from natural sources was confined to small-scale dyers, craftsmen, and a few small-scale exporters dealing with eco-friendliness as their motto. A small number of companies around the world are producing natural dyes for the textile industry. To mention a few, della Robbia, Milan, produces the extract of natural dyes under the Eco-Tex certifying system. Allegro Natural Dyes, United States, supplies natural dyes under the EcoColors label. They have also developed a mordant using a nontoxic aluminum and a biodegradable auxiliary substance. Similarly, Livos Pflanzenchemie Forschungs und Entwicklungs GmbH, Germany; Bleu de Pastel, France; and Rubia Pigmenta Naturalia, Netherlands, are involved in the production of natural dyes. India is one of the leading producers of natural dyes.

Natural dyes are going to be the next-generation dyeing method for the sustainability and safety of the ecosystems. They are chosen not only for their dyeing properties but for a combination of properties like UV protection and antibacterial and deodorizing properties. It is also used in textile printing on a small level. Consumers have become aware of the need for sun protection as increased exposure to UV radiation can lead to acceleration of skin aging and skin damage. Many researchers have found these evidences of getting a double-edged benefit of ultraviolet protection and coloration, for example, vegetable dyes extracted from Flos Sophorae for silk fabrics, weld dyes for flax and hemp fabrics, and eucalyptus leaf extract on silk fabrics. The deodorizing properties of fabrics dyed with various natural colors have shown to be around 50%–99%. From all the silk and wool fabrics that were natural dyed with *Cassia tora* L., coffee sludge (*Coffea arabica* L.), and pomegranate, the pomegranate results showed the highest deodorizing property of 99%. The same effects were seen when cotton, silk, and wool were naturally dyed with Amur cork tree, *Dryopteris crassirhizoma*, *Chrysanthemum boreale*, and *Artemisia* extracts. Wool showed the highest deodorizing property followed by silk and cotton. Natural dye powders have also been used along with thickeners to print on cotton, silk, wool, and flax. Natural dyes from alkanet and rhubarb have been printed by using the pigment printing technique with Meypro gum as thickener. Hand block prints and resist style of printing also use natural dyes. Chitosan from the shell of crabs also seems to improve the performance of natural dyes for printing.

Natural dyes produce soothing and soft shades that are not common in synthetic dyes. They have the potential to be used for certain specified areas to reduce pollution resulting from synthetic dyes. It can replace processes that are toxic and carcinogenic dyes and intermediaries that are polluting in nature (Rungruangkitkraj and Mongkholrattanasit 2012).

23.5.3.3 Clean Technologies

Any industrial or technical measures taken to reduce or eliminate the origin of the production of any pollution, waste, or nuisance and thereby save the natural resources, raw materials, and energy are termed as *clean* technologies. This objective can be implemented as the optimization of an existing process, process modification, or process change. There are basically four ways by which waste reduction can be accomplished—substitution and savings, integration of new technology or equipment for the reduction of pollution, internal recycling or recovery of materials or substances, and incorporation of a complete new system for radical innovations.

Some of the technologies used in textile production are as follows. The use of bifunctional reactive dyes for cotton has enabled high exhaustion, high reactivity with the fiber, and lower load of dye in the dye effluent. Two groups of bifunctional dyes, heterobifunctional and homobifunctional, are available in the market for dyeing cellulosic material. In continuous dyeing, there is liquor left in the pad trough as well as in the stock tank to be ready for running out of water before the dyeing is complete. Equipments have been developed to decrease the pad liquor volume to 15 or 10 L as the fabric comes to the end of the run.

Some approaches to minimize chemical usage are dyeing in low liquor ratio, right-first-time approach, process innovations in continuous dyeing, and use of low-salt reactive dyes. In reactive and vat dyeing systems, the material liquor ratio is usually 1:10. If this ratio is reduced to 1:5, then the pollution load will decrease by 40% and the auxiliaries used will also reduce. Optimization of process and systemization of production according to the recommendations of the manufacturer will help to do the right dyeing the first time. Spectrophotometer color matching would greatly help to decide on the correct process. Low-salt reactive dyes are available, which use 20 g salt per liter, while the conventional reactive dyes use 60–80 g/L. This helps to reduce the pollution of rivers and lakes by high salt concentration and also the aquatic life is also saved from extinction.

In pigment printing, emulsion thickener with kerosene or mineral turpentine was considered suitable for printing. These were reasons for fire hazards, health hazards, and air pollution. Synthetic thickeners based on polyacrylates have served as efficient alternatives to the highly polluting emulsion thickeners. In the area of wrinkle-free finishes, the cross-linking agents used were dimethylol dihydroxyethyleneurea (DMDHEU) of etherified DMDHEU. The release of formaldehyde is inevitable while finishing and subsequent storage. Carcinogenic effects and dermatitis can be avoided by the use of alternative cross-linking agents like polycarboxylic, maleic, and succinic acids. The hazards that have been caused to the ecology call for clean technologies that can be applied in many areas of textile production. They become eye openers for alternatives and process management as it has a great impact on the carbon footprint (UNCED 1992; Chavan 2011).

23.5.3.4 Recycling and Upcycling

The textile industry is an area that has abundant opportunities for recycling and upcycling. Recycling in textiles involves the reuse or reprocessing of clothing or waste cloth scraps from manufacture or any fibrous materials. Waste material comes during the production of yarn, fabrics, and products, for example, selvedge waste and waste after the cut plan in apparel industries. Recycled polyester is manufactured from used pet bottles (Layton 2009). Upcycling is to use the garment without diminishing its quality but may increase by the subsequent process. In the domestic setup, this sort of reuse occurs by the make-do and mend where a fabric that was intended for a particular use is redesigned to serve another function; otherwise, it would land up in a landfill. For example, old used jeans can be converted into durable bags or purses (Textile Environment Design).

Carbon footprint is a term that is commonly analyzed when sustainability of a product is taken into account. It computes the total amount of GHGs produced and is measured as the number of units of carbon dioxide emitted during two stages, namely, the primary and secondary stages. The primary footprints deal with the carbon dioxide emissions by direct usage from burning of fossil fuels like electricity usage and transportation; the secondary footprints include the indirect carbon dioxide emissions, which result from the manufacture, use, and cessation of products. The carbon dioxide emissions are harmful to the environment and living beings (Senthilkannan et al. 2011). The whole concept of sustainability is to plan the design, process, and manufacture of products that will emit negligible amounts of carbon dioxide from cradle to grave and that which can be biodegradable or serve as raw material for the next stage of product manufacture. Sustainable end products have their base in the selection of the fiber and the production techniques necessary to develop the fabric. If the production chain is sustainable right from the start to the end, the product is sure to pass the ecotest and claim to be sustainable as per the certification. All efforts are based on the quality of the raw materials, their origin, and the production technologies to make all products sustainable and hold the attention of the consumers.

23.6 FUTURE TRENDS

Sustainable textiles and fashion have become the buzz word and *green*, *social responsibility*, and *accountability* have been highlighted in the world of fashion production. This concept is further enhanced by eco-friendly and ethical collections from major retail chains, numerous popular designers, and mainstream fashion houses. Green textiles and fabrics were promoted early in the beginning of this century as people started realizing the drawbacks of the practices followed in apparel production. Today, all products under the eco-friendly label should be certified by governmental agencies or institutions like GOTS, OE, Bluesign, SA8000, and WRAP.

23.6.1 Trends in Sustainable Sourcing

23.6.1.1 Trends in Cotton and Natural Fibers

Many projects in India like Chetna and Aslli Sac are gaining ground as they devote to produce organic cotton and have scaled up the seed multiplication program for providing seeds for the cultivation of organic cotton. Cooperative movements have created a huge and long-lasting effect on farmers to bring about attitudinal changes in the minds of the farmers. The production of organic cotton has increased and there are many buyers who are interested in developing the ethical line of production. This has led to the vertical integration of production like growing of cotton, spinning, weaving, dyeing, finishing, and manufacturing all under one roof to increase the production efficiency and quality of fabrics. Retailers and buyers are getting directly involved in the production processes to help in sourcing and adding value to their products.

23.6.1.2 Trends in Silk and Luxury Fabrics

Ahimsa silk referred to as *peace silk* is produced by allowing the silkworms from the wild eri moth cocoons to emerge naturally, thus preventing the killing of 1500 silkworms per meter of silk manufactured. Innovative blends for luxury appeal is produced by blending silk with organic cotton, silk, and wool or with bio-based fibers like Modal and Tencel, creating new combinations in weight and textures. Nonwoven silks are being produced from hard silk waste, which is produced during the twisting and weaving stages of silk production. The reeling waste has been used regularly for spun silk yarn and noil yarn, but the hard waste has been utilized recently for nonwovens either by mechanical or adhesive bonding.

23.6.1.3 Trends in Innovative Fabrics, Dyeing, and Design

Bold and vibrant prints that unify natural and man-made world are being digitally printed on textiles by designers like Aimee Kent and The Social Studio (Figure 23.11). The designs are made to imbibe creativity and eco-friendliness in equal parts. Natural dyeing has become so popular and dyes are being extracted from almost all the available natural sources like food, trees, plants, minerals, animals, and other sources. The concept of warding off or treating diseases through clothing has become increasingly important with the production of herb-infused clothing. The wellness factor is brought into textiles known as *Ayurvastra* as a symbol of wellness, healing, fabric performance, and eco-friendliness.

23.6.1.4 Trends in Wool and Heavyweights

Under this category, dimensional weaves are created on woolen and heavyweight fabrics with a focus on 3D effects by means of reworking on the fabric surface and heavy embroidery. Checks or stripes are used as prints coupled with intricate techniques to enhance the qualities of the fabric. The latest trends are simple squares, checks, tartans, and tweeds for both men and women wear. Traditional "Panchachuli" designs of India that signify the five peaks of the Panchachuli mountain range are used as native designs for shawls, covers, and textiles and accessories (Panchachuliusa 2006).

FIGURE 23.11 Aimee Kent prints.

23.6.1.5 Trends in Components and Accessories

Great importance is given to accessories that portray traditional techniques and artisan craftsmanship. These accessories fetch high returns and are used for high-end products right from leather to buttons. Discarded waste is transformed to creative well-made accessories finding solutions to many problems (Ethical Fashion Forum 2013).

23.6.2 Trends in Sustainable Retailing

The concept of sustainability is changing the way retailing is done. This aspect has been the focal point on all areas of business that includes manufacturing, packaging, distribution, and marketing. Though environmental issues and planet stewardship were of great concern, more importance is being given to profits that are the lifeline of the business. The current trend is the opportunities seen by retailers to increase cash flow and profits by means of sustainable retailing. Today, sustainable retailing includes supplier management, brand management, and customer experience. Many researches reveal that sustainability and social responsibility are monitored throughout the supply chain. The main areas to be monitored for green retailing are retail supply chain, customer facing components, and facilities and infrastructure. Giving thought to innovative sustainable processes and solutions will enable the retailers to achieve a competitive edge, influence top and bottom lines in management, and give ideal customer experience to render satisfaction, acquisition, and retention (Pears 2006).

Retailers are looking at measures to incorporate sustainability by energy-saving and cost-saving measures. According to a recent supply chain management report, many retailers are involved in adopting energy-saving protocols by redesigning their supply chain network to improve the efficiency within the supply chain, implementing energy-efficient strategies like alternative fuels or sources of energy, or incorporating at least some criteria of sustainability in the supply chain management processes. These measures will serve to reduce energy costs and operating costs.

Good sustainable practices in green retailing will serve as a basis for boosting the brand value and improve customer loyalty apart from giving benefits from reducing overall costs. Energy usage reduction and its impact in lowering emissions would serve as a brand ambassador for the retail organization when the advantages of these two actions are clearly conveyed to the customer.

These actions would add the tag of *ethical organization*, which is an important brand differentiation in the global market. The customer loyalty to these organizations has been rated as very high when compared to the previous experience. Retail sustainability agendas are adopting energy management measures to work on the infrastructure and facilities to decrease the carbon footprint. A variety of alternative energy options like solar, wind, and fuel cell technology are being used to run all equipment in facilities and stores. The reducing aspect is extended to other areas like transport, labor, and fees for common facilities. Proper attention and planning of operations and technology improvements would help in reducing operational costs (Aberdeen Group 2012).

While dealing with sustainability in operations, many factors have to be taken into consideration. These include operational footprint, management of energy and waste, and integration of local and global economy. The carbon footprint for operations is estimated through a process of measurement, assessment, implementation of strategy, and reporting. This helps the retailer to realize the need for making changes in operation for continuous reductions in carbon footprint. Energy-saving techniques like installation of building automation systems that monitor energy and control temperature settings and alarms and help to identify energy-saving areas and replace mechanical systems with more efficient ones can be employed for managing energy efficiency. Corrective measures in refrigeration and lighting could save a lot of energy. Natural lighting is usually incorporated in areas to reduce the need for artificial light at daytime.

The essential features of waste management in retailing are material reduction, reuse, recycling, and disposal. The first thing is to minimize the volume of disposable material entering into the retail facility. Using reusable shopping bags and incentives for reuse is one way to avoid purchase of new shopping bags. Recycling is carried out for those materials that cannot be reused or eliminated. Retail organizations who are proactively involved in recycling efforts are saving in revenue as well as disposal costs. Retailers place value and work in unison with their stakeholders. The main stakeholder is the consumer, and the communication regarding sustainability will be through educational signage, onsite recycling stations, and buyback programs. Retailers often recognize terms like *green*, *sustainable*, *nontoxic*, and *recyclable*, which play an important role in influencing their buying decisions. From the community point of view, retailers constantly work on programs for the community to increase their self-esteem. Currently, such programs are directed toward sustainability objectives, and the term *corporate social responsibility* proves that the retailer is taking responsibility of the retailing services that are given to the consumers or community. The collaboration with community should be extended to nonprofit agencies, academic institutions, and governmental agencies, thereby portraying the importance of sustainability in retailing operations and supply chains. Retailing companies have linkages with nonprofit and governmental agencies like the Fair Labor Association and the Global Social Compliance Program (Siegel 2012).

23.6.3 Fashion Shows with a New Facelift

Fashion shows are earning a new thought and theme by turning its attention to the trendiest concepts of sustainable fashion. Fashion is no longer unethical and rigorous, neither on the environment nor on the organizing and participating team. Every part of the fashion show is dealt with a touch of sustainability, and these points are highlighted to the media as well as the world outside by various promotional strategies. Two cases of sustainability as concepts for fashion shows have been discussed below.

The annual vintage and fashion event *Undress Runways* is based on sustainable fashion with three categories: swimwear and lingerie, day wear and evening wear, and accessories. This show is focused on natural fibers, ethical production, food-dyed or natural-dyed garments, no-waste collections, and unique designs made from offcuts. The designers selected for the show are selected in terms of the extent of the sustainability aspect in the areas of sourcing, design, and manufacturing processes, which involves ethically made clothing, sustainable fibers, and minimal waste design. The whole show has been formulated on the basis of sustainability. Conversion of a car park into a dramatic stage, recycle inspired chairs for audience, organic cocktails, sustainable design exhibitions and field trip to a sustainable village are some of the highlights in implementing sustainability in fashion shows. Undress Runways, Australia, is dedicated to create an atmosphere for change and challenges the consumers to assess the sustainability aspects of their wardrobe. Knowledge is essential on *where the product comes from*, which indicates the origin of the product; *what it is made of*, which will include the contents of the product and the production methods; and *where it will end* in terms of the impact of the product on environment. These data may be asked to be presented in the tags accompanying the garments in retail stores (Siegel 2012).

The Ethical Fashion Show conducted in July 2014 in Berlin is based on the concept of ecology and ethics in design for innovative street and casual wear. A knowledge lounge for green fashion, panel discussions, and activities related to ethical fashion and the genuine B2B fair for professional fashion trade serve as the main platforms for the eco-fair fashion during the Berlin fashion week. In order to be part of the Ethical Fashion Show, Berlin, a company should comply on four criteria—sustainability, ecological aspects, social aspects, and transparency aspects. The standards include that 70% of the products exhibited by a company should fulfill the sustainability criteria specified by the organizers. In addition, an exhibitor should cover at least one ecological aspect and one social aspect, which will be verified by the transparency criteria. The ecological aspects express that fashion that is designed and manufactured to reduce environmental impacts from raw material and trim sourcing, production and consumer use, and disposal or reuse will be examined. With regard to social aspects, the company should design and manufacture fashion complying with the human rights and support for sustainable development. Transparency indicates clear communication of the sustainable strategy in order to control and monitor the sustainable development of its supply chains. The source also reveals that around 80 companies have registered to exhibit their products at this prestigious fashion show in Berlin (Hamar 2013).

23.6.4 IMPLEMENTATION OF SUSTAINABLE FASHION

Though many doctrines of sustainable fashion have been discussed, it is very difficult to implement this concept as there should be a basic change in all levels of the supply chain of fashion products. Education and experience would help the designer to work toward sustainability in accordance with the prevailing environment. However, certain barriers affect the implementation of sustainable design solutions. These include cost, education, and inexperience in sustainable design, material selection, and the consumer. Usually, sustainable designs are expensive as they are based on specific norms. Manufacturers and consumers want immediate savings instead of results in the long run as they do not want to spend time and money on research and development. It has also been found that many professionals are not exposed to sustainability principles during their basic education and in their professional work. Professionals have to be involved in many projects based on sustainability to acquire experience and continued professional development.

The most important step in sustainable fashion is material selection. While selecting an environment-friendly material, the designers are faced with limitation in the choice of raw material. The designers also find it hard to decipher the authenticity and reliability of information given by product suppliers and manufacturers due to the nontransparent nature of raw material suppliers and lack of certification of raw materials with regard to sustainability. In view of reducing cost, local vendors are approached and it may be difficult for them to maintain the required standards. Another most important barrier for the implementation of the sustainable feature is the attitude of the consumer that requires a tremendous change and cooperation for the successful execution of sustainable practices (Hankinson and Breytenbach 2012).

The barriers for implementation of sustainable fashion can be addressed by many means. Designers need to update their knowledge and take the help of a facilitator, consultant, or local resource centers for designing and developing sustainable designs. Continued professional development can be achieved by organizing/attending courses, workshops, and conferences in the context of sustainability. Government policies and regulations based on sustainability should be strictly enforced and the standards are to be raised to meet the global demands. Research and development of new products and sustainable processes are to be encouraged for reviewing and analyzing the impact of these products and processes on the environment and living beings. Use of rating tools would help to evaluate and improve the efforts invested in sustainable practices, and educating the consumer about the benefits of sustainability would help in bringing about a change in attitude.

The Ethical Fashion Programme by the International Trade Center, Switzerland; the Sustainable Clothing Action Plan by the Department for Environment, Food and Rural Affairs, United Kingdom; the Judicial Implementation of the Principles of Ecologically Sustainable Development in Australia and Asia, Sydney; and the Sustainable Development Strategy 2013–2016 by Employment and Social Development Canada are some of the action plans and organizations that are responsible for the execution of sustainable practices (ESDC 2006; Cipriani et al. 2011; DEFRA 2011). Most of these organizations work in a phased manner by setting targets and timelines for the enforcement of sustainability with regular evaluation and improvements. These programs start with baseline mapping of the impacts and interventions of sustainability across the clothing life cycle to find out the key impacts and the current improvement interventions. This is followed by action plans and processes for public understanding of sustainable clothing, reduction of negative environmental and social impacts, and maximizing reuse, recycling, and zero waste (Preston 2006).

All these efforts have led to a consciousness among manufacturers to undertake eco-friendly practices and consumers to look for certified organic or sustainable products. Organic fibers and materials are used as base materials to produce fabrics, and attention is focused in reducing the carbon footprint in all stages across the product life cycle. Awareness, media, education, and networking are aimed at transforming consumer trends and behavior, and market drivers have been created for sustainable clothing. Improving traceability along the supply chain and different tools for evaluation would help in better implementation and improvement of sustainable fashion.

Thus, sustainable design in fashion is a green concept and is at the forefront for the past decade and moving toward sustainability. During the past few years, the fashion industry has been faced with severe criticism about its environmental impact and carbon footprint. The reaction has been at the brand level where many organizations who are the driving forces of the leading brands have started establishing their own commitments and strategies at the retail level, and these efforts have moved on to an industry-wide scale. Many organizations like Sustainable Apparel Coalition and NRDC's Clean by Design Campaign are extending their services for sustainable production. Many industries have started taking part in the supply chain of the products and are addressing the environmental concerns right from the grass root level of the value chain. Sustainable design in fashion was largely focused on the quality of the raw material, but recently leading brands have developed indices to help the designers and product development teams to work in the right direction of choosing materials based on their environmental impacts along the entire value chain. The integration of technology, research and development, and product innovation will lead these efforts to create a sustainable environment for the health and safety of all living beings in this universe.

REFERENCES

Aberdeen Group. October 2012. Sustainable retailing today: 2012 green retailing report. http://www.retailsolutionsonline.com/solution/marketing-ops. (Accessed on March 7, 2014.)

Admin in Environment. August 27, 2012. The industrial revolution and its impact on our environment. http://eco-issues.com/TheIndustrialRevolutionandItsImpactonOurEnvironment.html. (Accessed on March 2, 2014.)

Allwood J.M., Lausen S.E., Malvedo de Rodriguez C., and Bocken N.M.P. 2006. Well dressed? The present and future sustainability of clothing and textiles in the United Kingdom. http://www.ifm.eng.cam.ac.uk/resources/sustainability/well-dressed/. (Accessed on March 4, 2014.)

Auterinen A.L. 2006. White biotechnology and modern technology processing. http://www.textileworld.com/Issues/2006/May-June/Dyeing_Printing_and_Finishing/White_Biotechnology_And_Modern_Textile_Processing. (Accessed on March 2, 2014.)

Bakari Mohamed E.K. 2014. Sustainability's inner conflicts: From ecologism to ecological modernization. *Journal of Sustainable Development Studies* 6 (1), 1–28. http://infinitypress.info/index.php/jsds/article/view/618/291. (Accessed on April 18, 2014.)

Ben-Eli M.U. 2006. Sustainability: The five core principles. http://www.sustainabilitylabs.org/page/sustainability-five-core-principles. (Accessed on March 3, 2014.)

Blackburn R.S. October 2009. Sustainable textiles: Life cycle and environmental impact. https://www.elsevier.com/books/sustainable-textiles/blackburn/978-1-84569-453-1. (Accessed on March 2, 2014.)

Borromeo L. September 23, 2013. Was sustainability on trend at London fashion week? http://www.theguardian.com/sustainable-business/sustainable-fashion-blog/sustainability-on-trend-london-fashion-week. (Accessed on January 4, 2014.)

Breyer M. September 11, 2012. 25 shocking fashion industry statistics. http://www.treehugger.com/sustainable-fashion/25-shocking-fashion-industry-statistics.html. (Accessed on January 4, 2014.)

Chavan R.B. July 11, 2011. Eco-friendly technologies for textile production. http://www.slideshare.net/nega2002/cleaner-production-technologies-for-textiles-iitd-dec10-12-2008-8563257. (Accessed on March 7, 2014.)

Cipriani S., Brown J., and Mukai C. July 18, 2011. The ethical fashion programme not charity, just work. http://www.intracen.org/workarea/downloadasset.aspx?id=51218. (Accessed on April 18, 2014.)

Crul M.R.M. and Diehl J.C. October 2005. Design for sustainability—A practical approach for developing economies. http://www.d4sde.org/manual/d4stotalmanual.pdf. (Accessed on March 7, 2014.)

Department for Environment, Food and Rural Affairs (DEFRA). 2011. Sustainable clothing action plan. http://archive.defra.gov.uk/environment/business/products/roadmaps/documents/summary-projects-sustain-clothing-ap.pdf. (Accessed on April 18, 2014.)

Dietz D. September 2, 2013. 8 easy steps to make fashion smarter and more sustainable. http://mic.com/articles/61857/8-easy-steps-to-make-fashion-smarter-and-more-sustainable. (Accessed on January 4, 2014.)

ESDC. 2006. Planning for a sustainable future. http://www12.hrsdc.gc.ca/servlet/sgpp-pmps-pub?lang=eng.

Ethical Fashion Forum. 2013. SOURCE Expo 2013: Trends in sustainable sourcing. http://source.ethicalfashionforum.com/article/source-expo-2013-trends-in-sustainable-sour. (Accessed on March 7, 2014.)

Ethical Fashion Forum (EFF). 2014. SOURCE: The global platform for sustainable fashion. http://www.ethicalfashionforum.com/About-the-SOURCE-platform. (Accessed on March 7, 2014.)

Friedman V. February 5, 2010. Sustainable fashion: What does green mean? http://www.ft.com/cms/s/2/2b27447e-11e4-11df-b6e3-00144feab49a.htm l#axzz3DXmAn YnU. (Accessed on March 2, 2014.)

Hamar E. October 1, 2013. What to expect at Undress Brisbane 2013. http://www.undressrunways.com/news/2014/8/18/what-to-expect-at-undress-brisbane-2013. (Accessed on March 5, 2014.)

Hankinson M. and Breytenbach A. 2012. Barriers that impact on the implementation of sustainable design. http://cumulushelsinki2012.org/cumulushelsinki2012.org/wp-content/uploads/2012/05/Barriers-that-impact-on-the-implementation-of-sustainable-design.pdf. (Accessed on April 18, 2014.)

Hawken P., Lovins A.B., and Lovins L.H. 2008. Natural capitalism creating the next industrial revolution. http://www.natcap.org/sitepages/pid20.php. (Accessed on April 17, 2014.)

Horn D. and Davis L. July 22, 2014. Sustainable design. http://www.gsa.gov/portal/content/104462. (Accessed on March 2, 2014.)

Hoskins T. October 29, 2013. Is fur trade sustainable? http://www.theguardian.com/sustainable-business/sustainable-fashion-blog/is-fur-trade-sustainable. (Accessed on 4 January 2014.)

Hough J. April 24, 2012. Campaign aims to highlight issues facing fashion industry through events. http://www.irishexaminer.com/ireland/campaign-aims-to-highlight-issues-facing-fashion-industry-through-events-191544.html. (Accessed on March 4, 2014.)

Image Credit. What are some of the ecofriendly fabrics? http://www.innovateus.net/earth-matters/what-are-some-ecofriendly-fabrics. (Accessed on March 7, 2014.)

Joy A., Sherry J.R., Venkatesh J.A., Wang J., and Chan R. 2012. Fast fashion, sustainability, and the ethical appeal of luxury brands, FT. doi:10.2752/175174112X13340749707123. http://www3.nd.edu/~jsherry/pdf/2012/FastFashionSustainability.pdf. (Accessed on March 2, 2014.)

Levi Strauss and Company. 2010a. Fashion futures 2025 (PPT). https://www.google.co.in/?gfe_rd=cr&ei=EZcZVPi0KKbV8gfqg4GYBw&gws_rd=ssl#q=Fashion+Futures+2025+(PPT). (Accessed on November 14, 2013.)

Levi Strauss and Company. 2010b. Fashion futures 2025. http://www.forumforthefuture.org/project/fashion-futures-2025/overview. (Accessed on November 14, 2013.)

Menezes E. and Choudhari M. December 14, 2011. Pre-treatment of textiles prior to dyeing. http://www.intechopen.com/books/textile-dyeing/pre-treatment-of-textiles-prior-to-dyeing. (Accessed on March 7, 2014.)

Netter S. November 14, 2013. Changing consumers' lifestyles towards more sustainable choices and behavior—Swedish trends. http://www.cbs.dk/node/259469. (Accessed on March 2, 2014.)

Nice Fashion. April 26, 2012. The nice consumer: Framework for achieving sustainable fashion consumption through collaboration. http://www.bsr.org/reports/nice-consumer-framework.pdf. (Accessed on March 3, 2014.)

Norwegian Fashion Institute. 2014. Climate challenges facing the clothing sector. http://www.google.co.in/webhp?sourceid=chrome-instant&ion=1&espv=2&ie=UTF-8#q=Climate%20challenges%20facing%20the%20clothing%20sector. (Accessed on March 4, 2014.)

Obregon C. 2012. Sustainable fashion: From trend to paradigm. Dissertation, Aalto University School of Arts, Design & Architecture, Helsinki, Finland. http://www.duoc.cl/cumulus2012/press/pdf-1/SUSTAINABILITY/ACADEMIC-PANEL/Carolina-Obregon.pdf. (Accessed on March 2, 2014.)

Organic Exchange (OE). Eco fibers. http://textileexchange.org/sites/default/files/eco_fibre.pdf. (Accessed on March 7, 2014.)

Panchachuliusa. 2006. In the beginning is the idea. http://www.panchachuliusa.com/index.html.

Parvathi C., Maruthavanan T., and Prakash T. 2009. Environmental impacts of textile industries. http://www.indiantextilejournal.com/articles/fadetails.asp?id=2420. (Accessed on January 4, 2014.)

Pears K.E. May 2006. Fashion re-consumption; developing a sustainable fashion consumption practice influenced by sustainability and consumption theory. School of Architecture and Design, RMIT University, Melbourne, Australia. http://researchbank.rmit.edu.au/eserv/rmit:6309/Pears.pdf. (Accessed on March 4, 2014.)

Periyasamy A.P. December 11, 2010. Eco-friendly in textile wet processing. http://www.slideshare.net/abiramtex/eco-friendly-in-textile-processing. (Accessed on March 7, 2014.)

Pinterest. Fish scales. http://www.pinterest.com/pin/114138171778234419/. (Accessed on April 16, 2014.)

Politowicz K. 2010. (D)urability day. http://www.tedresearch.net/research/detail/durability-day/. (Accessed on March 7, 2014.)

Pratt A., Borrione P., Lavanga M., and D'Ovidio M. 2012. International change and technological evolution in the fashion industry. http://www.andycpratt.info/andy_c_pratt/Research_Writing__Downloads_files/International%20Change%20and%20technological%20evolution%20in%20the%20Fashion%20Industry.pdf. (Accessed on March 4, 2014.)

Preston B.J. July 21, 2006. Judicial implementation of the principles of ecologically sustainable development in Australia and Asia. http://www.lec.justice.nsw.gov.au/agdbasev7wr/_assets/lec/m4203011721754/preston_judicial%20implementation%20of%20the%20principles%20of%20eologically%20sustainable%20development.pdf. (Accessed on April 18, 2014.)

Rodie J.B. September/October 2011. Eco-friendly raw material and fiber production are the first links in a sustainable textile manufacturing chain. http://www.textileworld.com/Issues/2011/September-October/Features/Fiber_First. (Accessed on March 7, 2014.)

Rungruangkitkrai N. and Mongkholrattanasit R. 2012. Ecofriendly of textiles dyeing and printing with natural dyes. In: *Abstracts of the RMUTP International Conference on Textiles & Fashion*, Bangkok, Thailand, July 3–4, 2012. http://textileconference.rmutp.ac.th/wp-content/uploads/2012/10/010-Eco-Friendly-of-Textiles-Dyeing-and-Print-with-Natural-Dyes.pdf. (Accessed on March 7, 2014.)

Senthilkannan M.S., Li Y., Hu J.Y., and Mok P.Y. 2011. Carbon footprint of shopping (grocery) bags in China, Hong Kong and India. *Atmospheric Environment* 45, 469–457. doi:10.1016/j.atmosenv.2010.09.054. http://www.sciencedirect.com/science/article/pii/S135223101000840X. (Accessed on March 18, 2014.)

Senthilkannan M.S. and Nithyanandhan R. 2008a. Modern textile preparatory processes using enzymes. http://www.fibre2fashion.com/industry-article/textile-industry-articles/modern-textile-preparatory-processes-using-enzymes/modern-textile-preparatory-processes-using-enzymes1.asp. (Accessed on March 17, 2014.)

Senthilkannan M.S. and Nithyanandhan R. February 2008b. Enzymatic application for bleach cleanup. http://www.indiantextilejournal.com/articles/FAdetails.asp?id=853. (Accessed on March 17, 2014.)

Siegel A. 2012. Sustainability in retail operations. http://www.retailsolutionsonline.com/doc/the-green-retailing-special-report-0001. (Accessed on March 7, 2014.)

Smith T. August 14, 2012. An introduction to sustainable fashion. http://www.rtcc.org/2012/05/11/an-introduction-to-sustainable-fashion/. (Accessed on March 2, 2014.)

Smithsonian Institution National Museum of National History. 2013. Biomimicry shark denticles. http://ocean.si.edu/ocean-photos/biomimicry-shark-denticles. (Accessed on March 7, 2014.)

Steph. November 11, 2012. Eco fabric: 14 strange and amazing textile innovations. http://webecoist.momtastic.com/2012/11/12/eco-fabric-14-strange-and-amazing-textile-innovations/. (Accessed on December 21, 2013.)

Textile Environment Design. 2014. Recycling and upcycling. http://www.tedresearch.net/media/files/Recycling.Upcycling.pdf. (Accessed on March 8, 2014.)

The Sustainable Angle. 2014. Fabric categories—Low impact leather. http://www.futurefabricsvirtualexpo.com/fabric-categories/. (Accessed on March 7, 2014.)

UNCED. 1992. Development of clean technology concept. http://www.wiley-vch.de/books/sample/3527320075_c01.pdf. (Accessed on March 8, 2014.)

Undress Runways Inc. 2014. Sustainability in fashion. http://undressrunways.com/tag/sustainable-fashion-2/. (Accessed on March 2, 2014.)

Layton, J. April 22, 2009. What are eco-plastics http://science.howstuffworks.com/environmental/green-tech/sustainable/eco-plastic1.htm. (Accessed on January 12, 2015.)

Xaxx J. 2014. Top ten ethical issues in a fashion business. http://smallbusiness.chron.com/top-ten-ethical-issues-fashion-business-21866.html. (Accessed on March 4, 2014.)

Index

A

Absorption method, 363
Acacia catechu, 338
Adidas DryDye technology, 487
Adjacent fabrics, quality control standards, 367
Ageing consumer and clothing, size and shape of, 245
    capturing, 251
    co-design process, 255
    global clothing chain, challenge, 252
Age-neutral marketing, 247
Aid by Trade Foundation, 427–428
Air emissions, 313–314
Air permeability, 361
Alkali-treated jute fiber structure, 99
Alkylphenol ethoxylates (APEOs), 325, 371
Alkylphenols (AP), 325
American Apparel and Footwear Association (AAFA), 421
The American Association of Textile Chemists and Colorists (AATCC), 354
American Society for Testing Materials (ASTM), 354
Ammonium polyphosphate (APP), 54
Amylases
    advantages, 84
    Bacillus amyloliquefaciens, 87–89
    bleaching cotton fabrics, 89
    combined desizing and bleaching, 90
    desizing, 87–89
    end products, 85–86
    enzymes, 84
    thermal and mechanical deactivation, 86–87
Analytic hierarchy process (AHP), 408
Analytic network process (ANP), 408
Antibacterial property, 338–339
Antifungal property, 339–340
Antimicrobial property, 338–340
Antimony and halogen-based chemicals, 55–56
Antioxidant property, 342
Apparel development methodologies, 99
Apparel industry
    apparel supply chains, 401–402
    Australia, 411–413
    global sourcing, 400–401
    incidents in apparel manufacturing, 402
    Smart Retailing Group, 411–413
    supplier selection
        decision making, 408
        economic criteria, 405–406
        environmental criteria, 406–407
        methods, 408
        social criteria, 406–407
    unethical practices, 402–403
Apparel supply chains, 399–400
    original brand manufacturing, 401–402
    original equipment manufacturing, 401–402
    supplier assessment (*see* Supplier assessment)
    types, 402
Aqueous agents, color fastness, 366
Aqueous extract, pH, 369
Aqueous liquid repellency, 362
Arnebia nobilis, 338
Artocarpus heterophyllus, 338
Arylamines, 321–323
Assisted shopping, 239
Association of Suppliers to the British Clothing Industry (ASBCI), 241–242
Atmospheric contaminants, color fastness, 367
Atmospheric pressure plasma (APP), 5–7
Atmospheric pressure plasma jet (AAPJ), 6–7
Atomic force microscope (AFM), 16
Audit frauds, 45–46
Auxochrome, 335
Azo dyes, 321–323

B

Baby boomers, 242
Bacillus amyloliquefaciens, 87–89
Bacillus subtilis, 338
Banana pseudostem sap (BPS), 52–53
    chemical analysis, 68, 70
    mechanism, 71
    thermal analysis, 66–67
Banned substances
    alkylphenol ethoxylates, 325
    alkylphenols, 325
    arylamines, 321–323
    azo dyes, 321–323
    chlorinated organic carriers, 326
    chlorinated phenol, 324
    dimethyl fumarate, 326
    flame retardants, 326
    formaldehyde, 323–324
    heavy metals, 325
    organotin compounds, 325
    perfluorooctane sulfonate, 326
    pesticides, 326–327
    phthalates, 324
    pH values, 326–327
Berberis vulgaris, 338
Best Ager market research, 247
Best available techniques (BATs), 496–497
Better Cotton Fast Track Program (BCTFP), 495
Better Cotton Initiative (BCI), 428, 495
Biodegradation, 489–490
Biodiversity protection, 303
Biological materials, 262
Biological oxygen demand (BOD), 173
Biomimetics, 299
Biomimicry, 478, 488–489

Biotechnology application
    cotton materials, 77–78
    desizing, scouring, and bleaching, 78
    enzymes, 81–84
    multienzyme scouring process
        cellulase and protease, 90
        characterization, 90
        enzyme treatment, 91–92
        multienzyme treatment, 92
        protease and pectinase concentrations, 90–91
    one-step desizing and bleaching, 87–90
    preparatory process, 78
    single-stage preparatory process (see Peroxide based single-stage process)
Bleaching agents, color fastness, 367
Bleaching cotton fabrics, 89
Blue Angel certification, 381
Bluesign system substances list (BSSL), 424
Bluesign® Technologies, 344, 424, 481
Boomerang, 457, 464–465
Borax and boric acid mixture, 54
BPS, see Banana pseudostem sap (BPS)
Business model, sustainable fashion, 453–456
Butane tetracarboxylic acid (BTCA), 52
Buyer-driven supply chains, 402

C

Carbon footprint, 420, 522; see also Greenhouse gas (GHG)
    definition, 145–146
    effects, 146–147
    energy intensity, 146
    global warming potential, 145
    natural fibers
        advantages, 150
        CO_2 emissions, 149
        cotton fiber products, 152–155
        energy consumption, 151
        energy sources, 150–151
        jute fiber and products, 156–157
        linen fiber products, 157–158
        value-added chain, 150
        wool fiber and products, 152–156
    strategies, 161–163
    synthetic fiber
        polypropylene (PP) bags, 158–160
        regenerated fibers, 160–161
    textile processes, 147–149
    types of, 141–142
Carbon neutral, 420
Carbon offsets, 386
Carcinogenic pesticides, 315
Care labeling, 368
Case-based reasoning (CBR), 408
CE, see Circular economy (CE)
Cellulosic textile, 11–12
Chemical modification, 99
Chemical oxygen demand (COD), 173
Chemical substitution, 320
Chlorinated organic carriers (COCs), 326
Chlorinated phenol, 324
Chlorine-based single-stage processes, see Peroxide based single-stage process

Chromophore, 335
Circular economy (CE), 262–263
Clean technologies, 521–522
Clevercare symbol, 483–484
Clothing
    design requirements for consumer, 258–260
    disposal, 277–278
    quality control standards, 372–373
    waste, 276–277
Clothing and fashion industry
    economic crisis, 208–209
    free-export zones, 209
    liberalization policies, 208
    textile and clothing industry, 207–208
Clothing field sustainability
    ageing consumer, 244
    ageing demographic, 242
    branding, 257
    changing retail practice, 235–236
    Chinese whispers, 234
    circular economy, 262–263
    co-design process, benefits of adopting, 261
    color, 255–256
    consumer decision making, 237–238
        confusing information, 240–242
        confusion with size and fit, 238–240
    fabrication, 256
    Finnish study, 244–247
    garment production, bringing to market, 264–266
    global structure, 233–235
    guidance to producers and consumers, 263–264
    low cost, 237
    new products to new consumer, 261–262
    older consumers, 242–244
    outdoor industry market research, 247
    perception, 236
    personal style, 255
    price, 257
    research findings, 257–260
    retail experience, 256–257
    size and shape, 255
    styling, 255
Clothing industry
    economic crisis, 208–209
    environmental communication, 377–378
    free-export zones, 209
    liberalization policies, 208
    Mauritius, 310
    social impacts (see Social impacts)
Clothing manufacturing and competitiveness
    architecture, 46–47
    environmental issues, 43–44
    global collaborative scheme, 46–47
    global value chains, 40–41
    implementation stage, 48
    labor conditions issue, 45–46
    operations, relational model, 47
    preferential trade agreements, 41–42
    relational mind-set, 47–48
    resource-based view, 40
    resource consumption, 39
    resource productivity issue, 44–45
    textile and clothing sectors, 39–40

theoretical frameworks, 47
theories and competitiveness, 42–43
transaction cost economics, 39–40
transparent supply chains, 46–47
Co-design process
  benefits of adopting, 261
  collaboration with users, 252
  functional clothing, 248–249
Cold/low-temperature/nonthermal plasma, 5
Collaboration group, 409
Collaborative fashion consumption, 456
Colorant characteristics, 368
Color fastness, quality control standards
  aqueous agents, 366
  atmospheric contaminants, 367
  bleaching agents, 367
  color and color differences measurement, 367
  colorant characteristics, 368
  dry-cleaning, 366
  heat treatments, 367
  light and weathering, 366
  principles, 365
  standard adjacent fabrics, 367
  vulcanization, 367
  washing and laundering, 366
Colors
  co-design process, 255–256
  and color differences, 367
  dye, 334–335
  dyeing mechanism, 336–338
  natural dye, 335–336
Community couture, 492–493
Comparative claims, 386
Constant rate of extension (CRE) tester, 356
Consumer-based quality attributes, 283
Consumer behavior; *see also* Clothing field sustainability; Fashion consumption
  clothing field
    ageing demographic, 242, 244
    confusing information, 240–242
    confusion with size and fit, 238–240
    decision making, 237–238
    design requirements, 258–260
    Finnish study, 244–247
    guidance, 263–264
    new products, 261–262
    older consumer, 242–244
  fashion field
    constant need, 284
    consumer values, 278–279
    current practices, 274–276
    environmental information, 279–281
    ethical issues, 281–282
    impact, 273–274
    impulse shopping, 276
    redirective fashion, 282–284
Consumption-related emotions, 272
Content Claim Standard (CCS), 392, 431
Control Union, 422
Corona discharge, 6
Corporate social responsibility (CSR), 399
  core elements, 213–214
  global scale, 213
  implementation, 215
  implications, 214–215
  requirements from stakeholders, 455
Cost-effective analysis (CEA), 218
Cotton fiber products, 152–155
Cotton Incorporated, 422
Cotton LEADS™, 422
Cotton Made in Africa (CmiA) Initiative, 427–428
Cotton Research and Promotion Act of 1966, 422
Cradle-to-cradle concept, 299
Cradle to Cradle Products Innovation Institute, 424–425
Crease recovery, 362
CSR, *see* Corporate social responsibility (CSR)
Cumulative energy demand (CED), 194
Curcuma domestica, 338

D

Data envelopment analysis (DEA), 408
Degradable claims, 386
Degumming with enzymes, 520
Deodorizing property, 340–341
Department for Environment, Food and Rural Affairs (DEFRA), 232
Desertification prevention, 303
Design for disassembly, 444
Design orientating scenario (DOS), 442
Desizing, scouring and bleaching (DSB)
  combination, 90
  desizing (*see* Amylases)
  process, 78
Dibutyltin (DBT), 325
Dichlorodifluoromethane (DCFM), 11
Dielectric barrier discharge (DBD), 6–7, 129
Dimethyl dihydroxy ethylene urea (DMDHEU), 54
Dimethyl fumarate (DMFU), 326
Do-it-yourself (DIY), 494
DOS, *see* Design orientating scenario (DOS)
Double transformation, 41–42
Dry-cleaning
  color fastness, 366
  quality control standards, 364–365
DSB, *see* Desizing, scouring and bleaching (DSB)
DyeCoo process, 487
Dyeing
  colors, 334–335
  mechanism, 336–337
  quality control standards, 370
  temperature, 336
Dyestuffs, 370

E

Eco-friendly fiber, 506, 517
Ecological footprint, 297
Eco-Management and Audit Scheme (EMAS), 380–381, 393–394
Economic supply chain networks, 402
Eco-promising, 381
  creating, 383–384
  current trends, 382
  guidelines, 383–384
  Nike's Considered Index, 382
  salient features, 382–383
  value-action gap, 383

Eco-testing
　ecolabeling, 343–344
　hazardous substances and mixtures, 315–316
　restricted substances list, 316–318
Electron spin resonance (ESR) analysis, 12
EMAS, *see* Eco-Management and Audit Scheme (EMAS)
Embodied energy, 302
End-of-life (EoL)
　business models, 457–458
　sustainable fashion consumption, 512
Energy dispersive x-ray (EDX) analysis, 14, 127
Energy efficiency, 482–484
ENERGY STAR, 431
Environmental claims, marketing, 384
　nontoxic claim, 387
　principles, 386
　refillable claims, 387
　renewable energy claims, 387
　source reduction claims, 387
　standard classification, 385–386
Environmental communication
　business dictionary, 376
　clothing industry, 377–378
　core messages, 376–377
　critical function, 376
　Danish textile industry, 376
　eco-promising, 381
　　creating, 383–384
　　guidelines, 383–384
　　Nike's Considered Index, 382
　　salient features, 382–383
　　value-action gap, 383
　textile industry, 377–378
　　FTC green guides, 379
　　product certification (*see* Product certification)
　　system certification (*see* System certification)
　　third-party certification, 379
Environmental design, *see* Sustainable design
Environmental impacts
　apparel and textile products, 167–168
　apparel manufacturing and distribution, 174–175
　consumer care
　　dry cleaning, 176
　　energy consumption and air pollution, 176
　　water, 175–176
　cotton fiber and yarn production
　　agrochemicals, 169–170
　　cotton yarn production, 171–172
　　water demands, 170–171
　end of life/disposal
　　environmental degradation, 176–177
　　solid waste, 177
　overview, 168–169
　polyester fiber and yarn production, 172
　textile production
　　polyvinyl/polyacrylic compounds, 172
　　quality of water, 173–174
　　water consumption, 173
Environmental issues
　information, consumption, 279–281
　sustainable development, 43–44
　sustainability, 295
Environmental labels and information schemes (ELIS), 384
Environmental management systems (EMS), 421

Environmental product declaration (EPD), 377–378
Environmental Protection Agency (EPA), 182, 430–431
Enzymes
　amylases, 84
　enzyme combinations, 83–84
　multienzyme scouring process, 90–92
　nitrogen-containing compounds, 83
　sustainable solutions, 81
　textile processing, 520–521
EoL, *see* End-of-life (EoL)
Escherichia coli, 338
Ethical consumption, 279
Ethical Fashion Forum (EFF), 503–504
Ethical Fashion Show, 526
Ethical hard-liners, 279
Ethylene glycol diacrylate (EGDA), 64
EU Ecolabel, 388–389
Eugenia caryophyllus, 338
European Committee for Standardization (CEN), 353–354
European Free Trade Association (EFTA), 353
European Outdoor Group (EOG), 422–423
European Waste Codes (EWC), 314

F

Fabrics
　dimensional changes, 364–365
　heavyweight, 523
　mechanical resistance, 358
　outdoor clothing design, 256
　structure and finishing, 361–364
　tests, 356–358
　trends, 523
Fairtrade Foundation, 428–429
Fairtrade International (FLO), 428–429
Fashion consumption
　Britain, 275
　consumerism, 283
　consumers constant need, 284
　consumer values, 278–279
　current practices, 274–276
　environmental and ethical issues, 281–282
　environmental impact, 273–274
　environmental information, 279–281
　impulse shopping, 276
　meaning, 272–273
　redirective fashion, 282–284
　slow/fast, 284–285
　vs. sustainability, 476–477
　transforming, 281
Fashion design for sustainability (FDS) model
　applying, 447–449
　design strategies, 448
　designer, 439
　development, 445
　improvements to garment, 449
　preparations for use, 445–447
　template, 446
Fashion design studio
　FDS model
　　applying, 447–449
　　development, 445
　　preparations for use, 445–447
　　template, 446

Index

global production, 439
integrating sustainable strategies, 449–450
life cycle of product, 443–444
scenario, 442–443
sustainable design strategy
    challenges, 449–450
    principles, 443–444
tools and checklists, 441–442
Fashion industry
  economic crisis, 208–209
  environmental issues, 508–509
  free-export zones, 209
  liberalization policies, 208
  metrics on impact, 510
  research and analysis phase, 440
  scenario, 507–508
  social issues, 509
  synthesis phase, 440
FDS model, *see* Fashion design for sustainability (FDS) model
Federal Trade Commission (FTC), 379, 386–387
Filippa K, 457, 461–462, 465
Flame-retardant finishing
  apparel and home textiles, 51–52
  banana pseudostem sap, 52–53
  curing operations, 52
  environment friendliness, 61
  flame-retardant formulations, 71–72
  limiting oxygen index, 52
  mechanism, 53
  nanotechnology, 61–63
  phosphorous, 52
  plant extract
    chemical analysis, 68–70
    global awareness, 65
    green-flame-retardant chemicals, 65–66
    mechanism, 71
    technologies, 65
    thermal analysis, 66–68
  plasma technology, 63–65
  technologies, 65
  thermoplastic synthetic fibers, 51
  traditional chemical finishing
    cotton, 54–56
    jute, 56–60
    synthetic fibers, 53–54
    thermoplastic fibers, 60
    wool, 60
Flo-Cert, 423, 428
Fluorocarbon precursor, hydrophobic textile
  helium–fluorocarbon (He/FC), 19
  nanoparticulate, 20–21
  2-(perfluorohexyl)ethyl acrylate (PFHEA), 20–21
  polyethylene terephthalate, 18
  polytetrafluoroethylene, 20
  SEM micrograph, 19–20
  silk fabric, 18
Formaldehyde, 323–324, 369
Fourier transform infrared (FTIR) spectroscopy, 11, 62
Free-export zones (FEZs), 209
Fugitive emissions, 313
Functional clothing, active ageing men and women
  co-design methodology, 248–249
  effective communication, 250
garment specification, 253–254
global clothing chain, 252
integration of technologies, 252–253
prototype co-design development, 252
size and shape, 251
user feedback, 251–252
work packages, 249

G

Garment industry
  Cambodia, 310
  product designer and architect, 253–254
  production
    bringing to market, 264–266
    lower-cost Asian countries, 272–274
Gas chromatography, 371
General environmental benefit claims, 386
Gentle Power Bleach, 490
GHG, *see* Greenhouse gas (GHG)
Global Organic Textile Standard (GOTS), 343–344, 380, 389–390
Global Recycle Standard (GRS), 431
Global supply chain, 233
Global value chains (GVCs), 40
Global warming potential (GWP), 145
Glucose oxidase, *see* Amylases
GOTS, *see* Global Organic Textile Standard (GOTS)
Gray market, 242
The Great Inequities, 293–295
The Great Unraveling, 292
The Great Warming, 292–293
GreenEarth Cleaning, 425
Green Garments, 470
Greenhouse gas (GHG)
  assessment, 142–143
  emission, 142, 290
  growth control, 300, 302
  sources, 143–145
  types, 143
Groz Beckert, 492
GWP, *see* Global warming potential (GWP)

H

Hazardous substances and mixtures, 315–316
Hazardous Substances and New Organisms Act, 323
Heat release rate (HRR), 62
Heat treatments, color fastness, 367
Hexabromocyclododecane, 60, 326
Higg Index, 433–434, 460
Hot/high-temperature/thermal plasma, 5
Hydrocarbons, 313
Hydrophobic textile
  definitions, 17
  dyed textiles and garments, 26–27
  fluorocarbon precursor
    helium–fluorocarbon (He/FC), 19
    nanoparticulate, 20–21
    2-(perfluorohexyl)ethyl acrylate (PFHEA), 20–21
    polyethylene terephthalate, 18
    polytetrafluoroethylene, 20
    SEM micrograph, 19–20
    silk fabric, 18

fundamental properties, 17
hydrocarbon precursor, 21–25
parameters, 16–17
silicone precursors, 25–26

I

Industrial ecology, 298–299
Industrial processes, 310–311
Industry affiliates, sustainable measures, 426
   American Apparel and Footwear Association, 421
   Bluesign® Technologies, 424
   Control Union, 422
   Cotton Incorporated, 422
   Cradle to Cradle Products Innovation Institute, 424–425
   European Outdoor Group, 422–423
   Flo-Cert, 423
   GreenEarth Cleaning, 425
   impact, 420
   input stream management, 424
   International Wool Textile Organization, 423
   Oeko-Tex, 423–424
   Zero Discharge of Hazardous Chemicals, 425, 427
Infrared (IR) dyeing and drying, 4
Input stream management, 424
Integrated pest management (IPM), 479
International Association of Natural Textile Industry (IVN), 344
International Standardization Organization (ISO), 352–353
International Wool Textile Organization (IWTO), 354, 423
Ion formation, 9
ISO 14001 environmental management systems, 381, 393–394
ISO 9001 quality management system, 394

J

Japan Organic Cotton Association (JOCA), 344
Jute-based apparels; *see also* Winter apparels
   apparel development methodologies, 99
   categories, 98–99
   characteristics, 99
   chemical modification, 99
   development methodologies, 99
   effect of washing, 109–110
   evaluation, 108–109
   finishing treatments, 103–107
   hollow polyester fabrics, 103–107
   jacket fabrics development, 102–103, 107
   natural fibers, 97–98
   polyester and cotton blended fabrics, 107–108
   warm garment, 103–107
   winter apparels, 101–102
   woollenized jute, 100–101
Jute fiber and products, 156–157

K

Klebsiella pneumonia (Kp) bacteria, 60
KRAV, 344

L

Labor conditions issue, 45–46
Laccifer lacca, 336–337
Lawsonia inermis, 338
LCA, *see* Life cycle assessment (LCA)
Least developed countries (LDCs), 45–46
Levi's life-cycle assessment, 483
Levi's Waterless™ denims, 487
Life cycle assessment (LCA)
   definition, 181–182
   fashion and sustainability, 441
   impact measure, 420, 423, 433
   life cycle impact assessment, 203
   limitations, 203–204
   natural fibers and textiles
      acidification potential, 187, 189
      comparative values, 194–195
      cotton and linen, 194–196
      cumulative energy demand, 194
      different fibers, 193–195
      emission of GHG, 188, 191
      energy demand, 188, 190
      energy distribution, 194
      eutrophication potential, 187, 189
      GHG emissions, 188–189
      global warming potential, 188–189
      life cycle emission of GHG, 188, 191
      modified cotton, 192
      primary energy, 188, 190
      raw silk, 194
      studies, 187–188
      types of cotton comparison, 191–192
      water consumption, 188, 190
      wool carpets, 193
   organization and standard methods, 184
   phases
      application, 183–184
      details, 182–183
      diagrammatic representation, 182
      different stages, 183
      holistic approach, 182
      interrelated components, 182
   regenerated cellulosic fibers
      different regenerated fibers, 197
      nonrenewable energy use, 196
      renewable energy use, 196
      viscose fiber consumption, 196
   SETAC, 181
   synthetic fibers and textiles
      assessment, 200–201
      CO_2 emission, 198–199
      energy consumption, 197–198
      energy requirement, 199–200
      GHG emissions, 198–199
      material resource requirement, 200–201
      nonrenewable energy resources, 197–198
      transport requirement, 199–200
   textile materials
      different phases, 184–185
      end of life cycle of clothing waste, 186, 188
      final share, 186, 188
      life cycle impacts, 185

Index

simplified flowchart, 185–186
systematic approach, 186–187
transportation and distribution, 186
textile recycling industry, 201–203
Limiting oxygen index (LOI), 52
jute, 56–57
mechanism, 53
nanotechnology, 61–62
thermal analysis, 66–67
traditional textiles, 53–54
wool, 60
Linear alkyl benzene sulfonates (LAS), 175
Linen fiber products, 157–158
Liquid chromatography, 371
LOI, see Limiting oxygen index (LOI)
Low cost of clothing, 237
Low-energy precision application (LEPA), 484
Low-impact materials, 481–482
Low-pressure plasma, 5–6

M

Maastricht Treaty, 353
Made-By®, 429–430
Marketing Practices Act, 384–385
Martindale abrasion tester, 359
Materials Sustainability Index (MSI), 433
Monobutyltin (MBT), 325
Montmorillonite (MMT), 62–63
Mordanting, 336–338
Multifiber arrangement (MFA), 400

N

Nanodispersed montmorillonite, 62
Nano-zinc oxide (ZnO), 52
National Organic Program, 379
National Resource Defense Council (NRDC), 432
Natural capitalism, 298
Natural dye–mordant complex, 337
Natural dyes, 521
    affinity, 335–336
    antimicrobial property
        antibacterial property, 338–339
        antifungal property, 339–340
    antioxidant property, 342
    chemical constitution, 336
    color, 345
    deodorizing property, 340–341
    eco-friendliness, 345
    environmental impacts of, 342–343
    extraction procedure, 345
    functional property, 345
    mordant and fiber, chemical bonding, 337–338
    sources, 345
    UV protection property, 341–342
Natural fibers, 479, 516–517, 523
Natural Step, 297
Near-sourcing, sustainable design, 495–496
Network resources, 42–43
Nike's Considered Index, 382
Nitrogen- and phosphorous-based chemicals, 54–55
NOMO Jeans, 468–469

Nonprofit organization (NPO), 435
    Aid by Trade Foundation, 427–428
    Better Cotton Initiative, 428
    CmiA Initiative, 427–428
    Environmental Protection Agency, 430–431
    Fairtrade Foundation, 428–429
    Higg Index, 433–434
    Made-By®, 429–430
    National Resource Defense Council, 432
    Sustainable Apparel Coalition, 433
    Textile Exchange, 430
    U.K. DEFRA, 432
Nontoxic claim, 387
Nonylphenol ethoxylates (NPEs), 175
Norm-activation theory, 280
NPO, see Nonprofit organization (NPO)
Nudie Jeans, 45, 461, 463–464

O

Oeko-Tex, 343–344, 423–424
Oeko-Tex 100 certification, 346
OEKO-TEX Standard 100, 380, 390–391
OEKO-TEX Standard 1000, 392–393
OE Standard, 431
Oil repellency, 362
Online retailing, 235–236
Optical brightening agent (OBA), 114, 120
Optical emission spectroscopy (OES), 10–11
Organic Content Standard (OCS), 392, 431
Organization for Economic Cooperation and
        Development (OECD), 44
Organotin compounds (OTCs), 325
Original brand manufacturing (OBM), 401–402
Original equipment manufacturing (OEM), 401–402
Outdoor clothing, 232
Outdoor industry
    guidance to producers and consumers, 263–264
    market research, 247
Oxidation, 9

P

PBT chemical, see Persistent bioaccumulative
        toxic (PBT) chemical
Pentachlorophenol, 324
Perfluorooctane sulfonate (PFOS), 326
Peroxide based single-stage process
    combined process, 78–79
    linear relation, 81
    perhydroxyl ions, 79
    peroxide–alkali process, 81
    radical mechanism, 80
    single-stage preparatory process, 81–82
    sodium chlorite affects, 79
    sodium chlorite–hydrogen peroxide, 81
Peroxide formation, 9
Persistent bioaccumulative toxic (PBT) chemical, 319
Pesticides, 326–327
pH
    aqueous extract, 369
    of dyeing, 336
    environmental impact, 313
    values, 326–327

Phthalates
  banned substances, 324
  quality control standards, 370
Pigments, quality control standards, 370
Plan–do–check–act (PDCA), 410
Planetary boundaries framework, 300
Plasma processing and finishing; *see also* Textile coloration
  atmospheric pressure plasma, 4–5
  chemical processing, 3–4
  desizing, 27–28
  environment-friendly
    advantages, 8
    application, 7–8
    heterogeneous reactions, 8–9
    substrate interaction, 9
    surface modification, 8
    value-added functionalities, 8
  flame-retardant finishing, 29–30
  generation and classification
    hot and cold temperature, 5
    low and atmospheric pressure, 5–7
  hydrophobic and superhydrophobic textile, 17
  hydrophobic textile (*see* Hydrophobic textile)
  method, 487
  optical properties and generation, 9–11
  sustainable textile, 30–31
  technological advancements, 4
  textile industry, 4
  types, 6
  water absorbency, 28–29
  wrinkle resistance treatment, 30
Point source emissions, 313
Pollution output, textile industry
  air emissions, 313–314
  emission during manufacturing, 312
  wastewater, 311–313
Pollution prevention, 318–319, 490–491
  chemical substitution, 320
  equipment modification, 321
  good operating practices, 321
  process modification, 320
  quality control of raw materials, 320
Pollution Prevention Act of 1990, 319–320
Polyamide (nylon 6) fabrics, 15–16
Polyester fiber, 16
Polyethylene terephthalate (PET), 18, 45
Polyhedral oligomeric silsesquioxanes (POSSs), 62
Polymerization, 9
Polypropylene (PP) bags, 158–160
Polytetrafluoroethylene (PTFE), 20
POTWs, *see* Publicly owned treatment works (POTWs)
Preparatory process, 336
Printing method, 363
Producer-driven supply chains, 402
Product certification, 380–381
    EU Ecolabel, 388–389
    GOTS, 389–390
    OEKO-TEX Standard 100, 390–391
    Textile Exchange, 391–292
Product ecology, 368
  alkylphenolethoxylates, 371
  dyes and pigments, 370
  flame retardants, 370–371
  formaldehyde, 369
  metals, 369
  pH of aqueous extract, 369
  phthalates, 370
Product life cycle, sustainable design, 513–514
Product-oriented PSS, 468
Product satisfaction, 467–469
Product service systems (PSS), 277, 460
  creation, 468
  fulfill consumers' cognitive needs, 470
  Green Garments, 470
  NOMO Jeans, 468–469
  offer change/fulfillment of fashion need, 469–470
Proteus vulgaris, 338
Prototype co-design development, 252
Pseudomonas aeruginosa, 338
PSS, *see* Product service systems (PSS)
Publicly owned treatment works (POTWs), 319

Q

Quality control standards for textiles
  care labeling, 368
  color fastness
    aqueous agents, 366
    atmospheric contaminants, 367
    bleaching agents, 367
    color and color differences measurement, 367
    colorant characteristics, 368
    dry-cleaning, 366
    heat treatments, 367
    light and weathering, 366
    principles, 365
    standard adjacent fabrics, 367
    vulcanization, 367
    washing and laundering, 366
  fabrics
    dimensional changes, 364–365
    mechanical resistance, 358
    structure and finishing, 361–364
    tests, 356–358
  fiber tests, 354–355
  raw materials, 320
  yarn tests, 355–356

R

Radical formation, 9
Radiofrequency (RF) drying, 4
Radio-frequency identification tags (RFIDs), 233
Raw materials, quality control, 320
Recombination, plasma–substrate interaction, 9
Recycled Claim Standard (RCS)
  content claims, 387
  standards, 431
  textile exchange, 392
Recycling
  sustainable design, 491–492
  sustainable fashion technology, 522
Regenerated fibers, 160–161
Research findings, clothing field sustainability, 257–260
Resource-based view (RBV), 40, 42–43
Resource productivity (RP), 39, 44–45
Responsible Down Standard (RDS), 392, 431

Index

Restricted substances list (RSL) testing, 316–318
Result-oriented PSS, 468
Retail practice, changing, 235–236
RSL testing, *see* Restricted substances list (RSL) testing
Rubbing, color fastness, 367

S

SAC, *see* Sustainable Apparel Coalition (SAC)
Safeguard water quality, 302
Safety of clothing, 372–373
SA8000 Standard, 394–395
Scanning electron microscopy (SEM), 104–106
Schwartz's theory, 280
Seam slippage, 360
Secondary ion mass spectrometer (SIMS), 13, 127
Service-and-flow model, 298
Silicon dioxide (SiO_2) layers, 63–64
Smart Retailing Group, 411–413
Social impacts
 corporate social responsibility (*see* Corporate social responsibility (CSR))
 cost–benefit analysis, 218
 cost-effective analysis, 218
 EarthCheck, 218
 life cycle analysis, 218–219
 measurement of
  benefits, 215
  broad approaches, 217
  dimensions, 216–217
  key elements, 216
  orientation, 217
  time frames, 217
 multicriteria appraisal, 218
 organizations
  critical issues, 220
  inspection and monitoring system, 222–223
  salient features, 220–222
 responsibilities, 224–225
 return investment, 218
 textile and clothing industry
  globalization, 209–210
  health and safety, 210
  raw materials, 211
  sweatshop, 209
  wages, 211
  worker's rights, 210
  work quality issues, 211–213
 tools, 217–220
 well-being valuation, 218
Social life cycle analysis (SLCA), 218–219
Social responsibility (SR)
 collaboration, 410
 supplier assessment, 400, 406
Society of Environmental Toxicology and Chemistry (SETAC), 181
Sodium silicate nonahydrate (SMSN), 59–60
Solvents (VOC), 326
Source reduction claims, 387
Spirality, quality control standards, 364
Standardization
 European Committee for Standardization, 353–354
 International Standardization Organization, 352–353
Staphylococcus aureus (Sa), 60, 338
Substantial waste, 234
Sun protective textile
 color, 119–120
 fiber and fabric properties, 119
 UV radiation, 118–119
Superhydrophobic textile, *see* Hydrophobic textile
Supplier assessment
 apparel industry
  apparel supply chains, 401–402
  global sourcing, 400–401
  unethical practices, 402–403
 continuous improvement, 410
 definition, 400, 403
 framework, 404
 relationship assessment program model, 404
 scorecard system, 403
 supplier development, 408
  collaboration, 410
  evaluation, 409–410
 supplier selection, 404–405
  decision making, 408
  economic criteria, 405–406
  environmental criteria, 406–407
  methods, 408
  social criteria, 406–407
Supplier development, 408
 collaboration, 410
 evaluation, 409–410
Supplier segmentation, 409
Supplier selection, apparel industry, 404–405
 decision making, 408
 economic criteria, 405–406
 environmental criteria, 405–407
 methods, 408
 social criteria, 405–407
Supply chains
 apparel supply chains (*see* Apparel supply chains)
 buyer-driven, 402
 producer-driven, 402
 sustainable fashion, 458–459
Supply chain transparency, 236
Surface cleaning, 9
Sustainable Apparel Coalition (SAC), 44, 263, 433, 460
Sustainable clothing action plan (SCAP), 476
Sustainable design
 best available techniques, 496–497
 business model, 453–456
 consumers, 233
 definition, 476
 fashion *vs.* sustainability, 476–477
 objectives, 505
 principles, 513
  biodegradable products, 489–490
  biomimicry, 488–489
  community couture, 492–493
  do-it-yourself and patchworks, 494
  energy efficiency, 482–484
  fair trade/ethical practices, 494–495
  low-impact materials, 481–482
  near-sourcing, 495–496
  pollution prevention, 490–491
  renewable sources, 482
  reuse and recycling, 491–492
  service substitution, 492

        sustainable fibers, 479
        waste management, 487–488
        water efficiency (see Water efficiency)
    production methods, 514–515
    product life cycle, 513–514
    science, 513
Sustainable fashion
    business model, 453–456
    collaborative consumption, 456
    communication strategies, 464–465
    concept, 502–503
    consumer-based eco-efficiency, 466, 470
    consumer satisfaction, 466
    consumption, 511–512
    definition, 505
    elements in apparel manufacturing, 506–507
    end-of-life, 457–458
    environment-friendly materials, 518–520
    Ethical Fashion Forum, 503–504
    Ethical Fashion Show, 526
    features, 504–505
    future trends, 523–526
    implementation, 526–527
    innovations, 459–460
    pillars, 503–504
    postretail business models, 457–458
    principles underlying framework, 511
    product satisfaction, 467–468
    product service systems, 468
        creation, 468
        fulfill consumers' cognitive needs, 470
        Green Garments, 470
        NOMO Jeans, 468–469
        offer change/fulfillment of fashion need, 469–470
    retailing, 524–525
    scenario, 514–516
    supply chain solutions, 458–459
    Swedish fashion design companies
        Boomerang, 464
        Filippa K, 461–462
        Nudie Jeans, 463–464
        Uniforms for the Dedicated, 462–463
    technology
        clean technologies, 521–522
        enzymes use in textile processing, 520–521
        natural dyeing, 521
        recycling and upcycling, 522
    World Global Sourcing Network, 504–505
    World Wide Fund for Nature, 506
Sustainable fiber program (SFP), Control Union, 422
Sustainable materials strategies, 44–45
Sustainable measures
    fibers, 479–480, 516–517
    impact, 420
    industry affiliates, 426
        American Apparel and Footwear Association, 421
        Bluesign® Technologies, 424
        Control Union, 422
        Cotton Incorporated, 422
        Cradle to Cradle Products Innovation Institute, 424–425
        European Outdoor Group, 422–423
        Flo-Cert, 423
        GreenEarth Cleaning, 425

        input stream management, 424
        International Wool Textile Organization, 423
        Oeko-Tex, 423–424
        Zero Discharge of Hazardous Chemicals, 425, 427
    NPO/governmental institutions, 435
        Aid by Trade Foundation, 427–428
        Better Cotton Initiative, 428
        CmiA Initiative, 427–428
        Environmental Protection Agency, 430–431
        Fairtrade Foundation, 428–429
        Higg Index, 433–434
        Made-By®, 429–430
        National Resource Defense Council, 432
        Sustainable Apparel Coalition, 433
        Textile Exchange, 430
        U.K. DEFRA, 432
Sustainable technology
    clean technologies, 521–522
    enzymes use in textile processing, 520–521
    natural dyeing, 521
    recycling and upcycling, 522
Sustainable Textile Production (STeP), 423–424
Swedish fashion design companies
    Boomerang, 464
    Filippa K, 461–462
    Nudie Jeans, 463–464
    Uniforms for the Dedicated, 462–463
Synthetic fibers, 479
    assessment, 200–201
    CO_2 emission, 198–199
    energy consumption, 197–198
    GHG emissions, 198–199
    material resource requirement, 200–201
    nonrenewable energy resources, 197–198
    transport requirement, 199–200
System certification, 380–381
    Eco-Management and Audit Scheme, 393–394
    ISO 14001 environmental management systems, 393–394
    ISO 9001 quality management system, 394
    OEKO-TEX Standard 1000, 392–393
    SA8000 Standard, 394–395

T

Technical materials, 262
2,3,5,6-Tetrachlorophenols, 324
Tetrakis hydroxymethyl phosphonium chloride (THPC), 54
Tetramethyldisiloxane (TDMS), 64
Textile and clothing (T&C) sectors, 39
Textile coloration
    cellulosic textile, 11–12
    dye exhaustion percentage, 14–15
    dye molecules, 14
    polyamide (nylon 6) fabrics, 15–16
    silk, 13–15
    textile exchange, 391–292, 430
    wool, 12–13
Textile industry
    banned substances (see Banned substances)
    eco-testing
        hazardous substances and mixtures, 315–316
        restricted substances list, 316–318

Index

environmental communication, 377–378
    FTC green guides, 379
    product certification (*see* Product certification)
    system certification (*see* System certification)
global structure, 233–235
industrial processes, 310–311
Mauritius, 310
pollution output
    air emissions, 313–314
    wastewater, 311–313
pollution prevention, 318–319
    chemical substitution, 320
    equipment modification, 321
    good operating practices, 321
    Pollution Prevention Act of 1990, 319–320
    process modification, 320
    quality control of raw materials, 320
processing flowchart, 310
Textile recycling industry, 232
Textile sustainability
    biodiversity protection, 303
    biomimetics, 299
    cradle-to-cradle, 299
    desertification prevention, 303
    ecological footprint, 297
    economic system goals, 296
    energy utilization, 302–303
    environmental system goals, 296
    greenhouse gases, growth control, 300, 302
    industrial ecology, 298–299
    natural capitalism, 298
    Natural Step, 297
    planetary boundaries, 300
    planetary pulse, 291
        The Great Inequities, 293–295
        The Great Unraveling, 292
        The Great Warming, 292–293
    safeguard water quality, 302
    social system goals, 296
    strategic vectors, 295–296, 301
    waste product management, 304
    Zero Emissions Research and Initiatives, 299–300
Textile waste, 276–277
Thermal insulation, 361
Thermogravimetric (TG) analysis, 59
Titanium dioxide (TiO_2), 125–128
Titanium tetra-isopropoxide (TTIP), 125
Total organic content (TOC), 313
Toxic Release Inventory (TRI) program, 319
Transaction cost economics (TCE), 40, 42
Transfer method, 363
Tributyltin (TBT), 325
Trichlorophenol, 324
Triethylenetetramine (TETA), 11
Trimethyl melamine (TMM), 54
Trust in networks, 43

U

U.K. Department for Environment, Food and Rural Affairs (U.K. DEFRA), 432
Ultraviolet protection factor (UPF), 113–114, 341–342
    color, 119–120
    evaluation, 116–118
    fiber and fabric properties, 119
    nanoparticles, 132–133
    natural polymer, 133–134
Ultraviolet (UV)-protective apparel textile; *see also* Sun protective textile
    evaluation
        sun protection factor, 116–117
        ultraviolet protection factor, 117–118
    impact, 115–116
    nanolignin and plant extracts, 133–134
    nanoparticles of natural polymers, 132–133
    nanotechnology, 122–123
        titanium dioxide (TiO_2), 125–128
        zinc oxide (ZnO), 123–125
    plant extract, 129–132
    plasma processing, 128–129
    radiation, 113
    traditional mechanism, 120–121
    ultraviolet protection factor, 113–114
    water-free plasma processing and finishing, 4
Ultraviolet rays (UVRs), 341
UN Global Compact study, 457
Uniforms for the Dedicated, 462–463, 465
Upcycling, sustainable fashion technology, 522
UPF, *see* Ultraviolet protection factor (UPF)
U.S. Environmental Protection Agency (U.S. EPA), 430–431
Use-oriented PSS, 468
UV protection of textiles, 364
UV rays (UVRs), *see* Ultraviolet rays (UVRs)

V

Value-action gap, 383
Vanity sizing, 238
Volatile organic compounds (VOCs), 173
Voluntary Product Environmental Profile (VPEP), 421
Vulcanization, color fastness, 367

W

Washing, fabrics
    color fastness, 366
    quality control standards, 364
Waste and Resources Action Programme (WRAP), U.K., 232
Waste product management, 304, 487–488
Water efficiency
    Adidas DryDye technology, 487
    Levi's Waterless™ denims, 487
    low-energy precision application, 484
    plasma processing method, 487
    water framework directive, 484
Water footprint, 420
Water framework directive (WFD), 484
Water-free plasma processing and finishing, *see* Plasma processing and finishing
Water-repellent textiles, *see* Hydrophobic textile
Wearable technology, 248–249, 262
Wet-cleaning, quality control standards, 364–365
Winter apparels, 101–102
Wool fiber and products, 152–156

Woollenized jute
    acrylic blended knitting yarn, 100–101
    hollow polyester, 100–101
    polyester, 100–101
    polypropylene blended yarn, 100
    wool blended yarn, 100
Work packages (WPs), 248
World Global Sourcing Network (WGSN), 504–505
World Trade Organization (WTO), 41–42
World Wide Fund for Nature, 506
Worldwide Responsible Accredited Production (WRAP), 476

X

X-ray diffraction (XRD), 62
X-ray photoelectron spectroscopy (XPS), 12

Y

Yarn tests
    formation, 310–311
    quality control standards, 355–356
    spinning waste, 314

Z

Zero Discharge of Hazardous Chemicals (ZDHC), 425, 427
Zero Emissions Research and Initiatives (ZERI), 299–300
Zero-waste fashion design, 488
Zinc oxide (ZnO), 123–125